The Adventure of Reason

The Adventure of Reason

Interplay between Philosophy of Mathematics and Mathematical Logic, 1900–1940

PAOLO MANCOSU

OXFORD
UNIVERSITY PRESS

Great Clarendon Street, Oxford OX2 6DP

Oxford University Press is a department of the University of Oxford.
It furthers the University's objective of excellence in research, scholarship,
and education by publishing worldwide.

Oxford is a registered trade mark of Oxford University Press
in the UK and in certain other countries

© in this volume Paolo Mancosu 2010

The moral rights of the author have been asserted

First published 2010
First published in paperback 2014

All rights reserved. No part of this publication may be reproduced,
stored in a retrieval system, or transmitted, in any form or by any means,
without the prior permission in writing of Oxford University Press,
or as expressly permitted by law, by licence or under terms agreed with the appropriate
reprographics rights organization. Enquiries concerning reproduction
outside the scope of the above should be sent to the Rights Department,
Oxford University Press, at the address above

You must not circulate this work in any other form
and you must impose this same condition on any acquirer

British Library Cataloguing in Publication Data
Data available

Library of Congress Cataloging in Publication Data
Data available

ISBN 978–0–19–954653–4 (Hbk)
ISBN 978–0–19–870151–4 (Pbk)

Ad Elena,
Compagna nell'avventura della vita

PREFACE

In the last ten years I have extensively investigated a variety of topics in the history of logic, the foundations of mathematics, and the philosophy of logic. The period of investigation goes roughly from the beginning of the twentieth century to the early 1940s. This period corresponds to the 'golden age' of mathematical logic including the discovery of Gödel's incompleteness theorems and the development of Tarskian semantics, to mention perhaps the two most well-known achievements. These and other results in mathematical logic both influenced and were influenced by contemporary developments in the philosophy of mathematics, the philosophy of logic, and general philosophy. While the general outline of such interactions are by now well known, a more nuanced picture of such developments can be gathered by exploiting, in addition to the published sources, the rich collection of materials found in several archives. The papers gathered here exploit untapped archival sources and thus make available a wealth of new information that deepens in significant ways our understanding of the above-mentioned areas. In pursuing this research, I worked at length in, among others, the Hilbert archive (Göttingen), the Bernays and Weyl archives (Zurich), the Carnap archive (Pittsburgh and Konstanz), the Neurath archive (Haarlem and Konstanz), the Tarski archive (Berkeley), the Woodger archive (London), the Behmann archive (Erlangen, now Berlin), the Mahnke archive (Marburg), the Kaufmann archive (Konstanz), and the Quine archive (Harvard).

While the results of this research were published separately, I am pleased now to bring these papers together in a single volume. The essays are closely linked by the fact that the subject matter is homogeneous and were written with a single major aim, namely that of reaching a deeper understanding of the interaction between developments in mathematical logic and the foundations of mathematics and logic from 1900 to 1940.

To the discerning reader, the developments described in these pages will justify the title of this book *The Adventure of Reason*. I borrowed this beautiful image from a letter from Becker to Mahnke, which also gives the title to Chapter 12. The adventure of reason reflects itself in the ambitious mathematical and philosophical aims that drive the constitution of mathematical logic, the philosophical and mathematical aims that it is made to serve, and the unexpected surprises lying in wait. That's the way I like to think of the ambitious goals by Russell and Whitehead, of Hilbert's program, and of Husserl's phenomenological ambitions, to name only a few major developments treated in the book. But just like Columbus had set sail to reach the Indies and found other unexpected—not less rich and exciting—lands, in the same way the adventure of reason in logic and the foundations of mathematics led to surprising new finds that dramatically reshaped the original goals.

The essays contained in this volume, all previously published except Chapters 17 and 18, fall naturally into the five parts that make up the book. The first part contains a lengthy account of the history of mathematical logic from the early part of the twentieth century to the middle of the 1930s. On account of its length, I recommend using this chapter as needed while reading the other chapters in the book. Part II contains seven chapters that deal with foundational issues in mathematics, namely the Russellian influence in Göttingen, constructivism, and Gödel's incompleteness theorems. The third part details a dramatic moment in the development of phenomenology through the three figures who, besides Husserl, had the best understanding of the new developments in logic, mathematics, and physics. I am referring to Hermann Weyl, Oskar Becker, and Dietrich Mahnke. In spite of their different solutions to the problem of the relationship between phenomenology and the exact sciences, they all agreed on the unprecedented strain put by the new scientific developments on the Husserlian phenomenological grounding of the sciences described in *Ideas*. Part IV focuses on the discussion of nominalism as a foundation for mathematics and science in the early stages of such discussions, that is during the meetings among Carnap, Quine, and Tarski in Harvard in 1940-41 and, more broadly in Quine's and Tarski's trajectory on such issues. Finally, part V both discusses the impact of Tarski's theory of truth and logical consequence in the philosophical milieu of the Vienna Circle and treats a number of issues related to the proper characterization of logical consequence in Tarski during that period. In an archival appendix, Chapter 18, I publish a hitherto unpublished lecture by Tarski, which contains important information on the notion of logical consequence and on the development of central notions in the semantics of theories formulated in higher order logic. I would like to thank professor Jan Tarski the son and Manager of Alfred Tarski's literary estate, for having generously granted me permission to publish this text.

Each part is prefaced by a summary of the contents of the articles and in addition contains a bibliographical update. These updates do not strive at completeness but I hope they are comprehensive enough to indicate the major new contributions that have appeared since the original publication of the articles. As a rule I have tried to reproduce the articles without major modifications. When I felt, as in the case of Chapter 2 and Chapter 16, that a substantial comment needed to be added, I have done so by appending an Addendum to those articles. Other, very minor, modifications are indicated by an insertion in the text within double brackets, as in [[modified from the original introduction]]. In a few cases I have removed inaccuracies from the originally published version and replaced reference to my published articles by the corresponding chapters of this book without flagging this explicitly. The articles are reproduced in their entirety with the only exception being Chapter 11 from which the first four sections have been removed. These sections are identical in content to the first four sections of Chapter 10, thus no loss in content ensues. The details of the original publications with permission to republish are contained in the Acknowledgments at the beginning of the book. The original acknowledgements, if any, appear as an uncued footnote at the bottom of the first page of the relevant chapter. Thus, I will here acknowledge only those scholars who helped me since conceiving this book. I would like to thank Peter Momtchiloff, the editor in Philosophy at OUP, Oxford, for having encouraged me to embark on this project starting with a conversation in Vancouver in November 2006 and for having followed the project through with the Press. I would like to thank Solomon Feferman, Marcus Giaquinto, Donald Gillies, and Daniel Isaacson for having been so supportive of the project from its very inception and for having provided many excellent suggestions on a variety of matters. Tom Ryckman and Richard Zach have helped me immensely with comments on the introductions and the bibliographical updates. Fabrizio Cariani has taken on, once again, the ungrateful task of resetting the original articles into a beautiful LaTeX format. His work was partially supported by a COR grant at U.C. Berkeley; but only partially and I am very grateful for his unfailing dedication to the project. In this connection, I would also like to thank Marco Dettweiler and Volker Peckhaus for having provided us with the source files of Chapters 11 and 12 which saved Fabrizio from the drudgery of resetting the papers in LaTeX. During the latest stages of the preparation of the book I was also fortunate to receive the help of Rebecca Millsop who saved me from the painful task of inserting an almost uncountable number of tiny corrections into the source files. It is also a pleasure to acknowledge my co-authors for Chapter 1 (Richard Zach and Calixto Badesa), Chapter 6 (Mathieu Marion), and Chapters 10 and 11 (Thomas Ryckman), for having allowed me to reprint our joint papers and, much more, for the intellectual stimulus and pleasure we shared while writing them.

Work on the final stages of the book has been made possible by a Guggenheim Fellowship and by a membership at the Institute for Advanced Study in Princeton in the Spring Term 2009. I cannot think of better conditions for bringing a book to completion than the ones I was fortunate to have. I am grateful to the Guggenheim Foundation and to the Institute for Advanced Study for their generous support and to the latter for having provided such ideal working conditions.

This book is dedicated to my wife Elena Russo, for having been close to me, throughout the composition of this book, in spirit, if not always in person.

Princeton,
Institute for Advanced Study,
March 27, 2009

ACKNOWLEDGMENTS

Chapter 1 was originally published in L. Haaparanta (ed.), *The Development of Modern Logic* (Oxford: Oxford University Press, 2009), pp. 318–470.

Chapter 2 was originally published in P. Mancosu (ed.), *From Brouwer to Hilbert* (Oxford: Oxford University Press, 1998), pp. 149–88.

Chapter 3 was originally published in *The Bulletin of Symbolic Logic*, 5(3) (1999), pp. 303–30. Reprinted with kind permission of the Association of Symbolic Logic.

Chapter 4 was originally published in *Synthese*, 137 (2003), pp. 59–101. Reprinted with kind permission from Springer Science + Business Media, ©2003 Kluwer Academic Publishers.

Chapter 5 was originally published in Wilfried Sieg, Richard Sommer, and Carolyn Talcott (eds.), *Reflections on the Foundations of Mathematics: Essays in Honor of Solomon Feferman*, Association for Symbolic Logic (2002), pp. 349–71. Reprinted by kind permission of the Association of Symbolic Logic.

Chapter 6 was originally published in Friedrich Stadler (ed.), *The Vienna Circle and Logical Empiricism: Re-Evaluation and Future Perspectives* (Dordrecht: Kluwer, 2003), pp. 171–88.

Chapter 7 was originally published in *History and Philosophy of Logic*, 20 (1999), pp. 33–45. Reprinted with kind permission of Taylor and Francis.

Chapter 8 was originally published in *Notre Dame Journal of Formal Logic*, 45(2) (2004), pp. 109–25. ©2004, Yale University. Reprinted by permission of the publisher, Duke University Press.

Chapter 9 was originally published in P. Mancosu (ed.), *From Brouwer to Hilbert* (Oxford: Oxford University Press, 1998), pp. 65–85.

Chapter 10 was originally published in *Philosophia Mathematica*, 10(3)(2002), pp. 130–202.

Chapter 11 was originally published in *Oskar Becker und die Philosophie der Mathematik*, ed. Volker Peckhaus (Munich: Wilhelm Fink Verlag, 2005) (Neuzeit & Gegenwart Philosophie in Wissenschaft und Gesellschaft), pp. 153–228. Reprinted by kind permission of Fink Verlag.

Chapter 12 was originally published in *Oskar Becker und die Philosophie der Mathematik*, ed. Volker Peckhaus (Munich: Wilhelm Fink Verlag, 2005) (Neuzeit & Gegenwart: Philosophie in Wissenschaft und Gesellschaft), pp. 229–43. Reprinted by kind permission of Fink Verlag.

Chapter 13 was originally published in *History and Philosophy of Logic*, 26 (2005), pp. 327–57. Reprinted with kind permission of Taylor and Francis.

Chapter 14 was originally published in *Oxford Studies in Metaphysics*, ed. Dean Zimmerman, vol. iv (2008), pp. 22–55.

Chapter 15 was originally published in Douglas Patterson (ed.), *New Essays on Tarski and Philosophy* (Oxford: Oxford University Press, 2008), pp. 194–224.

Chapter 16 was originally published in J. Ferreiros and J. Gray (eds.), *The Architecture of Modern Mathematics* (Oxford: Oxford University Press, 2006), pp. 209–37.

Chapter 18 contains the text of a lecture given by Tarski at Harvard in January 1940, transcribed by Paolo Mancosu. The original can be found in carton 15 of Tarski's archive at the Bancroft Library at U.C. Berkeley. Published by kind permission of the Bancroft Library and of Jan Tarski.

CONTENTS

PART I. MATHEMATICAL LOGIC, 1900–1935

Introduction — 2

1. The Development of Mathematical Logic from Russell to Tarski, 1900–1935 (with Richard Zach and Calixto Badesa) — 5

PART II. FOUNDATIONS OF MATHEMATICS

Introduction — 122

2. Hilbert and Bernays on Metamathematics — 125
   Addendum — 155
3. Between Russell and Hilbert: Behmann on the Foundations of Mathematics — 159
4. The Russellian Influence on Hilbert and His School — 176
5. On the Constructivity of Proofs: A Debate among Behmann, Bernays, Gödel, and Kaufmann — 199
6. Wittgenstein's Constructivization of Euler's Proof of the Infinity of Primes (with Mathieu Marion) — 217
7. Between Vienna and Berlin: The Immediate Reception of Gödel's Incompleteness Theorems — 232
8. Review of Gödel's *Collected Works*, Vols. IV and V — 240

PART III. PHENOMENOLOGY AND THE EXACT SCIENCES

Introduction — 256

9. Hermann Weyl: Predicativity and an Intuitionistic Excursion — 259
10. Mathematics and Phenomenology: The Correspondence between O. Becker and H. Weyl (with T. Ryckman) — 277
11. Geometry, Physics, and Phenomenology: Four Letters of O. Becker to H. Weyl (with T. Ryckman) — 308
12. "Das Abenteuer der Vernunft": O. Becker and D. Mahnke on the Phenomenological Foundations of the Exact Sciences — 346

PART IV. TARSKI AND QUINE ON NOMINALISM

    Introduction   358

13. Harvard 1940–1941: Tarski, Carnap, and Quine on a Finitistic Language of Mathematics for Science   361
14. Quine and Tarski on Nominalism   387

PART V. TARSKI AND THE VIENNA CIRCLE ON TRUTH AND LOGICAL CONSEQUENCE

    Introduction   412

15. Tarski, Neurath, and Kokoszyńska on the Semantic Conception of Truth   415
16. Tarski on Models and Logical Consequence   440
    Addendum   463
17. Tarski on Categoricity and Completeness: An Unpublished Lecture from 1940   469
18. Appendix: "On the Completeness and Categoricity of Deductive Systems" (1940)   485

Notes   493
Bibliography   571
Index   611

PART I

Mathematical Logic, 1900–1935

Introduction

SUMMARY

Chapter 1, written in collaboration with Richard Zach and Calixto Badesa, contains a history of mathematical logic in what might be called its 'golden period'. The reader who is mainly interested in the more philosophical sections of this book can use it as needed. However, I have put this article at the beginning of the book because most of the philosophical discussions contained in Parts II to V presuppose, in different degrees, the technical developments contained in Chapter 1. For instance, the reader interested in the material on Hilbert's program treated in Part II of the book would greatly benefit from reading in tandem itineraries 4 and 5 of Chapter 1; and the reader interested in the history of semantics, to add another example, would benefit from reading itineraries 1 and 8 of Chapter 1. On account of the scope and complexities of the material treated, Chapter 1 is divided into eight itineraries:

  I Metatheoretical Properties of Axiomatic Systems;
  II Bertrand Russell's Mathematical Logic;
  III Zermelo's Axiomatization of Set Theory and Related Foundational Issues;
  IV The Theory of Relatives and Löwenheim's Theorem;
  V Logic in the Hilbert School;
  VI Proof Theory and Arithmetic;
  VII Intuitionism and Many-Valued Logics;
  VIII Semantics and model-theoretic notions.

While Chapter 1 is structured, as much as possible, according to a chronological order of exposition, the reader should be able to use each itinerary independently of the others.

BIBLIOGRAPHICAL UPDATE

It is inevitable that such a long piece of work, written in the span of several years and whose publication was delayed by several years, suffers here and there from bibliographical omissions. While I cannot aim here at completeness, the following update should help the reader track down most of the relevant recent literature.

Itinerary I. Metatheoretical Properties of Axiomatic Systems. the most important publications here are Marchisotto and Smith (2007) on Pieri, and Hilbert's writings on the foundations of geometry published in Hilbert (2004).

Itinerary II. Bertrand Russell's Mathematical Logic. Here I refer to Linsky and Imaguire (2005) for recent essays on "On denoting" and to Linsky's forthcoming book on the second edition of *Principia Mathematica*. On Russell's completeness theorem for propositional logic in his 1906 article on the theory of implication see Milne (2008) which also contains much of relevance to logicians of the 'postulate' school treated in itinerary 1. On the history of type theory before 1940 see Kamareddine *et al.* (2002). For quantification theory in *Principia Mathematica* see Landini (2005). On Russell and the universalist conception of logic see Proops (2007).

Itinerary III. Zermelo's Axiomatization of Set Theory and Related Foundational Issues. Here the most important contribution is Ebbinghaus (2007) and Kanamori (2004) on Zermelo.

Itinerary IV. The Theory of Relatives and Löwehnheim's Theorem. On the Algebra of Logic see Peckhaus (2004a,b), Bondoni (2007), and recent entries in the *Stanford Encyclopedia of Philosophy*. More specific contributions to our understanding of Löwenheim include Löwenheim (2007), Thiel (2007), and von Plato (2007).

Itinerary V. Logic in the Hilbert School. On the history of combinatory logic see Bimbó (2008).

Itinerary VI. Proof Theory and Arithmetic. On anticipations of incompleteness by Emil Post see Mol (2006). On the Herbrand-Gödel correspondence see Sieg (2005). Wille (2008) discusses Ackermann's consistency proof. For Gentzen's life and work in proof theory see Menzler-Trott (2007) and von Plato (2008).

Itinerary VII. Intuitionism and Many-Valued Logics. Here see van Atten's entry on the history of intuitionism in the *Stanford Encyclopedia of Philosophy* and van Atten *et al.* (2008) on 100 years of intuitionism.

Itinerary VIII. Semantics and Model-Theoretic Notions. On Ajdukiewicz, see Betti (2008). On Lesniewski see Urbaniak (2008). On Tarski see the introduction to section V. On Sheffer see Scanlan (2000). On Gödel on completeness of first-order logic see von Plato (2004) and the rejoinder van Atten (2005b) and Moriconi (2006).

Finally, volume 5 of the *Handbook of the History of Logic* (Logic from Russell to Church), edited by Dov Gabbay and John Woods, Gabbay and Woods (2009), contains entries relevant to all eight itineraries. The list includes: Russell's logic; Meinongian Logic; The logic of Brouwer and Heyting; Skolem's logic; Herbrand's logic; The logic of Wittgenstein and Ramsey; Lesniewski's logic; Hilbert's logic; Hilbert's epsilon calculus and its successors; Gödel's logic; Tarski's logic; Post's logic; Gentzen's logic; Lambda calculus and Combinatory Logic; The logic of Curry and Church; Paradoxes in the 20th century.

1

THE DEVELOPMENT OF MATHEMATICAL LOGIC FROM RUSSELL TO TARSKI, 1900–1935 (WITH RICHARD ZACH AND CALIXTO BADESA)

THE following eight itineraries in the history of mathematical logic do not aim at a complete account of the history of mathematical logic during the period 1900–35. For one thing, we had to limit our ambition to the technical developments without attempting a detailed discussion of issues such as what conceptions of logic were being held during the period. This also means that we have not engaged in detail with historiographical debates which are quite lively today, such as those on the universality of logic, conceptions of truth, the nature of logic itself and so on. While of extreme interest, these themes cannot be properly dealt with in a short space, as they often require extensive exegetical work. We therefore merely point out in the text or in appropriate notes how the reader can pursue the connection between the material we treat and the secondary literature on these debates. Second, we have not treated some important developments. While we have not aimed at completeness, our hope has been that by focusing on a narrower range of topics our treatment will improve on the existing literature on the history of logic. There are excellent accounts of the history of mathematical logic available, such as, to name a few,

---

We have received comments on an earlier draft of this paper from Mark van Atten, José Ferreiros, Johannes Hafner, Ignasi Jané, Bernard Linsky, Enrico Moriconi, Chris Pincock, and Bill Tait. Their help is gratefully acknowledged.

Kneale and Kneale (1962), Dumitriu (1977), and Mangione and Bozzi (1993). We have kept the secondary literature quite present in that we also wanted to write an essay that would strike a balance between covering material that was adequately discussed in the secondary literature and presenting new lines of investigation. This explains, for instance, why the reader will find a long and precise exposition of Löwenheim's (1915) theorem but only a short one on Gödel's incompleteness theorem: Whereas there is hitherto no precise presentation of the first result, accounts of the second result abound. Finally, the treatment of the foundations of mathematics is quite restricted, and it is ancillary to the exposition of the history of mathematical logic. Thus, it is not meant to be the main focus of our exposition.[1]

Page references in citations are to the English translations, if available, or to the reprint edition, if listed in the bibliography. All translations are the authors', unless an English translation is listed in the references.

1.1 Itinerary I. Metatheoretical Properties of Axiomatic Systems

1.1.1 Introduction

The two most important meetings in philosophy and mathematics in 1900 took place in Paris. The First International Congress of Philosophy met in August and so did, soon after, the Second International Congress of Mathematicians. As symbolic, or mathematical, logic has traditionally been part of both mathematics and philosophy, a glimpse at the contributions in mathematical logic at these two events will give us a representative selection of the state of mathematical logic at the beginning of the twentieth century. At the International Congress of Mathematicians, Hilbert presented his famous list of problems (Hilbert 1900a), some of which became central to mathematical logic, such as the continuum problem, the consistency proof for the system of real numbers, and the decision problem for Diophantine equations (Hilbert's tenth problem). However, despite the attendance of remarkable logicians like Schröder, Peano, and Whitehead in the audience, the only other contributions that could be classified as pertaining to mathematical logic were two talks given by Alessandro Padoa on the axiomatizations of the integers and of geometry, respectively.

The third section of the International Congress of Philosophy was devoted to logic and history of the sciences (Lovett 1900–01). Among the contributors of papers in logic we find Russell, MacColl, Peano, Burali-Forti, Padoa, Pieri, Poretsky, Schröder, and Johnson. Of these, MacColl, Poretsky, Schröder, and Johnson read papers that belong squarely to the algebra of logic tradition. Russell read a paper on the application of the theory of relations to the problem of order and absolute position in space and time. Finally, the Italian school of Peano and his disciples—Burali-Forti, Padoa, and Pieri—contributed papers on the logical analysis of mathematics. Peano and Burali-Forti spoke on definitions, Padoa read his famous essay containing the "logical introduction to any theory whatever", and Pieri spoke on geometry considered as a purely logical system. Although there are certainly points of contact between the first group of logicians and the second group, already at that time it was obvious that two different approaches to mathematical logic were at play.

Whereas the algebra of logic tradition was considered to be mainly an application of mathematics to logic, the other tradition was concerned more with an analysis of mathematics by logical means. In a course given in 1908 in Göttingen, Zermelo captured the double meaning of mathematical logic in the period by reference to the two schools:

> The word "mathematical logic" can be used with two different meanings. On the one hand one can treat logic mathematically, as it was done for instance by Schröder in his Algebra of Logic; on the other hand, one can also investigate scientifically the logical components of mathematics. (Zermelo 1908a, p. 1)[2]

The first approach is tied to the names of Boole and Schröder, the second was represented by Frege, Peano, and Russell.[3] We will begin by focusing on mathematical logic as the logical analysis of mathematical theories, but we will return later (see itinerary IV) to the other tradition.

1.1.2 Peano's School on the Logical Structure of Theories

We have mentioned the importance of the logical analysis of mathematics as one of the central motivating factors in the work of Peano and his school on mathematical logic. First of all, Peano was instrumental in emphasizing the importance of mathematical logic as an artificial language that would remove the ambiguities of natural language, thereby allowing a precise analysis of mathematics. In the words of Pieri, an appropriate ideographical algorithm is useful as "an instrument appropriate to guide and discipline thought, to exclude ambiguities, implicit assumptions, mental restrictions, insinuations and other shortcomings, almost inseparable from ordinary language, written as well as spoken, which are so damaging to speculative research" (Pieri 1901, p. 381). Moreover, he compared mathematical logic to "a microscope which is appropriate for observing the smallest difference of ideas, differences that are made imperceptible by the defects of ordinary language in the absence of some instrument that magnifies them" (382). It was by using this "microscope" that Peano was able, for instance, to clarify the distinction between an element and a class containing only that element and the related distinction between membership and inclusion.[4]

The clarification of mathematics, however, also meant accounting for what was emerging as a central field for mathematical logic: the formal analysis of mathematical theories. The previous two decades had in fact seen much activity in the axiomatization of particular branches of mathematics, including arithmetic, algebra of logic, plane geometry, and projective geometry. This culminated in the explicit characterization of a number of formal conditions for which axiomatized mathematical theories should strive. Let us consider first Pieri's description of his work on the axiomatization of geometry, which had been carried out independently of Hilbert's famous *Foundations of Geometry* (1899). In his presentation to the International Congress of Philosophy in 1900, Pieri emphasized that the study of geometry is following arithmetic in becoming more and more "*the study of a certain order of logical relations*; in freeing itself little by little from the bonds which still keep it tied (although weakly) to *intuition*, and in displaying consequently the form and quality of *purely deductive, abstract and ideal science*" (Pieri 1901, p. 368). Pieri saw in this abstraction from concrete interpretations a unifying thread running through the development of arithmetic, analysis, and geometry in the nineteenth century. This led him to a conception of geometry as a hypothetical discipline (he coined the term "hypothetico-deductive"). In fact he goes on to assert that the primitive notions of any deductive system whatsoever "must be capable of arbitrary interpretations in certain limits assigned by the primitive propositions", subject only to the restriction that the primitive propositions must be satisfied by the particular interpretation. The analysis of a hypothetico-deductive system begins then with the distinction between primitive notions and primitive propositions. In the logical analysis of a hypothetico-deductive system it is important not only to distinguish the derived theorems from the basic propositions (definitions and axioms) but also to isolate the primitive notions, from which all the others are defined. An ideal to strive for is that of a system whose primitive ideas are irreducible, that is, such that none of the primitive ideas can be defined by means of the others through logical operations. Logic is here taken to include notions such as, among others, "individual", "class", "membership", "inclusion", "representation" and "negation" (383). Moreover, the postulates, or axioms, of the system must be independent, that is, none of the postulates can be derived from the others.

According to Pieri, there are two main advantages to proceeding in such an orderly way. First of all, keeping a distinction between primitive notions and derived notions makes it possible to compare different hypothetico-deductive systems as to logical equivalence. Two systems turn out to be equivalent if for every primitive notion of one we can find an explicit definition in the second one such that all primitive propositions of the first system become theorems of the second system, and vice versa. The second advantage consists in the possibility of abstracting from the meaning of the primitive notions and thus operate symbolically on expressions which admit of different interpretations, thereby encompassing in a general and abstract system several concrete and specific instances satisfying the relations stated by the postulates. Pieri is well known for his clever application of these methodological principles to geometrical systems (see Freguglia 1985; Marchisotto 1995). Pieri refers to Padoa's articles for a more detailed analysis of the properties connected to axiomatic systems.

Alessandro Padoa was another member of the group around Peano. Indeed, of that group, he is the only one whose name has remained attached to a specific result in mathematical logic, that is, Padoa's method for proving indefinability (see the following). The result was stated in the talks Padoa gave in 1900 at the two meetings mentioned at the outset (Padoa 1901, 1902). We will follow the "Essai d'une théorie algébrique des nombres entiers, précédé d'une introduction logique a une théorie déductive quelconque". In the Avant-Propos (not translated in van Heijenoort 1967a) Padoa lists a number of notions that he considers as belonging to general logic such as class ("which corresponds to the words: *terminus* of the scholastics, *set* of the mathematicians, *common noun* of ordinary language"). The notion of class is not defined but assumed with its informal meaning. Extensionality for classes is also assumed: "a class is completely known when one knows which individuals belong to it". However, the notion of ordered class he considers as lying outside of general logic. Padoa then states that all symbolic definitions have the form of an equality $y = b$ where y is the new symbol and b is a combination of symbols already known. This is illustrated with the property of being a class with one element. Disjunction and negation are given with their class interpretation. The notions "there is" and "there is not" are also claimed to be reducible to the notions already previously introduced. For instance, Padoa explains that given a class a to say "there is no a" means that the class not-a contains everything, that is, not-$a = (a$ or not-$a)$. Consequently, "there are a['s]" means: not-$a \neq (a$ or not-$a)$. The notion of transformation is also taken as belonging to logic. If a and b are classes and if for any x in a, ux is in b, then u is a transformation from a into b. An obvious principle for transformations u is: if $x = y$ then $ux = uy$. The converse, Padoa points out, does not follow.

This much was a preliminary to the section of Padoa's paper titled "Introduction logique a une théorie déductive quelconque". Padoa makes a distinction between general logic and specific deductive theories. General logic is presupposed in the development of any specific deductive theory. What characterizes a specific deductive theory is its set of primitive symbols and primitive propositions. By means of these, one defines new notions and proves theorems of the system. Thus, when one speaks of indefinability or unprovability, one must always keep in mind that these notions are relative to a specific system and make no sense independently of a specific system. Restating his notion of definition he also claims that definitions are eliminable and thus inessential. Just like Pieri, Padoa also speaks of systems of postulates as a pure formal system on which one can reason without being anchored to a specific interpretation, "for what is necessary to the logical development of a deductive theory is not *the empirical knowledge of the properties of things, but the formal knowledge of relations between symbols*" (1901, 121). It is possible, Padoa continues, that there are several, possibly infinite, interpretations of the system of undefined symbols which verify the system of basic propositions and thus all the theorems of a theory. He then adds:

> The system of undefined symbols can then be regarded as the *abstraction* obtained from all these interpretations, and the generic theory can then be regarded as the *abstraction* obtained from the *specialized theories* that result when in the generic theory the system of undefined symbols is successively replaced by each of the interpretations of this theory. Thus, by means of just one argument that proves a proposition of the generic theory we prove implicitly a proposition in each of the specialized theories. (1901, p. 121)[5]

In contemporary model theory, we think of an interpretation as specifying a domain of individuals with relations on them satisfying the propositions of the system, by means of an appropriate function sending individual constants to objects and relation symbols to subsets of the domain (or Cartesian products of the same). It is important to remark that in Padoa's notion of interpretation something else is going on. An interpretation of a generic system is given by a concrete set of propositions with meaning. In this sense the abstract theory captures all of the individual theories, just as the expression $x + y =$

$y + x$ captures all the particular expressions of the form $2 + 3 = 3 + 2$, $5 + 7 = 7 + 5$, and so on.

Moving now to definitions, Padoa states that when we define a notion in an abstract system we give conditions which the defined notion must satisfy. In each particular interpretation the defined notion becomes individualized, that is, it obtains a meaning that depends on the particular interpretation. At this point Padoa states a general result about definability. Assume that we have a general deductive system in which all the basic propositions are stated by means of undefined symbols:

We say that *the system of undefined symbols is **irreducible** with respect to the system of unproved propositions* when *no symbolic definition of any undefined symbol can be deduced from the system of unproved propositions*, that is, when we cannot deduce from the system a relation of the form $x = a$, where x is one of the undefined symbols and a is a sequence of other such symbols (and logical symbols). (1901, p. 122)

How can such a result be established? Clearly one cannot adduce the failure of repeated attempts at defining the symbol; for such a task, a method for demonstrating the irreducibility is required. The result is stated by Padoa as follows:

To prove that the system of undefined symbols is irreducible with respect to the system of unproved propositions it is necessary and sufficient to find, for any undefined symbol, an interpretation of the system of undefined symbols that verifies the system of unproved propositions and that continues to do so if we suitably change the meaning of only the symbol considered. (1901, p. 122)[6]

Padoa (1902) covers the same ground more concisely but also adds the criterion of compatibility for a set of postulates: "To prove the compatibility of a set of postulates one needs to find *an interpretation of the undefined symbols which verifies simultaneously all the postulates.*" (1902, p. 249) Padoa applied his criteria to showing that his axiomatization of the theory of integers satisfied the condition of compatibility and irreducibility for the primitive symbols and postulates.

We thus see that for Padoa the study of the formal structure of an arbitrary deductive theory was seen as a task of general logic. What can be said about these metatheoretical results in comparison to the later developments? We have already pointed out the different notion of interpretation which informs the treatment. Moreover, the system of logic in the background is never fully spelled out, and in any case it would be a logic containing a good amount of set-theoretic notions. For this reason, some results are taken as obvious that would actually need to be justified. For instance, Padoa claims that if an interpretation satisfies the postulates of an abstract theory, then the theorems obtained from the postulates are also satisfied in the interpretation. This is a soundness principle, which nowadays must be shown to hold for the system of derivation and the semantics specified for the system. For similar reasons the main result by Padoa on the indefinability of primitive notions does not satisfy current standards of rigor. Thus, a formal proof of Padoa's definability theorem had to wait until the works of Tarski (1934–35) for the theory of types and Beth (1953b) for first-order logic (see van Heijenoort 1967a, pp. 118–19, for further details).

1.1.3 Hilbert on Axiomatization

In light of the importance of the work of Peano and his school on the foundations of geometry, it is quite surprising that Hilbert did not acknowledge their work in the *Foundations of Geometry*. Although it is not quite clear to what extent Hilbert was familiar with the work of the Italian school in the last decade of the nineteenth century (Toepell 1986), he certainly could not ignore their work after the 1900 International Congress in Mathematics. In many ways Hilbert's work on axiomatization resembles the level of abstractness also emphasized by Peano, Padoa, and Pieri. The goal of *Foundations of Geometry* (1899) is to investigate geometry axiomatically.[7] At the outset we are asked to give up the intuitive understanding of notions like point, line, or plane and consider any three system of objects and three sorts of relations between these objects (lies on, between, congruent). The axioms only state how these properties relate the objects in question. They are divided into five groups: axioms of incidence, axioms of order, axioms of congruence, axiom of parallels, and axioms of continuity.

Hilbert emphasizes that an axiomatization of geometry must be complete and as simple as possible.[8] He does not make explicit what he means by completeness, but the most likely interpretation of the condition is that the axiomatic system must be able to capture the extent of the ordinary body of geometry. The requirement of simplicity includes, among other things, reducing the number of axioms to a finite set and showing their independence. Another important requirement for axiomatics is showing the consistency of the axioms of the system. This was unnecessary for the old axiomatic approaches to geometry (such as Euclid's) because one always began with the assumption that the axioms were true of some reality and thus consistency was not an issue. But in the new conception of axiomatics, the axioms do not express truths but only postulates whose consistency must be investigated. Hilbert shows that the basic axioms of his axiomatization are independent by displaying interpretations in which all of the axioms except one are true.[9] Here we must point to a small difference with the notion of interpretation we have seen in Pieri and Padoa. Hilbert defines an interpretation by first specifying what the set of objects consists in. Then a set of relations among the objects is specified in such a way that consistency or independence is shown. For instance, for showing the consistency of his axioms, he considers a domain given by the subset of algebraic numbers of the form $\sqrt{1+\omega^2}$ and then specifies the relations as being sets of ordered pairs and ordered triples of the domain. The consistency of the geometrical system is thus discharged on the new arithmetical system: "From these considerations it follows that every contradiction resulting from our system of axioms must also appear in the arithmetic defined above" (29).

Hilbert had already applied the axiomatic approach to the arithmetic of real numbers. Just as in the case of geometry, the axiomatic approach to the real numbers is conceived in terms of "a framework of concepts to which we are led of course only by means of intuition; we can nonetheless operate with this framework without having recourse to intuition". The consistency problem for the system of real numbers was one of the problems that Hilbert stated at the International Congress in 1900:

But above all I wish to designate the following as the most important among the numerous questions which can be asked with regard to the axioms: *To prove that they are not contradictory, that is, that a finite number of logical steps based upon them can never lead to contradictory results.* (1900a, p. 1104)

In the case of geometry, consistency is obtained by "constructing an appropriate domain of numbers such that to the geometrical axioms correspond analogous relations among the objects of this domain". For the axioms of arithmetic, however, Hilbert required a direct proof, which he conjectured could be obtained by a modification of the arguments already used in "the theory of irrational numbers".[10] We do not know what Hilbert had in mind, but in any case, in his new approach to the problem (1905b), Hilbert made considerable progress in conceiving how a direct proof of consistency for arithmetic might proceed. We will postpone treatment of this issue to later (see itinerary VI) and go back to specify what other metatheoretical properties of axiomatic systems were being discussed in these years. By way of introduction to the next section, something should be said here about one of the axioms, which Hilbert in his Paris lecture calls axiom of integrity and later completeness axiom. The axiom says that the (real) numbers form a system of objects which cannot be extended (Hilbert 1900b, p. 1094). This axiom is in effect a metatheoretical statement about the possible interpretations of the axiom system.[11] In the second and later editions of the *Foundations of Geometry*, the same axiom is also stated for points, straight lines and planes:

> (Axiom of completeness) It is not possible to add new elements to a system of points, straight lines, and planes in such a way that the system thus generalized will form a new geometry obeying all the five groups of axioms. In other words, the elements of geometry form a system which is incapable of being extended, provided that we regard the five groups of axioms as valid. (Hilbert 1902, p. 25)

Hilbert commented that the axiom was needed to guarantee that his geometry turn out to be identical to Cartesian geometry. Awodey and Reck (2002a) write, "what this last axiom does, against the background of the others, is to make the whole system of axioms categorical. ... He does not state a theorem that establishes, even implicitly, that his axioms are categorical; he leaves it ... without proofs" (11). The notion of categoricity was made explicit in the important work of the "postulate theorists", to which we now turn.

1.1.4 Completeness and Categoricity in the Work of Veblen and Huntington

A few metatheoretical notions that foreshadow later developments emerged during the early years of the twentieth century in the writings of Huntington and Veblen. Huntington and Veblen are part of a group of mathematicians known as the American postulate theorists (Scanlan 1991, 2003). Huntington was concerned with providing "complete" axiomatizations of various mathematical systems, such as the theory of absolute continuous magnitudes (positive real numbers) (1902) and the theory of the algebra of logic (1904). For instance, in 1902 he presented six postulates for the theory of absolute continuous magnitudes, which he claims to form a complete set. A *complete* set of postulates is characterized by the following properties:

1. The postulates are consistent;
2. They are sufficient;
3. They are independent (or irreducible).

By consistency he means that there exists an interpretation satisfying the postulates. Condition 2 asserts that there is essentially only one such interpretation possible. Condition 3 says that none of the postulates is a "consequence" of the other five.

A system satisfying conditions (1) and (2) we would nowadays call "categorical" rather than "complete". Indeed, the word "categoricity" was introduced in this context by Veblen in a paper on the axiomatization of geometry (1904). Veblen credits Huntington with the idea and Dewey for having suggested the word "categoricity". The description of the property is interesting:

Inasmuch as the terms *point* and *order* are undefined one has a right, in thinking of the propositions, to apply the terms in connection with any class of objects of which the axioms are valid propositions. It is part of our purpose however to show that there is *essentially only one* class of which the twelve axioms are valid. In more exact language, any two classes K and K' of objects that satisfy the twelve axioms are capable of a one-one correspondence such that if any three elements A, B, C of K are in the order ABC, the corresponding elements of K' are also in the order ABC. Consequently any proposition which can be made in terms of points and order either is in contradiction with our axioms or is equally true of all classes that verify our axioms. The validity of any possible statement in these terms is therefore completely determined by the axioms; and so any further axiom would have to be considered redundant. [Note: Even were it not deducible from the axioms by a finite set of syllogisms] Thus, if our axioms are valid geometrical propositions, they are sufficient for the complete determination of Euclidean geometry.

A system of axioms such as we have described is called *categorical*, whereas one to which it is possible to add independent axioms (and which therefore leaves more than one possibility open) is called *disjunctive*. (Veblen 1904, p. 346)

A number of things are striking about the passage just quoted. First of all, we are used to define categoricity by appealing directly to the notion of isomorphism.[12] What Veblen does is equivalent to specifying the notion of isomorphism for structures satisfying his 12 axioms. However, the fact that he does not make use of the word "isomorphism" is remarkable, as the expression was common currency in group theory already in the nineteenth century. The word "isomorphism" is brought to bear for the first time in the definition of categoricity in Huntington (1906–07). There he says that "special attention may be called to the discussion of the notion of isomorphism between two systems, and the notion of a sufficient, or categorical, set of postulates". Indeed, on p. 26 (1906–07), the notion of two systems being isomorphic with respect to addition and multiplication is introduced. We are now very close to the general notion of isomorphism between arbitrary systems satisfying the same set of axioms. The first use of the notion of isomorphism between arbitrary systems we have been able to find is Bôcher (1904, p. 128), who claims to have generalized the notion of isomorphism familiar in group theory. Weyl (1910a) also gives the definition of isomorphism between systems in full generality.

Second, there is a certain ambiguity between defining categoricity as the property of admitting only one model (up to isomorphism) and conflating the notion with a consequence of it, namely, what we would now call semantical completeness.[13] Veblen, however, rightly states that in the case of a categorical theory, further axioms would be redundant even if they were not deducible from the axioms by a finite number of inferences.

Third, the distinction hinted at between what is derivable in a finite number of steps and what follows logically displays a certain awareness of the difference between a semantical notion of consequence and a syntactical notion of derivability and that the two might come apart. However, Veblen does not elaborate on the issue.

Finally, later in the section Veblen claims that the notion of categoricity is also expressed by Hilbert's axiom of completeness as well as by Huntington's notion of sufficiency. In this he reveals an inaccurate understanding of Hilbert's completeness axiom and of its consequences. Baldus (1928) is devoted to showing the noncategoricity of Hilbert's axioms for absolute geometry even when the completeness axiom is added. It is, however, true that in the presence of all the other axioms, the system of geometry presented by Hilbert is categorical (see Awodey and Reck 2002a).

1.1.5 Truth in a Structure

These developments have relevance also for the discussion of the notion of truth in a structure. In his influential paper (1986), Hodges raises several historical issues concerning the notion of truth in a structure, which can now be made more precise. Hodges is led to investigate some of the early conceptions of structure and interpretation with the aim of finding out why Tarski did not define truth in a structure in his early articles. He rightly points out that algebraists and geometers had been studying "Systeme von Dingen" (systems of objects), that is, what we would call structures or models (on the emergence of the terminology, see itinerary VIII). Thus, for instance, Huntington in (1906–07) describes the work of the postulate theorist in algebra as being the study of all the systems of objects satisfying certain general laws: "From this point of view our work becomes, in reality, much more general than a study of the system of numbers; it is a study of any system which satisfies the conditions laid down in the general laws of §1."[14] Hodges then pays attention to the terminology used by mathematicians of the time to express that a structure A obeys some laws and quotes Skolem (1933) as one of the earliest occurrences where the expression "true in a structure" appears.[15]

However, here we should point out that the notion of a proposition being true in a system is not unusual during the period. For instance, in Weyl's (1910a) definition of isomorphism, we read that if there is an isomorphism between two systems, "there is also such a unique correlation between the propositions true with respect to one system and those true with respect to the other, and we can, without falling into error, identify the two systems outright" (Weyl 1910a, p. 301). Moreover, although it is usual in Peano's school and among the American postulate theorists to talk about a set of postulates being "satisfied" or "verified" in a system (or by an interpretation), without any further comments, sometimes we are also given a clarification that shows that they were willing to use the notion of truth in a structure. A few examples will suffice.

Let us look at what might be the first application of the method for providing proofs of independence. Peano in "Principii di geometria logicamente esposti" (1889) has two signs, 1 (for point) and $c \ \varepsilon \ ab$ (c is a point internal to the segment ab). Then he considers three categories of entities with a relation defined between them. Finally he adds:

> Depending on the meaning given to the undefined signs 1 and $c \ \varepsilon \ ab$, the axioms might or might not be satisfied. If a certain group of axioms is verified, then all the propositions that are deduced from them will also be true, since the latter propositions are only transformations of those axioms and of those definitions. (Peano 1889, pp. 77–78)

In 1900, Pieri explains that

> the postulates, just like all conditional propositions *are neither true nor false*: they only express conditions that can sometimes be verified and sometimes not. Thus for instance, the equality $(x+y)^2 = x^2 + 2xy + y^2$ is *true*, if x and y are real numbers and *false* in the case of quaternions (giving for each hypothesis the usual meaning to $+$, $\times$, etc.). (Pieri 1901, pp. 388–89)

In 1906, Huntington:

> The only way to avoid this danger [of using more than is stated in the axioms] is to think of our fundamental laws, not as axiomatic propositions about numbers, but as *blank forms* in which the letters a, b, c, etc. may denote any objects we please and the symbols $+$ and $\times$ any rules of combination; such a blank form will become a proposition only when a definite interpretation is given to the letters and symbols—indeed a true proposition for some interpretations and a false proposition for others... From this point of view our work becomes, in reality, much more general than a study of the system of numbers; it is a study of any system which satisfies the conditions laid down in the general laws of §1. (Huntington 1906–07, pp. 2–3)[16]

In short, it seems that the expression "a system of objects verifies a certain proposition or a set of axioms" is considered to be unproblematic at the time, and it is often read as shorthand for a sentence, or a set of sentences, being true in a system. Of course, this is not to deny that in light of the philosophical discussion emerging from non-Euclidean geometries, a certain care was exercised in talking about "truth" in mathematics, but the issue is resolved exactly by the distinction between axioms and postulates. Whereas the former had been taken to be true tout court, the postulates only make a demand, which might be satisfied or not by particular system of objects (see also on the distinction, Huntington 1911, pp. 171–72).

1.2 Itinerary II. Bertrand Russell's Mathematical Logic

1.2.1 From the Paris Congress to the Principles of Mathematics 1900–1903

At the time of the Paris congress, Russell was mainly familiar with the algebra of logic tradition. He certainly knew the works of Boole, Schröder, and Whitehead. Indeed, the earliest drafts of *The Principles of Mathematics* (1903; POM for short) are based on a logic of part-whole relationship that was closely related to Boole's logical calculus. He also had already realized the importance of relations and the limitations of a subject-predicate approach to the analysis of sentences. This change was a central one in his abandonment of Hegelianism[17] and also led him to the defense of absolute position in space and time against the Leibnizian thesis of the relativity of motion and position, which was the subject of his talk at the International Congress of Philosophy, held in Paris in 1900. However, he had not yet read the works of the Italian school. The encounter with Peano and his school in Paris was of momentous importance for Russell. He had been struggling with the problems of the foundation of mathematics for a number of years and thought that Peano's system had finally shown him the way. After returning from the Paris congress, Russell familiarized himself with the publications of Peano and his school, and it became clear to him that "[Peano's] notation afforded an instrument of logical analysis such as I had been seeking for years" (Russell 1967, p. 218). In Russell's autobiography, he claims that "the most important year of my intellectual life was the year 1900 and the most important event in this year was my visit to the International Congress of Philosophy in Paris" (1989, p. 12). One of the first things Russell did was to extend Peano's calculus with a worked-out theory of relations and this allowed him to develop a large part of Cantor's work in the new system. This he pursued in his first substantial contribution to logic (Russell 1901b, 1902b), which constitutes a bridge between the theory of relations, developed by Peirce and Schröder and Peano's formalization of mathematics. At this stage Russell thinks of relations intensionally, that is, he does not identify them with sets of pairs. The notion of relation is taken as primitive. Then the notion of the domain and co-domain of a relation, among others, are introduced. Finally, the axioms of his theory of relations state, among other things, closure properties with respect to the converse, the complement, the relative product, the union, and the intersection (of relations or classes thereof). He also defines the notion of function in terms of that of relation (however, in POM they are both taken as primitive). In this work, Russell treats natural numbers as definable, which stands in stark contrast to his previous view of number as an indefinable primitive. This led him to the famous definition of "the cardinal number of a class u" as "the class of classes similar to u". Russell arrived at it independently of Frege, whose definition was similar, but he was apparently influenced by Peano, who discussed such a definition in 1901 without, however, endorsing it. In any case, Peano's influence is noticeable in Russell's abandonment of the Boolean leanings of his previous logic in favor of Peano's mathematical logic. Russell now accepted, except for a few changes, Peano's symbolism. One of Peano's advances had been a clear distinction between sentences such as "Socrates is mortal" and "All men are mortal", which were previously conflated as being of the same structure. Despite the similar surface structure, the first one indicates a membership relation between Socrates and the class of mortals, whereas the second indicates an inclusion between classes. In Peano's symbolism we have $s \; \varepsilon \; \psi(x)$ for the first and $\phi(x) \supset_x \psi(x)$ for the second. With this distinction Peano was able to define the relation of subsumption between two classes by means of implication. In a letter to Jourdain in 1910, Russell writes:

Until I got hold of Peano, it had never struck me that Symbolic Logic would be any use for the Principles of mathematics, because I knew the Boolean stuff and found it useless. It was Peano's ε, together with the discovery that relations could be fitted into his system, that led me to adopt Symbolic Logic. (Grattan-Guinness 1977, p. 133)

What Peano had opened for Russell was the possibility of considering the mathematical concepts as definable in terms of logical concepts. In particular, an analysis in terms of membership and implication is instrumental in accounting for the generality of mathematical propositions. Russell's logicism finds its first formulation in a popular article written in 1901 where he claims that all the indefinables and indemonstrables in pure mathematics stem from general logic: "All pure mathematics—Arithmetic, Analysis, and Geometry—is built up of the primitive ideas of logic, and its propositions are deduced from the general axioms of logic" (1901a, 367).

This is the project that informed the *Principles of Mathematics* (1903). The construction of mathematics out of logic is carried out by first developing arithmetic through the definition of the cardinal number of a class as the class of classes similar to it. Then the development of analysis is carried out by defining real numbers as sets of rationals satisfying appropriate conditions. (For a detailed reconstruction see, among others, Vuillemin 1968; Rodriguez-Consuegra 1991; Landini 1998; Grattan-Guinness 2000). The main difficulty in reconstructing Russell's logic at this stage consists in the presence of logical notions mixed with linguistic and ontological categories (denotation, definition). Moreover, Russell does not present his logic by means of a formal language.

After Russell finished preparing POM, he also began studying Frege with care (around June 1902). Under his influence, Russell began to notice the limitations in Peano's treatment of symbolic logic, such as the lack of different symbols for class union and the disjunction of propositions, or material implication and class inclusion. Moreover, he changed his symbolism for universal and existential quantification to $(x)f(x)$ and $(Ex)f(x)$. He adopted from Frege the symbol $\vdash$ for the assertion of a proposition. His letter to Frege of June 16, 1902, contained the famous paradox, which had devastating consequences for Frege's system:

Let w be the predicate: to be a predicate that cannot be predicated of itself. Can w be predicated of itself? From each answer its opposite follows. Therefore we must conclude that w is not a predicate. Likewise there is no class (as a totality) of those classes which, each taken as a totality, do not belong to themselves. From this I conclude that under certain circumstances a definable collection does not form a totality. (Russell 1902a, p. 125)

The first paradox does not involve the notion of class but only that of predicate. Let Imp(w) stand for "w cannot be predicated of itself", that is, $\sim w(w)$. Now we ask: Is Imp(Imp) true or $\sim$Imp(Imp)? From either one of the possibilities the opposite follows. However, what is known as Russell's paradox is the second one offered in the letter to Frege. In his work *Grundgesetzte der Arithmetik* (Frege 1893, 1903), Frege had developed a logicist project that aimed at reconstructing arithmetic and analysis out of general logical laws. One of the basic assumptions made by Frege (Basic Law V) implies that every propositional function has an extension, where extensions are a kind of object.

In modern terms we could say that Frege's Basic Law V implies that for any property $F(x)$ there exists a set $y = \{x : F(x)\}$. Russell's paradox consists in noticing that for the specific $F(x)$ given by $x \notin x$, Frege's principle leads to asserting the existence of the set $y = \{x : x \notin x\}$. Now if one asks whether $y \in y$ or $y \notin y$ from either one of the assumptions one derives the opposite conclusion. The consequences of Russell's paradox for Frege's logicism and Frege's attempts to cope with it are well known, and we will not recount them here (see Garciadiego 1992). Frege's proposed emendation to his Basic Law V, while consistent, turns out to be inconsistent as soon as one postulates that there are at least two objects (Quine 1955a).[18]

Extensive research on the development that led to Russell's paradox has shown that Russell already obtained the essentials of his paradox in the first half of 1901 (Garciadiego 1992; Moore 1994) while working on Cantor's set theory. Indeed, Cantor himself already noticed that treating the cardinal numbers (resp., ordinal numbers) as a completed totality would lead to contradictions. This led him to distinguish, in letters to Dedekind, between "consistent multiplicities", that is, classes that can be considered as completed totalities, from "inconsistent multiplicities", namely, classes that cannot, on pain of contradiction, be considered as completed totalities. Unaware of Cantor's distinction between consistent and inconsistent multiplicities Russell in 1901 convinced himself that Cantor had "been guilty of a very subtle fallacy" (1901a, 365). His reasoning was that the number of all things is the greatest of all cardinal numbers. However, Cantor proved that for every cardinal number there is a cardinal number strictly bigger than it. Within a few months this conundrum led to Russell's paradox. In POM we find, in addition to the two paradoxes we have discussed, also a discussion of what is now known as Burali-Forti's paradox (Moore and Garciadiego 1981).

In POM Russell offered a tentative solution to the paradoxes: the theory of types. The theory of types contained in POM is a version of what is now called the simple theory of types, whereas the one offered in Russell (1908) (and *Principia Mathematica*, Whitehead and Russell 1910, 1912, 1913) is called the ramified theory of types (on the origin of these terms, see Grattan-Guinness 2000, p. 496). Russell's exposition of the theory of types (in 1903 as well as later) is far from perspicuous, and we will simply give the gist of it. The basic idea is that every propositional function $\phi(x)$ has a range of significance, that is, a range of values of x for which it can be meaningfully said to be true or false:

Every propositional function $\phi(x)$—so it is contended—has, in addition to its range of truth, a range of significance, that is, a range within which x must lie if $\phi(x)$ is to be a proposition at all, whether true or false. This is the first point in the theory of types; the second point is that ranges of significance form *types*, that is, if x belongs to the range of significance of $\phi(x)$, then there is a class of objects, the *type* of x, all of which must also belong to the range of significance of $\phi(x)$, however ϕ may be varied. (Russell 1903, p. 523)

The lowest type, type 0, is the type of all individuals (objects which are not "ranges"). Then we construct the class of all classes of individuals, namely, type 1. Type 2 is the class of all classes of classes of type 1, and so on. This gives an infinite hierarchy of types for which Russell specifies that "in $x \, \varepsilon \, u$ the u must always be of a type higher by one than x" (517). In this way $x \, \varepsilon \, x$ and its negation are meaningless and thus it is not possible for Russell's paradox to arise, as there are no ranges of significance, that is, types, for meaningless propositions. The other paradoxes considered by Russell are also blocked by the postulated criteria of meaningfulness. The presentation of the theory in POM is vastly complicated by the need to take into account relations and by a number of assumptions which go against the grain of the theory, for instance, that "$x \, \varepsilon \, x$ is sometimes significant" (525).

Russell, however, abandoned this version of the theory of types and returned to the theory of types only after trying a number of different theories. His abandonment of this theory is explained by the fact that the theory does not assign types to propositions and thus, as Russell pointed out to Frege (letter of September 29, 1902), this allows for the generation of a paradox through a diagonal argument applied to classes of propositions. His search for a solution to the paradoxes played a central role in his debate with Poincaré concerning impredicative definitions, to which we now turn.

1.2.2 Russell and Poincaré on Predicativity

In the wake of Russell's paradoxes, many more paradoxes were brought to light,[19] the most famous being Berry's paradox concerning the least ordinal number not definable in a finite number of words, Richard's paradox (see following discussion), and the König–Zermelo contradiction. The latter concerned a contradiction between König's "proof" that the continuum cannot be well ordered and Zermelo's (1904) proof that every set can be well ordered. Many more were added, and one finds a long list of paradoxes in the opening pages of Russell (1908). What the paradoxes had brought to light was that not every propositional function defines a class. Russell's paradox, for instance, shows that there is a propositional function, or "norm", $\phi(x)$ for which we cannot assume the existence of $\{ x : \phi(x) \}$. When trying to spell out which propositional functions define classes and which do not, Russell proposed in 1906 the distinction between predicative and nonpredicative norms:

> We have thus reached the conclusion that some norms (if not all) are not entities which can be considered independently of their arguments, and that some norms (if not all) do not define classes. Norms (containing one variable) which do not define classes I propose to call *nonpredicative*; those which do define classes I shall call *predicative*. (Russell 1906b, p. 141)

At the time Russell was considering various theories as possible solutions to the paradoxes and in the 1906 article he mentions three of them: the "no-classes" theory, the "zig-zag" theory, and the "limitation of size" theory. Accordingly, the Russellian distinction between predicative and nonpredicative norms gives rise to extensionally different characterizations depending on the theory under consideration. Russell mentions "simplicity" as the criterion for predicativity in the "zig-zag" theory and "limitation of size" in the "limitation of size" theory. In the case of the "no-classes" theory, no propositional function is predicative as classes are eliminated through contextual definitions. However, it is only with Poincaré's reply to Russell that we encounter the notion of predicativity that was at the center of their later debate.[20] Poincaré's discussion also takes its start from the paradoxes but rejects Russell's suggestion as to what should count as a predicative propositional function, on account of the vagueness of Russell's proposal. Poincaré suggested that nonpredicative classes are those that contain a vicious circle. Poincaré did not provide a general account, but he clarified the proposal through a discussion of Richard's paradox (Richard 1905). Richard's paradox takes its start by a consideration of the set E of all numbers that can be defined by using expressions of finite length over a finite vocabulary. By a diagonal process one then defines (by appealing explicitly to E) a new number N which is not in the list. As the definition of N is given by a finite expression using exactly the same alphabet used to generate E, it follows that N is in E. But by construction N is not in E. Thus N is and is not in E. Poincaré's way out was to claim that in defining N one is not allowed to appeal to E, as N would be defined in terms of the totality to which it belongs. Thus, according to Poincaré, reference to infinite totalities is the source of the nonpredicativity:

It is the belief in the existence of actual infinity that has given birth to these nonpredicative definitions. I must explain myself. In these definitions we find the word *all*, as we saw in the examples quoted above. The word *all* has a very precise meaning when it is a question of a finite number of objects; but for it still to have a precise meaning when the number of the objects is infinite, it is necessary that there should exist an actual infinity. Otherwise all these objects cannot be conceived as existing prior to their definition, and then, if the definition of a notion N depends on *all* the objects A, it may be tainted with the vicious circle, if among the objects A there is one that cannot be defined without bringing in the notion N itself. (Poincaré 1906, p. 294)

Poincaré was appealing to two different criteria in his diagnosis. On the one hand he considered a definition to be nonpredicative if the definiendum in some way involves the object being defined. The second criterion asserts the illegitimacy of quantifying over infinite sets.[21]

Russell, in "Les Paradoxes de la Logique" (1906a), agreed with Poincaré's diagnosis that a vicious circle was involved in the paradoxes, but he found Poincaré's solution to lack the appropriate generality:

I recognize, however, that the clue to the paradoxes is to be found in the vicious-circle suggestion; I recognize further this element of truth in M. Poincaré's objection to totality, that whatever in any way concerns *all* or *any* or *some* (undetermined) of the members of a class must not be itself one of the members of a class. In M. Peano's language, the principle I want to advocate may be stated: "Whatever involves an apparent variable must not be among the possible values of that variable." (Russell 1973, p. 198)

Russell's objection to Poincaré was essentially that Poincaré's proposal was not supported by a general theory and thus seemed ad hoc. Moreover, he pointed out that in many paradoxes infinite totalities play no role and thus he concluded that "the contradictions have no essential reference to infinity". Russell's position brought to light the coexistence of different criteria in Poincaré's notion of predicativity. However, what exactly the vicious circle principle amounted to remained vague also in Russell's work, which displayed several nonequivalent versions of the principle. We resume discussion of predicative mathematics in the section on set theory, and we move now to a discussion of the last element we need to discuss the ramified theory of types, the theory of denoting.

1.2.3 On Denoting

One of the key elements in the formalization of mathematics given in *Principia* is the contextual definition of some of the concepts appearing in mathematics. In other words, not every single mathematical concept is individually defined. Rather, there are concepts that receive a definition only in the context of a proposition in which they appear. The philosophical and technical tools for dealing with contextual definitions was given by the theory of denoting (Russell 1905; see de Rouilhan 1996; Hylton 1990). The theory of denoting allowed Russell to account for denoting phrases without having to assume that denoting phrases necessarily refer to an object. A characterization of denoting phrase is given by a list of examples. The examples include "a man, some man, any man, every man, all men, the present King of England, the present King of France". Whether a phrase is denoting depends solely on its form. However, whether a denoting phrase successfully denotes something does not depend merely on its form. Indeed, although "the present King of England" and "the present King of France" have the same form, only the first one denotes an object (at the time Russell is writing). Expressions of the form "the so-and-so", a very important subclass of denoting expressions, are called definite descriptions. Russell's theory consisted in parsing a definite description such as "the present King of France is bald" as "there exists a unique x such that x is King of France and x is bald". In this way "the so-and-so" is meaningful only in the context of a sentence and does not have meaning independently: "According to the view which I advocate, a denoting phrase is essentially part of a sentence, and does not, like most single words, have any significance on its own account." (Russell 1905; 1973, p. 113) It is hard to overestimate the importance of this analysis for the foundations of mathematics, as denoting phrases, and definite descriptions in particular, are ubiquitous in mathematical practice. In *Principia*, Russell and Whitehead will talk of "incomplete symbols" which do not have an independent definition but only a "definition in use", which determines their meaning only in relation to the context in which they appear. We are now ready to discuss the basic structure of the ramified theory of types.

1.2.4 Russell's Ramified Type Theory

Poincaré's criticism of impredicative definitions forced Russell and Whitehead to reconsider some of the work they had previously carried out. In particular, Poincaré had criticized the proof of mathematical induction (due to Russell) presented in Whitehead (1902). Poincaré found the definition of an inductive number as the intersection of all recurrent classes (i.e., a class containing zero and closed under successor) to be impredicative. Russell agreed with Poincaré's claim that a vicious circle is present in impredicative definitions and, as we mentioned, presented several theories as possible solutions for the problems raised by the paradoxes (Russell 1906b). Among the theories developed in this period, the substitutional theory (an implementation of the no-classes theory) has been recently subjected to detailed scrutiny (see de Rouilhan 1996, Landini 1998). However, these theories were eventually abandoned and it was the theory of types, as presented in (1908) and (1910), that became Russell's final choice for a solution to the paradoxes. Let us follow the exposition of Russell (1908) to convey the basic ideas of ramified type theory. Russell begins with a long list of paradoxes: Epimenides ("the liar paradox"), Russell's paradox for classes, Russell's paradox for relations, Berry's paradox on "the least integer not nameable in less than nineteen syllables", the paradox of "the least undefinable ordinal", Richard's paradox, and Burali-Forti's contradiction. Russell detects a common feature to all these paradoxes, which consists in the occurrence of a certain "self-reference or reflexiveness": "Thus all our contradictions have in common the assumption of a totality such that, if it were legitimate, it would at once be enlarged by new members defined in terms of itself." (Russell 1908, p. 155) Thus, the rule adopted by Russell for avoiding the paradoxes, known as the vicious circle principle, reads: "whatever involves *all* of a collection must not be one of a collection". Russell gives several formulations of the principle. A different formulation reads: "If, provided a certain collection had a total, it would have members only definable in terms of that total, then the said collection has no total" (Russell 1908, p. 155).[22]

Notice that the vicious circle principle implies that "no totality can contain members defined in terms of itself". This excludes impredicative definitions. However, Russell insists that the principle is purely negative and that a satisfactory solution to the paradoxes must be the result of a positive development of logic. This development of logic is the ramified theory of types. The second remark concerns the issue of when collections can be considered as having a total. By claiming that a collection has no total, Russell means that statements about *all* its members are nonsense. This leads Russell to a lengthy analysis of the difference between "any" and "all". For him the condition of possibility for saying something about all objects of a collection rests on the members of that collection as being of the same type. The partition of the universe into types rests on the intuition that to make a collection, the objects collected must be logically homogeneous. The distinction between "all" and "any" is expressed, roughly, by the use of a universally bound variable—which ranges over a type—versus a free variable, whose range is not bounded by a type.

In this way we arrive at the core of the ramified theory of types. Unfortunately, the exposition of the theory, both in (1908) and in *Principia*, suffers from the lack of a clear presentation.[23] We will not give a detailed technical exposition here, but only try to convey the gist of the theory with reference to the effect of the theory on the structuring of the universe into types. The distinction into types, however, can also be applied to propositions and propositional functions.

A type is defined by Russell as "the range of significance of a propositional function, that is, as the collection of arguments for which the said function has values" (Russell 1908, p. 163). We begin with the lowest type, which is simply the class of individuals. In (1908) the individuals are characterized negatively as being devoid of logical complexity, and hence as different from propositions and propositional functions. This is important to exclude the possibility that quantification over individuals might already involve a vicious circle. Type 1 will contain all the (definable) classes of individuals; type 2 all the (definable) classes of classes of individuals; and so on. What we have described is a form of the simple theory of types. This theory already takes care of some of the paradoxes. For instance, if x is an object of type n and y an object of type $n + 1$ it makes sense to write $x \in y$, but it makes no sense to write $x \in x$. Thus, in terms of class existence we already exclude the formation of problematic classes at the syntactic level by declaring that expressions of the form $x \in y$ are significant only if x is of type n, for some n, and y is of type $n + 1$. This significantly restricts the classes that can be formed.

However, the simple theory of types is not enough to guarantee that the vicious circle principle is satisfied. The complication arises due to the following possibility. One might define a class of a certain type, say, n, by quantifying, in the propositional function defining the class, over collections of objects which might be of higher type than the one being defined. It is thus essential to keep track of the way classes are defined and not only, so to speak, of their ontological complexity.[24] This leads to a generalized notion of **type** (boldface, to distinguish it from type as in the simple theory) for the ramified theory. Rather than giving the formal apparatus for capturing the theory, we will exemplify the main intuition by considering a few examples.

Type 0: the totality of individuals.

Type 1.0: the totality of classes of individuals that can be defined using only quantifiers ranging over individuals (**type** 0).

Type 2.0: the totality of classes of classes of individuals that can be defined by using only quantifiers ranging over objects of **type** 1.0 and **type** 0.

Type 2.1.0: the totality of classes of classes of individuals of **type** 1.0 that can be defined using only quantifiers ranging over elements in **type** 1.0 and in **type** 0.

And so on. Let us say that **type** 0 corresponds to order 0, **type** 1.0 to order 1, and that **type** 2.1.0 and **type** 2.0 are of order 2.

This system of **types** satisfies the vicious circle principle, as defining an object by quantifying over a previously given totality will automatically give a class of higher **type**. But this also implies that the development of mathematics in the ramified theory becomes unnatural. In particular, real numbers will appear at different stages of definition. For instance, given a class of real numbers bounded above, the least upper bound principle will, in general, generate a real number of higher **type** (as the definition of the least upper bound requires a quantification over classes containing the given class of reals). To provide a workable foundation for analysis, Russell is then forced to postulate the so-called axiom of reducibility. For its statement we need the notion of a *predicative propositional function* (notice that this notion of predicative is not to be confused with that which is at stake in impredicative definitions). A propositional function $\phi(x)$ is predicative if its order is one higher than that of its argument. To use the foregoing examples, **type** 1.0 and **type** 2.1.0 are predicative, but **type** 2.0 is not. The axiom of reducibility says that each propositional function is extensionally equivalent to a predicative function. Since predicative functions occupy a well-specified place in the hierarchy of **types**, the axiom has the consequence of rendering many of the **types** redundant, at least extensionally. Thus, to go back to our example, the axiom implies that all classes of **type** 2.0 are all extensionally equivalent to classes in **type** 1.0. The net effect of the axiom for the foundations of the real numbers is that it reestablishes the possibility of treating the reals as being all at the same level. In particular, the least upper bound of a class of reals will also be given, extensionally, at the same level as the class used in generating it. However, it has been often observed (most notably in Ramsey 1925), that the axiom of reducibility defeats the purpose of having a ramified hierarchy in the first place. Indeed, with the axiom of reducibility, the ramified theory is equivalent to a form of simple type theory.

1.2.5 The Logic of Principia

Russell and Whitehead's project consisted in showing that all of mathematics could be developed through appropriate definitions in the system of logic defined in *Principia*. One must distinguish here between the development of arithmetic, analysis, and set theory on the one hand and the development of geometry on the other hand. Indeed, for the former theories the axioms of the theory are supposed to come out to be logical theorems of the system of logic, thereby showing that arithmetic, analysis, and set theory are basically developments of pure logic. However, the logicist reconstruction of these branches of mathematics could only be carried out by assuming the axioms of choice ("the multiplicative axiom"), infinity, and reducibility among the available "logical" principles. This is one of the major reasons for the worries about the prospects of logicism in the twenties and thirties (see Grattan-Guinness 2000).

The situation for geometry, whose development was planned for the fourth volume of *Principia* (never published), is different. The approach there would have been a conditional one. The development of geometry in the system of logic given in *Principia* would have shown that the theorems of geometry can be obtained in the system of *Principia* under the assumption of the axioms of geometry. As these axioms say something about certain specific types of relations holding for the geometrical spaces in question, the development of geometry would result in conditional theorems of the logic of *Principia* with the form "if A then p", where A expresses the set of geometrical axioms in question and p is a theorem of geometry.

In both cases, the inferential patterns must be regulated by a specific set of inferential rules. The development of mathematical logic presented in part I of *Principia* (85–326) divides the treatment into three sections. Section A deals with the theory of deduction and develops the propositional calculus. Section B treats the theory of apparent variables (i.e., quantificational logic for types) and sections C, D, and E the logic of classes and relations. While the treatment is supposed to present the whole of logic, its organization already permits one to isolate interesting fragments of the logic presented. In particular, the axiomatization of propositional logic presented in section A of part I is the basis of much later logical work. Russell and Whitehead take the notion of negation and disjunction as basic. They define material implication, $A \supset B$, as $\sim A \vee B$. The axioms for the calculus of propositions are:

1. Anything implied by a true premiss is true.
2. $\vdash : p \vee p . \supset . p \vee q$

3. ⊢ :q . ⊃ .p ∨ q
4. ⊢ :p ∨ q . ⊃ .q ∨ p
5. ⊢ :p ∨ (q ∨ r) . ⊃ .q ∨ (p ∨ r)
6. ⊢ :.q ⊃ r . ⊃ :p ∨ q . ⊃ .p ∨ r

The sign "⊢" is the sign of assertibility (taken from Frege), and the dotted notation (due to Peano) is used instead of the now common parentheses. The only rule of inference is modus ponens; later Bernays pointed out the need to make explicit the rule of substitution, used but not explicitly stated in *Principia*. The quantificational part cannot be formalized as easily due to the need to specify in detail the type theoretic structure. This also requires checking that the propositional axioms presented remain valid when the propositions contain apparent variables (see Landini 1998 for a careful treatment).

Among the primitive propositions of quantificational logic is the following:

$$\vdash : \phi x . \supset .(\exists z).\phi z \tag{9.1}$$

About it, Russell and Whitehead say that "practically, the above primitive proposition gives the only method of proving 'existence-theorems': in order to prove such theorems, it is necessary (and sufficient) to find some instance in which an object possesses the property in question" (1910, 131). This is, however, wrong and it will be a source of confusion in later debates (see Mancosu 2002).

1.2.6 Further Developments

The present itinerary on Russell does not aim at providing a full overview of either Russell's development in the period in question nor of the later discussion on the nature of logicism. The incredible complexity of Russell's system and the wealth of still unpublished material make the first aim impossible to achieve here. As evidenced by the citations throughout this itinerary (limited to the major recent books), in the past decade there has been an explosion of scholarly work on Russell's contributions to logic and mathematical philosophy. Moreover, the history of logicism as a program in the foundations of mathematics in the 1920s would require a book on its own.[25] We thus conclude with a general reflection on the importance of *Principia* for the development of mathematical logic proper.

It is hard to overestimate the importance of *Principia* as the first worked-out example of how to reconstruct in detail from a limited number of basic principles the main body of mathematics (even though *Principia*, despite its length, does not even manage to treat the calculus in full detail). However, it became evident that a number of problematic principles—such as infinity, choice, and reducibility—were needed to carry out the reconstruction of mathematics within logic. These existential principles were not obviously logical, and in the case of reducibility seemed rather ad hoc. The further development of logicism in the twenties can be seen as an attempt to work out a solution to such problems. One possible solution was to simply reject the axiom of reducibility and accept that not all of classical mathematics could be obtained in the ramified theory of types. This was the strategy pursued by Chwistek in a number of articles from the early twenties. A second solution was offered by Ramsey's radical rethinking of the logicist project. Ramsey (1925) distinguished between mathematical and semantical antinomies. The former have to do with concepts of mathematics, which are purely extensional whereas the latter involve intensional notions, like definability, which do not belong to mathematics. By refusing to consider the semantical antinomies of relevance to mathematics, Ramsey was able to propose a simple theory of types that could account for classical mathematics and which he claimed took care of all the mathematical antinomies. This, however, came at the cost of excluding intensional notions from the realm of logic.

However, it can be said that despite their interest for the history of logicism, these developments did not, properly speaking, affect the development of mathematical logic for the period we are considering. What was the influence of *Principia* for developments in mathematical logic in the 1910s?

First of all, we have a number of investigations related to the propositional part of *Principia*. Among the results to be mentioned are Sheffer's (re)discovery (1913) of the possibility of defining all Boolean propositional connectives starting from the notion of incompatibility (Sheffer's stroke). Using Sheffer's stroke, Nicod (1916–19) was able to provide an axiomatization of the propositional calculus with only one axiom. This work was generalized in the early twenties in Göttingen by extending it to the quantificational part of the calculus. This development also marks the beginning of combinatory logic. A systematic analysis of the propositional part of *Principia* was also carried out in Bernays' *Habilitationsschrift* (1918). Much of this work required a metamathematical approach to logic, which was absent from *Principia* (on all this, see itineraries V and VIII).[26] *Principia* was also influential in the development of systems of logic that were strongly opposed to some of the major assumptions therein contained. In the 1910s the most important work in this direction was Lewis' development of systems of strict implication (Lewis 1918).

However, the major influence of *Principia* might simply be that of having established higher-order logic as the paradigm of logic for the next two decades. While it is true that first-order logic emerges as a (more or less) natural fragment of *Principia* (see itinerary IV) most logicians well into the thirties (Carnap, Gödel, Tarski, Hilbert–Ackermann) still considered higher-order logic the appropriate logic for formalizing mathematical theories (see Ferreiros 2001 for extensive treatment).

1.3 Itinerary III. Zermelo's Axiomatization of Set Theory and Related Foundational Issues

The history of set theory during the first three decades of the twentieth century has been extensively researched. One area of investigation is the history of set theory as a mathematical discipline and its influence on other areas of mathematics. A second important topic is the relationship between logic and set theory. Finally, much attention has been devoted to the axiomatizations of set theory, and even to the pluralities of set theories (naïve set theory, Zermelo, von Neumann, intuitionistic set theory, etc.). Here we focus on Zermelo's axiomatization.

1.3.1 The Debate on the Axiom of Choice

At the beginning of the century, set theory had already established itself both as an independent mathematical theory as well as in its applications to other branches of mathematics, in particular analysis.[27] In his address to the mathematical congress in Paris, Hilbert singled out the continuum problem as one of the major problems for twentieth-century mathematics. One of the problems that had occupied Cantor, which he was never able to solve, was that of whether every set is an aleph, or equivalently, whether every set can be well ordered. Julius König (1904) presented a proof at the third International Congress of Mathematicians in Heidelberg claiming that the continuum cannot be well ordered. A key step of the proof made use of a result by Felix Bernstein claiming that $\aleph_\alpha^{\aleph_\beta} = 2^{\aleph_\beta} \aleph_\alpha$. But after scrutinizing Bernstein's result in the wake of König's talk, Hausdorff (1904) showed that it holds only when α is a successor ordinal. Soon thereafter, Zermelo showed that every set can be well ordered (Zermelo 1904).[28] Let us recall that an ordered set F is well ordered if and only if every nonempty subset of it has a least element (under the ordering). Zermelo's proof appealed to

> the assumption that coverings γ actually do exist, hence upon the principle that even for an infinite totality of sets there are always mappings that associate with every set one of its elements, or, expressed formally, that the product of an infinite totality of sets, each containing at least one element, itself differs from zero. This logical principle cannot, to be sure, be reduced to a still simpler one, but is applied without hesitation everywhere in mathematical deduction. (Zermelo 1904, p. 141)[29]

Let M be the arbitrary set for which a well-ordering needs to be established. A covering γ for M in Zermelo's proof is what we would call a choice function, which for an arbitrary subset M' of a set M yields an element $\gamma(M')$ of M, called the distinguished element of M'. It is under the assumption of existence of such a covering that Zermelo establishes the existence of special sets called γ-sets. A γ-set is a set M_γ included in M which is well ordered and such that if $a \in M_\gamma$ and if $A = \{x : x \in M_\gamma \text{ and } x < a \text{ in the well ordering of } M_\gamma\}$, then a is the distinguished element of $M - A$ according to the covering γ. Zermelo then shows that the union of all γ-sets, L_γ, is a γ-set and that $L_\gamma = M$. Thus M can be well ordered.

Zermelo's proof immediately gave rise to a major philosophical and mathematical discussion.[30] The main exchange was published by the *Bulletin de la Société Mathématique de France* in 1905 and consisted of five letters exchanged among Baire, Borel, Lebesgue, and Hadamard (1905). Baire, Borel, and Lebesgue shared certain constructivist tendencies, which led them to object to Zermelo's use of the principle of choice, although in their actual mathematical practice they often made use (implicitly or explicitly) of Cantorian assumptions, including the principle of choice. For instance, Lebesgue's proof of the countable additivity of the measurable subsets of the real line relies on the principle of choice for countable collections of sets. Hadamard took a more liberal stand.

The debate began with an article by Borel, which appeared in *Mathematische Annalen* (Borel 1905). Borel claimed that Zermelo's proof had only shown the equivalence between the well-ordering problem for an arbitrary set M and the problem of choosing an arbitrary element from each subset of M. However, Borel did not accept this as a solution to the first problem, for the postulation of a choice function required by Zermelo was, if anything, even more problematic than the problem one began with. He found the application of the principle to uncountably many sets particularly problematic, but allowed for the possibility that the principle might be acceptable when we are dealing with countable collections of sets. Hadamard's reply to Borel's article defended Zermelo's principle. In the process of defending Zermelo's application of the principle, Hadamard also drew a few important distinctions. For instance, he distinguished between reasonings in which each choice depends on the previous ones (dependent choice) from Zermelo's principle, which postulated simultaneous independent choices. Moreover, he objected to Borel that he saw no essential difference between postulating the principle for a countable or an uncountable collection of sets. Finally, he also pointed at the fact that one had to distinguish between whether the choice could be made "effectively" or simply postulated to exist. He emphasized the essential difference between showing that an object (say, a function) exists, without however specifying the object, and actually providing a unique specification of the object. Hadamard claimed that whether one raises the first or the second problem essentially changes the nature of the mathematical question being investigated. The most radical position was taken by Baire, who defended a strong finitism and refused to accept one of the basic principles underlying Zermelo's proof. Indeed, he claimed that if a set A is given it does not follow that the set of its subsets can also be considered as given. And thus, he rejected that part of Zermelo's argument that allowed him to pick an element from every subset of the given set. Baire claimed that Zermelo's principle was consistent but that it simply lacked mathematical meaning. Lebesgue's point of view also emphasized the issue of definability of mathematical objects. He asked: "Can one prove the existence of a mathematical object without defining it?" He also defended a constructivist attitude and claimed that the only true claim of existence in mathematics must be obtained by defining the object uniquely. In the last of the five letters, Hadamard rejected the constructivist positions of Baire, Borel, and Lebesgue and claimed that mathematical existence does not have to rely on unique definability. He clearly set out the two different conceptions of mathematics that were at the source of the debate. On one conception, the constructivist one, mathematical objects are said to exist if they can be defined or constructed. On the other conception, mathematical existence is not dependent on our abilities to either construct or define the object in question. While allowing the reasonableness of the constructivist position, Hadamard considered it to rely on psychological and subjective considerations that were foreign to the true nature of mathematics.

The debate focused attention not only on the major underlying philosophical issues but also on the important distinctions that one could draw between different forms of the principle of choice. The positions of Baire, Borel, and Lebesgue on definability remained vague but influenced later work by Weyl, Skolem, and others.

Zermelo's proof was widely discussed and criticized. In the article "A new proof of the possibility of a well-ordering" (1908b), Zermelo gave a new proof of the well-ordering theorem, by relying on a generalization of Dedekind's chains, and gave a full reply to the criticism that had been raised against his previous proof (by, among others, Borel, Peano, Poincaré, König, Jourdain, Bernstein, and Schoenflies). We focus on Poincaré's objections.

Poincaré's criticism of Zermelo's proof occurred in his discussion (1906) of logicism and set theory. In particular, he had objected to the formation of impredicative sets which occur in the proof. Recall that in the final part of the first proof Zermelo defined the set L_γ as the union of all γ-sets, that is,

$$L_\gamma = \{ x : \text{for some } \gamma\text{-set } Y, x \in Y \}$$

According to Poincaré, this definition is objectionable because to determine whether an element x belongs to L_γ, one needs to go through all the γ-sets. But among the γ-sets is L_γ itself, and thus a vicious circle is involved in the procedure. Zermelo replied to Poincaré claiming that his critique would "threaten the existence of all of mathematics" (Zermelo 1908b, p. 198). Indeed, impredicative definitions and procedures occur not only in set theory but in the most established branches of mathematics, such as analysis:

> Now, on the one hand, proofs that have this logical form are by no means confined to set theory; exactly the same kind can be found in analysis wherever the maximum or the minimum of a previously defined "completed" set of numbers Z is used for further inferences. This happens, for example, in the well-known Cauchy proof of the fundamental theorem of algebra, and up to now it has not occurred to anyone to regard this as something illogical. (Zermelo 1908b, pp. 190–91)

Poincaré claimed that there was an essential difference between Cauchy's proof (in which the impredicativity is eliminable) and Zermelo's proof. This debate forced Poincaré to be more explicit on his notion of predicativity (see Heinzmann 1985) and contributed to Zermelo's spelling out of the axiomatic structure of set theory. After presenting the axioms of Zermelo's set theory we will return to the issue of impredicativity.

1.3.2 Zermelo's Axiomatization of Set Theory

Another set of objections that were raised against Zermelo's proof suggested the possibility that Zermelo's assumption might end up generating the set of all ordinals W and therefore fall prey to Burali-Forti's antinomy.[31] Zermelo claimed that a suitable restriction of the notion of set was enough to avoid the antinomies and that in 1904 he had restricted himself "to principles and devices that have not yet by themselves given rise to any antinomy" (Zermelo 1908b, p. 192). These principles were the subject of another article that contains the first axiomatization of set theory (Zermelo 1908c). Zermelo begins by claiming that no solution to the problem of the paradoxes has yielded a simple and convincing system. Rather than starting with a general notion of set, he proposes to distill the axioms of set theory out of an analysis of the current state of the subject. The treatment has to preserve all that is of mathematical value in the theory and impose a restriction on the notion of set so that no antinomies are generated. Zermelo's solution consists in an axiom system containing seven axioms. The main intuition behind his approach to set theory is one of "limitation of size", that is, sets which are "too large" will not be generated by the axioms. This is ensured by the separation axiom, which in essence restricts the possibility of obtaining new sets only by isolating (definable) parts of already given sets. Following Hilbert's axiomatization of geometry, Zermelo begins by postulating the existence of a domain $\mathfrak{B}$ of individuals, among which are the sets, on which some basic relations are defined. The two basic relations are equality (=) and membership ($\in$). For sets A and B, A is said to be a subset of B if and only if every element of A is an element of B. The key definition concerns the notion of definite property:

> A question or assertion $\mathfrak{E}$ is said to be *definite* if the fundamental relations of the domain, by means of the axioms and the universally valid laws of logic, determine without arbitrariness whether it holds or not. Likewise a "propositional function" $\mathfrak{E}(x)$, in which the variable term x ranges over all individuals of a class $\mathfrak{K}$, is said to be definite if it is definite for *each single* individual x of the class $\mathfrak{K}$. (Zermelo 1908c, p. 201)

This definition plays a central role in the axiom of separation (see the following) which forms the cornerstone of Zermelo's axiomatic construction. However, the notion of a propositional function being "definite" remained unclarified and Zermelo did not specify what "the universally valid laws of logic" are. This lack of clarity was immediately seen as a blemish of the axiomatization; it was given a satisfactory solution only later by, among others, Weyl and Skolem. Let us now list the axioms in Zermelo's original formulation.

Axiom I (Axiom of extensionality). If every element of a set M is also an element of N and vice versa, if, therefore, both $M \subseteq N$ and $N \subseteq M$, then always $M = N$; or, more briefly: Every set is determined by its elements. ...

Axiom II (Axiom of elementary sets). There exists a (fictitious) set, the null set, 0, that contains no element at all. If a is any object of the domain, there exists a set $\{a\}$ containing a and only a as element; if a and b are two objects of the domain, there always exists a set $\{a, b\}$ containing as elements a and b but no object x distinct from both. ...

Axiom III (Axiom of separation). Whenever the propositional function $\mathfrak{E}(x)$ is definite for all elements of a set M, M possesses a subset $M_\mathfrak{E}$ containing as elements precisely those elements x of M for which $\mathfrak{E}(x)$ is true. ...

Axiom IV (Axiom of the power set). To every set T there corresponds a set $\mathfrak{U}T$, the *power set* of T, that contains as elements precisely all subsets of T.

Axiom V (Axiom of the union). To every set T there corresponds a set $\mathfrak{S}T$, the *union* of T, that contains as elements precisely all elements of the elements of T. ...

Axiom VI (Axiom of choice). If T is a set whose elements are all sets that are different from 0 and mutually disjoint, its union $\mathfrak{S}T$ includes at least one subset S_1 having one and only one element in common with each element of T. ...

Axiom VII (Axiom of infinity). There exists in the domain at least one set Z that contains the null set as an element and is so constituted that to each of its element a

there corresponds a further element of the form $\{a\}$, in other words, that with each of its elements a it also contains the corresponding set $\{a\}$ as an element. (Zermelo 1908c, pp. 201–04)

Let us clarify how Zermelo's axiomatization manages to exclude the generation of the paradoxical sets and at the same time allows the development of classical mathematics, including the parts based on impredicative definitions. Previous developments of set theory operated with a comprehension principle that allowed, given any property $P(x)$, the formation of the set of objects satisfying $P(x)$, that is, $\{x : P(x)\}$. This unrestricted use of comprehension leads to the possibility of forming Russell's paradoxical "set" of all sets that do not contain themselves as elements, or the "set" of all ordinals W. However, the separation principle essentially restricts the formation of sets by requiring that sets be obtained, through some propositional function $P(x)$, as subsets of previously given sets. Thus, to go back to Russell's set, it is not possible to construct $\{x :\sim(x \in x)\}$ but only, for a previously given set A, a set $B = \{x \in A :\sim(x \in x)\}$. Unlike the former, this set is innocuous and does not give rise to an antinomy. In the same way, we cannot form the set of all ordinals but only, for any given set A, the set of ordinals in A. The paradoxes having to do with notions such as denotation and definability, such as Berry's or König's paradoxes, are excluded because the notions involved are not "definite" in the sense required for Axiom III. Zermelo's approach here foreshadows the distinction, later drawn by Ramsey (1925), between mathematical and semantical paradoxes, albeit in a somewhat obscure way. In his essay, Zermelo pointed out that the entire theory of sets created by Cantor and Dedekind could be developed from his axioms, and he himself carried out the development of quite a good amount of cardinal arithmetic.

To connect our discussion to the debate on impredicative definitions let us look more closely at the principles of Zermelo's system that allow the formation of impredicative definitions. We shall consider one classic example, namely, the definition of natural numbers according to Dedekind's theory of chains.

In *Was sind und was sollen die Zahlen?* (1888), Dedekind had given a characterization of the natural numbers starting from the notion of a chain. First he argued, in a notoriously fallacious way, that there are simply infinite systems (or sets), that is, sets that can be mapped one-one into a proper subset of themselves. Then he showed that each simply infinite system S contains an (isomorphic) copy of a K-chain, that is, a set that contains 1 and is closed under successor. Finally, the set of natural numbers is defined as the intersection of all K-chains contained in a simply infinite system. This is the smallest K-chain contained in S. From the logical point of view, the definition of the natural numbers by means of an intersection of sets corresponds to a universal quantification over the power set of the infinite system S. More formally, $N = \{X : X \subseteq S$ and X is a K-chain in $S\}$. Equivalently, $n \in N$ iff n is a member of all chains in S.

In Zermelo's axiomatization of set theory, the definition of N is justified by appealing to three axioms. First of all, the existence of an infinite simple system S is given through the axiom of infinity. By means of the power set axiom we are also given the set of subsets of S. Finally, we appeal to the separation axiom to construct the intersection of all chains in S.

It thus appears that the formalization of set theory provided by Zermelo had met the goals he set for himself. On the one hand, the notion of set was restricted in such a way that no paradoxical sets could arise. On the other hand, no parts of classical mathematics seemed to be excluded by its formalization. Zermelo's axiomatization proved to be an astounding success. However, there were problems left. Subsequent discussion showed the importance of the issue of definability, and further results in set theory showed that Zermelo's axioms did not quite characterize a single set-theoretic universe. This will be treated in the next section.

1.3.3 The Discussion on the Notion of "Definit"

One important contribution to the clarification of Zermelo's notion of "definit" came already in Weyl's "Über die Definitionen der mathematischen Grundbegriffe" (1910a). After reflecting on the process of "Logisierung der Mathematik", Weyl declares in this paper that from the logical point of view, set theory is the proper foundation of the mathematical sciences. Thus, he adds, if one wants to give general definitional principles that hold for all of mathematics, it is necessary to account for the definitional principles of set theory. First, he begins his definitional analysis with geometry. Relying on Pieri's work on the foundations of geometry, he starts with two relations, $x = y$ and $E(x, y, z)$. $E(x, y, z)$ means that y and z are equidistant from x. Then he adds that all definitions in Pieri's geometry can be obtained by closing the basic relationships under five principles:

1. Permutation of variables: if $\mathfrak{A}(x, y, z)$ is a ternary relation, so is $\mathfrak{A}(x, z, y)$.
2. Negation: if $\mathfrak{A}$ is a relation, then not-$\mathfrak{A}$ is also a relation.
3. Addition: if $\mathfrak{A}(x, y, z)$ is a ternary relation, then $\mathfrak{A}+(x, y, z, w)$ is a relation, which holds of x, y, z, w, iff $\mathfrak{A}(x, y, z)$ holds.
4. Subtraction: if $\mathfrak{A}(x, y, z)$ is a relation, then so is $\mathfrak{B}(x, y)$, which holds iff there exists a z such that $\mathfrak{A}(x, y, z)$
5. Coordination: if $\mathfrak{A}(x, y, z)$ and $\mathfrak{B}(x, y, z)$ are ternary relations, so is $\mathfrak{C}(x, y, z)$, which holds if and only if both $\mathfrak{A}(x, y, z)$ and $\mathfrak{B}(x, y, z)$ hold.

For Weyl, these definitional principles are sufficient to capture all the concepts of elementary geometry. In the later part of the article, Weyl poses the question: Can all the concepts of set theory be obtained from $x = y$ and $x \in y$ by closing under the definitional principles 1–5? Here his reply is negative. He claims that the fact that in set theory we have objects that can be characterized uniquely, such as the empty set, presents a situation very different from the geometrical one, where all the points are equivalent. He adds that the definitional principles 1–5 would have to be altered to take care of this situation. However the definitional principles still play an important role in connection to Zermelo's concept of "definit". After pointing out the vagueness of Zermelo's formulation of the comprehension principle he proposes an improvement:

"A *definite relation* is one that can be defined from the basic relationships = and ∈ by finitely many applications of our definitional principles modified in an appropriate fashion." (Weyl 1910a, p. 304) The comprehension principle is then stated not for arbitrary propositional functions, as in Zermelo, but in the restricted form for binary relationships: "If M is an arbitrary set, a an arbitrary object, and 𝔄 is a definite binary relationship, then the elements x of M which stand in the relationship 𝔄 to the object a constitute a set." (Weyl 1910a, p. 304) In a note to the text, Weyl also expresses his conviction that without a precise formulation of the definitional principles the solution of the continuum problem would not be possible. Weyl's attempt at making precise the notion of definite property is important because, despite a few remaining obscure points, it clearly points the way to a notion of definability based on closure under Boolean connectives and existential quantification over the individuals of the domain (definition principle 4). In *Das Kontinuum* (1918a), the analysis of the mathematical concept formation is presented as an account of the principles of combination of judgments with minor differences from the account given in (1910a). However, the explicit rejection of the possibility of quantifying over (what he then calls) ideal elements, that is, sets of elements of the domain, which characterizes Weyl's predicative approach in 1918, brings his approach quite close to an explicit characterization of the comprehension principle in terms of first-order definability.[32]

Two very important contributions to the problem of "definiteness" were given by Fraenkel (1922b) and Skolem (1922). The more influential turned out to be Skolem's account. Here is the relevant passage from Skolem's work:

A very deficient point in Zermelo is the notion "definite proposition". Probably no one will find Zermelo's explanations of it satisfactory. So far as I know, no one has attempted to give a strict formulation of this notion; this is very strange, since it can be done quite easily and, moreover, in a very natural way that suggests itself. (Skolem 1922, p. 292)

Skolem then listed "the five basic operations of mathematical logic": conjunction, disjunction, negation, universal quantification, existential quantification. His proposal is that "by a definite proposition we now mean a finite expression constructed from elementary propositions of the form $a ∈ b$ or $a = b$ by means of the five operations mentioned" (292–93). The similarity to Weyl's account is striking. Although Skolem does not mention Weyl (1910a), he was familiar with it, as he had reviewed it for the *Jahrbuch für die Fortschritte der Mathematik* (Skolem 1912).

One final point should be mentioned in connection to these debates on the notion of "definit". Weyl, already in (1910a), had pointed out that the appeal to a finite number of applications of the definitional principles showed that the notion of natural number was essential to the formulation of set theory, which however was supposed to provide a foundation for all mathematical concepts (including that of natural number). In *Das Kontinuum*, he definitely takes the stand that the concept of natural number is basic, and that set theory cannot give a foundation for it (Weyl 1918a, p. 24). Zermelo took the opposite stand. Analyzing Fraenkel's account of "definit" (1929), he rejected it on account of the fact that an explicit appeal to the notion of finitely many applications of the axiom was involved. But the notion of finite number should be given a foundation by set theory, which therefore cannot presuppose it in its formulation (see also Skolem 1929a).

Thus, two major problems emerged in the discussion concerning a refinement of the notion of "definit". The first concerned the question of whether set theory could be considered a foundation of mathematics. Both Skolem and Weyl (who had abandoned his earlier position) thought that this could not be the case. The second problem had to do with the choice of the formal language. Why restrict oneself to first-order logic as Skolem and Weyl were proposing? Why not use a stronger language? The problem was of course of central significance due to the relativization of set-theoretical notions that Skolem had pointed out in his 1922 paper (see itinerary IV). This topic would be worthy of further discussion, but we limit ourselves here to two observations. First of all, Skolem used the relativity of set-theoretic notions as an argument against considering set theory as a foundation of mathematics. Second, Zermelo proposed in "Über Stufen der Quantifikation und die Logik des Unendlichen" (1931) an infinitary logic with the aim of countering Skolem's position (which Zermelo called, disparagingly, "Skolemism").[33] As Ferreiros (1999, p. 363) argues, it was only after Gödel's incompleteness results that the idea of using first-order logic as the "natural" logical scaffolding for axiomatic set theory became standard.

1.3.4 Metatheoretical Studies of Zermelo's Axiomatization

In treating set theory as an axiomatic system, Zermelo had opened the way for a study of the metatheoretical properties of the system itself, such as independence, consistency, and categoricity of the axioms. It should be said from the outset that no real progress was made on the issue of consistency. A proof of the consistency of set theory was one of the major goals of Hilbert's program, but it was not achieved. Of course, much attention was devoted to the axiom of choice. The Polish set-theorist Sierpinski (1918) listed a long set of propositions which seemed to require the axiom of choice essentially, or which were equivalent to the axiom of choice. But was the axiom of choice itself indispensable, or could it be derived from the remaining axioms of Zermelo's system?[34] While this problem was only solved by the combined work of Gödel (1940) and Cohen (1966), an interesting result on independence was obtained by Fraenkel in (1922b). Fraenkel was able to show that the axiom of choice is independent of the other axioms of Zermelo's set theory, if we assume the existence of infinitely many urelements, that is, basic elements of the domain $\mathfrak{B}$ which possess no elements themselves. Unfortunately, the assumption of a denumerable set of urelements is essential to the proof, and thus the result does not apply immediately to Zermelo's system. Moreover, there were reasons to consider the assumption of urelements as foreign to set theory. Fraenkel himself (1922c) had criticized the possibility of having urelements as part of the domain $\mathfrak{B}$, posited at the outset by Zermelo, as irrelevant for the goal of giving a foundations of mathematics. The possibility of having interpretations of set theory with urelements, and others without, already suggested the inability of the axioms to characterize a unique model. Skolem (1922) (and independently also Fraenkel in the same year) also discusses interpretations of Zermelo's axioms in which there are infinite descending chains ... $\in M_2 \in M_1 \in M$, which he called a descending $\in$-sequence, a fact that had already been pointed out by Mirimanoff (1917).[35] A related shortcoming, which affects both the completeness and the categoricity of Zermelo's theory, is related to the inability of the theory to ensure that certain sets, which are used unproblematically in the practice of set theorists, actually exist. Skolem gives the following example. Consider the set M. By the power set axiom we can form $\mathfrak{U}(M)$, then $\mathfrak{U}(\mathfrak{U}(M))$, and so on for any finite iteration of the power set axiom. However, no axiom in Zermelo's set theory allows us to infer the existence of $\{M, \mathfrak{U}(M), \mathfrak{U}(\mathfrak{U}(M)), \ldots\}$. Skolem gives an interpretation that satisfies all the axioms of set theory, which contains M and all finite iterations of the power set of M, but in which $\{M, \mathfrak{U}(M), \mathfrak{U}(\mathfrak{U}(M)), \ldots\}$ does not exist. Both shortcomings, infinite descending chains and lack of closure at "limit" stages, pointed out important problems in Zermelo's axiomatization. The existence of infinite descending chains ran against the intuitive conception of set theory as built up in a "cumulative" way, and the lack of closure for infinite sets showed that genuine parts of the theory of ordinal and cardinal numbers could not be obtained in Zermelo's system. The latter problem was addressed by Skolem through the formulation of what came to be known as the replacement axiom:

> Let U be a definite proposition that holds for certain pairs (a, b) in the domain B; assume further, that for every a there exists at most one b such that U is true. Then, as a ranges over the elements of a set M_a, b ranges over all elements of a set M_b. (Skolem 1922, p. 297)

In other words, starting from a set a and a "definite" functional relationship $f(x)$ on the domain, the range of $f(x)$ is also a set. The name and an independent formulation, albeit very informal, of the axiom of replacement is due to Fraenkel (1922c). For this reason, Zermelo (1930, p. 29) calls the theory Zermelo–Fraenkel set theory. However, Fraenkel had doubts that the axiom was really needed for "general set theory". The real importance of the axiom became clear with the development of the theory of ordinals given by von Neumann, who showed that the replacement axiom was essential to the foundation of the theory.[36] Von Neumann (1923) gave a theory of ordinals in which ordinals are specific well ordered sets, as opposed to classes of equivalent well-orderings. This opened the way for a development of ordinal arithmetic independently of the theory of ordered sets. The definition he obtained is now standard and it was captured by von Neumann in the claim that "every ordinal is the set of the ordinals that precede it". The formalization of set theory he offered in (1925) is essentially different from that of Zermelo. Von Neumann takes the notion of function as basic (the notion of set can be recovered from that of function) and allows classes in addition to sets. This system of von Neumann was later modified and extended by Bernays and Gödel, to result in what is known as NBG set theory.[37] The central intuition is a "limitation of size" principle, according to which there are collections of objects which are too big (we now call them classes), namely, those that are equivalent to the class of all things. The difference between classes and sets is essentially that the latter but not the former can be elements of other sets or classes. A very important part of von Neumann (1925) consists in the axiomatic investigation of "models" of set theory. We will come back to this issue in itinerary VIII. Here it should be pointed out that von Neumann's technique foreshadowed the studies of inner models of set theory.

It is only with von Neumann that a new axiom intended to eliminate the existence of descending ∈-sequences (and finite cycles) was formulated (1925, 1928) (although Mirimanoff had foreshadowed this development by means of his postulate of "ordinariness" meant to eliminate "extraordinary" sets, that is infinite descending ∈-sequences). This was the axiom of well foundedness (von Neumann 1928, p. 498), which postulates that every (nonempty) set is such that it contains an element with which it has no element in common. The axiom appears in Zermelo (1930) as the *Axiom der Fundierung*:

Axiom of Foundation: Every (descending) chain of elements, each member of which is an element of the previous one, terminates with a finite index in an urelement. Or, equivalently: Every subdomain T (of a ZF-model) contains at least one element t_0 that has no element t in T. (Zermelo 1930, p. 31)

Thus by 1930 we have all the axioms that characterize what we nowadays call ZFC, namely, Zermelo–Fraenkel set theory with choice. However, the formulation given by Zermelo (1930) is not first-order, as it relies on second-order quantification in the statement of the axioms of separation and replacement. Even the second formulation of the axiom of foundation contains an implicit quantification over models of ZF.[38]

During the thirties, there were several competing systems for the foundations of mathematics, such as, in addition to Zermelo's extended system, simple type theory and NBG. It was only in the second half of the 1930s that the first-order formulation of ZFC became standard (see Ferreiros 1999, 2001).

1.4 Itinerary IV. The Theory of Relatives and Löwenheim's Theorem

1.4.1 Theory of Relatives and Model Theory

Probably the most important achievements of the algebraic tradition in logic are the axiomatization of the algebra of classes, the theory of relatives, and the proof of the first results of a clearly metalogical character. The origin of the calculus of classes is found in the works of Boole. De Morgan was the first logician to recognize the importance of relations to logic, but he did not develop a theory of relations. Peirce established the fundamental laws of the calculus of classes and created the theory of relatives.[39] Schröder proposed the first complete axiomatization of the calculus of classes and expanded considerably the calculus and the theory of relatives. This theory was the frame that made possible the proof by Löwenheim of the first metalogical theorem. "Über Möglichkeiten im Relativkalkül" (1915), the paper in which Löwenheim published these results, is now recognized as one of the cornerstones in the history of logic (or even in the history of mathematics) due to the fact that it marks the beginning of what we call *model theory*.[40]

The main theorems of Löwenheim's paper are (stated in modern terminology): (1) Not every first-order sentence of the theory of relatives is logically equivalent to a quantifier-free formula of the calculus of relatives (proved by Korselt in a letter to Löwenheim); (2) if a first-order sentence has a model, then it has a countable model; (3) there are satisfiable second-order sentences which have no countable model; (4) the unary predicate calculus is decidable, and (5) first-order logic can be reduced to binary first-order logic.

Nowadays, we use the term "Löwenheim–Skolem theorem" to refer to theorems asserting that if a set of first-order sentences has a model of some infinite cardinality, it also has models of some other infinite cardinalities. The mathematical interest of these theorems is well known. They imply, for example, that no infinite structure can be characterized up to isomorphism in a first-order language. Theorem 2 of Löwenheim's paper was the first one of this group to be proved and, in fact, the first in the history of logic that established a nontrivial relation between first-order formulas and their models.

Löwenheim's theorem poses at least two problems to the historian of logic. The first is to explain why the theory of relatives made it possible to state and prove a theorem which was unthinkable in the syntactic tradition of Frege and Russell. The second problem is more specific. Even today, Löwenheim's proof raises many uncertainties. On the one hand, the very result that is attributed to Löwenheim today is not the one that Skolem—a logician raised in the algebraic tradition—appears to have attributed to him. On the other hand, present-day commentators agree that the proof has gaps, but it is not completely clear which they are. We deal with these questions in the following pages.[41]

Schröder was interested in the study of the algebras of relatives. As Peirce and he himself conceived it, an algebra of relatives consists of a domain of relatives (the set of all relatives included in a given universe), the inclusion relation between relatives (denoted by $\subseteq$), six operations (union, intersection, complementation, relative product, relative sum, and inversion) and four distinguished elements called *modules* (the total relation, the identity relation, the diversity relation, and the empty class). Schröder's objective was to study these structures with the help of a calculus. He could have tried to axiomatize the calculus of relatives, but, following Peirce, he preferred to develop it within the theory of relatives. The difference between the theory and the calculus of relatives is roughly this. The calculus permits the quantification over relatives, but deals only with relatives and operations between them. The theory of relatives, on the other hand, is an extension which also allows the quantification over individuals. The advantage of the theory over the calculus is that the operations between relatives can be defined in terms of individuals and these definitions provide a simpler and more intuitive way of proving certain theorems of the calculus.

Neither Peirce nor Schröder thought that the theory of relatives was stronger than the calculus. Schröder in particular was convinced that all logical and mathematical problems could be addressed within the calculus of relatives (Schröder 1898, p. 253). So he focused on developing the calculus and viewed the theory as a tool that facilitated his task. Schröder did not address problems of a metalogical nature, in that he did not consider the relation between the formulas of a formal language and their models. Arguments or considerations of a semantic type are not completely absent from *Vorlesungen über die Algebra der Logik* (henceforth *Vorlesungen*), but they occur only in the proofs of certain equations, and so we cannot view them as properly metalogical.

Schröder posed numerous problems regarding the calculus of relations, but very few later logicians showed any interest in them, and the study of the algebras of relatives was largely neglected until Tarski. In his first paper on the subject, Tarski (1941) claimed that hardly any progress had been made in the previous 45 years and expressed his surprise that this line of research should have had so few followers.[42]

Schröder was not interested in metalogical questions, but the theory of relatives as he conceived it made it possible to take them into consideration. As a preliminary appraisal, we can say that in the theory of relatives two interpretations coexist: an algebraic interpretation and a propositional interpretation. This means that the same expressions can be seen both as expressions of an algebraic theory and as formulas of logic (namely, as well-formed expressions of a formal language which we may use to symbolize the statements of a theory to reflect its logical structure). We do not mean by this that the whole theory admits of two interpretations, because not all the expressions can be read in both ways, but the point is that some expressions do.

One way of viewing the theory of relatives that gives a fairly acceptable idea of the situation is as a theory of relations together with a partly algebraic presentation of the logic required to develop it.[43] The theory constitutes a whole, but it is important to distinguish the part that deals with the tools needed to construct and evaluate the expressions that denote a truth value (i.e., the fragment that concerns logic) from the one that deals specifically with relatives. So, to prove his theorem, Löwenheim had to think of logic as a differentiated fragment of the theory of relatives and delimit the formal language at least to the extent required to state and prove the theorem.

With the exception of the distinction between object language and metalanguage (an absence that needs emphasizing as it causes many problems in the proof by Löwenheim of his theorem), the basic components of model theory are found in one way or another in the theory of relatives. On the one hand, the part of the theory dealing with logic contains more or less implicitly the syntactic component of a formal language with quantification over relatives: a set of logical symbols with its corresponding propositional interpretation and a syntax borrowed from algebra. On the other, the algebraic interpretation supplies a semantics for this language in the sense that it is enough to evaluate the expressions of this language. In this situation, all that remains to be done to obtain the first results of model theory is, first, become aware that the theory does include a formal language and single it out; second, focus on this language and, in particular, on its first-order fragment; and third, investigate the relationship between the formulas of this language and the domains in which they hold. As far as we know, Löwenheim was the first in the history of logic to concentrate on first-order logic and investigate some of its nontrivial metalogical properties.

1.4.2 The Logic of Relatives

To understand Löwenheim's proof and the relationship between his paper and the theory of relatives, we need first to present the logic of relatives (i.e., the fragment of the theory that concerns logic).[44] In our exposition, we distinguish syntax from semantics, although such a distinction is particularly alien to the logic of relatives. Consequently, the exposition should not be used to draw conclusions about the level of precision found in Schröder or in Löwenheim.

Strictly speaking, relatives denote relations on the (first-order) domain and they are the only nonlogical symbols of the logic of relatives. However, as a matter of fact, in the writings of the algebraists the word *relative* refers both to a symbol of the language and to the object denoted by it. The only relatives usually taken into account are binary, on the assumption that all relatives can be reduced to binary.[45]

What we would call today *logical symbols* are the following: (a) indices; (b) module symbols: 1' and 0'; (c) operation symbols: $+$, $\cdot$ and $^-$; (d) quantifiers: Σ and Π; (e) equality symbol: $=$; and (f) propositional constants: 1 and 0.

Indices play the role of individual variables. As indices the letters h, i, j, k, and l are the most frequently used.

In the theory of relatives, the term *module* is used to refer to any of the four relatives 1, 0, 1', and 0'. The module 1 is the class of all ordered pairs of elements of the (first-order) domain; 0 is the empty class; 1' is the identity relation on the domain; and 0' is the diversity relation on the domain. In the logic of relatives, 1' and 0' are used as relational constants, and 1 and 0 are not viewed as modules but as propositional constants denoting the truth values.

There are six operations on the set of relatives: identical sum (union, denoted by $+$), identical product (intersection, denoted by $\cdot$), complement ($^-$), relative sum, relative product, and inversion. None of these operations belongs to the logic of relatives. The symbols corresponding to the first three operations are used ambiguously to refer also to the three well-known Boolean operations defined on the set $\{0, 1\}$. This is the meaning they have in the logic of relatives.

If i and j are elements of the domain and a is a relative or a module, then a_{ij} is a *relative coefficient*. For example, the relative coefficients of z in the domain $\{2, 3\}$ are z_{22}, z_{23}, z_{32}, and z_{33}. Relative coefficients can only take two values: the truth values (1 and 0). That is, if a_{ij} is a relative coefficient, then

$$a_{ij} = 1 \quad \text{or} \quad a_{ij} = 0.$$

Relative coefficients admit of a propositional interpretation: a_{ij} expresses that the individual i is in the relation a with the individual j. This interpretation allows us to regard relative coefficients as atomic formulas of a first-order language, but in the logic of relatives they are considered as terms.

If A and B are expressions denoting a truth value, so are $(A + B)$, $(A \cdot B)$, and $\overline{A}$; for example, $(a_{ij} + b_{ij})$, $(a_{ij} \cdot b_{ij})$, and $(\overline{a_{ij}})$ are meaningful expressions of this sort. Terms denoting a truth value admit a propositional reading when the symbols $+$, $\cdot$, and $^-$ occurring in them are viewed as connectives.

The symbols Σ and Π have different uses in the theory of relatives and they cannot be propositionally interpreted as quantifiers in all cases. We restrict ourselves to their use as quantifiers. If u is a variable ranging over elements (or over relatives) and A_u is an expression denoting a truth value in which u occurs, then

$$\sum_u A_u \quad \text{and} \quad \prod_u A_u$$

are, respectively, the sum and the product of all A_u, where u ranges over the domain (or over the set of relatives). From the algebraic point of view, these expressions are terms of the theory, because they denote a truth value. They also admit a propositional reading, Σ can also be interpreted as the existential quantifier and Π as the universal one. For example, $\Sigma_i \Pi_j z_{ij}$ can also be read as "there exists i such that for every j, i is in the relation z with j".[46]

The canonical formulas of the theory of relatives are the equations, that is, the expressions of the form $A = B$, where both A and B are terms denoting either a relative or a truth value. As a special case, $A = 0$ and $A = 1$ are equations.[47] The logic of relatives only deals with terms that have a propositional interpretation, that is, with terms denoting a truth value. A first-order term is a term of this kind whose quantifiers (if any) range over elements (not over relatives). In his presentation of the logic of relatives (1915), Löwenheim uses the word Zählausdruck (first-order expression) to refer to these terms and the word Zählgleichung (first-order equation) to refer to the equations whose terms are first-order expressions.[48] To move closer to the current terminology, in what follows we use the word "formula" for what Löwenheim calls Zählausdruck.

The set over which the individual variables range is the first-order domain (Denkbereich der ersten Ordnung) and is denoted by 1^1. The only condition that this domain must fulfill is to be nonempty. Schröder insists that it must have more than one element, but Löwenheim ignores this restriction. Relative variables range over the set of relations on 1^1. The second-order domain (Denkbereich der zweiten Ordnung), 1^2, is the set of all ordered pairs whose coordinates belong to 1^1. In this exposition we are using the word *domain* as shorthand for "first-order domain".

The current distinction between the individual variables of the object language and the metalinguistic variables ranging over the elements of the domain does not exist in the logic of relatives. From the moment it is assumed that an equation is interpreted in a domain, the indices play simultaneously the role of variables of the formal language and that of variables of the metalanguage. The canonical names of the elements of the domain are then used as individual constants having a fixed interpretation. Thus, the semantic arguments that we find in the logic of relatives are better reproduced when we think of them as arguments carried out in the expanded language that results from adding the canonical names of the elements to the basic language.

Interpreting an equation means fixing a domain and assigning a relation on the domain to each relative occurring in it. We can say that an *interpretation in a domain D* of an equation (without free variables) is a function that assigns a relation on D to each relative occurring in the equation. The interpretation of a relative z can also be fixed by assigning a truth value to each coefficient of z in D, because, in the theory of relatives, for every $a, b \in D$, $\langle a, b \rangle \in z$ if and only if $z_{ab} = 1$. Thus, an interpretation of an equation in a domain D can also be defined as an assignment of truth values to the coefficients in D of the relatives (other than $1'$ and $0'$) occurring in the equation.

The most immediate response to an equation is to inquire about the systems of values that satisfy it. This inquiry has a clear meaning in the context of the logic of relatives and it does not require any particular clarification to understand it. The equations of the logic of relatives are composed of terms which in a domain D take a unique value (1 or 0) for each assignment of values to the coefficients in D of the relatives occurring in them. An equation is satisfied by an interpretation $\mathfrak{I}$ in a domain if both members of the equation take the same value under $\mathfrak{I}$. There is no essential difference between asking if there is a solution (an interpretation) that satisfies the equation $A = 1$ and asking if the formula A is satisfiable in the modern sense.[49] In this way, in the logic of relatives semantic questions arise naturally, propitiated by the algebraic context. There is no precise definition of any semantic concept, but the meaning of these concepts is clear enough for the proof of theorems such as Löwenheim's.

1.4.3 Löwenheim's Theorem

The simplest versions of the Löwenheim–Skolem theorem can be stated as follows: For every first-order sentence A,

a. if A is satisfiable, then it is satisfiable in some countable domain;
b. if an interpretation $\mathfrak{I}$ in D satisfies A, there exists a countable subdomain of D such that the restriction of $\mathfrak{I}$ to the subdomain satisfies A.

Version b (the subdomain version) is stronger than version a (the weak version) and has important applications in model theory. Some form of the axiom of choice is necessary to prove the subdomain version, but not to prove the weak one.

All modern commentators of Löwenheim's proof agree that he proved the weak version, and that it was Skolem (1920) who first proved the subdomain version and further generalized it to infinite sets of formulas. By contrast, Skolem (1938, p. 455), a logician trained in the algebraic tradition, attributed to Löwenheim the proof of the subdomain version, and in our opinion, this attribution must be taken seriously. The fact that Löwenheim's proof allows two readings so at variance with each other shows patently his argument is far from clear.

As far as the correctness of the proof is concerned, no logician of Löwenheim's time asserts that the proof is incorrect or that it has major gaps. The only inconvenience mentioned by Skolem is that the use of fleeing indices complicates the proof unnecessarily.[50] Herbrand thought that Löwenheim's argument lacks the rigor required by metamathematics but considered it "sufficient in mathematics" (Herbrand 1930, p. 176). The most widely held position today is that the proof has some important gaps, although commentators differ as to precisely how important they are. Without actually stating that the proof is incorrect, van Heijenoort maintains that Löwenheim does not account for one of the most important steps. Dreben and van Heijenoort (1986, p. 51) accept that Löwenheim proved the weak version, but state that their reading of the proof is a charitable one. For Vaught (1974, p. 156), the proof has major gaps, but he does not specify what they are. Wang (1970, pp. 27 and 29) considers that Löwenheim's argument is "less sophisticated" than Skolem's in 1922, but does not say that it has any important gaps. Moore's point of view is idiosyncratic (see Moore 1980, p. 101; 1988, pp. 121–122). In his opinion, the reason Löwenheim's argument appears "odd and unnatural" to the scholars just mentioned is that they consider it inside standard first-order logic instead of considering it in the frame of infinitary logic.

This diversity of points of view makes manifest the difficulty of understanding Löwenheim's argument and at the same time the necessity to provide a new reading of it.

Theorem 2 of Löwenheim's paper is: "If the domain is at least denumerably infinite, it is no longer the case that a first-order fleeing equation is satisfied for arbitrary values of the relative coefficients." (Löwenheim 1915, p. 235)

A *fleeing equation* is an equation that is not logically valid but is valid in every finite domain. Löwenheim's example of a fleeing equation is:

$$\sum_l \prod_{i,j,h} (\bar{z}_{hi} + \bar{z}_{hj} + 1'_{ij})\bar{z}_{li} \sum_k z_{ki} = 0.$$

For the proof of the theorem, he assumes without any loss of generality that every equation is in the form $A = 0$. This allows him to go from equations to formulas, bearing in mind that "$A = 0$ is valid" is equivalent to "A is not satisfiable". Thus, Löwenheim's argument can also be read as a proof of the following.

Theorem *If a first-order sentence (a Zählausdruck) is satisfiable but not satisfiable in any finite domain, then it is satisfiable in a denumerable domain.*

Löwenheim's proof can be split into two lemmas. We state them for formulas (not for equations) and comment on their proof separately.

Lemma 1. *Every sentence of a first-order language is logically equivalent to a sentence of the form $\Sigma \Pi F$, where Σ stands for a possibly empty string of existential quantifiers, Π stands for a possibly empty string of universal quantifiers, and F is a quantifier-free formula.*

The central step in the proof of this lemma involves moving the existential quantifiers in front of the universal quantifiers, preserving logical equivalence. Löwenheim takes this step by applying the equality

$$\prod_i \sum_k A_{ik} = \sum_{k_i} \prod_i A_{ik_i}, \tag{1.1}$$

which is a notational variant of a transformation introduced by Schröder (1895, pp. 513–16). According to Löwenheim, Σ_{k_i} is an n-fold quantifier, where n is the cardinality of the domain (n may be transfinite).[51] For example, if the domain is the set of natural numbers, then

$$\sum_{k_i} \prod_i A_{ik_i} \tag{1.2}$$

can be developed in this way:

$$\sum_{k_1, k_2, k_3, \ldots} A_{1k_1} A_{2k_2} A_{3k_3} \ldots$$

Löwenheim warns, however, that this development of (1.2) contravenes the stipulations on language, even if the domain is finite.

Löwenheim calls terms of the form k_i *fleeing indices* (*Fluchtindizes*) and says that these indices are characterized by the fact that their subindices are universally quantified variables, but in fact, he also gives that name to the indices generated by a fleeing index when its universally quantified variables take values on a domain (k_1, k_2, k_3, ... in the example).

Schröder's procedure for changing the order of quantifiers is generally considered to be the origin of the concept of the Skolem function, and

$$\forall x \exists y A(x, y) \leftrightarrow \exists f \forall x A(x, fx)$$

as the current way of writing (1.1).[52] Even if we subscribed to this assertion, we should notice that neither Schröder nor Löwenheim associated the procedure for changing the order of quantifiers with the quantification over functions (as Goldfarb notes in 1979). Skolem did not make this association either. In addition, the interpretation of (1.2) in terms of *Skolem functions* does not clarify why Schröder and Löwenheim reasoned as they did, nor does it explain some of Skolem's assertions as this one: "But his [Löwenheim's] reasonings can be simplified by using the 'Belegungsfunktionen' (i.e., functions of individuals whose values are individuals)" (Skolem 1938, pp. 455–56). Finally, it is debatable whether fleeing indices are functional terms or not.

The usual way of interpreting Löwenheim's explanation of the meaning of (1.2) can be summarized as follows: (1.2) is a schema of formulas that produces different formulas depending on the cardinality of the domain under consideration; when the domain is infinite the result of the development is a formula of infinite length; in each case, (1.2) should be replaced by its development in the corresponding domain.[53] Against this interpretation the above-mentioned warning could be cited and also the fact that, strictly speaking, no step in Löwenheim's proof consists of the replacement of a formula by its development.

The main characteristic of fleeing indices is their ability to generate a different term for each element of the domain. If a is an element of the domain and k_i is a fleeing index, then k_a is an index. The terms generated by a fleeing index behave like any "normal" index (i.e., like any individual variable). Thus, Löwenheim can assert that k_a, unlike k_i, stands for an element of the domain.

In our view, Löwenheim's recourse to the development of quantifiers in a domain is a rather rough and ready way of expressing the semantics of formulas with fleeing indices. The purpose of the development of (1.2) is to facilitate the understanding of this kind of formulas. Today's technical and expressive devices allow us to express the meaning of (1.2) without recourse to developments. If for the sake of simplicity let us suppose that (1.2) has no free variables, then

(1.3) $\Sigma_{k_i} \Pi_i A_{ik_i}$ is satisfied by an interpretation $\Im$ in a domain D if and only if there is an indexed family $\langle k_a \mid a \in D \rangle$ of elements of D such that for all $a \in D$: A_{ik_i} is satisfied by $\Im$ in D when i takes the value a and k_i the value k_a.

This interpretation of (1.2) is what Löwenheim attempts to express and is all we need to account for the arguments in which (1.2) intervenes. Löwenheim (unlike Schröder) does not see (1.2) as a schema of formulas. The developments are informal explanations (informal, because they contravene the stipulations on language) whose purpose is to facilitate the understanding of quantification over fleeing indices. Löwenheim has no choice but to give examples, because the limitations of his conceptual apparatus (specifically, the lack of a clear distinction between syntax and semantics) prevents him from giving the meaning of (1.2) in a way analogous to (1.3). Many of Schröder's and Löwenheim's arguments and remarks are better understood when they are read in the light of (1.3). In particular, some of these remarks show that they did not relate quantification over fleeing indices with quantification over functions, because they did not relate the notion of indexed family with that of function.

In the proof of Lemma 1, Löwenheim aims to present a procedure for obtaining a formula of the form $\Sigma \Pi F$ logically equivalent to a given formula A. One of the most striking features of Löwenheim's procedure is that the order in which he proceeds is the opposite of the one we would follow today. First he moves the existential quantifiers of A in front of the universal ones, and then obtains the prenex form. This way of arriving at a formula of the form $\Sigma \Pi F$ introduces numerous, totally unnecessary complications. One of the most unfortunate consequences of the order that Löwenheim follows is that the prenex form cannot be obtained in a standard first-order language, because the formula that results from changing the order of the quantifiers will contain quantified fleeing indices. Thus, to obtain the prenex form we need equivalences that tell us how to deal with these expressions and how to resolve the syntactic difficulties that they present. Löwenheim ignores these problems.

The proof of the lemma presents some problems, but its first part, the one in which existential quantifiers are moved in front of universal ones, is an essentially correct proof by recursion. Löwenheim is not aware of the recursion involved, but his proof shows that he intuits the recursive structure of a formal language.

Lemma 2. *If $\Sigma \Pi F$ is satisfiable but not satisfiable in any finite domain, then it is satisfiable in a denumerable domain.*

First of all, Löwenheim shows with the aid of examples that for this proof we can ignore the existential quantifiers of $\Sigma \Pi F$. He notes that a formula of the form ΠF is satisfiable in a domain D if there exists an interpretation of the relatives occurring in F and an assignment of values (elements in D) to the free variables of F and to the indices generated by the fleeing indices when their subindices range over the domain. But this is precisely what it means to assert that $\Sigma \Pi F$ is satisfiable in D.

The proof proper begins with the recursive definitions of a sequence $(C_n, n \geq 1)$ of subsets of $C = \{1, 2, 3, \ldots\}$ and of some sequences of formulas as follows.

1. If ΠF is a sentence, $C_0 = \{1\}$. If $\{j_1, \ldots, j_m\}$ are the free variables of ΠF, then $C_0 = \{1, \ldots, m\}$. Let $\Pi F'$ be the result of replacing in ΠF the constant n ($1 \leq n \leq m$) for the variable j_n. Let F_1 be the product of all the formulas that are obtained by dropping the quantifiers of $\Pi F'$ and replacing the variables that were quantified by elements of C_0. For example, if $\Pi F = \Pi_i F(i, j_1, j_2, k_i)$ then, $C_0 = \{1, 2\}$ and

$$F_1 = F(1, 1, 2, k_1) \cdot F(2, 1, 2, k_2).$$

If F_1 has p fleeing indices, we enumerate them in some order from $m + 1$ to $m + p$. P_1 is the result of replacing in F_1 the individual constant n for the fleeing index t_n ($m + 1 \leq n \leq m + p$) and C_1 is the set of individual constants of P_1, that is, $C_1 = \{1, 2, \ldots, m, \ldots, m + p\}$. If ΠF and, therefore, F_1 has no fleeing indices, then $P_1 = F_1$ and $C_1 = C_0$. If in our example, the fleeing indices are enumerated from 2 onward in the order in which they occur in F_1, then

$$P_1 = F(1, 1, 2, 3) \cdot F(2, 1, 2, 4).$$

At this point Löwenheim makes the following claim:

Claim 2.1. *If P_1 is not satisfiable, then ΠF is not satisfiable.*

To determine whether P_1 is satisfiable or not, Löwenheim takes identity into account and considers all possible systems of equalities and inequalities between the constants that occur in P_1.[54] He implicitly assumes that we choose a representative of each equivalence class of each equivalence relation. Then, for each system of equalities between the constants of P_1, we obtain the formula resulting from

  i. replacing each constant of P_1 by the representative of its class; and
  ii. evaluating the coefficients of $1'$ and $0'$. This means that in place of $1'_{ab}$, we will write 1 or 0, depending on whether $a = b$ or $a \neq b$, and analogously for the case of $0'_{ab}$. Thus, each system of equalities determines the values of the relative coefficients of $1'$ and $0'$ and this allows us to eliminate these coefficients.

Because C_1 is finite, we obtain by this method a finite number of formulas:

$$P_1^1, P_1^2, \ldots, P_1^q.$$

Following Skolem's terminology (1922, p. 296), we use the expression *formulas of level 1* to refer to these formulas.

Löwenheim goes on by stating the following.

Claim 2.2. *If no formula of level 1 is satisfiable, then ΠF is not satisfiable.*

He could now have applied the hypothesis of the theorem to conclude that there are satisfiable formulas at level 1, but instead of doing so, he argues as follows: if no formula of level 1 is satisfiable, we are done; if some formula is satisfiable, we proceed to the next step of the construction.

2. Let F_2 be the product of all the formulas that are obtained by dropping the quantifiers of $\Pi F'$ and replacing the variables that were quantified by elements of C_1. Evidently, the fleeing indices of F_1 are also fleeing indices of F_2. Suppose that F_2 has q fleeing indices that do not occur in F_1. Enumerate these new fleeing indices in some order starting at $m + p + 1$. Now, P_2 is the result of replacing in F_2 each individual constant n for the corresponding fleeing index t_n ($m + 1 \leq n \leq m + p + q$), and C_2 is the set of individual constants of P_2, that is, $C_2 = \{1, 2, \ldots, m + p + q\}$. If ΠF and, therefore, F_1 has no fleeing indices, then $P_2 = P_1$ and $C_2 = C_1$. If in our example, the fleeing indices are enumerated from 4 onwards in the order in which they occur in F_1, then

$$P_2 = F(1,1,2,3) \cdot F(2,1,2,4) \cdot F(3,1,2,5) \cdot F(4,1,2,6).$$

As before, we take into account all possible systems of equalities between the elements of C_2, and for each of these systems, we obtain the formula resulting from replacing each constant by the representative of its class and from evaluating the coefficients of $1'$ and $0'$. Let the formulas obtained by this method (the formulas of level 2) be:

$$P_2^1, P_2^2, \ldots, P_2^r.$$

If no formula of level 2 is satisfiable, we are done; if any of them is satisfiable, we repeat the process to construct P_3, C_3, and the formulas of level 3. By repeatedly applying this method, we can construct for each $n \geq 1$, the formula P_n, the subset C_n and the associated formulas of level n.

We emphasize a number of points that will be important in the final part of the proof.

a. The number of formulas at each level is finite, since for each n, C_n is finite.

b. Let us say that a formula A is an *extension* of a formula B, if A is of the form $B \cdot B'$. Löwenheim assumes that for every n, F_{n+1} is an extension of F_n. Thus, if $n < m$, P_m is an extension of P_n, and each formula Q of level m is an extension of one and only one formula of level n. The relation of extension on the set of all formulas occurring at some level (the formulas P_n^r obtained from $P_1, P_2, \ldots$) is a partial order on the set of all formulas. This kind of partial order is what we today call a *tree*.

c. Because what we said about the formulas of level 1 goes for any $n > 1$ as well, the following generalization of Claim 2.2 can be considered as proven.

Claim 2.3. *If there exists n such that no formula of level n is satisfiable, then ΠF is not satisfiable.*

We now present the last part of Löwenheim's argument. We deliberately leave a number of points unexplained—points which, in our opinion, Löwenheim does not clarify. In the subsequent discussion we argue for our interpretation and explain all the details.

By the hypothesis of the theorem, there is an interpretation in an infinite domain D that satisfies $\Sigma \Pi F$ and, therefore, ΠF. As a consequence, at each level there must be at least one true formula under this interpretation and, therefore, the tree of formulas constructed by following Löwenheim's procedure is infinite. Among the true formulas of the first level which, we recall, is finite, there must be at least one has infinitely many true extensions (i.e., one that has true extensions at each of the following levels). Let Q_1 be one of these formulas. At the second level, which is also finite, there are true formulas which are extensions of Q_1 and also have infinitely many true extensions. Let us suppose that Q_2 is one of these formulas. In the same way, at the third level there must be true formulas which are extensions of Q_2 (and, therefore, of Q_1) and have infinitely many true extensions. Let Q_3 be one of these formulas. In this way, there is a sequence of formulas $Q_1, Q_2, Q_3, \ldots$ such that for each $n > 0 : Q_{n+1}$ is a true extension of Q_n. Consequently,

$$Q_1 \cdot Q_2 \cdot Q_3 \cdots = 1. \tag{1.4}$$

The values taken by the various kind of indices whose substitution gives rise to the sequence $Q_1, Q_2, Q_3, \ldots$ determine a subdomain of D on which ΠF has the same truth value as $Q_1 \cdot Q_2 \cdot Q_3 \cdots$. Because this subdomain cannot be finite, because ΠF is not satisfiable in any finite domain, we conclude that $\Pi F = 1$ in a denumerable domain. This ends the proof of the theorem.

Basically, this part of Löwenheim's argument is the proof of a specific case of what we know today as the infinity lemma proved later with all generality by Denes König (1926, 1927). The proof of this lemma requires the use of some form of the axiom of choice, but when the tree is countable (as in this case) any enumeration of its nodes allows us to choose one from each level without appealing to the axiom of choice. Since Löwenheim does not choose the formulas on the basis of any ordering, we can assume that he is implicitly using some form of the axiom of choice.

Modern commentators have seen in the construction of the tree an attempt to construct an interpretation of ΠF in a denumerable domain. van Heijenoort (1967a, p. 231) reads the final step in this way: "for every n, Q_n is satisfiable; therefore, $Q_1 \cdot Q_2 \cdot Q_3 \cdots$ is satisfiable". This step is correct but, as the compactness theorem had not been proven in 1915 and Löwenheim does not account for it, van Heijenoort concludes that the proof is incomplete. Wang considers that Löwenheim is not thinking of formulas but of interpretations. According to his reading, the tree that Löwenheim constructs should be seen as if any level n were formed by all the interpretations in D (restricted to the language of P_n) that satisfy P_n. The number of interpretations at each level is also finite, although it is not the same as the number of formulas that Löwenheim considers. Thus, when Löwenheim fixes an infinite branch of the tree, it should be understood that he is fixing a sequence of partial interpretations such that each one is an extension of the one at the previous level. The union of all these partial interpretations is an interpretation in a denumerable domain that satisfies P_n for every $n \in N$, and therefore ΠF.

The main difference between these readings of Löwenheim's argument and the foregoing version is that instead of constructing the sequence $Q_1, Q_2, Q_3, \ldots$ with satisfiable formulas or interpretations, we do so with formulas that are true under the interpretation that, by hypothesis, satisfies ΠF in D. Obviously, this means that we subscribe to the view that Löwenheim meant to prove the subdomain version of the theorem.

The aim of Löwenheim's proof is to present a method for determining a domain. The determination is made when all the possible systems of equalities are introduced.

In a way, it is as if the satisfiable formulas of a level n represented all the possible ways of determining the values of the constants occurring in P_n. Thus, when Löwenheim explains how to construct the different levels of the tree, what he means to be explaining is how to determine a domain on the basis of an interpreted formula; consequently, when the construction is completed, he states that he has constructed it.

In Löwenheim's view the problem of determining the system of equalities between numerals is the same (or essentially the same) as that of fixing the values taken by the summation indices of ΠF (the free variables, and the indices generated by the fleeing indices). Each system of equalities between the numerals of P_n is biunivocally associated to a formula of level n. The formulas of any level n *represent*, from Löwenheim's perspective, all the possible ways of determining the values taken by the numerals that occur in P_n and, in the last resort, the values taken by the indices replaced by the numerals (i.e. the free variables in ΠF and the indices generated when their fleeing indices range over the set of numerals occurring in P_{n-1}). Thus, any assignment of values to these indices is represented by a formula of level n. Now, if ΠF is satisfiable, at each level there must be at least one satisfiable formula. In the same way, if ΠF is true in a domain D, at each level there must be at least one true formula (in other words, for each n there exists an assignment of elements of D to the numerals of P_n that satisfies P_n, assuming that the relative coefficients are interpreted according to the interpretation that, by hypothesis, satisfies $\Sigma \Pi F$ in D). The infinite branches of the tree represent the various ways of assigning values to the summation indices of ΠF in a denumerable domain. The product of all the formulas of any infinite branch can be seen as a possible development of ΠF in a denumerable domain. This assertion is slightly inexact, but we think this is how Löwenheim sees it, and for this reason he claims without any additional clarification that for the values of the summation indices that give rise to the sequence Q_1, Q_2, $Q_3, \ldots$, the formula ΠF takes the same truth value as the product $Q_1 \cdot Q_2 \cdot Q_3 \cdots$. Thus, showing that the tree has an infinite branch of true formulas (in the sense just described) amounts, from this perspective, to constructing a subdomain of D in which ΠF is true, and this is what Löwenheim set out to do.

One of the reasons for seeing in Löwenheim's argument an attempt to construct an interpretation in a denumerable domain is probably that when it is seen as a proof of the subdomain version of the theorem, the construction of the tree appears to be an unnecessary complication. He could, it seems, have offered a simpler proof which would not have required that construction and which would have allowed him to reach essentially the same conclusion. Löwenheim reasons in the way he does because he lacks the conceptual distinctions required to pose the problem accurately. The meaning of ΠF and the relation between this formula and $\Sigma \Pi F$ cannot be fully grasped without the concept of assignment or at least without sharply distinguishing between the terms of the language and the elements they denote. From Löwenheim's point of view, the assumption that ΠF is satisfied by an interpretation in D does not imply that the values taken by the summation indices are fixed. All he manages to intuit is that the problem of showing that $\Sigma \Pi F$ is satisfiable is equivalent to the problem of showing that ΠF is satisfiable. He then proceeds essentially as he would with $\Sigma \Pi F$, but without the inconvenience of having to eliminate the existential quantifiers each time that a formula of the sequence $P_1, P_2, \ldots$ is constructed: he assumes that the nonlogical relatives (i.e., relatives other than $1'$ and $0'$) of ΠF have a fixed meaning in a domain D and proposes fixing the values of summation indices in a denumerable subdomain of D. This means that in practice Löwenheim is arguing as he would do if the prefix had the form $\Pi \Sigma$.

Löwenheim's strategy is then as follows: First he presents a procedure of a general nature to construct a tree of a certain type, and then (without any warning, and without differentiating between the two ideas) he applies the hypothesis of the theorem to the construction. The reason for the style that he adopts in the construction of the tree probably lies in his desire to make it clear that the technique he is presenting is applicable to any formula in normal form and not only to one that meets the conditions of the hypothesis. If the starting formula is not satisfiable, we will conclude the construction in a finite number of steps because we will reach a level at which none of the formulas is satisfiable; if the starting formula is satisfiable in a domain D, then, according to Löwenheim, this construction will allow us to determine a finite or denumerable subdomain of D in which it is satisfiable.

We must distinguish between what Löwenheim actually constructs and what he thinks is constructing. On the one hand, the tree (which he constructs) naturally admits a syntactic reading and can be viewed as a method of analyzing quantified formulas. This proof method was later used by Skolem, Herbrand, Gödel and more recently by Quine (though he related it with Skolem and not with Löwenheim) (Quine 1955b; 1972, 185 ff.). On the other hand, it is obvious that, contrary to Löwenheim's belief, the process of constructing the sequence $Q_1, Q_2, Q_3, \ldots$ does not represent the process of constructing a subdomain, because neither these formulas nor their associated systems of equalities can play the role of partial assignments of values to the summation indices. If we wanted to reflect what Löwenheim is trying to express, we should construct a tree with partial assignments rather than with formulas and modify his argument accordingly. Thus, Löwenheim's proof is not completely correct, but any assessment of it must take into account that he lacked the resources that would allow him to express his ideas better.

1.4.4 Skolem's First Versions of Löwenheim's Theorem

Although Skolem did not explicitly state the subdomain version until 1929, this was the version that he proved in 1920. At the beginning of this paper (1920, 254), Skolem asserts explicitly that his aim is to present a simpler proof of Löwenheim's theorem that avoids the use of fleeing indices. He then introduces what today we know as *Skolem normal form for satisfiability* (a prenex formula with the universal quantifiers preceding the existential ones), and then shows the subdomain version of the theorem for formulas in that form. This change of normal form is significant, because Löwenheim reasons as if the starting formula were in the form $\Pi \Sigma F$ (as remarked) and, therefore, the recourse to $\Pi \Sigma$ formulas seems to be the natural way of dispensing with fleeing indices. Skolem's construction of a countable subdomain is, in essence, the usual one. Let us suppose that $\Pi_{x_1} \ldots \Pi_{x_n} \Sigma_{y_1} \ldots \Sigma_{y_m} U_{x_1 \ldots x_n y_1 \ldots y_m}$ (his notation) is the $\Pi \Sigma$ formula that is satisfied by an interpretation $\mathscr{I}$ in a domain D. By virtue of the axiom of choice, there is a function h that assigns to each n-tuple $(a_1, \ldots, a_n)$ of elements in D the m-tuple $(b_1, \ldots, b_m)$ of elements in D such that $U_{a_1 \ldots a_n b_1 \ldots b_m}$ is satisfied by $\mathscr{I}$ in D. Let a be any element in D. The countable subdomain D' is the union $\bigcup_n D_n$, where $D_0 = \{a\}$ and for each n, D_{n+1} is the union of D_n and the set of elements in the m-tuples $h(a_1, \ldots, a_n)$ for $a_1, \ldots, a_n \in D_n$.

In 1922, Skolem proved the weak version of the theorem, which allowed him to avoid the use of the axiom of choice. The schema of Skolem's argument is as follows: (1) he begins by transforming the starting formula A into one in normal form for satisfiability which is satisfiable if and only if A is; (2) he then constructs a sequence of formulas which, in essence, is Löwenheim's $P_1, P_2, \ldots$, and, for each n, he defines a linear ordering on the finite set of (partial) interpretations that satisfy P_n in the set of numerals of P_n; and (3) after observing that the extension relation defined in the set of all partial interpretations is an infinite tree whose levels are finite, Skolem fixes an infinite branch of this tree; this branch determines an interpretation that satisfies A in set of natural numbers (assuming that A is formula without identity).

Skolem's (1922) seems similar to Löwenheim's in certain aspects, but the degree of similarity depends on our reading of the latter. If Löwenheim was attempting to construct a subdomain, the two proofs are very different: Each one uses a distinct notion of normal form, fleeing indices do not intervene in Skolem's proof, and, more important, the trees constructed in each case involve different objects (in Löwenheim's proof the nodes represent partial assignment of values to the summation indices, while in Skolem's the nodes are partial interpretations). These are probably the differences that Skolem saw between his proof and Löwenheim's. The fact is that in 1922 he did not relate one proof to the other. This detail corroborates the assumption that Skolem did not see in Löwenheim's argument a proof of the weak version of the theorem.

In 1964 Gödel wrote to van Heijenoort:

As for Skolem, what he could justly claim, but apparently does not claim, is that, in his 1922 paper, he implicitly proved: "Either A is provable or $\neg A$ is satisfiable" ("provable" taken in an informal sense). However, since he did not clearly formulate this result (nor, apparently, had he made it clear to himself), it seems to have remained completely unknown, as follows from the fact that Hilbert and Ackermann (1928) do not mention it in connection with their completeness problem. (Dreben and van Heijenoort 1986, p. 52)

Gödel made a similar assertion in a letter to Wang in 1967 (Wang 1974, p. 8). Gödel means that Skolem's (1922) argument can be viewed as (or can easily be transformed into) a proof of a version of the completeness theorem (see itinerary VIII). This is so because the laws and transformations used to obtain the normal form of a formula A, together with the rules employed in the construction of the sequence $P_1, P_2, \ldots$ associated with A and the rules used to decide whether a formula without quantifiers is satisfiable can be viewed as an informal refutation procedure. From this point of view, to say that P_n ($n \geq 1$) is not satisfiable is equivalent to saying that the informal procedure refutes it. Now, we can define what it means to be provable as follows:

1. A formula A is refutable if and only if there exists n such that the informal procedure refutes P_n;
2. A formula A is provable if and only if $\neg A$ is refutable.

An essential part of Skolem's argument is the proof of the following result.

Lemma 3. *If for every n, P_n is satisfiable, then A is satisfiable*

With the aid of the foregoing definitions the lemma can be restated as follows.

Lemma 4. *If A is not satisfiable, then A is refutable.*

This lemma (which is equivalent to Gödel's formulation: Either A is provable or $\neg A$ is satisfiable) asserts the completeness of the informal refutation procedure.[55]

Since the laws and rules used by Löwenheim in his proof can also be transformed into an informal refutation procedure (applicable even to formulas with equality), it is interesting to ask whether he proves Lemma 3 (for ΠF formulas). The answer to this question depends on our reading of his proof. If we think, as van Heijenoort and Wang do, that Löwenheim proved the weak version, then we are interpreting the last part of his argument as an (incomplete or unsatisfactory) proof of Lemma 3. Thus, if we maintain that Löwenheim proved the weak version, we have to accept that what Gödel asserts in the quotation applies also to Löwenheim as well. In our view, Löwenheim did not try to construct an interpretation, but a subdomain. He did not set out to prove Lemma 3, and, as a consequence, Gödel's assertion is not applicable to him.

1.5 Itinerary V. Logic in the Hilbert School

1.5.1 Early Lectures on Logic

David Hilbert's interests in the foundations of mathematics began with his work on the foundations of geometry in the 1880s and 1890s (Hilbert 1899, 2004). Although he was then primarily concerned with geometry, he was interested more broadly in the principles underlying the axiomatic method, and in Dedekind's work (1888). A number of factors worked together to persuade Hilbert around 1900 that a fundamental investigation of logic and its relationship to the foundation of mathematics was needed. These were his correspondence with Frege (1899–1900) on the nature of axioms and the realization that his formulation of geometry was incomplete without an axiom of completeness. They were manifest in his call for an independent consistency proof of arithmetic in his 1900 address, and in his belief that every meaningful mathematical problem had a solution ("no ignorabimus").

Although the importance of logic was clear to Hilbert in the early years of the 1900s, he himself did not publish on logic. His work and influence then consisted mainly in a lecture course he taught in 1905 and a number of administrative decisions he made at Göttingen. The latter are described in detail in Peckhaus (1990, 1994a, 1995b), and include his involvement with the appointment of Edmund Husserl and Ernst Zermelo at Göttingen.

Hilbert's first in-depth discussion of logic occurred in his course "Logical Principles of Mathematical Thought" in the summer term of 1905. The lectures centered on set theory (axiomatized in natural language, just like his axiomatic treatment of geometry), but in chapter V, Hilbert also discussed a basic calculus of propositional logic. The presentation is influenced mainly by Schröder's algebraic approach.

Axiom I   *If $X \equiv Y$ then one can always replace X by Y and Y by X.*

Axiom II   *From 2 propositions X, Y a new one results ("additively")*

$$Z \equiv X + Y$$

Axiom III   *From 2 propositions X, Y a new one results in a different way ("multiplicatively")*

$$Z \equiv X \cdot Y$$

The following identities hold for these "operations":

IV. $X + Y \equiv Y + X$
V. $X + (Y + Z) \equiv (X + Y) + Z$
VI. $X \cdot Y \equiv Y \cdot X$
VII. $X \cdot (Y \cdot Z) \equiv (X \cdot Y) \cdot Z$
VIII. $X \cdot (Y + Z) \equiv X \cdot Y + X \cdot Z$

There are 2 definite propositions 0, 1, and for each proposition X a different proposition $\overline{X}$ is defined, so that the following identities hold:

IX. $X + \overline{X} \equiv 1$
X. $X \cdot \overline{X} \equiv 0$
XI. $1 + 1 \equiv 1$
XII. $1 \cdot X \equiv X$.

(Hilbert 1905a, pp. 225–28)

Hilbert's intuitive explanations make clear that X, Y, and Z stand for propositions, + for conjunction, · for disjunction, ¯ for negation, 1 for falsity, and 0 for truth. In the absence of a first-order semantics, neither statement nor proof of a semantic completeness claim could be given. Hilbert does, however, point out that not every unprovable formula renders the system inconsistent when added as an axiom, that is, the full function calculus is not (what we now call) Post-complete.

1.5.2 The Completeness of Propositional Logic

Hilbert's work on the foundations of logic begins in earnest with a lecture course on the principles of mathematics he taught in the winter semester 1917/18 (1918c). These form the basis of Hilbert and Ackermann (1928) (see section 1.5.5 and Sieg 1999) and contain a wealth of material on propositional and first-order logic, as well as Russell's type theory. We focus here on the development of the propositional calculus in these lectures. Syntax and axioms are modeled after the propositional fragment of *Principia Mathematica* (Whitehead and Russell 1910). The language consists of propositional variables (*Aussage-Zeichen*) $X, Y, Z, \ldots$, as well as signs for particular propositions, and the connectives $\overline{\cdot}$ (negation) and $\times$ (disjunction). The conditional, conjunction, and equivalence are introduced as abbreviations. Expressions are defined by recursion:

1. Every propositional variable is an expression.
2. If α is an expression, so is $\overline{\alpha}$.
3. If α and β are expressions, so are $\alpha \times \beta$, $\alpha \rightarrow \beta$, $\alpha + \beta$ and $\alpha = \beta$.

Hilbert introduces a number of conventions, for example, that $X \times Y$ may be abbreviated to XY, and the usual conventions for precedence of the connectives. Finally, the logical axioms are introduced. Group I of the axioms of the function calculus gives the formal axioms for the propositional fragment (unabbreviated forms are given on the right, recall that XY is "X or Y".

1. $XX \rightarrow X$ $\overline{XX}\,X$
2. $X \rightarrow XY$ $\overline{X}(XY)$
3. $XY \rightarrow YX$ $\overline{XY}(YX)$
4. $X(YZ) \rightarrow (XY)Z$ $\overline{X(YZ)}((XY)Z)$
5. $(X \rightarrow Y) \rightarrow (ZX \rightarrow ZY)$ $\overline{XY}(\overline{ZX}(ZY))$

The formal axioms are postulated as correct formulas (*richtige Formel*), and we have the following two rules of derivation ("contentual axioms"):

a. Substitution: From a correct formula another one is obtained by replacing all occurrences of a propositional variable with an expression.
b. If α and $\alpha \rightarrow \beta$ are correct formulas, then β is also correct.

Although the calculus is very close to the one given in *Principia Mathematica*, there are some important differences. Russell uses $(2')$ $X \rightarrow YX$ and $(4')$ $X(YZ) \rightarrow Y(XZ)$ instead of (2) and (4). *Principia* also does not have an explicit substitution rule.[56] The division between syntax and semantics, however, is not quite complete. The calculus is not regarded as concerned with uninterpreted formulas; it is not separated from its interpretation. (This is also true of the first-order part, see Sieg 1999, B3.) Also, the notion of a "correct formula" which occurs in the presentation of the calculus is intended not as a concept defined, as it were, by the calculus (as we would nowadays define the term "provable formula" for instance) but rather should be read as a semantic stipulation: The axioms are true, and from true formulas we arrive at more true formulas using the rules of inference.[57] Read this way, the statement of modus ponens is not that much clearer than the one given in *Principia*: "Everything implied by a true proposition is true" (*1.1).

Hilbert goes on to give a number of derivations and proves additional rules. These serve as stepping stones for more complicated derivations. He proves a normal form theorem to establish decidability and completeness. In the new propositional calculus, however, Hilbert has to establish that arbitrary subformulas can be replaced by equivalent formulas, that is, that the rule of replacement is a dependent rule. He does so by establishing the admissibility of rule (c): If $\varphi(\alpha)$, $\alpha \to \beta$, and $\beta \to \alpha$ are provable, then so is $\varphi(\beta)$. With that, the admissibility of using commutativity, associativity, distributivity, and duality inside formulas is quickly established, and Hilbert obtains the normal form theorem just as he did for the first propositional calculus in the 1905 lectures. Normal forms again play an important role in proofs of decidability and now also completeness.

1.5.3 Consistency and Completeness

"This system of axioms would have to be called inconsistent if it were to derive two formulas from it which stand in the relation of negation to one another" (Hilbert 1918c, p. 150). Hilbert proves that the system of axioms is not inconsistent in this sense using an arithmetical interpretation. The propositional variables are interpreted as ranging over the numbers 0 and 1, $\times$ is just multiplication, and $\overline{X}$ is $1 - X$. One sees that the five axioms represent functions which are constant equal to 0, and that the two rules preserve that property. Now if a is derivable, $\overline{a}$ represents a function constant equal to 1, and thus is underivable.

Hilbert then poses the question of completeness in the syntactic sense for the propositional calculus in the following way:

Let us now turn to the question of *completeness*. We want to call the system of axioms under consideration complete if we always obtain an inconsistent system of axioms by adding a formula which is so far not derivable to the system of basic formulas. (Hilbert 1918c, p. 152)

This is the first time that completeness is formulated as a precise mathematical question to be answered for a system of axioms. Before this, Hilbert (1905a, p. 13) had formulated completeness as the question of whether the axioms suffice to prove all "facts" of the theory in question. The notion of completeness is of course related to the *axiom of completeness*. This axiom was missing from the first edition of *Grundlagen der Geometrie*, but was added in subsequent editions. Hilbert also added such an axiom to his axiomatization of the reals in (1900b); it states that it is not possible to extend the system of real numbers by adding new entities so that the other axioms are still satisfied. Following the formulation of the completeness axiom in Hilbert (1905a), we read:

This last axiom is of a general kind and has to be added to every axiom system whatsoever in some form. It is of special importance in this case, as we shall see. Following this axiom, the system of numbers has to be so that whenever new elements are added contradictions arise, regardless of the stipulations made about them. If there are things which can be adjoined to the system without contradiction, then in truth they already belong to the system. (Hilbert 1905a, p. 17)

The formulation of completeness can be seen to arise directly out of the completeness axioms of Hilbert's earlier axiomatic systems, only this time completeness is a theorem *about* the system instead of an axiom *in* the system. The completeness axiom stated that the domain cannot be extended without producing contradictions; the domain of objects is the system of real numbers in one case, the system of provable propositional formulas in the other.[58]

The completeness proof in the 1917/18 lectures itself is an ingenious application of the normal form theorem: Every formula is interderivable with a conjunctive normal form. As has been proven earlier in the lectures, a conjunction is provable if and only if each of its conjuncts is provable. A disjunction of propositional variables and negations of propositional variables is provable only if it represents a function which is constant equal to 0, as the consistency proof shows. A disjunction of this kind is equal to 0 if and only if it contains a variable and its negation, and conversely, every such disjunction is provable. So a formula is provable if and only if every conjunct in its normal form contains a variable and its negation. Now suppose that α is an underivable formula. Its conjunctive normal form β is also underivable, so it must contain a conjunct γ where every variable occurs only negated or unnegated but not both. If α were added as a new axiom, then β and γ would also be derivable. By substituting X for every unnegated variable and $\overline{X}$ for every negated variable in γ, we would obtain X as a derivable formula (after some simplification), and the system would be inconsistent.[59]

In a footnote, the result is used to establish the converse of the characterization of provable formulas used for the consistency proof: Every formula representing a function that is constant equal to 0 is provable. For, supposing there were such a formula which was not provable, then adding this formula to the axioms would not make the system inconsistent, by the same argument as in the consistency proof. This would contradict syntactic completeness (Hilbert 1918c, p. 153).

We have seen that the lecture notes to *Principles of Mathematics* 1917–18 contain consistency and completeness proofs (relative to a syntactic completeness concept) for the propositional calculus of *Principia Mathematica*. They also implicitly contain the familiar truth value semantics and a proof of semantic soundness and completeness. In his *Habilitationsschrift* (Bernays 1918), Bernays fills in the last gaps between these remarks and a completely modern presentation of propositional logic.

Bernays introduces the propositional calculus in a purely formal manner. The concept of a formula is defined and the axioms and rules of derivation are laid out almost exactly as done in the lecture notes. §2 of Bernays (1918) is titled "Logical interpretation of the calculus. Consistency and completeness." Here Bernays first gives the interpretation of the propositional calculus, which is the motivation for the calculi in Hilbert's earlier lectures (Hilbert 1905a, 1918c). The reversal of the presentation—first calculus, then its interpretation—makes it clear that Bernays is fully aware of a distinction between syntax and semantics, a distinction not made precise in Hilbert's earlier writings. There, the calculi were always introduced with the logical interpretation built in, as it were. Bernays writes:

The axiom system we set up would not be of particular interest, were it not capable of an important contentual interpretation.

Such an interpretation results in the following way:

The variables are taken as symbols for *propositions* (sentences).

That propositions are either true or false, and not both simultaneously, shall be viewed as their characteristic property.

The symbolic product shall be interpreted as the connection of two propositions by "or", where this connection should not be understood in the sense of a proper disjunction, which excludes the case of both propositions holding jointly, but rather so that "X or Y" holds (i.e., is true) if and only if at least one of the two propositions X, Y holds. (Bernays 1918, pp. 3–4)

Similar truth-functional interpretations of the other connectives are given as well. Bernays then defines what a provable and what a valid formula is, thus making the syntax-semantics distinction explicit:

The importance of our axiom system for logic rests on the following fact: If by a "provable" formula we mean a formula which can be shown to be correct according to the axioms [footnote in text: It seems to me to be necessary to introduce the concept of a provable formula in addition to that of a correct formula (which is not completely delimited) in order to avoid a circle], and by a "valid" formula one that yields a true proposition according to the interpretation given for any arbitrary choice of propositions to substitute for the variables (for arbitrary "values" of the variables), then the following theorem holds:

Every provable formula is a valid formula and conversely.

The first half of this claim may be justified as follows: First one verifies that all basic formulas are valid. For this one only needs to consider finitely many cases, for the expressions of the calculus are all of such a kind that in their logical interpretation their truth or falsehood is determined uniquely when it is determined of each of the propositions to be substituted for the variables whether it is true or false. The content of these propositions is immaterial, so one only needs to consider truth and falsity as values of the variables. (Bernays 1918, p. 6)

We have here all the elements of a modern discussion of propositional logic: A formal system, a semantics in terms of truth values, soundness and completeness relative to that semantics. As Bernays points out, the consistency of the calculus follows from its soundness. The semantic completeness of the calculus is proved in §3, along the lines of the footnote in Hilbert (1918c) just mentioned. The formulation of syntactic completeness given by Bernays is slightly different from the lectures and independent of the presence of a negation sign: It is impossible to add an unprovable formula to the axioms without thus making all formulas provable.[60] Bernays sketches the proof of syntactic completeness along the lines of Hilbert's lectures, but leaves out the details of the derivations.

Bernays also addresses the question of decidability. In the lecture notes, decidability was not mentioned, even though Hilbert had posed it as one of the fundamental problems in the investigation of the calculus of logic. In his talk in Zurich in 1917, he said that an axiomatization of logic cannot be satisfactory until the question of decidability by a finite number of operations is understood and solved (Hilbert 1918a, p. 1143). Bernays gives this solution for the propositional calculus by observing that

This consideration does not only contain the proof for the completeness of our axiom system, but also provides a uniform method by which one can decide after finitely many applications of the axioms whether an expression of the calculus is a provable formula or not. To decide this, one need only determine a normal form of the expression in question and see whether at least one variable occurs negated and unnegated as a factor in each simple product. If this is the case, then the expression considered is a provable formula, otherwise it is not. The calculus therefore can be completely trivialized. (Bernays 1918, pp. 15–16)

Consistency and independence are the requirements that Hilbert laid down for axiom systems of mathematics time and again. Consistency was established—but the "contributions to the axiomatic treatment" of propositional logic could not be complete without a proof that the axioms investigated are independent. In fact, however, the axiom system for the propositional calculus, slightly modified from the postulates in (*1) of *Principia Mathematica*, is not independent. Axiom 4 is provable from the other axioms. Bernays devotes §4 of the *Habilitationsschrift* to give the derivation, and also the interderivability of the original axioms of *Principia* (2′) and (4′) with the modified versions (2) and (4) in presence of the other axioms.

Independence is, of course, more challenging. The method Bernays uses is not new, but it is applied masterfully. Hilbert had already used arithmetical interpretations in (1905a) to show that some axioms are independent of the others. The idea was the same as that originally used to show the independence of the parallel postulate in Euclidean geometry: To show that an axiom α is independent, give a model in which all axioms but α are true, the inference rules are sound, but α is false. Schröder was the first to apply that method to logic. §12 of his *Algebra of Logic* (Schröder 1890) gives a proof that one direction of the distributive law is independent of the axioms of logic introduced up to that point (see Thiel 1994). The interpretation he gives is that of the "calculus of algorithms", developed in detail in appendix 4. Bernays combines Schröder's idea with Hilbert's arithmetical interpretation and the idea of the consistency proof for the first propositional calculus in Hilbert (1918c) (interpreting the variables as ranging over a certain finite number of propositions, and defining the connectives by tables). He gives six "systems" to show that each of the five axioms (and a number of other formulas) is independent of the others. The systems are, in effect, finite matrices. He introduces the method as follows:

In each of the following independence proofs, the calculus will be reduced to a finite system (a finite group in the wider sense of the word [footnote: that is, without assuming the associative law or the unique invertability of composition]), where for each element a composition ("symbolic product") and a "negation" is defined. The reduction is given by letting the variables of the calculus refer to elements of the system as their values. The "correct formulas" are characterized in each case as those formulas which only assume values from a certain subsystem T for arbitrary values of the variables occurring in it. (Bernays 1926, pp. 27–28)

We shall not go into the details of the derivations and independence proofs; see section 1.8.2.[61] Bernays' method was of some importance in the investigation of alternative logics. For instance, Heyting (1930a) used it to prove the independence of his axiom system for intuitionistic logic, and Gödel (1932b) was influenced by it when he defined a sequence of sentences F_n so that each F_n is independent of intuitionistic propositional calculus together with all $F_i, i > n$ (see section 1.7.1.7).[62]

1.5.4 Axioms and Inference Rules

In the final section of his *Habilitationsschrift*, Bernays considers the question of whether some of the axioms of the propositional calculus may be replaced by rules. This seems like a natural question, given the relationship between inference and implication: For instance, axiom 5 suggests the following rule of inference (recall that $\alpha\beta$ is Hilbert's notation for the disjunction of α and β):

$$\frac{\alpha \to \beta}{\gamma\alpha \to \gamma\beta}\ c,$$

which Bernays used earlier as a derived rule. Indeed, axiom 5 is in turn derivable using this rule and the other axioms and rules. Bernays considers a number of possible rules,

$$\frac{\alpha \to \beta \quad \beta \to \gamma}{\alpha \to \gamma}\ d \qquad \frac{\alpha\alpha}{\alpha}\ r_1 \qquad \frac{\alpha}{\alpha\beta}\ r_2 \qquad \frac{\alpha\beta}{\beta\alpha}\ r_3 \qquad \frac{\alpha(\beta\gamma)}{(\alpha\beta)\gamma}\ r_4$$

$$\frac{\varphi(\alpha\alpha)}{\varphi(\alpha)}\ R_1 \qquad \frac{\varphi(\alpha\beta)}{\varphi(\beta\alpha)}\ R_3,$$

and shows that the following sets of axioms and rules are equivalent (and hence, complete for propositional logic):

1. Axioms: 1, 2, 3, 5; rules: a, b.
2. Axioms: 1, 2, 3; rules: a, b, c.
3. Axioms: 2, 3; rules: a, b, c, r_1.
4. Axioms: 2; rules: a, b, c, r_1, R_3.
5. Axioms: $\overline{X}X$; rules: a, b, c, r_1, r_2, r_3, r_4.

Bernays also shows, using the same method as before, that these axiom systems are independent, and also the following independence results:[63]

6. Rule c is independent of axioms: 1, 2, 3; rules: a, b, d (showing that in (2), rule c cannot in turn be replaced by d).
7. Rule r_2 is independent of axioms: 1, 3, 5; rules: a, b, (thus showing that in (1) and (2), axiom 2 cannot be replaced by rule r_2).
8. Rule r_3 is independent of axioms: 1, 2; rules: a, b, c (showing similarly, that in (1) and (2), rule r_3 cannot replace axiom 3).
9. Rule R_3 is independent of axioms: $\overline{X}X$, 3; rules: a, b (showing that R_3 is stronger than r_3, since 3 is provable from R_3 and $\overline{X}X$).

10. Rule R_1 is independent of axioms: $\overline{X}X$, 1; rules: a, b (showing that R_1 is stronger than r_1, since 1 is provable from $\overline{X}X$ and R_1).

11. Axiom 2 is independent of axioms: $\overline{X}X$, 1, 3, 5; rules: a, b, and

12. Axiom 2 is independent of axioms: $\overline{X}X$; rules: a, b, c, r_1, R_3 (showing that in (5), $\overline{X}X$ together with r_2 is weaker than axiom 2).

The detailed study exhibits, in particular, a sensitivity to the special status of rules like R_3, where subformulas have to be substituted. The discussion foreshadows developments of formal language theory in the 1960s. Bernays also mentions that a rule (corresponding to the contrapositive of axiom 2), allowing inference of $\varphi(\alpha)$ from $\varphi(\alpha\beta)$ would be incorrect (and hence, "there is no such generalization of r_2").

Bernays' discussion of axioms and rules, together with his discussion of expressibility in the "Supplementary remarks to §2–3", shows his acute sensitivity for subtle questions regarding logical calculi. His remarks are quite opposed to the then-prevalent tendency (e.g., Sheffer and Nicod) to find systems with fewer and fewer axioms, and foreshadow investigations of relative strength of various axioms and rules of inference, for example, of Lewis' modal systems, or more recently of the various systems of substructural logics.

At the end of the "Supplementary remarks", Bernays isolates the positive fragment of propositional logic (i.e., the provable formulas not containing negation; here $+$ and $\rightarrow$ are considered primitives) and claimed that he had an axiomatization of it. He did not give an axiom system, but stated that it is possible to choose a finite number of provable sentences as axioms so that completeness follows by a method exactly analogous to the proof given in §3. The remark suggests that Bernays was aware that the completeness proof is actually a proof schema, in the following sense. Whenever a system of axioms is given, one only has to verify that all the equivalences necessary to transform a formula into conjunctive normal form are theorems of that system. Then completeness follows just as it does for the axioms of *Principia*.

In his next set of lectures on the "Logical Calculus" given in the winter semester of 1920 (Hilbert 1920a), Hilbert makes use of the fact that these equivalences are the important prerequisite for completeness. The propositional calculus we find there is markedly different from the one in Hilbert (1918c) and Bernays (1918), but the influences are clearly visible. The connectives are all primitive, not defined, this time. The sole axiom is $\overline{X}X$, and the rules of inference are:

$$\frac{X}{XY} \text{ b2}, \qquad \frac{\begin{array}{c} X \\ Y \end{array}}{X+Y} \text{ b3},$$

plus the rule (b4), stating: "Every formula resulting from a correct formula by transformation is correct." "Transformation" is meant as transformation according to the equivalences needed for normal forms: commutativity, associativity, de Morgan's laws, $\overline{\overline{X}}$ and X, and the definitions of $\rightarrow$ and $=$ (biconditional). These transformations work in both directions, and also on subformulas of formulas (as did R_1 and R_3).[64] One equivalence corresponding to modus ponens must be added, it is: $(X + \overline{X})Y$ is intersubstitutable with Y.

Anyone familiar with the work done on propositional logic elsewhere might be puzzled by this seemingly unwieldy axiom system. It would seem that the system in Hilbert (1920a) is a step backward from the elegance and simplicity of the *Principia* axioms. Adjustments, if they are to be made at all, it would seem, should go in the direction of even more simplicity, reducing the number of primitives (as Sheffer did) and the number of axioms (as in the work of Nicod and later Łukasiewicz). Hilbert was motivated by different concerns. He was interested not only in the simplicity of his axioms but in their efficiency. Decidability, in particular, supersedes considerations of independence and elegance. The presentation in Hilbert (1920a) is designed to provide a decision procedure which is not only efficient but also more intuitive to use for a mathematician trained in algebraic methods. Bernays' study of inference rules made clear, on the other hand, that such an approach can in principle be reduced to the axiomatics of *Principia*. The subsequent work on the decision problem is also not strictly axiomatic, but uses transformation rules and normal forms. The rationale is formulated by Behmann:

The form of presentation will not be axiomatic, rather, the needs of practical calculation shall be in the foreground. The aim is thus not to reduce everything to a number (as small as possible) of logically independent formulas and rules; on the contrary, I will give as many rules with as wide an application as possible, as I consider appropriate to the practical need. The logical dependence of rules will not concern us, insofar as they are merely of independent practical importance.... Of course, this is not to say that an axiomatic development is of no value, nor does the approach taken here preempt such a development. I just found it advisable not to burden an investigation whose aim is in large part the exhibition of new results with such requirements, as can later be met easily by a systematic treatment of the entire field. (Behmann 1922a, p. 167)

Such a systematic treatment, of course, was necessary if Hilbert's ideas regarding his logic and foundation of mathematics were to find followers. Starting in (1922c) and (1923a), Hilbert presents the logical calculus not in the form of *Principia* but by grouping the axioms governing the different connectives. In (1922c), we find the "axioms of logical consequence", in (1923a), "axioms of negation". The first occurrence of axioms for conjunction and disjunction seems to be in a class taught jointly by Hilbert and Bernays during winter 1922–23, and in print in Ackermann's dissertation (Ackermann 1924). The project of replacing the artificial axioms of *Principia* with more intuitive axioms grouped by the connectives they govern, and the related idea of considering subsystems such as the positive fragment, is Bernays'. In 1918, he had already noted that one could refrain from taking $+$ and $\rightarrow$ as defined symbols and consider the problem of finding a complete axiom system for the positive fragment. The notes to the lecture course from 1922–23 (Hilbert and Bernays 1923a, p. 17) indicate that the material in question was presented by Bernays. In 1923, he gives a talk titled "The role of negation in propositional logic", in which he points out the importance of separating axioms for the different connectives, in particular, giving axioms for negation separately. This emphasis of separating negation from the other connectives is of course necessitated by Hilbert's considerations on finitism as well. Full presentations of the axioms of propositional logic are also found in Hilbert (1928a), and in slightly modified form in a course on logic taught by Bernays in 1929–30. The axiom system we find there is almost exactly the one later included in Hilbert and Bernays (1934).

I. $A \to (B \to A)$
$(A \to (A \to B)) \to (A \to B)$
$(A \to (B \to C)) \to (B \to (A \to C))$
$(B \to C) \to ((A \to B) \to (A \to C))$

II. $A \& B \to A$
$A \& B \to B$
$(A \to B) \to ((A \to C) \to (A \to B \& C))$

III. $A \to A \vee B$
$B \to A \vee B$
$(B \to A) \to ((C \to A) \to (B \vee C \to A))$

IV. $(A \sim B) \to (A \to B)$
$(A \sim B) \to (B \to A)$
$(A \to B) \to ((B \to A) \to (A \sim B))$

V. $(A \to B) \to (\overline{B} \to \overline{A})$
$(A \to \overline{A}) \to \overline{A}$
$A \to \overline{\overline{A}}$
$\overline{\overline{A}} \to A$.[65]

Bernays (1927) claims that the axioms in groups I–IV provide an axiomatization of the positive fragment and raises the question of a decision procedure. This is where he first follows up on his claim in 1918 that such an axiomatization is possible.

1.5.5 Grundzüge der theoretischen Logik

Hilbert and Ackermann's textbook *Grundzüge der theoretischen Logik* (Hilbert and Ackermann 1928) provided an important summary of the work on logic done in Göttingen in the 1920s. Although (as documented by Sieg 1999), the book is in large parts a polished version of Hilbert's 1917–18 lectures (Hilbert 1918c), it is important especially for the influence it had in terms of making the work available to an audience outside of Göttingen. Both Gödel and Herbrand, for instance, became acquainted with the methods developed by Hilbert and his students through it.

In addition, *Grundzüge* contained a number of minor but significant improvements over (Hilbert 1918c). The first is a much simplified presentation of the axioms of the predicate calculus. Whereas Hilbert (1918c) listed six axioms and three inference rules governing the quantifiers, the formulation in Hilbert and Ackermann (1928) consisted simply in:

e. $(x)F(x) \to F(y)$,
f. $F(y) \to (Ex)F(x)$,

with the following form of the rule of generalization. If $\mathfrak{A} \to \mathfrak{B}(x)$ is provable, and x does not occur in $\mathfrak{A}$, then $\mathfrak{A} \to (x)\mathfrak{B}(x)$ is provable. Similarly, if $\mathfrak{B}(x) \to \mathfrak{A}$ is provable, then so is $(Ex)\mathfrak{B}(x) \to \mathfrak{A}$.

Another important part of *Grundzüge* concerns the semantics of the predicate calculus and the decision problem. The only publication addressing the decision problem had been Behmann (1922a); Bernays and Schönfinkel (1928) and Ackermann (1928a) appeared the same year as *Grundzüge* (although Bernays and Schönfinkel's result was obtained much earlier). Thus, the book was important in popularizing the decision problem as a fundamental problem of mathematical foundations. In a similar vein, although the completeness of the propositional calculus had been established already in 1918 by Bernays and in 1920 by Post, the Post-completeness and semantic completeness of predicate logic remained an open problem. Ackermann solved the former in the negative; this result is first reported in *Grundzüge*. It motivates the question of semantic completeness, posed on p. 68: "Whether the axiom system is complete at least in the sense that all logical formulas that are correct for every domain of individuals can be derived from it is still an unsolved question." This offhand remark provided the motivation for Gödel's landmark completeness theorem (see section 1.8.4).

1.5.6 The Decision Problem

The origin of the decision problem in Hilbert's work is no doubt his conviction, expressed in his 1900 address to the Paris Congress, that every mathematical problem has a solution:

This conviction of the solvability of every mathematical problem is a powerful incentive to the worker. We hear within us the perpetual call: There is the problem. Seek its solution. You can find it by pure reason, for in mathematics there is no *ignorabimus*. (Hilbert 1900a, p. 1102)

A few years later, Hilbert first explicitly took the step that this *no ignorabimus* should be reflected in the decidability of the problem of whether a mathematical statement is derivable from the axiom system for the domain in question:

So it turns out that for every theorem there are only *finitely many possibilities of proof*, and thus we have solved, in the most primitive case at hand, the old problem that it must be possible to achieve any correct result by a *finite proof*. This problem was the original starting point of all my investigations in our field, and the solution to this problem in the most general case[,] the proof that there can be no "ignorabimus" in mathematics, has to remain the ultimate goal.[66]

Hilbert's emphasis on the axiomatic method was thus not only motivated by providing a formal framework in which questions such as independence, consistency, and completeness could be given mathematical treatment, but so could the question of the solvability of all mathematical problems. In "Axiomatic Thought" (1918a, p. 1113), the problem of "decidability of a mathematical question in a finite number of operations" is listed as one of the fundamental problems for the axiomatic method.

Without a semantics for first-order logic in hand, it is not surprising that the formulation of the problem as well as the partial results obtained only made reference to derivability from an axiom system. For instance, as discussed, Bernays infers the decidability of the propositional calculus in this sense as a consequence of the completeness theorem. The development of semantics for first-order logic in the following years made it possible to reformulate the decision problem as a question of validity [*Allgemeingültigkeit*] or, dually, as one of satisfiability: "The decision problem is solved, if one knows a procedure which allows for any given logical expression to decide whether it is valid or satisfiable, respectively." (Hilbert and Ackermann 1928, p. 73) Hilbert and Ackermann (1928) call the decision problem the main problem of mathematical logic. No wonder it was pursued with as much vigor as the consistency problem for arithmetic.

1.5.6.1 The Decision Problem in the Tradition of Algebra of Logic

In the algebra of logic, results on the decision problem were obtained in the course of work on elimination problems. The first major contribution to the decision problem was Löwenheim's (1915) result. His theorem 4,"There are no fleeing equations between singular relative coefficients, not even when the relative coefficients of 1′ and 0′ are included as the only binary ones" (Löwenheim 1915, p. 243), amounts to the proposition that every monadic first-order formula, if satisfiable, is satisfiable in a finite domain. Recall from itinerary IV that a fleeing equation is one that is not valid but valid in every finite domain. If there are no fleeing equations between singular relative coefficients (i.e., monadic predicates), then every monadic formula valid in every finite domain is also valid.

It should be noted that both Löwenheim (1915) and Skolem (1919), who gave a simpler proof, state the theorem as a purely algebraic result. Neither draws the conclusion that the result shows that satisfiability of monadic formulas is decidable, indeed, this only follows by inspection of the particular normal forms they give in their proofs. In particular, the proofs do not contain bounds on the size of the finite models that have to be considered when determining if a formula is satisfiable.

Löwenheim (1915) proved a second important result, namely, that validity of an arbitrary first-order formulas is equivalent to a formula with only binary predicate symbols. This means that dyadic predicate logic forms a reduction class, that is, the decision problem for first-order logic can be reduced to that of dyadic logic. Löwenheim, of course, did not draw this latter conclusion, since he was not concerned with decidability in this sense. He does, however, remark that

Since, now, according to our theorem the whole relative calculus can be reduced to the binary relative calculus, it follows that we can decide whether an arbitrary mathematical proposition is true provided that we can decide whether a binary relative equation is identically satisfied or not. (Löwenheim 1915, p. 246)

A related result is proved in (Skolem 1920, theorem 1). A formula is in (satisfiability) Skolem normal form if it is a prenex formula and all universal quantifiers precede all existential quantifiers, that is, it is of the form

$$(\forall x_1)\ldots(\forall x_n)(\exists y_1)\ldots(\exists y_m)A(x_1,\ldots,x_n,y_1,\ldots,y_m).$$

Skolem's result is that for every first-order formula there is a formula in Skolem normal form that is satisfiable if and only if the original formula is. From this, it follows that the formulas in Skolem normal form are a reduction class as well.

1.5.6.2 Work on the Decision Problem after 1920

The word *Entscheidungsproblem* first appears in a talk given by Behmann to the Mathematical Society in Göttingen on May 10, 1921, titled "Entscheidungsproblem und Algebra der Logik".[67] Here, Behmann is very explicit about the kind of procedure required, characterizing it as a "mere calculational method", as a procedure following the "rules of a game", and stating its aim as an "elimination of thinking in favor of mechanical calculation".

The result Behmann reports on in this talk is that of his *Habilitationsschrift* (Behmann 1922a), in which he proves, independently of Löwenheim and Skolem, that monadic second-order logic with equality is decidable. The proof is by a quantifier elimination procedure, that is, a transformation of sentences of monadic-second order logic (with equality) into a disjunctive normal form involving expressions "there are at least n objects" and "there are at most n objects".

The problem was soon taken up by Moses Schönfinkel, who was a student in Göttingen at the time. In December 1922, he gave a talk to the Mathematical Society in which he proved the decidability of validity of formulas of the form $(\exists x)(\forall y)A$, where A is quantifier-free and contains only one binary predicate symbol (Schönfinkel 1922). This result was subsequently extended by Bernays to apply to formulas with arbitrary many predicate symbols (Bernays and Schönfinkel 1928). The published paper also discusses Behmann's (1922a) result and gives a bound on the size of finite models for monadic formulas, as well as the cases of prenex formulas with quantifier prefixes of the form $\forall^* A$, $\exists^* A$ and $\forall^*\exists^*$. In particular, it is shown there that a formula $(\forall x_1)\ldots(\forall x_n)(\exists y_1)\ldots(\exists y_m)A$ is valid iff it is valid in all domains with n individuals. In its dual formulation, the main result is that satisfiability of prenex formulas with prefix $\exists^*\forall^*$ (the Bernays-Schönfinkel class) is decidable. The result was later extended by Ramsey (1930) to include identity; along the way, Ramsey proved his famous combinatorial theorem.

The result dual to Bernays and Schönfinkel's first, namely, the decidability of satisfiability of formulas of the form $(\forall x)(\exists x)A$, was extended by Ackermann (1928a) to formulas with prefix $\exists^*\forall\exists^*$. The same result was proved independently later the same year by Skolem (1928); this paper as well as the follow-up (1935) also prove some related decidability results.

Herbrand (1930, 1931b) draws some important conclusions regarding the decision problem from his *theorème fondamental* (see following discussion) as well, giving new proofs of the decidability of the monadic class, the Bernays-Schönfinkel class, the Ackermann class, and the Herbrand class (prenex formulas where the matrix is a conjunction of atomic formulas and negated atomic formulas).

The last major partial solution of the decision problem before Church's (1936a) and Turing's (1937) proofs of the undecidability of the general problem was the proof of decidability of satisfiability for prenex formulas with prefix of the form $\exists^*\forall\forall\exists^*$. This was carried out independently by Gödel (1932a), Kalmár (1933), and Schütte (1934a, 1934b). Gödel (1933b) also showed that prenex formulas with prefix $\forall\forall\exists^*$ form a reduction class.[68]

1.5.7 Combinatory Logic and λ-Calculus

In the early 1920s, there was a significant amount of correspondence between Hilbert and his students (in particular, Bernays and Behmann) and Russell on various aspects of *Principia* (see Mancosu 1999a, 2003). One of the things Russell mentioned to Bernays was Sheffer's (1913) reduction of the two primitive connectives $\sim$ and $\vee$ of *Principia* to the Sheffer stroke. In 1920, Schönfinkel extended this reduction to the quantifiers by means of the operator $|^x$, where $\phi(x) |^x \psi(x)$ means "for no x is $\phi(x)$ and $\psi(x)$ both true". Then $(x)\phi(x)$ can be defined by $(\phi(x) |^y \phi(x)) |^x (\phi(x) |^y \phi(x))$. This led Schönfinkel to consider further possibilities of reducing the fundamental notions of the logic of *Principia*, namely, those of propositional function and variables themselves.

In a manuscript written in 1920, and later edited by Behmann and published (1924), Schönfinkel gave a general analysis of mathematical functions, and presented a function calculus based on only application and three basic functions (the combinators). First, Schönfinkel explains how one only needs to consider unary functions: A binary function $F(x, y)$, for instance, may be considered instead as a unary function which depends on the argument x, or, equivalently, as a unary function of the argument x which has a unary function as its value. Hence, $F(x, y)$ becomes $(fx)y$; fx now is the unary function which, for argument y has the same value as the binary function $F(x, y)$. Application associates to the left, so that $(fx)y$ can more simply be written fxy.

Just as functions in Schönfinkel's system can have functions as values, they can also be arguments to other functions. Schönfinkel introduces five primitive functions I, C, T, Z, and S by the equations

$$I x = x$$
$$(C x) y = x$$
$$(T \phi) x y = \phi y x$$
$$Z \phi \chi x = \phi(\chi x)$$
$$S \phi \chi x = (\phi x)(\chi x)$$

I is the identity; its value is always simply its argument. C is the constancy function: Cx is the function whose value is always x. T allows the interchange of argument places;

$T\phi$ is the function which has as its value for xy the value of ϕyx. Z is the composition function: $Z\phi\chi$ is the function which takes its argument, first applies χ, and then applies ϕ to the resulting value. The fusion function S is similar to composition, but here ϕ is to be thought of as a binary function $F(x, y)$: Then $S\phi\chi x$ is the unary function $F(x, \chi x)$.

So far this constitutes a very general theory of functions. In applying this to logic, Schönfinkel obtains an elegant system in which formulas without free variables can be written without connectives, quantifiers, or variables at all. In light of the reduction to unary functions, first of all relations can be eliminated; for example, instead of a binary relation $R(x, y)$ we have a unary function r from arguments x to functions that themselves take individuals as arguments, and whose value is a truth value. Then, instead of $|^x$, Schönfinkel introduces a new combinator, $U: U fg = fx \mid^x gx$—note that in the expression on the left the bound variable x no longer occurs. Together with the other combinators, this allows Schönfinkel to translate any sentence of even higher-order logic into an expression involving only combinators. For instance, $(f)(Eg)(x)\overline{fx \& gx}$ first becomes, using $|^x$:

$$[(fx \mid^x gx) \mid^g (fx \mid^x gx)] \mid^f [(fx \mid^x gx) \mid^g (fx \mid^x gx)]$$

Now replacing $|^x$ and $|^g$ by the combinator U, we get

$$[U(Uf)(Uf)] \mid^f [U(Uf)(Uf)]$$

To remove the last $|^f$, the expressions on either side must end with f; however, $U(Uf)(Uf) = S(ZUU)Uf$, and so finally we get $U[S(ZUU)U][S(ZUU)U]$.

Schönfinkel's ideas were further developed in great detail by Haskell Curry, who wrote a dissertation under Hilbert in 1929 (1929, 1930).[69]

Similar ideas led Church (1932) to develop his system of λ-calculus. Like Schönfinkel's and Curry's combinatory logic, the λ-calculus was intended in the first instance to provide an alternative to Russellian type theory and to set theory as a foundation for mathematics. Like combinatory logic, the λ-calculus is a calculus of functions with *application* (st) as the basic operation; like Curry, Church defined a notion of equality between terms using certain conversion relations. If t is a term in the language of the calculus with free variable x, the λ operator is used to form a new term $\lambda x.t$, which denotes a function with argument x. A term of the form $(\lambda x.t)s$ *converts* to the term $t(x/s)$ (t with all free occurrences of x replaced by s). This is one of three basic kinds of conversion; a term on which no conversion can be carried out is in *normal form*.

Unfortunately, as Kleene and Rosser (1935) showed, both Curry's and Church's systems were inconsistent and hence unsuitable in their original formulation to provide a foundation for mathematics. Nevertheless, combinatory logic and λ-calculus proved incredibly useful as theories of functions; in particular, versions of the λ-calculus were developed as systems of computable functions. In fact, Church's (1936b, 1936a) (negative) solution to the decision problem essentially involved the λ-calculus. Church (1933) and Kleene (1935) found a way to define the natural numbers as certain λ-terms $\bar{n}$ in normal form (Kleene numerals). The notion of λ-definability of a number theoretic function is then simply: A function f is λ-definable if there is a term t such that t applied to the

Kleene numeral $\bar{n}$ converts to a normal form which is the Kleene numeral of the value of $f(n)$. Church (1936b) showed that λ-definability coincides with (general) recursiveness and that the problem of deciding whether a term converts to a normal form is not general recursive. Church (1936a) uses this result to show that the decision problem is unsolvable.

1.5.8 Structural Inference: Hertz and Gentzen

Another important development in logic that originated in Hilbert's school was the introduction of sequent calculus and natural deduction by Gentzen. This grew out of the logical work of Paul Hertz. Hertz was a physicist working in Göttingen between 1912 and 1933. From the 1920s onward, he was also working in philosophy and in particular, logic. In a series of papers (Hertz 1922, 1923, 1928, 1929), he developed a theory of structural inference based on expressions of the form $a_1, \ldots, a_n \to b$. Hertz calls such expressions *sentences*; the signs on the left are the antecedents, the sign on the right the succedent. It is understood that in the antecedents each sign occurs only once. The two rules which he considers are what he calls *syllogism*:

$$\frac{\begin{array}{c} a_1^1, a_2^1, \ldots \to b^1 \\ a_1^2, a_2^2, \ldots \to b^2 \\ \vdots \\ a^1, a^2, \ldots, b^1, b^2 \to c \end{array}}{a_1^1, a_2^1, \ldots, a_1^2, a_2^2, \ldots, a^1, a^2 \to c}$$

and *direct inference*:

$$\frac{a_1, a_2, \ldots \to b}{a^1, a^2, \ldots, a_1, a_2, \ldots \to b}$$

In the syllogism, the premises on the left are called lower sentences, the premise on the right the upper sentence of the inference.

A set of sentences is called *closed* if it is closed under these two rules of inference. Hertz's investigations concern in the main criteria for when a closed system of sentences has a set of independent axioms—a concern typical for the Hilbert school. Hertz's other concern, and this is his lasting contribution, is that of proof transformations and normal forms. We cannot give the details of all these results, but a statement of one will give the reader an idea: A sentence is called *tautological* if it is of the form $a \to a$. An *Aristotelian normal proof* is one in which each inference has a nontautological upper sentence that is an initial sentence of the proof (i.e., not the conclusion of another inference). For instance, the following is an Aristotelian normal proof:

$$\frac{\dfrac{\dfrac{a \to b \qquad b \to c}{a \to c} \qquad c \to m}{a \to m} \qquad m, b \to d}{a, b \to d}$$

Hertz proves that every proof can be transformed into an Aristotelian normal proof.

Gentzen's first contribution to logic was a continuation of Hertz's work. Gentzen (1933b) shows a similar normal form theorem, as well as a completeness result relative to a simple semantics which interprets the elements of the sentences as propositional constants. A sentence $a_1, \ldots, a_n \to b$ is interpreted as: either one of the a_i is false or b is true. Gentzen's result is that if a sentence S follows from (is a tautological consequence of) some other sentences $S_1, \ldots, S_n$, then there is a proof of a certain normal form of S from $S_1, \ldots, S_n$.[70]

The basic framework of sentences and inferences, as well as the interest in normal form theorems, was contained in Gentzen's more important work on the proof theory of classical and intuitionistic logic. Gentzen (1934) extended Hertz's framework from propositional atoms to formulas of predicate logic. Sentences are there called *sequents*, and the succedent is allowed to contain more than one formula (for intuitionistic logic, the restriction to at most one formula on the right stands). Hertz's direct inference is now called "thinning;" there is an analogous rule for thinning the succedent: The antecedent and succedent of a sequent are now considered sequences of formulas (denoted by uppercase Greek letters). Thus, Gentzen adds rules for changing the order of formulas in a sequent and for contracting two of the same formulas to one. Syllogism is restricted to one lower sentence; this is the cut rule:

$$\frac{\Gamma \to \Theta, A \quad A, \Delta \to \Lambda}{\Gamma, \Delta \to \Theta, \Lambda}$$

To deal with the logical connectives and quantifiers, Gentzen adapts the axiom systems developed by Hilbert and Bernays in the 1920s by turning the axioms governing a connective into rules introducing the connective in the antecedent and succedent of a sequent. For instance, axiom group (III),

III. $A \to A \vee B$
$B \to A \vee B$
$(B \to A) \to ((C \to A) \to (B \vee C \to A))$,

results in the rules

OES: $\dfrac{\Gamma \to \Theta, A}{\Gamma \to \Theta, A \vee B} \quad \dfrac{\Gamma \to \Theta, B}{\Gamma \to \Theta, A \vee B}$ OEA: $\dfrac{A, \Gamma \to \Theta \quad B, \Gamma \to \Theta}{A \vee B, \Gamma \to \Theta}$

The rules, together with axioms of the form $A \to A$, result in the system **LK** for classical logic, and **LJ** for intuitionistic logic, where **LJ** is like **LK** with the restriction that each sequent can contain at most one formula in the succedent. The soundness and completeness of these systems is proved in the last section of the paper, by showing that they derive the same formulas as ordinary axiomatic presentations of Hilbert (1928a) and Glivenko (1929) (for the intuitionistic case).

Gentzen's main result (1934) is the *Hauptsatz*. It states that any derivation in **LK** (or **LJ**) can be transformed into one that does not use the cut rule; thus it is now also called the cut-elimination theorem. It has some important consequences: It establishes the decidability of intuitionistic propositional logic, and provides new proofs of the consistency of predicate logic as well as the nonderivability of the principle of the excluded middle in intuitionistic propositional calculus. Gentzen also proves an extension of the Hauptsatz, now called the midsequent theorem: Every derivation of a prenex formula in **LK** can be transformed into one that is cut-free and in which all propositional inferences precede all quantifier inferences. An important consequence of this theorem is a form of Herbrand's theorem (see section 1.6.4).

The second main contribution of Gentzen (1934) is the introduction of calculi of natural deduction. It was intended to capture actual "natural" reasoning more accurately than axiomatic systems do. Such patterns of reasoning are for instance the methods of conditional proof (to prove a conditional, give a proof of the consequent under the assumption that the antecedent is true) and dilemma (if a conclusion C follows from both A and B individually, it follows from $A \vee B$). In natural deduction then, a derivation is a tree of formulas. The uppermost formulas are assumptions, and each formula is either an assumption, or must follow from preceding formulas according to one of the rules:

$$\frac{A \quad B}{A \& B} \qquad \frac{A \& B}{A} \qquad \frac{A \& B}{B} \qquad \frac{A}{A \vee B} \qquad \frac{B}{A \vee B} \qquad \frac{A \vee B \quad C \quad C}{C} \, [A][B]$$

$$\frac{Fa}{\forall x \, Fx} \qquad \frac{\forall x \, Fx}{Fa} \qquad \frac{Fa}{\exists x \, Fx} \qquad \frac{\exists x \, Fx \quad C}{C} \, [Fa]$$

$$\frac{B}{A \supset B} \, [A] \qquad \frac{A \quad A \supset B}{B} \qquad \frac{\wedge}{\neg A} \, [A] \qquad \frac{A \quad \neg A}{\wedge} \qquad \frac{\wedge}{D}$$

In the rules, the notation $[A]$ indicates that the subproof ending in the corresponding premise may contain any number of formulas for the form A as assumptions, and that the conclusion of the inference is then *independent* of these assumptions. A derivation is a proof of A, if A is the last formula of the derivation and is not dependent on any assumptions.

1.6 Itinerary VI. Proof Theory and Arithmetic

1.6.1 Hilbert's Program for Consistency Proofs

The basic aim and structure of Hilbert's program in the philosophy of mathematics is well known: To put classical mathematics on a firm foundation and to rescue it from the attempted *Putsch* of intuitionism, two things were to be accomplished. First, formalize classical mathematics in a formal system; second, give a direct, finitistic consistency proof for this formal system. This project is first outlined in Hilbert (1922c) and received its most popular presentation in "On the infinite" (1926). The project has an important philosophical aspect, which we cannot do justice here (see Zach 2006). This philosophical aspect is the finitist standpoint—the methodological position from which the consistency proofs were to be carried out. At its most basic, the finitist standpoint is characterized as the domain of reasoning about sequences of strokes (the finitist numbers), or sequences of signs in general. From the finitist standpoint, only such finite objects which, according to Hilbert, are "intuitively given" are admissible as objects of finitist reflection; specifically, the finitist standpoint cannot operate with or assume the existence of completed infinite totalities such as the set of all numbers. Furthermore, only such methods of construction and inference are allowed that are immediately grounded in the intuitive representation we have of finitist objects. This includes, for example, definition by primitive recursion and induction as the basic method of proof. A consistency proof for a formal system, in particular, has to take roughly the following form: Give a finitist method by which any given proof in the formal system of classical mathematics can be transformed into one which by its very form cannot be a derivation of a contradiction, such as $0 = 1$. Such a finitist consistency proof not only grounds classical mathematics, but also can be taken as a reductio of one of the intuitionist's motivations, viz., that classical reasoning may lead to outright contradictions, since the finitist methods themselves are acceptable intuitionistically.

Hilbert envisaged the consistency proof for classical mathematics to be accomplished in stages of consistency proofs for increasingly stronger systems, starting with propositional logic and ending with full set theory. The crucial development that enabled Ackermann and von Neumann to give partial solutions to the consistency problem was the invention of the ε-calculus around 1922.[71] The ε-calculus is an extension of quantifier-free logic and number theory by term forming ε-operators: If $A(a)$ is a formula, then $\varepsilon_a A(a)$ is a term, intuitively, the least a such that $A(a)$ is true. Using such ε-terms, it is then possible to define the quantifiers by $(\exists a) A(a) \equiv A(\varepsilon_a A(a))$ and $(\forall a) A(a) \equiv A(\varepsilon_a \overline{A(a)})$. The axioms governing the ε-operator are the so-called transfinite axioms

$$A(a) \to A(\varepsilon_a(A(a))) \quad \text{and}$$

$$\varepsilon_a A(a) \neq 0 \to \overline{A(\delta \varepsilon_a A(a))}.$$

The first axiom allows the derivation of the usual axioms for $\exists$ and $\forall$; the second derives the induction axiom (δ is the predecessor function). The ε-substitution method used by Ackermann and von Neumann goes back to an idea of Hilbert: In a given proof, replace the ε-terms by actual numbers so that the result is a derivation of the same formula; then apply the consistency proof for quantifier-free systems.

1.6.2 Consistency Proofs for Weak Fragments of Arithmetic

Around 1900, Hilbert began championing the axiomatic method as a foundational approach, not only to geometry but also to arithmetic. He proposed the axiomatic method in contradistinction to the *genetic method*, by which the reals were constructed out of the naturals (which were taken as primitive) through the usual constructions of the integer, rational, and finally real numbers through constructions such as Dedekind cuts. In Hilbert's opinion, the axiomatic method is to be preferred for "the final presentation and the complete logical grounding of our knowledge [of arithmetic]" (Hilbert 1900b). The first-order of business, then, is to provide an axiomatization of the reals, which Hilbert first attempted in "Über den Zahlbegriff" (1900b). To complete the "logical grounding", however, one would also have to prove the consistency (and completeness) of the axiomatization. For geometry, consistency proofs can be given by exhibiting models in the reals; but a consistency proof of arithmetic requires a direct method. Hilbert considered such a direct proof of consistency the most important question that has to be answered for the axiomatization of the reals, and he formulated it as the second of his "Mathematical problems" (Hilbert 1900a). Attempts at such a proof were made in (Hilbert 1905b) and his course on "Logical principles of mathematical thought" (1905a). It became clear that a successful direct consistency proof requires a further development of the underlying logical systems. This development was carried out by Russell and Whitehead, and following a period of intense study of the *Principia* between 1914 and 1917 in Göttingen (see Mancosu 1999a; 2003), Hilbert renewed his call for a direct consistency proof of arithmetic in "Axiomatic thought" (1918a). This was followed by an increased focus on foundations in Göttingen. Until 1920, Hilbert seems to have been sympathetic to Russell's logicist approach, but soon became dissatisfied by it. In his course "Problems of mathematical logic", he explains:

Russell starts with the idea that it suffices to replace the predicate needed for the definition of the union set by one that is extensionally equivalent, and which is not open to the same objections. He is unable, however, to exhibit such a predicate, but sees it as obvious that such a predicate exists. It is in this sense that he postulates the "axiom of reducibility", which states approximately the following: "For each predicate, which is formed by referring (once or multiple times) to the domain of predicates, there is an extensionally equivalent predicate, which does not make such reference."

With this, however, Russell returns from constructive logic to the axiomatic standpoint. ...

The aim of reducing set theory, and with it the usual methods of analysis, to logic, has not been achieved today and maybe cannot be achieved at all. (Hilbert 1920b, pp. 32–33)

Precipitated by increasing interest in Brouwer's intuitionism and Poincaré's and Weyl's predicativist approaches to mathematics (Weyl 1918a, 1919a), and especially Weyl's (1921c) conversion to intuitionism, Hilbert finally formulated his own approach to mathematical foundations. This approach combined his previous aim of providing a consistency proof that does not proceed by exhibiting a model, or reducing consistency to the consistency of a different theory, with a philosophical position delineating the acceptable methods for a direct consistency proof. In the same course on "Problems of mathematical logic", he presented a simple axiom system for the naturals, consisting of the axioms

$$1 = 1$$
$$(a = b) \to (a + 1 = b + 1)$$

$(a + 1 = b + 1) \rightarrow (a = b)$

$(a = b) \rightarrow ((a = c) \rightarrow (b = c))$

$a + 1 \neq 1.$

An equation between terms containing only 1's and +'s is called *correct* if it is either $1 = 1$, results from the axioms by substitution, or is the end formula of a proof from the axioms using modus ponens. The system was later extended by induction, but for the purpose of describing the kind of consistency proof he has in mind, Hilbert observed that the axiom system would be inconsistent in the sense of deriving a formula and its negation iff it were possible to derive a substitution instance of $a + 1 = 1$. In this case, then, a direct consistency proof requires a demonstration that no such formula can be the end formula of a formal proof.

Thus we are led to make the proofs themselves the object of our investigation; we are urged toward a *proof theory*, which operates with the proofs themselves as objects.

For the way of thinking of ordinary number theory the numbers are then objectively exhibitable, and the proofs about the numbers already belong to the area of thought. In our study, the proof itself is something which can be exhibited, and by thinking about the proof we arrive at the solution of our problem.

Just as the physicist examines his apparatus, the astronomer his position, just as the philosopher engages in critique of reason, so the mathematician needs his proof theory, in order to secure each mathematical theorem by proof critique.[72]

This is the first occurrence of the term "proof theory" in Hilbert's writings.[73] This approach to consistency proofs is combined with a philosophical position in Hilbert's address in Hamburg in July 1921 (1922c), which emphasizes the distinction between the "abstract operation with general concept-scopes [which] has proved to be inadequate and uncertain", and contentual arithmetic which operates on signs. In a famous passage, Hilbert makes clear that the immediacy and security of mathematical "contentual" thought about signs is a precondition of logical thought in general, and hence is the only basis on which a direct consistency proof for formalized mathematics must be carried out:

As a precondition for the application of logical inferences and for the activation of logical operations, something must already be given in representation: certain extra-logical discrete objects, which exist intuitively as immediate experience before all thought. If logical inference is to be certain, then these objects must be capable of being completely surveyed in all their parts, and their presentation, their difference, their succession (like the objects themselves) must exist for us immediately, intuitively, as something that cannot be reduced to something else.... The solid philosophical attitude that I think is required for the grounding of pure mathematics—as well as for all scientific thought, understanding, and communication—is this: *In the beginning was the sign.* (Hilbert 1922c, pp. 1121–22)

Just as a contentual mathematics of number signs enjoys the epistemological priority claimed by Hilbert, so does contentual reasoning about combinations of signs in general.

Hence, contentual reasoning about formulas and formal proofs, in particular, contentual demonstrations that certain formal proofs are impossible, are the aim of proof theory and metamathematics. This philosophical position, together with the ideas about how such contentual reasoning about derivations can be applied to prove consistency of axiomatic systems—ideas outlined in the 1920 course and going back to 1905—make up Hilbert's program for the foundation of mathematics.

In the following two years, Hilbert and Bernays elaborate the research project in a series of courses and talks (Hilbert 1922a, 1923a, Hilbert and Bernays 1923b, Bernays 1922a). The courses from 1921–22 and 1922–23 are most important. It is there that Hilbert introduces the ε-calculus in 1921–22 to deal with quantifiers and the approach using the ε-substitution method as a proof of consistency for systems containing quantification and induction. The system used in 1922–23 is given by the following axioms (Hilbert and Bernays 1923b, pp. 17, 19):

1. $A \to B \to A$
2. $(A \to A \to B) \to A \to B$
3. $(A \to B \to C) \to (B \to A \to C)$
4. $(B \to C) \to (A \to B) \to A \to C$
5. $A \,\&\, B \to A$
6. $A \,\&\, B \to B$
7. $A \to B \to A \,\&\, B$
8. $A \to A \vee B$
9. $B \to A \vee B$
10. $(A \to C) \to (B \to C) \to A \vee B \to C$
11. $A \to \overline{A} \to B$
12. $(A \to B) \to (\overline{A} \to B) \to B$
13. $a = a$
14. $a = b \to A(a) \to A(b)$
15. $a + 1 \neq 0$
16. $\delta(a + 1) = a$.

Here, "$+ 1$" is a unary function symbol. In Hilbert's systems, Latin letters are variables; in particular, $a, b, c, \ldots$ are individual variables and $A, B, C, \ldots$, are formula variables. The rules of inference are modus ponens and substitution for individual and formula variables.

The idea of the consistency proof is this: Suppose a proof of a contradiction is available. (We may assume that the end formula of this proof is $0 \neq 0$.)

1. *Resolution into proof threads.* First, we observe that by duplicating part of the proof and leaving out steps, we can transform the derivation to one where each formula (except the end formula) is used exactly once as the premise of an inference. Hence, the proof is in tree form.

2. *Elimination of variables.* We transform the proof so that it contains no free variables. This is accomplished by proceeding backward from the end formula: The end formula contains no free variables. If a formula is the conclusion of a substitution rule, the inference is removed. If a formula is the conclusion of modus ponens it is of the form

$$\frac{\mathfrak{A} \qquad \mathfrak{A} \to \mathfrak{B}}{\mathfrak{B}'}$$

where $\mathfrak{B}'$ results from $\mathfrak{B}$ by substituting terms (*functionals*, in Hilbert's terminology) for free variables. If these variables also occur in $\mathfrak{A}$, we substitute the same terms for them. Variables in $\mathfrak{A}$ that do not occur in $\mathfrak{B}$ are replaced with 0. This yields a formula $\mathfrak{A}'$ not containing variables. The inference is replaced by

$$\frac{\mathfrak{A}' \quad \mathfrak{A}' \to \mathfrak{B}'}{\mathfrak{B}'}$$

3. *Reduction of functionals.* The remaining derivation contains a number of terms which now have to be reduced to numerical terms (i.e., standard numerals of the form $(\ldots(0+1)+\ldots)+1$). In this case, this is done easily by rewriting innermost subterms of the form $\delta(0)$ by 0 and $\delta(\mathfrak{n}+1)$ by $\mathfrak{n}$. In later stages, the set of terms is extended by function symbols introduced by recursion, and the reduction of functionals there proceeds by calculating the function for given numerical arguments according to the recursive definition.

To establish the consistency of the axiom system, Hilbert suggests, we have to find a decidable property of formulas [*konkret feststellbare Eigenschaft*] so that every formula in a derivation which has been transformed using the foregoing steps has the property, and the formula $0 \neq 0$ lacks it. The property Hilbert proposes to use is *correctness*. This, however, is not to be understood as truth in a model: The formulas still occurring in the derivation after the transformation are all Boolean combinations of equations between numerals. An equation between numerals $\mathfrak{n} = \mathfrak{m}$ is *correct* if $\mathfrak{n}$ and $\mathfrak{m}$ are equal, and the negation of an equality is correct if $\mathfrak{n}$ and $\mathfrak{m}$ are not equal.

If we call a formula which does not contain variables or functionals other than numerals an *"explicit [i.e., numerical] formula"*, then we can express the result obtained thus: Every provable explicit formula is end formula of a proof all the formulas of which are explicit formulas.

This would have to hold in particular of the formula $0 \neq 0$, if it were provable. The required proof of consistency is thus completed if we show that there can be no proof of the formula which consists of only explicit formulas.

To see that this is impossible it suffices to find a concretely determinable [*konkret feststellbar*] property, which first of all holds of all explicit formulas which result from an axiom by substitution, which furthermore transfers from premises to end formula in an inference, which however does not apply to the formula $0 \neq 0$. (Hilbert 1922b, part 2, 27–28)

This basic model for a consistency proof is then extended to include terms containing function symbols defined by primitive recursion, and terms containing the ε-operator. Hilbert's *Ansatz* for eliminating ε-terms from formal derivations is first outlined in the 1921–22 lectures and in more detail in the 1922–23 course.[74]

Suppose a proof involves only one ε-term $\varepsilon_a \mathfrak{A}(a)$ and corresponding *critical formulas*

$$\mathfrak{A}(\mathfrak{k}_i) \to \mathfrak{A}(\varepsilon_a \mathfrak{A}(a)),$$

that is, substitution instances of the transfinite axiom

$$A(a) \to A(\varepsilon_a A(a)).$$

We replace $\varepsilon_a \mathfrak{A}(a)$ everywhere with 0, and transform the proof as before by rewriting it in tree form ("dissolution into proof threads"), eliminating free variables, and evaluating numerical terms involving primitive recursive functions. Then the critical formulas take the form

$$\mathfrak{A}(\mathfrak{z}_i) \to \mathfrak{A}(0),$$

where $\mathfrak{z}_i$ is the numerical term to which $\mathfrak{f}_i$ reduces. A critical formula can now only be false if $\mathfrak{A}(\mathfrak{z}_i)$ is true and $\mathfrak{A}(0)$ is false. If that is the case, repeat the procedure, now substituting $\mathfrak{z}_i$ for $\varepsilon_a \mathfrak{A}(a)$. This yields a proof in which all initial formulas are correct and no ε terms occur.

If critical formulas of the second kind, that is, substitution instances of the induction axiom,

$$\varepsilon_a A(a) \neq 0 \to \overline{A(\delta \varepsilon_a A(a))},$$

also appear in the proof, the witness $\mathfrak{z}$ has to be replaced with the least $\mathfrak{z}'$ so that $\mathfrak{A}(\mathfrak{z}')$ is true.

The challenge is to extend this procedure to (a) cover more than one ε-term in the proof, (b) take care of nested ε-terms, and last (c) extend it to second-order ε's and terms involving them, that is, $\varepsilon_f \mathfrak{A}_a(f(a))$, which are used in formulations of second-order arithmetic. This was attempted in Ackermann's (1924) dissertation.

1.6.3 Ackermann and von Neumann on Epsilon Substitution

Ackermann's dissertation (1924) is a milestone in the development of proof theory. The work contains the first unified presentation of a system of second-order arithmetic based on the ε-calculus, a complete and correct consistency proof of the ε-less fragment (an extension of what is now known as primitive recursive arithmetic, PRA), and an attempt to extend Hilbert's ε-substitution method to the full system.

The consistency proof for the ε-free fragment extends a sketch of a consistency proof for primitive recursive arithmetic contained in Hilbert and Bernays' 1922–23 lectures. For primitive recursive arithmetic, the basic axiom system is extended by definitional equations for function symbols which define the corresponding functions recursively, for example,

$$\psi(0, \vec{c}) = \mathfrak{a}(\vec{c})$$
$$\psi(a+1, \vec{c}) = \mathfrak{b}(a, \psi(a, \vec{c}), \vec{c})$$

To prove consistency for such a system, the "reduction of functionals" step has to be extended to deal with terms containing the function symbols defined by evaluating innermost terms with leading function symbol ψ according to the primitive recursion specified by the defining equations. It should be noted right away that such a consistency proof requires the possibility of evaluating an arbitrary primitive recursive function, and as such exceeds primitive recursive methods. This means that Hilbert, already in 1922, accepted nonprimitive recursive methods as falling under the methodological, "finitary" standpoint of proof theory. Ackermann's dissertation extends this consistency proof by also dealing with what might be called second-order primitive recursion. A second-order primitive recursive definition is of the form

$$\phi_{\vec{b}_i}(0, \vec{f}(\vec{b}_i), \vec{c}) = \mathfrak{a}_{\vec{b}_i}(\vec{f}(\vec{b}_i), \vec{c})$$

$$\phi_{\vec{b}_i}(a + 1, \vec{f}(\vec{b}_i), \vec{c}) = \mathfrak{b}_{\vec{b}_i}(a, \phi_{\vec{d}_i}(a, \vec{f}(\vec{d}_i), \vec{c}), \vec{f}(\vec{b}_i))$$

The subscript notation indicates λ-abstraction; in modern notation the schema would more conspicuously be written as

$$\phi(0, \lambda \vec{b}_i.\vec{f}(\vec{b}_i), \vec{c}) = \mathfrak{a}(\lambda \vec{b}_i.\vec{f}(\vec{b}_i), \vec{c})$$

$$\phi(a + 1, \lambda \vec{b}_i.\vec{f}(\vec{b}_i), \vec{c}) = \mathfrak{b}(a, \phi(a, \lambda \vec{d}_i.\vec{f}(\vec{d}_i), \vec{c}), \lambda \vec{b}_i.\vec{f}(\vec{b}_i))$$

Second-order primitive recursion allows the definition of the Ackermann function, which was shown by Ackermann (1928b) to be itself not primitive recursive.

The first consistency proof given by Ackermann is for this system of second-order primitive recursive arithmetic. While for PRA, the reduction of functionals only requires the relatively simple evaluation of primitive recursive terms, the situation is more complicated for second-order primitive recursion. Ackermann locates the difficulty in the following: Suppose you have a functional $\phi_b(2, \mathfrak{b}(b))$, where ϕ is defined by

$$\phi_b(0, f(b)) = f(1) + f(2)$$

$$\phi_b(a + 1, f(b)) = \phi_b(a, f(b)) + f(a) \cdot f(a + 1)$$

Here, $\mathfrak{b}(b)$ is a term that denotes a function, and so there is no way to replace the variable b with a numeral before evaluating the entire term. In effect, the variable b is bound (in modern notation, the term might be more suggestively written $\phi(2, \lambda b.\mathfrak{b}(b))$). To reduce this term, we apply the recursion equations for ϕ twice and end up with a term like

$$\mathfrak{b}(1) + \mathfrak{b}(2) + \mathfrak{b}(0) \cdot \mathfrak{b}(1) + \mathfrak{b}(1) \cdot \mathfrak{b}(2).$$

The remaining $\mathfrak{b}$'s might in turn contain ϕ, for example, $\mathfrak{b}(b)$ might be $\phi_c(b, \delta(c))$, in which case the above expression would be

$$\phi_c(1, \delta(c)) + \phi_c(2, \delta(c)) + \phi_c(0, \delta(c)) \cdot \phi_c(1, \delta(c)) + \phi_c(1, \delta(c)) \cdot \phi_c(2, \delta(c)).$$

By contrast, reducing a term $\psi(3)$ where ψ is defined by first-order primitive recursion results in a term which does not contain ψ, but only the function symbols occurring on the right-hand side of the defining equations for ψ.

To overcome this difficulty, Ackermann defines a system of indexes of terms containing second-order primitive recursive terms and an ordering on these indexes. Ackermann's indexes are, essentially, ordinal notations for ordinals $< \omega^{\omega^\omega}$, and the ordering he defines corresponds to the ordering on the ordinals. He then defines a procedure to evaluate such terms by successively applying the defining equations; each step in this procedure results in a new term whose index is less than the index of the preceding term. Because the ordering of the indexes is wellfounded, this constitutes a proof that the procedure always terminates, and hence that the process of reduction of functionals in the consistency proof comes to an end, resulting in a proof with only correct equalities and inequalities between numerical terms (not containing function symbols).[75] This proof very explicitly proceeds by transfinite induction up to ω^{ω^ω}, and foreshadows Gentzen's (1936) use of transfinite induction up to ε_0. Ackermann was completely aware of the involvement of transfinite induction in this case, but did not see in it a violation of the finitist standpoint:

> The disassembling of functionals by reduction does not occur in the sense that a finite ordinal is decreased each time an outermost function symbol is eliminated. Rather, to each functional corresponds as it were a transfinite ordinal number as its rank, and the theorem that a constant functional is reduced to a numeral after carrying out finitely many operations corresponds to the other [theorem], that if one descends from a transfinite ordinal number to ever smaller ordinal numbers, one has to reach zero after a finite number of steps. Now there is naturally no mention of transfinite sets or ordinal numbers in our metamathematical investigations. It is however interesting, that the mentioned theorem about transfinite ordinals can be formulated so that there is nothing transfinite about it any more. (Ackermann 1924, pp. 13–14)

The full system for which Ackermann attempted to give a consistency proof in the second part of the dissertation consists of the system of second-order primitive recursive arithmetic together with the transfinite axioms:

1. $A(a) \to A(\varepsilon_a A(a))$ \qquad $A_a(f(a)) \to A_a((\varepsilon_f A_b(f(b))(a)))$
2. $A(\varepsilon_a A(a)) \to \pi_a A(a) = 0$ \qquad $A_a(\varepsilon_f A_b(f(b))(a)) \to \pi_f A_a(f(a)) = 0$
3. $\overline{A(\varepsilon_a A(a))} \to \pi_a A(a) = 1$ \qquad $\overline{A_a(\varepsilon_f A_b(f(b))(a))} \to \pi_f A_a(f(a)) = 1$
4. $\varepsilon_a A(a) \neq 0 \to \overline{A(\delta(\varepsilon_a A(a)))}$

The intuitive interpretation of ε and π, based on these axioms is this: $\varepsilon_a \mathfrak{A}(a)$ is a witness for $\mathfrak{A}(a)$ if one exists, and $\pi_a \mathfrak{A}(a) = 1$ if $\mathfrak{A}(a)$ is false for all a, and $= 0$ otherwise. The π functions are not necessary for the development of mathematics in the axiom system. They do, however, serve a function in the consistency proof, viz., to keep track of whether a value of 0 for $\varepsilon_a \mathfrak{A}(a)$ is a "default value" (that is, a trial substitution for which $\mathfrak{A}(a)$ may or may not be true) or an actual witness (a value for which $\mathfrak{A}(a)$ has been found to be true).

To give a consistency proof for this system, Ackermann first has to extend the ε-substitution method to deal with proofs in which terms containing more than one ε-operator (and corresponding critical formulas) occur, and then argue (finitistically), that the procedure so defined always terminates in a substitution of numerals for ε-terms which transform the critical formulas into correct formulas of the form $\mathfrak{A}(\mathfrak{t}) \to \mathfrak{A}(\mathfrak{s})$ (where $\mathfrak{A}$, $\mathfrak{t}$, and $\mathfrak{s}$ do not contain ε-operators or primitive recursive function symbols). To solve the first task, Ackermann has to deal with the various possibilities in which ε-operators can occur in the scope of other ε's. For instance, an instance of the transfinite axiom might be

$$\mathfrak{A}(\mathfrak{t}, \varepsilon_y \mathfrak{B}(y)) \to \mathfrak{A}(\varepsilon_x \mathfrak{A}(x, \varepsilon_y \mathfrak{B}(y)), \varepsilon_y \mathfrak{B}(y))$$

To find a substitution for $\varepsilon_x \mathfrak{A}(x, \varepsilon_y \mathfrak{B}(y))$ here, it is necessary to first have a substitution for $\varepsilon_y \mathfrak{B}(y)$. This case is rather benign, since the value for $\varepsilon_y \mathfrak{B}(y)$ can be determined independently of that for $\varepsilon_x \mathfrak{A}(x, \varepsilon y \mathfrak{B}(y))$. If $\varepsilon_y \mathfrak{B}(y)$ occurs in the term t on the left-hand side, the situation is more complicated. We might have, for example, a critical formula of the form

$$\mathfrak{A}(\varepsilon_y \mathfrak{B}(y, \varepsilon_x A(x))) \to \mathfrak{A}(\varepsilon_x \mathfrak{A}(x))$$

With an initial substitution of 0 for $\varepsilon_x \mathfrak{A}(x)$, we can determine a value for $\varepsilon_y \mathfrak{B}(y, \varepsilon_x \mathfrak{A}(x))$, that is, for $\varepsilon_y \mathfrak{B}(y, 0)$. With this value for $\varepsilon_y \mathfrak{B}(y)$, we then find a value for $\varepsilon_x \mathfrak{A}(x)$. This, however, now might change the "correct" substitution for $\varepsilon_x \mathfrak{A}(x)$, say, to n, and hence the initial determination of the value of the term on the left-hand side changes: We now need a value for $\varepsilon_y \mathfrak{B}(y, n)$.

The procedure proposed by Ackermann is too involved to be discussed here (see Zach 2003 for details). In short, what is required is an ordering of terms based on the level of nesting and of cross-binding of ε's, and a procedure based on this ordering which successively approximates a "solving substitution", that is, an assignment of numerals to ε-terms which results in all correct critical formulas. In this successive approximation, the values found for some ε-terms may be discarded if the substitutions for enclosed ε-terms change. A correct consistency proof would then require a proof that this procedure does in fact always terminate with a solving substitution. Unfortunately, Ackermann's argument in this regard is opaque.

The system to which Ackermann applied the ε-substitution method, as indicated, is a system of second-order arithmetic. Ackermann (and Bernays) soon realized that the proposed consistency proof had problems. Already in the published version, a footnote on p. 9 restricts the system in the following way: Only such terms are allowed in substitutions for formula and function variables in which individual variables do not occur in the scope of a second-order ε. Von Neumann clarified the restriction and its effect: In Ackermann's system, the second-order ε-axiom $A(f) \to \varepsilon_f A(f)$ does duty for the comprehension principle. In this system, the comprehension principle is $(\exists f)(\forall x)(f(x) = t)$, where t is a term possibly containing ε-terms. Under Ackermann's restriction, only such instances of the comprehension principle are permitted in which x is not in the scope of a second-order ε-operator; essentially this guarantees the existence of only such f's which can be defined by arithmetical formulas. Von Neumann (1927) also remarked that Ackermann's restriction makes the system predicative; it is roughly of the strength of the system ACA_0.

This alone restricts the consistency proof to a system much weaker than analysis; however, other problems and lacunae were known to Ackermann, one being that the proof does not cover ε-extensionality,

$$(\forall f)(A(f) \leftrightarrow B(f)) \to \varepsilon_f A(f) = \varepsilon_f B(f),$$

which serves as the ε-analog of the axiom of choice. Ackermann continued to work on the proof, amending and correcting the ε-substitution procedure even for first-order ε-terms. These corrections used ideas of von Neumann (1927), which was already completed in 1925. Von Neumann (1927) used a different terminology than Ackermann, and the precise connection between Ackermann's and von Neumann's proofs is not clear. Von Neumann's system does not include the induction axiom explicitly, because induction can be proved once a suitable second-order apparatus is available. Hence, the consistency proof for the first-order fragment of his theory does not deal with induction, whereas Ackermann's system has an induction axiom in the form of the second ε-axiom, and his substitution procedure takes into account critical formulas of this second kind. Another significant feature of von Neumann's proof is the precision with which it is executed: Von Neumann gives numerical bounds for the number of steps required until a solving substitution is found.[76]

Ackermann gave a revised ε-substitution proof, using von Neumann's ideas, and communicated it to Bernays in 1927. Both Ackermann and Bernays believed that the new proof would go through for full first-order arithmetic. Hilbert reported on this result in his lectures in Hamburg 1928 (1928a) (see also Bernays 1928d) and Bologna (Hilbert 1928b, 1929). Only with Gödel's (1930b, 1931) incompleteness results did it become clear that the consistency proofs did not even go through for first-order arithmetic. Bernays later gave an analysis of Ackermann's second proof (Hilbert and Bernays 1939) and showed that the bounds obtained hold for induction restricted to quantifier-free formulas, but not for induction axioms of higher complexity. Ackermann eventually, using ideas from Gentzen, gave an ε-substitution proof for full first-order arithmetic (1940).

1.6.4 Herbrand's Theorem

Herbrand's (1930) thesis "Investigations in proof theory" marks another milestone in the development of first-order proof theory. Herbrand's main influences in this work were Russell and Whitehead's *Principia*, from which he took the notation and some of the presentations of his logical axioms, the work of the Hilbert school, which provided the motivations and aims for proof theoretic research; and Löwenheim's (1915) and Skolem's (1920) work on normal forms. The thesis contains a number of important results, among them a proof of the deduction theorem and a proof of quantifier elimination for induction-free successor arithmetic (no addition or multiplication). The most significant contribution, of course, is Herbrand's theorem.

Herbrand's theorem shares a fundamental feature with Hilbert's approaches to proof theory and consistency proofs: Consistency for systems including quantifiers (ε-terms) is established by giving a procedure that removes quantifiers from a proof, reducing proofs containing such "ideal elements" to quantifier-free (essentially, propositional) proofs. Herbrand's theorem provides a general necessary and sufficient condition for when a formula of the predicate calculus is provable by reducing such provability to the provability of an associated "expansion" in the propositional calculus. The way such an expansion is obtained is closely related to obtaining a Skolem normal form of the formula. The Löwenheim–Skolem theorem reduces the validity of a formula in general to its validity in a canonical countable model. Skolem's and Löwenheim's methods, however, were semantic and used infinitary methods, both features that make it unsuitable for employment in the framework of Hilbert's finitist program. Herbrand's theorem can thus be seen as giving finitary meaning to the Löwenheim–Skolem theorem.

Let us now give a brief outline of the theorem. We follow Herbrand (1931b), which is in some respects clearer than the original (1930). Suppose A is a formula of first-order logic. For simplicity, we assume A is in prenex normal form; Herbrand gave his argument without making this restriction. So let P be $(Q_1 x_1)\ldots(Q_n x_n) B(x_1, \ldots, x_n)$, where Q_i is either $\forall$ or $\exists$, and B is quantifier-free. Then the Herbrand normal form H of A is obtained by removing all existential quantifiers from the prefix of A, and replacing each universally quantified x_i by a term $f_i(x_{j_1}, \ldots, x_{j_n})$, where $x_{j_1}, \ldots, x_{j_n}$ are the existentially quantified variables preceding x_i. Herbrand (1931a) calls this the elementary proposition associated with P, and f_i is the index function associated with x_i.

To state the theorem, we have to define what Herbrand calls *canonical domains of order* k. This notion, in essence, is a first-order interpretation with the domain being the term model generated from certain initial elements and function, and the terms all have height $\leq k$. (The height of a term is defined as usual: Constants have height 0, and a term $f_j(t_1, \ldots, t_k)$ has height $h + 1$ if h is the maximum of the heights of $t_1, \ldots, t_k$.) Herbrand did not use terms explicitly as objects of the domain, but instead considered domains consisting of letters, such that each term (of height $\leq k$) has an element of the domain associated with it as its value and such that if terms $t_1, \ldots, t_k$ have values $b_1, \ldots, b_k$, and the value of $f_i(b_1, \ldots, b_k)$ is c, then the value associated with $f(t_1, \ldots, t_k)$ is also c. A domain is *canonical* if it furthermore satisfies the condition that any two distinct terms have distinct values associated with them (i.e., the domain is freely generated from the initial elements and the function symbols). Last, a domain is of order k if each term of height $\leq k$ with constants only from among the initial elements has a value in the domain, but some term of height $k + 1$ does not.

The canonical domain of order k associated with P then is the canonical domain of order k with some nonempty set of initial elements and the functions occurring in the Herbrand normal form H of P. P is true in the canonical domain if some substitution of elements for the free variables in H makes H true in the domain, and false otherwise. Herbrand's statement of the theorem then is:

1. If [for some k] there is no system of logical values [truth value assignment to the atomic formulas] making P false in the associated canonical domain of order k, then P is an identity [provable in first-order logic].

2. If P is an identity, then there is a number k obtainable from the proof of P, such that there is no system of logical values making P false in every associated canonical domain of order equal to or greater than k. (Herbrand 1931b, p. 229)

By introducing canonical domains of order k, Herbrand has thus reduced provability of P in the predicate calculus to the validity of H in certain finite term models. If $H_1, \ldots, H_{n_k}$ are all the possible substitution instances of H in the canonical domain of order k, then the theorem may be reformulated as: (1) If $\bigvee H_i$ is a tautology, then P is provable in first-order logic; (2) if P is provable in first-order logic, then there is a k obtainable from the proof of P so that $\bigvee H_i$ is a tautology.

Herbrand's original proof contained a number of errors that were found by Peter Andrews and corrected by Dreben, Andrews, and Aanderaa (1963); Gödel had independently found a correction (see Goldfarb 1993; Andrews 2003 gives a detailed account of the discovery of the errors). Gentzen (1934) gave a different proof based on the midsequent theorem, which, however, only applies to prenex formulas and does not provide a bound on the size of the Herbrand disjunction $\bigvee H_i$. Another early complete and correct proof was given by Bernays (Hilbert and Bernays 1939) using the ε-calculus.

Herbrand was able to apply the fundamental theorem to give consistency proofs of various fragments of arithmetic, including the case of arithmetic with quantifier-free induction. The idea is to reduce the consistency of arithmetic with quantifier-free induction to induction-free (primitive recursive) arithmetic. This is done by introducing new primitive recursive functions that "code" the induction axioms used. The proof of Herbrand's theorem then produces finite term models for the remaining axioms, and consistency is established (Herbrand 1931a).

1.6.5 Kurt Gödel and the Incompleteness Theorems

Hilbert had two main aims in his program in the foundation of mathematics: first, a finitistic consistency proof of all of mathematics, and second, a precise mathematical justification for his belief that all well-posed mathematical problems are solvable, that is, that "in mathematics, there is no *ignorabimus*". This second aim resulted in two specific convictions: that the axioms of mathematics, in particular, of number theory, are complete in the sense that for every formula A, either A or $\sim A$ is provable,[77] and second that the validities of first-order logic are decidable (the decision problem). The hopes of achieving both aims were dashed in 1930, when Gödel proved his incompleteness theorems (1930b, 1931). The summary of his results (Gödel 1930b) addresses the impact of the results quite explicitly:

I. The system S [of *Principia*] is *not* complete [*entscheidungsdefinit*]; that is, it contains propositions A (and we can in fact exhibit such propositions) for which neither A nor $\overline{A}$ is provable and, in particular, it contains (even for decidable properties F of natural numbers) undecidable problems of the simple structure $(E x) F(x)$, where x ranges over the natural numbers.

II. Even if we admit all the logical devices of *Principia mathematica* ... in metamathematics, there does *not* exist a *consistency proof* for the system S (still less so if we restrict the means of proof in any way). (Gödel 1930b, pp. 141–43)

Soon thereafter, Church and Turing were able to show, using some of the central ideas in Gödel (1931), that the remaining aim of proving the decidability of predicate logic was likewise doomed to fail (Church 1936a,b, Turing 1937).

Gödel obtained his results in the second half of 1930. After proving the completeness of first-order logic, a problem posed by Hilbert and Ackermann (1928), Gödel set to work on proving the consistency of analysis (recall that according to Hilbert (1929), the consistency of arithmetic was already established). Instead of directly giving a finitistic proof of analysis, Gödel attempted to first reduce the consistency of analysis to that of arithmetic, which led him to consider ways to enumerate the symbols and proofs of analysis in arithmetical terms. It soon became evident to him that truth of number-theoretic statements is not definable in arithmetic, by reasoning analogous to the liar paradox. By the end of summer 1930, he had a proof that the analogous fact about *provability* is formalizable in the system of *Principia*, and hence that there are undecidable propositions in *Principia*. At a conference in Königsberg in September 1930, Gödel mentioned the result to von Neumann, who inquired whether the result could be formalized not only in type theory but already in first-order arithmetic. Gödel subsequently showed that the coding mechanism he had come up with could be carried out with purely arithmetical methods using the Chinese remainder theorem. Thus the first incompleteness theorem, that arithmetic contains undecidable propositions, was established. The second incompleteness theorem, namely, that in particular the statement formalizing consistency of number theory is such an undecidable arithmetical statement, was found shortly thereafter (and also independently by von Neumann).[78]

Let us now give a brief outline of the proof. The system P Gödel considers is a version of simple type theory in addition to Peano arithmetic. To carry out the formalization of predicates about formulas and proofs, Gödel introduces what is now known as "Gödel numbering". To each symbol of the system P a natural number is associated. A finite sequence of symbols a (e.g., a formula) can then be coded by $\Phi(a) = 2^{n_1} \cdot 3^{n_2} \cdots p_k^{n_k}$, where k is the length of the sequence, p_k is the kth prime, and n_i is the Gödel code of the ith symbol in the sequence. Similarly, a sequence of formulas (i.e., a sequence of sequences of numbers) with codes $n_1, \ldots, n_k$ is coded by $2^{n_1} \cdot 3^{n_2} \cdots p_k^{n_k}$.

To carry out the metamathematical treatment of formulas and proofs within the system, Gödel next defines the class of primitive recursive functions and relations of natural numbers (he simply calls them "recursive") and proves (theorems I–IV) that primitive recursive functions and relations are closed under composition, the logical operations of negation, disjunction, conjunction, bounded minimization, and bounded quantification. Using this characterization, he then shows that a collection of 45 functions can be defined primitive recursively. The functions are those necessary to carry out simple manipulations on formulas and proofs, or represent predicates about formulas and proofs. For instance, (31) is the function $Sb(x_y^v)$, the function the value of which is the code of a formula that results from the formula A (with code x) where every free occurrence of the variable with code v is replaced by the term with code y; (45) is the primitive recursive relation $x\,B\,y$, which holds if x is the code of a proof of a formula with code y. (46), finally is $Bew(x)$, expressing that x is the code of a provable formula with code x. $Bew(x)$ is not primitive recursive, because it results from $x\,B\,y$ by unbounded existential generalization: $Bew(x) \equiv (E\,y)y\,B\,x$. Gödel then proves (theorem V) that every recursive relation is numeralwise representable in P, that is, that if $R(x_1, \ldots, x_n)$ is a formula representing a recursive relation (according to the characterization of recursive relations given in theorems I–IV), then:

1. if $R(n_1, \ldots, n_k)$ is true, then P proves $Bew(m)$, where m is the code of $R(n_1, \ldots, n_k)$,

   and

2. if $R(n_1, \ldots, n_k)$ is false, then P proves $Bew(m)$, where m is the code of $\sim R(n_1, \ldots, n_k)$.

Then Gödel proves the main theorem.

Theorem VI *For every ω-consistent recursive class κ of* FORMULAS *there are recursive* CLASS SIGNS *r such that neither v Gen r nor Neg(v Gen r) belongs to Flg(κ) (where v is the* FREE VARIABLE *of r*. (Gödel 1931, p. 173)

Here κ is the recursive relation defining a set of codes of formulas to be considered as axioms, r is the code of a recursive formula $A(v)$ (i.e., one containing no unbounded quantifiers) with free variable v, v Gen r is the code of the generalization $(v)A(v)$ of $A(v)$, Neg(v Gen r) the code of its negation $\sim(v)A(v)$, and Flg(κ) is the set of codes of formulas that are provable in P together with κ. We may thus restate theorem IV somewhat more perspicuously thus: If P_κ is an ω-consistent theory resulting by adding a recursive set of axioms κ to P, then there is a formula $A(x)$ such that neither $(x)A(x)$ nor $\sim(x)A(x)$ is provable in P_κ. The requirement that P_κ is ω-consistent states that for no formula $A(x)$ does P_κ prove both $A(n)$ for all numerals n and $\sim(x)A(x)$; Rosser (1936) later weakened this requirement to the simple consistency of P_κ.

In the following sections, Gödel sharpens the result in several ways. First, he shows that (theorem VII) primitive recursive relations are arithmetical, that is, that the basic functions +, and × of arithmetic suffice to express all primitive recursive functions (this is where the Chinese remainder theorem is used). From this, theorem VIII follows, namely, that not only are there undecidable propositions of the form $(x)A(x)$ with A recursive (in particular, possibly using exponentiation x^y) but even with $A(x)$ arithmetical (i.e., containing only + and ×). Finally, in section 4, Gödel states the second incompleteness theorem.

Theorem XI *Let κ be any recursive consistent class of* FORMULAS; *then the* SENTENTIAL FORMULA *stating that κ is consistent is not κ*-PROVABLE; *in particular, the consistency of P is not provable in P, provided P is consistent (in the opposite case, of course, every proposition is provable).* (Gödel 1931, p. 193)

Although theorems VI and XI are formulated for the relatively strong system P, Gödel remarks that the only properties of P which enter into the proof of theorem VI are that the axioms are recursively definable, and that the recursive relations can be defined within the system. This applies, so Gödel, also to systems of set theory as well as to number theoretical systems such as that of von Neumann (1927).

Gödel's result is of great importance to the development of mathematical logic after 1930, but its most immediate impact at the time consisted in the doubts it cast on the feasibility of Hilbert's program. Von Neumann and Bernays immediately realized that the result shows that no consistency proof for a formal system of mathematics can be given by methods which can be formalized within the system—and since finitistic methods presumably were so formalizable in relatively weak number theoretic systems already, no finitistic consistency proofs could be given for such systems. This led Gentzen (1935, 1936), in particular, to rethink the role of consistency proofs and the character of finitistic reasoning; following him, work in proof theory has concentrated on, in a sense, *relative* consistency proof.

From [Gödel's incompleteness theorems] it follows that the consistency of elementary number theory, for example, cannot be established by means of *part* of the methods of proof used in elementary number theory, nor indeed by *all* of these methods. To what extent, then, is a genuine reinterpretation [*Zurückführung*] still possible?

It remains quite conceivable that the consistency of elementary number theory can in fact be verified by means of techniques which, in part, no longer belong to elementary number theory, but which can nevertheless be considered to be *more reliable* than the doubtful components of elementary number theory itself. (Gentzen 1936, p. 139)

Gentzen's proof uses transfinite induction on constructive ordinals $< \varepsilon_0$, and argues that these methods in fact are finitary, and hence "more reliable" than the infinitistic methods of elementary number theory.[79]

1.7 Itinerary VII. Intuitionism and Many-Valued Logics

1.7.1 Intuitionistic Logic

1.7.1.1 Brouwer's Philosophy of Mathematics

One of the most important positions in philosophy of mathematics of the 1920s was the intuitionism of Luitzen Egbertus Jan Brouwer (1881–1966).[80] Although our emphasis will be on the logical developments that emerged from Brouwer's intuitionism (as opposed to his philosophy of mathematics or the development of intuitionistic mathematics), it is essential to begin by saying something about his position in philosophy of mathematics. The essay "Intuitionism and Formalism" (1912b) contains many of the theses characteristic of Brouwer's approach. In it Brouwer discusses on what grounds one can base the conviction about the "unassailable exactness" of mathematical laws and distinguishes the position of the intuitionist from that of the formalist. The former, represented mainly by the school of French analysts (Baire, Borel, Lebesgue),[81] would posit the human mind as the source of the exactness; by contrast the formalist, by which Brouwer also means realists such as Cantor, would say that the exactness resides on paper. This rough and ready characterization of the situation, although objectionable, is very typical of Brouwer's style and perhaps contributed to the appeal of his radical proposal. Brouwer traces the origins of the intuitionist position back to Kant.[82] For Kant, time and space were the forms of our intuition, which shaped our perception of the world. He famously defended the idea that geometry and arithmetic are synthetic a priori. Brouwer only retains part of the Kantian intuitionism, in that he rejects the apriority of space but preserves that of time. The foundation of the Brouwerian account of mathematics is to be found in fact in the basal intuition of time:

The neo-intuitionism considers the falling apart of moments of life into qualitatively different parts, to be reunited only while remaining separate by time, as the fundamental phenomenon of the human intellect, passing by abstracting from its emotional content into the fundamental phenomenon of mathematical thinking, the intuition of the bare two-oneness. (Brouwer 1912a, p. 80)

The rest of mathematics is, according to Brouwer, built out of this basal intuition. Together with the emphasis on the centrality of intuition, Brouwer denigrates the use of language in mathematical activity and reserves to it only an auxiliary role. Talking about the construction of (countable) sets he writes:

And in the construction of these sets neither the ordinary language nor any symbolic language can have any other role than that of serving as a non-mathematical auxiliary, to assist the mathematical memory or to enable different individuals to build up the same set. (Brouwer 1912a, p. 81)

This is at the root of Brouwer's skeptical attitude toward a foundational role for formal work in logic and mathematics. Thus, the intuitionist position finds itself at odds with formalists, logicists, and Platonists, all guilty, according to Brouwer, of relying on "the presupposition of the existence of a world of mathematical objects, a world independent of the thinking individual, obeying the laws of classical logic and whose objects may possess to each other the 'relation of a set to its elements'". For this reason Brouwer criticized, among other things, the foundation of set theory provided by Zermelo and eventually produced (starting in 1916–17) his own intuitionist set theory. While in the realm of the finite there is agreement in the results (although not in the method) between intuitionists and formalists, the real differences emerge in the treatment of the infinite and the continuum. There is an important development in Brouwer's ideas here. Whereas in the 1912 essay he thought of real numbers as given by laws, later on (starting in 1917) he developed a very original conception of the continuum based on choice sequences.[83] This will lead him to the development of an alternative construction of mathematics, intuitionistic mathematics. Brouwer presented his new approach in two papers, titled "Foundation of set theory independent from the logical law of the excluded middle" (1918) and the companion paper "Intuitionist set theory" (1921). As already mentioned, the new approach to mathematics was characterized by the admission of "free choice" sequences, that is, procedures in which the subject is not limited by a law but can also proceed freely in the generation of arbitrary elements of the sequence. These sequences are seen as being generated in time and thus as "growing" or "becoming". This new conception of mathematics with the inclusion of free growth and indeterminacy goes hand in hand with one of the major claims of Brouwer's intuitionism, that is, the denial of the idea that mathematical entities and properties are always completely determined. The latter assumption is embodied, according to Brouwer, in the logical law of the excluded middle:

The use of the principle of the excluded middle is *not permissible* as part of a mathematical proof. It has only scholastic and heuristic value, so that the theorems which in their proof cannot avoid the use of this principle lack all mathematical content. (Brouwer 1921, p. 23)

Thus, for the intuitionist the only acceptable mathematical entities and properties are those that are constructed in thought; mathematical objects and properties do not have an independent existence. As a consequence, this leads to an abandonment of the unrestricted validity of the principle of the excluded middle and thus to a restriction of the available means of proof in classical mathematics. However, intuitionistic mathematics is not simply a subset of classical mathematics obtained by eliminating the excluded middle but rather a different development, due to the fact that the admission of "incomplete entities" such as free-choice sequences leads to a new and original theory of the mathematical continuum. One of the new concepts introduced by Brouwer is that of Species. This is the intuitionist equivalent of "property" in the classical setting. The constructive interpretation of property is presented by Brouwer in opposition to the principle of comprehension formulated by Cantor and in a restricted form by Zermelo. While in the classical setting any well-formed formula partitions the universe into the set of objects that satisfy the formula and those that do not, the new interpretation of property, or "Species", is obtained by limiting its domain to the entities whose constructions have already been achieved. However, the Species does not partition the already constructed entities into those that satisfy the Species and those that do not. An entity will belong to the Species if one can successfully carry out a proof that the constructed entity does indeed have the property in question (in Brouwer's terminology, "fitting in"). An entity will not belong if one can successfully carry out a construction that will show that the assumption of its belonging to the Species generates a contradiction. However, it is clear that the alternatives to a demonstration of "fitting in" can be twofold: either the demonstration of the absurdity of a "fitting in" or the absence of a demonstration either of "fitting in" or of its absurdity. The consequences of this strict interpretation of negation are that Brouwer has to produce a reconstruction of mathematics in which the principles of double negation and the principle of the excluded middle do not hold. The intuitionistic reconstruction of mathematics cannot be given here;[84] our focus is on the logical aspects of the situation.

1.7.1.2 Brouwer on the Excluded Middle

From the beginning of his publishing career, Brouwer gave pride of place to the mental mathematical activity and downplayed the foundational rôle of language and logic in mathematics. The system of logical laws is then seen as a mere linguistic edifice that at best can only accompany the communication of successful mathematical constructions. In 1908, Brouwer expresses doubts as to the validity of the principle of the excluded middle, since he claims that it is not the case that for an arbitrary statement S, we either have a proof of S or we have a proof of the negation of S. Of course, this already presupposes a constructive interpretation of the logical connectives. Issues about the excluded middle became central once Brouwer developed his new conception of mathematics based on the admissibility of "becoming" entities (such as choice sequences) and constructive properties (Species) for which, as we have seen, there is more than one alternative to the successful "fitting" of a constructed object to the Species. After the publication of "The Foundations of set theory independent of the logical principle of the excluded middle", which develops parts of mathematics without appeal to the excluded middle, he wrote a number of essays in which he analyzed the logic of negation implicit in the new reconstruction of mathematics. In "On the significance of the excluded middle in mathematics, especially in function theory" (1923b), Brouwer proposes a positive account of how we illegitimately move from the excluded middle on finite domains to infinite domains:

Within a specific finite "main system" we can always *test* (that is, either prove or reduce to absurdity) properties of systems ... On the basis of the testability just mentioned, there hold, for properties conceived within a specific finite main system, the *principle of excluded middle*, that is, the principle that for every system every property is either correct or impossible, and in particular the *principle of the reciprocity of the complementary species*, that is, the principle that for every system the correctness of a property follows from the impossibility of the impossibility of this property. (Brouwer 1923b, p. 335)

However, the validity on finite domains was arbitrarily extended to mathematics in general:

An a priori character was so consistently ascribed to the laws of theoretical logic that until recently these laws, including the principle of excluded middle, were applied without reservation even in the mathematics of infinite systems. (Brouwer 1923b, p. 336)

1.7.1.3 The Logic of Negation

In "Intuitionistic Splitting of the Fundamental Notions of Mathematics" (1923a), Brouwer for the first time engages in an analysis of the consequences of his viewpoint, in particular, his conception of negation as contradiction, for logic proper. He begins by pointing out that the

the intuitionist conception of mathematics not only rejects the principle of the excluded middle altogether but also the special case, contained in the principle of reciprocity of complementary species, that is, the principle that for any mathematical system infers the correctness of a property from the absurdity of its absurdity. (1923a, p. 286)

The rejection of the principle of the excluded middle is then argued by means of an example, which is paradigmatic of what are now called (weak) Brouwerian counterexamples.[85] Let k_1 be the least n such that there is a sequence 0123456789 appearing between the nth place and the $(n + 9)$th place of the decimal expansion of π, and let

$$c_n = \begin{cases} (-1/2)^{k_1} & \text{if } n \geq k_1 \\ (-1/2)^n & \text{otherwise.} \end{cases}$$

Then the sequence c_1, c_2, c_3, converges to a real number r. We define a real number g to be rational if one can calculate two rational integers p and q whose ratio equals g. Then r cannot be rational and at the same time the rationality of r cannot be absurd. This is because if r were rational we could compute the two integers, thereby solving a problem for which no computation is known (i.e., finding k_1). On the other hand, it is not contradictory that it be rational, because in that case k_1 would not exist and thus r would be 0, that is, a rational after all. In fact, the problem giving rise to the weak counterexample used by Brouwer has now been solved. But one can use other unsolved problems to generate similar counterexamples.

The counterexample shows that intuitionistically we cannot assert (until the problem is solved) "r is either rational or irrational", something that is of course perfectly legitimate from the classical point of view. However, the argument goes through only if one grants that the property of being rational requires the explicit computation of the integers p and q, which is of course not required in the classical setting. The consequences for the logic of negation are stated by Brouwer in the following principles:

1. Intuitionistically, absurdity-of-absurdity follows from correctness but not vice versa.
2. However, intuitionistically, the absurdity-of-absurdity-of absurdity is equivalent with absurdity.

As a consequence of these principles, any finite sequence of absurdity predicates can be reduced either to an absurdity or to an absurdity-of-absurdity.

It should be pointed out in closing this section that the notion of absurdity obviously involves the notion of a "contradiction" or "the impossibility of fitting in" or an "incompatibility". All these notions presuppose negation or difference, but Brouwer never spells out with clarity how to avoid the potential circularity involved here, although he refers to a primitive intuition of difference (not definable in terms of classical negation) in 1975 (73).

1.7.1.4 Kolmogorov

Kolmogorov's contribution to the formalization of intuitionistic logic and its properties date from "On the principle of the excluded middle" (1925), which however was not known to many logicians until much later, undoubtedly due to the fact that it was written in Russian. Thus, the debate that we describe in section 1.7.1.5 on the nature of Brouwer's logic, does not refer to Kolmogorov. In the introduction to his article, Kolmogorov states his aim as follows:

We shall prove that every conclusion obtained with the help of the principle of the excluded middle is correct provided every judgment that enters in its formulation is replaced by a judgement asserting its double negation. We call the double negation of a judgement its "pseudotruth". Thus, in the metamathematics of pseudotruth it is legitimate to apply the principle of the excluded middle. (Kolmogorov 1925, p. 416)

Kolmogorov's declared goal in the paper was to show why the illegitimate use of the excluded middle does not lead to contradiction. His results predate similar results by Gentzen (1933a) and Gödel (1933c), which are known as double negation interpretations or negative translations. Kolmogorov's points of departure are Brouwer's critique of classical logic and the formalization of classical logic given by Hilbert (1922c). He introduces two propositional calculi: $\mathfrak{B}$ and $\mathfrak{H}$.

Calculus $\mathfrak{B}$:

1. $A \to (B \to A)$,
2. $\{A \to (A \to B)\} \to (A \to B)$,
3. $\{A \to (B \to C)\} \to \{B \to (A \to C)\}$,
4. $(B \to C) \to \{(A \to B) \to (A \to C)\}$,
5. $(A \to B) \to \{(A \to \overline{B}) \to \overline{A}\}$.

Calculus $\mathfrak{H}$ is obtained by adding to $\mathfrak{B}$ the axiom

6. $\overline{\overline{A}} \to A$.

Rules of inference for both calculi are substitution and modus ponens.

It has been argued that Kolmogorov anticipated Heyting's formalization of intuitionistic propositional calculus (see section 1.7.1.6). This is almost true. The system $\mathfrak{B}$ (known after Johansson as the minimal calculus) differs from the negation-implication fragment of Heyting's axiomatization only by the absence of axiom

h. $A \supset (\overline{A} \supset B)$.

$\mathfrak{H}$ is equivalent to the formalization of classical propositional calculus given in Hilbert (1922c). We find in Kolmogorov also an attempt at a formalization of the intuitionistic predicate calculus, although he is not completely formal on this point. He regards as intuitive the rule "whenever a formula $\mathfrak{S}$ stands by itself [i.e., is proved], we can write the formula $(a)\mathfrak{S}$" (433; rule **P**) and states the following axioms:

I. $(a)\{A(a) \to B(a)\} \to \{(a)A(a) \to (a)B(a)\}$.
II. $(a)\{A \to B(a)\} \to \{A \to (a)B(a)\}$.
III. $(a)\{A(a) \to C\} \to \{(Ea)A(a) \to C\}$.
IV. $A(a) \to (Ea)A(a)$.

Adding to system $\mathfrak{B}$ the axioms I–IV and rule **P** would result in a complete system for intuitionistic predicate logic (Heyting 1930b) if axiom h and the following axiom,

g. $(a)A(a) \to A(a)$,

were also added. Kolmogorov considered axiom g to be true (see Wang 1967). He conjectured that $\mathfrak{B}$ is complete with respect to its intended interpretation ("the intuitively obvious" class of propositions), but he cautiously observed that "the question whether this axiom system is a complete axiom system for the intuitionistic general logic of judgments remains open" (422).

Whereas calculus $\mathfrak{B}$ corresponds, according to Kolmogorov, to the "general logic of judgments", calculus $\mathfrak{H}$ corresponds to the "special logic of judgments", since its range of application is narrower (it produces true propositions only when the propositional variables range over a narrower class of propositions). In section III of his paper, Kolmogorov individuates a class of judgments with the property that "the judgment itself follows [intuitively] from its double negation". Finitary judgments are of such type. Let $A^\bullet$, $B^\bullet$, $C^\bullet$, ... denote judgments of the mentioned kind. Then $\overline{\overline{(A^\bullet \to B^\bullet)}} \to (A^\bullet \to B^\bullet)$ and $\overline{\overline{A^\bullet}} \to A^\bullet$ are provable in $\mathfrak{B}$. Moreover, for every negative formula $\overline{A}$, $\mathfrak{B}$ proves $\overline{\overline{\overline{A}}} \to \overline{A}$. It is also shown that substitution for propositional variables, modus ponens, and the axioms of $\mathfrak{H}$ are all valid for this class of propositions. This shows that the system $\mathfrak{H}$ is intuitionistically correct if we restrict it to the class of judgments of the form $A^\bullet$. Thus, the domain for which the calculus $\mathfrak{H}$ is valid is the class of propositions that follow (intuitively) from their double negation, and this includes finitary statements and all negative propositions. This amounts to showing that all of propositional logic is included in intuitionistic propositional logic, if the domain of propositions is restricted to propositions of the form $A^\bullet$. In section IV, Kolmogorov introduces a translation from formulas of classical mathematics to formulas of intuitionistic mathematics:

> We shall construct alongside of ordinary mathematics, a "pseudomathematics" that will be such that to every formula of the first there corresponds a formula of the second and, moreover, that every formula of pseudomathematics is a formula of type $A^\bullet$. (Kolmogorov 1925, p. 418)

The translation is defined as follows: If A is atomic, then $A^* = \overline{\overline{A}}$; $\overline{A}^* = \overline{\overline{A^*}}$; and $(A \to B)^* = \overline{\overline{A^* \to B^*}}$. Thus, if $A_1, \ldots, A_k$ are axioms of classical mathematics (comprising the logical axioms), then we have $A_1, \ldots, A_k$ proves A in $\mathfrak{H}$ iff $A_1^*, \ldots, A_k^*$ proves A^* in $\mathfrak{B}$. The theorem is proved by showing that applications of substitution and modus ponens remain derivable in $\mathfrak{B}$ under the *-translation, using the results about double negations previously established. Moreover, the *-translations of the logical axioms are derivable in $\mathfrak{B}$.

Kolmogorov did not extend the result to predicate logic but the extension is straightforward. It should be pointed out that he asserts (IV, §5–6) that every axiom A of classical mathematics is such that A^* is intuitionistically true. But this would imply that all of classical mathematics is intuitionistically consistent, a result which is not established, for analysis and set theory, even to this day. However, as Wang remarks, "it seems not unreasonable to assert that Kolmogorov did foresee that the system of classical number theory is translatable into intuitionistic number theory and therefore is intuitionistically consistent" (Wang 1967, p. 415). We return to these results after describing the discussion on Brouwer's logic in the West.

1.7.1.5 The Debate on Intuitionist Logic

In 1926, Wavre published an article contrasting "logique formelle" (classical) and "logique empiriste" (intuitionist). This was, apart from Kolmogorov (1925), the first attempt to discuss systematically the features of "Brouwer's logic". Whereas classical logic is a logic of truth and falsity, "empirical" logic is a logic of truth and absurdity, where true means "effectively demonstrable" and absurd "effectively reducible to a contradiction". Wavre begins by listing similar principles between the two logics:

1. $((A \supset B) \& (B \supset C)) \supset (A \supset C)$.
2. From A and $A \supset B$, one can infer B.
3. $\neg(A \& \neg A)$.
4. $(A \supset B) \supset (\neg B \supset \neg A)$.

Among the different principles Wavre mentions the excluded middle and double negation. He then shows that $\neg A$ is equivalent, in empirical logic, to $\neg\neg\neg A$. Moreover he observed that in empirical logic the converse of (4) does not hold, unless B is a negative proposition. Much of Wavre's article only restated observations that were, implicitly or explicitly, contained in Brouwer (1923b). However, it had the merit of opening a debate in the *Revue de Metaphysique et de Morale* on the nature of intuitionistic logic which saw contributions by Wavre, Levy, and Borel. However, this debate did not directly touch on the principles of intuitionistic logic.[86] By contrast, Barzin and Errera (1927) claimed that Brouwerian logic was inconsistent, thereby sparking a long debate on the possibility of an intuitionistic logic, which saw contributions by Church, Levy, Glivenko, Khintchine, and others. Barzin and Errera incorrectly interpreted Brouwer's talk of undecided propositions (i.e., those for which there is neither an effective proof of their validity nor an effective proof of their absurdity) as claiming that there are propositions which are neither true nor false. These propositions are "tierce". Their aim was then to show that the admission of a "tierce" led to formal contradictions. They interpreted these "third" propositions not as a state of objective ignorance but rather as an "objective logical fact". They denoted "p is tierce" by p'. With this notation in place, they stated a principle of "quartum non datur": $p \vee \neg p \vee p'$ and claimed that Brouwer must accept it, if "tierce" is defined as being "neither true nor false". Finally, the equivalent of the principle of noncontradiction, which they claimed Brouwer must admit, is that no proposition can be true and false, or true and tierce, or false and tierce. Under these assumptions they claimed to show that one could prove the collapse of the truth values, that is, that in the calculus one could prove that every proposition that is true is also tierce, and every proposition that is tierce is also false. The proof is, however, inconclusive. First of all, there is a constant confusion between the object level and the metalevel of analysis; moreover, the proof makes use of principles that are classically but not intuitionistically valid.

Of the many replies to Barzin and Errera (1927), we discuss only Church's (1928).[87] In "On the law of the excluded middle" Church discussed, and rejected, the claims by Barzin and Errera by making essentially three points. First, he points out that the easiest alternative to a system that includes the law of the excluded middle is a system in which the excluded middle is not assumed "without assertion of any contrary principle". Thus, because this is a subsystem of the original one, no contradictions can be derived that could not be derived in the original system. To generate a contradiction we must admit a new principle that is not consistent with the law of the excluded middle. Second, one can drop the principle of the excluded middle and "introduce the middle ground between true and false as an undefined term" in which case it might be that "making the appropriate set of assumptions about the existence and properties of tierce propositions, we can produce a system of logic which is consistent with itself but which becomes inconsistent if the law of the excluded middle be added".[88] This possibility had already been proven by Łukasiewicz in developing many-valued logics (see later discussion), but Church does not mention Łukasiewicz. Third, the argument by Barzin and Errera fails because they introduce the tierce propositions by defining them as being neither true nor false, and this leads to an inconsistency. The argument by Barzin and Errera works only if one admits the faulty definition of a tierce (rather than leaving the notion undefined) and the principle of the excluded fourth, which again is defended using the faulty definition. Finally, Church argued that Barzin and Errera's argument is ineffective against those who simply drop the principle of the excluded middle, as "the insistence that one who refuses to accept a proposition must deny it can be justified only by an appeal to the law of the excluded middle".

1.7.1.6 The Formalization and Interpretation of Intuitionistic Logic

Glivenko (1928) contributed an article on intuitionistic logic in which he showed that Brouwerian logic could not admit a tierce. But of great technical interest is Glivenko (1929), which contains the following two theorems:

1. If a certain expression in the logic of propositions is provable in classical logic, it is the falsity of the falsity of this expression that is provable in Brouwerian logic.

2. If the falsity of a certain expression in the logic of propositions is provable in classical logic, that same falsity is provable in Brouwerian logic (Glivenko 1929, p. 301)

Although Glivenko's results do not yet amount to a translation of classical logic into intuitionistic logic, they certainly paved the way for the later results by Gödel and Gentzen (see Troelstra 1990; van Atten 2005a). By far the most important contribution in this period is the work of Heyting on the formalization of intuitionistic logic. Heyting's contributions were motivated by a prize question published in 1927 by the Dutch Mathematical Society on the formalization of the principles of intuitionism. Heyting was awarded the prize in 1928, but his result appeared in print only in 1930. Heyting (1930a) contains a formalization of the laws of intuitionistic propositional logic; (1930b) moves on to intuitionistic predicate logic and arithmetic; and finally, (1930c) investigates intuitionistic principles in analysis.

Heyting distilled the principles of intuitionistic logic by going through the list of axioms in *Principia Mathematica* and retaining only those that admitted of an intuitionist justification (letter to Becker, September 23, 1933; see Troelstra 1990; van Atten 2005a). The axioms for the propositional part were the following.

1. $A \supset (A \wedge A)$.
2. $A \wedge B \supset B \wedge A$.
3. $(A \supset B) \supset ((A \wedge C) \supset (B \wedge C))$.
4. $((A \supset B) \wedge (B \supset C)) \supset (A \supset C)$.
5. $B \supset (A \supset B)$.
6. $(A \wedge (A \supset B)) \supset B$.
7. $A \supset A \vee B$.
8. $A \vee B \supset B \vee A$.
9. $((A \supset C) \wedge (B \supset C)) \supset (A \vee B \supset C)$.
10. $\neg A \supset (A \supset B)$.
11. $((A \supset B) \supset (A \supset \neg B)) \supset \neg A$.

In the appendix, Heyting proves that all the axioms are independent, exploiting a technique used by Bernays for proving the independence of the propositional axioms of *Principia* (see section 1.5.3). Heyting (1930b) also gives an axiomatization for principles acceptable in intuitionistic first-order logic. He (1930a) only states the admissible principles and proved theorems from them but he was not explicit on the meaning of the logical connectives in intuitionistic logic. However, he (1930d) did provide an interpretation for intuitionistic negation and disjunction. The interpretation depends on interpreting propositions as problems or expectations:

> A proposition p like, for example, "Euler's constant is rational" expresses a problem, or better yet, a certain expectation (that of finding two integers a and b such that $C = a/b$), which can be fulfilled or disappointed. (Heyting 1930d, p. 307)

This interpretation is influenced by Becker's treatment of intuitionism in *Mathematische Existenz* (1927) where, appealing to distinctions found in Husserl's *Logical Investigations*, Becker distinguishes between the fulfillment of an intention (say a proof of "a is B"), the frustration of an intention (a proof of "a is not B") and the nonfulfillment of an intention (i.e., the lack of a fulfillment). Indeed, Heyting (1931) explicitly refers to the phenomenological interpretation and claims that "the affirmation of a proposition is the fulfillment of an intention" (1931, p. 59). He mentions Becker in connection with the interpretation of intuitionistic negation:

> A logical function is a process for forming another proposition from a given proposition. Negation is such a function. Becker, following Husserl, has described its meaning very clearly. For him negation is something thoroughly positive, viz., the intention of a contradiction contained in the original intention. The proposition "C is not rational" therefore, signifies the expectation that one can derive a contradiction from the assumption that C is rational. (Heyting 1931, p. 59)

Disjunction is interpreted as the expectation of a mathematical construction that will prove one of the two disjuncts. In Heyting (1934), it is specified that the mathematical construction fulfilling a certain expectation is a proof. Under this interpretation, $A \supset B$ signifies "the intention of a construction that leads from each proof of A to a proof of B". This interpretation of the intuitionistic connectives is now known as the Brouwer–Heyting–Kolmogorov interpretation. The presence of Kolmogorov stems from his interpretation of the intuitionistic calculus as a calculus of problems (1932). In this interpretation, for instance, $\neg A$ is interpreted as the problem "to obtain a contradiction, provided the solution of A is given". Although the two interpretations are distinct, they were later on treated as essentially the same, and Heyting (1934, p. 14) speaks of Kolmogorov's interpretation as being closely related to his.[89]

1.7.1.7 Gödel's Contributions to the Metatheory of Intuitionistic Logic

Glivenko's work had shown that classical propositional logic could be interpreted as a subsystem of intuitionistic logic, and thus be intuitionistically consistent. We have also seen that Kolmogorov (1925) implicitly claimed that classical mathematics is intuitionistically consistent. A more modest, but extremely important, version of this unsupported general claim was proved by Gödel and Gentzen in 1933. Gödel states:

> The goal of the present investigation is to show that something similar [to the translation of classical logic into intuitionistic logic] holds also *for all of arithmetic and number theory*, delimited in

scope by, say, Herbrand's axioms. Here, too, we can give an interpretation of the classical notions in terms of the intuitionistic ones *so that all propositions provable from the classical axioms hold for intuitionism as well.* (Gödel 1933c, pp. 287–89)[90]

Gödel distinguished the classical connectives from the intuitionistic connectives: $\neg, \supset, \vee, \wedge$ are the intuitionistic connectives; the corresponding classical connectives are $\sim, \rightarrow, \vee, \cdot$. Gödel's translation $'$ from classical propositional logic into intuitionistic logic is defined as follows: $p' = p$, if p is atomic; let $(\sim p)' = \neg p'$, $(p \cdot q)' = p' \wedge q'$; $(p \vee q)' = \neg(\neg p' \wedge \neg q')$; $(p \rightarrow q)' = \neg(p' \wedge \neg q')$.

He then shows that classical propositional logic proves a sentence A if and only if intuitionistic propositional logic proves the translation A'. The result is then extended to first-order arithmetic by first extending the translation to cover the universal quantifier so that $(\forall x\, P)' = \forall x\, P'$. Letting H' stand for intuitionistic first-order arithmetic and Z for first-order arithmetic (in Herbrand's formulation), then Gödel showed that a sentence A is provable in Z iff its translation A' is provable in H'.

From the philosophical point of view, the importance of the result consists in showing that under a somewhat deviant interpretation, classical arithmetic is already contained in intuitionistic arithmetic. Therefore, this amounts to an intuitionistic proof of the consistency of classical arithmetic. This result once and for all brought clarity into a systematic confusion between finitism and intuitionism, which had characterized the literature on the foundation of mathematics in the 1920s.[91] Gödel's result makes clear that intuitionistic arithmetic is much more powerful than finitistic arithmetic.

Two more results by Gödel on the metatheory of intuitionistic logic have to be mentioned. The first (1933a) consists in an interpretation of intuitionistic propositional logic into a system of classical propositional logic extended by an operator B ("provable", from the German *beweisbar*). It is essential that provability here be taken to mean "provability in general" rather than provability in a specified system. The logic of the system B turns out to coincide with the modal propositional logic S4. The system S4 is characterized by the following axioms:

1. $Bp \rightarrow p$,
2. $Bp \rightarrow (B(p \rightarrow q) \rightarrow Bq)$,
3. $Bp \rightarrow BBp$.

The translation † works as follows: Atomic sentences are sent to atomic sentences; $(\neg p)^\dagger = \sim Bp^\dagger$; $(p \supset q)^\dagger = Bp^\dagger \rightarrow Bq^\dagger$; $(p \vee q)^\dagger = Bp^\dagger \vee Bq^\dagger$; $(p \wedge q)^\dagger = p^\dagger \cdot q^\dagger$. Gödel showed that if A is provable in intuitionistic propositional logic, then $A^\dagger$ is provable in S4. This result was important in that it showed the connections between modal logic and intuitionistic logic and paved the way for the development of Kripke's semantics for intuitionistic logic, once the semantics for modal logic had been worked out.

One final result by Gödel concerns intuitionistic logic and many-valued logic. Gödel (1932b) proved that intuitionistic propositional logic cannot be identified with a system of many-valued logic with finitely many truth values. Moreover, he showed that there is an infinite hierarchy of finite-valued logics between intuitionistic and classical propositional logic.[92]

1.7.2 Many-Valued Logics

The systematic investigation of systems of many-valued logics goes back to Jan Łukasiewicz.[93] Łukasiewicz arrived at many-valued logics as a possible way out of a number of philosophical puzzles he had been worrying about. The first concerns the very foundation of classical logic, that is, the principle that every proposition p is either true or false. This he called the law of bivalence (1930, 53). The principle had already been the subject of debate in ancient times, and Aristotle himself expressed doubts as to its applicability for propositions concerning future contingents ("there will be a sea battle tomorrow"). The wider philosophical underpinnings of such debates had to do with issue of determinism and indeterminism, which Łukasiewicz explored at length (see, for instance, Łukasiewicz 1922). In all such issues, the notion of possibility and necessity are obviously central. Indeed, in his presentation of many-valued logic, Łukasiewicz motivates the system by a reflection on modal operators, such as "it is possible that p". The first presentation of the results goes back to two lectures given in 1920: "On the concept of possibility" (1920b) and "On three valued-logic" (1920a). Let us follow these lectures. In the first lecture, Łukasiewicz considers the relationship between the following sentences:

i. S is P.
ii. S is not P.
iii. S can be P.
iv. S cannot be P.
v. S can be non-P.
vi. S cannot be non-P (i.e., S must be P).

He distinguishes three positions that can be held with respect to the logical relationship between the above sentences:

a. If S must be P (vi), then S is P (i).
b. If S cannot be P (iv), then S is not P (ii).

When no further relationships hold between (i)–(vi), this corresponds to the point of view of traditional logic. The second position, corresponding to ontological determinism, consists of theses (a) and (b) plus the implications

c. If S is P (i), then S must be P (vi).
d. If S is non-P (ii), then S cannot be P (iv).

Finally, the third position, corresponding to ontological indeterminism, consists of (a), (b), and the implications

e. If S can be P (iii), then S can be non-P (v).

f. If S can be non-P (v), then S can be P (iii).

All these theses have, according to Łukasiewicz, a certain intuitive obviousness. However, he shows that if one reasons within the context of classical logic, there is no way to consistently assign truth values 0 and 1 to (i)–(vi) so that all of (a)–(f) will get value 1. However, this becomes possible if one introduces a new truth value, 2, which stands for "possibility". This gives rise to the need for the study of "three-valued logic".

In the second lecture, Łukasiewicz defines three-valued logic as a system of non-Aristotelian logic and defines the truth tables for equivalence and implication based on three values in such a way that the tables coincide with classical logic when the values are 1 and 0 but satisfy the following laws when the value 2 occurs. For the biconditional, one stipulates that the values for 02, 20, 21, and 12 is going to be 2; for the material conditional, the value is 1 for 02, 21, and 22 and it is 2 for 20 and 12. From the general analysis, it is also clear that for negation the following holds: If p is assigned value 2 then $\sim p$ is also 2.

While all tautologies of three-valued logic are tautologies of classical propositional (two-valued) logic, the converse is not true. For instance, $p \vee \sim p$ is not a tautology in three-valued logic, because if p is assigned the value 2, the value of $p \vee \sim p$ is also 2.

In Post (1921) we also find a study of many-valued logics. However, Post studies these systems purely formally, without attempting to give them an intuitive interpretation. It is perhaps on account of this fact that he was the first to develop tables for negation known as "cyclic commutation" tables. In the case of Łukasiewicz's system, negation is always defined by a "mirror" truth-table, that is, the value of negation is that of its opposite in the order of truth (the value of $\sim p$ is 1 minus the value of p). In the case of Post, the truth table for negation is defined by permuting the truth values cyclically. Here is a comparison of the tables for the two types of negations in three-valued logic:

Łukasiewicz		Post	
p	$\sim p$	p	$\sim p$
0	1	0	$\frac{1}{2}$
$\frac{1}{2}$	$\frac{1}{2}$	$\frac{1}{2}$	1
1	0	1	0

Post was motivated by issues of functional completeness and in fact one of the results (1921) is that the system of m-valued logic he introduces, with a "cyclic commutation" table for negation, and a disjunction table obtained by giving the disjunction the maximum of the truth values of the disjuncts, is truth-functionally complete. The table for negation, with values 1 to m, is as follows:

p	1	2	...	m
$\sim p$	2	3	...	1

Łukasiewicz generalized his work from three-valued logics to many-valued logics (1922). At first he looked at logics with n truth values and later he considered logics with $\aleph_0$ values. All these systems can be expressed as follows. Let n be a natural number or $\aleph_0$. Assume that p and q range over a set of n numbers from the interval $[0, 1]$. As usual at the time, let us standardize the values to be $k/(n-1)$ for $0 \leq k \leq n-1$ when n is finite and k/l ($0 \leq k \leq l$) when n is $\aleph_0$. Define $p \to q$ to have value 1 whenever $p \leq q$ and value $1 - p + q$ whenever $p > q$. Let $\sim p$ have value $1 - p$. If we select only 0 and 1 we are back in the classical two-valued logic. If we add to 0 and 1 the value $\frac{1}{2}$ we get three-valued logic. In similar fashion, one can create systems of n-valued logic. If p and q range over a countable set of values one obtains an infinite-valued propositional calculus. Many Polish logicians investigated the relationships between systems of many-valued logic (see Woleński 1989a). One of the first problems was to study how the sequence of logics L_n ($n > 1$) behaves. It was soon shown that all tautologies of L_n are also tautologies of L_2, but the converse does not hold. While $L_{\aleph_0}$ turns out to be contained in all finite L_n, the relationship between any two finite L_m and L_n is more complicated. Łukasiewicz and Tarski (1930) attribute to Lindenbaum the following result (theorem 19): For $2 \leq m$ and $2 \leq n$ (m, n finite) we have: L_m is included in L_n iff $n - 1$ divides $m - 1$. Among the early results concerning the axiomatization of many-valued logics one should mention Wajsberg (1931), which contains a complete and independent axiomatization of three-valued logic. However, the system is not truth-functionally complete. Słupecki (1936) proved that if one adds to the connectives $\supset$ and $\sim$ in three-valued logic, the operator T such that Tp is always $\frac{1}{2}$ (for $p = 1, 0$, or $\frac{1}{2}$), then the system is truth-functionally complete. To provide an axiomatization one needs to add some axioms for T to the axioms given by Wajsberg. Thus, the axiomatization provided by Słupecki is given by the following six axioms:

1. $p \supset (q \supset p)$.
2. $(p \supset q) \supset ((q \supset r) \supset (p \supset r))$.
3. $(\sim p \supset \sim q) \supset (q \supset p)$.
4. $((p \supset \sim p) \supset p) \supset p$.
5. $Tp \supset \sim Tp$.
6. $\sim Tp \supset Tp$.[94]

The axiomatizability of $L_{\aleph_0}$ was conjectured by Łukasiewicz in 1930, who put forth the (correct) candidate axioms, but a proof of the result was only given by Rose and Rosser (1958).

Let us conclude this exposition on many-valued logic in the twenties and the early thirties by mentioning some relevant work on the connection between intuitionistic logic and many-valued logic. We have seen that Gödel in 1932 showed that intuitionistic logic did not coincide with any finite many-valued logic. More precisely, he showed that no finitely valued matrix characterizes intuitionistic logic. Theorem I of Gödel (1932b) reads:

There is no realization with finitely many elements (truth values) for which the formulas provable in H [intuitionistic propositional logic], and only those, are satisfied (that is, yield designated truth values for an arbitrary assignment). (Gödel 1932b, p. 225)

In the process he identified an infinite class of many-valued logic, now known as Gödel logics. This is captured in the second theorem of the paper:

Infinitely many systems lie between H and the system A of the ordinary propositional calculus, that is, there is a monotonically decreasing sequence of systems all of which include H as a subset and are included in A as subsets. (Gödel 1932b, p. 225)

The previous result gave the first examples of logics that are now studied under the name of intermediate logics. One important result that should be mentioned in this connection was obtained by Jaśkowski (1936), who provided an infinite truth value matrix appropriate for intuitionistic logic.

1.8 Itinerary VIII. Semantics and Model-Theoretic Notions

1.8.1 Background

During the previous itineraries, we have come across the implicit and explicit use of semantic notions (interpretation, satisfaction, validity, truth, etc.). In this section we retrace, in broad strokes, the main contexts in which these notions occurred in the first two decades of the twentieth century. This will provide the background for an understanding of the gradual emergence of the formal discipline of semantics (as part of metamathematics) and, much later, model theory.

The first context we have encountered in which semantical notions make their appearance is that of axiomatics (see itinerary I). A central notion in the analysis of axiomatic theories is that of "interpretation", which of course has its roots in nineteenth-century work on geometry and abstract algebra (see Guillaume 1994; Webb 1995). The development of analysis, algebra, and geometry in the nineteenth century had led to the idea of an uninterpreted formal axiomatic system. We have seen that Pieri (1901) emphasized that the primitive notions of any deductive system "must be capable of arbitrary interpretations", with the only restriction that the primitive sentences are satisfied by the particular interpretation. The axioms are verified, or made true, by particular interpretations. Interpretations are essential for proofs of consistency and independence of the axioms. However, as we said, the semantical notions involved (satisfaction, truth in a system) are used informally. Moreover, all these developments took place without a formal specification of the background logic. With minor modifications from case to case, these remarks apply to Peano's school, Hilbert, and the American postulate theorists.

1.8.1.1 The Algebra of Logic Tradition

A second tradition in which semantic notions appear quite frequently is that of the algebra of logic. To this tradition we owe what is considered the very first important result in model theory (as we understand it today, i.e., a formal study of the relationship between a language and its interpretations). This is the Löwenheim–Skolem theorem. As stated by Skolem:

> In volume 76 of *Mathematische Annalen*, Löwenheim proved an interesting and very remarkable theorem on what are called "first-order expressions" [*Zählausdrücke*]. The theorem states that every first-order expression is either contradictory or already satisfiable in a denumerably infinite domain. (Skolem 1920, p. 254)

As we have already seen in itinerary IV, the basic problem is the satisfaction of (first-order) equations on certain domains. Domain and satisfaction are the key terminological concepts used by Löwenheim and Skolem (who do not talk of interpretations). However, all these semantical notions are used informally.

It can safely be asserted that the clarification of semantic notions was not seen as a goal for mathematical axiomatics. In 1918a, Weyl gestures toward an attempt at clarifying the meaning of "true judgment", but he does so by delegating the problem to philosophy (Fichte, Husserl). An exception here is Ajdukiewicz (1921), who was only accessible to those who read Polish. Ajdukiewicz stressed the issues related to a correct interpretation of the notions of satisfaction and truth in the axiomatic context. This was to leave a mark on Tarski, who was thoroughly familiar with this text (see section 1.8.7).

1.8.1.2 Terminological Variations (Systems of Objects, Models, and Structures)

Throughout the 1910s, the terminology for interpretations of axiomatic systems remains rather stable. Interpretations are given by systems of objects with certain relationships defined on them. Bôcher (1904) suggests the expression "mathematical system" to "designate a class of objects associated with a class of relations between these objects" (128). Nowadays, however, we speak just as commonly of models or structures. When did the terminology become common currency in axiomatics?

"Model", as an alternative terminology for interpretation, makes its appearance in the mathematical foundational literature in von Neumann (1925), where he talks of models of set theory. However, the new terminology owes its influence and success to Weyl's "Philosophy of Mathematics and Natural Science" (1927). In introducing techniques for proving independence, Weyl describes the techniques of "construction of a model [*Modell*]" (18) and described both Klein's construction of a Euclidean model for non-Euclidean geometry and the construction of arithmetical models for Euclidean geometry (or subsystems thereof) given by Hilbert.[95] Once introduced in the axiomatical literature by Weyl, the word "model" finds a favorable reception. It occurs in Carnap (1927, 2000 [1927–29], 1930), Kaufmann (1930), and in articles by Gödel (1930b), Zermelo (1929, 1930), and Tarski (1936b). The usage is, however, not universal. The word "model" is not used in Hilbert and Ackermann (1928) (but it is found in Bernays 1930b). Fraenkel (1928) speaks about realizations or models (353) as does Tarski (1936d). The latter do not follow Carnap in drawing a distinction between realizations (concrete, spatio temporal interpretations) and models (abstract interpretations). "Realization" is also used by Baldus (1924) and Gödel (1929).

As for "structure", it is not used in the twenties as an equivalent of "mathematical system". Rather, mathematical systems have structure. In *Principia Mathematica* (Whitehead and Russell 1912, part iv, *150 ff.) and then in Russell (1919, ch. 6) we find the notion of two relations "having the same structure".[96] In Weyl (1927, p. 21), two isomorphic systems of objects are said to have the same structure. This process will eventually lead to the idea that a "structure" is what is captured by an axiom system: "An axiom system is said to be monomorphic when exactly one structure belongs to it [up to isomorphism]" (Carnap 2000 [1927–29], 127; see also Bernays 1930b).

Here it should be pointed out that the use of the word "structure" in the algebraic literature was not yet widespread, although the structural approach was. It seems that "structure" was introduced in the algebraic literature in the early 1930s by Øystein Ore to denote what we nowadays call a lattice (see Vercelloni 1988; Corry 2004b).

1.8.1.3 Interpretations for Propositional Logic

A major step forward in the development of semantics is the clarification of the distinction between syntactical and semantic notions made by Bernays in *Habilitationsschrift* of 1918 (see itinerary V). We have seen that Bernays clearly distinguished between the syntax of the propositional calculus and its interpretations, a distinction that was not always clear in previous writers. This allowed him to properly address the problem of completeness for the propositional calculus. Bernays distinguished between provable formulas (obtainable from the axioms by means of the rules of inference) from the valid formulas (which yield true propositions for any substitution of propositions for the variables) and stated the completeness problem as follows: "Every provable formula is a valid formula and conversely." It would be hard to overestimate the importance of this result, which formally shows the equivalence of a syntactic notion (provable formula) with a semantic one (valid formula) (in section 1.8.4 we will look at the emergence of the corresponding notions for first-order logic). Post (1921) also made a clear distinction between the formal system of propositional logic and the semantic interpretation in terms of truth-table methods, and he also established the completeness of the propositional calculus (see section 1.8.3).

In this way, logic becomes an object of axiomatic investigation for which one can pose all the problems that had traditionally been raised about axiomatic systems. To get a handle on the problems, researchers first focused on the axiomatic systems for the propositional calculus and then moved on to wider systems (such as the "restricted functional calculus", i.e., first-order predicate logic). Here we focus on the metatheoretical study of systems of axiomatic logic rather than the developments of mathematical axiomatic theories (models of set theory, arithmetic, geometry, various algebraic structures, etc.).

1.8.2 Consistency and Independence for Propositional Logic

We have seen that the use of interpretations to provide independence results was exploited already in the nineteenth century in several areas of mathematics. Hilbert,

Peano and his students, and also the American postulate theorists put great value in showing the independence of the axioms for any proposed axiomatic system. Most of these applications concern specific mathematical theories. Applications to logic appear first in the tradition of the algebra of logic. For instance, in "Sets of independent postulates for the algebra of logic" (1904), Huntington studied the "algebra of symbolic logic" as an independent calculus, as a purely deductive theory. The object of study is given by a set K satisfying the axioms of what we would now call a Boolean algebra. Huntington provides three different axiomatizations of the "algebra of logic", of which we present the first, built after Whitehead's presentation in *Universal Algebra* (1898). Possible interpretations for the system are the algebra of classes and the algebra of propositions. Huntington claims originality in the extensive investigation of the independence of the axioms. The first axiomatization states the properties of a class K of objects on which are defined two operations, $\oplus$ and $\otimes$, satisfying the following axioms:

Ia. $a \oplus b$ is in the class whenever a and b are in the class;

Ib. $a \otimes b$ is in the class whenever a and b are in the class;

IIa. There is an element $\bigwedge$ such that $a \oplus \bigwedge = a$, for every element a;

IIb. There is an element $\bigvee$ such that $a \otimes \bigvee = a$, for every element a;

IIIa. $a \oplus b = b \oplus a$ whenever $a, b, a \oplus b$, and $b \oplus a$ are in the class;

IIIb. $a \otimes b = b \otimes a$ whenever $a, b, a \otimes b$, and $b \otimes a$ are in the class;

IVa. $a \oplus (b \otimes c) = (a \oplus b) \otimes (a \oplus c)$ whenever $a, b, c, a \oplus b, a \oplus c, b \otimes c, a \oplus (b \otimes c)$, and $(a \oplus b) \otimes (a \oplus c)$ are in the class;

IVb. $a \otimes (b \oplus c) = (a \otimes b) \oplus (a \otimes c)$ whenever $a, b, c, a \otimes b, a \otimes c, b \oplus c, a \otimes (b \oplus c)$, and $(a \otimes b) \oplus (a \otimes c)$ are in the class;

V. If the elements $\bigwedge$ and $\bigvee$ in postulates IIa and IIb exist and are unique, then for every element a there is an element $\bar{a}$ such that $a \oplus \bar{a} = \bigvee$ and $a \otimes \bar{a} = \bigwedge$;

VI. There are at least two elements, x and y, in the class such that $x \neq y$.

The consistency of the set of axioms is given by a finite table consisting of two objects, 0 and 1, satisfying the following:

$\oplus$	0	1		$\otimes$	0	1
0	0	1		0	0	0
1	1	1		1	0	1

The reader will notice that if we interpret $\oplus$ as conjunction of propositions and $\otimes$ as disjunction, we can read the table as the truth table for conjunction and disjunction of propositions (letting 0 stand for true and 1 for false). Similar tables are used by Huntington to prove the independence of each of the axioms from the remaining ones. In every case one provides a class and tables for $\oplus$ and $\otimes$ that verify all of the axioms but the one to be shown independent. For instance, IIIa can be shown to be independent by taking two objects 0 and 1 with the following tables.

$\oplus$	0	1		$\otimes$	0	1
0	0	0		0	0	0
1	1	1		1	0	1

All the axioms are satisfied, but $a \oplus b = b \oplus a$ fails by letting $a = 0$ and $b = 1$. Similarly for $a \otimes b$.

These techniques were not new and were already used in connection with the algebra of propositions by Peirce and Schröder. An application of this algebraic approach to the propositional calculus of *Principia Mathematica* was given by Sheffer (1913). Sheffer showed that one could study an algebra on a domain K with a binary K-rule of combination | satisfying the following axioms.

1. There are at least two distinct elements of K.
2. $a \mid b$ is in K whenever a and b are in K.
3. $(a \mid a) \mid (a \mid a) = a$ whenever a is an element of K and all the indicated combinations of a are in K.
4. $a \mid (b \mid (b \mid b)) = a \mid a$ whenever a and b are elements of K and all the indicated combinations of a and b are in K.
5. $(a \mid (b \mid c)) \mid (a \mid (b \mid c)) = ((b \mid b) \mid a) \mid ((c \mid c) \mid a)$ whenever a, b, and c are elements of K and all the indicated combinations of a, b, and c are in K.

Sheffer showed that this set of postulates implies Huntington's set by letting $\bar{a} = a \mid a$; $a \oplus b = (a \mid b) \mid (a \mid b)$ and $a \otimes b = (a \mid a) \mid (b \mid b)$. Conversely, by defining $a \mid b$ as $\bar{a} \otimes \bar{b}$, Huntington's set implies Sheffer's set of axioms. The application to *Principia* is now immediate. One can substitute a single connective $p \mid q$ defined as $\sim(p \vee q)$.

This work leads us to Bernays' (1918, 1926) studies of the independence of the axioms of the propositional fragment of *Principia*. Actually, Bernays was unaware of Sheffer's work until Russell mentioned it to him in 1920 (see Mancosu 2003). Bernays's (1926) formulation of the propositional logic ("theory of deduction") of *Principia* is given by

Taut. $\vdash : p \vee p . \supset . p$
Add. $\vdash : q . \supset . p \vee q$
Perm. $\vdash : p \vee q . \supset . q \vee p$
Assoc. $\vdash : p \vee (q \vee r) . \supset . q \vee (p \vee r)$
Sum. $\vdash :. q \supset r . \supset : p \vee q . \supset . p \vee r$

One also has rules of substitution and modus ponens.

The proof of independence of the axioms of the propositional calculus of *Principia*, with the exclusion of associativity, shown by Bernays to be derivable from the others, was given by appropriate interpretations in the style of the independence proofs we have looked at in the work of Huntington. However, one also has to show that the inference rules, and in particular modus ponens, preserve the right value. The technique is that of exhibiting "finite systems" consisting in the assignment of three or four finite values to the variables. One (or several) of these values are then singled out as distinguished value(s).

The proof of consistency of the calculus is given by letting propositions range over $\{0, 1\}$ and interpreting $\sim p$ as the numerical operation $1 - p$ and $p \vee q$ (disjunction) as the numerical operation $p \times q$. It is easy to check that the axioms always have value 0 and that substitution and modus ponens lead from formulas with value 0 to other formulas with value 0. This shows the calculus to be consistent, for were a contradiction provable, say $(p \& \neg p)$, then it would take the value 1.

The technique of proving independence of the axioms is similar (*Methode der Aufweisung*). Consider the axiom Taut. We give the following table with three values a, b, c with a distinguished value, say a.

$\vee$	a	b	c	$\sim$
a	a	a	a	b
b	a	b	c	a
c	a	c	a	c

It is easy to check that Add, Perm, and Sum always have value a, but not Taut as $(c \vee c) \supset c$ has value c ($\neq a$). Bernays also proved completeness by using the technique of normal forms (see section 1.5.3 for details on this and Bernays' independence proofs in 1918). Since Bernays' work did not appear in print until 1926, Post's paper (1921) contained the most advanced published results on the metatheory of the propositional calculus by the early 1920s. Similar results were also obtained by Łukasiewicz around 1924 (see Tarski 1983, p. 43).

1.8.3 Post's Contributions to the Metatheory of the Propositional Calculus

Post (1921) represents a qualitative change with respect to the previous studies of axiomatic systems for the propositional calculus by Russell, Sheffer, and Nicod. Post begins by explicitly stating the difference between proving results in a system and proving results about a system. He emphasizes that his results are about the system of propositional logic, which he takes in the version offered in *Principia* but regards it as a purely formal system to be investigated.[97] A basic concept introduced by Post is that of a truth table development. Post claims no originality for the concept, which he attributes to previous logicians. He denotes the truth value of any proposition p by $+$ if p is true and by $-$ if p is false.

The notion of truth table is then applied to arbitrary functions of the form $f(p_1, p_2, \ldots, p_n)$ of n propositions built up from $p_1, p_2, \ldots, p_n$ by means of arbitrary applications of $\sim$ and $\vee$. Because each of the proposition can assume either $+$ or $-$ as values, there are 2^n possible truth configurations for $f(p_1, p_2, \ldots, p_n)$. In general there will be 2^{2^n} possible truth tables for functions of n arguments. Let us call such truth tables of order n. Post proves first of all that for any n, to every truth table of order n there is at least one function $f(p_1, p_2, \ldots, p_n)$ which has it for its truth table. He then distinguishes three classes of functions: positive, negative, and mixed. Positive functions are those that always take + (this is the equivalent of Wittgenstein's propositional tautologies as defined in the *Tractatus* (1921,1922), say, $p \vee \sim p$, negative functions are those that always take − (say, $\sim(p \vee \sim p)$), and mixed are those functions those that take both +'s and −'s (e.g., $p \vee p$).

Post's major theorem then proves that a necessary and sufficient condition for a function $f(p_1, p_2, \ldots, p_n)$ to be a theorem of the propositional system of *Principia* is that $f(p_1, p_2, \ldots, p_n)$ be positive (i.e., all its truth values be +). In our terminology, $f(p_1, p_2, \ldots, p_n)$ is a theorem of propositional logic if and only if $f(p_1, p_2, \ldots, p_n)$ is a tautology. The proof makes use of the possibility of transforming sentences of the propositional calculus into special normal forms. Post emphasizes that the proof of his theorem gives a method both for deciding whether a function $f(p_1, p_2, \ldots, p_n)$ is positive and for actually writing down a derivation of the formula from the axioms of the calculus. Nowadays the property demonstrated by Post is called (semantic) completeness, but Post uses the word "completeness" in a different sense. He uses the word to discuss the adequacy of a system of functions to express all the possible truth tables (this is nowadays called truth-functional completeness). In this way he shows not only that through the connectives of *Principia* ($\sim$ and $\vee$) one can generate all possible truth tables but also that there are only two connectives which can, singly, generate all the truth tables. One is the Sheffer stroke, and the other is the binary connective that is always false except in the case when both propositions are false. The techniques used by Post are now standard, and we will not rehearse them here. Rather, we would like to mention another important concept introduced by Post. Post needed to introduce a concept of consistency for arbitrary systems of connectives (which therefore might not have negation as a basic connective). Because an inconsistent system brings about the assertion of every proposition, he defined a system to be inconsistent if it yields the assertion of the variable p (which is equivalent to the derivability of every proposition if the substitution rule is present). From this notion derives our notion of Post-completeness: A system of logic is Post-complete if every time we add to it a sentence unprovable in it, we obtain an inconsistent system. Post proved that the propositional system of *Principia* is thus both semantically complete and Post-complete.

Another powerful generalization was offered by Post in the last part of his article. There he defines m-valued truth systems, that is, system of truth values where instead of two truth values (+ and −), we have finitely many values. This development is, together with (Łukasiewicz 1920b), one of the first studies of many-valued logics (see itinerary VII).

One final point about Post. Although the truth table techniques he developed belong squarely to what we call semantics, this does not mean that Post was after an analysis of logical truth or a "semantics". Rather, his interest seems to have been purely formal and aimed at finding a decision procedure for provability (see Dreben and van Heijenoort 1986, p. 46).

To sum up: By 1921 the classical propositional calculus has been shown to be consistent, semantically complete, Post-complete, and truth-functionally complete. Moreover,

Bernays improved the presentation of the calculus given in *Principia* by showing that if one deletes associativity from the system, one obtains an axiomatic systems all of whose axioms are independent.

1.8.4 *Semantical Completeness of First-Order Logic*

With the work by Bernays (1918, 1926) and Post (1921), the notions of Post-completeness and semantic completeness had been spelled out with the required precision. After the recognition of first-order logic ("functional calculus" or "restricted functional calculus") as an important independent fragment of logic, due in great part to Hilbert's 1917–18 lectures and Hilbert and Ackermann (1928), the axiomatic investigation of first-order logic could also be carried out.

Chapter 3 of Hilbert and Ackermann (1928) became the standard exposition of the calculus. In section 9 of the chapter, Hilbert and Ackermann show that the calculus is consistent (by giving an arithmetical interpretation with a domain of one element). Then it is shown, crediting Ackermann for the proof, that the system is not Post-complete. To pose the completeness problem for first-order logic, it was necessary to identify the appropriate notion of validity [*Allgemeingültigkeit*]. This notion seems to be have been defined for the first time by Behmann (1922a). It turns out that Behmann's approach to the decision problem led to the notion of validity for first-order formulas (with variables for predicates) and for second-order formulas. This is well captured in Bernays' concise summary of the work:

In the decision problem we have to distinguish between a narrower and a wider formulation of the problem. The narrower problem concerns logical formulas of the "first-order", that is those in which the signs for all and exist (universal and existential quantifiers) refer only to individuals (of the assumed individual domain); the logical functions occurring here are variables, with the exception of the relation of identity ("x is identical with y"), which is the only individual [constant] relation admitted. The task consists in finding a general procedure which allows to decide, for any given formula, whether it is valid [*allgemeingültig*], that is whether it yields a correct assertion [*richtige Aussage*] for arbitrary substitutions of determinate logical functions.

One arrives at the wider problem by applying the universal and the existential quantifiers in connection to function variables. Then one considers formulas of the "second order" in which all variables are bound by universal and existential quantifiers, in whose meaning therefore nothing remains undetermined except for the number of individuals which are taken as given at the outset. For an arbitrary given formula of this sort one must now decide whether it is correct or not, or for which domains it is correct. (Bernays 1928b, pp. 1119–20)

A logical formula, in this context, is one that is expressible only by means of variables (both individual and functional), connectives, and quantification over individual variables, that is, there are no constants (see Hilbert and Ackermann 1928, p. 54). With this in place, the problem of completeness is posed by Hilbert and Ackermann as the request for a proof that every logical formula (of the restricted functional calculus) which is correct for every domain of individuals [*Individuenbereich*] be shown to be derivable from the axioms by finitely many applications of the rules of logical inference (68).[98]

Hilbert and Ackermann also posed the problem to show the independence of the axioms for the restricted functional calculus. Both problems were solved in 1929 by Kurt Gödel in his dissertation and published in "The completeness of the axioms of the functional calculus of logic" (1929, 1930a). The solution to the completeness problem is the most important one. As there exist already several expositions of the proof (Kneale and Kneale 1962, Dreben and van Heijenoort 1986) we can simply outline the main steps of the demonstration. Let us begin with the axioms for the system:

1. $X \vee X \to X$,
2. $X \to X \vee Y$,
3. $X \vee Y \to Y \vee X$,
4. $(X \to Y) \to (Z \vee X \to Z \vee Y)$,
5. $(x)F(x) \to F(y)$,
6. $(x)[X \vee F(x)] \to X \vee (x)F(x)$.

Rules of inference:

1. From A and $A \to B$, B may be inferred.
2. Substitution for propositional and functional variables.
3. From $A(x)$, $(x)A(x)$ may be inferred.
4. Individual variables (free or bound) may be replaced by any others (with appropriate provisos).

A valid formula (*allgemeingültige Formel*) is one that is satisfiable in every domain of individuals. Gödel's completeness theorem is stated as:

Theorem I *Every valid formula of the restricted functional calculus is provable.*

If a formula A is valid, then $\overline{A}$ is not satisfiable. By definition "A is refutable" means "$\overline{A}$ is provable". This leads Gödel to restate the theorem as follows:

Theorem II *Every formula of the restricted functional calculus is either refutable or satisfiable (and, moreover, satisfiable in the denumerable domain of individuals).*

Suppose in fact we have shown Theorem II. To prove Theorem I, assume that A is universally valid. Then $\overline{A}$ is not satisfiable. By Theorem II, it is refutable, that is, it is provable that $\overline{\overline{A}}$. Thus, it is also provable that A.

We can thus focus on the proof of Theorem II and, without loss of generality, talk about sentences rather than formulas. The first step of the proof consists in reducing the complexity of dealing with arbitrary sentences to a special class in normal form. The result is an adaptation of a result given by Skolem in 1920. Gödel appeals to the result (from Hilbert and Ackermann 1928) that for each sentence S there is an associated normal sentence S^* such that S^* has all the quantifiers at the front of a quantifier-free matrix, and it is provable that $S^* \leftrightarrow S$. Gödel then focuses on sentences that in addition to being in prenex normal form are such that the prefix of the sentence begins with a universal quantifier and ends with an existential quantifier. Let us call such sentences K-sentences.

Theorem III establishes that if every K-sentence is either refutable or satisfiable, so is every sentence. This reduces the complexity of proving Theorem II to the following.

Every K-sentence is either satisfiable or refutable. The proof is by induction on the degree of the K-sentence, where the degree of a K-sentence is defined by counting the number of blocks in its prefix consisting of universal quantifiers that are separated by existential quantifiers. The inductive step is quite easy (Theorem IV). The real core of the proof is showing the result for K-sentences of degree 1:

Theorem V *Every K-sentence of degree 1 is either satisfiable or refutable.*

Proof: Assume we have a K-sentence of degree 1 of the form

$$(P)M = (x_1)\ldots(x_r)(E\,y_1)\ldots(E\,y_s)M(x_1,\ldots,x_r,y_1,\ldots,y_s).$$

For the sake of simplicity, let us fix $r = 3$ and $s = 2$.

Select a denumerable infinity of fresh variables $z_0, z_1, z_2, \ldots$. Consider all 3-tuples of $z_0, z_1, z_2, \ldots$ obtained by allowing repetitions of the variables and ordered according to the following order: $\langle z_{k_1}, z_{k_2}, z_{k_3}\rangle < \langle z_{t_1}, z_{t_2}, z_{t_3}\rangle$ iff $(k_1 + k_2 + k_3) < (t_1 + t_2 + t_3)$ or $(k_1 + k_2 + k_3) = (t_1 + t_2 + t_3)$ and $\langle k_1, k_2, k_3\rangle$ precedes $\langle t_1, t_2, t_3\rangle$ in the lexicographic ordering. In particular, the enumeration begins with $\langle z_0, z_0, z_0\rangle$, $\langle z_0, z_0, z_1\rangle$, $\langle z_0, z_1, z_0\rangle$, and so on. Let $\mathbf{w}_n$ be the n-th triple in the enumeration.

We now define an infinite sequence of formulas from our original sentence as follows:

$M_1 = M(z_0, z_0, z_0; z_1, z_2)$

$M_2 = M(z_0, z_0, z_1; z_3, z_4) \& M_1$

$\vdots$

$M_n = M(\mathbf{w}_n; z_{2(n-1)+1}, z_{2n}) \& M_{n-1}$.

(Recall that our example works with $s = 2$).

Notice that the variables appearing after the semicolon are always fresh variables that have neither appeared before the semicolon nor in previous M_i's. Moreover, in each M_i except M_1 all the variables appearing before the semicolon have also appeared previously.

Now define $(P_n)M_n$ to be $(E\,z_0)(E\,z_1)\ldots(E\,z_{2n})M_n$. Thus, $(P_n)M_n$ is a sentence all of whose variables are bound by the existential quantifiers in its prefix.

With the above in place, Gödel proves (Theorem VI) that for every n, $(P)M$ implies $(P_n)M_n$. The proof, which we omit, is by induction on n and exploits the specific construction of the M_n's. The important point here is that the structure of the M_n's is purely propositional. Thus each M_n will be built out of functional variables $P_1(x_{p_1}, \ldots, x_{q_1})$, $\ldots$, $P_k(x_{p_k}, \ldots, x_{q_k})$ (of different arity) and propositional variables $X_1, \ldots, X_l$, (the elementary components, all of which are already in M) by use of "or" and "not". At this point we associate with every M_n a formula B_n of the propositional calculus obtained by replacing all the elementary components by propositional variables in such a way that to different components we associate different propositional variables. Thus, we can exploit the completeness theorem for the propositional calculus. B_n is either satisfiable or refutable.

Case 1. B_n is refutable. Then $(P_n)M_n$ is also refutable and so is

$$(x_1)\ldots(x_r)(E\,y_1)\ldots(E\,y_s)M(x_1,\ldots,x_r;y_1,\ldots,y_s).$$

Case 2. No B_n is refutable. Thus they are all satisfiable. Thus for each n, there are systems of predicates defined on the integers $\{0,\ldots,ns\}$ and truth values $t_0,\ldots,t_l$ for the propositional variables such that a true proposition results if in B_n we replace the P_i's by the system of predicates, the variables z_i by the natural numbers i, and X_i by the corresponding t_i.

Thus, for each M_n we have been able to construct an interpretation, with finite domain on the natural numbers, which makes M_n true. The step that clinches the proof consists in showing that since there are only finitely many alternatives at each stage n (given that the domain is finite) and that each interpretation that satisfies M_{n+1} makes true the previous M_n's, it follows that there is an infinite sequence of interpretations S_1, S_2, and so on such that S_{n+1} contains all the preceding ones. This follows from an application of König's lemma, although Gödel does not explicitly appeal to König's result. From this infinite sequence of interpretations it is then possible to define a system satisfying the original sentence $(x_1)\ldots(x_r)(E\,y_1)\ldots(E\,y_s)M(x_1,\ldots,x_r;y_1,\ldots,y_s)$ by letting the domain of interpretation be the natural numbers (hence a denumerable domain!) and declaring that a certain predicate appearing in M is satisfied by an n-tuple of natural numbers if and only if there is at least an n such that in S_n the predicate holds of the same numbers. Similarly the propositional variables occurring in M are given values according to whether they are given those values for at least one S_n. This interpretation satisfies $(P)M$.

This concludes the proof. Gödel generalizes the result to countable sets of sentences and to first-order logic with identity. The former result is obtained as a corollary to Theorem X, which is what we now call the compactness theorem: For a denumerably infinite system of formulas to be satisfiable, it is necessary and sufficient that every finite subsystem be satisfiable.[99]

1.8.5 Models of First-Order Logic

Although we have already discussed the notion of *Allgemeingültigkeit* in the presentation of the narrow functional calculus in Hilbert, it will be useful to go back to it to clarify how models are specified for such languages.

One first important point to notice is that both in Hilbert and Ackermann (1928) and in Bernays and Schönfinkel (1928), the problem of *Allgemeingültigkeit* is that of determining for logical expressions that have no constants whether a correct expression results for arbitrary substitution of values for the (predicate) variables. As a result, an interpretation for a logical formula becomes the assignment of a domain together with a system of individuals and functions. For instance $(x)(F(x) \vee \overline{F(x)})$ is, according to Bernays-Schönfinkel, *allgemeingültig* for every domain of individuals (i.e., by substituting a logical function for F one obtains a correct sentence). (Tarski 1933b, p. 199, n. 3) points out that what is at stake here is not the notion of "correct or true sentence in an individual domain a" because the central concept in Hilbert-Ackermann and Bernays-Schönfinkel is that of sentential functions with free variables and not that of sentence (Tarski implies that one can properly speak of truth of sentences only; this is also in Ajdukiewicz 1921). For this reason, Tarski says, these authors use *allgemeingültig*, as opposed to *richtig* or *wahr*. This is, however, misleading in that *richtig* and *wahr* are used by the above-mentioned authors all over the place. Tarski is nevertheless right in pointing out that when, for a specific individual domain, we assign an interpretation to F, say X (a subset of the domain), we are still not evaluating the truth of $(x)(F(x) \vee \overline{F(x)})$, because the latter expression is not a sentence as F is free in it.[100]

In Gödel's dissertation we find the following presentation of the notion of satisfaction in an interpretation:

> Let A be any logical expression that contains the functional variables $F_1, F_2, \ldots, F_k$, the free individual variables $x_1, x_2, \ldots, x_l$, the propositional variables $X_1, X_2, \ldots, X_m$, and otherwise, only bound variables. Let S be a system of functions $f_1, f_2, \ldots, f_k$ (all defined in the same universal domain), and of individuals (belonging to the same domain), $a_1, a_2, \ldots, a_l$, as well as propositional constants, $A_1, A_2, \ldots, A_m$.
>
> We say that this system, namely $(f_1, f_2, \ldots, f_k, a_1, a_2, \ldots, a_l, A_1, A_2, \ldots, A_m)$ satisfies the logical expression if it yields a proposition that is true (in the domain in question) when it is substituted in the expression. (Gödel 1929, p. 69)[101]

We see that also in Gödel's case the result of substituting objects and functions into the formula is seen as yielding a sentence, although properly speaking one does not substitute objects into formulas. Unless what he means is that symbols denoting the objects in the system have to be substituted in the formula. Lack of clarity on this issue is typical of the period.

1.8.6 Completeness and Categoricity

In the introductory remarks to his "Untersuchungen zur allgemeinen Axiomatik", written around 1927–29, Carnap wrote:

> By means of the new investigations on the general properties of axiomatic systems, such as, among others, completeness, monomorphism (categoricity), decidability [*Entscheidungsdefinitheit*], consistency and on the problems of the criteria and mutual relationships between these properties, it has become more and more clear that the main difficulty lies in the insufficient precision of the concepts applied. (Carnap 2000 [1927–29], p. 59)

Carnap's work remained unpublished at the time, except for the programmatic (1930), but the terminological and conceptual confusion reigning in logic had been remarked by other authors. Let us first pursue the development of the notions of completeness and categoricity in the 1920s and early 1930s.

Recall the notion of completeness found in the postulate theorists (see section 1.1.4): A complete set of postulates is one such that its postulates are consistent, independent of each other, and sufficient, where "sufficiency" means that only one interpretation is possible.

According to contemporary terminology, a system of axioms is categorical if all its interpretations (or models) are isomorphic. In the early part of the twentieth century it was usually mentioned, for example, that Dedekind had shown that every two interpretations of the axiom system for arithmetic are isomorphic. One thing on which there was already clarity is that two isomorphic interpretations make the same set of sentences true. We know today that issues of categoricity are extremely sensitive to the language and logic in which the theory is expressed. Thus the set of axioms for first-order Peano arithmetic is not categorical (an immediate consequence of the Löwenheim–Skolem theorem and/or of Gödel's incompleteness theorem) but second-order arithmetic is categorical (at least with respect to standard second-order models). This sheds light on some of the early confusions. One such confusion was the tendency to infer the possibility of incompleteness results from the existence of nonisomorphic interpretations. Consider Skolem (1922): "Since Zermelo's axioms do not uniquely determine the domain B, it is very improbable that all cardinality problems are decidable by means of these axioms."

As an example he mentions the continuum-problem.[102] The implicit assumption here is that if a system is not categorical, then there must be sentences A and $\neg A$ such that one of the interpretations makes A true and the other makes $\neg A$ true. That the situation is not as simple became clear only very late. In 1934, Skolem proved that there are nonisomorphic countable models of first-order Peano arithmetic which make true exactly the same (first-order) sentences. In later developments, the notion of elementary equivalent models was introduced to capture the phenomenon (see following discussion).

To gauge what the issues were surrounding a proper understanding of categoricity, let us look at how von Neumann deals with categoricity (1925). In the first part of his article, von Neumann discusses the Löwenheim–Skolem theorem, which shows that every set of first-order sentences which is satisfied by an infinite domain can also be satisfied in a denumerable domain. This immediately implies that no first-order theory which admits a nondenumerable interpretation can be categorical (in our sense). This should settle the problem of categoricity for the axioms being discussed by von Neumann. Indeed, von Neumann draws the right conclusion concerning the system of set theory: "We now know that, if it is at all possible to find a system S satisfying the axioms, we can also find such system in which there are only denumerably many I-objects and denumerably many II-objects." (von Neumann 1925, p. 409) Why then, in the following section (§6), does he discuss the issue of categoricity again? A careful reading shows that he is appealing to categoricity as nondisjunctiveness (see Veblen 1904), that is, an axiom system is categorical if it is not possible to add independent axioms to it.

An early attempt to provide a terminological clarification concerning different meanings of completeness is found in the second edition of *Einleitung in die Mengenlehre* (1923), where Fraenkel distinguishes between completeness in the sense of categoricity and completeness as decidability (*Entscheidungsdefinitheit*).[103] Both concepts of completeness are also discussed in Weyl (1927), but Weyl rejects completeness as decidability (for every sentence A, one should be able to derive from the axioms either A or $\neg A$) as a "philosopher's stone".[104] The only meaning of completeness that he accepts is the following: "The final formulation is thus the following: An axiom system is complete when two (contentual) interpretations of it are necessarily isomorphic." (Weyl 1927, p. 22) In this sense, he adds, Hilbert's axiomatization of geometry is complete.

In the third edition of *Einleitung in die Mengenlehre* (1928), Fraenkel adds a third notion of completeness, the notion of *Nichtgabelbarkeit* ("non-forkability"), meaning essentially that every two interpretations satisfy the same sentences (today we call this "semantic completeness"). Carnap (1927) claims that the first two notions are identical, and in 1930, he claimed to have proved the equivalence of all three notions (which he calls monomorphism, decidability, and nonforkability). The proofs were supposed to be contained in his manuscript "Untersuchungen zur allgemeinen Axiomatik", but his approach there is marred by his failure to distinguish between object language and metalanguage, and between syntax and semantics, and thus to specify exactly to which logical systems the proofs are supposed to apply (for an analysis of these issues see Awodey and Carus (2001); Carnap's unpublished investigations on general axiomatics are now edited in Carnap 2000 [1927–29]). Gödel, however, had access to the manuscript and, in fact, his 1929 dissertation acknowledges the influence of Carnap's investigations (as does Kaufmann 1930). Awodey and Carus (2001, p. 23) also point out that Gödel's first public mention of the incompleteness theorem in Königsberg in 1930 (see Gödel 1995, p. 29 and the introduction by Goldfarb) was aimed specifically at Carnap's claim. Indeed, when speaking of the meaning of the completeness theorem for axiom systems, he pointed out that in first-order logic monomorphicity (Carnap's terminology) implies (syntactic) completeness [*Entscheidungsdefinitheit*]. If syntactic completeness also held of higher-order logic then (second-order) Peano arithmetic, which by Dedekind's classical result is categorical, would also turn out to be syntactically complete. But, and here is the first announcement of the incompleteness theorem, Peano's arithmetic is incomplete (Gödel 1930a, pp. 28–30).

An important result concerning categoricity was obtained by Tarski in work done in Warsaw between 1926 and 1928. He showed that if a consistent set of first-order propositions does not have finite models, then it has a nondenumerable model (upward Löwenheim–Skolem). This shows that no first-order theory that admits of an infinite domain can be categorical (*kategorisch*). The result was mentioned publicly for the first time in 1934 in the editor's remarks at the end of Skolem (1934). A proof by Malcev stating that, under the assumptions, the theory has models of every infinite cardinality was published in (1936);[105] this result was apparently also obtained by Tarski in his Warsaw seminar (see Vaught 1974, p. 160). Other results that Tarski obtained in the period (1927–1929) include the result that a first-order theory that contains as an extralogical symbol "<" and that is satisfied in the order type ω is also satisfied in every set of order type $\omega + (\omega^* + \omega)\tau$, where ω^* is the reverse of the standard ordering on ω, and τ is an arbitrary order type. This was eventually to lead to the notion of *elementary equivalence*, defined for order types in the appendix to Tarski (1936b). This allowed Tarski to give a number of non-definability results. In the same appendix he shows that, using η for the order type of the rationals, every order type of dense order is elementarily equivalent to one of the following types: η, $1 + \eta$, $\eta + 1$, and $1 + \eta + 1$ (which are not elementary equivalent to each other). He thus concluded that properties of order types such as continuity or nondenumerability cannot be expressed in the language of the elementary theory of order. Moreover, using the elementary equivalence of the order types ω and $\omega + (\omega^* + \omega)$, he also showed that the property of well ordering is not expressible in the elementary theory of order (Tarski 1936b, p. 380).

One of the techniques investigated in Tarski's seminar in Warsaw was what he called the elimination of quantifiers. The method was originally developed in connection to decidability problems by Löwenheim (1915) and Skolem (1920). It basically consists in showing that one can add to the theory certain formulas, perhaps containing new symbols, so that in the extended theory it is possible to demonstrate that every sentence of the original theory is equivalent to a quantifier free sentence of the new theory. This idea was cleverly exploited by Langford to obtain, for instance, decision procedures for the first-order theories of linear dense orders without endpoints, with first but no last element and with first and last element (1927a) and for the first-order theory of linear discrete orders with a first but no last element (1927b). As Langford emphasizes (1927a), he is concerned with "categoricalness", that is, that the theories in question determine the truth value of all their sentences (something he obtains by showing that the theory is syntactically complete). Many such results were obtained afterward, such as Presburger's (1930) elimination of quantifiers for the additive theory of the integers and Skolem's (1929b) for the theory of order and multiplication (but without addition!) on the natural numbers. Tarski himself announced in 1931 to have obtained, by similar techniques, a decision procedure for elementary algebra and geometry (published however only in 1948). Moreover, he extended the results by Langford to the first-order theory of discrete order without a first or last element and to the first-order theory of discrete order with first and last element. This work is relevant to the study of models in that it allows the study of all the complete extensions of the systems under consideration and leads naturally to the notion of elementary equivalence between relational structures (for order types) that Tarski developed in his seminar. This work also dovetails with Tarski's "On certain fundamental concepts of metamathematics" (1930b), where for instance he proves Lindenbaum's result that every consistent set of sentences has a complete consistent extension. For reason of space, Tarski's contributions to metamathematics during this period cannot be discussed in their full extent, and we limit ourselves here to his definition of truth.[106]

Another important result concerning categoricity, or lack thereof, was obtained by Skolem (1933, 1934) (Skolem speaks of "complete characterizability"). The results we have mentioned so far, the upward and downward Löwenheim–Skolem theorems are consistent with the possibility that, for instance, there is only one countable model, up to isomorphism, for first-order Peano arithmetic. What Skolem showed was, in our terminology, that there exist countable models of Peano arithmetic that are not isomorphic. He constructed a model N^* of (classes of equivalence of) definable functions (hence the countability of the new model) which has all the constant functions ordered with the order type of the natural numbers and followed by non standard elements, which eventually majorize the constant functions, for instance, the identity function (for details see also Zygmunt 1973). Indeed, Skolem's result states that no finite (in 1933) or countable (in 1934) set of first-order sentences can characterize the natural numbers. The 1934 result implies that N^* can be taken to make true exactly those sentences that are true in N.

1.8.7 Tarski's Definition of Truth

The most important contribution to semantics in the early thirties was made by Alfred Tarski. Although his major work on the subject, "The concept of truth in formalized languages", came out in 1933 in Polish (1935 in German), Tarski said that most of the investigations contained in it date from 1929. However, the seeds of Tarski's reflection on truth were planted early on by the works of Ajdukiewicz (1921) and the lectures of Leśniewski.[107]

Tarski specifies the goal of his enterprise at the outset:

> The present article is almost wholly devoted to a single problem—*the definition of truth*. Its task is to construct—with reference to a given language—*a materially adequate and formally correct definition of the term "true sentence"*. (Tarski 1933b, p. 152)

A materially adequate definition is one that for each sentence specifies under what conditions it must be considered true. A formally correct definition is one that does not generate a contradiction and uses only certain concepts and rules specified in advance. One should not expect the definition to give a criterion of truth. It is not the role of the definition to tell us whether "Paris is in France" is true but only to specify under what conditions the sentence is true.

Tarski begins by specifying that the notion of truth he is after is the one embodied in the classical conception of truth, where a sentence is said to be true if it corresponds with reality. According to Tarski, the definition of truth should avoid appeal to any semantical concepts, which have not been previously defined in terms of nonsemantical concepts. In Tarski's construction, truth is a predicate of sentences. The extension of such a predicate depends on the specific language under consideration; thus the inquiry is to take the form of specifying the concept of truth for specific individual languages. The first section of the paper describes at length the prospects for defining truth for a natural language and concludes that this is a hopeless task. Let us see what motivates this negative conclusion. Tarski first proposes a general scheme of what might count as a first approach toward a definition of the expression "x is a true sentence":

x is a true sentence if and only if p. (∗)

Concrete definitions are obtained by substituting for "p" any sentence and for "x" the name of the sentence. Quotation marks are one of the standard devices for creating names (but not the only one). If p is a sentence, we can use quotation marks around p to form a name for p. Thus, a concrete example of (∗) could be

"It is snowing" is a true sentence if and only if it is snowing. (∗∗)

The first problem with applying such a scheme to natural language is that although (∗) looks innocuous, one needs to be wary of the possibility of the emergence of paradoxes, such as the liar paradox. Tarski rehearses the paradox and notices that at a crucial point one substitutes in (∗) for "p" a sentence, which itself contains the term "true sentence". Tarski does not see a principled reason that such substitutions should be excluded, however. In addition, more general problems stand in the way of a general account. First of all, Tarski claims that if one treats quotation mark names as syntactically simple expressions the attempt to provide a general account soon runs into nonsense. Therefore, he points out that quotation mark names have to be treated as complex functional expressions, where the argument is a sentential variable, p, and the output is a quotation mark name. The important fact in this move is that the quotation mark name "p" now can be seen to have structure. According to Tarski, however, even in this case new problems emerge, for example, one ends up with an intensional account, which might be objectionable (even if p and q are equivalent, their names, "p" and "q", will not be). This leads Tarski to try a new strategy by attempting to provide a structural definition of true sentence which would look roughly as follows:

A true sentence is a sentence which possesses such and such structural properties (i.e. properties concerning the form and arrangement in sequence of the single part of the expression) *or which can be obtained from such and such structurally described expressions by means of such and such structural transformations.* (Tarski 1933b, p. 163)

The major objection to this strategy is that we cannot, due to the open nature of natural languages, specify a structural definition of sentence, let alone of true sentence. Moreover, natural languages are "universal", that is, they contain such terms as "true sentence", "denote", "name", and so on, which allow for the emergence of self-reference such as the one leading to the liar antinomy. Tarski concluded:

If these observations are correct, *then the very possibility of a consistent use of the expression "true sentence" which is in harmony with the laws of logic and the spirit of everyday language seems to be very questionable, and consequently the same doubt attaches to the possibility of constructing a correct definition of this expression.* (Tarski 1933b, p. 165)

Thus, the foregoing considerations explain a number of essential features of Tarski's account. First, the account will be limited to formal languages. For such languages it is in fact possible to specify the syntactic rules that define exactly what a well-formed sentence of the language is. Moreover, such languages are not universal, that is, one can keep the level of the object language and that of the metalanguage (which is used to describe the semantic properties of the object language) separate. When we talk about theories specified in a certain language, then we distinguish between the theory and the metatheory, where the latter is used to study the syntactic and semantic properties of the former.

Tarski provides then the definition of truth for a specific language, that is, the calculus of classes, but the treatment is extended in the later sections of the essay to provide a definition of truth for arbitrary languages of finite type. One important point stressed by Tarski is that the definition of truth is intended for "concrete" deductive systems, specifically, deductive systems that are interpreted. For purely formal systems, Tarski claims that the problem of truth cannot be meaningfully raised.

The calculus of classes is a subtheory of mathematical logic that deals with the relationships between classes and the operation of union, intersection and complement. There are also two special classes, the universal class and the empty class. The intuitive interpretation of the theory that Tarski has in mind is the standard one with the individual variables ranging over classes of individuals. In the following we give an (incomplete) sketch of the structure of the language L of the calculus of classes (with only instances of the axioms) and of the metalanguage, ML, in which the definition of truth is given. It should be pointed out that Tarski does not completely axiomatize the metalanguage, which is presented informally, and he uses the Polish notation in his presentation.

The Language of the Calculus of Classes

Variables: $x_I, x_{II}, x_{III}, \ldots$

Logical constants: N [negation], A [disjunction], Π [universal quantifier]; relational constant: I [inclusion]

Expressions and formulas are defined as usual.

Logical axioms: $ANAppp$ [$\sim(p \vee p) \vee p$], etc.

Proper axioms: $\Pi x_1 I x_1 x_1$ [every class is included in itself];
$\Pi x_I x_{II} x_{III} A N I x_I x_{II} A N I x_{II} x_{III} I x_I x_{III}$ [transitivity of I], etc.

Rules of inference: substitution, modus ponens, introduction, and elimination of Π.

The Metalanguage

Logical constants: not, or, for all

Relational constants: $\subseteq$

Class theoretical terms: $\in$, individual, identical ($=$), class, cardinal number, domain, etc.

Terms of the logic of relations: ordered n-tuple, infinite sequence, relation, etc.

Terms of a structural descriptive kind: ng [for N]; sm [for A], un [for Π], v_k [the kth variable], $x \frown y$ [the expression that consists of x followed by y], etc. These form names of object-language expressions in the metalanguage.

Auxiliary symbols are introduced to give metatheoretical shorthands for whether an expression is an inclusion, a negation, a disjunction, or a universal quantification. They are: $x = \iota_{k,l}$ iff $x = (\text{in} \frown v_k) \frown v_l$, $x = \overline{y}$ iff $x = \text{ng} \frown y$; $x = y + z$ iff $x = (\text{sm} \frown y) \frown z$; $x = \bigcap_k y$ iff $x = (\text{un} \frown v_k) \frown y$.

Variables:

1. a, b [names for classes of an arbitrary character]
2. f, g [sequences of classes]
3. k, l, m, n [natural numbers and sequences of natural numbers]

4. t, u, w, x, y, z [expressions]

5. X, Y [sequences of expressions]

The Metatheory

Logical axioms: not (p or p) or p, etc.

Axioms of the theory of classes: for all a, $a \subseteq a$ etc.

Proper axioms: several axioms characterizing the notion of expression. Intuitively, this is the smallest class X containing ng, sm, $\cap$, ι, v_k, such that if x, y are in X then $x \frown y$ is in X.

With this in place, we can give names in ML to every expression in L. For instance, $NIx,x,,$ is named in ML by $((\text{ng} \frown \text{in}) \frown v_1) \frown v_2$ or $\overline{\iota_{12}}$. We can now define the notions of the following.

Sentential function (**Definition 10**): Sentential functions are obtained by the closure of expressions of the form ι_{ik} under negation, disjunction, and universal quantification.

Sentence: A sentential function with no free variables is a sentence.

Axioms: A sentence is an axiom if it is the universal closure of either a logical axiom or of an axiom of the theory of classes.

Theorems: A sentence is a theorem if it can be derived from the axioms using substitution, modus ponens, introduction, and elimination rules for universal quantifier.

With this machinery in place (all of which is purely syntactical), Tarski proceeds to give a definition of truth for the calculus of classes. The richness of the metalanguage provides us both with a name of the sentence and a sentence with the same meaning (a translation into the metalanguage) for every sentence of the original calculus of classes. For instance, to '$\Pi\, v,I\,v,v,$' in L corresponds the name $\cap_1 \iota_{11}$ and the sentence "for all a, $a \subseteq a$". The schema (*) should now be recaptured in such a way that for any sentence of the calculus of classes its name in the metalanguage appears in place of x and in place of p we have the equivalent sentence in the metalanguage:

$\cap_1 \iota_{11}$ is a true sentence if and only if for all a, $a \subseteq a$.

What is required of a satisfactory truth definition is that it contains all such equivalences in its extension. More precisely, let Tr denote the class of all true sentences and S the class of sentences. Then Tr must satisfy the following convention.

> **Convention T** *A formally correct definition of the symbol "Tr" formulated in the metalanguage, will be called an adequate definition of truth if it has the following consequences:*
>
> *a. all sentences which are obtained from the expression "$x \in Tr$ if and only if p" by substituting for the symbol "x" a structural-descriptive name of any sentence of the language in question and for the symbol "p" the expression which forms the translation of this sentence into the metalanguage;*

β. the sentence "for any x, if $x \in Tr$ then $x \in S$" (in other words, $Tr \subseteq S$). (Tarski 1933b, p. 188)

Ideally, one would like to proceed in the definition of truth by recursion on the complexity of sentences. Unfortunately, on account of the fact that sentences are in general not obtained from other sentences but rather from formulas (which, in general, may contain free variables), a recursive definition of "true sentence" cannot be given directly. However, complex formulas are obtained from formulas of smaller complexity, and here the recursive method can be applied. For this reason Tarski defines first what it means for a formula to be satisfied by given objects. Actually, for reasons of uniformity, Tarski defines what it means for an infinite sequence of objects to satisfy a certain formula. Definition of satisfaction (Definition 22): Let f be an infinite sequence of classes, and f_i the i-th coordinate. Satisfaction is defined inductively on the complexity of formulas (denoted by x, y, z).

Atomic formulas: f satisfies the sentential function $(\iota_{k,l})$ iff $(f_k \subset f_l)$.
Molecular formulas:

a. for all f, y: f satisfies $\overline{y}$ iff f does not satisfy y;

b. for all f, y, z: f satisfies $y + z$ iff f satisfies y or f satisfies z;

c. for all f, y, k: f satisfies $\bigcap_k y$ iff every sequence of classes that differs from f at most in the kth place satisfies the formula y.

This definition is central to Tarski's semantics, since through it one can define the notions of denotation (the name "c" denotes a, if a satisfies the propositional function $c = x$), definability, and truth. A closer look at the definition of satisfaction shows that whether a sequence satisfies a formula depends only on the coordinates of the sequence corresponding to the free variables of the formula. When the formula is a sentence, there are no free variables, and thus either all sequences satisfy it or no sequence satisfies it. Correspondingly, we have the definition of truth and falsity for sentences given in Definition 23: x is a true sentence iff x is a sentence and every infinite sequence of classes satisfies x. Tarski then argues that the definition given is formally correct and satisfies Convention T.

Among the consequences Tarski draws from the precise definition of the class of true sentences is the fact that the theorems of the calculus of classes are a proper subset of the truths of the calculus (under the intended interpretation).

Nowadays such definitions of satisfaction and truth are given by first specifying what the domain of the interpretation is, but Tarski does not do that. He speaks of infinite sequences of classes as if these sequences were taken from a universal domain. Indeed, on p. 199 of his essay, Tarski contrasts his approach with the relativization of the concept of truth to that of "correct or true sentence in an individual domain a". This is the approach, he points out, of the Hilbert school in Göttingen and contains his own approach as a special case. Of course, Tarski claims to be able to give a precise meaning of the notions (Definitions 24 and 27) that were used only informally by the Hilbert school.[108]

The remaining part of the essay sketches how to generalize the approach to theories of finite order (with a fixed finite bound on the types) and points out the limitations in extending the approach to theories of infinite order. However, even in the latter case Tarski establishes that "the consistent and correct use of the concept of truth is rendered possible by including this concept in the system of primitive concept of the metalanguage and determining its fundamental properties by means of the axiomatic method" (266).

By far the most important result of the final part of the essay is Tarski's celebrated theorem of the undefinability of truth, which he obtained after reading Gödel's paper on incompleteness.[109] Basically, the result states that there is no way to express $Tr(x)$ as a predicate of object languages (under certain conditions) without running into contradictions. In particular, for systems of arithmetic such as Peano Arithmetic, this says there is no arithmetical formula $Tr(x)$ such that $Tr(x)$ holds of a code of a sentence just in case that sentence is true in the natural numbers.

We have seen that Tarski emphasized that through the notion of satisfaction other important semantic notions, such as truth and definability, can be also defined. Thus, the work on truth also provided an exact foundation for (1930a) and (1931), on definable sets of real numbers and the connection between projective sets and definable sets, and to the general investigation on the definability of concepts carried out by Tarski in the mid-1930s.

One of the most important applications of the new semantic theory was the notion of logical consequence (1936d). Starting from the intuitive observation that a sentence X follows from a class of sentences K if "it can never happen that both the class K consists only of true sentences and the sentence X is false" (414), Tarski made use of his semantical machinery to give a definition of the notion of logical consequence. First he defined the notion of model. Starting with a class L of sentences, Tarski replaces all nonlogical constants by corresponding variables, obtaining the class of propositional sentences L'. Then he says:

An arbitrary sequence of objects which satisfies every sentential function [formula] of the class L' will be called a *model* or *realization* of the class L of sentences (in just this sense one usually speaks of an axiom system of a deductive theory). (Tarski 1936d, p. 417)

From this he obtains the notion of logical consequence: "The sentence X *follows logically* from the sentence of the class K if and only if every model of the class K is also a model of the sentence X." (Tarski 1936d, p. 417) There are several controversial issues concerning the exact interpretation of Tarski's theory of truth and logical consequence; these cannot be treated adequately within the narrow limits of this exposition.[110]

In any case, the result of Tarski's investigations for logic and philosophy cannot be overestimated. The standard expositions of logic nowadays embody, in one form or another, the definition of truth in a structure, which ultimately goes back to Tarski's article. Tarski's article marks also an explicit infinitistic attitude to the metatheoretical investigations, in sharp contrast to the finitistic tendencies of the Hilbert school. As a consequence the definition of truth is often nonconstructive. Often, but not always: In the particular case of the calculus of classes, Tarski shows that from the definition of truth one can also extract a criterion of truth; but he also remarks that this depends on the specific peculiarities of the theory and in general this is not so. Finally, Tarski's definition of truth and logical consequence have shaped the discussion of these notions in contemporary philosophy and are still at the center of current debates.

PART II

Foundations of Mathematics

Introduction

SUMMARY

Part II contains seven essays in the foundations of mathematics. The first two relate to Hilbert's program. Chapter 2 provides a presentation of Hilbert's program that is more sensitive to the diachronic changes in the program than had hitherto been the case while also setting Hilbert's program in the context of the philosophical exchanges Hilbert and Bernays had with their contemporaries. Chapter 4 explores an aspect of Hilbert's program that was completely ignored until recently, that is the influence of Russell's type theory and logicism on Hilbert and his school (Behmann, Bernays, Ackermann, Schönfinkel). This was considered to be an open problem in the literature. I was able to make progress on this aspect of Hilbert's program thanks to the discovery of the important dissertation work by Heinrich Behmann, a student of Hilbert's, who was primarily responsible for the reception of *Principia Mathematica* in Göttingen (see Chapter 3) and to the use of Hilbert's unpublished lectures in logic and the foundations of mathematics.

The work on constructivity (Chapters 5 and 6) is centered on two related episodes. The first (Chapter 5) concerns the problem of the relation of classical mathematics to intuitionistic mathematics. This problem was far from having a precise answer in the late twenties and in fact Felix Kaufmann formulated a conjecture to the effect that classical mathematics without the axiom of choice was intuitionistically acceptable. Heinrich Behmann gave a "proof" of this claim that was, at the time, accepted by Carnap as sound. However, Gödel was able to show (in a letter which I discovered and is now published in Gödel's *Collected Works*, vol. V) that the proof was wrong. While the proof did not stand, the debate (which involved Kaufmann, Behmann, Carnap, Bernays, and Gödel) provides a rich picture of a crucial moment in the development of the foundations of mathematics and contains some small gems such as the topological analysis of proofs given by Behmann. In Chapter 6, Marion and I were able to relate this debate to the only instance in which Wittgenstein attempted (successfully) the constructivization of a classical proof, viz. Euler's proof for the infinitude of primes. We use this analysis to claim, against much of the literature, that during the 'transitional period' Wittgenstein had a constructivist, more precisely Kroneckerian, position on mathematics.

The last two articles concern Gödel's incompleteness theorems. In Chapter 7, I studied the different reactions to Gödel's incompleteness theorem by Gödel and Bernays, who thought that Hilbert's program could be salvaged, and those of von Neumann and Herbrand, who saw the result as showing once and for all that Hilbert's program could not be carried out. The paper also contains interesting information on Carnap's reaction to the results. Chapter 8 is an essay review of volumes IV and V of Gödel's *Collected Works*. Here the aim is to study the heuristics that led to Gödel's incompleteness theorems by making use of the correspondence contained in volumes IV and V and to argue that knowledge of the correspondence deepens our understanding of various aspects of the theorems.

BIBLIOGRAPHICAL UPDATE

Of all the essays in this book, Chapter 2 was the first one to be written. Obviously, the bibliography represents the state of things up to 1996/7. Some of the developments since are taken into account in Chapter 1 and the following chapters in Part II of the book. The major developments in the years following the publication of Chapter 2 can be briefly summarized as follows. Concerning the historical development of Hilbert's program a number of contributions appeared shortly thereafter: Sieg (1999), Mancosu (1999a, 2003), and Zach (1999, 2003, 2004). All are characterized by the use of untapped archival resources and address both philosophical aspects and technical developments in Hilbert's program. Zach (2003) managed to show that in their metamathematical practice Hilbert and his school accepted forms of finitistic reasonings that go well beyond Primitive Recursive Arithmetic. At the historical level, this result challenges those formal characterizations of finitism in terms of Primitive Recursive Arithmetic. On this matter see the interesting discussion in Tait (2002) and the appendix to the reprint in Tait (2005) as well as Wille (2008), chapter 3. Let me also mention Zach (2006) as an excellent introduction to Hilbert's program and its historical development. Other aspects treated in Chapter 2 that have received further treatment in the literature concern the axiomatization program. First of all, one should mention Hallett (2008) and the first volume of the Hilbert Edition devoted to the foundations of geometry (Hilbert 2004). In addition, there has been very interesting scholarly work in the area of the axiomatization of physics. I refer the reader to Corry (2004a), Majer and Sauer (2006), and to Brading and Ryckman (2008); the reader will be able to track most of the additional relevant literature from those three entries. With regard to Hilbert's famous list of 23 mathematical problems see Gray (2000) and Yandell (2002). On the so-called 24th problem see Thiele (2003). Chapter 2 also touches upon the connections between Hilbert and Poincaré. A reference I had missed at the time is Folina (1992). In connection to Kronecker see now Kronecker (2001), Boniface (2004), Gauthier (2002), Edwards (2005), and Goldstein *et al.* (2007). On Paul Bernays' philosophical thought see also the introduction by Hourya Sinaceur to Bernays (2004). I have no further bibliographical recommendations for Chapters 3, 4, and 5. Concerning Wittgenstein's engagement with Euler's proof on the infinity of primes, treated in Chapter 6, see now also Lampert (2008).

On Wittgenstein's brand of constructivism during his transitional phase see Marion (2004, 2008). In the case of Gödel, I had to be conservative since there is a burgeoning literature on Gödel on philosophy of mathematics, phenomenology, etc. However, most of this concerns a different period of Gödel's philosophical reflection than the one addressed in Chapters 7 and 8. I will thus limit myself to the period leading to and immediately following the discovery of the incompleteness theorems. Here the best references are Tait's rich review of volumes IV and V of Gödel's *Collected Works* (Tait 2006) and Feferman (2008a). Moreover, Stadler's 1997 book is now available in English (Stadler 2003). In addition there are two scholarly editions in the making. The Hilbert Edition, undertaken by Springer, is projected to publish five volumes in German containing selections from the unpublished lecture notes preserved at the University of Göttingen. The only volumes published so far are Hilbert (2004) and (2009). There is also in preparation an edition of Bernays' foundational writings in English (http://www.phil.cmu.edu/projects/bernays/). This is scheduled to appear in 2010 for Open Court.

2

HILBERT AND BERNAYS ON METAMATHEMATICS

DAVID HILBERT (1862–1943) was one of the most important mathematicians of his generation. From 1886 to 1895 he taught at the university of Königsberg and then accepted an offer from Göttingen where he remained until his death. In addition to contributing to several branches of mathematics—number theory, analysis, and mathematical physics among others—he is also known to the broader philosophical community for his groundbreaking work in logic and the foundations of mathematics. The branch of mathematical logic known as proof theory began with his investigations, and some of the problems he proposed in the foundations of mathematics are still central to the field.[1] Hilbert's most extensive engagement in the foundations of mathematics took place in the 1920s. Through the help of his assistant Paul Bernays (1888–1977), who joined Hilbert in 1917 and worked with him throughout the 1920s and 1930s, Hilbert developed a far-reaching programme in the foundations of mathematics. Hilbert and Bernays rejected the revisionist approaches to mathematics defended by Brouwer and Weyl and aimed at a foundation of mathematics that would guarantee the reliability of mathematics without giving up any part of classical mathematics. This chapter is divided into 5 sections. The first section reviews the main contributions by Hilbert on foundational matters prior to the 1920s. The second section describes the new foundations of mathematics championed by Hilbert and Bernays. The third section describes the philosophical exchanges that opposed Hilbert and Bernays to other philosophers of mathematics. The fourth section attemps to spell out, if only in broad strokes, the specificity of Bernays' contributions to Hilbert's programme. Finally, section 5 gives an overview of the progress made by Hilbert and his school on the problem of consistency in the period under investigation.

2.1 Hilbert's Foundational Work Prior to the 1920s

Hilbert's thought on the foundations of mathematics reaches its more mature stage in the 1920s. However, his engagement with foundational issues goes back to the last decade of the nineteenth century. Hilbert's major published contributions to the foundations prior to the period we are examining are Hilbert (1899, 1900a,b, 1905b, 1918a). Characterizing this production is by no means an easy task. For example, Abrusci (1978) distinguishes three distinct stages in Hilbert's thought corresponding to the publications listed above. Peckhaus (1990, 1994a, 1995b) has also given an in-depth analysis of Hilbert's development from his early publications on the foundations of geometry to his work at the end of the 1910s. For our purposes it will be enough, appealing largely to the analyses of Abrusci and Peckhaus, to give an overview of this early production and emphasize the central aspects of the relationship between the previous research and the work by Hilbert and Bernays in the 1920s.

Hilbert's most famous contribution to the foundations of mathematics is the *Foundations of Geometry*, published in 1899.[2] The goal of the work is to investigate the field of geometry axiomatically. This led Hilbert to substantial revisions of the approach taken in the Euclidian *Elements*. For example, he gives up the separation of the first principles into axioms (*notiones communes*) and postulates.[3] The most fundamental distinction between Euclid's approach to geometry and Hilbert's, the old and the new axiomatics, consists, to speak with Bernays, in a "new methodological turn" in the conception of axiomatics.[4] When we open Euclid's *Elements*, we are presented with axioms and notions which are already interpreted; that is, when using the word 'point' or 'line', Euclid is referring to a specific object of geometrical intuition. The axioms are true of these objects. The axioms thus organize the body of geometrical knowledge by showing from which true propositions the rest of the geometrical truths can be derived by means of logical inference. Hilbert invites us to reach a higher level of generality. Rather than anchoring from the start the basic concepts (point, line, plane) to our pre-given geometrical understanding of those notions, he asks us to consider any three systems of objects and three sorts of relations between these objects (lie, between, congruent). The axioms of geometry then only state how these relationships relate the objects in question. As Bernays remarks, this led to the separation between the mathematical and the epistemological problems of axiomatics.

The important thing, then, about Hilbert's *Foundations of Geometry* was that here, from the beginning and for the first time, in the laying down of the axioms system, the separation of the mathematical and logical [spheres] from the spatial-intuitive [sphere], and with it from the epistemological foundation of geometry, was completely carried out and expressed with complete clarity. (Bernays 1922a, p. 192)

Of course, one should not rashly conclude that for Hilbert intuition plays no role anymore. Intuition is essential for suggesting which axioms we need to select in order to account for our geometrical knowledge, but having done that, we need to make an extra step by going beyond intuition to the consideration of any three systems of objects that might act as reference of the words 'point', 'line', and 'plane'. It might be worthwhile to emphasize this point, since the reproach has often been made to Hilbert's view on mathematics of reducing mathematics to a mere study of formalisms, a pure game with symbols. However, it is enough to peruse the 1919 lectures to realize how far this is from Hilbert's view of the matter.[5] Let us go back to the axiomatic method. In Hilbert's approach the axioms give an implicit definition of the relevant concepts. Let us quote Bernays again:

According to this conception, the axioms are in no way judgments that can be said to be true or false; they have a sense only in the context of the whole axiom system. And even the axiom system as a whole does not constitute the statement of a truth; rather, the logical structure of axiomatic geometry in Hilbert's sense, analogously to that of abstract group theory, is a purely hypothetical one. If there are anywhere in reality three systems of objects, as well as determinate relations betweeen these objects, such that the axioms of geometry hold of them (this means that by an appropriate assignment of names to the objects and relations, the axioms turn into true statements), then all theorems of geometry hold of these objects and relationships as well. Thus the axiom system itself does not express something factual; rather, it presents only a possible form of a system of connections that must be investigated mathematically according to its internal properties. (Bernays 1922a, p. 192)

According to Hilbert, an axiomatization of geometry must be complete and as simple as possible. The meaning of completeness at this stage is still vague, but what is likely meant by this requirement is that from the axioms we must be able to derive the theorems that correspond to the set of geometrical truths accepted in ordinary mathematics.[6] The requirement of simplicity includes, among other things, reducing the axioms to a finite number and showing the independence of the axioms. One more requirement for axiomatics is also the proof of the consistency of the axioms, that is, showing that they do not lead to a contradiction. This task was irrelevant to the old axiomatics, since axioms were assumed on account of their evidence and truth, and this was sufficient to ensure consistency. However, having given up reliance on an intuitive domain, the issue arises of proving that there is in fact an interpretation for the axiomatic system that makes the axioms true. To this end Hilbert exploited the possibility of interpreting the geometrical relations as arithmetical relations and thus showing that it is possible to interpret the geometrical axioms as truths about the real numbers. This provides a consistency proof relative to the consistency of the theory of real numbers (arithmetic, in a general sense). It is important to stress that Hilbert speaks of consistency in the sense that it is impossible to deduce from the axioms a contradiction in a finite number of steps.[7] Thus, the proof of consistency for geometry shows that a proof of a contradiction from the axioms of geometry turns, via a translation of geometry into arithmetic, into a proof of a contradiction in arithmetic.

As early as 1900 Hilbert focuses his energy on the axiomatization of arithmetic (in the sense of the theory of the real numbers) and states the important goal of giving a consistency proof for such a system. It seems that at this stage his motivation was to block the objections raised against the "existence" of the set of real numbers and of Cantor's transfinite numbers.[8] The origins of this programme can be traced back, according to Sieg (1990a), to the foundational investigations of the late nineteenth century (Kronecker, Dedekind) and in particular to Dedekind's proposed solution. In Hilbert (1905b), the paradoxes of set theory seem to loom larger in Hilbert's mind and this is certainly behind his idea of developing the theory of natural numbers and logic simultaneously. Both Abrusci and Peckhaus see a second phase of Hilbert's thought in this article. Peckhaus speaks of a "philosophical turn" motivated by the paradoxes and by Hilbert's correspondence with Frege on the foundations of geometry (see Peckhaus 1990, pp. 40–58; Sieg 1990a sees, however, a continuity between the two stages).

In a series of lectures given in 1905 in Göttingen, Hilbert explains, in analogy with his work in geometry, the new abstract approach to the axiomatic treatment of numbers as follows:

The numbers have become for us only a framework of concepts to which we are led of course only by means of intuition; we can nonetheless operate with this framework without having recourse to intuition. However, to ensure the applicability to the objects surrounding us, this conceptual system is constructed in such a way that it forms everywhere a complete analogy with the most trivial intuitions and with it to the facts of experience. (quoted in Peckhaus 1990, p. 60)[9]

The solution given to the problem of the foundations of geometry was only a relative consistency proof; in the case of arithmetic, to which the consistency of geometry was reduced, it was essential to find a direct proof.[10] The outline of an approach is found in Hilbert (1905b). Hilbert proposes here more concrete examples of how a direct consistency proof, that is one not based on reduction, should go. This strategy is in sharp contrast to those common in the late nineteenth century, for example, in Dedekind, which prove consistency by providing a model for the axioms (see Hallett 1995a). One should notice that Hilbert not only requests a consistency proof for the system of real numbers but also wants to push his programme to account for the notion of natural number. Indeed, he criticized Kronecker for not having carried out the foundational analysis far enough.

L. Kronecker, as is well known, saw in the notion of integer the real foundation of arithmetic; he came up with the idea that the integer—and, in fact, the integer as a general notion (parameter value)—is directly and immediately given; this prevented him from recognizing that the notion of integer must and can have a foundation. I would call him a dogmatist, to the extent that he accepts the integer with its essential properties as a dogma and does not look further back. (Hilbert 1905b, p. 130)

The main idea of a direct consistency proof, as delineated by Hilbert in 1905, can be summarized as follows. One begins with a finite (or denumerable) set of symbols and considers all the possible finite combinations from this set. The axioms serve the role of characterizing a group of expressions, and those derivable from them through specified inference rules, as being part of the entities. The other expressions in the language are called the non-entities. A proof of consistency for the system amounts to showing that nothing can be both an entity and a non-entity. Through the analysis of several examples

Hilbert showed that alternative, but related, techniques might be enlisted for achieving the goal. For instance, in an example concerning the notion of infinite set Hilbert shows that all the equations derivable from the axioms satisfy a certain property, whereas the non-entities satisfy the opposite property. However, the proof implicitly appeals to induction, in that one must show that the axioms give rise to equations having a certain property and that the inference rule applied to such equations yields equations that also have the property. But this is equivalent to an application of induction on the length of proofs. Poincaré was quick to seize on this point and to object to Hilbert's approach to the foundations, dubbing it as circular (see also Brouwer (1907, p. 93), and the section on Becker below).

In Hilbert (1905b) we are presented with the remarkable insight that we can "consider the proof itself to be a mathematical object." However, the article is unsatisfactory on account of its vagueness. There is no attempt to spell out the logic and the logical language underlying the mathematical theories—Hilbert only speaks of "the familiar modes of logical inference"—there is no real distinction between theory and metatheory, and most of the arguments are merely sketched. It is true that in his course of lectures in 1905 Hilbert presents an axiomatic system of logic in the style of Boole and Schröder, but he does not go beyond a system of propositional logic without quantification (see Peckhaus 1990, pp. 61–67, for a detailed analysis of the role of logic in Hilbert's work, see Hallett 1995a and Peckhaus 1995b). However, this paper marks two essential points of Hilbert's later approach to the foundations; first of all, the idea that logic and arithmetic have to be developed simultaneously; second, the idea that a consistency proof has to investigate the nature of proofs, an insight that will bear immense fruit in the more mature phase of Hilbert's thought.

Before we move to Hilbert's works in the 1920s we need to say something about Hilbert's development in the 1910s and the views he expressed in *Axiomatic Thought* (1918a). Determinant for Hilbert's approach during this period is the influence of Russell and Whitehead's *Principia Mathematica*. Hilbert sees the axiomatization of logic carried out in *Principia* as a success of the axiomatic method, comparable to that obtained by himself in geometry and by Zermelo in set theory.

But since the examination of consistency is a task thai cannot be avoided, it appears necessary to axiomatize logic itself and to prove that number theory and set theory are only parts of logic. This method was prepared long ago (not least by Frege's profound investigations): it has been most successfully explained by the acute mathematician and logician Russell. One could regard the completion of this magnificent Russellian enterprise of the axiomatization of logic as the crowning achievement of the work of axiomatization as a whole. (Hilbert 1918a, p. 1113)

Hilbert's fascination with the logicist programme did not last long. Soon after the talk of 1917, he returned to the idea of a simultaneous development of logic and mathematics (in opposition to Russell's construction of mathematics from logic). However, the logical work of Russell and Whitehead made it possible for Hilbert to overcome the limitations of his initial approach to the foundations of mathematics that were due to his lack of understanding of the important role that a formalization of logic had to play for carrying out his own programme of studying mathematical proofs as objects of mathematical investigation. At the end of the 1918 publication the goal is stated in clear terms:

All such questions of principle, which I characterized above and of which the question just discussed—that is, the question about decidability in a finite number of operations—was only the last, seem to me to form an important new field of research which remains to be developed. To conquer this field we must, I am persuaded, make the concept of specifically mathematical proof itself into an object of investigation, just as the astronomer considers the movement of his position, the physicist studies the theory of his apparatus, and the philosopher criticizes reason itself. (Hilbert 1918a, p. 1115)

Throughout the 1920s Hilbert would develop this programme by (a) trying to spell out the exact epistemological and foundational nature of his enterprise; and (b) by making Göttingen into a center for the development of mathematical logic. Among his students or collaborators we have to mention Ackermann, Bernays, and to some extent von Neumann. These men were instrumental in helping Hilbert develop his programme through the formal study of logic and, especially in the case of Bernays, by refining his philosophical perspective on the programme. Although the development of mathematical logic in the Hilbert school cannot be detached from the rest of the programme it would be out of place here to try to convey the variety of formalisms, results, and approaches pursued in mathematical logic by Hilbert and his students. Technical results will be mentioned later on only as needed to clarify the foundational nature of the programme (but see 2.5 (and Chapter 1)).

2.2 The New Foundation of Mathematics

By the early 1920s the situation in the foundations of mathematics had become more dire. Brouwer had begun an alternative development of set theory along intuitionistic lines, and Weyl had first presented an alternative foundation of analysis, in *The Continuum*, and in 1921 he had joined the intuitionistic camp. Thus, it is not surprising that Hilbert begins his "New Grounding of Mathematics"(1922c) with a polemic against Brouwer and Weyl. Indeed, it is clear that Hilbert was answering mainly Weyl's claims about a new foundational crises in mathematics than addressing Brouwer's overall intuitionistic philosophy of mathematics. Hilbert also does not distinguish clearly between the predicativist approach of *The Continuum* and the development of mathematics along intuitionistic lines. Both approaches are considered at once.

The second part of the paper is devoted to the new foundations of mathematics proposed by Hilbert. Hilbert's new approach involves three separate parts. One begins by providing an axiomatization by stages, of one specific area of mathematics or of all of mathematics. This also includes a complete formalization of the proofs by means of an axiomatic presentation of the logical calculus. However, according to views already present in the earlier foundational work, the axiomatic development of a theory requires a proof for the consistency of the axioms. In order to reply to Poincaré's objection concerning the possible circularity involved in such a strategy, Hilbert distinguishes three mathematical levels: ordinary mathematics, mathematics in the strict sense (or proper mathematics), and metamathematics.

I shall divide the treatment of the general features of Hilbert's approach to the foundations of mathematics into two sections. The first deals with his criticisms of other foundational enterprises. The second concerns the positive part of his programme.

Hilbert's criticisms of the other foundational schools

Kronecker. We have already mentioned that Hilbert in 1905 attacked Kronecker for having presupposed the notion of natural number without proper clarification in his foundational work. But what exactly was Kronecker's position? This is not an easy question to answer in detail. However, two theses can confidently be attributed to Kronecker:

> *Arithmetization thesis*: all of mathematics (with the exception of geometry) can be strictly arithmetized;
>
> *Decidability thesis*: only decidable definitions should be admitted in mathematics.

The second demand had very serious consequences for mathematical practice. It is well known that Kronecker rejected the Cantorian set-theoretical mathematics and the applications stemming from it. But he even rejected the arithmetization of analysis developed by Weierstrass on account of his use of infinite sets characterizing irrational numbers: Kronecker objected that the approach was unjustified in that there is no decision procedure for telling whether an infinite set characterizing a sequence defines an irrational or not.[11]

Hilbert met Kronecker early in his career during a trip he took to Berlin and other German universities. He recalls that as a young researcher he used to join some of his colleagues in the exercise of transforming "tranfinite proofs of mathematical theorems into finite terms, in accordance with Kronecker's paradigm" (Hilbert 1931, p. 268). He then added:

Kronecker only made the mistake of declaring the transfinite mode of inference to be inadmissible. He issued prohibitions against the transfinite mode of inference; in particular, according to him, one was not allowed to infer that, if a statement $A(n)$ does not hold for every integer n, then there must exist an integer n for which that statement is false. At the time, the whole of mathematics unanimously rejected his prohibition and went on to the business of the day. (Hilbert 1931, p. 268)

Hilbert referred to Kronecker in most of his foundational articles, and it seems that the Kronecker-Cantor opposition acted as a long standing motivation for his development of proof theory (see Reid 1970, p. 173). A careful reading of Hilbert's papers in chronological sequence shows a change of judgment with respect to Kronecker. In the 1922 article, which is certainly the most polemical against Brouwer and Weyl, there is not a clear cut distinction between the approaches by Brouwer and Weyl and those of Kronecker. However, In Hilbert (1931), Kronecker's position is identified, essentially, with Hilbertian finitism. This issue is related to the problem of the relationship between finitism and intuitionism, and I will come back to it below.[12]

Poincaré. Hilbert's polemic against Poincaré is not as sustained as the one against Kronecker, Weyl, and Brouwer. Poincaré had objected to Hilbert (1905b) that the consistency proof requires an appeal to induction and that therefore the whole procedure of giving a foundation for the theory of numbers by means of a Hilbertian consistency proof was circular. He also believed that the principle of induction was a synthetic a priori principle of the human intellect.[13] Hilbert replied to Poincaré (in Hilbert 1922c, 1928a, 1931). On Hilbert's account, Poincaré had not seen that there are two types of induction at play. Metamathematics uses a form of contentual induction on concretely given objects, whereas formal mathematics employs the full principle of induction. This issue will be discussed at greater length in the section on Becker and Hilbert below.[14]

Brouwer and Weyl. Hilbert's treatment of the objections raised by Brouwer and Weyl reflects quite clearly the historical development of the foundations of mathematics in the 1920s. In the 1922 article, which appears only one year after Weyl's conversion to intuitionism, Hilbert attacks Weyl and Brouwer together. He does not distinguish between the predicativist position of *The Continuum* and the intuitionistic views. Moreover, he does not make an exact distinction between Brouwer and Weyl's interpretation of intuitionism. However, Weyl will soon take his distance from intuitionism and from then on Brouwer becomes the main target of Hilbert's views. Nevertheless, even in the later stage many of the accusations that Hilbert attributed to Brouwer actually stem from Weyl's articles. For instance, in 1928 Hilbert characterizes intuitionism as claiming that all existential judgments are meaningless and replies to the accusation that his programme reduces everything to a game. This description of intuitionism corresponds better to Weyl's characterization than to Brouwer's.

There is no need here to rehearse in detail the objections raised by Weyl and Brouwer to classical mathematics (see van Stigt 1998). Hilbert recognizes the need for a secure foundation of mathematics, but he disagrees with Brouwer and Weyl when they claim that (constructive) evidence is the only guarantee in mathematics. Hilbert's plan is to show the admissibility of all of mathematics by establishing that the axiomatic systems for the various branches of mathematics cannot lead us to a contradiction. The proof of consistency will only make use of contentual reasoning, characterized by its evidence, but the mathematics expressed in those systems will not have the same evidential status of the metamathematical considerations. In other words, he agrees with Brouwer and Weyl that mathematics goes well beyond what can be founded by purely contentual reasonings, but this is not a reason to jettison those areas of mathematics that go beyond such contentual reasoning. By contrast, Hilbert wants to provide for these areas of mathematics, including analysis and set theory, a secure foundation by means of consistency proofs.

The logicists. Finally, a few words about Hilbert's position concerning the logicist programme. Although Hilbert will always maintain a distance from the main tenets of the various logicist programmes (Frege, Dedekind, Russell-Whitehead) his judgment of their achievement strongly contrasts with his negative assessment of the intuitionistic approach. In 1922 he criticized the shortcomings of the logicist schools but also recognized the positive achievements obtained. By their positive work in the foundations of analysis, Frege and Dedekind inaugurated a development that was to yield "the proper development of the so-called logical calculus, whose basic ideas prove more and more to be an indispensable tool in logico-mathematical investigations" (Hilbert 1922c, p. 202). This is what Hilbert cherished most in *Principia Mathematica*. *Principia* gave him the tools for a proper set up of his formalisms. Hilbert was, however, critical of *Principia* as a foundation for mathematics, and he objected to the axioms of infinity and reducibility ⟦see also Chapters 3 and 4⟧.[15] Let us now consider the more positive part of the programme.

The Positive part of the programme

Hilbert developed the main lines of his programme in a series of papers beginning with the 1922 "New Grounding of Mathematics" and ending with "The Grounding of Elementary Number Theory" (1931). *Grundlagen der Mathematik*, co-authored with Bernays, will provide a final presentation. During this period Hilbert changed views both on some key-issues (e.g., on the exact extent of finitary reasoning) and on the technical details of the formalisms used to carry out the programme. Following all such changes in detail would require more space than is suitable in this chapter. I shall thus mainly ignore the more technical developments ⟦on these, see Chapter 1⟧ in favor of a spelling out of the central concepts.[16]

After presenting the main lines of the programme presented in 1922–23, I shall concentrate on the following three issues: formalism, finitism, and the philosophical status of the a priori intuition that is at the basis of finitism.

The presentation given in 1922 proceeds in a twofold manner. Hilbert begins (p. 202) by giving a construction of (contentual) arithmetic.[17] This serves a double purpose. First of all, it shows the limits of the contentual development by showing the need to go beyond it, if one is to gain all of mathematics. Second, it functions as a paradigm for the formalistic development of all of mathematics (starting on p. 204). Hilbert begins by reflecting on the fact that Frege's and Dedekind's approach to the foundations of mathematics "has proved to be inadequate and uncertain" (p. 202). In clear contrast to Frege and Dedekind, Hilbert conceives of his approach as follows:

As we saw, abstract operation with general concept scopes and contents has proved to be inadequate and uncertain. Instead, as a precondition for the application of logical inferences and for the activation of logical operations, something must already be given in representation [*in der Vorstellung*]: certain extra-logical discrete objects, which exist intuitively as immediate experience before all thought. If logical inference is to be certain, then these objects must be capable of being completely surveyed in all their parts, and their presentation, their difference, their succession, (like the objects themselves) must exist for us immediately, intuitively, as something that cannot be reduced to something else. Because I take this standpoint, the objects [*Gegenstände*] of number theory are for me—in direct contrast to Dedekind and Frege—the signs themselves, whose shape [*Gestalt*] can be generally and certainly recognized by us—independently of space and time, of the special conditions of the production of the sign, and of insignificant differences in the finished product. The solid philosophical attitude that I think is required for the grounding of pure mathematics—as well as for all scientific thought, understanding and communication—is this: *In the beginning was the sign.* (Hilbert 1922c, p. 202)

The above quote raises several complex problems for the evaluation of Hilbert's programme. Let me mention three of them. First, what is at the basis of Hilbert's requirement that classical mathematics be justified? Second, what is the status of the appeal to an access to objects given in representation prior to any logical inference? Third, what exactly can be said about Hilbert's nominalism and formalism? I will address these three issues in what follows. The first one will lead us naturally into the topic of Hilbert's finitism. The second one probes the philosophical foundations of this finitism farther. Finally, the third section will try to clarify some of the philosophical underpinnings of Hilbert's view in connection to the objections raised by Aloys Müller and Leonard Nelson.

The need for a foundation of mathematics. In addition to the fact that proving the consistency of an axiom system was part of the requirements involved in pursuing axiomatized mathematics, there were also more pressing reasons for Hilbert's concern with proving the consistency of analysis and classical mathematics. We have seen that in 1922 one of these motivations was the failure faced by Frege and Dedekind in carrying out their programmes. These were not, however, the only worries Hilbert had concerning foundations. Weyl had recently attacked the usual procedures of analysis as containing a vicious circle, and later he joined Brouwer in attacking the reliability of classical logic. Hilbert implicitly accepts the correctness of the intuitionistic charge in his 1923. In that paper he raises the issue of the application of the tertium non datur to infinite totalities. Whereas the application of the classical laws of logic is perfectly safe in a finite context, the extension to infinite sets, according to Hilbert, is problematic and needs to be justified. Hilbert speaks in this context of "finite" and "transfinite" inferential procedures:

We therefore see that, if we wish to give a rigorous grounding of mathematics, we are not entitled to adopt as logically unproblematic the usual modes of inference that we find in analysis. Rather, our task is precisely to discover why and to what extent we always obtain correct results from the application of transfinite modes of inference of the sort that occur in analysis and set theory. The free use and full mastery of the transfinite is to be achieved on the territory of the finite! (Hilbert 1923a, p. 1140)

Let us consider an example. The validity of the disjunction $(a)A(a) \vee (E\,a)\neg A(a)$ is trivial for a finite domain; indeed, if the domain is finite we can run through the elements and, if we now abstract from issues of decidability, either find an a such that $A(a)$ is not satisfied, i.e. $\neg A(a)$, or not. Not so when the infinite is involved. Our search would only tell us that $(E\,a)\neg A(a)$ in case we run into one, but if we do not find one such element then we cannot assert, on account of the infinity of the domain, that $(a)A(a)$, since we are never done with our search.

But in mathematics these equivalences are customarily assumed, without further proof, to be valid for infinitely many individuals as well; and with this step we leave the domain of the finite and enter the domain of transfinite modes of inference. If we were constantly and blithely to apply to infinite totalities procedures that are admissible in the finite case, then we would open the floodgates of error. This is the same source of mistakes that we are familiar with from analysis. In analysis, we are allowed to extend theorems that are valid for finite sums and products to infinite sums and products only if a special investigation of convergence guarantees the inference; similarly we may not treat the infinite sums and products

$$A_1 \& A_2 \& A_3 \ldots$$

$$A_1 \vee A_2 \vee A_3 \ldots$$

as though they were finite, unless the proof theory we are about to discuss permits such as treatment. (Hilbert 1923a, pp. 1139–40)[18]

That is exactly Hilbert's goal. By adding to the formalism the "transfinite axioms", it is the goal of proof-theory to show that the generalized application of the logical laws discussed above will never lead to a contradiction. Hilbert compares his strategy to the postulation of ideal elements in mathematics. We now have to say something about this method.

The method of ideal elements Much has been written on the method of ideal elements. Hilbert's examples usually come from projective geometry (postulation of points and lines at infinity), algebra (postulation of the existence of n roots for an n-degree polynomial), algebraic number theory (ideals) and set theory (infinite cardinal and ordinal numbers). A good exposition of Hilbert's thoughts on the topic is contained in the 1919 lectures (see Hilbert 1992, chapter 9). In these lectures Hilbert begins by remarking that the opposition real-ideal has as one of its meanings the opposition actual–non-actual (merely thought). This might seem paradoxical, he adds, since in mathematics we are not dealing with actual things in the usual sense. However, mathematicians speak about mathematical existence, and Hilbert remarks that existence in the mathematical sense is always relative to a given system. When the system is extended by introducing new elements then, with respect to the original system, we can talk of ideal elements. However, the new system obtained might itself be extended, in which case all its elements are considered real and the new ones ideal. Thus the real-ideal distinction is relative to the systems. These extensions of the original system might occur either by an explicit construction, say when we introduce complex numbers as pairs of reals, or by postulating new axioms, for instance when we introduce as an axiom the fundamental theorem of algebra on the existence of n roots of an n-degree polynomial. In this second case the addition of the axiom must be shown to be consistent with the previous theory by a consistency proof. The advantages provided by the method of ideal elements are simplicity and generality. The recognition of the importance of the method of ideal elements played an important role in the foundational thought of Hilbert and Bernays. For example, Bernays in a review of Müller's book *Der Gegenstand der Mathematik* (1922) accuses Müller of *naiveté* on the issue of mathematical existence on account of his lack of understanding of the role of ideal elements in mathematics:

However, the treatment of the methodological questions of mathematics given here [Müller's book] cannot completely satisfy the mathematicians. The view that mathematical objects possess an ideal, timeless existence independent of all thought and that the method of the mathematical sciences is only determined by means of the characteristic feature of these objects may certainly serve for a first guidance as expression for an attitude that is fruitful and customary in the sciences. But simply to stop here seems, however, epistemologically too primitive; especially since Müller considers as object everything "that can become subject of a judgment". In fact, the mathematician knows that in his science an especially fruitful and continually applied procedure consists in the introduction of "ideal elements" that are introduced purely formally as subjects of judgments, and that, however, when detached from the statements in which they occur formally, are nothing at all. The proposition that the object is the product of the method holds widely in the realm of mathematics. (Bernays 1923b, p. 521)

This passage from Bernays clearly points at an instrumentalist reading of the method of ideal elements. They seem to possess only a linguistic reality and nothing else.

There is disagreement in the literature as to Hilbert's instrumentalism. Whereas most interpreters (Detlefsen, Giaquinto, Kitcher, among others) consider Hilbert as an instrumentalist, there are also those like Hallett who have been arguing otherwise (see Hallett 1990, pp. 233–43; cf. Resnik 1980, and Prawitz 1993, §5). Although I lean toward interpreting Hilbert as an instrumentalist, Hallet's arguments have also convinced me that the problem is a difficult one and that Hilbert and Bernays might not have been completely consistent in their positions. At the core of the instrumentalist reading of Hilbert are quotes such as the following:

To make it a universal requirement that each individual formula is interpretable by itself is by no means reasonable; on the contrary, a theory by its very nature is such that we do not need to fall back on intuition in the midst of some argument. What the physicist demands of a theory is that propositions be derived from laws ... solely by inferences, hence on the basis of a pure formula game, without extraneous considerations. Only certain consequences of the physical laws can be checked by experiment—just as in my proof theory only the real propositions are directly capable of verification. (Hilbert 1928a, p. 475)[19]

In the instrumentalist interpretation ideal elements function as the theoretical constructs in physics, which are only useful instruments in deducing hypotheses that can be empirically checked. Giaquinto (1983) draws some connections between Hilbert's programme and the empiricist revival in the foundations of physics in the first decades of the century. I believe that Giaquinto is right in concluding that "Hilbert rejected the destructive excesses of the empiricist revival"(p. 127). We can attempt to be more precise here by pointing out that Hilbert and Bernays strongly opposed the phenomenalism proposed by Mach. Mach's position is described by Hilbert, but not attacked, in the 1919 lectures in connection to the problem of ideal elements. Machian phenomenalism is intended to get rid of the notion of existence and get by merely with the concept of perceptual phenomenon:

From a certain philosophical standpoint, which only allows the contents of perception as real, all objects in the objective physical treatment of nature which we introduce for the purpose of ordering our perception-experiences in a surveyable [*übersichtlich*] way will be regarded as merely ideal forms. In particular, according to this view, anything we assume only by analogy with things known through perception is to be taken as just an ideal object, good examples being atoms or electrons. All of physical thought, and analogical thought in daily life, appears for this view as nothing other than an implicit application of the method of extension of a system through the addition of ideal elements. (Hilbert 1992, p. 98; translation from Hallett 1990, pp. 235–36)

Hallett rightly notices that here Hilbert is not espounding a view of his own. An explicit attack on Mach is contained in the 1924–25 lectures "On the Infinite" (Hilbert 1924–25, pp. 45–49; cf. Hilbert 1988, p. 80). The problem of the relationship to Mach's philosophy deserves a more thorough investigation but I will limit myself to a few remarks here. It was Weyl who pointed out the existence of a Machian tradition in the foundations of mathematics (Weyl 1924b, p. 452, Weyl 1925, p. 141, Weyl 1927, pp. 49–50). Weyl compared Brouwer's standpoint to that of Machian phenomenalism and Hilbert's approach to the creative urge evidenced by the atomic theory in physics.[20] Weyl, being "in the middle of the war of the factions", suggested that both attitudes are needed, in physics just as well as in the foundations of mathematics. The most explicit attack on Machian phenomenalism in the Hilbert school is contained in Bernays (1930a). Bernays' paper is extremely important for a proper interpretation of the epistemological background to Hilbert's finitism. As the Machian phenomenalism is insufficient to do justice to the development of modern physics, so is its version in the foundations of mathematics, Brouwer's intuitionism, inadequate to account for the developments of contemporary mathematics. Bernays proposed a modified Kantian framework that can account, according to him, for the non-phenomenal aspects of theoretical physics and, by means of Hilbert's proof theory, for the developments of contemporary mathematics. I will return to this paper in the section on the epistemological status of finitism.

It is impossible in this context to treat in more detail other important facets of the issues connected to the ideal elements, for instance, Hilbert's interpretation of the infinite, his distinction between real and ideal statements, and the similarities between the method of ideal elements and Kant's regulative use of reason. For more details on these issues see Detlefsen (1986, 1993b,a), Hallett (1990), Kitcher (1976), Majer (1991, 1993a,c).

Finitistic statements Hilbert distinguishes, in analogy with real and ideal elements, finitistic and transfinite (or ideal) statements. In the 1923 article the "transfinite axioms", which embody the classical rules for universal and existential quantifiers, are given in terms of a formulation that postulates the creation of "ideal" representatives for non-empty properties. Hilbert postulates the existence of a τ operator such that for any property $A(x)$ one can form $\tau A(x)$, where $\tau A(x)$ is such that if $A(x)$ holds of it, then for all objects x, $A(x)$ holds. Hilbert clarifies its meaning by means of the following example. Let $A(x)$ mean "to be corruptible". Then $\tau A(x)$ is a man with such moral integrity that if he turned out to be corruptible then all men are corruptible. Formally, we have the following transfinite axiom: $A(\tau A(x)) \rightarrow A(x)$. From 1925 on Hilbert will prefer to state the transfinite axioms in terms of the ϵ-operator: $A(x) \rightarrow A(\epsilon A(x))$. That is, if there is anything at all that satisfies $A(x)$ then $\epsilon A(x)$ does. For each $A(x)$, $\epsilon A(x)$ gives a representative for the class of objects satisfying $A(x)$, if the class is non-empty. The only important thing to keep in mind about the two operators is that from either one of the two axioms we can derive the classical quantificational laws for the universal and the existential quantifiers (modulo a redefinition of the quantifiers by means of the appropriate transfinite operator). However, the employment of the transfinite axioms must now be justified by a proof of consistency. But which statements count as finitistic and which count as transfinite (or ideal)? Hilbert's treatment in "On the Infinite" is the most extended on this issue, but this article is also notorious for the vagueness and unclarity of the explanations given by Hilbert.[21] Hilbert declines to investigate the extent of "finitistic" logic (see p. 379) since his goal is anyway that of giving a justification, by means of a consistency proof, for all of logic. However, it is clear that, at least in the domain of elementary number theory, numerical equations (e.g. $2 + 3 = 3 + 2$) are finitistic, and closure of such equations under negation and bounded universal or existential quantification also leads to finitistic statements. The transfinite is encountered when we cannot reduce an existential statement to a finite disjunction, which is often the case when we operate on infinite domains. Less clear is the case of the universal quantifier. Hilbert seems to think that statements of the form "for all numerals a, $a + 1 = 1 + a$" are finitistic. However, he does not read this as an infinite conjunction but "as a hypothetical judgment that comes to assert something when a numeral is given". Such statements cannot be denied finitistically, since they would give rise to existential propositions that cannot be reduced to finite disjunctions. This is why the excluded middle cannot be finitistically justified but its unrestricted use must be grounded by Hilbert's theory. From this passage it is also clear that for Hilbert the statement of consistency for a theory would be finitistic.

A synoptic view of the programme Let us summarize now what the Hilbertian approach looks like so far. According to Hilbert the formalization of the mathematical systems, including the formal rules of inference, is only the first part of his programme, to which he gives the name of proof theory. In addition to formalized mathematics, which proceeds purely formally, Hilbert requires, as part of his proof theory, a metamathematics which can make use of contentual reasoning:

In addition to this proper mathematics, there appears a mathematics that is to some extent new, a *metamathematics* which serves to safeguard it by protecting it from the terror of unnecessary prohibitions as well as from the difficulty of paradoxes. In this metamathematics—in contrast to the purely formal modes of inference in mathematics proper—we apply contentual inference; in particular, to the proof of the consistency of the axioms. (Hilbert 1922c, p. 212)

In the first approaches to the consistency problem (Hilbert 1900a,b, 1905b), Hilbert had not distinguished between the formalized theories and the metamathematical level of investigation. The present distinction allows Hilbert to reply to Poincaré's objections, which were taken over by Brouwer in 1907 and 1912, that a vicious circle is involved in the Hilbertian attempt to prove the consistency of arithmetic. In 1922 Hilbert recognizes that what is involved at the metamathematical level is only a small part of arithmetical reasoning and does not appeal to the full strength of the induction axiom. Moreover, the part of arithmetical reasoning used for metamathematical purposes is completely safe. Hilbert's foundational programme thus proceeds by requiring a formalization of arithmetic, analysis, and set theory (in the early 1920s the central concern seems to be analysis; later, number theory will take center stage) and associated with it a proof of consistency of such formalizations by means of proof theory. A proof of consistency will have to show, by appealing to contentual considerations that are completely unproblematic, that in the formalism in question it is never possible to derive the formula $a \neq a$, or alternatively it is not possible to prove both $a = b$ and $a \neq b$.

In 1922, after having given a construction of contentual arithmetic by means of substituting symbols for the numbers platonistically conceived, Hilbert remarked that the same strategy can be used for the construction of mathematics. Ordinary mathematics is formalized, its axioms also containing the formalized version of transfinite reasoning, and then one operates on this formalized mathematics by means of metamathematical considerations. These metamathematical considerations are, in opposition to the formal procedures contained in the formalized mathematics, provided with content [*inhaltlich*]. What has been gained by doing so? The formulas and proofs in a formalized system are, unlike the highly infinitistic objects invoked by mathematicians, finite combinations of symbols on which it is possible to operate with the certainty of reasoning required by Hilbert's strict finitistic demands. The consistency proof thus acts as a finitistic guarantee of the admissibility of the infinitistic objects considered in ordinary mathematics. Before we move on to discuss the "finite Einstellung", it is useful to discuss briefly the issue related to the formalization of ordinary mathematics and in particular the charge of formalism that has time and time again been leveled against Hilbert's programme.

Formalism The word *formalism* has often been used to characterize Hilbert's position in the philosophy of mathematics. There is no need to object to using this word to characterize Hilbert's position, but then one should clarify what is meant by this. As Detlefsen remarks,

It is this deeply motivated radical abstraction from meaning that, in all probability, is responsible for one of the worst misconceptions of Hilbert's formalism; namely, that according to which it says that mathematics is a "game" played with symbols. This at one time was, and may, I fear, still be, a common misconception of Hilbert's position. (Detlefsen 1993b, p. 299)

The lectures of 1919 leave no doubt that Hilbert never wanted to defend such a radical formalism. In order to get clear on the nature of Hilbert's formalism, it might be worthwhile to see how the label was attached to his position. In fact, whereas intuitionism was chosen by Brouwer to define his own position, Hilbert and Bernays did not refer to themselves as formalists. It was mainly Brouwer's "Intuitionistic Reflections on Formalism" (1928) that is responsible for the characterization of Hilbert's position as formalistic (within Hilbert's school use of the word *formalism* is found in Bernays (1930b) and von Neumann (1931)). Indeed, the layout of the article, beginning with a long list of references only by Hilbert and Brouwer, implied an identification of Hilbert's position with formalism. In the "first insight" Brouwer refers to the distinction between "the construction 'of a stock of mathematical formulae' (the Formalistic description of mathematics) and the intuitive (contentual) theory of this construction"(Brouwer 1928, p. 41). Thus, what is characteristic of formalism, according to Brouwer, is the identification of mathematics with a stock of mathematical formulas. Brouwer had used the word *formalism* before, notably in his article "Intuitionism and Formalism" (1912b). However, formalism in that article included not only Hilbert's approach (up to that point) but also Peano, Zermelo, and in general the development of set-theoretical mathematics. This broad characterization of formalism goes back to Brouwer's dissertation *On the Foundations of Mathematics* (1907), where Brouwer insists on the inadequacy of those positions (logistics, Cantorism, and Hilbertian axiomatics) on account of their "confusion between the act of constructing mathematics and the language of mathematics" (Brouwer 1975, p. 96).[22]

It is thus clear that Brouwer's conception of formalism was rather wide and it is therefore essential to point out exactly Hilbert's and Bernays' stand on the issue. I follow Detlefsen in distinguishing two versions of formalism in Hilbert. The earlier formalism, predating the 1905 paper, and the formalism of the 1920s. The earlier formalism amounts to "the idea that axiomatic geometry, and, indeed, axiomatic theories generally, are not to be regarded as sciences of a particular subject matter, but rather as "abstract sciences"—sciences, that is, which in some sense apply to a variety of different subject matters."[23] In the later formalism, Hilbert not only abstracts from the meaning of the mathematical primitives but also from the meaning of the logical terms. (Detlefsen 1993b, p. 299)

The formalistic position finds expression in several of the articles presented here. The new meaning of mathematics, "proper mathematics", is that of a "general theory of formalisms" or "a general theory of forms" (Bernays 1922a, p. 196). It is actually interesting to note that this very broad definition of mathematics is not found in Hilbert. Hilbert usually speaks of proper mathematics as a stock of provable formulas:

Everything that hitherto made up mathematics is now to be strictly formalized, so that *mathematics proper*, or mathematics in the strict sense, becomes a stock of provable formulae. (Hilbert 1922c, p. 211)

Having seen what is involved in the "proper mathematics" we can now move to the metamathematical level and in particular to the finitistic point of view.

"Die finite Einstellung" What are the contentual considerations that Hilbert wants to allow for the grounding of mathematics? As I already mentioned, the treatment in the 1922 paper proceeds by a double approach. Hilbert begins by developing contentual arithmetic by making use only of reliable procedures on symbols for the numbers. The approach is then generalized to include any system of formal symbols occurring in an axiom system (Hilbert 1922c, p. 202). Both in the case of the development of elementary arithmetic (which mainly deals with numerical equations) and in the development of the axiomatic systems, which also include logical inference, contentual thought operates on objects which are given in representation. In both the above mentioned cases these objects are symbols, either for the numbers or for the elements of the formalism. Let us recall what must be given to us prior to logical inference:

If logical inference is to be certain, then these objects must be capable of being completely surveyed in all their parts, and their presentation, their difference, their succession, (like the objects themselves) must exist for us immediately, intuitively, as something that cannot be reduced to something else. (Hilbert 1922c, p. 202)

This characterization is often repeated verbatim by Hilbert. It is in 1923 that we are given a slightly more precise characterization of what this finitism amounts to:

The elementary theory of numbers can also be obtained from these beginnings by means of "finite" logic and purely intuitive thought [*durch rein anschauliche Überlegungen*] (which includes recursion and intuitive induction for finite existing totalities); here it is not necessary to apply any dubious or problematical mode of inference. (Hilbert 1923a, p. 1139)

This is the only place in Hilbert's essays, where it is specified that in metamathematics we are only allowed induction and recursion on finite collections. A further clarification to the issue is given by the important commentary by Bernays:

This much is certain: We are justified in using the elementary ideas of sequence and ordering, as well as the usual counting, to the fullest extent. (For example, we can determine whether there are three occurrences of the sign → in a formula or fewer).

However, we cannot get by in this way alone; rather, it is absolutely necessary to apply certain forms of complete induction. Yet, by doing so we still do not go beyond the domain of the concretely intuitive.

In this regard, two types of complete induction are to be distinguished: the narrower form of induction, which relates only to something completely and concretely given, and the wider form of induction, which uses either the general concept of whole number or the operating with variables in an essential manner.

Whereas the wider form of complete induction is a higher form of inference whose justification constitutes one of the tasks of Hilbert's theory, the narrower form of inference belongs to the primitive intuitive mode of cognition and can therefore be applied as a tool of contentual inference.

As typical examples of the narrower form of complete induction, as they are used in the argumentations of Hilbert's theory, let us adduce the following two inferences:

1. If the sign + occurs at all in a concretely given proof, then in reading the proof one finds a place where it occurs for the first time.
2. If one has a general procedure for eliminating from a proof with a certain concretely describable property E the first occurrence of the sign Z, without the proof losing the property E in the process, then one can, by repeated application of the procedure, completely remove the sign Z from such a proof, without it losing the property E.

(Bernays 1922b, p. 221)

Does this distinction really answer Poincaré's objection? I will discuss the question in relation to Becker's objections to Hilbert's theory.[24]

Becker vs. Hilbert on metamathematical induction Oskar Becker wrote what is one of the most interesting books in the philosophy of mathematics in the 1920s, *Mathematische Existenz* (1927). He had began his career under Husserl with a dissertation on the concept of space. By the time he wrote *Mathematische Existenz* he had become heavily influenced by Heidegger. Indeed, his attempt to develop a philosophy of mathematics closer to the structure of *Dasein* led several authors to reject Becker's anthropologism in the philosophy of mathematics (see Geiger (1928), Becker (1928/29) and Cassirer (1929)). Becker's general standpoint in *Mathematische Existenz* was more favorable to the intuitionistic approach than to Hilbert's, although he attempted to extend the domain of what could be intuitively justified to transfinite set theory, by bringing in considerations related to the "transfinite Strukturkomplikationen des Bewußtseins". The similarity of this attempt to the infamous proof of the existence of an infinite set given by Bolzano and Dedekind left many puzzled (see Scholz 1928, p. 683). Another application of phenomenology to the philosophy of mathematics is his semantics for intuitionistic logic in terms of the Husserlian theory, contained in the sixth logical investigation, of fulfillment and frustration of intentional acts, that was to influence Heyting.

Becker's criticism of Hilbert's finitism occurs in §3 of the book (pp. 485–94). Becker's intention is that of vindicating Poincaré's original objections against Hilbert's proof of consistency, by showing that for such a proof to be obtained one has to go beyond the finite. The debate between Hilbert and Brouwer, he says, is a debate about the potential infinite [*das Endlose*].

Hilbert rejects the efforts of the intuitionists concerning the potential infinite as superfluous. For, on the one hand, from his point of view even the transfinite,—and thus even more the indefinite—can be "grounded" purely finitistically. On the other hand, an extension of intuition beyond the finite seems to him to be afflicted by a certain element of insecurity and on that account it should be avoided. (Becker 1927, p. 485)

So far the characterization seems accurate. However, it should be pointed out that clearly both Hilbert and Bernays put no limit on the possibility of considering objects of any finite complexity whatsover and of counting as far as it is required by any contentual consideration at hand. That is, there is no number a such that we cannot carry out contentual consideration on objects involving more than a symbols. Thus, in this sense they already admit a potential infinity, although it is true that at each stage we are working with finite objects. However, Becker's objection do not have to do with the arbitrary complexity of the objects under consideration. Rather, he criticizes Hilbert's distinction between the two forms of induction (explicitly given by Bernays in the above quote). The structure of a consistency proof, according to Hilbert, consists in assuming that a proof ends up with the end-formula $0 \neq 0$ and to show by reductio ad absurdum that this cannot be. This is where Becker objects. Now, there is no question that for each proof figure the above argument is finitistic. But, Becker insists, from this it does not follow that for all such proof figures the result hold:

This metamathematical proof is valid for any concretely given "proof figure" that is, of necessity, finite. For one then only needs the "narrower form" of complete induction. But from this in no way follows that it holds for all the proof figures which are in general possible. One ought to distinguish carefully between the two cases. (Becker 1927, p. 490)

He compares the situation to that of having to prove for specific numerals the binomial theorem for $(a + b)^n$ and to prove it for all n. In the previous case we only need the restricted form of induction. In the second case we need the full induction axiom. But the case of consistency is more like the latter case. Becker concluded by stating that for the proof of consistency one had to admit the potential infinite and the full axiom of induction in the metatheory.

Is Becker's charge justified? In Hilbert (1928a), probably unaware of Becker's book and addressing Poincaré's old argument, Hilbert still maintains that the restricted form of induction is sufficient for metamathematics:

> As my theory shows, two distinct methods that proceed recursively come into play when the foundations of arithmetic are established, namely, on the one hand, the intuitive construction of the integer as numeral (to which there also corresponds, in reverse, the decomposition of any given numeral, or the decomposition of any given array constructed just as a numeral is), that is, contentual induction, and, on the other hand, formal induction proper, which is based on the induction axiom and through which alone the mathematical variable can begin to play its role in the formal system. (Hilbert 1928a, pp. 472–73)[25]

It is not simple to determine, given the vagueness of Hilbert's and Bernays' statements on induction, what exactly the difference between the metamathematical and the mathematical induction amounts to or, as Becker alleges, whether metamathematics implicitly uses the full form of induction.[26]

It seems to me that the situation evidenced by Becker's critique can be summarized as follows. First of all, Hilbert and Bernays would have granted that the potential infinite is presupposed in their assumption that there is no bound to the complexity of the proof figures that might come under consideration. However, for each proof figure the form of induction used is finite. Thus, if I am working with a given proof figure which ends with $0 \neq 0$, I can certainly look for the first occurrence of the expression $0 \neq 0$ and there is no circularity here, since this appeal to the least number principle in the case of a finite proof figure is as harmless as the principle of induction on the finite proof figure. What is valid in Becker's charge is that Hilbert must restrain himself to the bare assertion of sentences of the form $A(n)$ for each numeral n, without being able, on pain of using an appeal to complete induction on an infinite totality, to conclude for all n, $A(n)$. In other words, the assertion of consistency, formulated as a universal statement of the form $(x)A(x)$, cannot be proved in the metatheory, only the instances $A(n)$ can be proved. This procedure would involve, according to Becker, an appeal to the potential infinity of the natural numbers.

One could be tempted to prove consistency by contradiction as follows: assume not $(x)A(x)$ then there exists an x such that $\neg A(x)$. But this move would not be allowed for Hilbert since this principle of logic is one of the principles that needs to be justified by the metatheory. And this last principle is in effect equivalent to the excluded middle on the infinite totality of the natural numbers in the form $(x)A(x)$ or $(Ex)\neg A(x)$ whose validity would involve the assumption that the natural number sequence is given in its totality. In conclusion, what Becker rightly objects to Hilbert is that in order to assert consistency in the form of a sentence of the form $(x)A(x)$ one needs to apply complete induction. What became clear with Gödel's second incompleteness theorem is that not even complete induction is sufficient to prove the consistency statement for a consistent theory that contains a minimal amount of arithmetic.

Intuitionism and finitism Becker's aim was to show that finitism had to assume the potential infinite, on account of the need to appeal to the principle of induction in its wider sense. This invites a more general question as to how Hilbert and his school saw the relationship between intuitionistic mathematics and metamathematics. I will attempt to establish two points. First, Hilbert and Bernays did extend their finitistic framework in the late 1920s and accepted the identification of finitistic and intuitionistic mathematics (thus, allowing induction on the natural numbers in the metatheory). Second, by the time of the *Grundlagen der Mathematik* they would argue again for the difference between intuitionism and finitism but not on the ground that finitism does not use induction on the natural numbers.

Establishing the first claim in the case of Hilbert requires a bit of textual reconstruction. It is perhaps best to begin with the uncontrovertial evidence within the Hilbert school. Von Neumann was the first in his 1927 article to identify finitism and intuitionism:

Here one must always distinguish clearly between two different forms of "proofs", that is between formalistic ("mathematical") proof within the formal system and the contentual ("metamathematical") proof about the system. While the former is an arbitrarily defined logical game, the latter is a chain of directly evident contentual insights. This "contentual proof" must thus proceed in the sense of the intuitionistic logic of Brouwer and Weyl. Proof-theory should construct, so to speak, classical mathematics on an intuitionistic basis and in this way reduce intuitionism ad absurdum. (von Neumann 1927, pp. 2–3)[27]

As for Hilbert's closest collaborator, Bernays, the evidence seems to be that around 1925 he thought finitism and intuitionism to be different. Van Dalen mentions a letter from Bernays to Hilbert, dated October 25, 1925, in which Bernays claims to have noticed a difference between the finitistic point of view and intuitionism (see van Dalen 1995, p. 163). This seems to imply that the distinction was not quite clear before, but need not imply that Hilbert and Bernays believed that intuitionism and finitism were the same thing. However, this latter position is defended in Bernays' articles from the late 1920s. For instance, in 1930 he identifies finitistic mode of thought and intuitionism (see 1930b, pp. 251–53) and in a comment to this paper written in 1976 he says:

First of all, as regards intuitionism, it was thought at first that the methodology of intuitionistic proofs coincided with that of Hilbert's "finitistic point of view." However, it has become clear that the methods of intuitionism go beyond the finitistic proof procedures intended by Hilbert. In particular, Brouwer makes use of the universal concept of contentual proof, to which the concept of "absurdity" is also connected, and which, however, is not made use of in finitistic inference. (Mancosu 1998a, p. 263)

This is exactly the distinction given in 1934.[28] Hilbert is never as explicit as this in his articles, although it is not hard to see that he actually held the position described by

Bernays (for example, in (1931) he identifies his position with that of Kronecker, which he had always associated with that of Brouwer. It is only in 1934 that Kronecker's position is associated with finitism and distinguished from intuitionism). In conclusion, throughout the 1920s Hilbert and Bernays never quite exactly spelled out what the boundaries of finitistic thought were supposed to be. Towards the end of the 1920s the general opinion, also within the Hilbert school, seems to have identified finitistic reasoning with intuitionism, and it was only in 1934 that Hilbert and Bernays argued that intuitionism goes substantially beyond finitistic thought. We now need to investigate what epistemological status Hilbert attributed to the postulation of a pre-logical faculty of representation.

The epistemology of finitism What is the epistemological status of the finitism defended by Hilbert? It is essential here to compare the different writings in order to characterize what I perceive as a change of emphasis during the 1920s. From the above quotes we see that in 1922 and 1923 Hilbert does not attempt to characterize the epistemological status of the intuition involved in metamathematics. Not surprisingly, it is Bernays who gives us more light on this philosophical problem. In 1922 Bernays sees Hilbert's project in the following terms:

An appeal to an intuitive grasp of the number series as well as to the multiplicity of magnitudes is certainly to be considered. But this could certainly not be a question of an intuition in the primitive sense; for certainly no infinite multiplicities are given to us in the primitive intuitive mode of representation. And even though it might be quite rash to contest any farther-reaching kind of intuitive evidence from the outset, we will nevertheless make allowance for that tendency of exact science that aims as far as possible to eliminate the finer organs of cognition [*Organe der Erkenntnis*] and to rely only on the most primitive means of cognition.

According to this viewpoint we will examine whether it is possible to ground those transcendent assumptions in such a way that only *primitive intuitive cognitions come into play*. On account of this restriction of the means of cognition, we cannot, on the other hand, demand of this grounding that it allow us to recognize as truths (in the philosophical sense) the assumptions that are to be grounded. Rather, we will be content if we succeed in proving the arithmetic built on those assumptions to be a possible (i.e., consistent) system of thought. (Bernays 1922b, pp. 215–16)

It is not easy to ascertain whether "primitive intuitive cognition" refers here to a Kantian pure intuition or whether, as the appeal to the reductionist tendency of the exact sciences suggests, the intuition mentioned here is an empirical one. I tend to read the above characterization of intuition as an empiricist one; that is, we are not dealing at this point with a Kantian pure intuition but rather with an empirical one (this agrees with Sinaceur's interpretation (1993, p. 260) of intuition in Hilbert, and Moriconi (1987, p. 104)). In 1926 Hilbert still speaks of "perceptual intuition" of the numerals and the formulas (1926, p. 379). However, I am aware that the above quotes could be consistent with a Kantian appeal to pure intuition. What is certain is that there is a striking change of emphasis between these articles from the early 1920s and the explicit appeal to Kant's pure intuition of the late 1920s. Let us reconstruct the steps of this development.

In 1923, while replying to the objections of Aloys Müller, Bernays still played the tune of the reductionist line:

> Hilbert's theory does not exclude the possibility of a philosophical attitude which conceives of the numbers as existing, non-sensible objects ... Nevertheless the aim of Hilbert's theory is to make such an attitude dispensable for the foundations of the exact sciences. (Bernays 1923a, p. 226)

In the process of replying to Müller, Bernays specified that for Hilbert "the number signs are also not created by thought. But this does not mean that they exist independently of their intuitive construction, to use this Kantian term quite appropriate here." (p. 226) This is the first place in which Kant is mentioned in this connection. At this point we have to distinguish between Hilbert and Bernays.

Hilbert mentioned Kant in this connection in "On the Infinite" (1926, p. 376). However, he did not proceed to make any claims on the nature of the finitistic metamathematical intuition. Rather he speaks of "perceptual intuition" (p. 379) and he posits this intuition as a condition for the possibility of any scientific thinking (see also Hilbert 1928a, pp. 464–65, where Kant is not mentioned). It should be remarked that in the 1922–23 lectures (Hilbert 1988) Hilbert appeals to Kant's theory of intuition as a foundation for the a priori in mathematics.[29] However, these lectures were not published, and it is only in (1931) that Hilbert provides in print a characterization of the finitistic intuition that includes it as a third realm between experience and pure thought.

> However, attentive reflection leads us to see that, besides experience and thought, there is yet a third source of knowledge. Even if today we can no longer agree with Kant in the details, nevertheless the most general and fundamental idea of the Kantian epistemology retains its significance: to ascertain the intuitive *a priori* mode of thought [*Einstellung*], and thereby to investigate the condition of the possibility of all knowledge. In my opinion, this is essentially what happens in my investigations of the principles of mathematics. The *a priori* is here nothing more and nothing less than a fundamental mode of thought [*Grundeinstellung*], which I also call the finite mode of thought: something is already given to us in advance in our faculty of representation: certain extra-logical concrete objects that exist intuitively as an immediate experience before all thought. (Hilbert 1931, pp. 1149–50)

As for Bernays we have seen that the reference to construction in intuition is already present in 1923. A clearly formulated position on the Kantian reading of the finitistic mode of thought is found in two articles by Bernays both coming out of papers he read in 1928 (Bernays 1928c, 1930a). In the paper on Nelson we read:

> There is, however, still another point of view from which Hilbert's grounding of mathematics is related to Nelson's philosophy: The "finitistic attitude" required by Hilbert as methodological basis must be characterized epistemologically as a form of pure intuition. For this "finitistic attitude" is, on the one hand, intuitive and, on the other hand, it goes beyond what can be actually experienced [*das eigentlich Erfahrbare*]. The demand for such an epistemic foundation is in itself still independent of the particular form of Hilbert's approach; it holds for any finite grounding of mathematics. However, what is characteristic for Hilbert's foundation is that here the finitistic standpoint is connected with the axiomatic grounding of the theoretical sciences. In this way the presuppositions of the finitistic attitude present themselves at the same time as conditions for the

possibility of the theoretical knowledge of nature, quite in the sense of the Kantian formulation of the problem. If this connection comes to be generally recognized then the possibility arises in this way that the leading thoughts of Kant's *Critique of Pure Reason* will come to life again in a new form, freed from the special forms of its historical relativity, from whose fetters theoretical science has freed us. (Bernays 1928c, pp. 144–45)

It is the explicit reference to Kant's pure intuition that is lacking from the articles of the early 1920s and that strongly suggests a development of views on the epistemological status of finitism. Indeed, in "Die Grundgedanken der Fries'schen Philosophie in ihrem Verhältnis zum heutigen Stand der Wissenschaft" (1930a) Bernays draws some more general epistemological consequences of the appeal to transcendental philosophy, and sees in it the bulkward against the dominant phenomenalism of the Machian school. Bernays described Fries' philosophy and that of his follower Nelson as being radically opposed to the phenomenalism characteristic of most philosophical directions in the first three decades of the century:

If we consider the different more recent philosophical theories, then we find that most of them are fundamentally opposed to transcendental idealism. In particular, this holds for the immanence philosophy of the Machian school—a philosophy which is not only widespread among scientists but almost absolutely predominant—which believes in general to be able to eliminate the concept of existence and to get by with the concept of phenomenon. According to this philosophy, there are basically no other forms of knowledge than perception, memory, the observation of the course of representations, and the comparison of contents of representation. (Bernays 1930a, p. 100)

Bernays also included in this tendency Schlick, Russell, Husserl, and Becker. In particular, the development in foundations has shown that "eine gewisse Art rein-anschaulicher Erkenntnis als Ausgangspunkt für die Mathematik genommen werden muß" (Bernays 1930a, p. 108). Unlike Kronecker, Weyl, and Brouwer, who follow Mach's reductionism, Hilbert's finitistic mathematics is used for a foundation of the rational element in mathematics. Bernays goes as far as to require for the philosophical completion of Hilbert's project the analogue of the transcendental deduction of the categories.[30]

In order better to clarify what is involved in Hilbert's finitistic attitude, and the nature of the "Kantian" reading of finitism, we must now look at the philosophical exchanges that forced Hilbert and Bernays to address the objections raised against the Hilbertian programme.

2.3 Philosophical Exchanges

Hilbert's views on the foundations of mathematics did not go unchallenged throughout the 1920s.[31] Challenges were raised from various philosophical viewpoints, as, for example the objections of Müller, working in the tradition of Rickert's philosophy, those of Nelson, who worked in the Kantian-Friesian tradition, and those of Becker, working broadly in the phenomenological tradition. The objections of Becker have already been mentioned. In this section I shall analyze the debate with Müller and then I will turn to Nelson.

Müller and Bernays

After the publication of Hilbert (1922c), the philosopher Aloys Müller published a critical piece on Hilbert's views entitled "Über Zahlen als Zeichen" (1923). Bernays replied in the same year (1923a), and in 1924 Müller published a long rejoinder in the *Annalen der Philosophie*, entitled "Über den Gegenstandscharakter der Zahlen (Eine Auseinandersetzung mit Hilbert)".[32]

In order to clarify Müller's objections to Hilbert, it is essential to say a few things about his philosophical standpoint. Müller followed the lead of Rickert's philosophy in starting from a dualistic epistemological theory, which opposes what can be perceived by means of the senses and what can only be grasped intellectually, and derives from it an ontology made up of four realms of actuality [*Wirklichkeit*]. He distinguishes real being [*reales Sein*], metaphysical superbeing [*metaphysisches übersein*], ideal being [*ideales Sein*], and the sphere of validity [*Gelten*]. According to Müller the objects of mathematics—numbers, figures, relations—belong to the ideal beings. These objects have a non-temporal reality and do not admit of causal relations.

It is within this general background that Müller's objections to Hilbert must be understood. In particular, the identification of numbers and signs found in Hilbert (1922c) was radically opposed to his conception of the actual existence of numbers. Bernays distinguished three main objections in Müller's article. First of all Müller objects to the use of the word *Zeichen* (sign) to indicate a meaningless symbol. Since signs are signs of something (playing on the *Zeichen-Bezeichnetes* relation) one should use the word *figure*. Bernays promptly granted the objection and indeed Hilbert, according to Bernays, from then on will use *Ziffer* instead of *Zahlzeichen*.[33]

The second and third objections are more interesting. They both aim at showing that in the formal construction of the number signs given by Hilbert a contentual aspect creeps in, so that a meaning is in fact somehow associated with the number signs. Müller argues that this happens both in the fact that the construction of the signs is given in the form of a serial arrangement, and moreover a spatial content is introduced when Hilbert speaks of numerical signs extending beyond other numerical signs. Although Bernays is quite right in rejecting the idea that a meaning is thereby associated with the number signs, Müller's objections have the merit of emphasizing that in the "construction" of the number signs we are making use of a certain amount of "intuition" both in the sense of an intuition involving sequential arrangements and a spatial intuition. This leads Bernays to grant that for the construction one needs "the idea of a determinate, concretely exhibitable, form of succession" and that, in order to justify Hilbert's claims about numerical signs extending beyond numerical signs, the spatial sense involved is not a metric sense. However, this seems to imply that a certain amount of spatial a-prioricity is built in the metamathematical intuition required for the construction of intuitive number theory and of formalized mathematics (Nelson will try to exploit this to show that one ought to stay close to Kant; see below). Finally, against Müller's attempt to show that Hilbert in other parts of the 1922 paper came closer to his conception—as when Hilbert assumes that certain objects must be given in representation before any logical thought—Bernays stresses that although Hilbert's attitude does not rule out the "possibility of a philosophical attitude which conceives of the numbers as existing, non-sensible objects", Hilbert's aproach consists exactly in avoiding this unreliable assumption that through the assumption of the simultaneous existence of the numbers would lead us beyond the immediately certain. Finally, Bernays proposes the following clarification:

However, the objects of intuitive number theory, the number signs, are, according to Hilbert, also not "created by thought". But this does not mean that they exist independently of their intuitive construction, to use this Kantian term quite appropriate here in this context. But the construction always only yields either a single determinate figure or a procedure for obtaining a further figure from a given one (e.g., by affixing "+1"). But this does not lead to the idea of a simultaneous existence. That the idea of the number series as a closed totality can be applied in mathematical inferences without danger of a contradiction, is precisely what is shown by Hilbert's proof theory. (Bernays 1923a, p. 226)

It seems to me that the above passage is extremely important for the clarification of a point that has come up time and time again in the literature. It is often stated that Hilbert's finitism excluded even recourse to the potential infinite (see Majer (1993a, p. 191): "therefore no infinity, whether actual or merely potential, belongs to the finite point of view"). I have already mentioned that Oskar Becker in his book *Mathematische Existenz* (1927) also reads Hilbert in the same way but then goes on to argue that, in order to be consistent, Hilbert must admit the potential infinite. It seems to me that above quote supports the interpretation I have given in the section on Becker to the effect that Hilbert and Bernays admit the potential infinite in the metatheory in the sense that there is no bound on the complexity of the objects which might be subjected to metamathematical consideration (see also Bernays 1930b, p. 247).

Kritische Philosophie und Mathematische Axiomatik

In analysing the epistemological status of the pure intuition founding the metamathematical investigations of Hilbert, we have pointed out its relation to Kant's notion of a priori intuition. We have seen that Hilbert and Bernays considered the original formulation of the transcendental aesthetic, with its postulation of the pure forms of space and time, irrelevant for a foundation of mathematics. They were willing, however, to postulate a modified form of the transcendental aesthetic which was in their eyes closer to scientific needs and at the same time, by rejecting the pure intuitions of space and time, would also reflect the dismissal of such postulations caused by the development of modern mathematics and physics. The close relationship to Kantianism also has its historical roots in the close connection that Hilbert and Bernays had with Leonard Nelson.

Leonard Nelson was active in Göttingen from 1904 to his early death in 1927. He was an advocate of a conception of "Philosophie als Wissenschaft", which he had endeavored to bring about by reviving the work of Jakob Friedrich Fries (1773–1843). He was the founder of the "Neue Fries'sche Schule" which counted among its close associates Gerhard Hessenberg, Otto Meyerhof, Kurt Grelling, and later Paul Bernays. The organ of the group was the *Abhandlungen der Fries'schen Schule*, the first issue of which was published in 1904.

In 1912–13 Nelson founded the Jakob Friedrich Fries-Gesellschaft, of which Paul Bernays was one of the co-founders. In 1917 Hilbert joined the association as a supporting member. Peckhaus has documented Hilbert's institutional engagement for Nelson, which was eventually successful in providing Nelson with an appointment as "Extraordinarius" at the University of Göttingen.

However close Hilbert and Bernays might have been to the idea of a scientific philosophy developed in strict connection to the exact sciences, they nonetheless disagreed with Nelson's "critical mathematics". Although many of Nelson's views go back to his early publications and were also foreshadowed in Hessenberg (see Peckhaus 1990 for details) it will be enough for us to refer to Nelson's 1928 article "Kritische Philosophie und Mathematische Axiomatik" and the two papers by Bernays (1928c, 1930a) that provide a critical assessment of Nelson's views from the standpoint of Hilbert's foundational approach.

Nelson proposed a revival of the Kantian-Friesian theory, according to which the axioms of mathematics are to be grounded by the a priori intuition of space and time. In his approach to the philosophy of mathematics, or, as he prefers, "critical mathematics", he distinguished two parts.[34] The first corresponds to the mathematical component of axiomatizing the discipline, and this corresponds roughly to Hilbert's contributions to axiomatizing various branches of mathematics. Moreover, one has to provide "eine kritische Deduktion der Axiome der Mathematik" that has the role of grounding epistemologically the axioms postulated in the various theories. From this standpoint Nelson considered mathematical axioms as a priori synthetic judgments grounded on a non-empirical, pure intuition (see Peckhaus 1990, p. 164). The article emphasizes the close similarities between Hilbert's approach to the foundations and "critical mathematics". To the systematic part in "critical mathematics" corresponds the setting up of "axioms systems"; to the specific epistemic investigation of the axioms corresponds in Hilbert's approach, according to Nelson, the metamathematical investigation of the systems in accordance with the "finitistic attitude". Overall, Nelson saw a strict parallel between the development of modern axiomatics (and the relative metatheoretical investigations into independence of the axioms, consistency, etc.) and the two parts of Friesian critical philosophy (regressive method, for the discovery of the axioms, and deduction, for the epistemic justification of the axioms). By doing so he was inaccurate in representing the Hilbertian project. In particular, according to this characterization of axiomatics, an axiomatic system would end up being a system of truths grounded in intuition, quite against many of the claims made by Hilbert and Bernays on this issue. Indeed, Nelson faced a strong opposition at the end of the talk from two of Hilbert's best collaborators, Bernays and Courant. Courant objected that "the usual concept of truth or validity should be restricted to metamathematics". Bernays pushed the point further. He claimed that posing the problem of consistency for an axiomatic system would be meaningless if we could rely on a foundation by means of an appeal to an a priori intuition of space and time. Bernays suggested, faithful to Hilbert's approach, that an a priori intuition can only be used in metamathematics and thus proposed that only a modification of Kant's approach which restricts the a priori intuition to the manipulation of surveyable symbols and combination of symbols could be preserved. However, the theory of the a priori intuition of space and time must be given up.[35]

Nelson provided a long reply to these objections. He tried to show that Courant and Bernays were forced to a sort of mathematical nihilism. By eliminating the epistemological issues from mathematics, they end up annihilating mathematics itself so that the original project for which metamathematics was created, that is, guaranteeing the certainty of the mathematical truths, becomes a sort of game with no real meaning. There is of course an interesting parallel here to the positions of Brouwer and Weyl, although Nelson does not share their aims. In addition, Nelson argued that the metamathematical intuition admitted by Bernays forces him to accept a spatial intuition as given a priori:

Those who dispute the epistemic character of the mathematical axioms deny in this way the existence of pure intuition as epistemic source for mathematical judgments... We cannot get by in metamathematics without intuition, as indeed is explicitly granted by Bernays in the discussion. But what kind of intuition is at issue here? The signs considered in metamathematics are extended entities, in fact, spatially extended entities given in spatial order. The intuition upon which metamathematics rests is thus the intuition of space. (Nelson 1928, p. 12)

Now, the argument proceeds, either this intuition of space is a priori or not. In the first case, Bernays would have to grant the validity of an a priori intuition of space, and thus the possibility of grounding mathematics through it. Otherwise, it must be empirical, but this would lead to the paradox of grounding the certainty of mathematical propositions on probabilities. Nelson concluded that the mathematical nihilism defended by Bernays and Courant rested on their philosophical preconceptions which consists in an "Angst vor der Erkenntnis a priori", and this bars the way to an appreciation of the most evident system of knowledge we have, the science of mathematics.

It should thus be clear from this exchange that the references to Kant found in Hilbert and Bernays do not immediately qualify their programme as standing in the Kantian tradition, or anyway that their departure from Kantianism was seen as a rather substantial one.

2.4 Paul Bernays: 1917–1930

In the above we have described the main features of Hilbert's approach to the foundations of mathematics by quoting often the works of Paul Bernays. Bernays had been called from Zurich to work as Hilbert's assistant in Göttingen in 1917. It is well known that Bernays played an important role in helping Hilbert develop his foundational programme and that he is in great part responsible for the writing of the *Grundlagen der Mathematik*. But what was Bernays' contribution to the programme in the 20s? I shall attempt to give a brief characterization of what emerges from the published articles. A more accurate assessment must await the study of Hilbert's lecture courses, preserved in Göttingen, and of Bernays' *Nachlaß*, preserved in Zurich.

If we exclude the more technical articles in logic (1926; 1927; 1928*d*), Bernays' contribution to the clarification of Hilbert's programme are centered around 1922–23 and 1928–30. It seems to me that Bernays' contribution, as it emerges from the published work, consists mainly in a more explicit discussion of the central philosophical topics surrounding Hilbert's programme. In this connection one should mention the following points:

(1) The epistemological connection between the empiricist "tendencies" of modern physics and Hilbert's programme developed in 1922*b*;

(2) The explicit definition of mathematics as a theory of abstract systems in 1922*a*;

(3) The sustained philosophical discussion with Müller in 1923*b*; 1923*a*.

(4) The explicit characterization of Hilbert's finitism in Kantian terms as an a priori intuition (see 1928*c*; 1930*a*; 1930*b*);

(5) The development of the distinction between logic and mathematics as based on two different types of abstraction, delineated in 1922*b* and then discussed at length in 1930*b*, and the importance attributed to reflection on the general features of intuitively given objects as essential for finitism (1930*b*) (cf. also Sieg 1990*b*, p. 178).

Furthermore, Bernays' articles were essential in clarifying a number of central points left unclear in Hilbert's publications. It is enough to think of the distinction between metatheoretical and formal induction contained in Bernays (1922*b*); one has to wait for Hilbert (1928*a*) to find the same clarity in Hilbert's published work (of course, this should not imply that Hilbert was not clear about this much earlier). All in all, Bernays' articles strike me as extremely useful for an introduction to Hilbert's thought on the foundations of mathematics, and on many occasions they display a philosophical sophistication that is lacking in Hilbert's work. This should not be surprising, for Bernays' solid training in philosophy (see Bernays 1910, 1913*a*,*b*) was one of the factors that influenced Hilbert in deciding to use Bernays as his collaborator in the foundations of mathematics.

2.5 Technical Results

It is unfortunately not possible within the limited context of this chapter to give an idea of the variety of technical formalisms and investigations carried out throughout the 1920s by Hilbert and his school. For example, in 1922 Hilbert does not make use of an operator for negation and opts for the introduction of an inequality $\neq$ along with the equality $=$. In 1923 the formalisms contain the negation operator. Another instance concerns the use of the universal quantifier (see Goldfarb 1979). It appears in 1922 as basic, but in 1923 it is obtained through the use of the τ operator and later on by means of the ϵ operator.

Recent work in this area, carried out by Sieg (see 1996 for a statement of the results and Sieg 1999), has shown that the 1917/18 lectures, written by Bernays, contain essentially all of the material presented in Hilbert-Ackermann (1928). Rather than follow in detail this development, let me simply summarize the global results obtained within Hilbert's school in the 1920s concerning the consistency problem.

The emphasis in the 1922/23 publications is on the formalization and consistency of analysis. The positive result of these two articles can be seen as establishing a finitistic proof of consistency for symbolisms that do not contain quantified variables (or transfinite operators). These systems are, however, too weak for what is needed in mathematical practice, and Hilbert's hints for how to deal with quantifiers, or transfinite operators, are only based on simple examples.

In 1924 Wilhelm Ackermann in a dissertation written under Hilbert's supervision believed to have achieved, by developing Hilbert's previous attempts, the consistency proof for the complete formalism of analysis (thus including the transfinite operators). However, during the revision of the proofs of his article, he realized that the proof goes through only if a modification is made to the formalism. In essence he obtains a consistency proof for a fragment of arithmetic, which restricts the induction axiom to formulas not containing bound variables. Among the improvements reflected in Ackermann's dissertation is the use of a unique formalism for the formalization of analysis, as opposed to the sequence of formalisms introduced in Hilbert's previous articles.

The next important step was von Neumann (1927). In this important article von Neumann improved on the techniques given by Ackermann, but he still failed to abtain the desired proof of consistency for analysis (for more details see van Heijenoort 1967a, p. 489). However, these partial results created an atmosphere of excitement in Hilbert and his co-workers. Indeed Hilbert on several occasions during these years spoke as if accomplishing the consistency proof for analysis was only a matter of working out a few combinatorial details (see Hilbert 1928a, p. 479, 1929, p. 229).

Hilbert's aims in the late 1920s are well described in his 1929 article. He concentrates on four general problems. The first problem is the extension of Ackermann's and von Neumann's techniques for the elimination of the transfinite operator in connection with function variables (this would give the required extension of the consistency proof to all of analysis). The second problem mentions an extension of the above problem so as to include more powerful set-theoretical principles. The third problem concerns the completeness of number theory. It is interesting that Hilbert uses interchangeably two properties that we have learned to distinguish, categoricity and syntactic completeness. Finally, the fourth problem raised the issue of proving the completeness of first-order logic.

As is well known the results by Gödel in 1930 and 1931 had a profound impact on the programme of research described by Hilbert in 1929. In 1930 Gödel proved the completeness of first order logic. Furthermore, in 1931 he established the incompleteness of elementary number theory. The result had the most significant consequences for Hilbert's problems. In particular, through Gödel's incompleteness theorems it became clear that, at least in their original formulation, the first three problems mentioned by

Hilbert in the 1928 address to the Mathematical Congress in Bologna (Hilbert 1929) could only admit of a negative solution.

Much more should be said about the results obtained during the late 1920s. Hilbert's introduction of the ω-rule (see Hilbert 1931), the works by Herbrand and other logicians, and the monumental *Grundlagen der Mathematik* should be mentioned in this connection. However, this story cannot be told here.[36]

Addendum to Chapter 2
(Added March 24, 2009)

I have little to add to the majority of the essays contained in the book. However, I would like to comment on the oldest of them, namely Chapter 2. I still stand by the presentation of Hilbert's program given there. In particular, I still claim that there is a shift in the epistemological foundations of the program between an empirical (or phenomenistic) conception of finitistic intuition and the Kantian notion of pure intuition. While some of the essential points supporting my case are already made in Chapter 2, I would like to summarize and expand on the evidence in favor of my interpretation. I am moved to this position by four considerations, which I shall presently discuss.

First of all, an enormous change in emphasis exists between the articles written by Hilbert and Bernays in the early part of the twenties and in the later articles' explicit reference to Kant's pure intuition in characterizing the program. In particular, in the early articles intuition is never characterized as "pure" [*rein*] in contrast to what we find in the later articles. The textual evidence for this is given in Chapter 2 and I will not rehearse it here.

Second, the appeal to the reductionist tendencies of the exact sciences in Bernays (1922b) points to a methodological phenomenalism. In Chapter 2, I point out that in 1922 and 1923 Hilbert does not attempt to characterize the epistemological status of the kind of intuition appealed to in metamathematics. Rather, it is Bernays who provides substantial clarifications with respect to this problem. According to Bernays' description of the program, the axiomatic construction of arithmetic (taken here to include also analysis and set theory) proceeds, analogously to that of geometry, by assuming at the outset a system of objects with determinate relational properties that satisfy the axioms. This postulation of a given set of entities satisfying the axioms is an existential assumption. Bernays speaks here of it as a "transcendent" assumption. In the epistemological literature of the time "transcendent" is opposed to "immanent" and these are key words, for instance, in the debates concerning the proof of the existence of the external world and of realism in general (see Freytag 1902, 1904; see also Schlick 1918). Indeed, the subject was topical at the time since the Machian school (including Avenarius and Petzoldt) defended a philosophy according to which the concept of (transcendent) existence was to be eschewed and all philosophical reflection should be limited to immanent contents of consciousness. Bernays' approach seems to follow the same methodological guidelines:

An appeal to an intuitive grasp of the number series as well as to the multiplicity of magnitudes is certainly to be considered. But this could certainly not be a question of an intuition in the primitive sense; for, certainly no infinite multiplicities are given to us in the primitive intuitive mode of representation. And even though it might be quite rash to contest any farther reaching kind of intuitive evidence from the outset, we will nevertheless make allowance for that tendency of exact science that aims as far as possible to eliminate the finer organs of cognition [*Organe der Erkenntnis*] and to rely only on the most primitive means of cognition. According to this viewpoint we will examine whether it is possible to ground those transcendent assumptions in such a way that only primitive intuitive cognitions *come into play*. On account of this restriction of the means of cognition we cannot, on the other hand, demand of this grounding that it allow us to recognize as truths (in the philosophical sense) the assumptions that are to be grounded. Rather, we will be content if we succeed in proving the arithmetic built on those assumptions to be a possible, i.e. consistent, system of thought. (Bernays 1922b, pp. 215–16)

There is no trace in the lectures notes from 1920 or 1921–22 given by Hilbert of a discussion of these methodological issues. This epistemological interpretation of the situation seems to be original with Bernays. Although Bernays does not refer to any specific school, the characterization of the Machian school he provides in a lecture delivered in 1928 (Bernays 1930a) seems to correspond accurately:

It is especially the immanence philosophy of the phenomenalism of Mach's school which is widespread among researchers in the exact sciences (indeed, it dominates almost absolutely), which believes to be able to eliminate the notion of existence in general and to get by with the notion of phenomenon. According to this philosophy there is fundamentally no other kind of knowledge as perceiving, remembering, following the sequence of representations and comparing the contents of representations. (Bernays 1930a, p. 100)

While the above description in 1928 occurs in the context of a defense of a version of transcendental idealism, it is remarkable that in 1922 Bernays seems to align himself with the goals of the Machian school. However, he makes it very clear that this is a methodological position. One does not reject out of hand the possibility of further-reaching types of intuition, such as the intuition of infinite collections or of continua; rather one tries to get by without that "transcendent" assumption. This is also quite clearly expressed against Müller in 1923:

Hilbert's theory does not exclude the possibility of a philosophical attitude that conceives of the numbers as existing, nonsensible objects ... Nevertheless the aim of Hilbert's theory is to make such an attitude dispensable for the foundations of the exact sciences. (Bernays 1923a, p. 226)

The third consideration stems from Behmann who, in his 1918 thesis written under Hilbert's supervision (see Chapters 3 and 4), bases the construction of mathematics on empirical perception because all that involves higher functions of cognition (such as abstract thought) might cause paradoxes. While only a reading of Chapters 3 and 4 provides the full picture, I claim that some of Hilbert's key epistemological passages on "intuition" are strikingly similar to those of Behmann. Discussing the non-viability of grounding arithmetic by assuming at the outset the collection of natural numbers, Behmann says:

Since, due to the nature of our subject, we cannot even presuppose the simplest logical concepts in their traditional use as given, we must thus initially mistrust all abstract thinking. But since we need some kind of starting point, it must be such that our abstract thinking is not yet involved, and whose existence is not—unlike the existence of numbers—based only on the possibility of thought, and thus cannot be distorted by thought. The only things satisfying this requirement, however, are now those that can be known directly without the aid of thought, i.e. solely through sense perception, which thus provide thought with the prerequisite material for its possibility. (So far, almost all false metaphysics suffered from the effort of wanting to create the objects of thought through thinking. As everywhere, however, here too the Archimedean point necessarily lies outside.) Thus, our initial assumption is that it is permissible to take as given objects of experienced reality (individuals) whose existence precedes all thought. (With this, of course, we consider ourselves justified in completely ignoring any possible distortion through perception.) We thus regard these individuals as already given and directly at the disposal of our thought. (Behmann 1918, pp. 44–45)

Now Hilbert:

as a precondition for the application of logical inferences and for the activation of logical operations, something must already be given in representation [in der Vorstellung]: certain extra-logical discrete objects, which exist intuitively as immediate experience before all thought. Hilbert (1922c, E.t., p. 202).

As I said, I am struck by the similarity.

The fourth and final consideration concerns Fraenkel who, in the second edition of his *Einleitung in die Mengenlehre* in 1923, devotes a few pages to the new Hilbertian approach. In that book he emphasizes that the new Hilbertian approach to the foundations of mathematics excludes pure intuition [reine Anschauung], for instance the intuition of the series of natural numbers as advocated by Poincaré (and we can certainly add the intuition of the spatial continuum):

Rather the roots of the new procedure should lie exclusively in primitive, direct intuitive knowledge [Erkenntnisse] without the use of deeper tools of logic. This limitation, of course, does not present a sharp program but it can be made, up to a certain degree, more precise through the enumeration of the admissible intuitive considerations; in any case "pure intuition" in the finer sense should be excluded, e.g. that of the totality of the natural numbers placed by Poincaré at the top of mathematics, as well as, in general, any intuition of infinite sets. (Fraenkel 1923, p. 236)

That this characterization found the approval of Bernays is confirmed by the fact that in the introduction of his book Fraenkel thanks Bernays for "several valuable observations" on the pages from which the above quotation is taken.

What can we conclude from this? There is no question that here Bernays is after a characterization of finitism from the epistemological point of view that would block the type of Kantianism associated with Poincaré. Poincaré had objected to Hilbert's first attempts at consistency proofs (Hilbert 1905b) that they were circular since they had to rely on induction and implicitly assume the totality of natural numbers, which indeed Poincaré thought should be assumed as given a priori. The assumption of the totality of natural numbers is also part and parcel of Weyl's *Das Kontinuum*, whereas it is rejected by Brouwer (and the intuitionist Weyl of 1921c). The appeal to the contemporary tendency of exact science would then seem to put Bernays in the immanentist camp of Mach and the phenomenalists but within a few years Bernays and Hilbert will explicitly interpret their immanentism as a modified form of Kantianism. This appears clearly already in lectures given by Hilbert in 1923. In a series of lectures given in Hamburg in the summer of 1923 Hilbert writes:

> There have been philosophers, and Kant is the classical representative of this tendency, that claim that in addition to logic and experience we also have a certain a priori knowledge of reality. Now I grant that already in the construction of theoretical edifices certain a priori intuitions are needed and that these intuitions are always at the basis of the emergence of our knowledge. I believe that mathematical knowledge is based on a sort of intuition [*anschaulicher Einsicht*] and that even in the construction of number theory a certain intuitive a priori conception is necessary. In this way, Kant's most general idea retain their importance; namely, the philosophical problem of determining that intuitive conception a priori and with it the conditions of possibility of every possible conceptual knowledge and, at the same time, of every experience. This is what has essentially happened in my theory of the foundations of mathematics. (Hilbert 1923b, pp. 424–25 of lecture 3)[1]

In the passages following the above, Hilbert goes on to discard the Kantian proposal in its original terms by rejecting the a priori forms of space and time. Nonetheless he accepts a "finitistic" a priori, as the foundation of the mathematical sciences. And from this point on (see the textual evidence given in Chapter 2) the appeal to Kant remains a constant in Hilbert and Bernays' foundational writings.

3

BETWEEN RUSSELL AND HILBERT: BEHMANN ON THE FOUNDATIONS OF MATHEMATICS

3.1 Logic and Foundations of Mathematics in Göttingen from 1910 to 1921

Recent work on Hilbert's program has focused, among other things, on the development of logic in Hilbert's school and on the philosophical underpinnings of the program. Sieg (1999) and Moore (1997) have investigated the development of first-order logic in Hilbert's 1917–18 lectures, Zach (1999) has given an in-depth analysis of the propositional calculus in Hilbert's school from 1918 to 1928, and Mancosu (1998c) has investigated the philosophical context of Hilbert's approach to the foundations of mathematics. The *Habilitationsschrift* by Bernays (1918) and Hilbert's 1917–18 lectures (Hilbert 1918c) represent the starting point of these important developments. However, these lectures were not the product of a sudden reawakening of interest in logic and the foundations of mathematics within the Göttingen mathematical community. Rather, they were prepared by previous work in foundational issues which began around 1914. Central to these efforts was, in addition to Hilbert's charismatic role, the work of Heinrich Behmann and, starting in 1918, that of Bernays. Another factor which should not

---

I would like to thank Volker Peckhaus, Christian Thiel, Peter Bernhard and Richard Zach for comments and for making it possible to access and reproduce some of the materials contained in the Behmann Archive in Erlangen. I am grateful to an anonymous referee for his comments, which helped me sharpen a number of issues raised in the paper. I am also grateful to the curators of the following collections for their help: Russell archive at McMaster University, Hamilton; Bernays *Nachlaß*, ETH Zürich; Hugo Dingler-*Nachlaß*, Aschaffenburg; Hilbert *Nachlaß*, Göttingen. I would finally like to thank the Wissenschaftskolleg zu Berlin for having provided ideal conditions for work on the first draft of this paper during the academic year 1997–98.

be underestimated is the presence in Göttingen in the early twenties of two Russian mathematicians, Schönfinkel and Boskovitz, who were also engaged in foundational work. Schönfinkel is well known for his article "Über die Bausteine der mathematischen Logik," all of whose contents go back to a talk he gave in Göttingen in 1920.[1] Moreover, the paper by Bernays and Schönfinkel (1928) goes back to work by Schönfinkel in 1922 (see Bernays and Schönfinkel 1928, p. 350). We know little about Boskovitz but he is, together with Behmann, thanked in the second edition of *Principia Mathematica* (Whitehead and Russell 1910, henceforth PM) for his help in correcting various oversights found in the first edition.[2] It is during the period from 1918 to 1922 that the major ideas and some results about metatheoretical investigations, such as completeness and decidability, are formulated and established with the appropriate degree of logical clarity.

If we look at the development of logic and the foundations of mathematics in Hilbert's thinking and in Göttingen in general in the early part of the 1910s we find very little of interest. At the time Hilbert seems to have been very busy with problems in the foundations of physics. Moreover, the Göttingen Mathematical Society does not list in its very active Colloquium any talks on the foundations of logic and mathematics in the period 1910–13. The situation changes suddenly in 1914.[3] A list of the lectures given in logic and foundations of mathematics at the Colloquium between 1914 and 1921, as reported yearly in the *Jahresbericht der Deutschen Mathematiker-Vereinigung*, gives an idea of the increasing interest in logic and foundations in Göttingen from 1914 onwards.

> February 17, 1914, D. Hilbert, Axiome der ganzen Zahlen
> December 1, 1914, H. Behmann, Über mathematische Logik (see Behmann 1914)
> December 18, 1914, F. Berstein and K. Grelling, Über mathematische Logik. Ergänzungen und genauere Ausführungen zum Referat vom 1. Dezember
> February 16, 1915, F. Bernstein, Über J. Königs "Neue Grundlagen der Logik, Arithmetik und Mengenlehre"
> November 7, 1916, E. Zermelo, Über einige neuere Ergebnisse in der Theorie der Wohlordnung
> November 14, 1916, H. Behmann, Die Russell-Whiteheadsche Theorie und die Paradoxien (see Behmann 1918)
> July 3, 1917, H. Behmann, Die Russell-Whiteheadsche Theorie und die Grundlagen der Arithmetik (see Behmann 1918)
> July 10, 1917, H. Behmann, Die Russell-Whiteheadsche Theorie und die Grundlagen der Arithmetik (Schluß) (see Behmann 1918)
> July 17, 1917, F. Bernstein, Geschichte der Mengenlehre
> July 24, 1917, F. Bernstein, Geschichte der Mengenlehre (Schluß)
> July 31, 1917, D. Hilbert, Referat über seine Vorlesungen über Mengenlehre (see Hilbert 1918b)
> November 20, 1917, P. Bernays, Weyl über die Grundlagen der Analysis (see Bernays 1917)

November 27, 1917, D. Hilbert, Über axiomatisches Denken (see Hilbert 1918a)

December 7, 1920, M. Schönfinkel, Elemente der Logik (see Schönfinkel 1924)

February 1 and 8, 1921, R. Courant and P. Bernays, Über die neuen arithmetischen Theorien von Weyl und Brouwer

February 21, 1922, D. Hilbert, Eine neue Grundlegung des Zahlbegriffes (see Hilbert 1922c)

May 10, 1921, H. Behmann, Das Entscheidungsproblem der mathematischen Logik (see Behmann 1922a)

December 6, 1921, P. Bernays and M. Schönfinkel, Das Entscheidungsproblem im Logikkalkül

Even a cursory look at the above list of talks should convince the reader that Behmann was a central player in the development of logic and the the foundations of mathematics in Hilbert's circle. I will try to spell out how knowledge of *Principia Mathematica* was acquired in Göttingen, thereby shedding light on the important problem of the relationship between Russell's logicism and Hilbert's program. After presenting a short account of Behmann's early career I will proceed to investigate his role as *trait d'union* between Russell's logicism and Hilbert's program.[4]

3.2 Heinrich Behmann's Early Career

Heinrich Behmann was born in 1891. He attended the Realgymnasium in Vegesack (Bremen), graduated in 1909 and subsequently enrolled at the University of Tübingen, where he spent two semesters to study mathematics and physics. Later he moved to Leipzig, where he spent three semesters, and eventually to Göttingen in 1911. In 1914, Behmann volunteered for military duty. He was wounded in Poland in May 1915, which required a prolonged stay in a hospital. He resumed his studies in Göttingen in 1916. On June 5, 1918 he successfully defended his dissertation *Die Antinomie der transfiniten Zahl und ihre Auflösung durch die Theorie von Russell und Whitehead*. His advisor was David Hilbert. During his studies he had attended courses by, among others, Hilbert, Landau, Nelson, Reinach, and Weyl. From 1919 to 1921 he worked on his *Habilitationsschrift* (Behmann 1922a). On July 9, 1921 he completed the Habilitation and obtained the *venia legendi*. From 1921 to 1925 he was active as Privatdozent at the institute for mathematics in Göttingen, where Hilbert and Bernays were also active. Behmann left Göttingen in 1925 following a call to the University of Wittenberg (Halle).[5]

3.3 Behmann's 1914 Lecture on *Principia Mathematica* in Göttingen

We have seen from the above list of lectures that Behmann spoke on Russell's system in 1916 and 1917. However, his lecture from December 1, 1914 "Über mathematische Logik" was also advertised in the *Jahresberichte* as an introduction to the system of PM

("Der Vortrag bringt einige neue Gedanken aus den in den 'Principia mathematica' von Russell und Whitehead niedergelegten logischen Untersuchungen"). One would of course like to know what aspects of PM were emphasized in this first public introduction of Russell's work to the mathematical community in Göttingen. Fortunately, our curiosity can be satisfied. The lecture is extant in its entirety in the Behmann *Nachlaß* although it does not appear in the list of the documents contained in the *Nachlaß* published in (Haas and Stemmler 1981). I will here stress only two aspects of this lecture that seem to be quite relevant for later developments in Hilbert's program. Let me also remark that on December 18, 1914 Bernstein and Grelling gave a follow-up talk about PM at the Colloquium in Göttingen.

Behmann began the lecture by stressing the fact that the term "Mathematische Logik" is ambiguous and can be used to characterize two different traditions. The first tradition consists in a general construction of logic by mathematical means ("Mathematik der Logik") and is associated with the names of Boole, Schröder, and partly Peano. The second tradition ("Logik der Mathematik") analyzes the role played by logic in the construction of mathematics. Behmann mentions Bolzano, Frege, and Russell as representative of this direction of work. Although PM belongs squarely in the second tradition it also accounts for results developed in the first tradition. Thus, for Behmann, PM is the first unified account of these two traditions.

Behmann moves on to introduce the main concepts of individual, proposition, and propositional function. As I will come back to these issues when analyzing Behmann's dissertation let me simply remark that individuals are defined as "everything which in reality is directly given to us, thus not, say, things like sets or similar conceptual entities [*Gedankenbilde*]." He then goes on to discuss the concept of propositional function, the vicious circle principle, and type theory. The word "type" is translated as *Stufe* and Behmann asserts that he does this in reference to Frege's use of concepts of first and second order. After discussing the complications due to the ramified constructions, Behmann wonders whether the solution proposed by PM is not too drastic. It is true, he says, that the paradoxes are avoided but at the same time many propositions, which seem perfectly innocuous, are in this way excluded an declared meaningless, e.g. "All propositions are either true or false." Concerning the paradoxes, whose solution "must be seen as a vital question for logic and especially for arithmetic," Behmann goes on to describe Russell's solution with the slogan "There are no classes at all." It is clear that Behmann has read more than PM, as he speaks of several unsuccessful previous attempts by Russell to solve the paradoxes and then mentions that the slogan given above captures in a nutshell what in the literature is called the "no-class theory." In this connection Poincaré is also quoted.

There are now two points in the remaining part of the lecture which deserve emphasis. The first concerns an explication of the no-class theory in terms of ideal elements in mathematics. Behmann uses the example of postulation of points at infinity in projective geometry. After expounding the conceptual difficulties encountered by the student presented with such concepts for the first time, Behmann goes on to dispel the puzzlement as follows:

Nevertheless, the correct solution of all these geometrical inconsistencies, once it is given, is not difficult to comprehend. To wit, it simply states that the ideal elements are not objects of geometry in the proper sense, but are first of all only words. As such they are parts of ways of speaking which one uses in geometry to bring a certain class of propositions into a form which is as simple as possible.[6]

Behmann then asserts that one only needs to generalize the above reflections on projective geometry to arrive at the no-class theory. As terms purporting to refer to points at infinity can always be eliminated from a sentence in which they occur, likewise can words like "class," "totality" etc. be eliminated from sentences in which they occur in favor of equivalent sentences in which these words do not occur.

It will be obvious to the reader that the role of ideal elements in mathematics, so much emphasized by Hilbert in his foundational work in the 1920s, is here brought to the fore with greatest clarity in connection to an explication of the main theses of *Principia*. There is no need to follow in detail Behmann's explication of how the strategy is supposed to work. More important is to stress another point of connection to Hilbert's work that emerges from this talk. This relates the axiom of reducibility—which claims that any class definable by a propositional function can be defined by a propositional function of the lowest level (i.e., a predicative function)—to completeness axioms in Hilbert's sense:

The axioms thus states that in this one order there already are enough functions to define all possible classes; it can therefore be viewed as a kind of completeness axiom for the predicative functions.[7]

We will see that Hilbert was particularly interested in this intepretation of the axiom of reducibility.

Behmann gave three more talks on the system of *Principia* between 1916 and 1917. We do not have the texts of these talks but we can conjecture that they were preparatory work to his 1918 dissertation to which we now turn.

3.4 Behmann's Dissertation

Behmann's dissertation *Die Antinomie der transfiniten Zahl und ihre Auflösung durch die Theorie von Russell und Whitehead* was written in 1918. The original is still preserved at the Niedersächsische Universitätsbibliothek in Göttingen. In 1922 Behmann published an account of the main contents of the dissertation in the *Jahrbuch der Mathematisch-Naturwissenschaftlichen Fakultät* with the same title (see Behmann 1922b). The original dissertation is 351 pages long whereas the article is only 10 pages long. It is clear that much of the contents of the dissertation could thus only be mentioned in the article and much was glossed over.

The dissertation aims in the first place at giving an introduction to the system of PM. However, as not all of PM could be treated, Behmann selected the topic of the foundations of the natural numbers including the developments of Cantorian set theory. The topic is developed by using the solution of the antinomies of transfinite numbers—by means of the system developed by Russell and Whitehead—as a unifying theme.

We have seen that interest in the system of PM was alive in Göttingen since 1914 and one should consider that Hilbert himself was quite attracted by the logicist solution to the foundations of mathematics and for a while seems to have considered the logicist solution valid. Moore (1997, p. 85) and Sieg (1999, pp. 3, 11) have no doubts that Hilbert defended a logicist point of view in 1917–18 although he later became disenchanted with Russell's solution (see Hilbert 1918a, p. 412, trans. p. 1113 for a strong praise of Russell's achievements). During this period Hilbert supervised Behmann's dissertation and also encouraged Bernays to write a *Habilitationsschrift* on topics related to the logical system of PM (see Bernays 1918). However, whereas Bernays' 1918 Habilitation centers on logic, Behmann's works skips over the logical parts of *Principia* to focus instead on the mathematical developments more immediately related to the construction of arithmetic and of Cantorian set theory. Behmann also took, at least up to 1922, the logicist solution to the foundations of mathematics to be successful.[8]

Behmann's dissertation remained unpublished due to the difficult situation of the postwar period. A letter from Behmann to Russell dated August 8, 1922, gives an account of the reasons that prompted Behmann to work on PM. He had discovered his passion for logic through PM but the difficulty of the exposition had led him to the idea of writing an introduction to it.[9]

Then Hilbert suggested a more focused approach to the topic:

Prof. Hilbert then proposed to me, as a theme of dissertation, not to treat the whole work as such, but rather to make clear the particular way by which the plainly most serious among the antinomies of the Theory of Aggregates, that concerning the transfinite cardinal and ordinal number, is avoided by the logical theory of the Principia Mathematica. So I tested the theme in the way suggested by Prof. Hilbert, adding critical comments wherever I believed it necessary or desirable.

But unhappily, on account of the want of paper, which was a consequence of the War and the Revolution strongly felt in our public life, there was no possibility of having a book printed "having a character extremely scientific and addressed to a small class of readers"—as the publisher remarked—, so that I am very sorry to not to be able to send you a copy of it.

Russell did actually get to see a copy of the work, through the Göttingen library, and judged it "a valuable and important piece of work."[10]

In order to give an overview of the contents of the dissertation it might be useful to list the contents of the six chapters that make it up. The first chapter (pp. 1–43) considers the formulation of the problem and the way to its solution. The second chapter (pp. 44–89) treats sentences and functions up to the first order. The third chapter (pp. 90–147) is devoted to functional logic in general. The fourth chapter (pp. 148–225) treats of the logic of classes. Chapter five (pp. 226–83) develops cardinal arithmetic. Finally chapter six (pp. 284–351) discusses the general results and more general philosophical questions. Of course, given the length of the work it is not possible to give a detailed account of its contents. In Behmann (1922b), Behmann emphasized the technical differences between his own exposition and that given in PM. In my analysis I will not emphasize the technical reconstruction but rather the philosophical outlook defended by Behmann in this period. I believe that Behmann's dissertation sheds additional light on the approach to the foundations of mathematics held by Hilbert and his co-workers during this period.

3.5 Abstract Terms and Context Principle

Behmann begins his dissertation by posing the problem of certainty in science. The development of arithmetic and the discovery of Cantorian set theory have shown that also within arithmetic there is room for doubt. Together with the mathematical fruitfulness came also a number of contradictions which could not be solved by the available logic and mathematics. However, the freedom originated by the Cantorian approach to mathematics should not be limited on accounts of the doubts originated by the paradoxes.

These antinomies, according to Behmann, are of different sorts (p. 3) and affect mathematics in different ways. The paradox of number [*Paradox der Zahl*] lies at the heart of mathematics. Rather than rejecting Cantorian mathematics one ought to analyze the mathematics of the finite and in particular the concept of number (p. 4). A solution in this direction has been given by symbolic logic (or mathematical logic) and in particular by PM. Given the complexity of PM, Behmann proposes to give an exposition of the system by developing only what is necessary to show how the paradox of number can be solved in this theory (p. 6). In the cardinal form the paradox is nothing else but Cantor's paradox of the cardinality of the universal set. In its ordinal form it resembles the Burali-Forti paradox. Behmann's emphasis is on the cardinal version of the paradox, and this is motivated by pedagogical reasons. At this point Behmann introduces the relevant set-theoretic notions. In this connection he announces that although he is not dealing with an axiomatic project he will use the axiom of choice as sparingly as possible. After presenting various cardinal and ordinal forms of the antinomies Behmann declares that the goal of the investigation is to free mathematics from the antinomies and that this should not be done by first constructing a pure logic but rather he will try to ground arithmetic so that the logical forms allowable will become clear through this foundational work (pp. 31–32). §5 (pp. 33–39) contains some of the most interesting methodological reflections. Antinomies arise, says Behmann, when two sentences stand in contradiction. In arithmetic—and Behmann explicitly includes also transfinite set theory under this name—such sentences contain concepts such as number and set. However, the concepts themselves cannot give rise to antinomies or contradictions. Contradictions only arise when the concepts are used in sentences. This leads Behmann to state a form of context principle:

As a result, it would be a mistake from the start to try to investigate the abstract concepts and relations of arithmetic in themselves alone without bearing in mind the fact that all such terms mean something only in the context of the *sentence*, and that even the best-constructed concepts are of no use to us and cannot protect us from contradiction until we know how to *apply* them properly.[11]

This goes hand in hand with giving the concept of sentence pride of place in the foundational investigation:

... thus, as the latter consideration shows, we must now, from the outset, place the concept of *proposition* in the center of all our investigations. This means we may never regard concepts occurring in propositions—like "set," "number," "statement," "space," "force," etc.—in isolation, as independent logical objects, but always in the context of the propositions in which they can meaningfully appear.[12]

Behmann does not use the term 'context principle', he speaks of *Erkenntniszusammenhang* (epistemic context). The project now consists in developing a consistent system of sentences in which arithmetic can be developed.[13]

The last section of chapter 1 presents the six assumptions that will be used in the remaining part of the work (p. 40):

1. Axiom of individuals;
2. Axiom of true and false sentences;
3. Axiom of negation and disjunction;
4. Axiom of the variation of the constants and generalization;
5. Axiom of individual identity;
6. Axiom of reducibility.

In the next section I will spell out how the first four axioms are appealed to in Behmann's reconstruction.

3.6 Individuals, Propositional Functions, and Classes

It is customary to develop arithmetic by taking for granted the system of natural numbers. Given the possibility of the emergence of the antinomies this approach is, according to Behmann, barred for foundational research. Moreover, procedures involving abstract objects are also not to be allowed:

Since, due to the nature of our object, we cannot even presuppose the simplest logical concepts in their traditional use as given, we must thus initially mistrust *all* abstract thinking. But since we need some kind of starting point, it must be such that our abstract thinking is not yet involved, whose existence is not—unlike the existence of numbers—based only on the possibility of thought, and thus cannot be falsified by thought. The only things satisfying this requirement, however, are now those that can be recognized directly without the aid of thought, i.e., solely through sense perception, which thus provide thought with the prerequisite material for its possibility. (So far, almost all false metaphysics suffered from the effort of wanting to create the objects of thought through thinking. As everywhere, however, here too the Archimedean point necessarily lies outside.) Thus, our *initial assumption* is that it is permissible to *take as given objects of experienced reality* (individuals) *whose existence precedes all thought.* (With this, of course, we consider ourselves justified in completely ignoring any possible distortion through perception.) We thus regard these individuals as already given and directly at the disposal of our thought.[14]

It is essential to remark that for Behmann reliance on empirical reality provides a consistency proof for the system:

Since—as we may here undoubtedly assume—the objective world, i.e. the totality of the individuals with all their properties and relations, certainly constitutes a consistent domain in the end, then clearly *every* proposition that has only individuals as objects, and which thus does not presuppose the existence of other things, must either correspond to or contradict the actual case, and thus fulfill the principles of contradiction and of the excluded middle.[15]

What is essential for Behmann is that these individuals can be considered as existent and they serve as primitive material to our thinking (p. 46). Moreover, Behmann asserts (p. 47) that these individuals have to be considered as a fixed domain that cannot be altered.[16] By means of properties grasped empirically [*empirisch aufzufassenden Eigenschaften und Wechselbeziehungen der Individuen*] Behmann defines the elementary propositions [*einfache Aussagen*] as those in which a direct perception is expressed.[17]

In order to be able to say something about these individuals it is necessary to use "elementary propositions." This leads to the second assumption: it must be possible to make consistent elementary propositions concerning individuals. More precisely, elementary propositions are characterized by containing only individuals and their objective properties. By contrast, abstract concepts such as "proposition," "property," "set," "number," and "existence" cannot occur in them. Moreover, universal and existential quantification cannot occur in them because that would contain an implicit appeal to sets. In Behmann's characterization an elementary proposition is such that by varying the name of individuals we always get a meaningful sentence. This is not true were we to allow abstract objects ("the sentence is green" is simply meaningless). The elementary propositions are, as soon as they are about specific individuals, either true or false. However, Behmann says that one should not say that propositions exist but only that the individuals with their empirical properties exist (p. 49). Talk of the collection of all elementary propositions cannot lead, according to Behmann, to antinomies as a proposition that refers to this collection is not elementary (p. 50).

In §8 Behmann states that the elementary propositions are closed under negation and disjunction. Implication and conjunction are defined in the usual way. We can also consider an elementary proposition without thinking of any specific individual. This leads to generalization: from fa to fx, and to the notion of propositional function $f\hat{x}$. We are at this point still not justified in treating these propositional functions as objects of discourse. The fourth assumption thus leads to assert that given any elementary proposition it must be allowed to vary at will the individuals in order to construct the associated elementary propositional function. Moreover, it must be allowed to create new propositional functions or propositions by means of existential and universal generalization. The propositional functions, resp. propositions, thus obtained are called first-order functions or first-order propositions (p. 60). It is important to remark that propositions about propositional functions are only apparently about them and should be paraphrased as being about the individuals. Behmann clarifies that parts of sentences such as "no object," "proposition," "set," "function," "numbers" should be interpreted only as formal components of propositions. Let us simply give two quotes as paradigms for the many passages in which Behmann insists that all entities not reducible to the individuals are not existent.[18] On p. 106, by making use of the comparison with the differential calculus, Behmann reasserts that propositional functions are not real:

> In the same sense [as the differential dx], however, propositional functions, like all abstracta, are not real objects, but create only the deceptive appearance of object-hood by providing certain resting points for abstract thought moving at a great distance from its real, concrete objects.[19]

When we move to classes we find the same strategy of deontologization of anything which is not an individual or an empirical fact, a strategy which is of course the Russellian "no-class theory":[20]

> The classes—and by the same token, incidentally, the numbers—are thus, as we earlier hinted, nothing else than figures of speech extremely useful for ease and clarity in presenting arithmetic, but can nonetheless become quite problematical as soon as one takes them seriously and, in violation of their nature, takes them for the names of objects.[21]

It should be clear from the above quotes that the general philosophical project defended by Behmann consists in reducing appeal to abstract entities to a formal construction with no ontological import. When we are clear about this it also becomes possible, within certain contexts, to treat the abstract objects *as if* they were individuals.[22]

I will now move to the last, and most philosophical, part of the dissertation.

3.7 The Philosophical Foundations of Arithmetic

Experience of concrete objects as the foundation of arithmetic

In the last chapter of the dissertation Behmann comes back to his approach to the foundations of arithmetic described in the earlier part of the dissertation and engages in a number of interesting philosophical reflections. The system presented should of course recover the full power of ordinary arithmetic but his account of numbers in terms of the role played by numerals in a sentence should not be interpreted as formalism:

> the primary advantage of the theory proposed here is that our arithmetic in its entire development is in no way merely a playing with signs, but has from the beginning an objective content.[23]

In particular, Behmann stresses that whereas formalism cannot account for the applicability of mathematics to reality, in his approach this is taken care of by the very construction of the system. The request that all of mathematics refers in the last analysis to an empirical reality can of course raise issues concerning the problem of whether we can have only arithmetical statements to which an experience could correspond. On this view, not only can any statement involve only finitely many entities but indeed there must be a certain finite bound on the number of entities in any possible experience. The only reply given by Behmann is that accepting this limitation would make it impossible to obtain a complete and simple theory. In any case, when we generalize beyond experience by means of axioms we do indeed lose the possibility of checking every statement directly but, on the other hand, every single theorem is such that every possible experience must conform to it.

Concerning the appeal to concrete objects as the foundations of arithmetic Behmann provides a historical sketch that shows how the Russellian position is the natural development of a sequence of events in the foundations of mathematics. He mentions Dedekind's reduction of the irrationals to cuts of rationals and Kronecker's arithmetical standpoint. Kronecker criticized contemporary analysis and did not realize that the concept of number could be reduced to that of set. This latter step was taken by Frege. However, Frege's development led to contradictions and it was Russell who eliminated the notion of set in favor of that of concrete individuals.[24] The antinomies show that sets and relations, propositions, and properties cannot be the starting point:

Since all abstract things—at least to the degree that they can enter into any relationship with arithmetic—are thus already excluded, all that remains are the concrete objects of experienced reality.[25]

Behmann's position thus ends up emphasizing the non-existence of the natural numbers except as a purely formal part of sentences:

Arithmetic in truth does not have its own objects at all, but assumes rather no other object-hood than the other sciences do (which need not be imagined as real, but as at least possible), that is that of experience. Since we may have every right to assume that every reasonable statement about this reality conforms to the law of contradiction, the consistency in arithmetic is thus fundamentally ensured.[26]

However, the attempt to provide objectivity for arithmetic is often at odds with statements which sound like a straightforward instrumentalist position:

we grasp the numbers as mere formal components of propositions ... In this sense, the numbers, too, are nothing other than a means of easily deriving and expressing specific bits of knowledge; but aside from this purpose, in themselves, they are nothing at all.[27]

Of course, when engaged in practical arithmetical work we do not need to go back to the empirical intuitions that form the objective content of our reflections. But the relation of abstract thought to concrete experience is best analyzed in terms of the problem of the apriority of arithmetic.

The apriority of arithmetic

In §51 of the dissertation Behmann tries to escape the almost inevitable conclusion that arithmetic becomes in this way a posteriori knowledge. He asks himself: how can one reconcile the apriority of arithmetic with such emphasis on its natural connection to experience? For Behmann the question becomes equivalent to that of whether the essence of a priori knowledge consists in having no relationship to experience and in not presupposing the latter as a condition of possibility for its significance. In other words, if we make experience an essential condition of possibility of arithmetic how can we preserve its aprioricity? Behmann holds that mathematical knowledge is synthetic a priori. According to him the essential point is that "synthetic a priori judgments are only possible in that they bring to expression the form of a possible experience" (p. 339). His claim is that the essential feature of a priori knowledge consists not in having actual experiences of a certain sort or other but rather that every possible experience must accord to it. To strengthen his position he also gives a long list of quotes from Kant and Schopenhauer to the effect that a priori knowledge is essentially tied to the possibility of experience. Consider for instance the following quotation by Schopenhauer:

> Since, as has been shown, the concepts borrow their material from intuition, and since as a result the entire construction of our world of thought is based on the world of intuition, we must thus be able, though perhaps via intermediary steps, to trace each concept back to the intuitions from which it itself directly derives or back to the concepts from which it is abstracted: i.e. we must be able to exhibit it with intuitions that relate to the abstractions as examples. These intuitions thus provide the true content of all our thought, and wherever they are lacking we have not had true concepts in mind, but mere words.[28]

Behmann goes as far as claiming that Schopenhauer had provided with his insights the philosophical foundation for the type-theoretical project presented in PM. Of course, Behmann is aware of the rather paradoxical fact that he has given an account in terms of synthetic a priori knowledge provided by Russell's construction whereas Russell himself was, like the majority of logicist writers, quite anti-Kantian. One final thing should be emphasized about Behmann's attempt to put together empirical content and aprioricity of mathematics. The account makes no mention of pure intuition which for Kant was essential for claimimg the synthetic a priori nature of mathematics. In Behmann's case the synthetic part is given by empirical experience. However, for aprioricity his claims only amount to the fact that accord with experience and applicability are not in conflict with the aprioricity of mathematics. Unfortunately, what is lacking in Behmann is a positive account to the effect that mathematics is not empirical and indeed synthetic a priori.

Mathematical objects as fictions: Relationship to Vaihinger's philosophy of the As If

It could be considered remarkable, in light of the above claims, that when actually developing number theory and analysis one needs concern oneself with experience and the real meaning of numerical formulas and inferences as little as one would with the "internal construction of a calculator which has proved to be reliable."[29] Mathematics strikes us normally as a field of a priori [*immanente*] truths and this calls for an explanation. The concept which is here required, according to Behmann, is that of fiction. He claims that it would make no sense to go back to the experiential content of arithmetical equations every time. Rather, it is a need of thought economy [Denkersparnis] to interpret mathematical statements (e.g., $5 + 7 = 12$) as expression of a relation between fictional objects. This justifies the normal approach of mathematicians in starting with the number system and leaving the analysis of the real meaning of the arithmetical formulas to the business of the foundations of arithmetic. It is this point of view of economy of thought that also explains the formalist and platonist tendencies of many mathematicians and philosophers. The first reduces numbers to their symbols and the second hypostatizes their existence. Both of them are however unjustified as a foundational account of arithmetic, although practically, according to Behmann, they can be accounted for by this need for thought economy. When we admit these fictional objects for thought-convenience we proceed as in practical life when we use checks, which are a fictional substitute for a real sum of money. The concept of fiction was of course much discussed during this period. One of the major philosophical developments of the decade had been the publication of Hans Vaihinger's *Philosophie des Als Ob* (Vaihinger 1911). However, the book had not been received favorably by mathematicians. Behmann found the dismissal of the book to be explained by a number of blunders and infelicities in Vaihinger's discussion of issues pertaining to mathematics. However, the fact that in mathematical practice mathematicians have always made use of fictions calls, according to Behmann, for a reevaluation of Vaihinger's ideas in the foundations of mathematics. He thus accepts Vaihinger's idea that numbers are fiction, although with an important specification:

Note: it is not the number that is a fiction—in the sense explained earlier, the number is rather a merely formal component of arithmetical propositions—but rather: we *avail* ourselves of a fiction when we (unnecessarily) attribute to it an object-hood not appropriate to it.[30]

The mistake in Vaihinger's position, according to Behmann, had been that of considering fictions as necessary parts of arithmetical statements and inferences. Finally, Behmann agrees with Vaihinger that a fiction according to its essence contains in itself a contradiction. In fact, Behmann says, as soon as one goes beyond a certain realm of thought in which the fiction is applicable, one ends up with contradictions.

In conclusion, Behmann's approach to the foundations of mathematics can be summarized as follows. Empirical perception grounds the objectivity and consistency of mathematics. Number statements have to be considered as formal objects which acquire a meaning only by being traced back to their empirical source. However, mathematics is a priori and certain. Moreover, the practical need for thought economy justifies a philosophy of the As If for the practicing mathematician. One is allowed to hypostatize the numbers and treat them as fictions, although in reality, i.e. independently of their role as formal components of sentences, the numbers do not exist.

3.8 Behmann's Approach to the Foundations and Its Relationship to Hilbert's Foundational Views

We have already mentioned that Hilbert was attracted to the logicist solution for a short period. However, his new approach to the foundations of mathematics in the early twenties seems to mark a dramatic new beginning. Sieg (1999, p. 12) laments the gap in historical understanding concerning the Russellian influence on Hilbert. It seems to me that through Behmann's dissertation and his earlier lectures it is possible to see some elements of continuity between the logicist approach and Hilbert's program as it was presented in the early twenties. I claim that on some issues Hilbert was at this stage still in agreement with Russell's and Behmann's general outlook. But let me begin by first pointing out continuities with the earlier foundational work by Hilbert. Hilbert gave a positive evaluation of Behmann's achievements. This is found in manuscript in the *Promotion* files for 1918 at the University of Göttingen

Göttingen, 1 February 1918,

The original aim and intent of the treatise submitted by Behmann is to introduce the world of thought of the discipline nowadays known as symbolic logic, which, in the hands of a number of eminent mathematicians and logicians, has become an important part of epistemology, and which has received its most mature treatment and presentation in recent years in the large-scale work "Principia Mathematica" by Russell and Whitehead. While symbolic logic for a long time seemed nothing more than a superficial formalistic development of Aristotle's theory of inference figures, Russell was the first to achieve definite successes in applying symbolic logic to difficult epistemological questions. At the very top of Russell's theory—as highest axiom of thought—stands the so-called axiom of reducibility. This axiom, including the related theory of types of Russell, poses extraordinary difficulties for its comprehension. To remove these, Behmann turns to the axiom of completeness I have introduced to arithmetic, which is not only logically similar to the axiom of reducibility, but is even internally materially related to it. By bringing out these connections, Behmann succeeds not only in clarifying the axiom of reducibility, but in surpassing Russell in applying Russell's theory to a particular deep problem—the solution of the antinomy of transfinite number. His result is essentially: all transfinite axiomatics is by its nature something incomplete, the set-theoretical concepts of Cantor are however strictly admissible. Every meaningfully posed set-theoretical problem retains meaning and is therefore capable of a solution. The presentation takes pains to presuppose no specific mathematical knowledge; it is thus accessible also to the non-mathematician. I hope to find a publisher for Behmann's work since—it seems to me—it fills a present mathematical and philosophical need. I propose to accept just 2 printed sheets as dissertation.

Grade
Hilbert[31]

We see that Hilbert strongly emphasizes the similarities between his arithmetical completeness axiom and the axiom of reducibility. Indeed, there is a note in Hilbert's hand in the margin of Behmann's official copy of the dissertation where, in connection with Behmann's discussion of the axiom of reducibility, we read "Mein arithmetisches 'Vollständigkeitsaxiom' citiren!" (see p. 157).

Let us now move on to the connection between Behmann's dissertation and Hilbert's mature program of the 1920s.[32] I have already emphasized that the issue of ideal elements is raised quite clearly by Behmann in 1914 and treated at length in his dissertation of 1918. Hilbert's detailed treatment of ideal elements is first found in the 1919–20 lectures "Natur und mathematisches Erkennen" (Hilbert 1992). I should emphasize that I am not making here a priority claim on behalf of Behmann. Rather, I am simply interested in pointing out that within Hilbert's circle extensive discussion of ideal elements seems to have begun in connection to a discussion of *Principia*. This is of course consistent with the presence in Hilbert's early writings of the notion of ideal element, which was clearly central to nineteenth-century mathematics (e.g., Dedekind). What seems to me more original in Behmann, and strictly tied to Russell's no-class theory, is the connection between the context-principle and ideal elements. We find traces of this idea in a review of Müller's work *Der Gegenstand der Arithmetik* written by Bernays in 1923:

> In fact, the mathematician knows that in his science an especially fruitful and continually applied procedure consists in the introduction of "ideal elements" that are introduced purely formally as subjects of judgments, and that, however, when detached from the statements in which they occur formally, are nothing at all. (Bernays 1923b, p. 521)

Once again, I am not concerned with priority issues but rather with highlighting, through Behmann's work, ideas which seem to have been shared by Hilbert and his immediate co-workers.

Let us now consider Behmann's emphasis on the importance of individuals as the only objective grounding of the edifice of PM. Behmann sees in this the only way to appeal to concrete objects without relying on abstract objects. In Hilbert we also find the idea that in order to overcome the dangers represented by abstract operations or thinking we must have extra-logical objects given to us before all thought:

> As we saw, abstract operation with general concept-scopes and contents has proved to be inadequate and uncertain. Instead, as a precondition for the application of logical inferences and for the activation of logical operations, something must already be given in representation [*in der Vorstellung*]: certain extra-logical discrete objects, which exist intuitively as immediate experience before all thought. If logical inference is to be certain, then these objects must be capable of being completely surveyed in all their parts, and their presentation, their difference, their succession, (like the objects themselves) must exist for us immediately, intuitively, as something that cannot be reduced to something else.[33]

I see here, in the acceptance of a class of concrete individuals as given before abstract thought, an element of continuity between Russell's (and Behmann's) conception and Hilbert's mature foundational thought. Notice that for Hilbert, just as for Russell and Behmann, the individuals are not susceptible of further analysis. Hilbert goes on to say:

> Because I take this standpoint, the objects [*Gegenstände*] of number theory are for me—in direct contrast to Dedekind and Frege—the signs themselves, whose shape [*Gestalt*] can be generally and certainly recognized by us—independently of space and time, of the special conditions of the production of the sign, and of insignificant differences in the finished product. The solid philosophical attitude that I think is required for the grounding of pure mathematics—as well as for all scientific thought, understanding and communication—is this: In the beginning was the sign.

At this point we see in what one of the novelties of Hilbert's approach consists. Rather than appealing to the totality of Russellian individuals he simply singles out a subclass of them: the signs.[34] On this basis he then goes on to provide his account of mathematics.

Notice that even in Hilbert's case we have a process of de-ontologization of mathematics quite analogous to that found in Russell and emphasized in Behmann. One can provide a foundation of mathematics without making the problematic assumption that the numbers—conceived as independently existing entities—exist as abstract objects. Of course, there is also a major difference. Hilbert puts great emphasis on the surveyability of the objects and in general on the combinatorial (finitistic) operations that we must be able to carry out on them. This demand, which is the true core of Hilbert's finitism, is absent in Behmann's account of the individuals. There is one last issue I would like to raise in connection to the nature of Hilbert's finitism and a possible relation to Behmann's work. We have seen that for Behmann the individuals are given through empirical perception. What is Hilbert's position on this issue? This is a complicated issue. In Mancosu (1998c, pp. 168–71) I have argued that the intuition of concrete objects postulated in Hilbert in the early 20s as a precondition for logical thought should be understood as empirical perception and not as a form of Kantian a priori intuition. I also argued that it was only in the late twenties that Hilbert and Bernays started appealing to a form of pure intuition as the grounding for finitism. Of course, this is not the place to rehearse the evidence for my claim, which relies of a detailed textual reading of Hilbert's and Bernays' texts from the early twenties.[35] [[On this matter, see the Addendum to Chapter 2 in this volume.]] If my analysis of the nature of finitist intuition in the early twenties is correct then this would point to an interesting convergence of views on the nature of intuition between Hilbert and his student Behmann.

There is of course much more that could be said about specific connections between Behmann's work and Hilbert's program. My intention here was only to prepare the ground for further analyses. For instance, one could develop the connection between Behmann and Hilbert in a different direction. In 1922 Behmann wrote an important *Habilitationsschrift* on the decision problem [*Entscheidungsproblem*]. This was connected to ideas formulated in Hilbert's Paris talk in 1900. After the formalization of logic given in PM the project could finally be tackled at a technical level. And Behmann is, together with Bernays and Schönfinkel, one of the most engaged in looking for a solution to the problem, referred in the twenties as "the main problem of mathematical logic." The history of the attempts to solve the decision problem in the 1920s still awaits a devoted historian but we know the cast of actors: Behmann, Hilbert, Schönfinkel, Bernays, Herbrand, Ramsey, Ackermann. It is in a lecture by Behmann "Entscheidungsproblem und Algebra der Logik," delivered in Göttingen on May 10, 1921, that the main ideas for this set of investigations are first spelled out in Hilbert's school. Behmann introduces the term *Entscheidungsproblem* and speaks about it as an *"übermathematisches Problem"* (Behmann 1921, p. 2).[36] Although a detailed analysis of the technical results is here out of the question I would like to quote from this lecture to indicate how important themes which are associated with Hilbert's school seem to occur here for the first time and foreshadow important developments such as the analysis of decidability by means of computability theory:

As is well known, symbolic logic can be *axiomatized*, that is, it can be reduced to a system of relatively few basic formulas and basic rules, so that *proving theorems* appears as a *mere calculating procedure*. One has merely to write new formulas next to given ones, where the rules specify what can be written in every case. Proving has assumed the *character of a game*, so to speak. It is just like in chess, where one transforms *the given position into a new one* by moving one of one's own pieces and removing one of the opponent's pieces if appropriate, and where moving and removing must be allowed by the rules of the game.[37]

But this comparison makes it blatantly clear that the *standpoint of symbolic logic* just outlined *cannot suffice* for our problem. For that standpoint tells us, like the rules of chess, *only what one may do but not what one should do*. In both cases, what one should do remains a matter of *inventive* thinking and fortunate *combination*. We require much more: not just every single allowed operation, but the *process of calculation* itself is specified by rules, in other words, we require the *elimination of thinking in favor of mechanical calculation*. When a logical mathematical proposition is presented, the procedure we ask for must give complete instructions on how *to ascertain by a deterministic computation after finitely many steps* whether the given proposition is true or false. I would like to call the problem formulated above the *general decision problem*.

It is of fundamental importance for the character of this problem that *only mechanical calculations* according to given instructions, without any thought activity in the stricter sense, are admitted as tools for the proof. One could, if one wanted to, speak of *mechanical* or *machinelike* thought (Perhaps one could later let the procedure be carried out by a machine).[38]

Behmann then went on to describe his work as a reevaluation and a unification under a new viewpoint of the tradition of the algebra of logic (Boole, De Morgan, Schröder, Peirce) emphasizing at the same time that none of the mathematicians active in that tradition had formulated the *Entscheidungsproblem* although they had as one of their chief goals the construction of algorithms in algebra for the solution of equations. However, Schröder's notation is considered completely unusable [*vollständig unbrauchbar*] and thus Behmann frames his presentation in the notation of Russell and Whitehead with some modifications.

In conclusion, we have seen that Behmann's work contains several ideas which are connected to Hilbert's foundational views. Of course, one would need to delve deeper into the connection between Behmann and Hilbert and the role Behmann played in reevaluating the tradition of the algebra of logic within Hilbert's school. However, this could not be achieved within the scope of this paper. I will have reached my goal if I have convinced the reader that Behmann's work up to 1921 contains a wide variety of aspects which are central for a proper understanding of the context within which Hilbert's program developed and that they deserve the attention of the historian of logic and the foundations of mathematics.

4

THE RUSSELLIAN INFLUENCE ON HILBERT AND HIS SCHOOL

4.1 Introduction

In the summer of 1914 Norbert Wiener was studying in Göttingen. He had gone there following Russell's advice. He attended lectures by Hilbert in mathematics and by Husserl in phenomenology. However, his impression of the level of foundational studies in Göttingen was rather bleak:

Symbolic logic stands in little favor in Göttingen. As usual, the mathematicians will have nothing to do with anything so philosophical as logic, while the philosophers will have nothing to do with anything so mathematical as symbols. For this reason, I have not done much original work this term: it is disheartening to try to do original work where you know that not a person with whom you talk about it will understand a word you say. (Wiener to Russell, June/early July 1914, Russell archive, McMaster)

In his reply of July 12, 1914 Russell wrote:

I am interested in the account of your work in Göttingen, but I am disappointed about symbolic logic; I had hoped that the mathematicians there had begun to see the point of it. (Russell to Wiener, Russell archive, McMaster).

The above exchange cannot fail to surprise the reader who grew up with the idea of Göttingen as the hotbed of foundational activity in Germany in the first thirty years of the

---

I would like to thank Wilfried Sieg, Chris Pincock and Richard Zach for many useful comments. Many thanks to the curators of the following archives for their help in making available to me the unpublished material used in this paper: Russell archive, McMaster University, Hamilton; Dingler archive, Hofbibliothek, Aschaffenburg; Behmann archive, Erlangen; Weyl archive, ETH, Zurich; Bernays archive, ETH, Zurich; Hilbert archive, Universitätsbibliothek Göttingen; Zermelo archive, Freiburg, I.B; Paul Hertz archive, University of Pittsburgh.

twentieth century. After all, it was in Göttingen that the first chair of mathematical logic was established (in 1907) and which could count at some point or other during 1900–14 among its active faculty people like D. Hilbert, E. Husserl, F. Bernstein,[1] K. Grelling,[2] L. Nelson,[3] H. Weyl,[4] E. Zermelo,[5] and visitors such as L. Chwistek.[6]

In what follows I will try to reconcile Wiener's report with the status of foundational research in Göttingen by providing an account of the fortunes of foundational work in mathematics in Göttingen until the early twenties. My first aim in doing so will be to clarify the influence exercised by Russell's thought in Göttingen after the publication of the monumental *Principia Mathematica* until Hilbert's first statement of his program in 1922. I will show that after a period of intense foundational work, culminating with the departure of Zermelo and Grelling in 1910, we witness a reemergence of interest in foundational issues towards the end of 1914, that is only a few months after Wiener's visit to Göttingen. It is this second period of foundational work that is my specific interest. However, I will begin by saying something about how Russell's work influenced the Göttingen researchers before the appearance of *Principia Mathematica*.

4.2 The Russellian Influence from the *Principles* to *Principia* (1903–1910)

In the book *Hilbertprogramm und kritische Philosophie*, V. Peckhaus has given a detailed analysis of the relationship between Hilbert's foundational work (up to 1917) and the philosophical and mathematical activity in Göttingen. At various points of the exposition he touches on Russell's influence in Göttingen, especially during the period 1903–10.

When we look at Hilbert's published and unpublished works during this period, the Russellian influence can mainly be detected at the level of the role played by the antinomies in accounting for Hilbert's foundational shift in 1905 and the following work on the foundations of set theory.[7] This influence can be adduced to account, at least partially, for some of the major features of Hilbert's foundational work after 1905 (Peckhaus 1995b, pp. 72–78):

1. Inclusion of mathematical logic as a mathematical subject in the curriculum;
2. Inclusion of set theory in the axiomatic program;
3. Inclusion of logic in the axiomatic program;
4. Development of logical calculi.

However, the Russellian influence at this stage should not be overemphasized. In fact, there is nothing at all on Russell in Hilbert's unpublished lecture notes of 1905 and Hilbert is basically silent on foundational questions until 1917. Moreover, the lectures on mathematical logic given by Zermelo in Göttingen in 1908 do not build on Russell and rely mostly on the symbolisms of Peano, Frege, and Schröder.

Nonetheless, the discussion of Russell's paradox was alive during this period. For one thing, knowledge of Russell's work was certainly disseminated by the polemic with

Poincaré in the *Revue de Metaphysique et de Morale*. Zermelo often refers to this exchange in his published work. In 1905, W. H. Young (who lived in Göttingen from 1899 to 1908) talked about Russell's *Principles of Mathematics* to the Mathematisches Kolloquium. Moreover, several philosophers and mathematicians tied to Hilbert—let us mention Zermelo, Nelson, and Grelling—extensively dealt with the antinomies.[8] In particular, we should mention here the work by Nelson and Grelling "Bemerkungen zu den Paradoxieen von Russell und Burali-Forti" (1908) which contains a discussion of Russell's solution by means of the no-class theory formulated in Russell (1906a).[9]

Grelling found Russell's work of great value and in 1909 he wrote to Russell—whom he had met in Rome in 1908—asking for permission to translate the *Principles*.[10] Meanwhile, Chwistek was visiting Göttingen and he also wrote to Russell in 1909 sending him an essay which was based on Russell's 1908 paper on the theory of types.[11] The departure of Zermelo, Grelling, and Chwistek from Göttingen in 1910 was to leave a gap in foundational activities until late 1914.

4.3 Logic and Foundations of Mathematics in Göttingen from 1910 to 1917

If we look at the development of logic and the foundations of mathematics in Hilbert's thinking and in Göttingen in general in the early part of the 1910s we find very little of interest. At the time Hilbert seems to have been very busy with problems in the foundations of physics (see Corry 1997). Moreover, the Göttingen Mathematical Society does not list in its very active Colloquium any talks on the foundations of logic and mathematics in the period 1910–13. The situation changes suddenly in 1914.[12]

In a letter to Dingler at the end of 1914 Hilbert reports a discussion on *Principia* which took place in the Mathematisches Kolloquium in Göttingen and mentions his closest collaborators on foundational issues:

Just before the break we had there [Mathematical Colloquium] a thorough discussion on the big three volume work by Russell, which was extremely lively and instructive. In particular Mr Grelling, Hertz and Bernstein are those here that are interested in the epistemological aspects of mathematics. (Hilbert to Dingler, December 26, 1914)[13]

Kurt Grelling was for a short time back in Göttingen in 1914. I have already mentioned Grelling's correspondence with Russell and I should add that in 1910 he wrote to Russell from Berlin for some clarifications on parts of *Principia*. On that occasion he mentioned that he was writing an essay on the theory of types. There are however no publications by Grelling on Russell until the end of the 1920s.[14] Paul Hertz was mainly interested in physics but in the 1920s did logical work of importance foreshadowing later developments by Gentzen.[15] Finally, Bernstein had been a doctoral student of Hilbert (1901) and was interested in set theory.[16]

In the list of lectures from 1914 to 1917 of the Mathematisches Kolloquium we catch a glimpse of the renewed interest in foundational questions:

February 17, 1914, D. Hilbert, *Axiome der ganzen Zahlen*
December 1, 1914, H. Behmann, *Über mathematische Logik* [see Behmann 1914]
December 18, 1914, F. Berstein and K. Grelling, *Über mathematische Logik. Ergänzungen und genauere Ausführungen zum Referat vom 1. Dezember*
February 16, 1915, F. Bernstein, *Über J. Königs "Neue Grundlagen der Logik, Arithmetik und Mengenlehre"*
November 7, 1916, E. Zermelo, *Über einige neuere Ergebnisse in der Theorie der Wohlordnung*
November 14, 1916, H. Behmann, *Die Russell-Whiteheadsche Theorie und die Paradoxien* [see Behmann 1918]
July 3, 1917, H. Behmann, *Die Russell-Whiteheadsche Theorie und die Grundlagen der Arithmetik* [see Behmann 1918]
July 10, 1917, H. Behmann, *Die Russell–Whiteheadsche Theorie und die Grundlagen der Arithmetik (Schluß)* [see Behmann 1918]
July 17, 1917, F. Bernstein, *Geschichte der Mengenlehre*
July 24, 1917, F. Bernstein, *Geschichte der Mengenlehre (Schluß)*

At the time of Hilbert's letter to Dingler, Heinrich Behmann was still a graduate student.[17] However, he was soon to become a central player in the development of logic and the foundations of mathematics in Hilbert's circle. I will try to spell out how knowledge of *Principia Mathematica* (PM) was acquired in Göttingen, thereby shedding light on the important problem of the relationship between Russell's logicism and Hilbert's program. I shall begin with Behmann's first lecture to the Mathematisches Kolloquium.

4.4 Behmann's 1914 Lecture on *Principia Mathematica* in Göttingen

The first official presentation of the system of *Principia Mathematica* in Göttingen was given by Behmann on December 1, 1914. The lecture "Über mathematische Logik" was advertised in the *Jahresbericht* as "offering some new thoughts from the logical investigations set out in the '*Principia Mathematica*' by Russell and Whitehead" ("Der Vortrag bringt einige neue Gedanken aus den in den 'Principia mathematica' von Russell und Whitehead niedergelegten logischen Untersuchungen"). Since the text of this lecture is extant in the Behmann *Nachlaß* we can get an accurate idea of what aspects of *Principia* were emphasized in this first presentation to the mathematical community in Göttingen. I should also point out that on December 14, 1914, Bernstein and Grelling gave a follow up talk on *Principia*.

Behmann begins by focusing on what is meant by mathematical logic and points out that the expression 'mathematical logic' can be used to talk about two different traditions. The first tradition constructs logic by means of mathematical techniques. This is the tradition of algebra of logic which pursues a 'mathematics of logic' [*Mathematik der Logik*] and has as its foremost representatives Boole, Schröder, and partly Peano. The second tradition is associated to the names of Bolzano, Frege, and Russell. Its goal is to analyze the role played by logic in the construction of mathematics. The distinction was quite standard in Göttingen; it is for instance emphasized at the very outset of Zermelo's unpublished 1908 lectures in mathematical logic. Behmann points out that although Russell belongs squarely in the second tradition his work also accounts for the results obtained in the "algebra of logic" tradition. Thus, in his opinion, *Principia Mathematica* represents the first unification of these separate traditions in mathematical logic. Behmann emphasizes the importance of using a symbolic language and remarks that the approach taken in *Principia* is axiomatic, resting as it does on primitive propositions and primitive ideas.

The specific description of the Russellian project begins with the presentation of the concepts of individual, proposition and propositional function. Behmann then presents the type theoretic solution to the paradoxes by describing the vicious circle principle and the functional hierarchy. Individuals are defined as "everything that in reality is directly given to us, thus not, say, things like sets or similar conceptual entities [*Gedankengebilde*]". This characterization is important in that it already foreshadows the nominalist interpretation of the Russellian project given by Behmann in his dissertation.

The description of the concept of propositional function emphasizes the difficulty of mastering the multifarious contexts in which it occurs. Moreover, dealing with propositional functions has often led to paradoxes. Behmann uses the liar paradox to motivate Whitehead and Russell's vicious circle principle and the type theoretic solution offered in *Principia*. He also points out that the type theoretic structure is not linear but rather complex, due to the appearance of orders within types, and that forming a clear idea of what this functional hierarchy looks like is one of the greatest difficulties one encounters in trying to understand *Principia*.

As a critical point, Behmann wonders whether the restriction offered in *Principia* for avoiding the paradoxes might not be too drastic, since sentences like 'All sentences are either true or false' are rejected as meaningless due to the type theoretic constraints. The motivation for such a move is that the expression 'all sentences' presupposes the totality of all sentences. But a sentence that presupposes this totality cannot belong to the totality, otherwise it would presuppose itself. However, Behmann thinks that the objection can be removed by reinterpreting the sentence as 'if p is a sentence, then p is either true or false' and "from this one sees that the sentence in reality does not represent a sentence about 'all sentences', i.e., the totality of sentences, but rather it has the individual sentences as objects". It should be remarked that this interpretation of the universal quantifier is found in the writings of Weyl, Bernays and Hilbert in the twenties as a way to show that universal quantification need not commit us to infinite totatilities. Whereas common sense would push us in the direction of accepting the above sentence as meaningful, type theory forbids us from doing so. Behmann takes no position on the issue, which he takes as a major open problem.

The seriousness of the paradoxes however is displayed not by the liar but by the set-theoretic paradoxes, such as Russell's and Burali-Forti's, the latter of which has become a 'Schreckgespnst' of set theory. The solution of these paradoxes must be seen as a

"Lebensfrage der Logik und von allem der Arithmetik". The solution given by Russell is captured in the claim that "there are no sets at all". This is in essence, Behmann continues, what is known as the no-class theory. It is irrelevant here to discuss whether this is a faithful reading of *Principia*. More important for our purposes is to emphasize the explication provided by Behmann of the no-class theory. This is done by comparing the role of classes in *Principia* with that of ideal elements in other branches of mathematics. Behmann uses the example of points at infinity in projective geometry. The use of points at infinity is misleading if one thinks that points at infinity denote objects. However, no problems can arise if the points at infinity are taken for what they are, i.e., mere words:

Nevertheless, the correct solution of all these geometrical inconsistencies, once it is given, is not difficult to comprehend. To wit, it simply states that the ideal elements [*die idealen Elemente*] are not objects of geometry in the proper sense, but are first of all only words. As such they are parts of ways of speaking which one uses in geometry to bring a certain class of propositions into a form which is as simple as possible.[18]

Thus, for instance, the sentence "the lines a and b intersect at infinity" means no more and no less than "the lines a and b are parallel". And in exactly the same way to every sentence mentioning ideal elements corresponds a logically equivalent sentence in which ideal elements are not mentioned. ("In genau derselben Weise läßt sich nach bestimmten Vorschriften jedem Satz über uneigentliche Elemente ein logisch vollkommen gleichwertiger an die Seite stellen, in dem die uneigentlichen Elementen überhaupt nicht vorkommen.") In this way the whole question about the existence of the ideal elements as geometrical objects is completely bypassed.

By generalizing the above reflections one obtains the explication of Russell's and Whitehead's solution, which consists in treating classes as the ideal elements in projective geometry.

Given Hilbert's emphasis on ideal elements in the 1920s, it is striking to see that an extensive discussion of the role of ideal elements in Hilbert's circle appeared for the first time in connection to an explanation of the basic intuition of *Principia*. I will not emphasize Behmann's description of how the strategy is supposed to work but I will mention one more point in Behmann's talk that is related to Hilbert's early foundational work. This relates the axiom of reducibility—which states that any class definable by a propositional function can be defined by a propositional function of the lowest level (i.e., a predicative function)—to completeness axioms in Hilbert's sense:

The axiom thus states that in this one order there already are enough functions to define all possible classes; it can therefore be viewed as a kind of completeness axiom for the predicative functions.[19]

Hilbert was particularly interested in this interpretation of the reducibility axiom, which indeed will be proposed again in Behmann's 1918 dissertation to which we will return in Section 4.7.

4.5 Hilbert on Type Theory and *Principia Mathematica*

We have seen that interest in the system of PM was alive in Göttingen since 1914. This influence reaches its peak around 1917–18. First of all one should consider that Hilbert himself was quite attracted to the logicist solution to the foundations of mathematics and for a while seems to have considered the logicist solution valid. Moore (1997, p. 85) and Sieg (1999, pp. 3 and 11), have no doubts that Hilbert defended a logicist point of view in 1917–18 although he later became disenchanted with Russell's solution. In "Axiomatisches Denken" we read:

> But since the examination of consistency is a task that cannot be avoided, it appears necessary to axiomatize logic and to prove that number theory and set theory are only parts of logic. This method was prepared long ago (not least by Frege's profound investigations); it has been most successfully explained by the acute mathematician and logician Russell. One could regard the completion of this magnificent Russellian enterprise of the axiomatization of logic as the crowning achievement of the work of axiomatization as a whole. (Hilbert 1918a, p. 412, Ewald 1996, p. 1113)

During this period Hilbert supervised Behmann's dissertation Behmann (1918) and also encouraged Bernays to write a *Habilitationsschrift* on topics related to the logical system of PM Bernays (1918). However, whereas Bernays' 1918 Habilitation centers on logic, Behmann's work skips over the logical parts of *Principia* to focus instead on the mathematical developments more immediately related to the construction of arithmetic and of Cantorian set theory. Behmann also took, at least up to 1922, the logicist solution to the foundations of mathematics to be successful.[20]

But before looking in detail at Bernays' and Behmann's work on PM let us try to point out the extent of the Russellian influence on Hilbert's texts. In the 1905 and 1910 lectures there are only generic discussions of the paradoxes and one cannot talk of an explicit Russellian influence. For the period before the 1917/18 lectures the situation is well summarized by Sieg:

> The notes for lectures Hilbert gave before the winter term 1917/18, even for those of the immediately preceding summer term 1917, do not contain any reference to *Principia Mathematica* nor any hint of a Russellian influence. There is only one exception I discovered; in his lectures Probleme und Prinzipien der Mathematik given in the winter term 1914/15, Hilbert mentions Russell and remarks briefly that type theory contains something true, but that it has to be deepened significantly. Here is a real gap in our historical understanding... (Sieg 1999, p. 14)

Sieg also published the correspondence between Hilbert and Russell concerning a possible invitation for Russell to come and lecture in Göttingen. The visit never materialized. Hilbert's fascination with the logicist program seems to have lasted only until 1920. In the lectures given in the summer semester 1920 Hilbert expressed serious doubts concerning the possibility of carrying out the logicist project:

> The goal or reducing set theory, and by this also the usual methods of analysis, to logic is not achieved as of today and it is perhaps not achievable. (Hilbert 1920b, p. 33)[21]

Of course, this is not to say that in this way the Russellian influence on Hilbert disappeared. In 1922 after criticizing the shortcomings of the logicist program Hilbert was quick to acknowledge his debt to the works of the logicists "for the proper development of the so-called logical calculus, whose basic ideas prove more and more to be an indispensable tool in logico-mathematical investigations" (Hilbert 1922c, p. 202). This position towards Russell and logicism remains constant in all published works by Hilbert in foundations of mathematics in the 1920s.

At this point we have to analyze the 1917/18 lectures by Hilbert, which contain the first explicit discussion of type theory in Hilbert's lectures. Zach (1999) has shown, by a detailed comparison of Bernays' notes taken during the lectures and the final typescript of the 1917/18 lectures, that the section on type theory is actually due to Bernays. Accordingly, the next section is devoted to Bernays.

4.6 Bernays on Type Theory and *Principia Mathematica*

Although we may assume that Bernays must have been familiar with Russell's works while a student in Göttingen (1908–12)—perhaps through Nelson and Grelling and his strong ties to the Neue Fries'schen Schule—his detailed acquaintance with *Principia Mathematica* seems to date from the time of his writing up the 1917/18 lectures. Indeed, in a letter to Russell dated April 8, 1920 he writes:

As you probably know Professor Hilbert—whose assistant I have the honour to be—has been working for some years quite actively on problems of mathematical logic. I have been spurred to a more precise study of these questions, and especially of your theory of mathematical logic, by a course which Professor Hilbert has delivered on this matter during the Winter [Semester] 1917/18.[22]

Much later in his life, in an interview on August 27, 1977 (quoted in Zach 1999, p. 347), Bernays also pointed out that the Russellian treatment of generality was not to his taste and for that reason the 1917/18 lectures have a formalization of the quantifier axioms without free variables. However, one of the axiomatizations of the propositional calculus offered in the 1917/18 lectures is closely related to that of *Principia*. It is thus clear that the reception of the Russellian work in Göttingen was far from uncritical. I will now discuss the section on type theory in the 1917/18 lectures and then move on to Bernays' achievements in his *Habilitationsschrift*, which deals with the propositional logic of *Principia Mathematica*.

4.6.1 Prinzipien der Mathematik *(WS 1917/1918)*

The importance of this set of lectures for the history of mathematical logic has been recently emphasized by Sieg (1999, 2000) and Moore (1997). I will only deal with them from the point of view of the presentation of type theory they contain. As I mentioned, Zach has shown that the part on type theory of the 1917/18 lectures (corresponding to pp. 219–45) were not part of the oral delivery and thus were added by Bernays in his preparation of the final text. He also argued that Bernays' *Habilitationsschrift* was written after the 1917/18 lectures.

For the goal of formalizing logical inference one could be satisfied with the propositional and predicate logic developed on pp. 63–188 of the lectures. It is only when the foundations of mathematics come into play that more needs to be done and the extension of the calculus becomes important:

> Not only do we want to develop individual theories from their principles in a purely formal way, but we also want to investigate the foundations of the mathematical theories and examine what their relation to logic is and how far they can be built up from purely logical operations and concepts; and for this purpose the logical calculus is to serve as an auxiliary tool.[23]

The extension in question will allow the use of propositions and function variables.[24] A few examples introduce quantification over propositions and functions. For instance, Bernays gives the second-order formalization of mathematical induction. This amounts to treating propositions and functions (predicates) "as if they were individuals". Another example is the definition of identity by means of second order quantification (p. 191). One of the consequences of treating functions and sentences as objects is that now there arises the possibility of having functions of propositions and functions of functions. The necessity of introducing functions of functions is explained in reference to the concept of number. Bernays specifies that the numbers are not objects but rather properties of concepts, i.e., predicates of predicates.

> The numbers appear accordingly as properties of predicates and in our calculus a number is represented as a certain function of a variable predicate. The importance of this representation of the numbers consists in the fact that the functions of predicates which constitute the numbers can be expressed completely by means of the logical symbols, and in this way it becomes possible to incorporate number theory into logic. (Hilbert 1918c, pp. 192-93)[25]

Since for the mathematician it is more common to think of numbers as properties of sets, the lectures proceed to show that considering numbers as properties of predicates amounts to the same thing.[26] By means of the correspondence between sets of sets and functions of predicates the lectures discuss how to translate set theory into the formal language introduced earlier. This is done by appealing to two examples: the union of a set of sets and the intersection of a set of sets.

After this the rules for second-order quantificational inference are introduced and several examples of derivations are provided (pp. 200–08). At this point Bernays mentions that the extension of the calculus would satisfy our needs were it not for the emergence of logical paradoxes. Bernays discusses in detail Russell's paradox of self-predication, a version of the liar paradox and the paradox of the least number which cannot be named by any human being in the 20th century. The logical paradoxes show that something has gone wrong in the extension of the calculus.[27]

The diagnosis of what has gone wrong is that a vicious circle is taking place in the extension of the calculus:

> In the original method of the function calculus we have assumed from the beginning a system, or several systems (sorts), of objects as given. The manipulation of variables (especially with

the parentheses [universal quantifier]) received its logical meaning through the relationship to such totalities of objects. The extension of the calculus consisted in considering the propositions, predicates and relations as types of objects and accordingly in allowing symbolical expressions whose logical interpretation required a reference [*Bezugnahme*] to the totality of propositions or to the totality of functions. This way of proceeding is objectionable in that those expressions which receive their content only through reference to the totality of propositions or functions are in turn included in the [totality of] propositions or functions. On the other hand, in order to be able to relate to the totality of propositions or functions we must consider the propositions and functions as determined from the start. Thus, there is here a type of logical circle and we have grounds for the assumption that this circle constitutes the cause for the emergence of the paradoxes.[28]

A theory of *Stufen* is then introduced as the only way to preserve the possibility of quantifying over propositions and functions without allowing the objectionable totalities of propositions and functions. At the first level we only quantify over objects; this gives us the theory of the first level. At the second level we quantify over objects of the first level and the functions and propositions that occur in the theory of the first level; moreover, we consider the relationships that hold between objects of this extended domain. And so on for every n. Formally the idea can be implemented by the introduction of indexes to propositions and function signs, which indicate the corresponding domain of application of the signs. Then Bernays gives a set of rules for logical inference in this calculus (simply an extension of his rules for first and second order logic, but with a restriction of the second-order calculus). Bernays then shows how the paradoxes are avoided in this framework.

Now the question becomes whether this restriction to types might not be too strong, i.e., whether essential inferential patterns required for the foundation of mathematics might not have been excluded by the new restrictions.[29] A discussion of identity and of Cantor's theorem (for each level n) in this context confirms that the restriction imposed by this theory of types does not allow for certain importance inferences in our new calculus.[30] In order to overcome these limitations the axiom of reducibility is introduced:

For every functional expression which occurs in the calculus of types there is an equivalent predicative expression.[31]

The discussion of the methodological significance of the axiom is of great interest. Before looking at it in detail let me conclude the exposition of the lectures. Bernays goes on to show the advantages of the assumption of the reducibility axiom by formalizing, among others, the following theorems:

1. Cantor's theorem;
2. Every bounded set of rationals has a least upper bound (at the same level).

The lectures conclude by claiming that the introduction of the reducibility axiom is the appropriate tool for the development in the logical calculus of the foundations of mathematics.[32]

Let us go back to the discussion of the methodological significance of the assumption of the axiom of reducibility. Whereas the type theoretic construction—without the reducibility axiom—will only allow functions which can be explicitly defined (or constructed), the introduction of the reducibility axiom requires that we consider certain predicates and relations as existing in themselves so that their manifold depends neither on an actually given definition nor in general on our definitional possibilities:

> Such a way of proceeding seems at first very strange but it is unavoidable in so far as we aim at developing the foundations of set theory and analysis from the system of functions of our calculus. For, so long as we insist that every function must be given explicitly, the functions of our calculus can only constitute a countable totality and thus cannot suffice to represent the uncountable totalities which occur in those mathematical disciplines.[33]

This clarification of the methodological meaning of the reducibility axiom and the lengthy development of the proof of the least upper bound principle in the lectures leave no doubt that this part of the lectures must be seen as a reply to Weyl's project in *Das Kontinuum*. Indeed, the text of Bernays' lecture on Weyl's predicative system given in Göttingen in 1917 gives us the right categories for interpreting the methodological significance of the reducibility axiom according to Bernays.

4.6.2 Bernays' Lecture on Weyl (November 20, 1917)

On November 20, 1917, Bernays reported on Weyl's predicative project of *Das Kontinuum* to the Göttingen Mathematical Society. I will not give here a detailed account of this lecture but only mention some of the central key concepts that are relevant to the previous section. Bernays begins by stating that Weyl criticizes analysis on account of a vicious circle present in its procedures. The vicious circle consists in defining real numbers by quantifying over the totality of real numbers:

> The result of such a definition depends on what real numbers there are; on the other hand, the result of the definition contributes to determine what real numbers there are.[34]

In order to evaluate Weyl's criticisms it is important, according to Bernays, to look at the several methodological standpoints that one might possibly take with respect to the foundations of analysis. These alternative standpoints, according to Bernays, turn out to be the same in mathematical applications. In particular, the alternative standpoints are not affected by whether we think of real numbers as given by numerical fundamental sequences or by Dedekind cuts. However, the difference consists in the logical interpretation [*logische Deutung*] of the alternative procedures. Bernays describes three procedures:

1. The first standpoint consists in accepting as a real number anything that is given by a cut (say by the condition $x^3 < 2$). The problem with this first method is that it does not delimit at the outset the domain of the real numbers ("der Begriff der reellen Zahlen wird nicht 'bestimmt' umgrenzt"). For this reason we truly have a vicious circle here, since real numbers are defined by partitions which in turn are defined by reference to what real numbers [partitions] exist possessing a specified property. But, according to Bernays, one does not always follow this standpoint. Rather one follows standpoint 2.

2. Here one simply assumes that the concept of partition of rational numbers (resp. of a fundamental sequence of rational numbers) is precisely delimited with respect to

content and extension. In this case the definition of real numbers as a partition of rational numbers which possesses the cut property (resp. as a monotone bounded sequence of rational numbers) contains no circularity. But Weyl, Bernays adds, rejects this procedure on different grounds than the one adduced for stategy 1. Weyl's argument here is that the concept of partition or of sequence is dependent on the concept of set. The paradoxes of set theory have shown that relying on the concept of set is problematic. Thus, Weyl concludes, the concept of set of rational numbers is also unreliable. The objection by Bernays here is that from the fact that the concept of set is unreliable it does not follow that the concept of set of natural numbers is also unreliable. Moreover, mathematical experience tells against the Weylian argument. No area in mathematics has been so deeply investigated as real analysis. And in all these investigations not a shadow of a discordance has ever been found.[35] What the paradoxes of set theory show, Bernays concludes, is that we do not understand with sufficient clarity the grounds of this harmony. And it does not suit the demand of mathematical rigor to have a theory like analysis based on assumptions whose correctness is not fully understood.

3. The third possibility for grounding analysis is the axiomatic one. This is not open to the objections raised in strategies 1 and 2 but Weyl does not consider it. One characterizes the system of the real numbers as a system of things that are connected by several relations. One then characterizes the properties of this domain by axioms. In place of the definition of real numbers by Dedekind cuts (or fundamental sequences) we have the cut-axiom or the equivalent completeness axiom (in connection with the Archimedean axiom). The fundamental sequences of the cuts serve then to define individual real numbers but not the concept of real number. Rather, a real number is now what belongs to the system of things. This type of foundation is common to several mathematical disciplines:

The same axiomatic foundation is present in comprehensive application when number theory, analysis, and set theory are developed from a homogeneous set of things. Such a construction [*Aufbau*] is found on the one hand in Zermelo's set theory in which the basic relationship represents the membership of a thing to a set and on the other hand in Russell's logic, which also presents an axiom system with the predicative relationship as basic relationship. (We have to abstract here from the usual contentual understanding of these relationships) (Although Russell does not have the pure axiomatic tendency).[36]

This third way of proceeding is thus unobjectionable although the task of a consistency proof still remains.

The reason why, in conclusion, Weyl is against the axiomatic method is that he attempts a constructive development of mathematics. He would like, Bernays adds, pure mathematics to create the objects of which it treats and not to make statements about an unknown system of objects. The last part of the lecture explains this opposition by characterizing the two methods as 'constructive' and 'existential' and provides a clarification by appealing to geometry. In other lectures and publications by Hilbert and Bernays these are called the 'genetic' and 'axiomatic' methods.

4.6.3 The Methodological Significance of the Reducibility Axiom (Lectures from SS 1920, WS 1921/1922 and Grundzüge der theoretischen Logik (1928))

The reducibility axiom, by postulating a set of entities and relationships independent of construction, defies the constructive approach and leads us back to the existential or axiomatic approach. This shows that the construction of analysis, through the assumption of the reducibility axiom, given in the 1917/18 lectures was consciously an axiomatic one. In later lectures the same starting point leads, as Sieg (1999) points out, to an explicit rejection of logicism.

In the lectures given by Hilbert in SS 1920 (again available in notes written up by Bernays and Schönfinkel) Hilbert goes over the topic of reducing talk of sets to talk of predicates, as it occurs in Russell, and notes that in principle one does not need, axiomatically speaking, to postulate at the outset all the predicates but that in fact one can get by with postulating only one predicate for a class of equivalent predicates. Indeed, Hilbert claims, that's exactly Russell's procedure (Hilbert 1920b, p. 29).

After going over again the reduction of set theory to predicate theory—by showing how to express the power set and union set by means of predicates—Hilbert claims that if all we wanted to do was to transform set-theoretic terminology into predicate terminology we could stop here. However, the transformation of set theory into predicate theory was done with a specific goal in mind, that is reducing the system of axiomatic set theory to logic. But here is where we meet difficulties. Indeed, the predicate versions of the power set and the union set require an existential quantification over the set of predicates. Now in set theory the existential quantification always refers to a given domain B, postulated at the start. In logic one could also think of the domain of predicates as being a domain; but this domain, according to Hilbert, cannot be postulated as being given at the outset. Rather, the predicates must first be constructed by means of logical operations and only later—through the rules of logical construction—the domain of predicates becomes definite. Thus, Hilbert concludes that in the rules for the logical construction of the predicates reference to the domain of predicates cannot be allowed on pain of a vicious circle (Hilbert 1920b, p. 31).

Thus, Hilbert agrees with Russell and Weyl that there is a difficulty here. Of course, he recognizes that Russell and Weyl take different ways out. Russell postulates by means of the reducibility axiom the existence of a predicate satisfying the existential claims, as they appear for instance in the union set. But in this way "Russell returns from constructive logic to the axiomatic standpoint" (Hilbert 1920b, p. 32). By contrast, Weyl follows a constructive method which however is unable to account for analysis. Hilbert concludes that

The goal of reducing set theory, and by this also the usual methods of analysis, to logic is not achieved as of today and it is perhaps not achievable. (Hilbert 1920b, p. 33)[37]

His conclusion is that whether or not mathematics can be reduced to logic the axiomatic method cannot be replaced and maintains its value. Indeed, even Brouwer's and Weyl's investigations can be seen as axiomatic investigations on how a certain part of analysis can be developed from a weaker system of assumptions (Hilbert 1920b, p. 34).

Whereas the SS1920 lectures still left open whether or not the reduction of mathematics was still possible by the time of 1921/22 lectures (pp. 69–100) the tone is downright negative. Indeed, in introducing on p. 76 the discussion of Russell and Whitehead, Hilbert already mentions what the conclusion will be: "It is shown that also here the goal of reducing arithmetic to logic is not obtained".[38]

Two types of problems emerge in attempting the reduction. The first one has do with the fact that the domain of individuals must be postulated to be infinite, otherwise the predicate for the numbers will turn out to be equivalent once they go beyond the (finite) cardinality of the domain of individuals. If one postulates the domain to be infinite at the outset then one has to give up the prospect of a logical foundation for the existence of an infinite set (Hilbert 1922b, p. 88). But this problem could be set aside if at least one could get analysis out of the logical theory (plus the assumption of an infinite set). The development of the foundations of real analysis follows now that given in the 1917/18 lectures. This leads to a discussion of the least upper bound principle and of the axiom of reducibility. In favor of the plausibility of the axiom it has to be said that it serves at several key points in the construction of mathematics and not just in the special case of the least upper bound principle. However, assuming the axiom of reducibility brings about a complete change in the presuppositions of our logical theory. The construction of the real numbers for instance begins with a set of individuals (say, rational numbers) and certain basic relationships on them. The further relationships and predicates should be derived constructively out of these. But when we come to apply the axiom of reducibility we have no guarantee that the relationships postulated at the outset satisfy its application. So, in each particular case we have to make sure that in addition to the basic relationships we also add all those needed to satisfy the axiom of reducibility. But how is this to be done in a logical-constructive fashion?[39] There is no other way than by postulating that the axiom of reducibility is indeed satisfied. But in this way we give up the constructive attitude and we are back to the axiomatic standpoint, "for a reduction [of arithmetic and analysis] to logic is given only in name".[40] The axiom of reducibility is plausible but the same plausibility can be ascribed to the inferences of analysis we were trying to ground in the first place. Thus, Hilbert concludes that transfinite logic is not able to give a safe foundation for arithmetic. Out of this impasse Hilbert was to develop his new foundations of mathematics.

Let me make a remark on how these reflections find their way in the first edition of *Grundzüge der theoretischen Logik* by Hilbert and Ackermann (1928). The exposition of the extended functional calculus and type theory found in chapter IV of the book depends not only on the 1917/18 lectures but follows closely also the 1921/22 lectures. For instance, the requirement that the domain be infinite (p. 88), if number theory is to be developed, is taken straight out of the later lectures and is absent from the 1917/18 lectures. Moreover, *Grundzüge* also mentions the axiom of infinity and connects the topic to Russell's *Introduction to Mathematical Philosophy* (German translation in 1923), and this explicit appeal to the axiom of infinity in connection to Russell was absent in both lectures. It is actually interesting to note the changes which occur in the exposition of 1928 but I will limit myself to the topic of reducibility. Although the presentation follows closely that given in the 1917/18 lectures (there is no corresponding part on type theory in the 1921/22 lectures) there are some interesting changes. First of all, Russell and Whitehead are quoted at the outset as the authors of the theory being discussed (surprisingly, this acknowledgement is missing in the 1917/18 lectures).[41] Second, whereas the 1917/18 lectures only spoke of *Stufen* now the development refers explicitly to the "Typentheorie oder den Stufenkalkül von Whitehead und Russell" (p. 99).[42] Finally, the final reflections on the reducibility axiom come from the 1921/22 lectures and reproduce its argument faithfully. However, the conclusion is sharpened. If we postulate that the collection of functions of the first level already includes all functions then developing a type theory would be, Hilbert and Ackermann conclude, "an unnecessary complication" (this might reflect the influence of Ramsey 1925). Thus, as in the 1921–22 lectures Hilbert's new foundational investigations are hinted at as a way to develop mathematics without the complications due to the reducibility axiom.

4.6.4 Bernays' Habilitationsschrift *(1918) and Correspondence with Russell*

Bernays was of course involved in all the philosophical reflections that we have described above. Moreover, he also made positive contributions to the axiomatic investigation of *Principia Mathematica*. They are contained in his *Habilitationsschrift* of 1918 (partially published in 1926). Of these results he informed Russell in correspondence. These technical achievements are well known and described elsewhere (see Kneale and Kneale 1962, and Zach 1999). I will present only some aspects of the correspondence with Russell and point out the main results.

1. Bernays showed the derivability of the associativity axiom in *Principia* from the other axioms of the propositional calculus.[43]

2. He also showed, by means of the technique of value assignments, the independence of the remaining axioms of *Principia*. He gives detailed proofs to Russell in a letter dated 19.3.21, which includes an attachment of 11 pages containing the proofs.[44]

3. The conclusion of the letter from 19.3.21 also points to another influence of Russell on the development of logic in Göttingen. Russell had informed Bernays about Sheffer's stroke. Bernays in turn informed the mathematical community in Göttingen about this development and as a consequence Schönfinkel developed in 1920 the predicate calculus by means of a single sign, $/^x$ (where $fx/^x gx$ means $(x).\neg f(x) \vee \neg g(x)$).[45] This was the beginning of combinatory logic. It is interesting here to note that in a letter from Behmann to Russell, Behmann credits Bernays with the idea of looking at $(x)..f(x). \vee .\neg p(x)$.[46]

4. Finally, I want to conclude this section on Bernays by pointing out that in a letter from Boskovitz to Russell, dated July 3, 1923, containing a list of corrections to the first edition of *Principia Mathematica*, Boskovitz mentions that Bernays had put him to

work on the issue of how far one can develop the foundations of set theory without the axiom of reducibility.[47]

Both Behmann and Boskovitz were thanked for their list of corrections in the preface to the second edition of *Principia*. However, in light of the above correspondence it is surprising that Russell seems to imply in the second edition of *Principia* that Sheffer's stroke was the most important development in logic to take place since the first edition.

4.7 Behmann's Dissertation

Behmann's work on *Principia* had been interrupted by the war but upon his return from the field he was able to successfully finish his dissertation *Die Antinomie der transfiniten Zahl und ihre Auflösung durch die Theorie von Russell und Whitehead* (1918). The original is still found at the Niedersächsische Universitätsbibliothek in Göttingen. This voluminous work, totalling 351 pages in two volumes, aimed at providing an introduction to the system of *Principia*. As I have devoted an entire article to Behmann's dissertation (see the previous Chapter) I will only give a succinct account here and draw attention to some aspects of the work that I had not previously emphasized. Writing to Russell on August 8, 1922, Behmann recounts how he had come to work on the topic. His passion for logic was sparked by reading *Principia*. However, he found the work very difficult to master and thus conceived of the idea of writing a general introduction to it.[48]

It was Hilbert who encouraged Behmann to focus on the set-theoretical antinomies and their solution in *Principia*:

Prof. Hilbert then proposed to me, as a theme of dissertation, not to treat the whole work as such, but rather to make clear the particular way by which the plainly most serious among the antinomies of the Theory of Aggregates, that concerning the transfinite cardinal and ordinal number, is avoided by the logical theory of the *Principia Mathematica*. So I tested the theme in the way suggested by Prof. Hilbert, adding critical comments wherever I believed it necessary or desirable. But unhappily, on account of the want of paper, which was a consequence of the War and the Revolution strongly felt in our public life, there was no possibility of having a book printed "having a character extremely scientific and addressed to a small class of readers"—as the publisher remarked—, so that I am very sorry to not to be able to send you a copy of it.

Russell did actually get to see a copy of the work, through the Göttingen library, and judged it "a valuable and important piece of work".[49]

As a consequence of Hilbert's suggestion, Behmann decided to develop the treatment by using as a unifying theme the solution to Cantor's paradox of the set of all sets, thus focusing on the foundations of arithmetic including the Cantorian theory of cardinal numbers. He also decided not to develop the theory of logic to its full extent but only insofar as needed for the foundations of cardinal arithmetic.

The dissertation is divided into six chapters. Chapter one is introductory. The next three chapters deal with propositions, functional logic, and the theory of classes. Chapter five (pp. 226–83) develops cardinal arithmetic. Finally, chapter six (pp. 284–351) discusses the general results and more general philosophical questions. I will only emphasize some aspects of the dissertation, which in my opinion counts as one of the most important documents concerning the assimilation of the system of *Principia* in Göttingen.

4.7.1 Abstract Terms and Context Principle

The discovery of Cantor's set theory came from the investigation into the grounds for our belief in the truth of arithmetic. The domain of arithmetic was extended from the finite to the infinite but this conquest of new mathematical realms did not come without a cost. Indeed, Behmann adds, Cantor himself was aware that set theory was tangled in inextricable contradictions that the available tools of logic and mathematics were insufficient to solve. The freedom gained by the Cantorian extension of mathematics should not be relinquished on account of the doubts caused by the paradoxes and by returning to the narrow limits of a finite arithmetic. Rather, one ought to investigate whether the old arithmetic might not already carry within it the seeds of the contradictions, which only came to light in the more extended domain. In this situation, Behmann continues, the only way out is to reconstruct arithmetic on the basis of a critical analysis of the concept of number (p. 4). A powerful tool to carry out the task is offered by the mathematical logic developed by Russell and Whitehead in *Principia*. After the introduction of the basic set-theoretic notions, Behmann asserts that the project of eliminating the antinomies from mathematics will be pursued not by first grounding a pure logic but rather the foundation will be effected in such a way that the allowable logical inferences will become clear in the construction of arithmetic itself (pp. 31–32).

In Section 5 we encounter some important programmatic statements. The goal is to construct a system of propositions which will be powerful enough to express all arithmetical propositions and such that from the propositions that are recognized as true it is not possible to derive a contradiction. Antinomies arise not on account of the concepts taken in isolation but rather as a consequence of the way in which these concepts are used in sentences. For this reason it is essential to consider concepts always in the context of the sentences in which they appear:

As a result, it would be a mistake from the start to try to investigate the abstract concepts and relations of arithmetic in themselves alone without bearing in mind the fact that all such terms mean something only in the context of the *sentence*, and that even the best-constructed concepts are of no use to us and cannot protect us from contradiction until we know how to *apply* them properly.[50]

The concepts of arithmetic, which for Behmann also includes set theory, contain concepts such as "set" and "number". For this reason the notion of proposition becomes central to the foundation project:

Thus, as the latter consideration shows, we must now, from the outset, place the concept of *proposition* in the center of all our investigations. This means we may never regard concepts occurring in propositions—like 'set', 'number', 'statement', 'space', 'force', etc.—in isolation, as independent logical objects, but always in the context of the propositions in which they can meaningfully appear.⁵¹

The project now consists in constructing a consistent system of sentences in which arithmetic can be developed.⁵² In this connection, Behmann points out that the notion of proposition is taken as primitive in his project although, strictly speaking, propositions should be considered in connection to propositional acts and the latter in relationship to epistemic acts. Another restriction is that only the logic necessary for the reconstruction of arithmetic will be developed and not all logically valid propositions. As in the 1914 lecture, Behmann disagrees with Russell and Whitehead and holds the sentence 'an arbitrary proposition is either true or false' to be perfectly meaningful. However, it will not occur in his system, as it is not needed for the purposes at hand, although he claims that from a purely logical point of view nothing can be objected to it. Concerning the nature of the logicist project, Behmann remarks that the consistency of the axioms of Euclidean geometry was obtained by a reduction to the consistency of arithmetic but that this process until then did not seem applicable to the axioms of arithmetic. However, the new developments have shown that "arithmetic has found in the logic of propositions the domain of knowledge which is at its basis" (p. 38). The last section of chapter 1 presents the six assumptions that will be used in the remaining part of the work (p. 40):

1. Axiom of individuals;
2. Axiom of true and false sentences;
3. Axiom of negation and disjunction;
4. Axiom of the variation of the constants and generalization;
5. Axiom of individual identity;
6. Axiom of reducibility.

I will not follow in detail the reconstruction but rather emphasize the nominalist interpretation of *Principia* given by Behmann and then say something about his position on the axioms of reducibility, choice, and infinity.

4.7.2 Individuals, Propositional Functions, and Classes

Arithmetic is usually developed by starting at the outset with the natural numbers considered as a system of things given once and for all. Assuming the existence of numbers in this way is, according to Behmann, not allowable in a foundational project, as one would assume with the natural numbers also all the possibilities of contradiction hidden in the concept. For the same reason, no abstract concept can be taken as a starting point of the reconstruction of arithmetic:

Since, due to the nature of our subject, we cannot even presuppose the simplest logical concepts in their traditional use as given, we must thus initially mistrust *all* abstract thinking. But since we need some kind of starting point, it must be such that our abstract thinking is not yet involved, and whose existence is not—unlike the existence of numbers—based only on the possibility of thought, and thus cannot be distorted by thought. The only things satisfying this requirement, however, are now those that can be known directly without the aid of thought, i.e., solely through sense perception, which thus provide thought with the prerequisite material for its possibility. (So far, almost all false metaphysics suffered from the effort of wanting to create the objects of thought through thinking. As everywhere, however, here too the Archimedean point necessarily lies outside.) Thus, our *initial assumption* is that it is permissible to *take as given objects of experienced reality* (individuals) *whose existence precedes all thought*. (With this, of course, we consider ourselves justified in completely ignoring any possible distortion through perception.) We thus regard these individuals as already given and directly at the disposal of our thought.[53]

A certain amount of idealization enters into the postulation that individuals are considered as completely independent of one another (p. 47) and as fixed (on p. 153 Behmann says that they are spatially distinct and temporally unalterable).[54] Behmann also declines to give a definition of the concept of individual. The only essential thing for him is that the individuals are the only things to which "a thought-independent existence can be ascribed". While refusing to engage in a philosophical clarification of the logical meaning of 'existence', Behmann specifies that by saying that the individuals exist he only means that "these objects of thought are logically prior to thought or, more precisely, that we are moving in the realm of that thought that might presuppose these things as given" (p. 46). Individuals are however of no use unless we can make assertions about them. This leads Behmann to define an 'elementary proposition' [*einfache Aussage*] as one in which an empirical property of an individual (or an empirical relation between individuals) is expressed.[55] It should be pointed out that although Russell and Whitehead used statements expressing empirical properties of empirical individuals as instances of elementary propositions by no means they committed themselves to the much stronger theory propounded by Behmann.

With the notion of elementary proposition begins the construction of the system of sentences for the development of arithmetic, which is the ultimate goal of the reconstruction. The second assumption then postulates that elementary propositions are unobjectionable. What is meant by this is spelled out by means of Cantor's distinction between consistent and inconsistent totalities. A consistent domain, Behmann explains, is one such that the grasp of the totality is not presupposed in the grasp of any of its members. In the case of the totality of propositions we have an inconsistent domain since the proposition 'All propositions are true', which belongs to the totality, essentially presupposes a grasp of the totality to which it belongs. This problem does not arise with elementary propositions, as they are 'objective' propositions about empirical individuals and their empirical properties and thus cannot refer to abstract concepts such as 'set' or 'class' (or 'proposition', 'property', 'number', and 'existence'). For the same reason, the universal and existential quantifiers do not occur in elementary propositions, as the use of the quantifiers presupposes the existence of given totalities over which they range.

An elementary proposition is either true or false depending on whether or not it corresponds to the actual state of affairs [dem wirklichen Tatbestand entspricht oder widerspricht]. The class of elementary propositions is also closed under negation and disjunction (Section 8). In this way we see how the constraints on the construction of arithmetic are supposed to operate. First of all we need classes of propositions which form consistent totalities in Cantor's sense. Talk of the collection of all elementary propositions cannot lead, according to Behmann, to antinomies as a proposition that refers to this collection as a whole is not elementary (p. 50). Thus, the collection of elementary proposition is consistent in Cantor's sense. Moreover, we need to show that what is taken as true will never lead to a contradiction. At the first stage of the construction the consistency problem is thus discharged on the consistency of the empirical reality. Thus, later in the dissertation, we read:

Since—as we may here undoubtedly assume—the objective world, i.e., the totality of the individuals with all their properties and relations, certainly constitutes a consistent domain in the end, then clearly every proposition that has only individuals as objects, and which thus does not presuppose the existence of other things, must either correspond to or contradict the actual case, and thus fulfill the principles of contradiction and of the excluded middle.[56]

Indeed, the same strategy will be pursued at the different levels of the functional hierarchy (as I point out in the conclusion a fuller study should be devoted to this aspect). In order to achieve this Behmann will be forced to offer a deflationary account of propositional functions, classes, and propositions. Thus, talking about first order propositional functions he states that propositions about propositional functions are only apparently about them and should be paraphrased as being about the individuals.

Behmann's position can be summarized in the claim that the parts of sentences such as 'no object', 'proposition', 'set', 'function', 'numbers' should be interpreted only as formal components of propositions ("Derartige formale Bestandteile sind nach unsere Auffassung nun aber auch Aussagen, Funktionen, Mengen und Zahlen", p. 62). Thus, whatever is not reducible to the individuals has no ontological import.[57] For instance, in an extended comparison with the notion of differential Behmann states that propositional functions are not real:

In the same sense [as the differential dx], however, propositional functions, like all abstracta, are not real objects, but create only the deceptive appearance of object-hood by providing certain resting points for abstract thought moving at a great distance from its real, concrete objects.[58]

The Russellian no-class theory is of course the background for Behmann's position on classes:[59]

The classes—and by the same token, incidentally, the numbers—are thus, as we earlier hinted, nothing else than figures of speech extremely useful for ease and clarity in presenting arithmetic, but can nonetheless become quite problematical as soon as one takes them seriously and, in violation of their nature, takes them for the names of objects.[60]

It should be clear from the above quotes that the general philosophical project defended by Behmann consists in reducing appeal to abstract entities to a formal construction with no ontological import. When we are clear about this it also becomes possible, within certain contexts, to treat the abstract objects *as if* they were individuals.[61]

In the last part of the dissertation Behmann presents his philosophical position more thoroughly. This can be summarized as follows. Empirical perception grounds the objectivity and consistency of mathematics. Number statements have to be considered as formal objects which acquire a meaning only by being traced back to their empirical source. However, mathematics is a priori and certain. Moreover, the practical need for thought economy justifies a philosophy of the *Als Ob* for the practicing mathematician. One is allowed to hypostatize the numbers and treat them as fictions, although in reality, i.e., independently of their role as formal components of sentences, the numbers do not exist (see the previous Chapter for more details).

4.7.3 Choice, Reducibility, and Infinity in Behmann's Dissertation

Let us briefly look at Behmann's position on the three problematic postulates of *Principia*. Behmann explicitly rejects in their case the use of the word axiom, as this should apply to a necessary and immediately plausible truth [*notwendige und unmittelbar einleuchtende Wahrheit*]. But the postulates of choice, reducibility, and infinity are far from satisfying those conditions.

Concerning Zermelo's choice postulate [*Auswahlpostulat*] Behmann points out (p. 12) that following *Principia* he will not make use of it unless this is absolutely necessary. In this way he hopes to forestall possible criticisms of his reconstruction on this count. The postulate is used for instance to prove that every cardinal number is an aleph (§42 p. 279). In §49, devoted to axiomatic problems, Behmann spells out that the reasons for doubting Zermelo's postulate have to do with the fact that the postulate ascribes to the mind the possibility of carrying out something which as a matter of fact cannot be carried out in thought. In order to clarify the obscure nature of the postulate it is thus important to investigate it axiomatically. Behmann points out the topical nature of such an investigation. In recent times new mathematical problems have emerged for which there is no clue as to which path to pursue to decide them or even whether they can be decided on the basis of the known axioms. One of the problems mentioned by Behmann is: given a definite real number, say e^{π}, is it necessarily decidable whether it is rational or irrational, resp. algebraic or transcendent? And here we encounter the notion of completeness for an axiom system:

We arrive quite naturally at the question of whether the given axiom system has the completeness property, that is whether any rational question is capable of being answered in the intended sense.[62]

As an example from set theory, Behmann mentions the continuum problem. Until we either prove that $a_1 = 2^{a_0}$ or that $a_1 = 2^{a_0}$ [his symbols] generates a contradiction when added to the axioms of set theory, the question remains undecided. Then Behmann conjectures that "it might even be possible to prove the undecidability [of the continuum problem] or even the undecidability of the decidability problem" (p. 325).[63]

We have here one of the earliest occurrences of the notion of completeness in its syntactic sense and a statement of what will become the general 'Entscheidungsproblem' (the problem is first stated in Behmann 1921).

Behmann also thought that it would be interesting to develop type theory without the axiom of reducibility (p. 326) in order to measure the exact effect of the axiom on the system. However, since basic results, such as Cantor's theorem, depend on it essentially this was not something he felt could be pursued in his introductory project. Behmann interprets the axiom of reducibility along the lines of a completeness axiom in Hilbert's sense. In this sense the axiom asserts that it is impossible to add any new element to the class of predicative functions of an individual which is not already contained in it. Hilbert was very interested in this interpretation of the axiom. In his evaluation of Behmann's achievements, preserved in the *Promotion* files for 1918 at the University of Göttingen, he writes:

At the very top of Russell's theory—as highest axiom of thought—stands the so-called axiom of reducibility. This axiom, including the related theory of types of Russell, poses extraordinary difficulties for its comprehension. To remove these, Behmann turns to the axiom of completeness I have introduced into arithmetic, which is not only logically similar to the axiom of reducibility, but is even internally materially related to it. By bringing out these connections, Behmann succeeds not only in clarifying the axiom of reducibility, but in surpassing Russell in applying Russell's theory to a particular deep problem—the solution of the antinomy of transfinite number. His result is essentially: all transfinite axiomatics is by its nature something incomplete, the set-theoretical concepts of Cantor are however strictly admissible. Every meaningfully posed set-theoretical problem retains meaning and is therefore capable of a solution.[64]

Indeed, there is also a note in Hilbert's handwriting in the margin of Behmann's official copy of the dissertation where, in connection with Behmann's discussion of the axiom of reducibility, we read "Mein arithmetisches "Vollständigkeitsaxiom" citiren!" (see p. 157) Behmann's most extensive discussion of the axiom of reducibility occurs in §29 of the dissertation (pp. 196–202). In this section Behmann argues that the axiom of reducibility cannot be a necessary truth as there are domains on which it is false. He proves this by using as the domain of individuals the natural numbers and then by a standard diagonal argument he shows that it is always possible to diagonalize out of the class of predicative functions (a countable class) and define (in the system) a new class of integers which is not coextensive with any predicatively defined class.

Finally, concerning the axiom of infinity, Behmann finds himself forced to introduce it as a postulate when developing arithmetic in the fifth part of the dissertation. This axiom allows him to postulate an infinity of individuals and thus provide a general construction of the arithmetic of *all* finite numbers. Of course, this creates a problem with Behmann's general approach to the project as the axiom of infinity cannot be justified by appeal to experience.[65] It is then on grounds of convenience that the axiom is assumed in the development, as without it no uniform development of arithmetic could be given (pp. 319–21). But what can be said about the consistency of the axiom of infinity with the rest of the system? This is touched upon in section 49 where problems of axiomatic nature are discussed. Here Behmann proposes in outline a strategy which calls for reinterpreting sentences of the system in such a way that if a sentence in the original system plus the axiom of infinity is provable then the new interpretation would yield a sentence provable in the system without the axiom of infinity. Since the interpretation sends inconsistent pairs of statements into inconsistent pairs of statements this would show the consistency of the axiom of infinity (p. 323). That the proof of such a statement might involve surveying all possible proofs is implied by the reflection, immediately following the description of the strategy, to the effect that in type theory one cannot talk about all possible proofs but only about all possible proofs of a certain order. Although there is obviously some connection, I think a more detailed study is needed here to see how this strategy relates to Hilbert's consistency program.

4.8 Conclusion

In the above I focused on the reception of Russell's system in Göttingen between 1910 and 1922. During this period Hilbert and his co-workers took very seriously the Russellian proposal of reducing mathematics to logic. In particular, the axiomatization of logic provided by Russell was the starting point for the formalization of mathematics, which was central also to Hilbert's program. Although Hilbert and his school eventually rejected the logicist solution to the foundations of mathematics there are a number of problems that still deserve full investigation. The first consists in analyzing in detail how much continuity there is between *Principia* and the 1917/18 lecture notes "Prinzipien der Mathematik" (Sieg 2000, p. 12, note 33). Another desideratum is a thorough study of how consistency is argued for by Behmann in the construction by stages of the Russellian system and to relate his approach to consistency to Hilbert's later program. Finally, I have argued in the previous Chapter that there are elements of continuity between Behmann's work on Russell and Hilbert's mature program.[66] For instance, Hilbert's conception of 'ideal' statements in number theory as analogous to ideal elements in mathematics is obviously related to Behmann's general strategy of treating propositions containing abstract notions along the lines of sentences containing points at infinity in geometry. Moreover, Behmann's emphasis on individuals as given prior to any abstract thought finds a striking analogue in Hilbert's emphasis in the twenties that "as a precondition for the application of logical inferences and for the activation of logical operations, something must already be given in representation [*in der Vorstellung*]: certain extra-logical discrete objects, which exist intuitively as immediate experience before all thought" (Hilbert 1922c, E.t., p. 202).[67] And this is obviously related to Hilbert and Bernays' emphasis on relying on primitive forms of cognition as a starting point for the construction of arithmetic.[68] These connections between Russell's logicism and Hilbert's mature program should be pursued in more detail.

Although there is still much more to be done in understanding the role played by *Principia* in the development of logic and the foundations of mathematics in Hilbert's school I believe that we now have the milestones laid out for future excursions.

5

ON THE CONSTRUCTIVITY OF PROOFS: A DEBATE AMONG BEHMANN, BERNAYS, GÖDEL, AND KAUFMANN

To Sol, on the occasion of his 70th birthday[1]

5.1 Introduction

The question of whether proofs by contradiction can be eliminated in favor of direct proofs has occupied the minds of many philosophers and mathematicians for centuries. Aristotle, Arnauld, Kant, Bolzano, and many others up to our day have tackled this topic.[2] In the present paper I present the details of a debate centered around the relationship between constructivity and indirect proofs which took place in 1930 and involved some of the most able logicians and philosophers of mathematics of the century, namely Heinrich Behmann, Paul Bernays, Rudolf Carnap, Kurt Gödel, and Felix Kaufmann. The debate never made its way into print but it is still accessible in its entirety in documents preserved in the Kaufmann archive in Konstanz, the Bernays archive at the ETH in Zurich, and the Behmann archive in Erlangen. Related materials are also found in the Carnap archive

---

[1] I would like to thank the curator of the Felix Kaufmann archive in Konstanz, Dr Martin Endress, the curators of the Behmann archive in Erlangen, Prof. C. Thiel and Dr V. Peckhaus, the curator of the Bernays *Nachlaß* in Zurich, Dr Yvonne Vögeli, for their help and kindness. In addition, many thanks to John Dawson Jr., Johannes Hafner, Richard Zach, and an anonymous referee for several helpful suggestions. Finally, it is a pleasure to acknowledge that the research leading to this paper was carried out during my tenure as fellow of the Wissenschaftskolleg zu Berlin in 1997–98.

in Pittsburgh. The Kaufmann archive in Konstanz contains, among other things, a large folder entitled *Zur Frage der Konstruktivität von Beweisen* which is of great interest to the historian of logic and the foundations of mathematics. Felix Kaufmann was an associate of the Vienna Circle and he was in constant contact with, among others, Carnap, Gödel, Hahn, Waismann, Hempel, and Behmann. Most of the material we will discuss is found in this folder but supplementary materials from the other archives are used to fill in the picture when needed.[3]

5.2 A Conjecture by Kaufmann

On October 24, 1930 Heinrich Behmann sent a letter to Paul Bernays, assistant editor of *Mathematische Annalen*, submitting a manuscript for publication entitled *Zur Frage der Konstruktivität von Beweisen*. He explained to Bernays the circumstances from which the paper originated:

> [The paper] concerns the solution of a problem Herr Kaufmann suggested to me during my stay in Vienna in September. Although I do not share the constructivist point of view I still deem such investigations important so that the practical import of the constructivist principles will not be overestimated and consequently great parts of mathematics be, unjustifiably, put into doubt.[4]

He also added "Herr Carnap has already checked the proof and he agrees with it in every respect."[5]

Behmann had visited Kaufmann for 11 days in September 1930 from September 9 to September 20.[6] Neurath had first suggested to Behmann in 1929 that he should deliver a talk for the Schlick circle.[7] Indeed, on the occasion of his visit to Vienna, Behmann delivered a lecture at the invitation of the Schlick circle entitled *Widersprüche in Mathematik und Logik* (September 15, 1930). This took place just after the famous Königsberg meeting.

During his visit Kaufmann had proposed a problem on the constructivity of existence proofs to which Behmann proposed a solution just a few days after leaving Vienna.[8] The problem raised by Kaufmann was, roughly, the following. Is it possible to prove that all proofs of existence claims, with the exception of those obtained by means of the axiom of choice, can be shown to rely on an implicit exhibition of an instance? Kaufmann's conjecture was that this would in fact be the case. The consequences for the debate on the foundations of mathematics would have been remarkable. A proof of the conjecture would have implied that most of classical mathematics—the only exception being those parts relying on the axiom of choice—is already "constructive." Before we look at Behmann's solution let us try to be more precise on the nature of Kaufmann's conjecture. Kaufmann's position is expressed in his *The Infinite in Mathematics*, which was published in 1930. Since the thesis is somewhat unclear, let me quote the entire relevant passage:

> Our train of thought is this: if we make perfectly clear to ourselves what the criteria are for a mathematical assumption being absurd, we recognize that they lie in the fact that two signs for natural numbers presupposed as synonymous turn out, at the end of a proof, not to be synonymous. Every contradiction therefore relates to certain natural numbers; that it arises at all

means that it occurs at certain points and in exhibiting it we must therefore implicitly determine those points. That this determination does not always show in mathematical proofs themselves must be traced back to the abbreviations in the formation of mathematical concepts (symbolism); but a completely articulated theory of proof, as aimed at by Hilbert, would have to bring it into the open. If so, the illusion that there are pure existence proofs would vanish and Brouwer's demand for constructivity throughout would have been met; but it is precisely the fulfilment of this demand that will show that whenever the question is that of the existence of finite bounds, the only question that arises in the building up of mathematics (apart from set theory, to be analysed later), a divergence between absurdity of absurdity and mathematical existence did not even 'originally' exist, so that the results of existence proofs in classical mathematics will be perfectly justified from Brouwer's point of view as well, leaving aside differences in terminology.

However, this does not hold for the pseudo-existence proofs in the doctrine of the non-denumerable infinite, which, as we shall see, do not have a constructive basis. (Kaufmann 1978, pp. 57–58)

If we abstract for a moment from the appeal to Hilbert's proof theory in the motivation of Kaufmann's conjecture the overall thesis appears to consist of two main parts:

(a) There is no distinction between constructive existence proofs and general existence proofs in the domain of "countable" mathematics;

(b) Such a distinction arises as soon as the "uncountable" makes its appearance.

It should be noted that the thesis displays a certain amount of ambiguity; the boundaries between countable and uncountable mathematics are not spelled out in formal terms; moreover, it is not clear how the proof of such a claim should be carried out. But the general nature of the claim is clear enough. It is a question of showing that all existence claims proved in those parts of classical mathematics that do not rely on non-constructive existence axioms, such as the axiom of choice, are constructive existence claims, i.e. can be shown to be grounded on the exhibition of an instance for the claim. Kaufmann did not even hint at an argument for establishing this conjecture but he spoke about it to an able logician, Heinrich Behmann, who shortly thereafter provided a "proof" of the conjecture.

Behmann sent a paper containing a proof of Kaufmann's conjecture to Kaufmann on October 6, 1930 requesting that the paper be shown to Carnap, Waismann and others. Moreover, in the same letter, Behmann discussed the opportunity ("wegen der Wichtigkeit der Sache") of publishing the result in a mathematical journal such as *Mathematische Annalen* or *Mathematische Zeitschrift*. Kaufmann's reply to Behmann showed complete agreement with the solution proposed by Behmann and contained two additions which Kaufmann proposed to add as an appendix to Behmann's paper.

5.3 Behmann's Typescript and Its Contents

Let us now consider Behmann's typescript. Although it was never published it survives in many archives (in Zurich, Konstanz, Erlangen, Pittsburgh). Kaufmann had made copies for Waismann, Carnap, Gödel, Wittgenstein, and Hahn. Moreover, a copy was sent to

Bernays and it is still found among Bernays' papers at the ETH in Zurich. The latter version of the paper shows signs of reworking that are absent from the original copy sent to Kaufmann. Occasionally, I will remark on the difference between the different versions of the typescript in the notes.[9]

The first section of the paper is entitled "A conjecture [*Vermutung*] by Kaufmann".[10] Behmann begins by discussing the contemporary constructive tendencies and their restriction of classical mathematics in methods of proof and, consequently, in their theorems. This takes place in the form of three restrictions:

(1) Elimination of non-constructive existence axioms (in particular the axiom of choice)

(2) Elimination of non-predicative existence proofs (in as far as they are not already excluded by (1))

(3) Elimination of concepts of higher type (that is, restriction to the so-called *engeren Funktionenkalkül*)

Behmann has no doubts that the first restriction is an actual restriction, for it was well known that there are statements of classical mathematics that would not hold without the axiom of choice.[11]

Concerning the restriction in (2) Behmann observes that existence proofs can be obtained either directly by exhibition [*Aufweisung*] or indirectly, by means of a proof by contradiction. (2) would thus constitute an essential restriction only if there are existence proofs proved by contradiction that cannot be made constructive. Equivalently, (2) represents a restriction if there are proofs of existence claims which, although not relying on non-constructive existence axioms, can only be proved by contradiction.[12]

Kaufmann's conjecture is exactly that an indirect existence proof, provided it satisfies (1), is also constructive:

> Herr Kaufmann has recently formulated an interesting conjecture by means of which the aforementioned question would be answered negatively. He is of the opinion that an indirect proof, in so far as it satisfies condition (1), is also in essence constructive; more precisely, that it relies on a hidden instance, which through an appropriate analysis of the demonstration can automatically [*zwangsläufig*] be made explicit.[13]

I hasten to warn the reader that the use of the word "non-predicative" in condition (2) was an oversight by Behmann, which in fact generated confusion in the exchange with Bernays. Behmann simply meant "non-constructive" rather than "non-predicative". Let me also add a word about condition (3). In other parts of the Behmann-Kaufmann correspondence we find many discussions as to whether quantification over properties can be eliminated without loss of any parts of classical mathematics. This was not a new problem as Weyl's system *Das Kontinuum* (1918a) was in fact an attempt to develop mathematics without assuming quantification over subsets of the domain but only over individuals of the domain. One of the topics of discussion between Behmann and Kaufmann was whether the theory of real numbers and theorems such as Weierstrass' theorem on the convergence of bounded monotone sequences can be proved within a framework which does not assume quantification over classes or predicates. However, I will not enter into this issue as it is not relevant to our main concern.[14]

Behmann claimed to have proved Kaufmann's conjecture and the rest of the paper offers a proof of the claim.

5.4 Behmann's Theorem: Every Indirect Proof Can Be Transformed into a Direct One

The second section of the paper is entitled *Die Struktur des mathematischen Beweises*. Despite the promising title Behmann's treatment of the topic is extremely sketchy and somewhat confusing. The source of the confusion lies in the fact that there is no specification of a language or of the rules of inference. Thus, the notion of proof is not given by a derivation system and it is left completely unspecified. Moreover, in the course of the presentation Behmann uses certain diagrammatic configurations to represent—at different times—proofs, sentences, and inferences. At the source of this lack of clarity is the fact, I claim, that Behmann tacitly identifies inference and logical implication. I will come back to this after we have gone through Behmann's presentation.

Behmann begins by representing [*darstellen*] direct and indirect proofs [*Beweisführungen*] through certain schematic diagrams [*Schemata*]. Let a, b, c, etc. be sentences used for the proof of a certain proposition d.

A direct proof is represented by

An indirect proof is represented by

Behmann then remarks that in passing from a direct to an indirect proof the negation of one of the assumptions becomes the conclusion and the negation of the conclusion one of the assumptions.

The above schemata stand for proofs. However, Behmann goes on to claim that the same schemata hold for each single inferential step of the corresponding proof. In the case of a direct proof each inferential step can be seen as leading from a set of premises

$a, b, c, \ldots$ (taken to be true on account of the fact that they are obtained in the course of a direct proof) to a proposition d, considered as the conclusion of the inferential step. In the case of an indirect proof $\neg d, c, b$ will stand for the assumptions and $\neg a$ for the conclusion of the inferential step. In short, Behmann is claiming that the global structure of direct and indirect proofs is replicated at the local level for each inferential step.[15]

Behmann now describes a procedure that transforms one diagram into another by using negations and reversing the direction of the arrows. In order to make this step more comprehensible he provides a new interpretation of the "arrow" symbolism, which is now interpreted as a representation of sentences, in particular of disjunctions. To see how this works let us begin with some easy configurations.

The first configuration stands for the disjunction $p \vee q \vee r$. But if we reverse the direction of the arrow, this means that the negation of the formula goes into the disjunction. Thus in the second configuration we have $\neg p \vee \neg q \vee r$, i.e. $(p \& q) \supset r$. In short, one can reverse the direction of the arrow when one wants to substitute the occurrence of a formula with its negation. The third configuration will be read as $p \supset (q \vee r)$. Behmann denotes the arrows pointing inside as "condition-arrows" and those pointed outwards "conclusion-arrows".

Any proof can now be seen, according to Behmann, as a system of such configurations [*Sternfiguren*]. Under the condition—commonly assumed in proof theory, says Behmann—that each single occurrence of the same sentence in the proof is considered independent of other occurrences (and thus is proved afresh every time it is introduced) we have the following characterization of a proof structure:

At the bottom stands (with arrows pointing downwards) the configuration of the last inference of the proof. Insofar as the sentences related by the arrows are not the assumptions of the whole proof there is an associated configuration with its arrows. [...] Thus with the exception of the assumptions and of the conclusion every single occurrence of a proposition stands at the beginning and at the end of exactly one arrow.[16]

The scheme has the structure of a tree—the terminology is Behmann's—with the conclusion as the root, and the immediately connected configurations are the branches. At the end of the branching we find the assumptions of the proof, i.e. the leaves [*Zweigende*]. Behmann specifies that all arrows are necessarily open and that there cannot be a closed polygon of arrows. He then goes on to assert that if one abstracts from the direction of the arrows, the proof structure can be looked at as if any of the leaves of the proof were in fact the conclusion. He says:

It is now intuitively evident [anschaulich evident] that within the described structure as such, if one abstracts from the direction of the arrows, the position of the conclusion is, from a topological point of view, equivalent to that of the assumptions, so that one can think of the [proof-]schema as if it were constructed in exactly the same way from any leaf [*Zweigende*]. If one turns the direction of the associated configuration in such a way that the leaf in question is directed downwards (i.e. the arrow is directed upwards) then it is again true that to those configurations, which contain neither the assumptions nor the conclusion of the proof, is associated exactly one configuration whose arrows we can think as reversed etc.[17]

I find remarkable the idea of applying to proofs, even if at such an elementary level, graph-theoretic and topological considerations. This might very well be the first time such talk occurs.[18]

We have seen that the characterization of proofs given by Behmann uses a diagrammatic device that can be interpreted in three different ways, depending on whether we take the arrow configurations to stand for proofs, inferences, or disjunctive propositions. How can the three interpretations be reconciled? The only reading that makes sense of this is to interpret Behmann as claiming that to every inference there corresponds a disjunctive tautological sentence (or, equivalently, a tautological implication). Consequently, the whole proof corresponds to a tautological sentence (see section five for a concrete example). To see that this assumption cannot be granted to Behmann it is enough to think that while the claim is true for derivations in propositional logic, it is certainly not the case for inferences in first-order logic. Indeed, some of the inferential rules in first order logic, such as the rule of universal generalization, do not correspond to tautological implications.

We now move to section three of the paper where Behmann outlines his strategy for transforming an indirect proof into a direct one. Without loss of generality assume that the negation of the proposition to be proved appears only at one end of the proof structure (this can be obtained by making the relevant transformations, according to the preceding explanations). One then follows these steps:

(1) Eliminate the negation sign of the conclusion sentence and reverse the direction of the arrow making it into a conclusion arrow;

(2) Keep reversing all the arrows connecting the conclusion to the negation of the assumption which is the desired conclusion.

Example:

Assume that d is an established theorem and that $\neg d$ is the last step needed for inferring g by contradiction. Then:

Step 1.

```
       a       b
        ↘     ↙
         ↓
  ¬g     e
   ↘    ↙
     ↓
    ¬f      c
      ↘    ↙
        ↓
       ¬d
```

Step 2.

```
       a       b
        ↘     ↙
         ↓
   g ←   e
     ↘  ↙
     ↓
    ¬f      c
      ↘    ↙
        ↓
       ¬d
```

Step 3.

```
       a       b
        ↘     ↙
         ↓
   g ←   e
     ↘  ↙
     ↑
     f      c
      ↘    ↙
        ↓
       ¬d
```

Step 4.

```
       a         b
        ↘       ↙
         ↓
     g       e
      ↘    ↙
         ↑
         f       c
                ↙
         ↑
         d
```

If the negation of the claim to be proved occurs several times in the proof as a leaf then one has to apply the transformation for all the leaves in question. The claim thus appears as end-inference in several places of the direct proof obtained by transforming the indirect proof. The whole procedure boils down to reversing arrows and taking the negation of the formulas whose arrows are being reversed. How is this going to establish Kaufmann's conjecture?

Since in the absence of a non-constructive existence axiom every existence proposition can only be proved directly by providing an instance—or so does Behmann claim—then at some stage of the proof there must be an occurrence of the form $fa \supset \exists x fx$. If the arrow in question goes in the reverse direction in the original proof then the inference has the form $(x)\neg fx \supset \neg fa$. According to Behmann this proves the claim. The rest of the paper goes on to show in the specific case of Euler's proof of the infinitude of primes how to transform the indirect proof into a direct one. We need not delve into the example since we will show that Behmann's argument is faulty. In order to do this we will attempt to clarify some of the hidden presuppositions of Behmann's argument by working out in detail an example of a formal derivation in the propositional system of *Principia Mathematica*.

5.5 The Propositional Axioms of *Principia Mathematica* and Behmann's Argument

In order to gain a better insight into Behmann's argument I am going to supplement his general exposition by looking at one specific system, the propositional system of *Principia Mathematica*, and a specific example of a derivation.[19] In order to keep the similarity with the exposition given by Behmann the axioms of the system in question and the rule of inference will be expressed using disjunction and negation as primitives. The axioms of the system are:

(1) $\neg(p \vee p) \vee p$
(2) $\neg p \vee (p \vee q)$
(3) $\neg(p \vee q) \vee (q \vee p)$
(4) $\neg(\neg p \vee q) \vee (\neg(r \vee p) \vee (r \vee q))$

Rules of inference:

(R1) Substitution
(R2) Modus ponens: From p and $\neg p \vee q$ infer q.

A proof of $p \vee \neg p$ can be given as follows (without explicit mention of the appropriate substitutions in the axioms):

(a) $\neg(\neg(p \vee p) \vee p) \vee [\neg(\neg p \vee (p \vee p)) \vee (\neg p \vee p)]$ (by axiom 4)
(b) $\neg(p \vee p) \vee p$ (by axiom 1)
(c) $\neg(\neg p \vee (p \vee p)) \vee (\neg p \vee p)$ (by R2)
(d) $\neg p \vee (p \vee p)$ (by axiom 2)
(e) $\neg p \vee p$ (by R2)
(f) $\neg(\neg p \vee p) \vee (p \vee \neg p)$ (by axiom 3)
(g) $p \vee \neg p$ (by R2)

In Behmann's setup this would have the following tree structure:

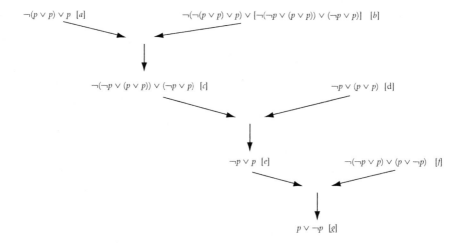

To this tree structure corresponds a propositional tautology, constructed as follows:

$$(([a]\&[b]) \supset [c]) \& (([c]\&[d]) \supset [e]) \& (([e]\&[f]) \supset [g])$$

Let us remain within the boundaries of this formalization of propositional logic. Notice that in this system we proceed from truths to truths. However, following Behmann's idea, we could now turn the arrows of the graph and obtain a scheme for an indirect proof, where (∗) denotes the negation of the result to be proved.

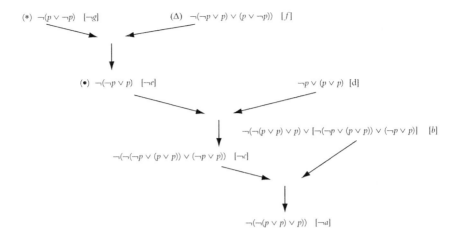

We see that the end-sentence is a negation of an axiom and thus would be the pivotal point of a proof by contradiction. The result of the transformation once again gives rise to a propositional tautology:

$$(([\neg g]\&[f]) \supset [\neg e])\&(([\neg e]\&[d]) \supset [\neg c])\&(([\neg c]\&[b]) \supset [\neg a])$$

However, it is also important to notice that although the transformation proposed by Behmann sends a graph into a graph which preserves logical consequence, we do not get a proof after the transformation has been carried out. For instance, in our system we have no rule that would allow us to go from (∗) and (Δ) to (•). However, under the assumption of completeness of the calculus, a proof can in principle be found for each of the steps. Likewise, if we start with a proof by contradiction the transformation will not in general lead us to a proof.

We thus see how in the propositional case Behmann could move between the several interpretations of his arrow diagrams without trouble. However, things get more complicated as soon as we move to the first-order calculus, where the Behmannian assumption that each inference rule corresponds to a true disjunction does not hold anymore. Consider the usual rule for universal generalization: from $A(x)$ infer $(x)A(x)$. Assume that the rule is applied as an inferential step in an indirect proof. Behmann's strategy for obtaining a direct proof would call for a reversal of the arrows of implication. But this amounts to confusing the rule of inference with the formula $A(x) \supset (x)A(x)$. In his argument Behmann made two assumptions, one implicit and the other explicit. The implicit assumption is that only the inferential step concerning the introduction of the existential quantifier is relevant for the success of the transformation of every indirect proof into a direct proof. But, as I pointed out above, there are other quantificational steps, which do not correspond to implications, involved in a first-order demonstration that cause problems.

The explicit assumption concerns the introduction of the existential quantifier. According to Behmann, in the absence of non-constructive existence axioms the only way to introduce $\exists x f(x)$ is by first obtaining $f(a)$ (or by the equivalent rule: from $(x)\neg f x$ infer

¬fa). From the reactions to Behmann's proof we will discover the roots of the explicit assumption and its problematic nature.

5.6 Kaufmann, Carnap, and Gödel on Behmann's Proof

The reception of the proof in Vienna was marked by acceptance of its validity. Kaufmann wrote back on October 9, 1930: "Thank you very much for your letter and the expected proof. It seems to me absolutely conclusive and of paradigmatic clarity"[20] The letter continues by saying that copies of the essay are being made for Hahn, and Waismann (intended for Wittgenstein). At the same time Kaufmann asks permission on behalf of Carnap to show the essay to Gödel. The rest of the letter discusses the appropriateness of trying to publish the essay either in *Mathematische Zeitschrift* or *Mathematische Annalen*. On account of the importance of the essay Kaufmann said he would not talk to Hahn about possible publication in the *Monatshefte für Mathematik und Physik* (which, ironically, will publish Gödel's paper on incompleteness just a few months later). Another point raised in the letter concerns the similarities between a constructivization of Euler's proof for the infinitude of primes recently given by Wittgenstein and the one contained in the essay. [[See Chapter 6]] Finally, Kaufmann also adds two appendices that he would like to publish together with the essay by Behmann. In the first, Kaufmann tries to point out the philosophical importance of Behmann's result for the dispute over the foundations of mathematics. The second appendix concerns the issue of whether quantification over classes or predicates, can be eliminated and answers the question positively. Part of the later correspondence between Kaufmann and Behmann delves more into this issue but I will not follow this up. Let me just report the passages where Kaufmann spells out the philosophical importance of Behmann's result:

In my opinion, the claim proved by Mr. Behmann and put forward in my book [*The Infinite in Mathematics and its Elimination*] on p. 66 f decides (as was set out there on pp. 59 ff) the methodological dispute between formalism and neo-intuitionism (Brouwer). In the first place this follows for those propositions which are proved without the help of the axiom of choice. For the mathematical core of Brouwer's teachings consists in posing the following trilemma:

(1) A mathematical existence claim E, e.g., the occurrence of a sequence 0, 1, 2, 3, ..., 9, in the decimal expansion of π, is verified by direct exhibition.

(2) The absurdity of E is proved, i.e., $\neg E$ is proved for all numbers.

(3) The absurdity of (2), thus the absurdity of the absurdity of (1) is proved without it being possible to give a construction (1).

The possibility of (3) would mean that one could not infer the validity of an existence claim from the absurdity of the absurdity of that existence claim. On the assumption of this possibility rests Brouwer's thesis that the law of the excluded middle is not valid for infinite domains. This thesis coincides, as Brouwer himself has stressed (see e.g., Intuitionistische Mengenlehre, Jahresber. d. Deutsch. Math. Ver., Bd 28, pp. 203–208, 1919), with his negation of the definiteness of decision of arithmetic and analysis.

So the basis for this claim disappears for all propositions which are proved without the help of the axiom of choice.

We obtain the following result:

The intuitionistic requirement of universal constructivity is legitimate; but it appears to be implicitly satisfied in the so-called existence proofs of classical mathematics, and so its realization does not entail a reduction of the assets of mathematics.[21]

The latter claim is qualified later by Kaufmann by remarking that the claim holds with the exception of results depending on the axiom of choice; in particular, it does not hold for results not reducible to countable cardinalities. It is thus clear why Behmann's result generated such attention. If true, it would have shown that classical mathematics was, to an unsuspected extent, in essence intuitionistic. It should be remarked that Kaufmann's characterization of the philosophical nature of the project emphasizes the goal of providing a positive contribution to the debate between formalism and intuitionism. In this connection, then, it should be mentioned that Behmann's "theorem" can be seen as one of the many contributions to the clarification of the relationship between classical and intuitionistic mathematics, such as those of Kolmogorov, Glivenko, Gödel and others. I am referring to the negative translations (or double negation interpretations) aimed at interpreting fragments of classical mathematics into intuitionistic mathematics. Indeed, as presented in the above quote, Behmann's theorem would have provided a translation from classical (countable) mathematics into intuitionistic mathematics such that if in classical mathematics we have $\exists x A(x)$ then intuitionistic mathematics proves $A^*(t)$, where $*$ stands for some suitable translation. As a consequence, if intuitionistic mathematics proves $\neg\neg \exists x A(x)$, it would also prove $A^*(t)$ for some term t.

In his reply of October 12, Behmann declines the suggestion to publish Kaufmann's remarks with his essay, motivating his choice on strategic grounds by a desire not to alienate acceptance of his result on account of ideological resistance, and states that he is submitting the essay to Bernays for publication in *Mathematische Annalen*. On the same day Kaufmann writes to Behmann to announce that Carnap and Waismann find the proof absolutely compelling:

Today I had a longish, very nice discussion with Mr. Carnap; he is very happy about the successful proof, which he finds—just as I expected—completely transparent and he recognizes its immense importance. The same is true of Mr. Waismann with whom I spoke on the phone today.[22]

Let us now look at Gödel's contribution to this debate.

In a letter dated October 9, 1930 Kaufmann asks permission on Carnap's behalf to send a copy of the paper to Gödel.[23] A handwritten letter from Kaufmann to Gödel, which informs Gödel of some corrections to the proof made by Behmann (in a letter to Kaufmann dated October 11, 1930) is still to be found in Kaufmann's papers. The letter is a handwritten draft and it is dated October 14, 1930.[24] Carnap and Gödel probably met to discuss the issue on October 16, 1930 and Carnap made a record of the discussion in his diary.[25] Dawson (1997, p. 73) mentions an entry in Carnap's diary (dated October 16, 1930) which "recorded Gödel's refutation of a claim by Heinrich Behmann that every

212   Foundations of Mathematics

existence proof could be made constructive". Dawson rightly relates Gödel's refutation to Brouwer's doubt as to "whether any formal system could comprise all of intuitionistic mathematics". A fuller account of Gödel's objections can be gathered from a letter from Kaufmann to Behmann. Indeed, just three days after meeting Carnap Gödel went to see Kaufmann and expressed his doubts concerning Behmann's result.

Vienna, October 19, 1930.

Dear Mr Behmann,
Dr Gödel is with me right now and he raises the following objections against my constructivity thesis and your proof of it:

He claims to be able to construct examples where clearly an existence claim is proved although one could not give a construction. The simplest example mentioned by Mr Gödel—which may serve in principle as representative for all the remaining examples—is the following:

Let a one-to-one mapping between the natural numbers and certain rational numbers from the interval (0, 1) be given. Then one can prove in the usual fashion that the given sequence of rational numbers has an accumulation point, although it would in general not be possible to specify it. Let us consider an example where one certainly cannot give an accumulation point! A number is called a Goldbach number if all even numbers smaller than it are sums of two prime numbers. The sequence of rational numbers is now defined as follows; the natural number n is mapped to $\frac{1}{n}$ if n is a Goldbach number, and to the number $1 - \frac{1}{n}$ if n is not a Goldbach number. Then one can prove that this sequence has an accumulation point (and indeed that it is rational), namely either 0 or 1. However, without a solution of Goldbach's problem no accumulation point can be specified.

Mr Gödel has also talked with Mr Carnap about it who finds the objection very plausible.

I would be very grateful if you could let me know your reaction to this objection and in the meantime I will also rack my brains about it.

Mr Gödel also adds that although in this case it is a question of a disjunction between *two* possibilities, this is not essential; the disjunction between infinitely many cases could also remain undecided.

I am yours with warmest regards
Felix Kaufmann[26]

The letter was completed by a *Nachtrag* which gave the immediate reaction of Kaufmann to the result. Kaufmann's letter to Behmann was written together with Gödel and as such it represents an important testimony to Gödel's activities during this period. Indeed, the copy of the letter in Behmann's archive contains a handwritten note by Gödel on the side ("Die besten Grüße, Ihr erg. Kurt Gödel"). Gödel's objection can be restated as follows. Working within the background of classical mathematics let $E(x)$ stand for *x is even* and $P(x)$ for *x is prime*. Define $G(n)$ to be $(\forall x < n)(E(x) \supset \exists y z (P(y) \& P(z) \& x = y + z))$ Let

$$a(n) = \begin{cases} \frac{1}{n}, & \text{if } G(n); \\ 1 - \frac{1}{n}, & \text{if } \neg G(n). \end{cases}$$

$a(n)$ converges to 0 if Goldbach's conjecture is true; it converges to 1 if Goldbach's conjecture is false. We thus have a sentence of classical mathematics that says that there exists an x such that $a(n)$ converges to x. The proof of the last claim does not rely on a proof by contradiction. The argument starts from the premise that either Goldbach's conjecture is true or it is false. In either case it concludes that $a(n)$ converges. We thus have a direct proof for an existence claim, which however, does not rely on a specific instance of the claim, for as a matter of fact we have established neither that $a(n)$ converges to 0 nor that $a(n)$ converges to 1. This shows that there are ways to introduce an existence statement in a direct proof for which we have no instance. It should be pointed out that the counterexample provided by Gödel shows the influence of the Brouwerian counterexamples on Gödel's thinking.

In his answer to Kaufmann dated October 21, 1930, Behmann relativizes the impact of the objection by saying: "First of all, it must be said that the undoubtedly very interesting objection by Mr Gödel in any case does not affect my proof as it stands; it affects indeed the presupposition of my proof that every direct existence proof rests on the exhibition of an instance [auf einer Aufweisung]."[27] However, it was exactly that unquestioned assumption that made Behmann's result of interest. Moreover, as we have seen, it is not the case that even granting that assumption the proof goes through. In the same letter Behmann also gives examples similar to the one provided by Gödel. In subsequent correspondence between Behmann and Kaufmann the discussion about the proof is continued but nothing of great importance emerges, except the realization that the claim had not been proved conclusively. Kaufmann also suggested that a revised version of the paper might include a discussion of the recently published article by Hölder, who claimed to have shown that proofs by contradiction were not eliminable in favor of direct ones (see Hölder 1929, 1930). However, the realization that much more work needed to be done if the claim were to be conclusively established came with the criticisms raised by Bernays to which we now turn.

5.7 Bernays' Objections to Behmann's Proof

The correspondence between Bernays and Behmann on the subject of constructive proofs is preserved in the Bernays archives at the ETH. We saw that Behmann had submitted the paper to Bernays on October 24, 1930. Bernays replies on October 31 from Göttingen and raises a number of objections to the work. He begins by stating that he is not convinced that Behmann's result succeeds in showing the eliminability of non-predicative existence proofs. As mentioned earlier talk of non-predicativity was caused by a mistake in Behmann's orginal formulation of his second requirement. Bernays goes on to state that he has no qualms with the main lemma on the transformation of indirect into direct proofs. However, this goal had already been achieved by Löwenheim in a paper read to the Berlin Mathematical Society.[28] But this result does not allow one to conclude that anything follows for the elimination of non-predicative existence theorems. In particular, the opposition set up by Behmann between "direct through instantiation or indirect" is misleading in that indirect proof procedure is not what characterizes non-predicative proofs. Bernays then goes on to discuss the example of Euler's proof of the

infinitude of primes. However, Bernays argues that also in this case the essential point of the transformation from a non-predicative to a predicative proof has nothing to do with the issue of direct vs. indirect proof. Many mathematicians, he remarks, have conjectured that it is in general true that every non-predicative proof can be transformed into a predicative one but, he adds, so far there is no proof of this claim. He diagnosed the difficulty of giving such a proof as follows: "The difficulty of the proof lies in the variety of forms in which the impredicative moment can be introduced."[29]
Suppose we infer, for instance,

$\exists x \, B x$
from
$(x) \exists y \, A x y \supset \exists x \, B x,$
$\exists x (y) \neg A x y \supset \exists x \, B x,$
and
$(x) \exists y \, A x y \vee \exists x (y) \neg A x y.$

Another problem is definition by cases such as

$$\varphi(n) = \begin{cases} 0, & \text{if } (x) A x n; \\ 1, & \text{if } \exists x \neg A x n. \end{cases}$$

We see, as Gödel did in his counterexample, that Bernays makes use of definition by cases based on the law of the excluded middle.[30] Bernays also says that of course one could object that such definitions use the "incomplete symbol" "the x such that" (indicated by ιx) and that this symbol is in principle eliminable. But the proof, says Bernays, is not easy at all and it has given him quite a lot of trouble although he has just succeeded in the elimination of the ι-operator in the case of the first-order predicate calculus.[31] Problems only get worse when the ι-operator is used in connection with recursive definitions or with quantification over higher variables. The last part of the letter concerns the problem of the elimination of second-order variables and a few comments by Bernays to the effect that Kaufmann's results on elimination of second-order variables from theorems such as Weierstrass' theorem on monotone convergence are not original and are found in Weyl's *Das Kontinuum*. The letter ends with the advice to Behmann to avoid the use of imprecise terms such as "constructive" and "instance" and replace them with sharp logical notions. Finally, he asks Behmann why he keeps at a distance from Hilbert's proof theory, which can be the only framework adequate for treating with precision the problems Behmann is addressing.

Behmann's reply is dated November 1, 1930. First of all Behmann regrets having used the word non-predicative in that he realizes this led to misunderstandings. All he meant in point (2) of his essay was "non-constructive". He clarifies that by "instance" [*Aufweisung*] he means the replacement of an $\exists x f(x)$ by a suitable $f(a)$. By "constructive" he means that for every "there is" a corresponding "instance" is given. Behmann went on to say that it was far from his intentions to ignore the developments of proof theory but that he was not as familiar with the details of proof theory as Bernays was, as he never had occasion before to make use of proof theory.

The most important remark concerns the assumption that every direct existence proof, which used only constructive axioms, already contains an instance. He claims to have taken this from *Principia Mathematica* (p. 20) and adds that the other Viennese take this result for granted. The relevant passage in *Principia* reads:

An asserted proposition of the form "$(E x). f x$" expresses an "existence-theorem," namely "there exists an x for which $f(x)$ is true." The above proposition gives what is in practice the only way of proving existence-theorems: we always have to find some particular y for which $f y$ holds, and thence to infer "$(E x). f x$". If we were to assume what is called the multiplicative axiom, or the equivalent axiom enunciated by Zermelo, that would, in an important class of cases, give an existence-theorem where no particular instance of its truth can be found. (Whitehead and Russell 1910, p. 20)

Behmann recognized that, after Bernays' criticism, this claim could not be accepted as evident and that also the meaning of constructive axiom had to be clarified.

The final reply by Bernays came on November 11, 1930. Bernays acknowledges that if one reads "constructible" for "non-predicative" (in Behmann's essay) then Behmann's claim is more plausible. But still one must provide a proof along the following lines:

It must be shown that in the case of the derivability of a formula E prefixed by unnegated universal and existential quantifiers, we can also derive a formula E^* which is obtained from E in that to each existential there is a corresponding instance; and indeed it must be shown that the derivation of E^* can be carried out in such a way that no existential quantifiers or negated universal quantifiers occur in it.[32]

Bernays then gave a precise axiom for the introduction of the ι-operator and advised Behmann to follow Hilbert-Ackermann's text (1928) for the details of formalization.

This concludes the exchange between Bernays and Behmann and the debate on the constructivity of proofs. Behmann saw the strength of Bernays' objections and wrote to Kaufmann:

Following this it appears questionable whether publication of my article, in particular of the proof of replaceability of every indirect proof by a direct proof, is worthwhile at the moment. At the least it should not just be assumed that every direct existence proof contains an exhibition [of an instance], rather, this would have to be introduced explicitly as a mere conjecture. (November 2, 1930)[33]

The essay was never published and when asked in 1934 by a Japanese student, J. Hirano, whether the essay had been published, Behmann replied that it had not on account of the fact that he had never seen quite clearly into the matter:

A few years ago I completed an article "On the question of constructivity of proofs," but never published it. For the most part, my aim was to show that indirect existence proofs, just like direct ones, are constructive, as long as they do not use the axiom of choice, i.e., they contain explicit exhibition of a thing of the kind in question. An important step to that conclusion was the proof that indirect proofs are nothing but transformations of direct ones.

Prof. Bernays, to whom I had sent the manuscript for consideration, then wrote to me that this latter proof had been given some time before by Löwenheim, who presented it to the mathematicians in Berlin, but never published it. Furthermore, he gave a detailed critique of the article as a whole, which I did not follow in all the details and I therefore do not know in how far it was justified; it may be that he was somewhat misled by a mistake of mine ("not predicative" instead of "non-constructive"). Therefore I was in any case unsure whether it was reasonable to publish. (Behmann to Hirano, December 8, 1934)[34]

5.8 Conclusion

We have seen that Behmann's paper has several shortcomings, both technical and conceptual, and that his principal claim is simply false. The aims and ideas underlying his paper, together with the discussions with Bernays and Gödel, however, provide interesting additions to the history of notions such as "constructive proof" and "proof transformation." Behmann's idea of a tree-like proof configuration and of a proof transformation proceeding by local transformations in the proof structure is, although vaguely formulated, one of the first times that the structure of a formal proof is made a subject of investigation. We find similar ideas already in Hilber's work on proof theory in the early 1920s, where derivations are transformed from linear into tree-like form, but it is only with Gentzen's work on proof theory in 1934 that the notion of tree-like derivation with branches and leaves is as central a topic as in Behmann's paper. The paper's title, "On the Constructivity of Proofs," emphasizes the notion of constructivity. Behmann's aim was to show how one could extract an instance from a proof—direct or indirect—of an existence claim. His argument rested in part on the false assumption—taken over from Russell—that any direct proof of an existential rests on the proof of an instance. Although the argument fails, Behmann's effort and the ensuing discussion is an interesting record of the state of thought about constructivity around 1930. The basic idea, i.e., that it should be possible to extract concrete information from a proof of an existence statement, and the conditions for when this is possible, is certainly the most important aspect of the debate. Again, Hilbert's work on proof theory bears some resemblance here—the epsilon-substitution method provided a way to replace quantified formulas by instances in a proof. The clearest early examples pointing in similar directions are, however, Herbrand's theorem and the closely related midsequent theorem by Gentzen. Both results establish that it is possible to extract from a proof of an existence statement in the predicate calculus a finite set of terms one of which is a witness to the existential quantifier. Whereas Behmann's result wanted to provide one instance from a proof of an existential in a classical theory, Herbrand's and Gentzen's results show that one can obtain at least a disjunction of instances in predicate logic, which is a propositional tautology, from the proof of an existential. Although more limited in scope than Behmann's original intentions, Herbrand's and Gentzen's results laid the foundation for more sophisticated results for arithmetical theories, such as the no-counterexample interpretation and witnessing theorems in bounded arithmetic.

6

WITTGENSTEIN'S CONSTRUCTIVIZATION OF EULER'S PROOF OF THE INFINITY OF PRIMES (WITH MATHIEU MARION)

> Es war eben auch hier die Hand eines Schleifers notwendig, um den Glanz der Edelsteine Eulers voll herauszuarbeiten.
>
> L. Kronecker

6.1 Introduction

Ever since Georg Kreisel ended his review of Wittgenstein's *Remarks on the Foundations of Mathematics* (1959) by saying that "it seems to me to be a surprisingly insignificant product of a sparkling mind" (1959, p. 158) there has been a strong presumption that Wittgenstein did not know much about mathematics in general and about issues pertaining to the foundations of mathematics in particular, and that his remarks on these topics were marred by this lack of proficiency in mathematical matters. One may indeed ask how much mathematics did Wittgenstein know or, in other words, how much mathematics was he able to do? For a long time, it has been generally believed that Wittgenstein held a peculiar foundational stance, 'strict finitism'. This is certainly not the case today. Indeed, one would nowadays tend to approve of Gordon Baker and Peter Hacker, when they write that "his philosophy of mathematics does not defend a form of 'strict finitism', depsychologized 'intuitionism' or 'constructivism'" and his purpose was "not to take sides in the debates between rival schools of mathematicians, but rather to question the presuppositions which provided the framework of their debates" (Baker and Hacker 1985, p. 345).

We will discuss a *mathematical* proof found in Wittgenstein's *Nachlaß*, a constructive version of Euler's proof of the infinity of prime numbers. Although it does not amount to much, this proof allows us to see that Wittgenstein had at least some mathematical skills. At the very least, the proof shows that Wittgenstein was concerned with mathematical practice and it also gives further evidence in support of the claim that, after all, he held a constructivist stance, at least during the transitional period of his thought (1929–33).

6.2 Behmann, Kaufmann, and Wittgenstein

On October 24, 1930, Heinrich Behmann, a student of Hilbert at Göttingen then teaching mathematics at the University of Wittenberg-Halle,[1] submitted a paper to *Mathematische Annalen* entitled "Zur Frage der Konstruktivität von Beweisen" (1930). Behmann visited Vienna from the 9th to the 20th of September, 1930, to deliver a talk for Schlick's circle. During his stay he learned about a conjecture by the Viennese philosopher, Felix Kaufmann, according to which, roughly, all existence theorems of classical mathematics which do not rely on non-constructive existential axioms, such as the axiom of choice, are in effect constructive existence theorems. In other words, it was conjectured by Kaufmann that one should be able to show that all such existence theorems rely on an implicit exhibition of an instance. Kaufmann had made the conjecture in a book that had appeared in the spring of 1930, *Das Unendliche in der Mathematik und seine Ausschaltung* (1930, pp. 66–67; 1978, pp. 57–58).[2] In his paper, Behmann proposed a 'proof' of this conjecture. Criticisms from Paul Bernays, who was at the time an assistant-editor for the *Annalen*, and from Kurt Gödel forced him to withdraw his paper, which remained unpublished. In its first section, entitled "A Conjecture by Kaufmann", Behmann presents three constructivist restrictions on methods of proofs (and, consequently, on admissible theorems):

(1) Elimination of non-constructive existential axioms (in particular the axiom of choice)
(2) Elimination of non-predicative existential proofs (in so far as they are not already excluded by (1))
(3) Elimination of concepts of higher type (that is, restriction to the so-called "restricted functional calculus")[3]

(The use of the expression "non-predicative" in (2) was an oversight on Behmann's part; it should read "non-constructive".) There is no point in discussing here restriction (3), as it does not bear on the topic of this paper. Behmann clearly was of the opinion that restriction (1) is a genuine one, i.e., there are indeed theorems of classical mathematics that cannot be proved without use of the axiom of choice. Now, restriction (2) would also be a genuine restriction if there were indirect existential proofs (i.e., proofs by contradiction or *reductio ad absurdum*) that cannot be turned into direct ones or if there were proofs of existential claims which, although *not* relying on non-constructive existential axioms, can only be proved by contradiction. Kaufmann's conjecture is to the effect that this is not the case, in other words that all indirect existential proofs satisfying (1) are also constructive. As Behmann put it in his paper:

> Mr. Kaufmann has recently formulated an interesting conjecture [. . .] He is of the opinion that an indirect proof, in so far as it satisfies condition (1), is also in essence constructive, more precisely, that it relies on a hidden instance, which through an appropriate analysis of the demonstration can automatically be made explicit.[4]

The truth of Kaufmann's conjecture would have implied that existential proofs of classical mathematics that do not rely on non-constructive existential axioms, would implicitly satisfy the requirement of constructivity. This, in turn, would imply that the requirement of constructivity would not entail a restriction of current mathematics. Although Kaufmann shared Brouwer's rejection of (non-denumerable parts of) set theory, he believed that classical mathematics was already constructive and therefore that the elimination of the non-denumerable infinite would leave "mathematics as it is", so to speak.

Gödel himself provided within a few days a counterexample that refuted Behmann's claim that he had proved Kaufmann's conjecture.[5] (However, Gödel's objection to Behmann's proof does not use his incompleteness theorems.) Nevertheless, Behmann's unpublished typescript is interesting for a number of reasons. In it, he presented a general method by means of which one could turn, allegedly, a non-constructive existence proof into a constructive one. The techniques used for the proof involve one of the earliest uses of graph-theoretical and topological considerations in the study of proofs.[6] Behmann's method can also be seen as related to proof-theoretical methods developed later by Hilbert & Bernays, Herbrand, and Kreisel. Of peculiar interest to us here, however, is the fact that Behmann presented, in the last section of his paper, an application of his method to the constructivization of a well-known existence proof presented by Leonhard Euler in his *Introductio in analysin infinitorum* (1748, I, 235). There exist many proofs of the infinity of prime numbers,[7] starting with Euclid's famous proof.[8] Euler proved the theorem by *reductio ad absurdum* relying on the equation:

$$\sum_{n=1}^{\infty} \frac{1}{n} = \prod_{p} \frac{1}{1 - \frac{1}{p}} \qquad (A)$$

where p ranges over primes.[9] The left-hand side is the harmonic series. Euler had proved that it diverges, i.e., one can say that the sum is infinite. Euler's proof works as follows: assume $p_1, p_2, \ldots, p_n$ are all the prime numbers. From the equation

$$\sum_{n=1}^{\infty} \frac{1}{n} = \prod_{p_k} \frac{1}{1 - \frac{1}{p_k}}$$

one obtains

$$(1 + \frac{1}{p_1} + \frac{1}{p_1^2} + \ldots) \cdot (1 + \frac{1}{p_2} + \frac{1}{p_2^2} + \ldots) \cdots (1 + \frac{1}{p_n} + \frac{1}{p_n^2} + \ldots) =$$

$$= \frac{p_1}{p_1 - 1} \cdot \frac{p_2}{p_2 - 1} \cdots \frac{p_n}{p_n - 1} = 1 + \frac{1}{2} + \frac{1}{3} + \ldots$$

Thus, $\sum_{n=1}^{\infty} \frac{1}{n}$ is finite, which is a contradiction.[10]

However, as Leopold Kronecker pointed out in his *Vorlesungen über Zahlentheorie* (1901, p. 270), equation (A) is actually invalid; he rectified Euler's proof using:

$$\sum_{n=1}^{\infty} \frac{1}{n^z} = \prod_p \frac{1}{1 - \frac{1}{p^z}} \qquad (A^*)$$

for $z > 1$. If there were only a finite number of primes p, the product would remain finite as z approaches 1, while the sum increases indefinitely. Kronecker also showed how to define an interval from m to n, in the left-hand sum, of which it is shown that it contains at least one prime, however great m is taken (1901, pp. 270–71), thus providing a constructive addition to Euler's proof. In this connection he made a very important remark:

This proof, as Euclid's proof, can also be cast in a form which yields an interval $(m \ldots n)$ that contains a new prime number $p > m$ no matter how big m is. Only then is the last and highest requirement satisfied which has to be imposed on a rigorous mathematical proof.[11]

Euler's proof is of great importance for results about the infinity of primes in general arithmetical progression, e.g., for Dirichlet's theorem for $mz + n$, with m and n relatively prime.[12] The constructivization of Euler's proof is thus far from being a mere exercise for idle minds. At all events, what is of specific interest to us, however, is the fact that, in a footnote right at the beginning of the section of his paper dealing with the constructivization of Euler's proof, Behmann wrote:

I simultaneously learned about a transformation [*Überführung*] of this proof into a constructive form by Mr. Wittgenstein and about Kaufmann's conjecture.[13]

The attribution of a proof to Wittgenstein is quite an astounding claim and we should like to look closely into it. As it turns out, even this application of Behmann's method to Euler's proof is not satisfactory, but it remains of interest to us because of the light it sheds on Wittgenstein's thinking about mathematics.

On October 6, before submitting his paper, Behmann had therefore written to Kaufmann:

Since I have no memory of the details of Wittgenstein's completion of Euler's proof that Mr. Waismann had presented, I would very much like to find out to which degree my transformation [*Umwandlung*] of the proof, which I worked out independently on the basis of the general scheme that I developed, corresponds to Wittgenstein's. At any rate, it would be necessary to reach an agreement with Mr. Wittgenstein, when I shall have to mention this point, which, for reasons of fairness, can hardly be ignored; it seems best to do it through the mediation of Mr. Waismann.[14]

Answering Behmann's request, Kaufmann wrote reassuringly on the 12th that:

Concerning the proof that you have given as an example of your general theorem and that of Wittgenstein, Mr. Waismann considers that they are indeed related but that your footnote appropriately deals with the relationship between them, all the more so since this result is of fundamental importance to your general considerations. He does not consider himself, however, to be in a position to reach a final decision on this point and will, in accordance to your request, forward your proof to Wittgenstein and inform him about its circumstances. At all events, there is no reason to delay publication. Perhaps I should ask you to modify your note to the effect that you had learned about Wittgenstein's transformation [*Überführung*] into a constructive form of Euler's proof about prime numbers prior to devising your own constructivization of the proof, while not mentioning in this context that you had previously learned about my conjecture, since the latter is already in my book.[15]

As a result of Kaufmann's request, Behmann reworded the footnote as follows in the final version of his paper:

The fact that I recently learned about a transformation into a constructive form of this proof by Mr. Wittgenstein motivated me to devise the general procedure that I described above. Kronecker had already given in his *Vorlesungen über Zahlentheorie* a constructive completion (not, however, a transformation [*Umformung*]) of Euler's proof.[16]

One may infer from these quotations that Behmann learned about Kaufmann's conjecture and Wittgenstein's proof during his short stay in Vienna and that he learned about the latter talking to Waismann. It does not seem that Wittgenstein actually took part in any discussions involving Behmann, Kaufmann and Waismann; the latter seems to have acted as a go-between. It is also certainly clearly implied that Wittgenstein had his own constructive version of Euler's proof, which closely resembled Behmann's.

This evidence prompted a search for Wittgenstein's proof. The section on "Euler's Proof" which is printed in *Philosophical Grammar* (1974, pp. 383–86) does not contain a proof but it ends with some calculations that are hard to interpret at first blush. The book *Philosophical Grammar* was edited by one of Wittgenstein's literary executors, Rush Rhees, who used as a basis the so-called *Big Typescript*, which is a voluminous typescript of 768 pages produced by Wittgenstein in 1933. Anthony Kenny (1984) has shown the extent of Rhees' editorial intervention but a comparison of the section on Euler's proof with the corresponding section in the recently published *Big Typescript* (2000, pp. 427–29) shows that they are virtually similar. However, most of the remarks that form the *Big Typescript* originated in manuscripts for the years 1929–31 that are by now published as the first five volumes of the *Wiener Ausgabe*. It turns out that the remarks that form the section on Euler's proof in the *Big Typescript* and the *Philosophical Grammar* originated in one of these manuscripts, namely MS 108, which is entitled "IV. Philosophische Betrachtungen",[17] and can be dated from the summer and autumn of 1930—therefore around the time of Behmann's visit to Vienna.[18] In this manuscript, now published as part of volume 2 of the *Wiener Ausgabe* (1994),[19] the strange calculations that end the section on Euler's proof are continued and one can actually see the point of these, as it is clear that Wittgenstein is calculating a bound below which one can find a new prime number greater than all the elements of any given set of primes $p_1, \cdots, p_m$, thus providing a constructivization of Euler's proof.

It appears that both Wittgenstein and Behmann used as a point of departure Euler's original proof, based on the invalid equation (A) and not the corrected version provided by Kronecker, although Behmann learned about it at one stage,[20] since he mentions it in the footnote that we already quoted. It is unlikely that either of them used Euler's original version but it is impossible to tell from which presentation of the proof they worked. We shall now look at both proofs.

6.3 The Transformation of Euler's Proof in Behmann

Behmann starts from Euler's invalid equation:

(A) $$\sum_{n=1}^{\infty} \frac{1}{n} = \prod_p \frac{1}{1 - \frac{1}{p}}$$

Assuming that (B) the number of primes is finite, say $p_1, \ldots, p_m$, one obtains that

(C) $$\prod_{n=1}^{m} \frac{1}{1 - \frac{1}{p_n}} = (1 + \frac{1}{2} + \frac{1}{2^2} + \ldots)(1 + \frac{1}{3} + \frac{1}{3^2} + \ldots) \cdots$$
$$\cdots (1 + \frac{1}{p_m} + \frac{1}{p_m^2} + \ldots) = 1 + \frac{1}{2} + \frac{1}{3} + \ldots$$

This implies (D) that $\sum_{n=1}^{\infty} \frac{1}{n}$ is finite, which is a contradiction, since Euler has already shown that the harmonic series diverges. One should notice that the last equality of (C) is justified by the unique factorization theorem. That is every $\frac{1}{n}$ must be obtained from the previous product. By uniqueness each $\frac{1}{n}$ will only appear once. The above proof is non-constructive since it does not tell us how to find the next prime beyond p_m.

Behmann's project was to show that, by a transformation of the proof into a direct one, one could extract a bound for finding another prime number not contained in $p_1, \ldots, p_m$. Although we will not present Behmann's general strategy in detail, the following outline should suffice to get the gist of the transformation. According to Behmann's own conventions the proof could be schematized thus:

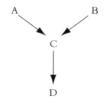

One can, for ease, read the arrows as indicating implications. According to Behmann's strategy we obtain a direct proof by using the negation of D and reversing the arrows of implication from D to C and also negating C and reversing the arrow of implication from C to B. In short the direct proof would look as follows:

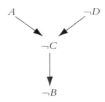

One should note here that ¬B states that there are infinitely many primes.

Thus, the direct proof of the result of the infinitude of prime numbers would proceed in three steps: from the fact that the harmonic series diverges (¬D) and equation (A) one would obtain

$$\forall m \; \exists n \; (\prod_{i=1}^{m} \frac{p_i}{p_i - 1} < \sum_{i=1}^{n} \frac{1}{i}) \tag{¬C}$$

And, therefore, there exist infinitely many prime numbers (¬B). However, this single transformation does still not give us the required instance for the existential. For this reason, Behmann claims that this information must be hidden in the statement that $\sum_{n=1}^{\infty} \frac{1}{n}$ diverges. Indeed, the statement can be re-expressed as saying that

$$\forall r \in \mathcal{R} \; \exists n \in \mathcal{N} \; (r < \sum_{i=1}^{n} \frac{1}{i}).$$

This, says Behmann can be shown easily by rewriting the harmonic series as

$$1 + \frac{1}{2} + (\frac{1}{3} + \frac{1}{4}) + (\frac{1}{5} + \frac{1}{6} + \frac{1}{7} + \frac{1}{8}) + (\frac{1}{9} + \ldots + \frac{1}{16}) + \ldots$$

and noticing that each expression in parenthesis is $> \frac{1}{2}$.

The fact that every real number r can be majorized through the addition of sufficiently many sums s of which the first is 1, the second $\frac{1}{2}$ and all the others greater than $\frac{1}{2}$, can easily be shown by choosing the number of summands to be $s = [2r]$. This will certainly suffice for $r \geq \frac{1}{2}$. (Here, $[r]$ gives the greatest natural number smaller or equal to r.) We thus need to sum up $[2r]$ members of the transformed harmonic series to go beyond r, or, what amounts to the same thing, $2^{[2r-1]}$ elements of the original harmonic series, (i.e. choose $n = 2^{[2r-1]}$).[21] In this way one obtains a general proof of

$$\forall r \in \mathcal{R} \; (r \geq \frac{1}{2} \rightarrow r < \sum_{i=1}^{2^{[2r-1]}} \frac{1}{i}).$$

In turn, by substituting the product $\prod_{i=1}^{m} \frac{p_i}{p_i-1}$ (denoted by $\prod_m$) for r one obtains $\prod_m < \sum_{i=1}^{2^{[2\prod_m -1]}} \frac{1}{i}$. The proof is concluded by noticing that the right-hand side of the inequality with finitely many summands must contain at least one summand that cannot be obtained by multiplying out the factors on the left hand side, i.e., a summand whose denominator must be divisible through a prime greater than p_m, if one decomposes into prime factors all numbers between p_m and $2^{[2\prod_m -1]}$.

Two things should be pointed out about the above transformation. First, although Behmann wants to use this theorem as an example of how the general strategy for turning indirect existential proofs into direct proofs is supposed to work, it should be remarked that the constructivization of Euler's proof does not follow from syntactic manipulations but rather by refining one of the initial assumptions. However, in general we have no guarantee that for every assumption of the form ∀∃ we would be able to provide such refinement. On the contrary, we know that in general this is not possible. Secondly, unlike Wittgenstein's approach to the problem, Behmann seems to be already knowledgeable about which bounds are going to work for the situation at hand. Indeed, later on he points out that he could have transformed the proof by using the better inequality

$$\sum_{i=1}^{n} \frac{1}{i} > \int_{0}^{n} \frac{dx}{x+1} = log(n+1)$$

which gives a bound for the next prime under $e^{\prod_m}$ where e is the natural logarithm.

6.4 Back to Wittgenstein

The section on "Euler's proof" in *Philosophical Grammar* begins with a straightforward reflection on Euler's proof:

From the inequality

$$1 + \frac{1}{2} + \frac{1}{3} + \frac{1}{4} + \ldots \neq (1 + \frac{1}{2} + \frac{1}{2^2} + \frac{1}{2^3} + \ldots) \cdot (1 + \frac{1}{3} + \frac{1}{3^2} + \ldots)$$

can we derive a number ν which is still missing from the combinations on the right-hand side? Euler's proof that there are infinitely many prime numbers is meant to be an existence proof, but how is such a proof possible without a construction?

$$\sim 1 + \frac{1}{2} + \frac{1}{3} + \ldots = (1 + \frac{1}{2} + \frac{1}{2^2} + \frac{1}{2^3} + \ldots) \cdot (1 + \frac{1}{3} + \frac{1}{3^2} + \ldots)$$

The argument goes like this: The product on the right is a series of fractions $\frac{1}{n}$ in whose denominators all multiples of the form $2^\nu 3^\mu$ occur; if there were no numbers besides these, then this series would necessarily be the same as the series $1 + \frac{1}{2} + \frac{1}{3} + \ldots$ and in that case the sums also would necessarily be the same. But the left-hand side is ∞ and the right-hand side only a finite number $\frac{2}{1} \cdot \frac{3}{2} = 3$, so there are infinitely many fractions missing in the right-hand series, that is, there are on the left-hand side fractions that do not occur on the right. (Wittgenstein 1974, p. 383; 1994, pp. 321–22)

Again, it is clear that Wittgenstein discusses Euler's equation (A). As he points out a bit further, if I extend the sum on the left-hand side until it is greater than any number which would be the result of the right hand product for prime numbers up to any given one, this part of the sum "must contain a term that doesn't occur in the right hand series, for if the right hand series contained all those terms it would yield a larger and not a smaller sum" (Wittgenstein 1974, p. 385; 1994, p. 324).

As was pointed out, there is a passage in Wittgenstein's manuscripts for the year 1930 where we find the following computations, which were neither reproduced in the *Big Typescript*, nor in *Philosophical Grammar*. This passage should be read in conjunction with the calculations at the end of the section on Euler's proof in both of these:

Thus, a sufficient condition for $\frac{1}{n} + \frac{1}{n+1} + \ldots + \frac{1}{n+\nu} \geq 1$, is that $\nu \geq 3n - 1$. If I represent the beginning of the series $1 + \frac{1}{2} + \frac{1}{3} + \ldots$ of the ordered segments that are ≥ 1, then the first of these segments goes from

	1	to	3
the second from	4	to	15
the third from	16	to	63
the mth		to	$4^m - 1$

The sum of the series $1 + \frac{1}{2} + \frac{1}{3} + \ldots$ developed up to 4^m thus surely exceeds m. Thus, we have

$$1 + \frac{1}{2} + \frac{1}{3} + \ldots + \frac{1}{4^m} > (1 + \frac{1}{2} + \frac{1}{2^2} + \ldots) \cdot (1 + \frac{1}{3} + \frac{1}{3^2} + \ldots) \cdots (1 + \frac{1}{m} + \frac{1}{m^2} + \ldots)$$

It is thus the case that in the first 4^m whole numbers, there must be one which is not divisible by any of the first m numbers. (Wittgenstein 1994, p. 325)

We will reproduce Wittgenstein's calculations in the appendix and limit ourselves here to a synthesis of the results. First of all, Wittgenstein solves the following general problem: if one starts from $\frac{1}{n}$, for arbitrary n, how far does one have to go for the harmonic series to sum up to a number ≥ 1, i.e., to determine ν such that $\frac{1}{n} + \ldots + \frac{1}{n+\nu} \geq 1$. He works out a bound through algebraic manipulations (see appendix) whereby the condition is $\nu \geq \frac{[3n^2-(n+2)]}{(n+3)}$. And since $\frac{[3n^2-(n+2)]}{(n+3)} < 3n - 1$, he thus concludes that from any point $\frac{1}{n}$ of the harmonic series one only needs to go to $\frac{1}{[n+(3n-1)]} = \frac{1}{4n-1}$ in order for that segment of the harmonic series to sum up to a number greater than 1. Applying the result to the harmonic series from its start (i.e., $n = 1$) Wittgenstein determines that $m \leq 1 + \frac{1}{2} + \ldots + \frac{1}{4^m}$. This is not as good a bound as the one used by Behmann (2^{2m-1}), although it is closely related since Wittgenstein's evaluation, $2(2^{2m-1}) = 2^{2m} = 4^m$, requires always two times as many steps as Behmann's. This is due to the fact that Wittgenstein's condition on ν gives him a very slow start when trying to majorize any number m. We have to assume that the full passage from MS 108 gives us a good idea of the proof by Wittgenstein that was referred to by Behmann, Kaufmann and Waismann in their exchange of letters. Therefore, Waismann's claim (quoted above in Kaufmann's letter dated 12.x.1930) that the two proofs are related seems to be justified.

6.5 Wittgenstein on Mathematical Existence

Wittgenstein's critical discussion of Euler's original proof and his constructive version of it are, if not of much mathematical interest, of great significance to the interpretation of his philosophy: *this is the sole instance of an extended discussion of a specific existential proof and the sole instance of a 'proof' in the whole of his writings* and, as such, they cannot be bypassed. Wittgenstein's comments on Euler's proof are extremely critical:

> Here again we have that remarkable phenomenon that we might call proof by circumstantial evidence in mathematics—something that is absolutely never permitted. It might also be called a proof by *symptoms*. The result of the summation is (or is regarded as) a symptom that there are terms on the left that are missing on the right. The connection between the symptom and what we would like to have proved is a *loose* connection. That is, no bridge has been built, but we rest content on *seeing* the other bank.
>
> All the terms on the right-hand side occur on the left, but the sum on the left-hand side is ∞ and the sum of the right-hand side is only a finite number, *so there must* ... but in mathematics nothing *must* be except what *is*.
>
> The bridge has to be built. (Wittgenstein 1974, p. 384; 1994, p. 322; 2000, p. 427)

These critical comments amount to a straightforward rejection of the validity of existential proofs: he calls them "proofs by circumstantial evidence" and says that they are "absolutely never permitted"; alternatively, he calls them "proofs by symptoms" but claims that "there are no symptoms" in mathematics. He uses the metaphor of the bridge and likens existential proofs to resting content with seeing the other bank, while he requests that "the bridge has to be built". A further critical remark in the same section can help us to understand Wittgenstein's more general thoughts about existential proofs:

> We might also put the question thus: if you had only this proof, what would you bet on it? If we discovered the primes up to N, could we later go on for ever looking for a further prime number—since the proof guarantees that we will find one?—Surely that is nonsense. For "if we only search long enough" has no meaning. (That goes for existence proofs in general). (Wittgenstein 1974, p. 384; 1994, p. 323; 2000, p. 428)

Not only is this reasoning perfectly in line with Wittgenstein's proof, which provides an interval within which one will find a new prime number, but it also squares perfectly well with the very reasoning of Brouwer and Weyl which is at the basis of their rejection of the Law of Excluded Middle.[22] Moreover, it bears an obvious relation to Kronecker's demand that an existential proof must be accompanied by a bound for the existential if it has to satisfy the demands of mathematical rigor.

Wittgenstein's critical discussion of Euler's original proof and his constructive version of it are also extremely useful because they allow us to infuse meaning into his more general pronouncements about mathematics. (From an exegetical point of view, to ignore or to bypass these remarks and the proof would simply amount to plain dishonesty.) For example, one finds in the very passage of MS 108 that we have been referring to, the following remark, which encapsulates Wittgenstein's central thought about foundations of mathematics:

> In mathematics *everything* is algorithm and *nothing* is meaning; even when it doesn't look like that because we seem to be using *words* to talk *about* mathematical things. Even these words are used to construct an algorithm. (Wittgenstein 1974, p. 468; 1994, p. 321; 2000, p. 494)

A similar remark was uttered by Wittgenstein, during a meeting with Schlick and Waismann in June 1930:

> I believe that mathematics, once the conflict about its foundations has come to an end, will look just as it does in elementary school where the abacus is used. The way of doing mathematics in elementary school is absolutely strict and exact. It need not be improved upon in any way. Mathematics is always a machine, a calculus. The calculus does not describe anything [...] A calculus is an abacus, a calculator, a calculating machine; it works by means of strokes, numerals, etc. (Wittgenstein 1979, p. 106)

And remarks to the same effect made their way into the typescript of 1931 which was edited under the title *Philosophical Remarks*:

> Let's remember that in mathematics, the signs themselves *do* mathematics, they don't describe it. The mathematical signs *are* like the beads of an abacus. (Wittgenstein 1975, §157)

> ... we can't describe mathematics, we can only do it. (And that of itself abolishes every "set theory".) (Wittgenstein 1975, §159)

"In mathematics *everything* is algorithm": it is obvious that the rejection of (nonconstructive) existential proofs is entailed by this very Kroneckerian thesis.

Perhaps it is worth pointing out here another element contained in these remarks, i.e. the idea that there are no descriptions in mathematics. According to Wittgenstein it is a mistake to think that a foundational theory (set theory or *Principia Mathematica*) could be a description, hence a 'theory'. As he would put it *"Calculation with letters is not a theory"* (1979, p. 136); "what is calculus must be separated off from what attempts to be (and of course cannot be) *theory*" (1974, p. 468; 2000, p. 494). This is the true ground of Wittgenstein's *limited* anti-theoretical and anti-foundational stance, a stance that has been entirely misunderstood by those who claim that he held no 'theses' at all.[23]

6.6 Viennese Finitism

What about Wittgenstein's reaction to Kaufmann's conjecture? If there is any discussion of it in Wittgenstein's *Nachlaß*, it remains to be found. A few remarks are nevertheless in order. To begin with, although it seems from the available evidence that Wittgenstein never met with Behmann, it is also clear that he was not kept out of the loop. A copy of the last version of Behmann's paper was sent to Wittgenstein, who must have known about Kaufmann's conjecture. There are at any rate traces left of contacts between Wittgenstein and Kaufmann. For example, in a letter to Kaufmann dated April 6, 1930, Wittgenstein wrote:

Thank you very much for sending me your book. Sadly, I am too busy to be able to read it. I hear from Mr. Waismann that our points of view coincide on many points.[24]

There is also the incredible story of a suitcase bearing the name "Kaufmann" that was apparently found in Wittgenstein's lodgings in Cambridge after his death and which Rush Rhees mistakenly sent to Walter Kaufmann, only later realizing that the owner must have been Felix Kaufmann. Later, Rhees unsuccessfully tried to recover the suitcase; one will never know its contents.[25]

It is most probable that Wittgenstein would have found Kaufmann's attempt at grounding arithmetic with help of Husserl's phenomenology uncongenial. However, Kaufmann's finitism is not very far from Wittgenstein's own *Standpunkt*, which also should be seen as a form of finitism.[26] It should be recalled here that, just prior to Behmann's visit to Vienna, Waismann had just participated in the famous meeting at Königsberg organized by the *Wiener Kreis* and Hans Reichenbach's *Gesellschaft für empirische Philosophie*, by giving a paper at a symposium on foundations of mathematics which included presentations by Rudolf Carnap on logicism, Arend Heyting on intuitionism, and Johann von Neumann on formalism.[27] Waismann's paper, entitled "Über das Wesen der Mathematik. Der Standpunkt Wittgensteins" (Waismann 1982), was not ready to be included alongside the others in the 1931 issue of *Erkenntnis* devoted to the Königsberg meeting; it appeared only in 1982. It is clear from Wittgenstein's attitude prior to the meeting that he took the opportunity to have his views publicized very seriously. There was, therefore, such a thing as Wittgenstein's *Standpunkt*. Waismann's presentation of his ideas, although rather faithful, was not formal. As Carnap pointed out in the discussion that was to be published in the same issue of *Erkenntnis*, the *Standpunkt Wittgensteins* had not yet reached a stage of development at which it could be evaluated (Hahn et al. 1931, p. 141). It never reached that stage, either in Waismann's or in Wittgenstein's own work. We are now left with the task of reconstructing this *Standpunkt Wittgensteins* and the above discussion of his remarks on the constructivization of Euler's proof should be seen as a contribution to this task. That Kaufmann's finitism is not very far from Wittgenstein's own *Standpunkt* is supported by the above-quoted letter, where Wittgenstein writes: "I hear from Mr. Waismann that our points of view coincide on many points". Therefore, something like Kaufmann's conjecture could not have been very far from Wittgenstein's mind. This would go a long way to explain his interest in constructivizing Euler's proof. But, before one might be tempted to speak of an indigenous, Viennese form of finitism, it should be pointed out that Kaufmann's conjecture implied that classical mathematics was already constructive and that Brouwer's critique was to leave it essentially untouched. Wittgenstein did not believe this. In his manuscripts for the years 1930–31, we find claims to the contrary, e.g.:

What will distinguish mathematicians of the future from those of today will really be a greater sensitivity, and that will—as it were—prune mathematics; ... (Wittgenstein 1974, p. 381; 1996, p. 23; 2000, p. 426)

It is clear that Wittgenstein had no illusions about the effects of philosophical clarity on mathematics or about the consequences of his *Standpunkt*, parts of which was the rejection of pure existential proofs.

6.7 Appendix

Wittgenstein's goal is to determine what bound is necessary for v such that $\frac{1}{n} + \ldots + \frac{1}{n+v} \geq 1$. He begins by transforming the left-hand side into

$$\frac{\frac{n}{n} + \ldots + \frac{n}{n+v}}{n}$$

This, in turn, can be written as

$$\frac{1 + \frac{n}{n+1} + \ldots + \frac{n}{n+v}}{n}$$

At this point, in order to work out a bound, Wittgenstein's divides the above sum into a first part up to $2n - 1$ ($= n + (n - 1)$) and a second part. Moreover, since

$$1 + \frac{n}{n+1} + \ldots + \frac{n}{n+(n-1)} = 1 + (1 - \frac{1}{n+1}) + (1 - \frac{2}{n+2}) + \ldots$$
$$+ (1 - \frac{n-1}{n+(n-1)})$$

We have

$$\frac{1 + \frac{n}{n+1} + \ldots + \frac{n}{n+v}}{n} = \frac{[1 + (1 - \frac{1}{n+1}) + (1 - \frac{2}{n+2}) + \ldots + (1 - \frac{n-1}{n+(n-1)})]}{n} +$$
$$+ \frac{\frac{n}{2n} + \frac{n}{2n+1} + \frac{n}{2n+2} + \ldots + \frac{n}{n+v}}{n}$$

We now consider the expression in square brackets. The 1's sum up to n and the difference

$$-(\frac{1}{n+1} + \frac{2}{n+2} + \ldots + \frac{n-1}{n+(n-1)})$$

is such that we can bound every one of its terms by a multiple of $\frac{1}{n+1}$. Indeed,

$$\frac{1}{n+1} \leq \frac{1}{n+1}$$

$$\frac{2}{n+2} = \frac{1}{n+2} + \frac{1}{n+2} \leq \frac{1}{n+1} + \frac{1}{n+1} = \frac{2}{n+1}$$

And in general,
$$\frac{\kappa}{n+\kappa} = \underbrace{\frac{1}{n+\kappa} + \ldots + \frac{1}{n+\kappa}}_{\kappa-times} \leq \underbrace{\frac{1}{n+1} + \ldots + \frac{1}{n+1}}_{\kappa-times} = \frac{\kappa}{n+1}$$

Thus,
$$\frac{1}{n+1} + \frac{2}{n+2} + \ldots + \frac{n-1}{n+(n-1)} \leq \frac{1}{n+1} + \frac{2}{n+1} + \ldots + \frac{n-1}{n+1}$$
$$\leq \frac{1}{n+1} \cdot (1 + 2 + \ldots + (n-1))$$
$$\leq \frac{1}{n+1} \cdot \frac{n \cdot (n-1)}{2}$$

As for the second part of the sum
$$\frac{n}{2n} + \ldots + \frac{n}{n+\nu}$$

Wittgenstein looks at the sum obtained from it by replacing each summand by $\frac{n}{n+\nu}$. Since there are $n + \nu - (2n - 1) = (\nu - n + 1)$ many summands, this new sum amounts to $(\nu - n + 1) \cdot \frac{n}{n+\nu}$. Also, it is clearly less than the sum from which it is obtained. Now the question becomes: Knowing a bound on

$$1 + \frac{n}{n+1} + \cdots + \frac{n}{n+(n-1)}$$

how big must we choose ν so that

$$\frac{[n - \frac{1}{2}n(n-1) \cdot \frac{1}{n+1}] + (\nu - n + 1) \cdot \frac{n}{n+\nu}}{n} \geq 1$$

The left hand side is equal to

$$1 - \frac{n-1}{2n+2} + \frac{\nu - n + 1}{n + \nu}$$

Thus,
$$1 - \frac{n-1}{2n+2} + \frac{\nu - n + 1}{n + \nu} \geq 1$$

iff
$$-\frac{n-1}{2n+2} + \frac{\nu - n + 1}{n + \nu} \geq 0$$

iff
$$-(n+\nu)(n-1) + (2n+2)(\nu - n + 1) \geq 0$$

iff
$$-n^2 - \nu n + n + \nu + 2n\nu + 2\nu - 2n^2 - 2n + 2n + 2 \geq 0$$

iff

$$nv + 3v - 3n^2 + 2 + n \geq 0$$

iff

$$v(n+3) - 3n^2 + 2 + n \geq 0$$

iff

$$v(n+3) \geq 3n^2 - (n+2)$$

iff

$$v \geq \frac{3n^2 - (n+2)}{n+3}$$

The final observation consists in noticing that $3n - 1$ majorizes $\frac{[3n^2-(n+2)]}{n+3}$ for all $n \geq 1$. Remark: in Wittgenstein (1994, p. 325), the symbol '$\Rightarrow$' should be read as '$\geq$'. In Wittgenstein (1974, p. 386) and 2000, p. 429 some of the equality signs should be '$\geq$'.

7

BETWEEN VIENNA AND BERLIN: THE IMMEDIATE RECEPTION OF GÖDEL'S INCOMPLETENESS THEOREMS

7.1 Introduction

"What is going on with Mr Gödel? I hear all kinds of exciting things but I cannot make out what this is all about."[1] So wrote Heinrich Scholz to Rudolf Carnap on April 16, 1931. The rumor of Gödel's groundbreaking discoveries was spreading quickly and Gödel's work was immediately perceived as having tremendous importance for the debate on the foundations of mathematics.[2]

Through the work of Dawson (1990, 1997) and Köhler (1991) we are offered much information about the reception of Gödel's theorems in the immediate aftermath of Gödel's announcement of his revolutionary result at the conference in Königsberg which took place from the 5th to the 7th of September 1930.[3] Through the study of the correspondence with Bernays, Herbrand, and von Neumann, which is to be published

---

I would like to thank John W. Dawson, Solomon Feferman, Wilfried Sieg, and Richard Zach for many useful suggestions on a previous draft of this paper. A special thanks goes to the archivists of the Rudolf Carnap *Nachlaß* and of the Hans Reichenbach *Nachlaß* in Konstanz (Dr Brigitte Uhlemann), of the Heinrich Behmann *Nachlaß* in Erlangen (Prof. Christian Thiel and Dr Volker Peckhaus), of the Felix Kaufmann *Nachlaß* in Konstanz (Dr Martin Endress) and of the Bernays *Nachlaß* in Zurich (Frau Yvonne Vögeli) for their help. I am also very grateful to the Wissenschaftskolleg zu Berlin for having provided ideal conditions for work during the year 1997–8.

in the fourth volume of Gödel's *Collected Works* [see Gödel 2003b,a], we are also well informed about the reaction of these logicians to Gödel's results. The main outline of these exchanges is discussed in Dawson (1997, pp. 68–75). We learn there that Carnap was probably the first one to learn about the results on August 26, 1930 during a conversation at the Café Reichsrat in Vienna. Feigl was apparently also there and Waismann joined the group later that afternoon. After the very modest remark by Gödel at the Königsberg conference, more precisely on September 7, 1930, John von Neumann was quick in realizing the importance of the discovery announced by Gödel. Indeed, he discovered that one of the consequences of Gödel's statement was that arithmetic could not prove its own consistency, that is, the second incompleteness theorem. However, just before receiving von Neumann's letter stating the result Gödel had already sent his article, which also included a statement of the second incompleteness theorem, for publication (October 23, 1930). John von Neumann decided to leave the priority of the discovery to Gödel. As for Bernays, the correspondence shows clearly that he at first had some trouble in understanding some of the technical details in Gödel's proofs. More importantly, the correspondence shows that Gödel was fully conscious of the undefinability of truth. As pointed out in Sieg (1994, p. 103) Herbrand learned about Gödel's results in Berlin through von Neumann in November 1930. Inspired by Gödel's results, Herbrand wrote a long letter to Gödel whose contents were to inspire Gödel to formulate the notion of general recursive function.[4]

The first public presentation of the incompleteness results was given on January 15, 1931 in front of the Schlick circle. Minutes of the event in the hand of Rose Rand are preserved in the Carnap archives. These minutes have recently appeared in Stadler (1997, pp. 278–80) and are translated here in section 7.3 with the original German in note. The following week Gödel presented his results in Menger's Colloquium. Later in the year Zermelo took issue with Gödel's results in private correspondence. This exchange has been studied by Dawson (1985) and Grattan-Guinness (1979) (cf. Taussky-Todd 1987).

In the following I would like to supplement the information given above by making use of several documents from different archives containing material relevant to the history of the foundations of mathematics for the period in question. I will do this by presenting the relevant documents with a little commentary when needed. Section 7.2 will give extracts of the correspondence between Hempel and Kaufmann concerning the first reactions in Berlin to Gödel's results. Section 7.3 reproduces the discussion within the Schlick circle. Section 7.4 contains some of the correspondence between Carnap, von Neumann, and Reichenbach concerning the publication of the proceedings of the Königsberg meeting. The letters speak against Dawson's assertion that Carnap had not understood Gödel's results immediately. What seems to emerge from the correspondence concerning the publication of the lectures from the Königsberg meeting is that the results were perceived as so devastating that the lectures should only reflect the pre-Gödelian state of things. Finally, the last section will give yet another instance of the reaction of a logician, Leon Chwistek, to his first reading of Gödel's results.

7.2 Kaufmann and Hempel

Dawson (1997) quotes Hempel saying, in an interview with Richard Nollan on March 24, 1982, that John von Neumann had given an announcement of Gödel's second incompleteness result in his lectures on proof theory. Dawson is however tentative about the exact datation of the event ("presumably in the fall of 1930", p. 70). A fuller report emerges from a letter sent from Hempel to Kaufmann dated December 13, 1930:

> Shortly before the beginning of the last Colloquium [i.e. Reichenbach's] we were strongly impressed by the news given by Mr von Neumann that Mr Gödel managed to prove the undecidability of the consistency of mathematics. Of course, we find this report of the utmost interest and I have just yesterday asked Mr Carnap to ask Mr Gödel to send a copy of his work to Berlin. If it arrives in time we intend to organize a special session before Christmas to discuss the work. It would be very nice if you could give Mr Gödel a quiet word of moral encouragment, if a presentation of his proof already exists.[5]

This report differs from that given by Hempel in 1982 in being richer but also in specifying that the announcement was made for the first time in connection to Reichenbach's Colloquium and not in the lectures on proof theory by von Neumann.[6]

Kaufmann's reply to Hempel's letter shows that knowledge of Gödel's results were well known in Vienna. Moreover, Kaufmann held a phenomenological point of view concerning the completeness of number theory which he will also express in the discussion following Gödel's lecture to Schlick circle (see section 7.3). In essence, Kaufmann argued that the consistency proof for arithmetic could be obtained by means of an "intellectual insight". On December 19, 1930 he wrote to Hempel:

> I have called Mr Gödel on the phone immediately after receiving your letter. He has promised to send to Mr von Neumann the galley proofs of his article on the critique of the formalist investigations on consistency, which will presumably appear at the end of January in Monatshefte für Mathematik. You already know my ideas on the matter. I am of the opinion that one can obtain the consistency of mathematics directly, with all the obviousness that one could strive for and achieve, directly from Peano's axioms if only one transforms the fifth axiom [the axiom of induction] in the form that I have indicated, but that any attempt to justify operating with uncountable infinities by means of a consistency proof is hopeless.[7]

It does seem that Gödel's result confirmed Kaufmann's skepticism towards the possibility of justifying uncountable set theory, but that he did not immediately perceive the possible consequences for his own approach to the consistency of arithmetic. This will also emerge from the next document.

7.3 The Discussion of Gödel's Result at Schlick's Circle

The discussion following the first public presentation of the results is preserved in minutes of the event written by Rose Rand.[8] The minutes confirm that Gödel, unlike von Neumann, was resisting use of the incompleteness results to claim that Hilbert's program had failed and that he thought intuitionism was not affected by his results. Moreoever, we see the sharp objection he gave to the phenomenological claims made by Kaufmann. The reader should keep in mind that the discussion on intuitionistic proofs is here made with the tacit assumption that the class of intuitionistic proofs and that of finitistic proofs coincide. For the relationship between intuitionism and finitism in this period see Mancosu (1998a, pp. 167–68). In the hope of providing a more coherent text I have interpolated in square brackets a translation which departs from the original document.

Schlick circle

Minutes for January 15, 1931

On consistency and decidability in axiomatic systems Discussion on Mr Gödel's lecture.

Kaufmann asks how things stand with respect to the decidability of sentences of a subsystem.

Gödel replies that in as far as it can be proved this proof must make use of means that cannot be formalized within the subsystem. This also agrees, he continues, with his proofs.

In reply to a question of Hahn, Gödel recalls once more the main ideas of his proof for the impossibility of a consistency proof. If one adds the statement of consistency for a system to the system itself—and this addition can be carried out formally—then in this extended system a sentence which was undecidable in the original system will become decidable. Consequently, the consistency of a system cannot be shown in the system itself. In reply to a question of Schlick Gödel formulates the conjecture by Mr von Neumann: if there is at all a finite proof of contradiction [consistency] then this can also be formalized. Thus Gödel's proof implies the impossibility of a proof of contradiction [consistency] in general.

Hahn asks about the application to the axiomatic system by Heyting:

Gödel: Heyting's system is narrower than Russell's system. If it is o-consistent [omega-consistent] then there will also be undecidable statements in it.

Hahn points out that one of the basic ideas of the proof that "there is no meaningful totality of what is constructible simpliciter" plays, since Cantor's diagonal proof, a decisive role in set theory.

Gödel remarks that the application of this very thought also puts into question whether the totality of all intuitionistically unobjectionable proofs finds a place in *one* formal system. This is, he continues, the weak point in von Neumann's argument.

Kaufmann asks how things are with, say, the consistency of sentences that have no pair of concepts in common or, say, the Peano axioms. There is a first, there is a last number [no last number].

Gödel replies that in a consistency proof it is not a question of the concepts as such. Here it is not at all a question of a consistency proof in the contentual sense.

To an objection by Kaufmann that contentual consistency proofs are not ruled out Gödel clarifies: that such "insights" are not proofs at all in the sense of a formalistic theory.

Neumann asks whether there are systems so simple that the concrete form of the undecidable sentence can be given in a transparent way.[9]

Gödel replies that this depends on the system in which one represents it. He recalls the decisive trick of his procedure. The isomorphic mapping of the inferential figures to sequences f_2 of sequences f_1 of numbers which makes it possible to formulate provability internally in the first place. If $S(f_2)$ denotes, e.g., an inference figure and $l(f_2)$ the length of the corresponding [inferential] chain then the derivability of f_1 is written as

$$Bewf_1 \equiv (\exists f_2)\{S(f_2) \& f_2[l(f_2)] = f_1\}$$

One can rest satisfied with this or proceed and replace the symbol S by its definition.

Hahn draws attention to the book by Lusin "Sur les ensembles analytiques". Lusin carefully distinguishes higher classes in the existence proofs for the Borelian sets according to whether the diagonal procedure applies or not.

Moreover Hahn asks:

whether it would be possible to eliminate the diagonal procedure from Gödel's proof.

In response, Gödels states that the undecidable formula given by him is actually constructible. Its content is finitary just like that of Goldbach's or Fermat's theorem.

Finally to a remark by Kaufmann Gödel expresses his opinion: that Intuitionism according to Brouwer's conception is not touched by his work because it does not want to be contained in any formal system.

7.4 Von Neumann, Carnap, and Reichenbach

As I mentioned earlier the correspondence between John von Neumann and Gödel is going to be published in the fourth volume of Gödel's *Collected Works* [see Gödel 2003b,a]. I would like to reproduce here three letters, one from Carnap to von Neumann, one from von Neumann to Carnap, and one from von Neumann to Reichenbach which give an idea of how they dealt with Gödel's results when faced with the issue of the publication of the proceedings of the Königsberg conference. Moreover, if we take Carnap's statements at face value, and we have no reason not to do so, they do show that even before the Königsberg meeting he was fully aware of the dramatic impact that Gödel's results had for the foundations of mathematics as conceived in the twenties. As I mentioned in the introduction this goes against the reading of the situation given by Dawson, who claims that Carnap had at first not understood Gödel's ideas:

In any case, Gödel did confide his discovery to Carnap prior to the discussion at Königsberg, as we know from *Aufzeichnungen* in Carnap's *Nachlass*. Specifically, on 26 August 1930, Gödel met Carnap, Feigl, and Weismann at the Cafe Reichsrat in Vienna, where they discussed their travel plans to Königsberg. Afterward, according to Carnap's entry for that date, the discussion turned to "Gödels Entdeckungen: Unvollständigkeit des Systems der PM; Schwierigkeit des Widerspruchsfreiheitbeweises." Three days later another meeting took place at the same Cafe. On that occasion, Carnap noted "Zuerst [before the arrival of Feigl and Waismann] erzählt mir Gödel von seiner Entdeckungen." Why then at Königsberg did Carnap persist in advocating consistency as a criterion of adequacy for formal theories? That he might have done so just to provide an opening for Gödel seems hardly credible. It seems much more likely that he simply failed to *understand* Gödel's ideas. (As it happens, a subsequent note by Carnap dated 7 February 1931, after the appearance of Gödel's paper, provides confirmation: "Gödel hier. Über seine Arbeit, ich sage, dass sie doch sehr schwer verständlich ist") (Dawson 1990, p. 86, 1984, p. 115, 1997, p. 69).

My reading of the situation is that although Carnap surely had difficulties with the technical details of the proof—just like Bernays—the importance of Gödel's result was immediately clear to him. There is another explanation for his position in Königsberg and it is given in Carnap's reply to von Neumann's letter.

Berlin,[10] June 6, 1931
Dear Mr Carnap,
Many thanks for your letter and your suggestions.

In my contribution I have referred to the more detailed bibliography contained in Weyl's handbook article[11], without giving any further literature. I have done this mainly because I did not want to give value judgments by means of my choice. There are some programmatic publications by Hilbert in which he claims that certain things have been proven or almost proven while this is in fact not even approximately the case (continuum problem and so on). Therefore I would like neither to quote these articles nor to correct or ignore them, and I therefore believe that the best course of action is to include a general reference to the handbook article.

Concerning your suggestion on mailing the off-prints [of the lectures by Heyting, Carnap, and von Neumann] together I would like to say the following: I have known from Gödel (whom you also know?) about his results in Königsberg and later in correspondence. These theorems (which have since then appeared) show that certain logical and mathematical statements, such as for instance the consistency of analysis, are unprovable in certain formal-logical systems. In my opinion, however, they also show that it is also impossible contentually, for I am of the opinion that all contentual inferences are reproducible in *one* specific formal system (which I do not want to describe more precisely but for which Gödel's theorems hold). Thus I am today of the opinion that

1. Gödel has shown the unrealizability of Hilbert's program
2. There is no more reason to reject intuitionism (if one disregards the aesthetic issue, which in practice will also for me be the decisive factor)

Therefore I consider the state of the foundational discussion in Königsberg to be outdated, for Gödel's fundamental discoveries have brought the question to a completely different level. (I know that Gödel is much more careful in the evaluation of his results, but in my opinion on this point he does not see the connections correctly). I have often spoken with Reichenbach whether it makes any sense, under these circumstances, to publish my lecture. Had I had to give the lecture four weeks later it would have sounded fundamentally different. We agreed finally that it should be written down as the description of a certain outdated state of things.* [* I would like to emphasize: *nothing* in Hilbert's aims is *wrong*. Could they be carried out then it would follow from them absolutely what he had claimed. But they cannot in fact be carried out, this I know only since September 1930]

As you will be able to gather from these lines I would not otherwise send out a single off-print of my lecture. I do not want, however, to interfere with a joint action and I agree to your proposal and to the list you suggested.

With best regards, Johann von Neumann
Berlin W 10, Hohenzollernstrasse 23

[Prof. Dr Rudolf Carnap[12]
Vienna XIII/5 Stauffergasse 4]
Vienna, July 11, 1931

Dear Mr von Neumann,

I understand quite well the doubts you have against the publication of your Königsberg lecture. I had similar doubts not only about the publication but already about the delivery of my lecture. However, I have taken the standpoint that here it was not an issue of presenting a final solution or to present my personal conception of the different points but rather to inform on the basic thought from which logicism takes its start and on the possibilities and difficulties for its implementation in so far as they can be surveyed at the moment. My own conception is not that of Russell but rather rests on a combination of Russell's and Hilbert's ideas: therefore it will also be affected quite strongly by Gödel's results. For publication the description of the state of the problems at the point of the Königsberg conference has however value at least so long as the consequences which derive from the new results for the different conceptions are not yet clearly surveyable.

The references to the literature in Weyl are rather scanty. Would it not perhaps be more appropriate in its place or in addition to refer to Fraenkel's *Set Theory* (3rd ed.)[13] which gives extensive references to the literature up to 1928.

With best regards, R.C.

We have no record of von Neumann's conversations with Reichenbach. However, a letter from von Neumann to Reichenbach, concerning the issue of the publication in *Erkenntnis* of von Neumann's Königsberg lecture, gives more information about von Neumann's reaction to Gödel's results. The letter does not specify where von Neumann is writing from and it is undated but it must have been written around Summer 1931.

Dear Professor Reichenbach,[14]

Here enclosed I am sending the report you asked for. Since the manuscript had already been completed by the time your letter arrived it does not have the desired title. Please modify the title so that it fits your arrangement. By the way, I have decided not to mention Gödel since the view that there is still a certain hope for proof theory has found some advocates: among others Bernays and Gödel himself. Of course, in my opinion this view is mistaken but a discussion of this question would lead out of the given context. Therefore I would like to speak about this on another occasion.

Sincerely yours, J. v. Neumann

P.S. Please forgive me for not having sent the long-promised off-prints but they are now in Berlin. I feel very ashamed and I will send them as soon as I come back to Berlin.

7.5 Chwistek

Several logicians working in the late twenties and early thirties had of course to reckon with the consequences of Gödel's results for their foundational thinking. I have already mentioned the exchange between Zermelo and Gödel on this issue. As an example of another reaction to Gödel's results let me now adduce the case of Leon Chwistek. On July 26, 1931, Chwistek wrote to Kaufmann from Lwov:

There is much talk here about Mr Gödel's work. I would like very much to have this work. I would be very grateful if you could convince Mr Gödel to send me his work [...] Gödel's result seems to me very welcome for the nominalist standpoint. First of all one sees that Hilbert's method is in no way essential. Moreover, that simple type theory does not lead at all to a pure logic but rather to an ontological system, something I have been emphasizing for a long time. However, Gödel's result seems to me impossible in pure type theory and thus in classical mathematics. The introduction of the concept "all properties of x" in metamathematics, where practically one has to operate with properties of higher order is a clear misunderstanding. Even if it does not lead to antinomies it is nonetheless clear that it must lead to results whose value is problematic.[15]

The next letter confirms that Chwistek had received copies of Gödel's works and raises some objections to Gödel's results.

Dear Colleague,[16]

Thank you very much for your kindness in obtaining Dr Gödel's work for me. Please extend my thanks to Dr Gödel.

Dr Gödel's method has caused me some doubts, first of all since it does not seem possible to carry it out purely symbolically. The problem concerns the theoretical semantics that I am developing and which leads to the hierarchy of types of expressions. I still have doubts whether it would not be possible to construct an antinomy in Dr Gödel's system. One can indeed instead of class-signs speak of the *meaningful class-signs*, which is also granted by Dr Gödel himself (p. 174 of [Gödel 1931]). Thus we can instead of formula (1) [p. 175] use the formula $n \in K \equiv \overline{[R(n);n]}$ and consider the class-sign $R(q)$ so that $[R(q);q]$ says, when interpreted contentually, that n belongs to K. It is then easy to show that the sentence $[R(q);q]$ is contradictory.

I would be very grateful if you could tell me how things stand with this problem.

I am now busy with the development of a rigorous symbolical system of theoretical semantics and metamathematics in which the class symbol $f\{x\}$ is reduced to the substitution schema (*abcd*).

Best regards, L. Chwistek.

Dear Colleague,[17]

In my last letter I have raised some doubts concerning Dr Gödel's work that have completely disappeared after a more attentive study of the problem. I had at first thought that there was a tacit introduction of non-predicative functions which makes the use of a rigorous symbolical procedure impossible and I even feared the possibility of an antinomy. Now I see that this is out of the question. The method is truly wonderful and it fits pure type theory thoroughly. Please communicate this rectification to Mr Carnap.[18] I will soon prepare the article for *Erkenntnis*.[19] I hope that the article "Neue Grundlagen" 2nd Communication will appear soon.[20]

Best regards, L. Chwistek

8

REVIEW OF GÖDEL'S COLLECTED WORKS, VOLS. IV AND V

8.1 Introduction

When I arrived at Stanford as a graduate student in 1984 I immediately heard about the projected publication of Gödel's *Collected Works* (henceforth CW). My interest was piqued not only by hearing Solomon Feferman talk about the project but also through my friendship with one of the early collaborators, a young man from Japan, Tadashi Nagayama, who was an expert on Gabelsberger, the special stenography used by Gödel and many other German and Austrian academicians at the time to take notes or draft lectures and letters. I often wondered: how does someone from Japan end up being an expert on Gabelsberger? Perhaps at the time I was naïve enough to think that the major obstacle to such an edition was the transcription of such documents. Twenty years afterwards and with five volumes of Gödel's *Collected Works* on my desk it has become apparent to me what a gigantic effort the publication of these five volumes has been. The Gödel *Nachlaß* at Princeton had to be catalogued, a selection of what would go into the volumes had to be determined, experts on and off the editorial board had to be commissioned to write introductions and so on to other, and not less daunting, problems of choices of typography, layout, textual annotations, etc.

Following the original plan, all the published writings were reprinted in Vol. I (1986) and Vol. II (1990), accompanied by facing translations into English for German originals. Beyond these, the remaining volumes were to contain "Gödel's unpublished manuscripts, lectures, lecture notes, and correspondence, as well as extracts from his scientific notebooks" (CW, Vol. I, 1986, preface). Indeed, Volume III (1995) contained a comprehensive selection of unpublished essays and lectures, while the present Volumes IV and V are devoted to an equally comprehensive selection from the correspondence. However, the plan to publish extracts from the scientific notebooks had to be abandoned, even though extensive transcriptions had been made from them, because the task of publishing them in a coherent form became forbidding. This is material for future scholars to pursue.

The two new volumes give us a broad and representative selection of the most important correspondence in which Gödel engaged throughout his life. In total, fifty individual correspondents are represented. The criteria followed for the selections are described by the editors as follows: "In all cases our criterion for inclusion was that letters should either possess intrinsic scientific, philosophical or historical interest or should illuminate Gödel's thoughts or his personal relationship with others." (CW, Vol. IV, p. v) The edition itself contains, in addition to the letters also the corresponding calendars. Furthermore, volume V contains a full inventory of Gödel's *Nachlaß*, an extremely useful tool for any future research on Gödel. The list of correspondents is impressive. Represented are, among others, Bernays, Boone, Carnap, Church, P. Cohen, Herbrand, Menger, A. Robinson, Tarski, von Neumann, Wang, and Zermelo. But the editors were wary of just going after the big names. Indeed, they included correspondence with lesser known figures and sometimes even completely unknown persons (at least in academic circles). Interesting in this connection are also items of more personal correspondence, such as the letters to his mother, in which Gödel discusses his opinions on, among other things, a number of religious topics.

8.2 The Correspondents

The editors have wisely decided to publish the correspondence not in chronological order but rather alphabetically according to the correspondent's name. When the letters are not in English we are presented with the German text and a facing translation into English. Following the format of Volumes I–III, each exchange with a correspondent is prefaced by an introduction written by one of the editors or by a specialist commissioned especially for the occasion. The list of collaborators for the project is impressive and it goes without saying that these, often lengthy, introductions are outstanding pieces of scholarship and a great help to the reader who is thereby informed as to the context and importance of the exchange. The alphabetical presentation and the relevant introduction allow for easier reading as many topics are pursued from letter to letter with the same correspondent.

We find broadly four types of correspondents:

(1) Academics (this represents the majority of the selections)

(2) Relatives (Gödel's mother, Marianne)

(3) Editors (A. Angoff, T. Honderich, P. Schilpp and others)

(4) Occasional correspondents who solicited Gödel's help on various issues related to his work or career (C. Reid, Grandjean and others)

Before discussing in more detail some of the correspondence, I would like to point out that it not only enlightens us about Gödel's life and work but also gives us precious information about the correspondents themselves in some of whose cases, for example, Herbrand, we have very little else left.

We do, of course, know quite a bit about Gödel's life and work, in particular through the books by Wang (1974, 1987) and Dawson (1997) among others.[1] Wang had consulted the Gödel *Nachlaß* early on and having recently reread his book *Reflections on Kurt Gödel* I was actually surprised at how much he was able to take in of this vast collection. However, Wang's treatment of Gödel's opinions has to be approached cautiously and thus the correspondence is a more trustworthy source of information (on Wang on Gödel see Parsons 1998). And in any case, no summary can do full justice to the original texts now in front of us.

Gödel was obviously a very reserved person on personal matters. The correspondence in Volumes IV and V does not tell us much about, say, his relationship with his wife, or with the academic community in Princeton, or about his health problems, or his views on contemporary world affairs. Such topics might surface from time to time (for instance, Gödel discusses various types of drugs with Bernays, CW, Vol. IV, pp. 298–301) but overall we do not get much. Incidentally, even the letters to his mother published here do not so much discuss his feelings or personal matters but rather discuss more intellectual matters. Of course, since I have not seen the complete bulk of the correspondence (totaling 245 letters from Gödel to his mother; five of them published (excerpts) in Volume IV) it is quite possible that the selection here was determined by the intrinsic intellectual interest of the topics discussed.[2] In these letters Gödel shows a keen interest in religion. As John Dawson informs us in the lucid introduction to the exchange, Gödel, who as a youth attended Protestant schools in Brno, was very critical of the religious instruction he had received. He seemed to have despised organized religion while holding on to the importance of religion per se. He characterized his position as theistic but not pantheistic (see the unsent reply to Grandjean, CW, Vol. IV, p. 448). While the correspondence *says* little about the more personal aspects of Gödel's life, it certainly *shows* much about him as a person and a scholar. Gödel was meticulous and cautious. While these are virtues, especially in a logician, in the case of Gödel he was so almost to a fault. This led to feelings of exasperation and in some cases rage on the part of the affected correspondents.[3] The correspondence is a real window into this aspect of Gödel's personality. Examples of such are the finicky discussion with Behmann about a minor correction of a statement by Dubislav to be published in *Erkenntnis*, the stalling tactics with Heyting about a projected joint book on foundations or with Schilpp for the essay on Carnap, and his correspondence with Nagel, Angoff and Follett concerning the publication of Nagel and Newman's "Gödel's Proof" (1959). Additional examples could easily be provided. In some cases, it was obvious intellectual dissatisfaction that stopped Gödel (as in the case of the failure to deliver to Schilpp the essay for the volume on Carnap, "Is mathematics syntax of language"; or the English translation of a revised version of "On a hitherto unutilized extension of the finitary standpoint", long promised to Bernays and published only posthumously); in other cases, such as the ones mentioned above with Nagel, Angoff and Follett, personal idiosyncrasies played a role too.

Gödel's caution was especially evident in the case of philosophical topics, as in the two cases just mentioned. While he had very strong opinions (for instance about the negative effect of "anti-metaphysical" or "anti-platonist" attitudes for logic, science and culture in general) he rarely expressed them publicly. However, the correspondence is more explicit about these issues. Famously, he claimed that what stopped Skolem or anyone else from discovering the completeness theorem was the finitist bias shared by all researchers in logic in the 1920s. On December 7, 1967, discussing issues related to Skolem's work, he wrote to Wang:

> This blindness (or prejudice, or whatever you might call it) of logicians is indeed surprising. But I think the explanation is not hard to find. It lies in a widespread lack, at that time, of the required epistemological attitude towards metamathematics and toward non-finitary reasoning... I might add that my objectivistic conception of mathematics and metamathematics in general, and of transfinite reasoning in particular, was fundamental also to my other work in logic. (CW, Vol. V, pp. 397–98; already in Wang 1974, p. 8)

It is nothing short of paradoxical that the person with whom Gödel shared this criticism of the "finitistisches Vorurteil" (CW, Vol. V, p. 422), that is, Ernst Zermelo, was to show very little understanding of Gödel's achievements.[4] This particular commitment to finitism seems to have been only one facet of what Gödel characterized as a prejudice of the time, which could perhaps be negatively characterized as the rejection of "Platonism" in philosophy, religion, and the sciences. Instances of such prejudice were mechanism in biology, nominalism in the philosophy of mathematics and the claim that there is no mind separate from matter (see Wang 1974, p. 326). The correspondence gives us many examples of Gödel's position and how it influenced even the history of his publications or lack thereof. For instance, concerning his essay on Carnap, Gödel wrote to Schilpp:

> However, I feel I owe you an explanation why I did not send my paper earlier. The fact is that I have completed several different versions, but none of them satisfies me. It is easy to allege very weighty and striking arguments in favor of my views, but a complete elucidation of the situation turned out to be more difficult than I had anticipated, doubtless in consequence of the fact that the subject matter is closely related to, and partly identical with, one of the basic problems of philosophy, namely the question of the objective reality of concepts and their relations. On the other hand, ~~in view~~ because of widely held prejudices, it may do more harm than good to publish half done work. (Gödel to Schilpp, Feb. 3, 1959, Vol. V, p. 244)

As for the consequences in religion of these biases, in a letter to his mother he remarks that "90% of contemporary philosophers see their principal task to be that of beating religion out of men's heads, and in that way they have the same effect as the bad churches" (CW, Vol. IV, pp. 436–37).

8.3 Some Novel Aspects of Gödel's Thought Which Emerge from the Correspondence

One obvious question to ask is whether the correspondence gives us novel information about Gödel's thought and work. Thus, the question I would like to address is: *What more do we know that we could not have known without the correspondence?*

It would obviously be useless within the scope of a review to try to answer this question by giving a survey of the range of topics treated with so many different correspondents. It suffices to say that Feferman's clear and informative introduction to the long term correspondence with Bernays runs to thirty-nine pages and the aim there is only to highlight the major topics of discussion between the two scholars. In the light of this I propose to mention in this section two aspects, of the many which could be selected, of Gödel's thought about which the correspondence provides novel information. Then the rest of the review will focus on one specific issue only, that is what the correspondence contributes to our knowledge of the context of Gödel's most important and best known result, that is the 1931 incompleteness theorem (this is shorthand for first and second incompleteness theorems).

Let me thus briefly indicate the importance of the correspondence in enlightening Gödel's thought on philosophy and finitism.

Philosophy

It is well known that, after 1943, philosophical interests came to dominate Gödel's thinking. Writing in 1966 to Bernays he expresses thanks "for the wishes concerning my philosophical investigations. For they have been my principal interest for a long time" (Gödel to Bernays, May 22, 1966, Vol. IV, p. 253) Concerning Gödel on philosophy the correspondence is surprising in two ways. First of all, for what it contains. The correspondence with Günther, wonderfully introduced by Parsons, shows that Gödel had quite a substantial interest in post-Kantian idealistic philosophy (Fichte, Hegel, Schelling). In the correspondence with Bernays we find extensive discussions of, among others, Fries, Nelson, Wittgenstein, and Hegel. In some cases the comments are quite biting: "As for Wittgenstein's book on the foundations of mathematics, I also read parts of it. It seemed to me at the time that the benefit created by it may be mainly that it shows the falsity of the assertions set forth in it." The footnote adds: "and in the *Tractatus*. (the book itself really contains very few assertions)". (Gödel to Bernays, Oct. 30, 1958, Vol. IV, p. 161)

However, the two volumes of correspondence contain very little on the two philosophers Gödel spent most time reading and thinking about, Husserl and Leibniz. This is where we wish we had more by way of transcriptions of the notebooks and probing deeper in this connection is a task left for future researchers.

Finitism

Gödel's position on finitism underwent several changes. As Feferman points out in his introduction to the Gödel–Bernays exchange, the correspondence allows us to gauge the extent to which Gödel views concerning the upper bound of finitary reasoning remained "unsettled". While in the 1931 paper (in a remark to be discussed below), Gödel seems to entertain the possibility that finitistic reasoning might outstrip the resources of Peano Arithmetic (**PA**), later comments indicate that Gödel would consider finitary reasoning as contained in **PA**. From previously available material, it would have seemed safe to conclude that for Gödel finitistic reasoning could in fact be captured in systems much weaker than Peano Arithmetic, quite possibly in Primitive Recursive Arithmetic. However, the correspondence with Bernays shows that Gödel was much taken by Kreisel's 1960 characterization of finitistic proof by means of autonomous progressions of formal systems of ordinal logics (in the sense of Turing). In 1961 Gödel writes to Bernays: "I had interesting discussions with Kreisel. He now really seems to have shown in a mathematically satisfying way that the first ϵ-number is the precise limit of what is finitary. I find this result very beautiful, even if it will perhaps require a phenomenological substructure in order to be completely satisfying" (CW, Vol. IV, p. 193). And again to Bernays in 1967: "I am now convinced that ϵ_0 is a bound on finitism, not merely in practice but also in principle, and that it will also be possible to prove that convincingly" (CW, Vol. IV, p. 255). The issue of what exactly is the extent of finitism re-emerges also in later correspondence concerning Bernays' proof of transfinite induction up to ϵ_0 for the second edition of *Grundlagen der Mathematik* (1968–70). This led to a discussion of whether free choice sequences are to be included in finitary mathematics. Other essential information on the topic of Gödel's views on finitism is also to be gained from other correspondence, such as the exchange with Herbrand, von Neumann (see below), and others.

Let us now move to the question of what the correspondence contributes to our knowledge of the context of Gödel's incompleteness theorem. I will survey in succession:

(1) what Gödel says about the heuristics of the theorem;

(2) a number of comments he makes related to the proof;

(3) the immediate impact of the theorem (wonderfully documented by the correspondence with, among others, Bernays, Herbrand, and von Neumann);

(4) finally, a few things about Gödel's own interpretation of the lasting philosophical significance of the result.

I will assume that the reader has encountered the incompleteness theorem before. I should also state here that throughout the review I refer to parts of the correspondence that might have been familiar to, and used in publications by, those researchers who had access to parts of the Gödel correspondence prior to their publication in the *Collected Works*. But this does not detract from the fact that the information I will refer to became known through the correspondence, which is now made available to the general public. For this reason I will only refer briefly in notes to articles in the literature where the correspondence was already exploited.

8.4 The Heuristic Path to the Theorem: Truth and Provability

As we have seen, it was his unorthodox epistemological attitude that Gödel identified as the condition of possibility for his groundbreaking results on completeness. In the case of the incompleteness theorem the key to the result was again to focus on a notion of which philosophers and logicians were skeptical, that is, the notion of truth. He writes this explicitly to Wang (CW, Vol. V, p. 398). To the same attitude he credits his work on the consistency of the axiom of choice with the remaining axioms of set theory in contrast to similar developments in Hilbert (see letter to van Heijenoort, July 8, 1965, Vol. IV, p. 324).

Let us begin with a description of the heuristics which led to the incompleteness theorem given in an unsent letter to Yossef Balas, a master degree student at the University of Northern Iowa.[5] The letter dates from around 1970:

I have explained the heuristic principle for the construction of propositions undecidable in a given formal system in the lectures I gave in Princeton in 1934 ... The occasion for comparing truth and demonstrability was an attempt to give a relative model-theoretic consistency proof of analysis in arithmetic. This leads almost by necessity to such a comparison. (Gödel to Balas, undated, Vol. IV, p. 10)

A crossed-out paragraph connects nicely with remarks I made in the previous section on the prejudices of the time:

However in consequence of the philosophical prejudices of our times 1. nobody was looking for a relative consistency proof because i[t] was considered axiomatic that a consistency proof must be finitary in order to make sense 2. a concept of objective mathematical truth as opposed to demonstrability was viewed with greatest suspicion and widely rejected as meaningless. (Gödel to Balas, undated, Vol. IV, p. 10)

An interesting question here is: How did Gödel manage not to fall prey to what he called the prejudices of the time? This is something that the correspondence does not clear up for us. He writes to Grandjean (CW, Vol. IV, p. 448) that one important philosophical influence was Heinrich Gomperz. Wang (1987, p. 22) seems to locate the source of Gödel's Platonism in his discovery of Plato through Gomperz's lectures. But the correspondence is silent on this. I think this is a point that would be interesting to investigate more thoroughly.[6]

Let us conclude with a the description of the heuristics for the discovery of the incompleteness results given by Gödel to Balas:

⟨For an arithmetical model of analysis is nothing else but an arithmetical ϵ-relation satisfying the comprehension axiom $(\exists n)(x)[x \in n \equiv \phi(x)]$. by an arithmetical Now, if in the latter "$\phi(x)$" ⟨ is replaced ⟩ by "$\phi(x)$ is provable", such an ⟨$\epsilon-$⟩ relation can easily be defined. Hence, if truth were equivalent to provability, we would have reached our goal. However, (and this is the decisive point) it follows from the correct solution of the semantic paradoxes i.e., the fact that the concept of "truth" of the propositions of a language *cannot be expressed* in the same language, while provability (being an arithmetical relation) *can*. Hence true $\neq$ provable.⟩ (Gödel to Balas, undated, vol. IV, p. 10)

This suggests that the undefinability of truth (a theorem usually attributed to Tarski) was the key fact in the heuristics leading to the incompleteness theorem.[7] Of course, since these lines were written in 1970 we want to make sure that they are not just a "rational reconstruction" of what happened forty years earlier. The best evidence we have to support Gödel's account in this connection comes from a letter to Bernays dated

April 2, 1931 where Gödel discusses at length the definition of truth for a first-order system Z in a second-order system S. Gödel writes that

> Simultaneously and independently of me (as I gathered from a conversation), Mr. Tarski developed the idea of defining the concept "true proposition" in this way (for other purposes, to be sure). (Gödel to Bernays, Vol. IV, p. 97)

Concerning this passage, Feferman remarks that

> the specific definition of W [the set of true sentences of arithmetic, PM] that [Gödel] describes depends on the fact that every element of the standard model of Z is denoted by a numeral; the more general definition of truth for languages of other structures given by Tarski in terms of satisfaction is not noted by Gödel. (CW, Vol. IV, p. 45)

Gödel appeals to the undefinability of truth within a formal system of arithmetic also in his explanations to Zermelo on October 12, 1931:

> In connection with what has been said, one can moreover also carry out my proof as follows: the class W of correct formulas *is never* coextensive with a class sign of that same system (for the assumption that that is the case leads to a contradiction). The class B of provable formulas *is* coextensive with a class sign of that same system (as one can show in detail); consequently B and W cannot be coextensive with each other. But because $B \subseteq W$, $B \subset W$ holds, i.e. there is a correct formula A that is not provable. Because A is correct, not-A is also not provable, i.e., A is undecidable. This proof has, however, the disadvantage that it furnishes no construction of the undecidable statement and is not intuitionistically unobjectionable. (Gödel to Zermelo, October 12, 1931, Vol. V, pp. 427 and 429, his emphasis)

We thus see why the reasoning that led to the discovery of the theorem was later removed from the final presentation of the proof in 1931, where Gödel does not prove the undefinability of arithmetical truth. While he did not share in the supposed prejudice of the time he realized that his argument would be open to objection had he made use of a proof that was not "intuitionistically unobjectionable". The reader should keep in mind that in 1931 "intuitionistically unobjectionable" was taken to mean "finitistically unobjectionable" (see Mancosu 1998a, pp. 167–68). In a postscript to a letter to van Heijenoort written on February 22, 1964, Gödel remarks again on the motivations that led him to the incompleteness theorem:

> Perhaps you were puzzled by the fact that I once said an attempted relative consistency proof for analysis led to the proof of the existence of undecidable propositions and another time that the heuristic principle and the first version of the proof were those given in Sect. 7 of my 1934 Princeton lectures. But it was precisely the relative consistency proof which made it necessary to formalize either "truth" or "provability" and thereby forced a comparison of the two in this respect. (Gödel to van Heijenoort, February 22, 1964, Vol. V, p. 313)

It was thus the comparison between truth and provability that was the heuristic key to the theorem. And while some of this could have been already gathered from the published version of the 1934 lectures (in Davis (1965, pp. 39–74); and now in CW, Vol. I), the correspondence adds considerable information on the issue.

8.5 The 1931 Presentation

Concerning the specific details of the incompleteness paper we learn something new from the correspondence with van Heijenoort. The 1931 article shows the existence of independent arithmetical statements from a theory P which consists of a system of axioms for arithmetic with a simple theory of types as background logic. The simple theory of types also has two essentially mathematical axioms needed to include Peano Arithmetic, namely axiom I.3, induction in second order form, and axiom IV, full comprehension. The individual variables of the system range over the class of individuals, which is here identified with the class of natural numbers. However, Gödel could have proceeded by simply developing arithmetic within the simple theory of types (with the axiom of infinity). Why did he choose the former approach? One reason is given in a letter dated August 14, 1964. Gödel says:

I identified the individuals of PM with the integers in order to obtain a system every proposition of which has a well-defined meaning in classical mathematics and, therefore, viewed from the standpoint of classical mathematics, must be either true or false. The question of completeness is of philosophical interest only for systems which satisfy some requirement of this kind. (Gödel to van Heijenoort, Vol. V, p. 316)

And in the following letter (August 15, 1964) a second motivation is adduced:

On rereading my letter of Aug. 14 I find that in suggestion 2. ad M(4) I have given the wrong impression that what I say there was my only reason for adjoining Peano's axioms. Another reason, of course, was the simplification of the proofs which ⟨results⟩ from it. In fact, I believe that either one of these two considerations would have been sufficient by itself. However, if the second one had been my only reason, I could have omitted the axiom of complete induction, thereby admitting other individuals besides the integers. (p. 317)

As Goldfarb remarks in his useful introduction to the Gödel-van Heijenoort correspondence, Gödel's "remarks here are not echoed in any other known writing of his" (CW, Vol. V, p. 304)

8.6 The Impact of Gödel's Incompleteness Theorem

After the announcement of his first incompleteness theorem at Königsberg in September 1930, the news of Gödel's epoch-making result traveled fast. Nöbeling informs Menger (CW, Vol. IV, p. 85), Courant and Schur inform Bernays (CW, Vol. IV, p. 81), von Neumann tells Herbrand (CW, Vol. V, p. 15; see also Hempel to Kaufmann, December 13, 1930 in Mancosu (1999b)) who in turn tells Behmann (CW, Vol. IV, p. 39). In Mancosu (1999b) [[Chapter 7]] I documented the rapid spread of the news among several logicians and philosophers of mathematics. However, knowing of the forthcoming edition of the correspondence, I did not discuss the correspondence between Gödel and Bernays, Herbrand[8] and von Neumann[9] on the relevance of Gödel's incompleteness theorem for Hilbert's program, an issue which had already been briefly summarized in Wang (1987, pp. 43 and 84–91), and Dawson (1997, pp. 68–75). Gödel in his 1931 paper states that his results

do not contradict Hilbert's formalistic viewpoint. For this viewpoint presupposes only the existence of a consistency proof in which nothing but finitary means of proof is used, and it is conceivable that there exist finitary proofs that cannot be expressed in the formalism of P. (CW, Vol. I, p. 195)

Gödel will eventually abandon this viewpoint.[10] For instance in March 1966 he writes to Constance Reid that

Hilbert's scheme for the foundations of mathematics remains highly interesting and important in spite of my negative results. What has been proved is only that the *specific epistemological* objective which Hilbert had in mind cannot be obtained. This objective was to prove the consistency of the axioms of classical mathematics on the basis of evidence just as concrete and immediately convincing as elementary arithmetic. (Gödel to Reid, March 22, 1966, Vol. V, p. 187, his emphasis)

But this is not the way things stood in the immediate aftermath of the discovery and publication of the result. Whereas von Neumann and Herbrand saw in Gödel's results a definitive defeat for Hilbert's program, Bernays and Gödel were more cautious. Let us recall that after the unassuming remark made by Gödel at the Königsberg meeting stating what amounts to the first incompleteness theorem (September 7, 1930), John von Neumann very quickly realized the implications of the result Gödel had announced. Soon afterwards von Neumann was able to prove that "the consistency of mathematics is unprovable", that is what we now call the second incompleteness theorem. He informed Gödel of the discovery:

I have recently concerned myself again with logic, using the methods you have employed so successfully in order to exhibit undecidable properties. In doing so I achieved a result that seems to me to be remarkable. Namely, I was able to show that the consistency of mathematics is unprovable. (von Neumann to Gödel, November 20, 1930, Vol. V, p. 337)

Here is how von Neumann states the result:

In a formal system that contains arithmetic one can express, following your considerations, that the formula $1 = 2$ cannot be the end-formula of a proof starting with the axioms of this system—in fact, this formulation is a formula of the formal system under consideration. Let it be called W. In a contradictory system any formula is provable, thus also W. If the consistency [of the system] is established intuitionistically, then it is possible, through a "translation" of the contentual intuitionistic considerations into the formal [system], to prove W also. (On account of your result one might possibly doubt such a "translatability". But I believe that in the present case it must obtain, and I would very much like to learn your view on this point). Thus with unprovable W the system is consistent, but the consistency is unprovable. I showed now: W is always unprovable in consistent systems, i.e., a putative effective proof of W can certainly be transformed into a contradiction. (von Neumann to Gödel, November 20, 1930, Vol. V, p. 337)

Von Neumann concludes his letter by asking Gödel to express his point of view on the issue of the "translation" of "contentual intuitionistic" considerations into formal proofs.

Moreover, he asked when Gödel's article would appear and whether he could get a copy of the proofs so as to present his result in agreement with Gödel's presentation. Finally, he informed Gödel that the mathematician E. Schmidt considered his result "to be the greatest logical discovery in a long time" (CW, Vol. V, p. 337)

However, before von Neumann wrote this to Gödel, the latter had already sent (October 23, 1930) the incompleteness article for publication. In it Gödel had also stated the second incompleteness theorem. Gödel's reply to von Neumann's letter seems to have been lost but on 29 November, 1930 von Neumann thanks Gödel for his letter and says:

> Many thanks for your letter and your reprint. As you have established the theorem on the unprovability of consistency as a natural continuation and deepening of your earlier results, I clearly won't publish on this subject. (von Neumann to Gödel, November 29, 1930, Vol. V, p. 339)

Two points in this letter deserve special emphasis. The first concerns the possibility of translating intuitionistic proofs into formalistic proofs:

> I believe that every intuitionistic consideration can be formally copied, because the "arbitrarily nested" recursions of Bernays-Hilbert are equivalent to ordinary transfinite recursions up to the appropriate ordinals of the second number class. This is a process that can be formally captured, unless there is an intuitionistically definable ordinal of the second number class that could not be defined formally—which is in my view unthinkable. Intuitionism clearly has no finite axiom system, but that does not prevent its being a part of classical mathematics that does have one. (von Neumann to Gödel, November 29, 1930, Vol. V, p. 339)

One has to keep in mind that the discussion on intuitionistic demonstrations is here carried out with the assumption that intuitionism and finitism are equivalent (it will only be in 1933 that the two will definitely be shown to be non equivalent).

Von Neumann drew from Gödel's result the conclusion that the so-called *Grundlagenfrage* could only be answered negatively:

> Thus, I think that your result has solved negatively the foundational question: there is no rigorous justification for classical mathematics. (von Neumann to Gödel, November 29, 1930, Vol. V, p. 339)[11]

Gödel replied sending two letters and the proofs of the 1931 article but the letters have been lost. The last letter from von Neumann to Gödel dealing with questions of incompleteness was written on January 12, 1931. It contains several interesting points:

1. Von Neumann claims to have a decision procedure for deciding the provability or unprovability of sentences built by means of Boolean operations (conjunction, negation) and the predicate $B(x)$, "provable". As pointed out by Sieg in his introduction (CW, Vol. V, p. 332) this seemingly anticipates a solution to Friedman's 35th problem, given in print by Boolos in 1976.
2. Apparently Gödel had communicated some observations on ω-consistency; in particular, he claimed that ω-consistency at a certain level could be inferred from consistency at the next higher level.

3. Von Neumann expressed disagreement with Gödel's position on the formalization of intuitionistic proofs. Gödel had claimed (in one of the two lost letters) that it is not at all clear that every intuitionistic proof can be captured in a formal system. Von Neumann replied:

"Clearly I cannot prove that every intuitionistically correct construction of arithmetic is formalizable either in A or M or even in Z. [A = first order arithmetic; M = second order arithmetic; Z = von Neumann's axiomatization of set theory] for intuitionism is undefined and undefinable. But is it not a fact, that not a single construction is known that cannot be formalized in A, and that no living logician is in the position of naming such [a construction]? Or am I wrong and you know an effective intuitionistic arithmetic construction whose formalization in A creates difficulties? If that, to my great surprise, should be the case, then the formalization should certainly work in M or Z!" (von Neumann to Gödel, January 12, 1931, Vol. V, p. 343)[12]

4. Finally, von Neumann suggested a simplification for the proof of the second incompleteness theorem as presented by Gödel.

This concludes the correspondence between the two scholars centering on the incompleteness theorem.

Another scholar who shared von Neumann's evaluation of the foundational situation in the light of Gödel's results was Jacques Herbrand. I will not, for reasons of space, discuss this part of the correspondence but only refer to Sieg's insightful introduction to the Gödel-Herbrand correspondence in which he spells out the relevance of the exchange both for the issue of the scope of finitism and for the consequences of Gödel's theorems for Hilbert's program. The former topic will lead the reader to explore one facet of the emergence of computability theory through Herbrand's influence on Gödel's notion of general recursive function. The latter topic nicely ties in with the other exchanges Gödel was having with von Neumann and Bernays.

The issue of the effect of Gödel's theorems on Hilbert's program was of course a central one in the discussion with Bernays. The correspondence with Bernays on this topic is much more extended and more technical than the one with von Neumann and thus I will only emphasize a few points.

In a letter dated December 24, 1930 Bernays asks Gödel whether he can send him the proofs of the article on incompleteness, of which he got wind through Courant and Schur. Gödel replies on December 31 sending the proofs of the article. On January 18, 1931, Bernays replies saying that he has received the proofs on January 14. What comes next is very important for the issue of whether the ω-rule in Hilbert (1931) was introduced by Hilbert as a remedy to Gödel's incompleteness. Bernays writes:

Your results have moreover a special topical interest for me that goes beyond their general significance, in that they cast light on an extension of the usual framework for number theory recently undertaken by Hilbert. (Bernays to Gödel, January 18, 1931, Vol. IV, p. 83)

Bernays specifies that by number theory he means first-order arithmetic. Then he states the extension proposed by Hilbert:

Hilbert's extension now consists in the following rule: If $A(x_1, ..., x_n)$ is a *recursive* formula (according to your designation), which might be shown, finitarily, to yield a numerical identity for arbitrarily given numerical values $x_1 = z_1, x_2 = z_2, ..., x_n = z_n$, then the formula $(x_1)...(x_n)A(x_1, ..., x_n)$ can be used as an initial formula (i.e., as an axiom). (Bernays to Gödel, January 18, 1931, Vol. IV, p. 83)

According to Bernays, Hilbert has shown that if $(x_1)...(x_n)A(x_1, ..., x_n)$, with $A(x_1, ..., x_n)$ quantifier-free, can be shown to be consistent with number theory by means of finitistic considerations, then $(x_1)...(x_n)A(x_1, ..., x_n)$ is provable in number theory augmented by the new rule. Bernays also claimed that the consistency of the new rule followed by the techniques for proving consistency known from the works of Ackermann and von Neumann (as shown by a student, R. Schmidt). Now Bernays draws his conclusion for what Gödel's results mean for Hilbert's program. Gödel has shown that there are (recursive) formulas $A(x_1, ..., x_n)$ such that all their numerical instances can be proved in number theory but number theory does not prove $(x_1)...(x_n)A(x_1, ..., x_n)$. Thus, Bernays concludes, there are statements that are finitistically justifiable but unprovable in number theory. The consequence to be drawn, according to him, is that if a formal system can be finitistically shown to be consistent, then there is a finitistic sentence that cannot be finitistically justified in the formal system. By contrast, if one assumes with von Neumann that all the finitistic considerations are already included in number theory then one needs to conclude that a finitistic consistency proof of number theory is impossible:

Thus if, as von Neumann does, one takes it as certain that any and every finitary consideration may be formalized within the framework of the system P [the system of Principia]—like you, I regard that in no way as settled—one comes to the conclusion that a finitary demonstration of the consistency of P is impossible. (Bernays to Gödel, January 18, 1931, Vol. IV, p. 87)

However, Bernays shows that he is unhappy with Hilbert's proposed extended system. He considers it inelegant to have both the axiom of induction and the new ω-rule. This leads him to propose a new rule, R, from which he can derive both the axiom of induction and Hilbert's new rule. R is formulated as follows:

If $A(x_1, ..., x_n)$ is a (*not necessarily recursive*) formula in which only $x_1, ..., x_n$ occur as free variables and which, through the substitution of any numerical values whatever in place of $x_1, ..., x_n$, is transformed into a formula such as is derivable from the formal axioms and the formulas already derived, the formula $(x_1)...(x_n)A(x_1, ..., x_n)$ may be adjoined to the domain of the derived formulas. (Bernays to Gödel, January 18, 1931, Vol. IV, p. 89)

For this rule R, Bernays claims to have a sketch of a consistency proof along the lines of those given for number theory.

Notwithstanding Feferman's opinion to the contrary (see Vol. IV, p. 44, note 1), in my view the above exchange (and a detailed study of the chronology surrounding the publication of Hilbert's 1931 article on the ω-rule) is strong evidence that the ω-rule was not introduced by Hilbert as a reaction to Gödel's incompleteness theorems.[13] However, I do agree with Feferman that if one discards his preferred explanation then one needs to give an account of what might have led Hilbert to entertain the extension. At the moment no such account is on offer.

Incidentally, it is only in this correspondence with Gödel that Bernays begins worrying about the formalizability in number theory of an alleged finitistic demonstration of the consistency of number theory (see letters from Bernays to Gödel in April and May 1931; CW, Vol. IV, pp. 91–105). The demonstration in question is referred to by Hilbert in the 1928 paper "The foundation of mathematics" and is attributed to Ackermann. This is not the 1924 proof contained in Ackermann's dissertation but rather a different proof by means of ϵ-substitution proposed by Ackermann in 1927 (see Zach 2003, p. 242). Bernays considers the proof correct and worries about which parts of the proof, in light of Gödel's theorem, cannot be formalized in Z (first-order number theory). He ends up singling out certain forms of nested recursion as the culprit. But this is somewhat puzzling because von Neumann in previous correspondence with Bernays had already pointed out that Ackermann's 1927 proof had a gap (see Zach 2003, p. 243). Moreover, from our point of view it is surprising that Bernays thinks that nested recursions are not formalizable in Z, since they are. But evidently that was not clear to Bernays at the time and Bernays' position is understandable given that Gödel had only explicitly shown the primitive recursive functions to be be representable in Z.

8.7 The Philosophical Relevance of the Incompleteness Theorem According to Gödel

On November 1942 Gödel wrote to Schilpp that "the meddling of scientist[s] into philosophy has so often proved useful for both". (CW, Vol. V, p. 219) That is certainly the case for Gödel's own pronouncements on various philosophical topics. Concerning Gödel's evaluation of the lasting importance of the incompleteness theorems the correspondence confirms what was already published in Wang (1974) and the claims made by Gödel in the Gibbs lecture (1951) now published in Volume III. I will however quote here an extremely nice and succinct formulation of how Gödel saw the philosophical consequences of his result for general philosophy and Hilbert's program in particular. In a letter to Leon Rappaport written on August 2, 1962 Gödel wrote:

Nothing has been changed lately in my results or their philosophical consequences, but perhaps some misconceptions have been dispelled or weakened. My theorems only show that the *mechanization* of mathematics, i.e. the elimination of the *mind* and of *abstract* entities, is impossible, if one wants to have a satisfactory foundation and system of mathematics.

I have not proved that there are mathematical questions undecidable for the human mind, but only that there is no *machine* (or *blind formalism*) that can decide all number theoretical questions (even of a very certain special kind).

Likewise it does not follow from my theorems that there are no convincing *consistency* proofs for the usual mathematical formalisms, notwithstanding that such proofs must use modes of reasoning not contained in those formalism. What is practically certain[1] is that there are for the classical formalisms, no conclusive combinatorial consistency proofs (such as Hilbert expected to give), i.e. no consistency proofs that use only concepts referring to finite combinations of symbols and not referring to any infinite totality of such combinations. (Gödel to Rappaport, August 2, 1962, Vol. V, p. 176)

In note 1 Gödel specified: "No formal proof has yet been given because the concept of a combinatorial proof, although intuitively clear, has not yet been precisely defined". This last quote could be the starting point for tracing in the correspondence Gödel's position(s) on the extent of finitism, the philosophical import of the consistency proof given by the Dialectica interpretation, his notion of abstract entity and so on. The reader will have to discover these and many other wonderful topics by going through the correspondence. The above was only meant to give a sample of the rewards to be expected by delving into the volumes.

8.8 Conclusion

The completion of the publication of Gödel's *Collected Works* marks an epochal moment in our appreciation of the career of one of the most brilliant minds of the 20th century. The two volumes of the *Collected Works* containing the correspondence provide us with important and essential information about Gödel's life and intellectual achievements. But they also shed light on several important aspects of the history of logic and philosophy in the twentieth century. The correspondence refines and extends our knowledge of the technical and philosophical issues related to constructivism, proof theory, model theory, recursion theory, and set theory. Moreover, it provides information on other areas of Gödel's thought such as physics, general philosophy, and theology. Through it one can retrace the important debates and the profound thoughts that led to, and originated from, some of the most important logical results of the twentieth century. Thanks to the introductions by the editors (Dawson, Feferman, Goldfarb, Parsons, and Sieg) and those of the collaborators (Beeson, Fenstad, Kanamori, Linnebo, Machover, and Malament) the reader can immediately access the correspondence with the required historical, philosophical, and technical background. While I referred to some of the introductions in my review it is impossible to convey the full richness of content contained in them and the fact that these introductions are uniformly of superb quality and of the highest level of scholarship. The translations are accurate and readable at the same time. The editorial apparatus is precise without being daunting. The volumes are beautifully produced in T$_E$Xand contain a rich collection of photographs. The two volumes are expensive but one hopes that OUP will soon bring out a paperback edition, as it has done for Vols. I–III.

I can only conclude by saying that the completion of the publication of Gödel's *Collected Works* is an extraordinary achievement of the highest intellectual importance.

PART III

Phenomenology and the Exact Sciences

Introduction

SUMMARY

Part III concerns the relation between phenomenology and the exact sciences during the twenties. This is still a largely unexplored area and one of the goals of these papers has been that of showing how rich that discussion was and how fruitful it is to investigate this topic without stopping only at Husserl's contributions. In particular the investigations center on Hermann Weyl, Oskar Becker, and Dietrich Mahnke. Becker's exchange with Weyl is of pivotal importance as Husserl saw in Weyl, the mathematician, and in Becker, the philosopher, who, perhaps in tandem, might provide an adequate phenomenological foundation for mathematics and physics. Chapter 9 could belong to Part II but I think it works well here by providing some of the background to Hermann Weyl's engagement with phenomenology. Chapters 10 and 11 (co-authored with Thomas Ryckman) investigate the major point of rupture between the claims of phenomenology to be able to ground each cognitive act in intuition and the fact that both mathematics and physics had by the 1920s shown the limits of such ambitions. This realization led Becker and Weyl to abandon the Husserlian form of phenomenology presented in *Ideas* and to embrace either a new form of phenomenology (as in Becker's mantic phenomenology) or, as in Weyl's case, a symbolic construction of the world. The outcome is a large fresco of historical, philosophical, and technical problems facing Becker and Weyl in their attempt to square phenomenology and the sciences. Ryckman and I planned (and at first submitted) Chapters 10 and 11 as a unified treatment of these issues but on account of length we had to split the original into two separate articles. I am delighted to have the opportunity to reunite them. The only thing missing from the original articles is the transcription of Becker's letters to Weyl (Weyl's letters to Becker have not survived) that were provided in the respective appendices. For reasons of space, it seemed unsuitable to include them here again. The interested reader will be able to access them by going back to the original publications. In addition, I omitted the first four sections of Chapter 11, for they are identical to the first four sections of Chapter 10.

Chapter 12 explores similar topics by looking at the correspondence between Mahnke and Becker. Dietrich Mahnke was a student of Hilbert and Husserl in Göttingen. The study details the opposition, within the context of an account of the exact sciences, between a phenomenological approach faithful to Husserl's *Ideas* (Mahnke) and the more Heideggerian (anthropological) approach defended by Becker. The correspondence between Mahnke and Becker has been published in Becker (2005).

BIBLIOGRAPHICAL UPDATE

There has been a flurry of activity both on Hermann Weyl and on Oskar Becker since the articles in this part were written. Here I will only mention the literature not already cited in the articles. However, I will not take into account the rich literature on Husserl's philosophy of mathematics or on phenomenological foundations of logic, mathematics, and physics from a contemporary point of view (connections to Gödel etc.; see van Atten (2007, 2009), Boi (2007), and Tieszen (2005) for recent books in this connection), which is not a topic addressed in the articles of part II. Let's begin with Weyl. On Weyl and the continuum see van Atten *et al.* (2002), Bell (2000), Folina (2008), Scholz (2000). The recent literature on Weyl's philosophy of mathematics and physics has also emphasized, in addition to the Husserlian influence, the important role of Fichte on Weyl's thought. On these matters see Bell (2004), Feist (2002, 2004), Pollard (2005). Especially rich for Weyl's thought on foundations of geometry and physics are Ryckman (2003, 2009), Scholz (2004, 2005, 2006), and Sieroka (2007, 2009). For a recent overview of predicativity see Feferman (2004b). As a consequence of the yearly meetings in Hagen devoted to the thought of Oskar Becker, organized by Anneliese Gethmann-Siefert, we also have many new contributions to Becker's philosophy of mathematics. They are all contained in the two volumes Gethmann (2002) and Peckhaus (2005c). On Becker and geometry see Janich (2002); on Becker and intuitionism see Gethmann (2002), van Atten (2005c,a). On Becker and Zermelo see Peckhaus (2005a). On Becker's criticism of formalist conceptions of existence see Peckhaus (2005b). On more general aspects of Becker's position in philosophy of mathematics see Thiel (2005), Giugliano (2005), Poser (2005), Emrich (2005), and Wille (2005). As for Mahnke the reader is referred to Becker (2005) for a publication of his correspondence with Becker and to de Risi (2009) for a study of his correspondence with Reichenbach.

9

HERMANN WEYL: PREDICATIVITY AND AN INTUITIONISTIC EXCURSION

HERMANN WEYL'S role in the debate on the foundations of mathematics in the 1920s cannot be overestimated. Although Weyl is not as often mentioned as Hilbert or Brouwer in the popular accounts, he played an important role in sparking off the debate on foundational issues (Weyl 1921c). Furthermore, his ideas on predicativity, as expressed in *Das Kontinuum* (1918), continue to exert a lasting influence on contemporary work in predicative analysis and set theory.

Weyl (1885–1955) took his doctorate with Hilbert in Göttingen in 1908. In 1913 he became Professor at the Eidgenössische Technische Hochschule in Zurich. In 1930 he succeeded Hilbert in Göttingen, and after 1933 he moved to Princeton as professor at the Institute for Advanced Study until his retirement in 1951.

Weyl was a universal mathematician. He contributed to number theory, algebra, geometry, analysis, mathematical physics, logic and epistemology. In the field of philosophy of mathematics and logic Weyl published two books and several articles. The books are *Das Kontinuum* (1918) and *Philosophy of Mathematics and Natural Sciences* (1927, 1949).[1]

9.1 Weyl's Contribution to the Philosophy of Mathematics up to the Late 1920s

Weyl's contribution to the philosophy of mathematics in the second and third decade of the twentieth century can for simplicity be divided into four phases. Weyl begins his foundational work with a *Habilitation* on the definition of mathematical concepts. The main problem addressed by Weyl is that of making precise the vague idea of definite property found in Zermelo's work on the axiomatization of set theory (Zermelo 1908c).

In doing so he provided an answer that in large part anticipates that given by Skolem in 1922, the latter being the one that is now commonly accepted. Weyl's point of view is still in large measure "classical". In any case we do not find the explicit criticisms to the foundations of mathematics that are to be found in his later work.

Weyl's second phase, and the most creative one, coincides with the publication of the book *Das Kontinuum* (1918). In this book Weyl criticizes set theory and classical analysis as "a house built on sand" and proposes a remedy to the uncertainty that in his opinion affected a great part of classical mathematics. In my analysis of the book I will distinguish two moments. In the destructive part Weyl proposes a number of criticisms of contemporary analysis and set theory, but also to some of the accepted reconstructions of classical mathematics, such as Zermelo's set theory and Russell's theory of types. In the positive part Weyl develops his own arithmetical continuum, or a Weylean number system, in which large portions of analysis can be carried out.

Just a few years after the publication of *Das Kontinuum*, Weyl discovers the new foundationalist proposal championed by Brouwer, that is, intuitionism. This was the most radical proposal in the foundations of mathematics: Most radical because it urged an abandonment of the logical principle of the excluded middle for infinite totalities, and the abandonment of most of infinitistic mathematics. The costs to be paid were high: The intuitionistic reconstruction of mathematics had to sacrifice a great deal of classical mathematics. However, this did not frighten Weyl who spoke of Brouwer as "*die Revolution*" (incidentally, this led Ramsey to talk of Brouwer and Weyl as Bolscheviks, which in turn sparked Wittgenstein's characterization of Ramsey as a "bourgeois philosopher"). Weyl joined Brouwer's critique of classical mathematics, and abandoned his previous position, mainly on epistemological grounds.

Finally, the last 5 years of the 1920s yielded an attempt on the part of Weyl to try to find a middle ground between intuitionism and Hilbert's programme. Hilbert was conducting a personal dispute with Brouwer (which eventually culminated in the *Annalen* affair, see van Stigt 1990) and he could not easily swallow that the best of his students had joined Brouwer's camp in the area of foundations. Weyl attempts in this second part of the 1920s to do justice to Hilbert's programme but his allegiance remained with constructivism. I will also refer to the philosophical background (Husserl's phenomenology), which formed the constant underpinning of Weyl's intuitionistic tendencies.

At the end of the 1920s, and in particular with the discovery of Gödel's incompleteness theorems, the foundational crisis inugurated by Weyl's article in 1921 lost some of its urgency, and in 1955 Weyl himself spoke of it as a paper written in a bombastic style which reflected the atmosphere of an excited time (i.e. the time following World War I). Let us now explore the details of Weyl's contribution to the philosophy of mathematics.

9.2 Set Theory and Impredicative Definitions[2]

In the Concluding Remarks to Chapter 1 of *The Continuum*, Weyl gave an appraisal of the steps that led to his position in the philosophy of mathematics. The first stage of his philosophical apprenticeship was dominated by the set-theoretical approach, which through the work of Cantor and Dedekind had drastically changed the face of mathematics.[3]

Let us recall Dedekind's contribution. Dedekind gave, by means of set-theoretical techniques, a thoroughgoing justification of analysis, and thus of irrational numbers, in his booklet *Continuity and Irrational Numbers* (1872). In this work irrational numbers are defined as entities corresponding to the cuts in the field of rational numbers. Dedekind's justification of the notion of irrational number on the notion of set of rational numbers satisfying certain properties, namely that of being a cut, presupposed, however, the notion of *rational number* and that of (infinite) set. It was also Dedekind's belief that the notion of number in general could be characterized by appealing to basic logical concepts. This he attempted to show in his work on *Was sind und was sollen die Zahlen* (1888), which presents a foundation of the natural numbers based on his theory of chains, that is, sets with specific properties. The reduction of analysis to logic (containing a large amount of what we classify as set theory) seemed to have been achieved once and for all. However, problems began to emerge. The process of reduction of arithmetic to logic had in fact used at various stages a number of problematic notions, or at least as problematic as the notions that had to be grounded, for example, the notion of infinite set, the notion of a set of all objects of thought, and a number of problematic procedures, the so-called impredicative definitions.

Well known is also Frege's attempt to provide a logicistic foundation for arithmetic and the great difficulties which he encountered in carrying out the project. Frege had assumed that for any property $P(x)$ it made sense to talk about the course of values of $P(x)$ as a totality. More formally, and anachronistically, Frege postulated that given any $P(x)$, one could speak of the totality of objects satisfying $P(x)$, that is,

$$\exists X (x \in X \leftrightarrow P(x))$$

Russell's paradox showed that even at this very basic level one could run into problems. He considered the following property $P(x) = x \notin x$ and showed that the set X such that $x \in X \leftrightarrow x \notin x$ (which is supposed to exist according to Frege's postulation) gives rise to an antinomy, that is $X \in X \leftrightarrow X \notin X$.[4]

The following years witnessed an attempt to take care of this situation by means of different strategies. The most important ones are those related to the names of Zermelo and Russell.

Zermelo, following Hilbert's axiomatization of geometry, offered an axiomatization of set theory, which assumed only the existence of those sets whose definition could be given through a "definite propositional function" [*definite Klassenaussage*]. More precisely: if $P(x)$ is a "definite propositional function" and Y is a set already given then

$$\exists X (x \in X \leftrightarrow x \in Y \text{ and } P(x)) \qquad \text{(Axiom of separation)}$$

Here is Zermelo's original formulation:

Axiom III. (Axiom of separation [*Axiom der Aussonderung*]). Whenever the propositional function $C(x)$ is definite for all elements of a set M, M possesses a subset M_C containing as elements precisely those elements x of M for which $C(x)$ is true. (Zermelo 1908c, p. 202)

Unfortunately, Zermelo did not specify what it means for a *Klassenaussage* to be *definit*.

Russell developed a theory of types in which the type of self-referential situation evidenced by Russel's paradox could not arise. In particular, impredicative definitions were excluded. One of the consequences of this was a rather awkward reconstruction of mathematics. In particular one had to deal with real numbers of different levels. In the attempt to avoid these undesired consequences, Russell introduced the notorious Axiom of Reducibility, which states that for any set defined at some level n there is already an extensionally equivalent set at level 1.

Since impredicative definitions will play an important role later on, it is worthwhile to spell out how they appear in a classical setting and how Zermelo's set theory allows these types of definition.

One of the very first occurrences of the distinction between predicative and impredicative definitions goes back to (Poincaré 1906, p. 307). According to Poincaré, a definition is predicative "only if it excludes all objects that are dependent upon the notion defined". Poincaré rejected the use of impredicative definitions, and the ban against impredicative definitions was also maintained by Russell. As the latter put it: "No totality can contain numbers defined in terms of itself (*vicious circle principle*)" (Russell 1908). The history of the various, and sometimes conflicting, formulations of the notion of impredicative definition and of the vicious circle principle cannot be recounted here (see Thiel 1972, Chihara 1973, Heinzmann 1985). The discussion on impredicative definitions became topical in the debates following Zermelo's proof that every set can be well-ordered (Zermelo 1904). As is well known, Zermelo had given a proof of this claim by making use of the axiom of choice. In the proof he had to consider the so-called γ sets in order to create a new γ set L_γ given by the following condition:

$$L_\gamma = \{x : \text{ for some } \gamma \text{ set } Y, x \in Y\}$$

Since L_γ is itself a γ set we are faced with a non predicative definition. The vicious circle involved in this procedure, according to Poincaré, is that in order to determine whether $x \in L_\gamma$ we have to overview the totality of γ sets to see whether x belongs to at least one of them. But since L_γ itself is a γ set, this would involve a vicious circle. Zermelo's reply to Poincaré's critique, which he thought would "threaten the existence of all of mathematics" (Zermelo 1908b, p. 198), was to point out the widespread use of such definitions not just in the realm of abstract set theory, as, for example, Dedekind's theory of chains, but also in the realm of analysis:

Now, on the one hand, proofs that have this logical form are by no means confined to set theory; exactly the same kind can be found in analysis wherever the maximum or the minimum of a previously defined "completed" set of numbers Z is used for further inferences. This happens, for example, in the well-known Cauchy proof of the fundamental theorem of algebra, and up to now it has not occurred to anyone to regard this as something illogical. (Zermelo 1908b, pp. 190–91)[5]

In defending the legitimacy of impredicative definitions Zermelo also insisted that the object being defined is not "created" through such a determination. As we shall see, this is exactly the crucial philosophical issue concerning impredicative definitions.

Let us see what principles of Zermelo's set theory allow the formation of impredicative definitions. We shall consider two classic examples, Dedekind's theory of chains and the construction of the least upper bound for a bounded set of real numbers.

In 1888 Dedekind proceeded to the characterization of the natural numbers in two steps. First he showed, by a fallacious argument, that there exists a simply infinite system (or set), that is, a system that can be mapped one-to-one into a proper subset of itself. Then he showed that each simply infinite system contains (an isomorphic copy of) a K-chain, that is, a set that contains 1, and such that whenever x belongs to it, its successor also belongs to it. Finally, the set of natural numbers is characterized, up to isomorphism, with the intersection of all K-chains contained in a simply infinite set. From the logical point of view, an intersection corresponds to the definition of a set by means of a universal quantification where the quantifier ranges over the power set of the simply infinite system.

In Zermelo's axiomatization of set theory such a characterization of the natural numbers is permissible on the following grounds. First of all, Zermelo rejects Dedekind's fallacious proof for the existence of a simply infinite system and postulates an axiom of infinity:

Axiom VII. (Axiom of Infinity). There exists in the domain at least one set Z that contains the null set as element and is so constituted that to each of its elements a there corresponds a further element of the form $\{a\}$, in other words that with each of its elements a it also contains the corresponding set a as an element. (Zermelo 1908c, p. 204)

Two more axioms enter into the proof for the existence of the natural numbers. The first is the power set axiom:

Axiom IV. (Axiom of the power set). To every set T there corresponds another set UT, the power set of T, that contains as elements precisely all subsets of T. (Zermelo 1908c, p. 203) [U stands for subsets [*Untermenge*]]

The other axiom is the separation axiom which allows the creation of the intersection set of a set of sets (Zermelo 1908c, p. 202). Thus Zermelo establishes the existence of the natural numbers by the following three steps. The axiom of infinity provides a simply infinite system Z. The power set axiom allows him to construct the power set of Z. Finally, by means of the separation axiom he can take the intersection of all chains contained in Z. It is essential to realize that from a logical point of view the application of the separation axiom involves a universal quantification over the set of subsets of Z.

A similar situation arises in the case of the least upper bound principle for sets of real numbers. The principle is essential to the development of analysis along classical lines and says that if the set of elements of a set A is bounded above by a certain number r, then there is a least number b such that all elements of A lie below b. In Zermelo's set theory the principle is justified as follows: The set of natural numbers and the set of rational numbers are constructed in the usual fashion. We then consider the power set of the rational numbers and by separation we construct the set of Dedekind cuts on the rationals. This is identified with the set of real numbers. It is easy to show that the reals have the same power as the power set of the rationals and the power set of the natural numbers.

Suppose we are now given a set of real numbers S which is bounded above. Since the reals are identified with the Dedekind cuts on the rationals, S is a set of sets of rationals. The least upper bound b is simply the union of the set of sets in S, that is $b = \bigcup\{X : X \in S\}$. In Zermelo's set theory this can be obtained in two ways, either by making use of the union axiom (axiom V) or more perspicuosly by defining b as follows, $b = \{r \in Q : \exists Y (Y \in S \text{ and } r \in Y)\}$. Y here ranges over all subsets of the rationals, thus over b itself.

In terms of second order arithmetic the situation can be logically characterized as follows: Since there is a one-to-one mapping between natural and rational numbers and a one-to-one mapping between the reals and the power set of the natural numbers, we can identify S with a subset of the power set of the natural numbers, say S^*. Then b would coincide with a set of natural numbers given by $b = \{n : \exists Y (Y \in S^* \text{ and } n \in Y)\}$ where the range of quantification of Y is the power set of the natural numbers.

The two examples show that impredicative definitions of the sort that Poincaré objected to are allowed in Zermelo's set theory. Their justification relies essentially on the axioms of separation, infinity, and power set. In both examples these axioms allow us to rise above a certain given domain and quantify over the sets of subsets of this domain. In the first case the natural numbers are obtained by quantifying over the set of subsets of an infinite set Z; in the case of the least upper bound we quantify over the set of subsets of the rational numbers.

We are now ready to deal more carefully with Hermann Weyl's position.

9.3 The Continuum

Weyl's first contribution to the area of the foundations of mathematics is a paper entitled "Über die Definitionen der mathematischen Grundbegriffe" published in 1910. The main result of the paper, the determination of the principles of concept formation, are embodied in his work *The Continuum*. Weyl's main goal in this article is to provide an answer to a problem raised by Zermelo's axiomatization of set theory, that is, how are we to understand the notion of definability contained in the separation axiom.[6] He achieved this by characterizing the notion of definite property. This project, which in reference to Husserl he considered in terms of the formulation of a "pure syntax of relations", constitutes the first building block of Weyl's approach in *The Continuum*. In order to fix precisely the role of this pure syntax of relations, let me quickly list the main sections of *The Continuum*. The booklet is divided into two chapters. The first chapter is titled "Set and Function. (Analysis of the Mathematical Concept-Formation)", while the second has the title "Number Concept and Continuum (Foundations of Infinitesimal Analysis)". Whereas the first chapter gives a general theory of the formation of concepts in mathematics—a theory that, as we shall see, contains many original theses—the second part provides a reconstruction of analysis according to the principles stated in the first part.

The first chapter is in turn divided into two parts, a logical part and a mathematical part. Those who are familiar with mathematical logic know that a (formal) theory is always given by first stating the logical system underlying it and then specifying the mathematical axioms of the theory. This can help us understand Weyl's organization of Chapter 1, although we have to be careful and look at the treatment more closely.

Weyl begins his treatment by spelling out the central notions of property, relation, existence, judgment, state of affairs, etc. A judgment makes a claims about a state of affairs [*Ein Urteil behauptet einen Sachverhalt*]. Judgments can only be expressed by meaningful propositions, and states of affairs can only correspond to true judgments. However, Weyl does not intend to carry out a detailed analysis of these concepts, which he delegates to philosophy. What is central for Weyl is the precise determination of which principles are used to form judgments, that is, *the principles of judgment combination*. The definition takes the form of an inductive definition.

Base Case: Basic (given) properties and relations $E(x)$, $P(x, y)$, etc., give rise to simple or basic judgment schemas. Among them is also the relation $x = y$.

Inductive step: If P and Q are judgment schemas and $a, b, c, \ldots$ are objects of the appropriate type for the variables contained in P and Q, then:

1. (Negation) $\overline{P}$, the negation of P, is also a judgment schema;
2. (Identification of variables): If $P(x, y)$ is a judgment schema, so is $P(x, x)$;
3. (Conjunction) P and Q is a judgment schema;
4. (Disjunction) P or Q is a judgment schema;
5. (Filling in): If $P(x, y)$ is a judgment schema, then $P(a, b)$ is a judgment;
6. (Existential quantification): $\exists x \, P(x, y)$ is a judgment schema. [Weyl uses an asterisk * to denote existential quantification: $P(*, y)$ means $\exists x \, P(x, y)$].

The above definition of judgment construction differs from contemporary logical treatments in that normally now one first defines what the terms of the theory are, and then the substitution principle (5) is stated in terms of replacing a variable by a term. Weyl, as we have seen, talks about the filling in of a variable by an object.

These principles of judgment formation reflect the procedure used in practice by the mathematicians to formulate properties and to make claims about the mathematical structures at hand.

Finally, Weyl defines the notion of tautology, contradiction, logical consequence, logical equivalence, and states the general principles underlying the setting up of an axiomatic theory. One starts from axioms and deduces the theorems by means of logical inferences.[7]

The mathematical part begins in Section 1.4 with a treatment of the notion of set. Sets can be defined in two different ways: either by listing their members individually (and this can only be done for finite sets) or by providing a property that defines them. Thus sets are presented as extensions of definable properties: To every property P on a domain D corresponds a set S_P such that for a in D, $a \in S_P$ iff $P(a)$. Equality for sets is defined extensionally: Two sets S and S' are equal iff for all a in D, $a \in S$ iff $a \in S'$. More generally, relations $R(x_1, \ldots, x_n)$ give rise to multidimensional sets $S_R = \{(a_1, \ldots, a_n) : R(a_1, \ldots, a_n)\}$.

The one-dimensional and multidimensional sets over a domain D are considered by Weyl as ideal objects. The process of their formation is called by Weyl the *mathematical process*. Thus one begins with a basic domain D (usually the set N of natural numbers) and constructs over this domain the ideal objects by means of the mathematical process. This logical conception of set stands in opposition to the Platonist conception of set. Weyl draws two consequences already at this stage: (1) There is no notion of set independent of given basic domain and relations; and (2) the attempt to give a foundation for natural numbers by means of the notion of set is deeply misconceived:

> A set-theoretic treatment of the natural numbers such as that offered in Dedekind (1888) may indeed contribute to the systematization of mathematics; but it must not be allowed to obscure the fact that our grasp of the basic concepts of set theory depends on a prior intuition of iteration and of the sequence of natural numbers. (Weyl 1994, p. 24)

The centrality of the category of natural number is spelled out in Section 1.5. Central to the understanding of the natural numbers is the relation $F(x, y)$ which says that y is the immediate successor of x. Among the basic facts about this structure Weyl mentions the following:

1. Each number has a unique successor;
2. 1 is the only number which is not the successor of a preceding number;
3. Each number $\neq 1$ has a unique predecessor;
4. Mathematical induction.

The natural numbers owe their special privilege also to the fact that every mathematical discipline—geometry, topology, and algebra, among others—already presupposes them as given. When the natural numbers are assumed as given then we must extend the principles of definitions by the addition of the principle of iteration (see below).

Section 1.6 spells out the ontological and epistemological boundaries of Weyl's approach to analysis. Weyl defines here a general "operation domain" consisting of one or more basic categories of objects together with basic properties and relations. The category of natural numbers with the successor relation F is called the *absolute* operation domain and is always assumed to be included in the given operation domain. Now by the mathematical process described above we can form new ideal elements over this operation domain (that is, definable sets), and then by unlimited iteration go on to form new ideal elements over the previous ones. Thus we introduce sets of higher and higher levels, and this leads, according to Weyl, to impredicative definitions. Weyl describes the process of formation of ideal elements by levels. At level 1 we have all the sets (and functions) that can be formed starting from the relation over the given domains by closing under the definition principles. At level 2 we also have the relation $\in$ holding between a set and its elements. If we now reiterate the construction principles arbitrarily, we obtain all sets (and functions) of level 2. Notice that this would involve existential quantifications over sets of level 1. We proceed in the same way to form sets of level n. It should be remarked that at each level we also obtain new subsets of the original domain. In other words, at level 1 we only obtain the definable sets of level 1; more subsets of the domain (not definable at the first level) can be obtained at further stages. The main difference between level 1 and the successive levels is that in order to obtain the sets belonging to level $n > 1$, one must quantify over ideal objects, that is, sets of level i for $1 \leqslant i < n$. Now if one does not pay attention to the distinction of levels, then one ends up in a vicious circle:

> If, failing to bear in mind the distinctions between levels, one chose to speak here of a relation whose existence is linked to there being a relation such that ... —one would trap oneself in an endless circle, in absurdities and contradictions entirely analogous to Russell's well-known paradox involving the set of all sets which are not members of themselves. (I maintain, and I will presently show in some detail, that our current version of analysis spins in such circles constantly). (Weyl 1994, pp. 29–30)

However an analysis stratified by levels, with reals of all levels, is artificial and useless. Weyl's way out is to restrict attention to the so called *"engeres Verfahren"* that is, to restrict attention to the sets and functions of the first level ("It seems natural to restrict this application of the existence concept to objects of the basic categories" (p. 30)). This rejection of quantification over sets (and properties) is related to Poincaré's and Russell's vicious circle principle. I will come back to this.

Let me finish describing how this first half of the book concludes. In the second part of Section 1.6 Weyl explains what is to be intended by a function of a real variable. Of course, the technical complexity consists in showing what is meant by this notion when everything ought to be definable. In this context real numbers are special sets of rational numbers. Finally, for these functions Weyl specifies in Section 1.7 the principles of substitution and iteration. Whereas substitution is straightforward, things are more problematic in the case of iteration[8] (see Feferman 1988a). Finally, Section 1.8 offers a summary of the development contained in the first part of the book.

Having given up the most basic principles of analysis, like the least upper bound principle for bounded sets of reals, do we have any hope of developing a reasonable theory of analysis over the Weyl continuum? Weyl's great intuition was to see that although one has to give up the least upper bound (l.u.b.) principle for sets of reals, for most applications it is really sufficient to use the l.u.b. principle for sequences of reals and that this holds at level 1. In other words, given a sequence bounded above by a real number, the least upper bound is definable by means of an existential quantification, where the existential quantifier ranges over the natural numbers.

The reconstruction of analysis according to Weylean principles is carried out in the second part of *The Continuum*. Weyl develops his number system starting from the natural numbers. Rational numbers are represented by quadruples of natural numbers (m, n, p, q). The real numbers are then introduced as lower Dedekind sections in the set of rational numbers and hence may be treated as sets of 4-tuples of natural numbers. It is then shown that the operations of + and × are arithmetically definable. This allows Weyl to show that real polynomials are also arithmetically definable functions on the Weyl Continuum. The rest of the development consists in showing that the theory of analysis for continuous functions (of course, continuous functions definable within the Weylean framework) can be developed in Weyl's system. What cannot be developed? Weyl states that the l.u.b. principle for bounded sets of reals, as well as the Heine-Borel theorem for intervals, fail. However, he was not able to justify these claims logically. Indeed, one needs much deeper results about definability in order to prove those claims, and those results were not available to Weyl. For a more technical discussion the reader should consult Feferman (1988a), which presents an axiomatization of Weyl's theory, as presented in *The Continuum*, thereby providing the proper logical setting for investigating definability and proof-theoretical results.

9.4 The Discussion on the Vicious Circle Principle

Although Weyl's proposal on how to restrict mathematics to what can be defined over the ground categories without making use of quantification over ideal objects is clear enough and indeed workable, one ought to look closer at Weyl's reasons for banning impredicative definitions. Having given an overview of the development contained in the first part of the book, it is now time to consider more carefully the foundational grounds on which the Weylean project rests. To this end we need to look very carefully at Section 1.6, wherein Weyl begins to spell out the consequences of his approach to the construction of analysis. Let us assume that a real number a is a set of rational numbers satisfying the following properties:

1. If r is an element of a, so is every rational number r', for which $r - r'$ is positive;
2. For every $r \in a$ there is a rational $r*$ such that $r* \in a$ and $r * -r$ is positive;
3. a is nonempty, but it is not the set of rational numbers.

Now in classical analysis we have not only sets of rationals (individual real numbers) but also sets of real numbers, functions between such, etc. If one does not follow the "engeres Verfahren", then one ends up with an analysis organized by levels; that is, we would have real numbers (and functions) of level 1, 2, 3, etc. We could, of course, ignore the distinction of levels and act as if everything occurs at the same level. This is in fact the case in classical analysis, and it is the cause, according to Weyl, of the vicious circles present in analysis. Weyl mentions a number of examples of vicious circles present in mathematics: the construction of the upper bound of a bounded set of real numbers;

Dedekind's theory of chains; Zermelo's theory of finite sets. In 1925 he also adds a new example (see Weyl 1925, p. 130 and the discussion Hölder 1926). All these constructions sin against the vicious circle principle; that is, no element can be defined with reference to the totality to which it belongs. We have already seen in the previous section how this takes place in the classical context.

However, a closer look at Weyl's exposition shows the lack of a real argument. Weyl simply poses the alternative: Either allow impredicative definitions and end up in "absurdities and contradictions entirely analogous to Russell's well-known paradox", and this is the case of classical analysis, or accept a "hierarchical" version of analysis that is, however, "artificial and useless". Many mathematicians felt that it was Weyl who had introduced an "artificial" critique to the principles of analysis. In order to get a clearer picture, it is useful to discuss the reception of Weyl's criticism of analysis in the 1920s.

Immediately after the publication of *The Continuum*, Weyl published "The circulus vitiosus in the current foundations of analysis" (1919a), which is an excerpt from a letter to the mathematician O. Hölder. This article contains some novelties. In it Weyl gives an argument for why he thinks that the concept of "property of natural numbers" is not extensionally determined, that is does not form, unlike the natural numbers, a complete collection. The same argument is repeated in 1921 and 1925. It goes as follows: Assume any sphere **k** of properties of natural numbers is given as a completed totality, that is, as extensionally determinate. Assume also that A is a property of properties of natural numbers. Then we can always define a new property that lies out of the sphere **k**. Weyl defines P_A to be such that $x \in P_A$ iff there is a "**k**-property" $X \in A$ such that $x \in X$. Weyl then claims that the property P_A most certainly differs in sense from every **k**-property. He does not deny that it might be extensionally equivalent to a **k**-property. However, this is in general highly unlikely:

Clearly, however, it is extraordinarily unlikely that it is possible, in an exact way, to set down an extensionally determinate concept "**k**-property" such that each property P_A, whose definition involves the *totality* of **k**-properties as indicated above, is extensionally equivalent to a **k**-property. In any case, *not even the shadow of a proof* of such a possibility exists; but precisely this proof would have to be effected in order for the assertion of the least upper bound's existence to *receive a sense in all cases and be universally true*. (Weyl 1919a, p. 112)

It seems to me that the only thing Weyl establishes with his argument is a skeptical suspicion for the grounds on which we assume that the set of sets of properties of natural numbers is a completed totality. But this is a far cry from pointing out a vicious circle in the foundations. Indeed Weyl has no argument that would convince a classical mathematician committed to the power set axiom that there is something wrong in analysis. Let us look at some of the reactions to Weyl's claim of circularity in analysis.

Hilbert replied to Weyl's alleged circle on a number of occasions. The most complete discussion of this problem in the Hilbert school is contained in a manuscript in the handwriting of Bernays preserved in the Hilbert *Nachlaß*.[9] Hilbert (or Bernays) ⟦later evidence actually points to Bernays as the author of this talk⟧ grants that there is a standpoint in the foundations of analysis that leads to vicious circles. This standpoint proceeds by progressive determination of real numbers without delimiting a previous domain of real numbers.[10] This corresponds to the standpoint described in Hilbert (1922c) as based on "the principle of constructivity". By contrast, mathematicians normally take the standpoint that the concept of partition of rational numbers is exactly delimited.[11] To this conception Weyl objects, the manuscript continues, that the concept of partition involves that of set, and set-theoretical developments have led to paradoxes. Thus the general concept of set of rational, or even natural, numbers is inadmissible. To this objection the Hilbertian reply is that it is not cogent. Just because there have been problems with the general notion of set one cannot infer that there is something wrong with the notion of a "set of natural numbers" (cf. Hilbert 1922c, p. 199). Indeed the theory of analysis does not at all display the "chaos" alleged by Weyl. However, if one has qualms about such assumptions, one can always take the standpoint of the axiomatic method. Overall the manuscript reflects quite accurately the line taken by Hilbert against Weyl contained in the articles by Hilbert of the early 1920s.

Objections to Weyl's claims came also from Hölder (1926). Hölder argued that the construction of the least upper bound in analysis did not conform to the examples presented by Weyl. The discussion rests on an example of vicious circle given in Weyl (1925). However, Hölder defended Weyl's general claims concerning the impossibility of constructing *The Continuum* arithmetically, a position he had already put forth in 1892. Hölder's construction of the least upper bound rests on the assumption that an infinite set of real numbers must be given by a law, and thus for each such number (considered as a Dedekind cut) it is the case that a rational x either belongs to it or not. Thus one effects the construction the upper bound b of a bounded set of real numbers by remarking that one has to quantify only over an extensionally definite totality, that is, the sequence of real numbers given by a law. The constructivism defended by Hölder is therefore very close to Weyl's own approach in *The Continuum* except that Hölder starts from an unclarified concept of law, just as Weyl did in 1921, whereas Weyl had appealed to explicitly defined construction principles. Moreover, since Hölder thinks that such constructive procedures have been followed by mathematicians all along, he also concludes that "the 'foundational crisis' of mathematics of which recently even the philosophical journals speak has been greatly exaggerated" (Hölder 1926, p. 147).

The topic of the construction of the least upper bound was also taken up by Ramsey in an important article written in 1926 entitled "The foundations of mathematics". In that paper Ramsey proposes to allow impredicative definitions by espousing a strong form of platonism. Carnap in 1931 ("Die logizistische Grundlegung der Mathematik") summarized the situation as follows:

Ramsey (1926) outlined a construction of mathematics in which he corageously tried to resolve this difficulty by declaring the forbidden impredicative definitions to be perfectly admissible. They contain, he contended, a circle but the circle is harmless, not vicious. Consider, he said, the description "the tallest men in this room". Here we describe something in terms of a totality to which it itself belongs. Still no one thinks this description inadmissible since the person described already exists and is only singled out, not created, by the description. Ramsey believed that the same considerations applied to properties. The totality of properties already exists in itself... For these reasons Ramsey allows impredicative definitions. (Carnap 1931, pp. 49–50)

Carnap objected to Ramsey's solution because of the strong form of platonism involved which led him to dub Ramsey's approach as a "theological mathematics". Carnap was, however, sympathetic to Ramsey's attempt to allow impredicative definitions and proposed his own solution to the problem. I do not need to go into his theory here. Finally, let me mention that in Kaufmann (1930) the vicious circle principle in the Weylean formulation is accepted starting from the assumption that is makes no sense to accept the totality of all subsets of natural numbers to be given all at once.[12]

9.5 Weyl's Conversion to Intuitionistic Mathematics

In the years following the publication of *The Continuum*, Weyl familiarized himself with the works of Brouwer. Weyl was deeply impressed with the works of the Dutch mathematician and his intuitionistic foundational viewpoint.[13] This led him to abandon his previous approach to foundational matters and join the intuitionistic camp. The manifesto of his new faith was Weyl (1921c). Written, as he later admitted, in a rather bombastic style—a style that revealed the excitement of the post-World War I period—these lectures have a twofold interest. On the one hand, they were instrumental in giving a clear exposition of the intuitionistic approach.[14] On the other hand, they represent Weyl's own approach to intuitionism. He was not in agreement with Brouwer on various points, and thus one should not identify his position with that of Brouwer tout court. Brouwer himself was quite aware of that. In a short handwritten note containing comments on Weyl (1921c) (preserved in the Brouwer Archief) we read:

Limiting arithmetic and analysis to general statements about numbers and free-becoming sequences. This restriction of mathematics to mathematical entities and species of the lowest order is totally unjustified. This clearly refers to p. 109, where he dismisses my theory of species as meaningless, and it shows that in the end Weyl only half understands what intuitionism is about. (Mancosu 1998a, p. 122)

While referring the reader to van Stigt (1998), I will concentrate on Weyl's article and attempt simply to point out the most remarkable differences between the two viewpoints.

Weyl begins by embedding the debates on the foundations of mathematics in a wider epistemological framework. According to his reconstruction the attempts to deal with the continuum—and this is the central object of debate in the foundations—have always alternated between an atomistic theory of the continuum and a "becoming" theory of the continuum. Speaking of his own attempt in the continuum, which was still written according to the atomistic viewpoint, he says:

Through this conceptual restriction, an ensemble of individual points is, so to speak, picked out from the fluid paste [*fliessender Brei*] of the continuum. The continuum is broken up into isolated elements, and the flowing-into-each-other of its parts is replaced by certain conceptual relations between these elements, based on the "larger-smaller" relationship. This is why I speak of the *atomistic conception of the continuum*. (Weyl 1921c, p. 91)[15]

This point of view also characterized classical analysis. One is presented here with statical theories of the continuum. The continuum is given as something completed, contained in itself. At the opposite extreme is Brouwer's position in which the real numbers and the continuum are given only as developing: *Das Kontinuum als Medium freien Werdens*. And it is exactly on epistemological grounds that Weyl decided to join Brouwer:

> So I now abandon my own attempt and join Brouwer. I tried to find a solid ground in the impending dissolution of the state of analysis (which is in preparation, even though still only recognized by few) without forsaking the order upon which it is founded, by carrying out its fundamental principle purely and honestly. And I believe I was successful—as far as this is possible. For, *this order is in itself untenable*, as I have now convinced myself, and Brouwer—that is the revolution! (Weyl 1921c, pp. 98–99)

The first part of Weyl (1921c) gives a survey of the atomistic theory of the continuum developed in Weyl (1918a). The second part moves on to the theory of the continuum as medium of free becoming. This second part is divided into four sections. Section I presents the basic ideas; section II spells out the concept of function; section III deals with mathematical propositions, properties, and sets. Finally, section IV develops in more detail the new theory of the continuum. Of course, it would be impossible to follow in detail Weyl's development here, and thus I will limit the exposition to some of the central topics. A good starting point is Weyl's own summary of his relationship to Brouwer's own approach.

> As far as I understand, I no longer completely concur with Brouwer in the radical conclusions which are drawn here. After all, he immediately begins with a general theory of functions (the name "set" is used by him to refer to what I call here *functio continua*); he looks at properties of functions, properties of properties, etc. and applies the identity principle to them. (I am unable to find a sense for many of his statements). From Brouwer I borrowed (1) the basis that is essential in every respect, namely, the idea of the developing sequence and the doubt in the principium tertii exclusi, and (2) the concept of the functio continua. I am responsible for the concept of the functio mixta and the conception I summarize in the following three theses: (1) The concept of a sequence alternates, according to the logical connection in which it occurs, between "law" and "choice", that is, between "Being" and "Becoming"; (2) universal and existential theorems are not judgments in the proper sense; they do not make a claim about a state of affairs, but they are judgment instructions and judgment abstracts, respectively; (3) arithmetic and analysis merely contain general statements about numbers and freely developing sequences; there is no general theory of functions or sets of independent content! (Weyl 1921c, pp. 109–10)

Let us try to spell out what is contained in the above quote. Weyl begins §1 of section II with the definition of a real number as an infinite sequence of nested (dual) intervals. Such intervals can be identified with integer characters, and thus it becomes of central importance to investigate sequences of natural numbers. Weyl then distinguishes three possibilities for generating sequences. The first, corresponding to the viewpoint of classical analysis, is given through arbitrary infinitely many choices and then considers the result of the choices as a completed totality. For Weyl, this notion is "absurd and untenable". The two further possibilities are both to be allowed. The first consists of specifying a sequence by means of a law. One can consider such sequences as given. The second acceptable possibility consists of giving a sequence by free acts of choice. However, in this case the sequence cannot be considered as given but only as developing. (see p. 94). Whereas a law corresponds to an individual real number, the continuum "does not dissolve into a set of real numbers as finished beings; we rather have a *medium of free becoming*"(p. 94). In connection to the law/free-choice opposition Weyl motivates his doubts about the validity of the excluded middle. In the continuum he had interpreted existential statements as meaningful if the existential quantifier could be shown to range over an extensionally definite totality. However, the concept of sequence is not extensionally definite, and in the new context an existential statement of the form "there is a σ such that $E(\sigma)$", where $E(\sigma)$ is a property that is meaningful for the domain of number sequences, has meaning only if a law σ has been given by a construction such that $E(\sigma)$. By contrast, the universal statement of the form "for all σ, $\sim E(\sigma)$" has a meaning that refers to any developing sequence (and not just a law). According to this interpretation of the universal and the existential quantifiers, "it would be absurd to think of a complete disjunction in this context". This is why the excluded middle must be doubted. In this context one must remark that Weyl claims originality in having pointed out that the notion of sequence oscillates depending on the logical context (see quote above).

The most interesting part of Weyl's discussion of the excluded middle occurs in reference to the excluded middle on the natural numbers. Here Weyl again claims to differ from Brouwer (see his point 2 in the quote given above) and proposes an interpretation of the meaning of universal and existential theorems not as jugments but as judgment abstracts and judgment instructions:

An *existential statement*—say, "there exists an even number"—is not at all a judgment *in the strict sense, which claims a state of affairs*. Existential states of affairs are empty inventions of logicians. "2 is an even number": This is an actual judgment expressing a state of affairs; "there is an even number" is merely a judgment abstract gained form this judgment. If knowledge is a precious treasure, then the judgment abstract is a piece of paper indicating the presence of a treasure, yet without revealing at which place. Its only value can be to drive me on to look for the treasure. The piece of paper is worthless as long as it is not realized by an underlying judgment like "2 is an even number". Indeed, in the context of number sequences and the laws which determine them in infinitum, we have already said: if we have suceeded in constructing a law with property E, then we are justified in making the claim that there are laws of kind E. Only the *successful* construction can provide the justification for this; the mere *possibility* is out of the question.... The general statement "every number has property E—for example, for every number m, $m + 1 = 1 + m$"—is equally not an actual judgment, but rather a general *instruction for judgments*. (Weyl 1921c, pp. 97–98)

This interpretation of the existential and universal sentences as judgment abstracts and instructions, respectively, is decisive, in Weyl's view, for the final overcoming of the foundational theory espoused in *The Continuum*. This interpretation of quantified sentences is mentioned by Weyl as one of the main points of originality in his approach. There is here a certain ambiguity in Brouwer's reaction to this claim. In a note on the manuscript he says: "I wholly disagree with (2) and (3)". However, in a different note on this claim Brouwer says "Referring to general and existence theorems as resp. 'judgment instructions' and 'judgments abstracts'. This is only a matter of name and certainly does not reflect any lacking insight on my part."(Mancosu 1998a, p. 122)

Finally, concerning the third claim of originality, Brouwer recognized that Weyl was much more radical in his restriction of the objects of mathematics. Weyl rejects a general theory of sets, or even functions. Moreover, whereas Weyl claims originality for his concept of functio mixta, Brouwer claimed that the concepts of functio mixta and functio discreta (see section II, §2 of Weyl (1921c)) were already contained in his notion of functio continua.

A more thourough analysis of these differences between Brouwer and Weyl would force us to introduce many technical details and cannot be carried out here. Fortunately, we can refer the reader to van Dalen (1995), which contains an explicit discussion of these technical topics and more biographical information on the relationship between Brouwer and Weyl.

9.6 The Fourth Phase: Weyl "in the middle of the war of the factions"

We have seen that Hilbert was very much against the standpoint defended by Weyl, both in *The Continuum* and in his 1921 paper, Hilbert feared the influence of such mathematicians as Poincaré, Brouwer, and Weyl on the further development of mathematics. In particular, the situation with Weyl was made difficult by the fact that Weyl was certainly the most oustanding student of Hilbert. Hilbert certainly did not appreciate the critique that his foundational approach reduces everything to a game of formulas (see Weyl 1924b, p. 449; 1925, p. 136, and Hilbert 1928a, p. 475 where, however, only Brouwer is mentioned). However, from 1925 Weyl already attempts to take a middle stand between Hilbert and Brouwer. In a letter from Bernays to Hilbert (25.10.25) Bernays, reporting on a meeting with Weyl in Zurich said: "In particular we discussed your foundational theory, to which, as you know, Weyl is now no longer opposed." (quoted in van Dalen 1995, p. 163) The subsequent publications (Weyl 1925, 1927, 1928a) are marked by a certain attempt at finding a middle course between the opposite camps. In particular, Weyl stresses three points:

(1) The difference between Hilbert and the intuitionists is not as deep as Hilbert wants to claim.[16] Both agree that contentual thought is not sufficient to reconstruct classical mathematics.

(2) Brouwer has the merit of having raised the issue of the limits of contentual thought [*inhaltliches Denken*]. However, starting from 1925, Weyl expresses a certain dissatis-

faction with the complexity and the sacrifices imposed by the intuitionistic reconstruction of mathematics.

> With Brouwer, mathematics gains the highest intuitive clarity; his doctrine is idealism in mathematics thought to the end. But, full of pain, the mathematician sees the greatest part of his towering theories dissolve in fog. (Weyl 1925, p. 136, cf. 1928a, p. 484)

(3) The formalistic approach defended by Hilbert can still be fulfilled with meaning if one interprets the formalisms in analogy with theories in theoretical physics:

> Without doubt, if mathematics is to remain a serious cultural concern, then some sense must be attached to Hilbert's game of formulae; and I see only one possibility of attributing it (including its transfinite components) an independent intellectual meaning. In theoretical physics we have before us the great example of a [kind of] knowledge of completely different character than the common or phenomenal knowledge that expresses purely what is given in intuition. While in this case every judgment has its own sense that is completely realizable within intuition, this is by no means the case for the statements of theoretical physics. In that case it is rather the system as a whole that is in question if confronted with experience. (Weyl 1925, p. 140; cf. 1924b, p. 451; 1927, pp. 49–50, 1928a, p. 484)

Weyl will exphasize more and more the interplay between intuition and formalism both in physics and mathematics.[17] In 1928 Weyl draws a broader philosophical conclusion from this state of affairs:

If Hilbert's view prevails over intuitionism, as appears to be the case, then I see in this a decisive defeat of the philosophical attitude of pure phenomenology, which thus proves to be insufficient for the understanding of creative science even in the area of cognition that is most primal and most readily open to evidence—mathematics. (Weyl 1928a, p. 484)

The relationship of mathematics to phenomenology was a central concern for Weyl. Whereas Hilbert and Bernays found their philosophical inspiration in the Kantian and neo-Kantian tradition, Weyl was from his doctoral days a serious reader of Husserl, with whom he had direct contact when he was in Göttingen. Husserl's influence is visible in the care with which Weyl treated issues like the relationship between intuition and formalization (see *The Continuum*, part II, §6), the connection between his construction postulates and the idea of a pure syntax of relations, the appeal to a *Wesensschau* etc.(see Weyl 1918a, p. 2). Husserl had read Weyl's work *Das Kontinuum* and *Raum, Zeit, Materie* and found them close to his own view of a philosophically informed treatment of scientific matters (see van Dalen 1984, Tonietti 1988). In particular, Husserl praised Weyl's attempt to develop a philosophy of mathematics on the original ground of logico-mathematical intuition.[18] When Weyl, in the quote given above, pondered the possible defeat of the phenomenological attitude in the foundations of mathematics, he was referring to the possible insufficiency of a foundation of mathematics "*auf den Urboden logisch-mathematischer Intuition*". In any case, despite the pessimism expressed in the last quote, Weyl was never to lose interest in the constructivist approaches to mathematics and in phenomenology (Weyl 1985, 1955).

Weyl's influence is visible in Becker 1923, 1927 and Kaufmann (see Kaufmann 1930, Gillies 1980), both involved in a phenomenological treatment of the formal sciences. In addition to the philosophical influence mentioned above, Weyl's ideas, especially those expressed in *The Continuum*, have had a lasting success in more technical areas of the foundations of mathematics. Through the work of Lorenzen, Feferman and others, the scope of predicative mathematics has been extended considerably and predicative formal systems still hold a prominent role in proof theory.[19]

10

MATHEMATICS AND PHENOMENOLOGY: THE CORRESPONDENCE BETWEEN O. BECKER AND H. WEYL (WITH T. RYCKMAN)

10.1 Introduction

It has been observed that 'phenomenological reflections on the sciences are not widely known and, insofar as they are known, usually not well received.' (Kockelmans 1989, p. 365) In this and a companion paper [Mancosu and Ryckman 2005, chapter 11] we aim at making available one of the most important episodes in the relationship between

---

While writing this paper we have incurred many debts. First of all, we would like to thank Richard Zach for having transcribed the Becker-Weyl correspondence and having set the paper into LaTeX. Interpreting Becker's *Kurrent* is a daunting task but he successfully cracked the code. Without his careful work, including suggestions on the contents, this paper would never have become a reality. We are also very grateful to Frau Astrid Becker, daughter of Oskar Becker, for answering our queries on her father's *Nachlaß* and for having granted permission to publish the correspondence. Frau Dr Yvonne Vögeli of the *Wissenschaftshistorische Sammlungen* of the ETH Zurich was most corteous and helpful during the first author's visit to the ETH in 1997. She subsequently helped us in securing further material and the agreement of the ETH to our publication of the correspondence.

The first author would also like to thank the Hellman Family Faculty Fund for sponsoring his visit to Zurich in the summer 1997. Moreover, a research assistantship from the Committee on Research at U. C. Berkeley (1998–99) and a Faculty Grant (1999–2000) were instrumental in funding Richard Zach's work mentioned above.

The first occasion for bringing together some of the topics which make up this paper, and its companion, was a mini-symposium entitled 'Hermann Weyl: Mathematics, Physics, and Philosophy' (Berkeley, April 1999). We are very grateful to the Townsend Center and the Department of Philosophy for co-sponsoring the event and to the other speakers (Solomon Feferman and Richard Tieszen) and the audience for comments.

phenomenology and science, namely the encounter between Oskar Becker and Hermann Weyl on mathematics, physics, and phenomenology in 1923–26. Our hope is that a transcription of these richly philosophical letters from Becker to Weyl, together with the presentation of a detailed account of the mathematical, physical, and philosophical background, will provide a better understanding and appreciation of the historical and theoretical association between phenomenology and science. This chapter focuses on that part of the correspondence, the last two letters from Becker to Weyl, dealing with issues in philosophy of mathematics. Chapter 11 treats the rest of the correspondence, the first four letters from Becker to Weyl, concerned with problems in the philosophy of geometry and classical field physics. Although this division of material does not do justice to the unity of concerns that informed the work of Weyl and Becker, it is however forced on us by the rather lengthy nature of the correspondence—totalling thirty-five pages in our transcription 〚not included in this volume〛—and by the variety of topics touched upon in it. We hope that readers will be able to read both chapters and thus gain a unified view of the philosophical exchange between Becker and Weyl.

The first author found the six letters of Becker to Weyl while working, in the summer of 1997, at the ETH archives in Zurich. His opinion of the major importance of these documents was shared by the second author, and thus we conceived of the project of editing the letters and writing a substantial introduction to them.

The chapter is organized as follows. Sections 10.2 and 10.3 sketch the careers of Weyl and Becker in the years prior to, and immediately following, the correspondence, noting the significant relationship of each to Husserlian phenomenology. Section 10.4 details archival information about the letters and chronicles what is known about the personal contact of Weyl and Becker. In section 10.5, on overview is given of the transformation in phenomenological philosophy underway during this period. Section 10.6 then presents detailed commentary and analysis of the the last two letters, concerned with the foundational matters of Becker (1927).[1]

10.2 Weyl's Early Career and His Relationship to Phenomenology

Hermann Weyl (1885–1955), one of the most outstanding mathematicians of the century, studied in Munich and Göttingen.[2] He received his doctorate under Hilbert in 1908 writing on the topic of integral equations (Weyl 1908). In 1910 he wrote his *Habilitationsschrift* on ordinary differential equations with singularities (Weyl 1910b). He then taught in Göttingen as *Privatdozent* from 1910 to 1913. In 1913 he accepted a position at

---

We are also much indebted to Frau Dr Brigitte Uhlemann, curator of the Oskar Becker *Nachlaß*, at the University of Konstanz. She has helped us on numerous occasions, going always beyond the call of duty. While collecting unpublished correspondence of Oskar Becker we also received the help of Prof. A. Troelstra (Amsterdam), Prof. D. van Dalen (Brouwer Archief, Utrecht), Dr M. Endress (Felix Kaufmann *Nachlaß*, Konstanz), Dr U. Bredehorn (Mahnke *Nachlaß*, Marburg), Dr W. Hagenmaier (Zermelo *Nachlaß*, Freiburg i. B.). Mr Richard Wilhelm Ackermann kindly replied to our requests of information concerning his father's *Nachlaß*. Finally, we would like to thank Mark van Atten, Giuseppe Longo, Christian Thiel, Richard Tieszen, and an anonymous referee, for careful comments.

the ETH in Zurich where he stayed until his return to Göttingen in 1930 (see Frei and Stammbach 1992). During this period Weyl wrote several books, each of which rapidly attained the status of a classic of twentieth-century mathematics, physics, or philosophy: *Die Idee der Riemannschen Fläche* (1913); *Das Kontinuum* (1918a); *Raum, Zeit, Materie* (1918c, 1923a); *Philosophie der Mathematik und Naturwissenschaft* (1927); *Gruppentheorie und Quantenmechanik* (1928b).

While Weyl was often viewed by physicists as an interloper in their domain, repeatedly plaguing them with the necessity of acquiring new mathematics, his contributions to physics have proven to be fundamental. His 1918 generalization of Riemannian geometry spurred developments both in differential geometry and in theoretical physics. The resulting theory of gravitation and electromagnetism initiated the geometrical unified field theory program that Einstein pursued until his death in 1955. In addition, it gave birth to the idea of 'gauge invariance' which, reinterpreted by Weyl himself in 1929 as pertaining to a factor of complex phase and not of scale, has become an integral part of the conceptual framework of quantum field theories. Moreover, Weyl (and independently, Eugene Wigner) pioneered the application of the powerful methods of group theory to the new quantum mechanics.

In addition to his outstanding productivity in mathematics and physics, Weyl also played a central role in the debates on the foundations of mathematics. After his predicativist proposal of 1918, presented in *Das Kontinuum*, he publicly joined Brouwer's intuitionism in 1921 in his article 'Über die neue Grundlagenkrise in der Mathematik' (1921c). However, by 1924 he was already leaning towards an abandonment of the intuitionistic perspective in favor of a 'symbolic construction of the world.'[3] As we shall see in the main part of the paper, the abandonment of intuitionism was tied to a negative reassessment of the prospects of a phenomenological foundation of mathematics. This was a dramatic change for Weyl, who had been very close to phenomenology up to that point.

Weyl's exposure to phenomenology goes back to his graduate student days between 1904 and 1908. The information we have is however scant. In the curriculum vitae appended to his dissertation Weyl says he attended lectures by Husserl (Weyl 1908). However, it is, as far as we know, unknown which lectures he attended. But we do know that he was away from Göttingen from April 1905 to April 1906. Thus, he could not have been in the seminar on philosophy of mathematics given by Husserl in the Summer Semester of 1905, which was attended by Born, Hellinger, and R. König. Husserl was however the chair of Weyl's dissertation examination committee in 1908 (Schuhmann 1977, p. 113). In the much later account (Weyl 1955), Weyl claimed that it was around 1912—at the time of meeting his wife Helene Joseph (who specifically came to Göttingen to be a student of Husserl in 1911, see Weyl (1996)—that he owed to Husserl's influence a liberation from his previous positivistic allegiances.

Indeed, the works written after this period bear a strong imprint of Husserl's influence. This is visible in the care with which Weyl treated issues like the relationship between intuition and formalization (see *Das Kontinuum*, part II, §6 and the discussion in 10.6.2 below), the connection between his construction postulates and the idea of a pure syntax of relations, the appeal to a *Wesensschau*, etc. In the preface to *Das Kontinuum*, Weyl declares that concerning the epistemological side of logic he agrees with the conceptions that underlie the *Logische Untersuchungen* (and which, he adds, have been placed in the context of a general philosophy in *Ideen I*). Husserl had read Weyl's work *Das Kontinuum* and *Raum, Zeit, Materie* and found them close to his own view of a philosophically informed treatment of scientific matters (see van Dalen (1984), Tonietti (1988)). In particular, Husserl praised Weyl's attempt to develop a philosophy of mathematics on the 'original soil [*Urboden*] of logico-mathematical intuition'.[4] Moreover, Husserl emphasized to Weyl in private correspondence that his works were being read very carefully in Freiburg and had had an important effect on the new phenomenological investigations, in particular those of his assistant-to-be Oskar Becker.

In addition to the several passages in Weyl's major works (see also Tonietti 1988, van Dalen 1984, Tieszen 2000, and da Silva 1997) there are two sources which confirm Weyl's allegiance to phenomenology in this period. The first is a letter written to Husserl on the occasion of Husserl's gift of the second edition of the *Logische Untersuchungen* to Weyl and his wife:

You have made me and my wife very happy with the last volume of the *Logical Investigations*; and we thank you with admiration for this present. I have managed to acquaint myself with the contents only now (if such a superficial study such as the one I was able to achieve so far can at all be called 'acquaintance'). Despite all the faults you attribute to the *Logical Investigations* from your present standpoint, I find the conclusive results of this work—which has rendered such an enormous service to the spirit of pure objectivity in epistemology—the decisive insights on evidence and truth, and the recognition that 'intuition' [*Anschauung*] extends beyond sensual intuition, established with great clarity and concisiveness [...] I also was upset by the ridiculous remarks of Schlick concerning phenomenology, all the more so since his book—regrettably, but understandably—finds great resonance among the leading theoretical physicists.[5]

The preface of the second edition of *Logische Untersuchungen* (1913, 1921) contains a dismissal of the critical comments that Schlick had raised in his 1918 *Allgemeine Erkenntnislehre* against Husserl's phenomenology.

Soon thereafter Weyl reviewed Schlick's book in four dense pages of the *Jahrbuch über die Fortschritte der Mathematik* (published in 1923/24). Most of the review gives an objective presentation of Schlick's approach to the problem of knowledge and reality but in two paragraphs Weyl expressed his strong rejection of Schlick's 'semiotic' view of cognition:

According to Schlick the essence of knowledge [*Erkenntnisprozess*] is exhausted by the above. He himself describes his conception as semiotic. It is incomprehensible to the reviewer how anybody who has ever striven for insight can be satisfied with this. It is true that Schlick speaks of acquaintance [*Kennen*] (in opposition to knowing [*Erkennen*]) as a mere intuitive grasping of the given; but he says nothing of its structure, nor does he say anything about the grounding connections between the given and the meanings by which it is expressed. If he ignores intuition to such an extent, in so far as it extends beyond the mere sensory experienceable, then he rejects the evidence outright, which is however the only original source of all insight. (Weyl 1923/24, p. 60)

It should be noticed how faithfully Husserlian these criticisms are. In particular, Weyl emphasizes the need for an analysis of the structure of the given, and calls for the recognition that intuitive evidence is much more encompassing than empirical evidence. Husserl was very pleased to have such a prominent mathematician on his side and highly praised Weyl's work. We'll come back to the relationship between Husserl and Weyl in section 10.4.

10.3 Becker's Early Career and His Relationship to Phenomenology

Among all the early phenomenologists interested in science—including Hans Lipps, Moritz Geiger, Alexandre Koyré, Felix Kaufmann, and Dietrich Mahnke—Oskar Becker (1889–1964) was the most knowledgeable in mathematics and physics. He had arrived at phenomenology from mathematics. In a letter to Hilbert, written from Freiburg i. B. in 1930, he mentions his mathematical studies:

> I studied mathematics for 12 semesters with Hölder and Herglotz in Leipzig and then got the doctorate in this area with a thesis on axiomatic geometry (1914). After 4 years of military duty I then turned to philosophy and obtained my Habilitation here with Husserl in 1922. Even in most recent years I have had many exchanges, epistolary as well as in person, with mathematicians, including some from your inner circle (W. Ackermann and J. von Neumann), moreover with H. Weyl and here, of course, often with E. Zermelo in whose seminar I have even lectured several times on problems in the theory of transfinite ordinals.[6]

The dissertation, written in 1914, was titled *Über die Zerlegung eines Polygons in exklusive Dreiecke auf Grund der ebenen Axiome der Verknüpfung und Anordnung* (Leipzig, 1914, xii + 71 pp.). He then devoted himself to philosophy and wrote a *Habilitationsschrift*, under Husserl's direction, on the phenomenological foundations of geometry and relativity theory: *Beiträge zur phänomenologischen Begründung der Geometrie und ihre physikalischen Anwendungen* (Becker 1923). Husserl was impressed with the results and saw in Becker the philosopher who could join technical competence with an understanding of his new work on the constitution of nature (as developed in investigations that now make up the second volume of *Ideen*). Becker became Husserl's assistant in 1923. In 1927 he published his major work *Mathematische Existenz*. The book was strongly influenced by Heidegger's investigations. Becker had in fact attended many of the seminars offered by Heidegger in Freiburg before the latter's departure for Marburg. The book was sharply criticized, especially by Moritz Geiger, Heinrich Scholz, and Ernst Cassirer (see Geiger 1928, Scholz 1928, Cassirer 1929). This led Becker to reply in polemic fashion (see Becker 1928/29, 1930b). From 1928 to 1930 Becker was on the editorial board of Husserl's *Jahrbuch*, on which however he had served informally as editor since becoming Husserl's assistant. In 1928 he was promoted *Ausserordentlicher Professor* in Freiburg where he stayed until 1931, at which point he took up a chair in philosophy at the University of Bonn.

Although there is now a Becker *Nachlaß* in Konstanz, there is very little in it regarding the period before the Second World War. In a personal communication, Frau Astrid Becker, daughter of Oskar Becker, has confirmed that most of that material was destroyed during the war. This is a great loss since Becker corresponded with several leading philosophers, mathematicians and physicists of the period. The list includes: Wilhelm Ackermann,[7] Arend Heyting,[8] David Hilbert,[9] Felix Kaufmann,[10] J. von Neumann,[11] Hans Reichenbach,[12] Dietrich Mahnke,[13] Abraham Fraenkel,[14] Hermann Weyl,[15] and Ernst Zermelo.[16] A number of Becker's letters can still be found in several archives around the world. However, most of the letters sent to him—except, of course, in the case where a copy was kept by the sender—are considered to be lost. From the point of view of the history of the relationship between phenomenology and science, the most important are surely the letters exchanged with Reichenbach and Weyl.

10.4 The Interaction between Becker and Weyl

In a letter to Weyl dated April 9, 1922, Husserl wrote:

Dr Becker's *Habilitation*, which has just been completed and submitted to the Faculty, shows how intensively interested my Freiburg circle is in your work. I have studied it thoroughly and I have reviewed it very favorably. It is nothing less than a synthesis of Einstein's and your discoveries with my phenomenological investigations on nature. In original and exhaustive considerations, it attempts to show that Einstein's theory, but *only* when it is completed and supported through your investigations on infinitesimal geometry, represent those forms of the 'structural lawfulness' of nature (in contrast to the specific 'causal' lawfulness of nature), which must be required as necessary on deepest transcendental-constitutive grounds: thus the one which (according to its form) is the only possible and ultimately understandable. What will Einstein say when it is shown that nature requires a relativity-theoretical structure on the a priori grounds of phenomenology and not on positivistic principles, and that only in this way a completely understandable, and ultimately exact, science is possible.

Dr Becker also found it necessary in the first part of his work to enter into the general fundamental questions concerning the theorization of vague experiential data, with its vague continuity, and to sketch a constitutive theory of the continuum (rational apprehension of the vague continuum through limits and approximations). In that part too, he aims at showing that the Brouwer-Weyl theories are the only ones appropriate for the definite and necessary requirements of an investigation of the constitutive-phenomenological sources.

All this must make you happy. Too bad that Zurich is completely unreachable for us Germans; otherwise I would already have sent you Dr Becker, deeply knowledgeable both in mathematics and phenomenology, from whom, in his own way, you would have certainly received much stimulation.

I was very sorry that you have given the important paper, which at the time you promised me for the Jahrbuch, to the *Mathematische Zeitschrift*.[17]

It is evident from this letter how highly Husserl considered Weyl's work and how important it was for the cause of phenomenology to have a mathematician of the stature of

Weyl on his side.[18] Through this letter, Weyl learns for the first time of the work of Becker. Husserl's satisfaction with Becker's work is evident, and the letter indicates a strong recommendation. Weyl must have been thus well disposed to interact with the newly-credentialled Dr Becker. Becker approached Weyl by letter on April 12, 1923. Becker had sent his dissertation, which had just appeared in print, to Weyl and in the letter he thanks Weyl for the inspiration he received by reading Weyl's work on the foundations of mathematics, relativity, and the theory of space. This was not mere courtesy.[19] According to Becker, it is Weyl's work that makes a phenomenological foundation of geometry possible.

> It is my conviction that your conception of the continuum problem as well as of the structure of space and time is just the one that makes it possible to give a complete phenomenological foundation of geometry (in the sense of 'world geometry') and in my work I have tried to give the outline of such a phenomenological foundation.[20]

This was the beginning of an exchange that lasted for four years. Before getting into matters discussed in the correspondence, we should say a bit more about the correspondence itself. The correspondence is preserved at the Wissenschafthistorische Sammlungen of the ETH in Zurich in the Weyl *Nachlass*. It consists of six letters from Becker to Weyl:

1. Freiburg, April 12, 1923 (Hs 91.470)
2. Freiburg, April 25, 1923 (Hs 91.471)
3. Freiburg, June 27, 1923 (Hs 91.472)
4. Freiburg, October 10, 1924 (Hs 91.473)
5. Freiburg, July 2–7, 1926 (Hs 91.474)
6. Kötteritzsch, August 16, 1926 (Hs 91.475)

As we said above, none of the letters from Weyl to Becker are extant. However, two fortunate factors help us in our understanding of what Weyl's letters must have contained. First of all, Becker is a very expansive letter-writer, with the result that it is usually possible to reconstruct the points raised by Weyl from Becker's careful replies. Often, he quotes Weyl verbatim, so that we get a very good sense of what Weyl was after. Second, Becker quoted entire passages from several of Weyl's letters in his correspondence with Dietrich Mahnke, an important phenomenologist with interests also in the foundations of the exact sciences.[21] Accordingly, the correspondence sheds light not only on Becker but also on important aspects of Weyl's foundational thinking during this period.

The first four letters from Becker to Weyl deal mainly with problems related to Becker's *Habilitationsschrift* on transcendental-phenomenological foundations of physical geometry with reference to classical physics and to the general theory of relativity. The last two letters relate to topics treated by Becker in *Mathematische Existenz*. These letters mark a strong contrast to the previous four. It seems that Weyl, an editor of *Symposion*, had suggested that Wilhelm Benary, publisher of *Symposion*, solicit from Becker a contribution for the forthcoming issue of the journal.[22] Becker sent a paper which contained the essence of a phenomenological foundation of the transfinite. There must have been some misunderstanding at the editorial level, whose nature remains unclear from the correspondence. What is not in doubt, however, is that Weyl was extremely unhappy with Becker's piece, which—he claimed—'would discredit the name of phenomenology among the concrete sciences'. This was no small accusation to level against one of Husserl's foremost disciples. Thus, the two last letters are much more confrontational in character than the first four but they also bring to light, in crisper form, the significant differences among the 'classic', eidetic, style of phenomenological analysis, Weyl's most recent position in the foundations of mathematics and natural science, and Becker's own approach to phenomenological investigations.

Becker and Weyl eventually met in person. On October 10, 1924 Becker writes that he would have liked to visit Weyl in Zurich ('*Es würde mir eine grosse Freude sein, Sie persönlich kennen zu lernen*'). At the latest they met in 1927. In a letter to Mahnke (August 21, 1927) Becker reports a recent conversation with Weyl.[23]

Although the last part of the correspondence was far from amiable, Becker avoided any polemic with Weyl in *Mathematische Existenz* and gratefully thanks Weyl:

What is said at the end of the work (in §6cIV) on the idea of a symbolic mathematics which interprets nature owes much to the decisive stimulation of the newest philosophical writings of H. Weyl and moreover to his kind epistolary clarifications. (Becker 1927, p. 444)

There is no more documented contact between Becker and Weyl after 1927. In order to facilitate the understanding of the context of Weyl's and Becker's attitude to phenomenology the next section provides a brief overview of the changes in the phenomenological movement during the period. This will be instrumental in understanding the specific approach to phenomenology defended by Becker and the reasons for Weyl's skepticism, starting in 1924, of a phenomenological foundation of the sciences, and mathematics in particular.

10.5 The Phenomenological Background.

In his first letter to Weyl, Becker positions himself in the phenomenological spectrum of the time:

What is important here is that my philosophical starting point (in agreement with the Freiburg direction of phenomenology in opposition to the Munich-Cologne direction) is the principle of transcendental idealism, from which the fundamental problem of the phenomenological constitution of nature arises. I believe that the same idealistic conception also constitutes the background of your theory of the continuum and of your 'pure infinitesimal geometry'.[24]

The opposition between the two schools mentioned in the letter represents the major fracture in the phenomenological movement in the 1910s and early 20s. Husserl's *Logische Untersuchungen* had seen an enthusiastic response on the part of several young philosophers. The initial circle of phenomenologists around Husserl, J. Daubert, A. Pfänder, M. Scheler, A. Reinach, and later Moritz Geiger, Hedwig Conrad-Martius and others, found the *Logische Untersuchungen* to be a source of inspiration. For them phenomenology was to direct the 'gaze of the conscious Ego' at objective entities in general. This attitude towards phenomenology was soon to come into conflict with the new developments of Husserl's thought. When *Ideen zu einer reinen Phänomenologie und phänomenologischen Philosophie I* came out in 1913 many phenomenologists refused to follow Husserl in the new transcendental direction it outlined. In particular, they saw the new investigations on transcendental constitution as a return to idealism or psychologism. Husserl felt quite betrayed by the inability of his students to follow the new paths he was opening.[25] With the death of Reinach (who taught in Göttingen) in the war in 1917, and Husserl's move to Freiburg in 1916, the most significant geographical poles of the phenomenological movements become Freiburg and Munich. In Freiburg, Husserl developed the analyses initiated in the first volume of *Ideen* and further pursued his investigations on the transcendental constitution of the natural, animal, and intellectual realm (see *Ideen II* and *III*). Heidegger, and later Becker, were among his assistants during that period. In Munich and Cologne, an opposing 'realist' school formed, represented by Pfänder, Geiger, and Scheler.

The principle of transcendental idealism is, in Becker's characterization, the fundamental starting point of Husserl's most recent work. The principle is supposed to capture the 'transcendental constitution of all ontological essences in pure consciousness' developed by Husserl in the last sections of *Ideen I*. Here is how Becker puts the matter in 1923:

> According to the basic principle of transcendental idealism, an object only 'is' and a state of affairs only 'subsists' ('holds') in as much as and so far as it can exhibit itself in consciousness with the degree and the type of evidence that is characteristic for it. Everything 'transcendent' constitutes itself in pure consciousness, 'reality' inheres in it only through rationally motivated 'reality theses' [*Wirklichkeitsthesen*]. (Becker 1923, p. 394)

The basic consequence of this principle is that it is dogmatism to speak of realities that are in principle inaccessible to consciousness. Indeed, Becker at one point reformulates the principle of transcendental idealism as that according to which 'one cannot say of any state of affairs that it holds, if one does not have a way in principle to decide whether it holds or not' (Becker 1923, p. 414). The reader might well wonder on what ground the principle is defended. Here Becker admits frankly that 'the foundations of our construction are received and are not put up for discussion' (Becker 1923, p. 387). As such, he thinks of the project as a bridge between the central philosophical theses of Husserlian phenomenology, whose major achievements he takes for granted, and the most recent research in the foundations of mathematics and physics. On the basis of these major philosophical tenets (and the related Husserlian notions of definite manifold, material and formal ontology, etc.) Becker attempts to provide a transcendental a priori grounding of Euclidean geometry, the geometry of the actual space of sensory intuition that is phenomenologically constituted in step-by-step fashion from lower levels of spatial and pre-spatial forms of intuition. In this way, in virtue of its constitutive origin, Becker claims to account for the peculiar double nature of geometry, as related both to pure reason and to the sensible world.

As he clearly indicates, he considers his account to square both with the Brouwer-Weyl[26] theory of the continuum, and with Weyl's extension of Riemannian geometry and its application to general relativity.

With some justification, Becker considers Weyl to be a supporter of the principle of transcendental idealism. (Becker 1923, p. 544) adduces the following passage from Weyl's *Das Kontinuum* as evidence of Weyl's commitment to the principle of transcendental idealism:

> Therefore, points and sets of points can be defined only relative to (i.e., as functions of) a coordinate system, never absolutely. (The coordinate system is the unavoidable residue of the annihilation of the ego in that geometrico-physical world which reason sifts from the given using 'objectivity' as its standard—a final scanty token in this sphere that existence is only given and can only be given as the intentional content of the processes of consciousness of a pure, sense-giving ego.)[27]

This is a striking passage indeed (see Chapter 11 for a full analysis) and it fits perfectly well with the contents of Husserl's *Ideen*. In *Mathematische Existenz* the principle, also called accessibility principle [*Zugangsprinzip*], is stated as follows:

> To each object [*Gegenständlichkeit*] there is (in principle, i.e., independently of 'technical' complications) an access. (Becker 1927, p. 502)

Whereas the principle of transcendental idealism per se is compatible with some forms of realism, in *Mathematische Existenz* it is given a strong idealistic connotation, which prejudges its use in deciding the battle between Hilbert and Brouwer (in favor of the latter). Becker, as in his previous work, interprets the principle in light of the Husserlian theory of constitution, according to which for each objectivity one has to provide the constitution in consciousness corresponding to it. However, he seemingly slips into an extreme form of idealism by claiming for example that 'ultimately only phenomena "exist"' (p. 527) and that 'accessibility is part of the phenomena' (p. 502).[28]

Three more points will be useful to put the discussion between Becker and Weyl in context. *Mathematische Existenz* is strongly influenced by Heidegger's new investigations on the facticity of *Dasein*. This leads Becker to pose the problem of mathematical existence within the confines of human existence:

> The factual life of mankind [...] is the ontical foundation also for the mathematical. (Becker 1927, p. 636)

This standpoint in philosophy of mathematics—dubbed at the time as 'anthropologism' (see Geiger 1928)—also leads Becker to find the origin of mathematical abstractions in *concrete* aspects of human life. In this he is—like Heidegger, of course—critical of the Husserlian style of phenomenological analysis which rests, according to Becker, on a form of 'combinatorial phantasy' without a 'concrete, that is, not purely epistemological, motivation' (Becker 1927, p. 543).

We will see how this anthropological current plays a role in Becker's analysis of the transfinite. The second point concerns the role of philosophy. Becker criticizes the neo-Kantians for according to philosophy only the role of a servant of science. The phenomenology defended by Becker reclaims for philosophy the right to criticize, and to provide appropriate foundations for, the sciences (see Becker 1927, p. 808). Finally, it turns out that, having become critical of Husserl's abstract phenomenological considerations, Becker is also not fully satisfied with Heidegger's hermeneutical phenomenology of *Dasein*. This leads to his proposal, first in the correspondence with Weyl and later in *Mathematische Existenz*, of a 'mantic phenomenology', whose role is to be that of interpreting, or divining, the inner structure of nature. We are now ready to enter into the details of the last two letters from Becker to Weyl.

10.6 The Last Two Letters: Phenomenology and Foundations of Mathematics

10.6.1 Overview of The Last Two Letters

We begin with a brief summary of the content of the letters. For further details and context, reference is made to the developments below.

The letter of July 2–7, 1926. Almost two years had elapsed since Becker had corresponded with Weyl on the foundations of geometry. The correspondence between the two scholars resumes in 1926 with a letter from Weyl to Becker. The letter was in the form of a referee report on a paper that the publisher of *Symposion*, Wilhelm Benary, had solicited from Becker. The invitation to contribute a paper for *Symposion*, as transpires from the correspondence, stemmed from Weyl, who belonged to the editorial board of the journal. Becker had submitted a paper, no longer extant, on the phenomenological foundation of the transfinite. Weyl's comments on the paper were so negative that Becker decided to withdraw it from publication. Weyl had suggested to Becker that he leave out the reflections on the transfinite process but Becker objected that, were he to do that, nothing would be left of the paper. Moreover, he insisted that factually, he had nothing to retract. The strongest remark made by Weyl was that Becker's style of investigation would 'discredit phenomenology among the concrete sciences'. Becker addresses in his reply the factual criticism made by Weyl and sends him a part of his new work *Mathematische Existenz*, quite clearly the part of the book related to the foundation of the transfinite process.

The rest of Becker's letter can be divided into two parts. The first concerns a number of clarificatory remarks relating to points made by Weyl. The discussion focuses on four topics: iterated representation; iterated reflection; remarks on the paradox of the 'set' W (Burali-Forti's paradox); remarks on Becker's criticisms of Hilbert's and Ackermann's consistency proofs. We will return to this in section 10.6.3.

The second part leaves matters of detail to look at the more general differences between Becker's approach and Weyl's. Becker claims that the source of the disagreement stems from a different conception of the relationship between mathematics and philosophy. He claimed that until recently this relationship saw philosophy as *ancilla mathematicae*. The neo-Kantians Cohen and Natorp are singled out as paradigm examples of this attitude. Implicitly, Becker assimilates, at least on this point, Weyl's position to theirs. However, Becker continues, things are different nowadays and philosophy must claim the right to pursue its 'ontological' questions also in the domain of mathematics. And the project must not be confused with an attempt at providing mathematics with a foundation in the Hilbert style, so that the mathematician might safely construct his theories without worrying about ontology. Rather, philosophy must pose to mathematics objective-ontological problems. To the general objection that it is arbitrary what one decides to take as intuitive, Becker—declaring himself an intuitionist—strongly disagrees, claiming that the phenomenon of iteration of intentionality requires the Cantorian theory of the transfinite, just as the discovery of the 'reality' of the electrical field by Faraday requires its appropriate formulation by means of the vector and tensor calculus. Thus, Becker is critical of 'symbolic' mathematics exactly because it simply ignores all ontological questions, but these questions still remain even when mathematics as an individual science '*ihren sicheren Gang geht*'. As we shall see in section 10.6.2, Weyl's insistence on symbolic mathematics corresponds to a new stage in his foundational thinking, characterized exactly by the abandonment of the intuitionistic position defended in Weyl (1921c). Becker wrote this letter under the general perception that a number of misunderstandings were responsible for the differences of opinion. But it is only in the next letter that Becker realizes that Weyl had undergone a major change in his philosophical thinking and that the roots of the disagreements were indeed very deep.

The letter of August 16, 1926. Weyl had replied to the previous letter and also sent back the manuscript of the fragment of *Mathematische Existenz* Becker had sent. Becker had meanwhile read Weyl's booklet *Philosophie der Mathematik und Naturwissenschaft* (Weyl 1927), which had just appeared, and had come to realize 'that the sharp rejection of my views does not rest on a mere misunderstanding of my presentation, but rather on a deep difference of opinions'. He points out how radically different Weyl's position of 1921 is from the one defended in the new booklet. Becker realizes that this change of position coincides also with a turn away from phenomenology in favor of a 'constructive philosophy', denoted also with the names of 'metaphysics' or 'metalogic', terms probably stemming from Weyl's letter. For this reason, Becker prefers to focus on major points of disagreements rather than on technical points. The first general issue concerns the right of philosophy to investigate mathematics. Once again, as transpires from the following quote, Weyl had been rather hard on Becker's claims:

I do not want to appeal to the claim of philosophy to its own domain of investigation and even to a right to critique the problems and methods of the individual sciences, in order to 'cover up' my mathematical 'wrongs'. (Original German in Mancosu and Ryckman 2002, p. 186)

This leads Becker to restate his position concerning the relationship between science and philosophy, once again in open polemic against the neo-Kantian position, by claiming the right for philosophy and mathematics to exercise a factual criticism of one another:

I think it therefore makes good sense to require that philosophy and mathematics should have the free right to mutual objective criticism. We cannot possibly return today to the neo-Kantian position, where one let philosophy pant behind the positive sciences with the surprised cry: 'How is this (this sacrosanct, wonderful science) possible!?'[29] (Original German in Mancosu and Ryckman 2002, p. 186)

Moving then to Weyl's specific criticisms of his 'objective' grounding of transfinite set theory Becker divides the objections into a mathematical and an almost-philosophical one. The philosophical objection is discussed first. Weyl had claimed that he saw neither the necessity nor the exhibitability of a special transfinite 'horizon' in consciousness. In his reply, Becker defends his phenomenologico-methodological approach:

Concerning your criticism of my attempt to ground an 'objective' [*sachlich*] theory of the transfinite, it seems to me to fall into two parts, one mathematical and one quasi-philosophical. You write that you see neither the necessity nor the possibility of exhibiting a special (transfinite) consciousness-horizon. On this point I must decisively object. Unfortunately, phenomenological descriptions have different persuasive force for different people and there is no way to make them compelling. But I claim to have exhibited this consciousness-horizon very clearly and I can also appeal to Husserl's agreement with me on this matter. (Original German in Mancosu and Ryckman 2002, p. 186)

A succinct description of his project is also given with the remark that Weyl's previous foundational stand was more in accord than the present one with Becker's own ontological endeavours:

My efforts to provide an objective interpretation of the transfinite process have thus the purpose of developing the ontological character of the great Cantorian ideas. This seems to me to be an extremely important philosophical problem. And I believe that in the past you would have also considered it as such, before the idea of a symbolic mathematics led you to consider all objective questions to be uninteresting, in so far as their solution seemed not necessary for a consistency proof.(Original German in Mancosu and Ryckman 2002, p. 187)

What is the relationship between the phenomenological analysis and the mathematical theory of the transfinite? The allegedly transfinite structure of pure intentional consciousness has a double role to play vis-à-vis the mathematical theory of the transfinite. On the one hand, the former is an application of the latter, just as a mathematical theory can be applied in physics. On the other hand, the 'matter' to which the mathematical theory is applied is not an empirical entity (whether physical or psychological) but rather the phenomena of transcendental ('pure') consciousness, and indeed the phenomena which play a decisive role in the constitution of the mathematical theory in transcendental consciousness (see section 10.6.3.3 for details). Becker remarks that the phenomenological exhibition of the transfinite structure of consciousness can only contribute to the *clarification* of the ontological meaning of the mathematical theory but does not push the mathematical theory itself further.

With this we arrive at the mathematical objections raised by Weyl. Here the discussion focuses on three topics:

1. Weyl's requirement that the transfinite objects be 'finitely describable'.
2. A problem concerning critical numbers.
3. The relationship between generality of consistency proofs and the idea of hypothetical generality (see section 10.6.3.4 for details).

The following section contains a remark on Weyl's opinion on the debate between Brouwer and Hilbert on the foundations of mathematics. Weyl had clearly expressed his dissatisfaction with intuitionism and Becker says he understands Weyl's inclination towards symbolic mathematics 'stemming as it does from the regret of the mathematician "to see his towering edifice dissolve in fog"'.[30] Weyl drew further consequences from the alleged failure of intuitionism. In particular, he claimed that the victory of formalism over intuitionism implied that phenomenology could no longer aspire to play the role of basic philosophical science (we will come back to Weyl's position on intuitionism and phenomenology in sections 10.6.2.1 and 10.6.2.2). Becker disagreed. According to him, Weyl judges the situation in this way because he ignores the new developments in phenomenology, which in different ways have abandoned the 'maxim of immanent description'. These new approaches to phenomenology, he continues, have moved to a different type of phenomenological analysis, which strives to reach beyond the phenomenal level.

I cannot agree with you when you claim that this victory of formalism destroys the position of phenomenology as philosophical fundamental science. On the other hand, you probably must judge the situation this way since the most recent development of phenomenology is likely hardly known to you. (Original German in Mancosu and Ryckman 2002, p. 190)

Becker then goes on to describe the new and exciting developments in phenomenology that had occurred in the preceding years (Husserl, Pfänder, Heidegger, Scheler, and himself). He then describes his own approach to phenomenology, mantic phenomenology (see section 10.6.3.5 for details).

Remarkably, in reflecting on his own approach to phenomenology, mantic phenomenology, Becker realizes the importance that symbolic thinking ('i.e., without an ultimate grasp') plays in it and connects his own approach to Weyl's new position in favor of symbolic mathematics. This convergence of points of view is in fact spelled out in the last part of *Mathematische Existenz*. In this connection Becker reports another interesting remark made by Weyl in his previous letter:

Your thought that in the course of its (recent) history physics is moving more and more away from the intuitively-graspable and that the time has come, when the last rope is cut that keeps it at the 'phenomenal shore', the recognition of formal ontology, is certainly momentous. Perhaps you are right, but for the moment I cannot yet make up my mind. In any case the problem of the boundary between mathematical phenomena and transphenomena is incredibly important and difficult. (Original German in Mancosu and Ryckman 2002, pp. 191–92)

The letter is concluded by a number of specific remarks on Weyl (1927), which need not detain us here.

The importance of the last two letters for an understanding of Weyl's and Becker's philosophies of mathematics cannot be overestimated. Becker realizes that Weyl has shifted his position on basic issues such as intuitionism and phenomenology. Moreover, Becker himself is led to see that there are also affinities between his mantic phenomenology and Weyl's symbolic mathematics. In order to put the two letters in context we will devote section 10.6.2 to Weyl's change of position in the middle of the 1920s, emphasizing especially his thoughts on intuitionism and phenomenology. Then in section 10.6.3 we will describe Becker's proposed phenomenological foundation of the transfinite and discuss some of the technical issues raised in the letters. We will conclude with a description of Becker's own brand of phenomenology and its relation to Weyl's symbolic construction of the world.

10.6.2 *Weyl's Foundational Thinking in the Period 1910–1930*

Weyl's contribution to the philosophy of mathematics in the second and third decades of the twentieth century can be divided into four phases.[31] Weyl begins his foundational work with an article on the definition of mathematical concepts (Weyl 1910a). The main problem addressed by Weyl is that of making precise the vague idea of definite property found in Zermelo's work on the axiomatization of set theory (Zermelo 1908c). In doing so he provided an answer which in large part anticipates that given by Skolem in 1922, the latter being the one that is now commonly accepted. Weyl's point of view is still in large measure 'classical'. In any case we do not find here explicit criticisms to the foundations of mathematics of the sort that are to be found in his later work.

Weyl's second phase, and the most creative one, coincides with the publication of the book *Das Kontinuum* (1918a). In this book Weyl criticized set theory and classical analysis as 'a house built on sand' and proposes a remedy to the uncertainty that in his opinion affected a great part of classical mathematics. The book can be divided into two parts. In the first part, Weyl proposes a number of criticisms to contemporary analysis and set theory, but also to some of the accepted reconstructions of classical mathematics, such as Zermelo's set theory and Russell's theory of types. In the second part, Weyl develops his own arithmetical continuum, or Weylean number system, in which large portions of analysis can be carried out.[32]

Just a few years after the publication of *Das Kontinuum*, Weyl discovered the new foundationalist proposal championed by Brouwer, i.e., intuitionism. Weyl was deeply impressed with the works of the Dutch mathematician and his intuitionistic foundational viewpoint.[33] Perhaps the first evidence of such fascination we have is contained in a letter to Bernays dated January 9, 1920:

Moreover, I am again working on the foundations of analysis. A meeting with Brouwer in the summer has given a new impetus to the matter. I am modifying my standpoint substantially. Brouwer is a devil of a fellow and a wonderfully intuitive person. Spending a few hours with him made me feel positively elated. I will envy you if you get him to take Hecke's [old] position in Göttingen.[34]

The encounter with Brouwer led him to abandon his previous approach to foundational matters and join the intuitionistic camp. The manifesto of his new faith was (Weyl 1921c). These lectures have a twofold interest. On the one hand, they were instrumental in giving a clear exposition of an intuitionistic approach.[35] On the other hand, they represent Weyl's own approach to intuitionism. He was not in agreement with Brouwer on various points and thus one should not identify his position with that of Brouwer *tout court*.[36] Weyl's intuitionist phase lasted only until 1924. In the next section, we will point out the reasons for Weyl's dissatisfaction with intuitionism.

Finally, the last five years of the 1920s show an attempt on the part of Weyl to do justice to Hilbert's programme and in doing this he changed his philosophical position substantially.

We will focus on the following topics connected to the fourth phase of Weyl's foundational reflection:

1. Weyl's abandonment of intuitionism (§10.6.2.1);
2. The connection between phenomenology and intuitionism (§10.6.2.2);
3. Weyl's 'symbolic construction of the world' (§10.6.2.3).

10.6.2.1 Weyl's Abandonment of Intuitionism

Hilbert was very much against the standpoint defended by Weyl, both in *Das Kontinuum* and in his Weyl (1921c). He feared that the influence of such mathematicians as Poincaré, Brouwer, and Weyl would stifle the further development of mathematics. However, already by 1925 Weyl was moving closer to Hilbert's position. In a letter from Bernays to Hilbert (October 25, 1925) Bernays, reporting on a meeting with Weyl in Zurich, said: 'in particular we discussed your foundational theory, to which, as you know, Weyl is now no longer opposed' (quoted in van Dalen 1995, p. 163).

Whereas Weyl's intuitionistic phase has been the object of detailed analysis,[37] there is very little in the secondary literature about Weyl's new position in the middle of the twenties. Why did Weyl give up the intuitionist approach? A careful reading of his writings on the subject from 1924 to 1931 seems to point to two independent reasons. The first one concerns the inability of intuitionist mathematics to go very far in recapturing much of classical mathematics. Although Weyl had himself contributed to the constructivization of some central theorems, such as the fundamental theorem of algebra, he expresses a definite dissatisfaction with the complexity and the sacrifices imposed by the intuitionistic reconstruction of mathematics.[38]

With Brouwer, mathematics gains the highest intuitive clarity; his doctrine is idealism in mathematics thought to the end. But, full of pain, the mathematician sees the greatest part of his towering theories dissolve in fog.[39]

However, more was bothering Weyl. It was the fact that intuitionistic mathematics could not account for the mathematics used in physics and according to him 'it is the function of mathematics to be at the service of the natural sciences'. In *Die Stufen des Unendlichen*

Weyl makes clear that the reason to go beyond intuitionistic mathematics are not internal to mathematics:

> If one considers mathematics on its own then one should restrict oneself with Brouwer to the evident truths [*einsichtigen Wahrheiten*], in which the infinite enters only as an open field of possibilities; one cannot find any motive that pushes beyond that. (Weyl 1931a, pp. 17–18; cf. Weyl 1932, p. 82)

What leads us to accept the non-intuitionist part of mathematics is its decisive role in physics. In this connection it is important to return to the passage in the last letter from Becker to Weyl where Becker said that he perfectly understood Weyl's inclination towards symbolic mathematics on account of the inability of intuitionistic mathematics to recapture a great deal of classical mathematics. Fortunately, the passage of Weyl's letter to which Becker was reacting was reported verbatim by Becker in his letter to Mahnke in 1926:

> For me the Hilbert-Brouwer conflict has of course quite a central significance. I am too much of a mathematician to be able to ignore that in practice Brouwer's mathematics is not what we need, and it will not prevail. In this respect I believe in history and I am a pious child of this world, in that, together with Hilbert, I find that success is what is decisive. *If Hilbert prevails over Brouwer, this means for me at the same time that with it phenomenology as a fundamental science is condemned* .... For me theoretical construction—which does not rest in the end on any intuitive foundation as absolute unsurpassable limit—is beginning to emerge as the main epistemological problem. I have become skeptical of the attempt to replace through phenomenology, the intuitively given, the metaphysical support that is the foundation in Leibniz.[40]

Becker replied to these passages arguing that it would be useful to know exactly the limitations of intuitionism for our knowledge of nature:

> Your main remark on the significance of the Brouwer-Hilbert conflict has interested me very much, especially in connection with your new article in the 'Handbook'. I understand your inclination towards symbolic mathematics very well, stemming as it does from the regret of the mathematician 'to see his towering edifice dissolve in fog'. It seems to me almost certain that in the public opinion of the mathematicians Hilbert, or presumably a semi-renewal of the old 'existential absolutism', will prevail. In general, I would not think very much of this public opinion, which always prefers mediocrity. However, I cannot close my eyes to the lack of success of the intuitionists; negative success in the sense that they probably have to give up all of modern physics in the end. This however is in dire need of proof, that is one would absolutely have to know the limits of intuitive mathematics. No axiomatic investigation seems to me more important and urgent than this one. (Original German in Mancosu and Ryckman 2002, p. 189)

That the connection to the issue of whether intuitionism could account for natural science was raised by Weyl in his letter is confirmed by Becker's comment to Mahnke immediately after the quoted passage:

> We are facing here a certain crisis of the phenomenological method itself. If Weyl is right that Brouwer's intuitionism cannot support theoretical physics, then phenomenology in Husserl's

'classical' sense must also be seen as incapable of securing the modern form of knowledge of nature—and making it completely understandable.[41]

One important point raised by Becker is that it would be important to study the limit of intuitionistic mathematics, in particular with reference to the problem of whether intuitionistic mathematics is sufficient for the needs of natural science. This is still a topical issue in logic and philosophy of mathematics. For instance, Feferman (1988a, 1993b) has argued that predicative mathematics is sufficient for all the needs of physics and there are several investigations concerning whether some important theorems of mathematics appearing in physics (such as Gleason's theorem in quantum mechanics) can be constructivized.[42]

We have thus reached the important point of connection between Weyl's abandonment of intuitionism and his abandonment of the hope that phenomenology could account for scientific activity in its entirety. To avoid misunderstandings, it should be pointed out that by speaking of 'abandonment', or 'rejection', of phenomenology and intuitionism we do not mean to imply that Weyl lost interest in what could be done intuitionistically or what could be grounded phenomenologically. It is rather that he abandons the hope that intuitionism and phenomenology will be sufficient to account for our mathematics and science in their entirety.

10.6.2.2 The Connection between Intuitionism and Phenomenology in Weyl's Foundational Thought

We should say from the outset that we cannot here rehearse the major claims of Husserl's phenomenology and compare them specifically to Weyl's passages.[43] Rather, we are interested in explaining why, for Weyl, the fortunes of intuitionism and phenomenology (as a fundamental science) are strictly connected. Moreover, in his comments to Mahnke, Becker indicates his agreement with this confluence, in so far as phenomenology is taken in its 'classical', eidetic, sense.

Eidetic phenomenology refers here to Husserl's account of phenomenology, in particular the one offered in *Ideen I*. To understand what follows we have to recall briefly some facts about Husserl's classification of the eidetic sciences. In §3 of *Ideen I* Husserl distinguishes two essentially different sorts of intuitions, 'intuition of something individual' ('experiencing') and 'eidetic intuition' ('intuition of essences'). Different eidetic sciences correspond to the different essences that can be presented in 'eidetic intuition'. Husserl considers pure mathematics as a science founded on the intuition of essences (*Ideen I*, §7). There are however differences among mathematical disciplines. Husserl distinguishes between formal and material eidetic sciences. The first group of sciences belongs to the complex of 'formal-ontological disciplines', which study the 'essence of anything objective whatsoever' (*Ideen I*, §8). Formal ontology includes, among other sciences, arithmetic, pure analysis, the theory of multiplicities, and (parts of) formal logic. The material eidetic sciences concern specific 'regions'. Thus, geometry is concerned with the specific 'region' dealing with the essence of 'spatial form' (*Ideen I*, §9).

Husserl also distinguishes exact and descriptive eidetic sciences. The former are the sciences of exact essences, such as all those found in the mathematical disciplines, whereas the latter deal with non-exact, vague, 'morphological' essences (*Ideen I*, §74). The most important example of a descriptive eidetic science is phenomenology itself, i.e., the '*descriptive* eidetic doctrine of transcendentally pure mental processes' (*Ideen I*, §75).

The important point is that genuine science requires 'as the foundation of all proofs, immediately valid judgments which derive their validity from *originally presentive intuitions*' (*Ideen I*, §19). All eidetic sciences are ultimately founded on the intuition of essences:

> Immediate 'seeing', not merely sensuous, experiential seeing, but seeing in the universal sense as an originally presentive consciousness of any kind whatever, is the ultimate legitimizing source of all rational assertions. (*Ideen I*, §19)

Let us now return to Weyl. Weyl's skepticism on the adequacy of phenomenology for the foundations of science became public in 1928 with words very close to those reported in Becker's 1926 letter to Mahnke:

> If Hilbert's view prevails over intuitionism, as appears to be the case, *then I see in this a decisive defeat of the philosophical attitude of pure phenomenology*, which thus proves to be insufficient for the understanding of creative science even in the area of cognition that is most primal and most readily open to evidence—mathematics. (Weyl 1928a, p. 88; van Heijenoort 1967a, p. 484)

But this is not at all obvious and requires an explanation. The first thing to notice is that there is no obvious connection between Weyl's phenomenological leanings and his espousal of intuitionism that might help us here. Neither the correspondence nor Weyl (1921c) give us any clue that Weyl's conversion to Brouwer was due to a set of considerations related to phenomenology. Weyl (1921c) does appeal to epistemological considerations but they have to do with the opposition between 'static' ('atomistic') and 'becoming' theories of the continuum. The former type is exemplified by classical and predicative analysis, the latter by the intuitionist theory of the continuum as a 'medium of free becoming'. Weyl emphasizes that one of the gains of Brouwer's theory is to narrow the chasm between the intuitive and the mathematical continuum but, as we said, no explicit appeal to Husserl's phenomenology is offered by Weyl as motivation for his intuitionist conversion.

We thus have to reconstruct, from the scattered statements in Weyl's publications, what he thought about the connection between intuitionism and phenomenology and why, once he convinced himself of the shortcomings of intuitionistic mathematics, he alleged the insufficiency of phenomenology. Our reconstruction of Weyl's train of thought will rely on making explicit what seems to be an unspoken assumption in Weyl, i.e., the claim that, as far as mathematics is concerned, the resources of phenomenological *Anschauung* do not go beyond those of intuitionistic *Anschauung*.[44]

We should also warn the reader that Weyl often uses the notion of 'intuition' informally, without providing an explicit philosophical characterization. However, he does specify in (Weyl 1924b, note 19, p. 451) that ' "Intuition" [*Anschauung*] will not be restricted here to sensible intuition, but rather denotes any presentive act [*jeden gebenden Akt*].' The terminology is clearly Husserlian. We will return to this issue after we have given the reconstruction of the Weylean argument.

Weyl holds that intuitionism is characterized, among others, by the following features:

1. Each sentence of intuitionistic mathematics is meaningful and its meaning is fulfillable in intuition (Weyl 1927, p. 49);
2. Intuitionistic mathematics consists of contentual, intuitive truths (Weyl 1927, p. 45).

Thus, under the assumption, never explicitly stated by Weyl, that

3. Intuitionistic *Anschauung* is the only relevant form of intuition when it comes to mathematics (i.e., there are no other forms of phenomenological intuition that could phenomenologically ground parts of mathematics which go beyond intuitionistic mathematics)

and

4. Classical mathematics far outstrips the limits of intuitionistic *Anschauung* (Weyl 1927, p. 45), one can conclude that:
5. Classical mathematics cannot be grounded by intuition in its entirety, i.e., 'pure phenomenology proves to be insufficient for the understanding of creative science even in the area of cognition that is most primal and most ready open to evidence—mathematics' (Weyl 1928a, p. 484).

Let us remark first of all that in order for the argument to go through we need not specify the exact nature of intuitionistic *Anschauung*.[45] Since classical mathematics contains many statements, including Hilbert's transfinite statements, which cannot be interpreted 'intuitively' or 'contentually', it follows that much of classical mathematics cannot be grounded on intuition, 'eidetic' or otherwise. In a nutshell, for Weyl a phenomenological foundation of mathematics rests 'on intuitive foundations as an unsurpassable limit', but mathematics cannot be reduced to what is intuitively accessible. For Weyl, as we will see in the next section, there is a metaphysical step beyond what is accessible to intuition, a step epistemically characterized as belief, which is constitutive of mathematics and physics in their 'symbolical construction of the world'.[46]

The real problem with the Weylean argument is assumption 3. A defender of Husserl's phenomenology could well argue that the resources of Husserlian phenomenology and the forms of intuition available in it far surpass the limited resources of intuitionistic *Anschauung*. Just how much classical mathematics can be accounted for phenomenologically is still a matter of philosophical debate, which certainly cannot be settled here.[47] It is enough to point out that Weyl is implicitly making a rather large assumption that can be resisted already on the basis of Husserl's position.

Becker also seems to accept the problematic assumption we have pointed out in Weyl. We have already seen that Becker, writing to Mahnke, seems to accept as cogent the Weylean argument that if intuitionistic mathematics cannot support theoretical physics then phenomenology, in its classical 'eidetic' form, cannot account for our knowledge of nature. Becker goes further and claims that in mathematics the situation is such as to warrant the recognition of 'formations' [*Gebilde*] going beyond formal ontology[48] in the

Husserlian sense. On August 22, 1926, Becker writes to Mahnke and points out that one needs to distinguish between three types of mathematical formations:

1. Formations which are accessible to the (idealized) sensible intuition; examples include Euclidean geometrical formations in their application to intuitive space.
2. Formations which are accessible to categorial ('eidetic') intuition; these include non-euclidean geometrical formations which can only be intuited arithmetically-analytically; moreover, all the 'objectivities' of Brouwer's 'intuitive' mathematics.
3. Formations which are not accessible through any intuition but are nonetheless consistent. The example here is given by Hilbert's transfinite elements.

Becker specifies to Mahnke that formations of the third sort do not belong to the Husserlian 'formal ontology' but only to that of consistent posits [*widerspruchsfreien Gesetztheiten*], whose proper realm is not truth but consequence.[49] The recognition of a realm of formations that is not accessible, even in principle, to any intuition represents a radical departure from Husserl's phenomenological approach and in fact Becker, in *Mathematische Existenz*, characterizes the distinction between (b) and (c) as new (see p. 483). We thus see that also Becker ends up pointing out the insufficiency of formal ontology to account for mathematics, for Hilbert's transfinite elements are objectivities that are not accessible to formal ontology.[50] We will return to this issue in section 10.6.3.1.

In conclusion, we have seen that both Becker and Weyl accept the limitation of classical 'eidetic' phenomenology for accounting for mathematics as a whole. However, their positions rest on a very problematic assumption concerning the resources of Husserlian phenomenology in the area of mathematics, that is its limitation to intuitionistic *Anschauung*. And, as we have already pointed out, there are reasons to doubt this assumption from a Husserlian perspective.

Returning to Weyl, it should be pointed out that he does not reject intuitionism and phenomenology in the sense that they have no role to play. They do play a role in accounting for those parts of mathematics and physics, which are accessible to intuition. However, Weyl is convinced that phenomenology and intuitionism leave unaccounted for those parts of mathematics (and natural science) which do not fall within 'the light of intuition', and he turns this conviction into a *reductio ad absurdum* of the adequacy of intuitionism and phenomenology for a foundation of mathematics.

10.6.2.3 The Symbolic Construction of the World

Weyl's opening towards the foundational position defended by Hilbert still left him with the problem of what meaning to give to ideal statements. Given that contentual thought is not sufficient to reconstruct classical mathematics, how can meaning be given to the non-contentual part of mathematics? According to Weyl, the formalistic approach defended by Hilbert can still be filled with meaning if one interprets the formalisms in analogy with theories in theoretical physics:

Without doubt, if mathematics is to remain a serious cultural concern then some sense must be attached to Hilbert's game of formulae; and I see only one possibility of attributing to it (including its transfinite components) an independent intellectual meaning. In theoretical physics we have before us the great example of a [kind of] knowledge of a completely different character from the commonly intuitive or phenomenal knowledge which expresses purely what is given in intuition. While in this case every judgment has its own sense which is completely realizable within intuition, this is by no means the case for the statements of theoretical physics. In that case it is rather the system as a whole which is in question if confronted with experience. [...] If Hilbert is not just playing a game of formulae, then he aspires to a theoretical mathematics in contrast to Brouwer's intuitive one. But where is that transcendent world carried by belief, at which its symbols are directed? I do not find it, unless I completely fuse mathematics with physics and assume that the mathematical concepts of number, function, *etc.* (or Hilbert's symbols), generally partake in the theoretical construction of reality in the same way as the concepts of energy, gravitation, electron, etc.[51]

Thus, meaning can be attributed to mathematical and physical theories only holistically. They can never be tested statement by statement but it is only the theory as a whole that gets tested.[52] Similarly, in mathematics one strives for consistency without attempting to give a contentual interpretation to all the transfinite parts. Let us quote one more passage from the section on 'Symbolic Mathematics' in Weyl (1927), which nicely summarizes many of the themes we have touched upon:

[I]t is the function of mathematics to be at the service of the natural sciences. The propositions of theoretical physics, however, certainly lack that feature which Brouwer demands of the propositions of mathematics, namely, that each should carry within itself its own intuitively comprehensible meaning. Rather, what is tested by confronting theoretical physics with experience is the physical system as a whole. It seems that we have to differentiate carefully between phenomenal knowledge or insight—such as expressed in the statement: 'This leaf (given to me in a present act of perception) has this green color (given to me in the same perception)'—and theoretical construction. Knowledge furnishes truth, its organ is 'seeing' in the widest sense. Though subject to error, it is essentially definitive and unalterable. Theoretical construction seems to be bound only to one strictly formulable rational principle, that of concordance, [...] which in mathematics, where the domain of sense data remains untouched, reduces to consistency; its organ is creative imagination.[53]

In order to introduce Becker's work on the transfinite it will be important to recapitulate how Weyl, according to his symbolical construction of the world, sees the situation in mathematics. During the middle of the twenties, Weyl sees the philosophy of mathematics as proceeding according to three stages:[54]

1. The stage of naïve realism; during this stage one is unaware of the transition from the given to the transcendent;
2. the second stage reduces all truth to the intuitively given;
3. the stage where we go beyond the given and attempt to reach the transcendent but only by means of symbolic thought.

It will certainly be useful to the reader to recast the above in more concrete terms. The dialectic displayed in the above three stages can be brought closer to home by considering the foundations of set theory. The first stage consists of the naïve hypostatization of set-theoretical entities as given out there, independently of the human mind. Cantorian realism is paradigmatic of this stage. The second stage is intuitionistic set theory. Sets become a construction of the human mind. In the third stage, we go beyond what is intuitively accessible to us by means of ideal statements. This corresponds to Hilbertian formalism. The transcendent reality is not intuited but symbolically represented. According to Weyl, Fichte had come close to this third stage but in the end fell prey to the 'mystical error of expecting the transcendent ever to fall within the lighted circle of our intuition'.[55]

According to Weyl, thus, it would be however a significant mistake to attempt to give these transfinite elements an intuitive interpretation. In the correspondence with Becker we see Weyl accusing Becker exactly of that mistake.

10.6.3 Becker's Phenomenological Grounding of the Transfinite

As mentioned in section 3, we do not have the original typescript submitted by Becker for publication in *Symposion*. However, it is clear that the paper contained the essence of the phenomenological grounding of the transfinite presented in *Mathematische Existenz*. This is not the place to give a full exposition of the structure and aim of *Mathematische Existenz*, a complex book of 370 pages. For our purposes we can limit ourselves to the opening reflections (pp. 439–810 in the *Jahrbuch* pagination) on formalism and intuitionism and, more specifically, to the phenomenological grounding of the transfinite presented in section §5.a (pp. 521–69). These sections of the book, plus the appendixes and the conclusion, contain all the topics touched upon in the correspondence with Weyl. Detailed blow-by-blow reconstruction of the exchange would increase the length of the exposition unduly. Consequently, we will only focus on the major points and refer the reader to the central sections of *Mathematische Existenz*, for the passages which are closely related to the discussion of some of the technical points.

10.6.3.1 The Ontological Inaccessibility of Hilbert's Ideal Elements

The phenomenological characterization given by Becker of intuitionism and formalism puts emphasis on the fact that intuitionism proceeds according to the logic of truth, whereas formalism proceeds according to the logic of consequence. The distinction is spelled out as follows:

The basic distinction consists in the fact that the logic of truth is related to states of affairs of formal-ontological nature and strives to intuit (of course, categorially) these facts themselves insightfully, i.e., to grasp them in their original givenness. [...] By contrast, the 'logic of consequence' (and with it also Hilbert's mathematics—but not his metamathematics!) is not directed, strictly speaking, at objects (which indeed have to be somehow accessible, in some way phenomena) but rather to mere 'posits [*Gesetztheiten*]', which in their inner structure are impenetrable. (Becker 1927, p. 509)

The reason why Hilbert's mathematics, *i.e.*, formalized mathematics, cannot aspire to fulfillment in intuition is spelled out in the analysis of the role of ideal elements in Hilbert's foundational approach. Becker distinguishes between the old use of ideal elements, such as complex numbers, and the use of the transfinite machinery in Hilbert's program. Complex numbers were eventually provided with a fulfillment in intuition—given by the geometrical interpretation—and thus shown to be both consistent and accessible to a categorial intuition. However, for Hilbert's transfinite axioms things are different. The reader should have in mind here something like Hilbert's formalization of set theory given in Hilbert (1926). A metamathematical consistency proof might show the consistency of the 'posits [*Gesetztheiten*]' formulated in the transfinite axioms, but the transfinite component of Hilbert's formalism is, according to Becker, *in principle* not fulfillable by a categorial intuition (see also the discussion of Becker in §10.6.2.2). The reason is that the transfinite component is obtained by analogy with the laws holding for finite sets and when interpreted contentually the transfinite axioms presuppose the givenness of actually infinite sets. But actually infinite sets cannot be explicitly given to consciousness:

> For the objects of the transfinite axioms, the transfinite sets, are simply *not thinkable* in the sense of the formal-ontological possibility; they are *not* categorially seeable [*erschaubar*]; they are merely empty posits [*Gesetztheiten*]. They cannot, *by any means*, become *realities*. Indeed, even 'consistency' has no 'contentual' meaning at all; it does not mean, in particular, the *conditio sine qua non* of the formal-ontological *dabilitas*. The *dabilia* of formal ontology, of course, must not exhibit any contradiction, but first and foremost they must contain something which is apprehensible in formal-ontological categories—and the transfinite sets do not. (Becker 1927, pp. 515–16)

Becker's criticism of Hilbert's treatment of ideal elements was also shared by Weyl. In a letter to Mahnke (written in September 1927) Becker reports an interesting conversation he had with Weyl on the subject of ideal elements in Hilbert's foundational approach:

> I spoke in August with Weyl who told me that he was still in complete agreement with my critique of Hilbert. [He said that] he himself asked Hilbert several times about the meaning of his (H.'s) 'ideal statements' and that Hilbert had 'regularly turned his head away and changed the topic of discussion'. Nonetheless, despite the complete ontological incomprehensibility of Hilbert's 'mathematics', Weyl considers a symbolical physical knowledge to be possible, which he considers to be freely creative, related to artistic activity.[56]

We have seen in the previous section how Weyl thought one could provide the Hilbertian formalism with meaning. It is now time to turn to Becker's own proposal for a phenomenological account of the transfinite.

10.6.3.2 The Set-Theoretical Background

Becker begins his treatment by going back to Cantor's philosophical pronouncements concerning the infinite. He notices that in some passages Cantor seems to imply that the series of transfinite ordinals has a close relationship with that of the natural numbers, and thus could be interpreted as being closely connected to the idea of potential infinity. Of course, Becker grants that many passages by Cantor actually emphasize the actual infinity of the transfinite ordinals and cardinals. However, Becker's tendency is to distinguish in Cantor an original intuition, phenomenologically valid, from the later developments requiring an explicit appeal to the actual infinite.

Becker emphasizes the two stages in Cantor's development of the theory of transfinite ordinals. The first stage goes back to the treatment given by Cantor in his *Mannigfaltigkeitslehre* in 1883. Here, Cantor gives a development of transfinite ordinals according to generation principles. In the second stage of his thought, espoused in the *Beiträge* of 1895 and 1897, Cantor criticizes the previous theory as inadequate and gives a new foundation of the transfinite numbers, starting from the notion of well-ordered set. This stage is, according to Becker, philosophically problematic on account of its appeal to the actual infinite.

The reader should then keep in mind that it is not Becker's goal to give a foundation for set theory as it is ordinarily understood. In other words, he is not striving to achieve a phenomenological foundation of the ordinals as well-ordered transfinite *sets*. Rather, he is after the transfinite process of ordinal generation, where the ordinals are understood as pure order numbers and not as set-theoretical entities. Becker's phenomenological foundation of the transfinite process will then show how to interpret the generating principles of Cantor's first ordinal theory as indexes of the level-characteristics of acts of transcendental consciousness. Let us begin with Cantor's original generation of the ordinals and then we will move to the phenomenological interpretation.

Cantor (1883) begins with the natural numbers 1, 2, 3, ... and then adds a new ordinal number ω, which follows all the previous ones. However, this is not the last number since the sequence can now be extended to $\omega + 1$, $\omega + 2$, $\omega + 3$, etc. The general construction of the ordinals is regulated by the following generating principles (following Hallett 1984, p. 49):

1. If α is an ordinal number (whether finite or transfinite) then there is a new ordinal number $\alpha + 1$ which is the immediate successor of α;

2. Given any unending sequence of increasing ordinal numbers there is a new ordinal number following them all as their 'limit.'

We are so used to this procedure, by many years of familiarity with set theory that, according to Becker, we do not see anymore how extraordinary the move is, even into the first step of the transfinite, ω. Of course, it is not enough to have trust in the consistency of the formalism. Consistency can give at most the condition of possibility for an ontological fulfilment but it achieves nothing from an 'ontological' point of view:

Intentions which are in principle unfulfilable may well be completely consistent. Actual existence—in the *ontological* sense—requires *phenomenal* givenness, 'access'. After all, ultimately only phenomena 'exist'. (Becker 1927, p. 527)

Becker's goal is thus to show that transfinite numbers can be given in consciousness, as phenomena. And since he rejects the accessibility of actually infinite sets, he aims at grounding the transfinite ordinals by means of concrete phenomenological experiences that relate to potential infinity. Thus, all of Becker's examples of concrete transfinite processes of consciousness will try to achieve two goals. First, show that these processes emerge naturally in concrete situations. Second, avoid carefully the possibility of falling back into admitting the actual infinite and arguing that the transfinite processes involved in the examples can be carried out by appealing to dynamic, potential, processes. In short, the transfinite is to be justified by appealing to potential infinity. The discussion with Weyl will show that Becker is prepared to qualify his claim significantly, in that he is not sure how much of the transfinite is really grounded on phenomenological considerations.

It should be remarked, although we will not delve into this, that Becker's emphasis on concrete examples is a mark of the fact that his notion of ontology is very close to Heidegger's notion (Becker 1927, p. 621). Correspondingly, Becker also criticizes Husserl's phenomenological analyses for being at times only related to the combinatorial possibilities of 'phantasy' and not to the concrete phenomena.[57]

10.6.3.3 The Phenomenological Fulfillment of the Transfinite Process

Iterated representation. Let us begin with examples of iterated representation (they immediately come up in the discussion with Weyl and are also discussed first in *Mathematische Existenz*; see pp. 538–541 and 792–794). Consider the following geometrical figure in which the squares are iterated infinitely many times: Each successive square can be seen as a representation of the previous one. In this way one reaches arbitrary finite levels of representation. But a transfinite level is still not given:

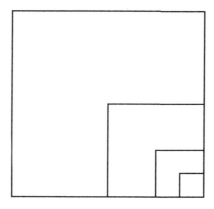

But when one thinks of a picture [of the type given above] as pictured once again, and this picture as once again pictured, then one obtains pictures of level ω, $\omega + 1$, $\omega + 2$, *etc*. In this way one has actually given an intuitive representation, at least for the first transfinite numbers ω, $\omega + 1$, $\omega + 2, \ldots$, (Becker 1927, p. 540)

Weyl's objection to this procedure, as it emerges from the letter, was based on a geometrical diagram that contained the ω-th step of the process within the figure itself. However, Becker remarks that the ω-th step is represented by the entire figure and not by something within the figure. Becker recognizes that there is a difference between the specific finite levels and the ω-th level. Thus, Weyl's discussion of the first example already points at one of the problematic aspects of Becker's construction. Becker is trying to resist the idea of having a step ω inserted in the iteration at the end of all the finite steps. But what does it mean now to say that the ω-th step is represented by the figure containing all the finite levels, given that only finitely many levels can actually be pictured? In *Mathematische Existenz* Becker gives the following clarification:

> It is clear that in an actual drawing only a finite (and indeed a very small) number of pictures can be nested into each other [...] But it is nevertheless the case that the first nestings make the infinite process evident according to its pure ideal possibility. This 'ideal' infinite complication of picture nestings is represented *symbolically* in one of the described figures. And this, in a certain sense, *symbolic* representation is what the further (ω-th, [$\omega + 1$]-th, *etc.*) representational procedure is applied to. This 'symbolic' [representation] is, however, not abstract-conceptual in any way; rather it is intuitive. (Becker 1927, p. 541)

In replying to Weyl's first objection Becker also points out that he prefers examples which are non-mathematical, for often what is mathematically easiest is not necessarily what is phenomenologically most accessible. Finally, he points out that, although many of the examples he is using are known, the novelty of his treatment concerns the connection to Husserl's intentionality and structuring of acts as treated by Husserl in §100 of *Ideen*.

Iterated reflection. Indeed, the following examples of transfinite iteration of consciousness have to do with iterated reflection. Talking about noematic levels, Husserl—in §101 of *Ideen*—says:

> To every noematic level there *belongs a characteristic appropriate to that level* as a kind of index with which each thing characterized manifests itself as belonging to its level—whether it would otherwise be a primary object or one lying in some line or other of the reflective regard. For *indeed to every level belong possible reflections at that level*, so that, *e.g.*, with respect to remembered things at the second level of remembering, ⟨there are⟩ reflections on perceivings of just these things belonging to the same level (thus presented at the second level). (Husserl 1982, p. 247)

What Becker wants to do is to extend the Husserlian characterization of levels, which for Husserl is always limited to finite iterations, to transfinite iterations. The example given by Becker is summarized as follows:

> If we examine the process of iteration of reflection on itself carefully, *i.e.*, entirely concretely, then we see the following:
> I reflect on the reflection I just carried out, then again on this reflection of second level just carried out, and so on. After I have carried out a determinate finite number of nested reflections, say n, I rise—by *stopping* with the continuation of the more and more deeply nested reflection acts—to a level of contemplation, which lies above all these infinitely nested acts. At this level, the ω-th, the object (the theme or the matter) of my present reflection act, is this entire endless sequence of the reflections carried out previously. (Becker 1927, p. 546)

Weyl's objection to (something like) the above seems to have been the following. Before one can reflect on an infinite sequence one must have it as a whole [*ein Ganzes*]. But this can only be done through a 'law'.

We must admit that we do not fully grasp the connection between the objection and Becker's lengthy reply. However, Becker's complex and lengthy discussion of the objection can somehow be summarized by saying that there is an essential difference between a law such as $\phi(n)$ and its argument variable n. But the development of the number sequence, represented by the variable n, must be reducible to the basic intuition of the temporal sequence 'always one more'. If, as Weyl claims, a law is needed in order to be able to grasp an infinite sequence, this would also be true of the fundamental intuition of the 'always one more'. But this would be contradictory as a law expresses how the terms of the law are related to the basic intuition of the 'always one more'. In *Mathematische Existenz* the issue is discussed, with the same example used by Weyl, by distinguishing between functional and argumental laws (pp. 346–47). Becker concludes by saying that the 'law' $\phi(\ )$ giving the infinite sequence is, from the intentional point of view, a finite structure. Infinity appears only in the variable. The temporal sequence is for Becker a 'basic phenomenon' which cannot be reduced to something else. The aim of his work on the transfinite numbers is to show a similar transfinite basic phenomenon. Also in the case of the transfinite basic phenomenon, Becker claims that it is not a matter of a law. Rather, the development of the transfinite is in many ways more similar to that of Brouwer's free choice sequences. Becker clarifies this point by mentioning that the extension of the transfinite to higher and higher transfinite numbers (as in Mahlo's work) is not obtained through a mechanical rule but requires inventiveness. However, this extension is always obtained by means of finitely many symbols and Becker concludes that the transfinite process is, to use Husserlian terminology, 'a system of internal horizons, but at each obtained stage of finite complication'.

In *Mathematische Existenz* he is more explicit:

> One has to distinguish carefully the structure *of the contemplation of* the nesting of the intentionalities and the structure of this nesting itself. The first structure is of *finite* complication whereas the second is of *transfinite* complication. (Becker 1927, p. 547)

The conclusive part of Becker's answer to Weyl's second objection is then that these higher horizons are not laws, just as the basic temporal sequence is not a law, but can serve as basis for a transfinite law $\phi(\beta)$.

The 'Set' W. Concerning the third objection, Weyl seems to have asked whether one does not fall prey to the paradox of the largest ordinal number, or the 'set' containing all ordinals, possibly by means of a reflection that oversees all the previous ones. Here Becker replies that there is no possibility of overseeing all the possible extensions of the transfinite sequence, which thus resembles a free choice sequence in the sense of Brouwer.[58]

Becker and Weyl on Hilbert's Consistency Proofs. Finally, Weyl's fourth objection seems to be directed against Becker's criticism of Hilbert's consistency proofs. Becker revived Poincaré's objections that a consistency proof must appeal to the principle of induction, and that by so doing finitism is committed to the potential infinite and mathematical induction. Here Weyl's objection is close to the one used by Hilbert to defend his approach. He distinguishes between two interpretations of a universally quantified sentence. In the first interpretation one thinks of the universal quantifier as standing for an infinite conjunction. In the second the universal quantifier is read hypothetically, 'each time something of the appropriate type is given, then ...'. One argues hypothetically that each given proof figure cannot give rise to a contradiction. Becker's reply to this is that the infinity is already contained in the 'each', since all the possible proofs of an axiom system have to be taken into account, and there are infinitely many of those. He thus rejects Weyl's way out of the situation.[59]

What Counts as Intuition. In the final part of the letter Becker replies to Weyl's criticism that the 'transfinite complications of pure consciousness' are a 'vague idea'. Here Becker retorts that a logicist such as Frege, intent on grounding the natural numbers on logical laws—the only exact laws, Frege would claim—could accuse Weyl in the same way for assuming the basic number sequence as basic phenomenon on which all of mathematics should be developed. Becker reiterates that the transfinite must be grounded phenomenologically by appealing to reflection processes and that the formal concordance of the iteration of reflection with the investigations of Cantor and Hilbert is a confirmation of the validity of his approach. Moreover, Becker claims that as a phenomenon the transfinite iteration of reflection is just as determined as the usual indefinite horizon of the 'always one more', only more complex and more difficult to access.

Finally, one more objection by Weyl is mentioned. The reflection process leads to more and more emptiness [*ins immer Leerere*]. To this Becker replies by mentioning that phenomenology is used to many investigations into 'empty' phenomena and that in any case Husserl himself is in agreement with Becker's results.

10.6.3.4 The Phenomenological Grounding of the Transfinite in the Second Letter

The last letter reflects Weyl's reaction after reading the fragment of *Mathematische Existenz* and Becker's reaction after reading *Philosophie der Mathematik und Naturwissenschaft*. Many of the same topics discussed in the previous letter are touched upon again. Let us make a few comments on the technical discussion. The next section will deal with Becker's idea of mantic phenomenology.

We have seen in the description of the letter that Becker discusses three mathematical objections by Weyl. The first topic concerns the mathematical standing of the theory of the transfinite. The context of the letter seems to suggest that in his previous letter Weyl had made the remark that the first Cantorian theory (given in the *Mannigfaltigkeitslehre*) is too narrow for the needs of mathematics. On the other hand, the second Cantorian theory (given in the *Beiträge*) uses the actual infinite and the *tertium non datur*. Moreover, Brouwer's and Hilbert's attempts to ground the transfinite do not even manage to ground the second number class (i.e., all the countable ordinals).[60] Becker expresses his hope that a reinterpretation of the theory of the well-ordered sets will open the way for an intuitionist account of at least a part of the theory. Weyl also seems to have suggested that perhaps the second number class is like Cantor's absolute infinite. Becker replies he is not sure whether this is the case and he conceives of the possibility that even the third number class could be accessible to phenomenological reflection. It is at this point that Becker mentions Weyl's insistence on the criterion of 'finite describability'. Becker replies that this notion is difficult to grasp and in any case remarks that the construction of the transfinite he is envisaging requires only a finite system of horizons built upon one another.

The most interesting point raised by Weyl concerns the issues of 'critical numbers', i.e., numbers such as ε_0 and ε_1, which satisfy the equation $\omega^{\varepsilon_0} = \varepsilon_0$ and $\varepsilon_0^{\varepsilon_1} = \varepsilon_1$, respectively. Becker speaks of Weyl's hypothesis but it is not clear what the hypothesis was. We can however find a clarification in a note of *Mathematische Existenz*. The main text (Becker 1927, pp. 567–68) refers exactly to the problem of whether every ordinal of the second number class is finitely describable, where this means not that each single ordinal should be so describable—this would be impossible for cardinality considerations—but that only a finite set of new symbols is necessary to reach an arbitrary ordinal of the second number class by means of Cantor's generation principles (given in his *Mannigfaltigkeitslehre*). Becker notices that Brouwer's reconstruction of set theory only gives a systematic construction only up to the second ε number, ε_1, but that the possibility remains open that the transfinite process will simply terminate at a countable stage, at 'an upper bound, which is at first hidden but that eventually emerges, once one tries to determine it exactly'. The note following this passage credits Weyl for the observation:

> In a letter Professor Weyl has drawn my attention to this problem. Each single definitional system, which determines larger transfinites on the basis of already defined transfinites, eventually breaks down. (Phenonenon of 'catching up' or of 'critical numbers'. Example: ε, where $\omega^\varepsilon = \varepsilon$. It is possible [according to Weyl] that there is a limit 'within the second number class' below which already *all* possible definitional systems break down. (Becker 1927, p. 568, fn. 3)

It must be said that Becker found this objection quite important as often he qualifies his claims by remarking that his phenomenological foundations reaches for sure only up to the first ε number, i.e., ε_0 (Becker 1927, p. 569).[61]

The third issue goes back to the distinction between number variables and hypothetical universal judgments and their relevance for consistency proofs. It concerns a particular formalization given in von Neumann (1927). However, Becker remains unconvinced.[62]

10.6.3.5 Becker on Mantic Phenomenology

Let us say a few concluding words on Becker's own approach to phenomenology. Becker's description of mantic phenomenology in the letter to Weyl plays three roles. First of all, it serves to point out to Weyl that new types of phenomenology are emerging which might be able to account for the sciences and thus that Weyl's pessimism about the prospects of phenomenology might be unwarranted. Second, he points out the centrality of the role of symbolism also in his own reflection and thus shows how Weyl's symbolical construction of the world and mantic phenomenology are trying to give a solution to the same set of concerns. Finally, Becker claims that just as symbolical mathematics emerges from having pushed intuitionism to its limits, so mantic phenomenology emerges from the limits of the old eidetic phenomenology. Thus, in a rather interesting way, Becker seems to have reached a position that closely parallels Weyl's. And indeed, as he admits in his letter and more formally in *Mathematische Existenz* (p. 444), it was Weyl's influence that led him to clarify to himself the new mantic phenomenology.

After listing a number of new directions in phenomenological research, which include the investigations of Husserl, Pfänder, and Scheler, Becker refers to his own investigations. Spurred by the influence of Heidegger's work on '*Geist*' and '*historisches Dasein*', Becker had investigated, in a series of lectures given in the previous semester, the concept of '*Natur*' in the widest sense, i.e., as inorganic, organic, and primitive-spiritual [*primitivseelische*]. He has followed the foundations of this concept not only at the phenomenal level but also at the transphenomenal level. These combined investigations on '*Geist*' and '*Natur*' thus give rise to two important directions in phenomenology, the hermeneutical direction (Heidegger) and the mantic direction (Becker). With the word 'mantic' Becker refers to the fact that there is a divinatory aspect related to any attempt to understand '*Natur*'. Here he also points at the fact that hermeneutical phenomenology does not leave the realm of the immanent, whereas mantic phenomenology tries to reach the transphenomenal, i.e., what lies beyond the phenomena. But why this need to go beyond the phenomena? This is because, as he explains in clearer terms in *Mathematische Existenz*, the upshot of the ontological investigations on mathematics still leaves something unaccounted for. The development of modern physics shows that Hilbert's symbolical, transfinite mathematics, reveals the harmony of the world.

> Thus, the empty game of symbols has a mysterious relationship with the metaphysical structure of the world. What is the meaning and the ground of that relationship? (Becker 1927, p. 767)

It is this question that shows the problem of a divination of a relationship between the phenomena, which is completely different from the task of accounting for the formal constitutions of the phenomena. In this way mathematics appears as a divinatory science which, by means of symbols, allows us to go beyond what is accessible. From this point of view, Becker finds Weyl's observations on contemporary physics and the symbolical construction of the world quite close to his own concerns. Indeed, just as for Weyl symbolic mathematics is the stage that follows the demise of intuitionism, so for Becker mantic phenomenology will have to replace the older 'eidetic' phenomenology. And this sets the stage for a problem which joins Weyl's symbolic construction of the world and Becker's mantic phenomenology: where does the phenomenal stop and the transphenomenal begin?[63]

Added in proof: The following volume, which also contains an article by the first author (PM), came out too late for its contents to be properly acknowledged in our discussion: A. Gethmann-Siefert and J. Mittelstraß, eds., *Die Philosophie und die Wissenschaften. Zum Werk Oskar Beckers* (Fink-Verlag, Munich, 2002).

11

GEOMETRY, PHYSICS, AND PHENOMENOLOGY: FOUR LETTERS OF O. BECKER TO H. WEYL (WITH T. RYCKMAN)

Introduction [Modified from the Original Introduction]

This chapter is the second installment of a study devoted to the relationship between phenomenology and science in the 1920s. The subject is obviously a large theme and our attention has been focused on the interaction between Oskar Becker and Hermann Weyl in 1923–26, an interchange that took the form of a correspondence between the

In this chapter I omit the first four sections of the original publications for they are identical to sections 10.1 to 10.4. I also omit the transcription of the correspondence which was given in appendix to the original article. As a consequence endnote – in this chapter corresponds to note $n + 25$ in the original publication. While writing this paper we have incurred many debts. First of all, we would like to thank Richard Zach for having transcribed the Becker-Weyl correspondence. Interpreting Becker's *Kurrentschrift* is a daunting task but he successfully cracked the code. Without his careful work, including suggestions on the contents, neither this paper, nor Mancosu and Ryckman (2002), would have become a reality. We are also very grateful to Frau Astrid Becker, daughter of Oskar Becker, for answering our queries on her father's *Nachlaß* and for having granted permission to publish the correspondence. Frau Dr Yvonne Vögeli of the *Wissenschaftshistorische Sammlungen* of the ETH (Zurich) was most courteous and helpful during the first author's visit to the ETH in 1997. She subsequently helped us in securing further material and the agreement of the ETH to our publication of the correspondence.

The first author would also like to thank the Hellman Family Faculty Fund for sponsoring his visit to Zurich in the summer 1997. Moreover, a research assistantship from the Committee on Research at U.C. Berkeley (1998–99) and a Faculty Grant (1999–00) were instrumental in funding Richard Zach's work mentioned above.

The first occasion for bringing together the topics which make up this paper was a mini-symposium entitled "Hermann Weyl: Mathematics, Physics, and Philosophy" (Berkeley, April 1999). We are very grateful to the Townsend Center and the Department of Philosophy for co-sponsoring the event and to the other speakers (Solomon Feferman and Richard Tieszen) and the audience for comments.

two scholars. Six letters from Becker to Weyl are still preserved in the Weyl *Nachlaß* at the ETH in Zurich. Although we had planned at the beginning to publish in a single piece the transcription of the letters (due to Richard Zach) [[here omitted]] and our commentary, the length of the project forced us to divide the treatment into two parts. While not ideal for accounting for the unity of concerns that motivated Becker and Weyl this division is not artificial. By subject matter and chronology the letters fall into two groups. The first four, from 1923–24, are primarily concerned with issues in physics and geometry, in particular issues arising in Weyl's response to Becker's *Habilitationsschrift* (1923) on the necessary phenomenological grounding of Euclidean geometry attempted there. The other two letters, from 1926, deal with issues related to the foundations of mathematics. In Mancosu and Ryckman (2002), we published a transcription of the last two letters with a detailed commentary on the relevant background in foundations of mathematics [[see Chapter 10]]. Here, we publish the first four letters [[omitted in this chapter]] together with our introduction detailing the relevant background in physics, geometry, and philosophy.

The chapter is divided into three sections. Section 11.1 provides a synopsis of the first four letters of Becker to Weyl. Section 11.2 treats in detail the related mathematical and phenomenological context of Weyl's work on geometry and relativity physics that inspired this work of Becker. Finally, we consider, in section 11.3, the argument of Becker's monograph in the light of the background laid out in section 11.2.

11.1 Overview of the Four Letters on Geometry, Physics, and Phenomenology

We begin the main part of our discussion with a synopsis of the content of these four letters of Becker to Weyl. For further details and context, reference is made to the individual discussions of §§ 11.2 and 11.3.

The letter of 12 iv 23

This first letter to Weyl accompanies the *Habilitationsschrift* (Becker 1923; see §11.3 below) to which Husserl had alerted Weyl in his letter of 9 April 1922 (see §10.3 above). The letter, as well as the introductory remarks of the *Habilitationsschrift*, conveys a warm appreciation of Weyl's influence, expressing Becker's "deeply felt thanks for the *decisive* scientific stimuli which I have received from your writings on the foundation of mathematics and on relativity theory and the theory of space." Indeed, it is Weyl's work that has first "made possible a complete phenomenological foundation for geometry

---

We are also much indebted to Frau Dr Brigitte Parakenings, curator of the Oskar Becker *Nachlaß*, at the University of Konstanz. She has helped us on numerous occasions, going always beyond the call of duty. While collecting unpublished correspondence of Oskar Becker we also received the help of Prof. A. Troelstra (Amsterdam), Prof. D. van Dalen (*Brouwer Archief*, Utrecht), Dr M. Endress (Felix Kaufmann *Nachlaß*, Konstanz), Dr U. Bredehorn (Mahnke *Nachlaß*, Marburg), Dr W. Hagenmaier (Zermelo *Nachlaß*, Freiburg i. B.). Mr Richard Wilhelm Ackermann kindly replied to our requests for information concerning his father's *Nachlaß*. Finally, we would like to thank Dr Mark van Atten and Professors Michael Friedman and Richard Tieszen for careful comments on an earlier incarnation of this paper.

(in the sense of 'world geometry')" (see 11.2.1 and 11.2.3). Concordance with Weyl, in Becker's estimation, extends beyond the inspiration provided by Weyl's substantial technical contributions to these subject matters. Indeed, Becker states his belief that "the same idealistic basic conception" underlies both Weyl's continuum theory (see Chapter 10) and his "purely infinitesimal geometry" (see 11.2.1). This "philosophical starting point" is identified as "the principle of transcendental idealism [...] out of which arises the fundamental problem of the phenomenological constitution of nature" (see 10.4). The decisive indicator of this fundamental agreement in philosophical perspective is identified as lying in Weyl's designation of a coordinate system as the "residuum of Ego annihilation" ("*Residuum der Ichvernichtung*", see 11.2.2).

Becker concedes that his work contains "nothing new, on the mathematical side". But he does claim to have established a historical antecedent for the conception of a "guiding field" [*Führungsfeld*], Weyl's term for the combined gravitational-inertial field of general relativity. Mathematically, the "guiding field" is a manifold with an affine connection; physically, trajectories of "force-free" neutral test particles are represented by the spacetime geodesics of the connection.[1] While Einstein did not take up Weyl's terminology, the conception is fully in accord with the theory of general relativity and it was strategically employed by Weyl (in the fourth (1921) and fifth (1923) editions of *Raum-Zeit-Materie*) to treat the vexed problem of the relativity of motion.[2] Becker thinks he has found an ancestor of the "guiding field" in works of mechanics going back to the middle of the 19th century:

No one has recognized Andrade as precursor of Einstein in the conception of the 'guiding field' that comprises gravitation and inertia together. I mean that Andrade's concept 'the natural course of things', [...] is directly identical with the concept of 'guiding field'. Perhaps a similar conception can be found in Reech, whom Andrade cites.[3]

It will emerge that Weyl, having looked into the texts, does not accept that there is a "direct identity" with Andrade's non-dynamical pre-relativistic concept (see the synopsis of the letter of 10 x 24, below). Becker concludes by asking several favors. He has heard from Brouwer's students that Brouwer in a lecture the previous winter had developed the differential and integral calculus without the logical law of the excluded middle. Yet nothing seems to have appeared in print. Does Weyl know anything of this? In addition, Becker naturally expresses a "great interest in (Weyl's) Spanish lectures on the problem of space" and he inquires whether a German edition will appear, or whether Weyl knows of a German public library possessing the Spanish edition. As will appear in the fourth letter, Weyl later posts Becker a copy of the German edition (Weyl 1924a). Finally, Becker asks for an offprint of Weyl (1920a) on statistical and causal laws of nature that appeared in a hard-to-find Swiss medical weekly (see 11.2.4).

The letter of 25 iv 23

In responding to Becker's first letter, Weyl erroneously omitted the first page of his reply and enclosed instead a page written to a Spanish scholar. Apparently Weyl had written that he would send Becker his work on the problem of space when it is available (see 11.3).

After informing Weyl of his mistake, Becker expresses his gratitude for Weyl's acceptance of his *Habilitationsschrift* and promises to respond to the issues raised regarding it upon receipt of the missing part of the letter:

to my great joy I see that you have very kindly accepted my writing and I thank you heartily for this. I will first enter into a factual discussion of the points touched on by you after I am in possession of your entire letter.

The letter of 27 vi 23

In this letter, Becker provided the promised detailed reply to queries Weyl raised to his *Habilitationsschrift*. Although Weyl's objections must be inferred, it seems probable they were directed upon Becker's "transcendental-phenomenological grounding" of the a priori necessity of both the three-dimensionality and the Euclidean metrical structure of "phantom (intuitive) space". Becker responds as follows: With regard to three-dimensionality, Becker himself is not satisfied, but with respect to Euclidean structure, he still maintains that there is a "compelling justification". In his monograph (see 11.3.2) this justification has two separate components, ontological and transcendental phenomenological. Becker refers to his discussion at pp. 482–84, which, however, he fears "will not make a strong impression on a mathematician"; the passage concerns the "ontological determination" that space, as an optimal *"principium individuationis,"* must have Euclidean metrical structure. On the other hand, Becker also reiterates his transcendental-phenomenological justification of the necessary Euclidean nature of space on which he seems to stake a greater confidence. Becker appears to yield some ground to Weyl, stating that his justification should proceed on the basis of Weyl's "Pythagorean metric determination" and without any reference to the Helmholtz-Lie solution to the "problem of space" involving the condition of "free mobility". At the same time, however, Becker holds ("I now maintain") that Weyl's group-theoretic treatment of the "problem of space", which distinguishes between the "nature" and the "orientation" of the metric, abides the same fundamental phenomenological distinction between rotational and translational (radial) motions that he originally identified. See the more detailed comments in 11.3.2 below.

The letter of 10 x 24

Weyl has sent several off-prints, including apparently one of Weyl (1923b). It seems that he has also sent two monographs, the German language version of his Madrid and Barcelona lectures (Weyl 1924a), and a reprint of two articles appearing that year in the journal *Die Naturwissenschaften* (Weyl 1924c). Becker again mentions Andrade and Reech as precursors to Weyl's "guiding field" characterization of purely gravitational-inertial motion in general relativity.[4] He deems Weyl's comments regarding the "rotation of the star compass"[5] as very noteworthy; they are a physical illustration of the directional bundle of rays proceeding from a point-eye in Becker's phenomenological grounding of Euclidean space (see 11.3.2). Becker states that Weyl's recent further investigations of the Euclidean rotation group (in the first off-print mentioned above) are of great interest, however, they are too complicated to bring to intuitive or visualizable representation. This situation prompts Becker to wonder whether another, less mathematically complex, basis for Weyl's "Pythagorean metric determination" might not be found, and he asks Weyl's indulgence for the sketch that follows of one such attempt. Becker's idea, citing a contemporary text of affine differential geometry,[6] is based on an affine invariant (an "affine sphere", the surface generated by the tip of a rotating "unit vector") rather than Weyl's group theoretical characterization of the rotation of a "volume-true vector body", a congruence and hence, metrical concept. But it is questionable whether his sketch is developed or coherent enough to merit discussion, and in any event, Becker does not see how this method of procedure will yield the "fundamental fact" of infinitesimal geometry, that the metric of a Riemannian manifold uniquely determines the affine connection. This is a potentially fatal flaw (see 11.2.3).

The most interesting passage of this letter concerns a lengthy response to Weyl's speculative considerations on matter in Weyl (1924c) that have "awakened in me a desire to bring out of them the purely a priori". Such a project (of *"Apriorisierung"*), Becker concedes, always runs a danger since it is anchored in the specific physical ideas of a given epoch, usually the present. In any event, Becker deems very important Weyl's "monadic" analogy of the electron and the Ego (see 11.2.4) and he states his opinion regarding the a priori determinate structure of the Ego's phenomenal forms of representation, space and time. Presumably with reference to Weyl's suggestive picture of the Ego tracing out its world line in relativistic spacetime where "nothing happens", this is a structure in which the smallness of the three-dimensional space (surrounding that Ego) stands in vivid contrast to the vast unbounded extension of the time dimension. There follows an extraordinary musing on the linkage between scientific theories and their historical, social and psychological context.

The starting point is a reflection on the "objective standpoint" possible within this phenomenological conception of the Ego, infinitesimal in relation to the world, and to history, which requires the unboundedness of past time. But, Becker asks, does such a representation of the Ego, or even of a geometry whose elements are its world lines, reflect anything more than the "historical consciousness of our time"? Is it not philosophically naïve to think that conceptions of objectivity can transcend their historical epoch? Intriguingly he then observes, with arresting directness, that the reasons for the recent turn of physics to indeterminism can not be accountable as due to an increase in objective knowledge of the physical world:

that we have today taken the leap to become indeterminists in physics, certainly doesn't lie in the increase of our objective knowledge of the physical world!

This is a remarkable statement, an almost classic case of a genre considered in Paul Forman's controversial essay on Weimar culture and causality (Forman 1971, see 11.2.4). Becker's ruminations go even further than relativizing objectivity to historical epoch; speculative scientific theories might also be illuminated by reference to their individual creators and their differing modes of "life-fulfillment". For example, Laplacian determinism ("so grotesque to us today") is accounted "a late baroque idea". "Limiting myself to the present", he goes on to ask, "doesn't the De Sitter cosmology have a different *ethos* than the Einstein cosmology?"[7] Of course, Becker is aware of the highly uncertain character of this mode of thinking in which, he cites Shakespeare's Falstaff, "reasons are cheaper than blackberries". But it is clear, he continues, that identifiable tendencies of such a linkage between theory and context are even more prevalent in speculative scientific theories (such as Weyl's "agent theory of matter", or in the Einstein or De Sitter cosmologies) than in those that have a narrow tie to experiment. All of this is now rather suddenly brought to bear on the above-mentioned project of *"apriorization"*. For it may turn out that a particular theory is the only one possible: this would mean that the *"apriorization"* cannot be only formal, a "boring axiomatics". Rather at its core must be a "material a priori" that alone will satisfy the phenomenological demand for *"Evidenz"*. Becker ends the discussion with a cautionary remark to the effect that he doesn't know what Weyl would say to all of this, but that he has the feeling such a line of thought is not strange to Weyl.

The letter concludes with what the next few years would prove to be an overly optimistic forecast of the future prospects of phenomenology in foundations of mathematics. Becker sees the new philosophy, latent within the domain of pure mathematics, as now emerging; the best examples are the views of Brouwer and Weyl on existence in mathematics and on the continuum. He states his belief that the same intellectual root (*geistige Wurzel*) as is currently transforming epistemology into phenomenology is also at work in the transformation of the classical (*"Cantor-Dedekind-Weierstrass'schen"*) theory of the continuum into that of Brouwer and Weyl. In support of this claim, he notes that even Hilbert, though defending "classical" mathematics, in fact employs "dialectical" methods, in particular, the method of seeking consistency proofs, that seems in accord with the modern "dialectical" tendency in philosophy. However this may be, it is clear that this letter bears witness to the transformation of Becker's own understanding of phenomenology marked by the "anthropological" and existentialist orientation of *Mathematische Existenz*, which, he notes in the closing paragraph, is still some time away from completion.

11.2 The Phenomenological Context of Weyl's Work on Geometry and General Relativity

In this section, we bring to light the phenomenological context of Weyl's innovations in differential ("infinitesimal") geometry and in relativity theory that Becker has credited as having had a "decisive" influence on his own *Habilitationsschrift*.

1918 was Weyl's *annum mirabilis*. Within the first six months of that year, Weyl published both his monograph on the continuum (Weyl 1918a) and the first edition of *Raum-Zeit-Materie* (1918), the first comprehensive treatise on general relativity, extolled by Einstein as a "symphonic masterpiece" wherein "every word has its relation to the whole."[8] While *RZM* was still in press, in the early Spring of 1918, Weyl produced the first geometrical unification of gravitation and electricity (Weyl 1918b,d). All of these achievements reveal the influence of Husserl's transcendental phenomenological idealism. These philosophical motivations are most clearly articulated in Weyl's classic monograph on general relativity. In the light of its self-contained mathematical treatment of relativity theory (and other mathematical gems, e.g., a new derivation of the Schwarzschild solution for the static centrally symmetric gravitational field that Einstein (1922, p. 94) considered "particularly elegant"), the book's densely philosophical "Introduction" surely comes as a surprise. Readers, who believed they had in their hands what the journal *Nature* termed, on its appearance in English, "the standard treatise on the general theory of relativity", could only have been puzzled by such passages as:

> The actual world [*wirkliche Welt*], each of its components and all their determinations, is, and can only be, given as intentional objects of conscious acts. The conscious experiences that I have are absolutely given—just as I have them.
>
> The immanent is absolute, that is, it is exactly what it is just as I have it, and I can eventually bring this, its essence [*Wesen*] to givenness [*Gegebenheit*] before me in acts of reflection.
>
> The given to consciousness [*Bewußtseins-Gegebene*] is the starting point in which we must place ourselves in order to comprehend the sense and the justification of the posit of reality [*Wirklichkeitsetzung*].
>
> "Pure consciousness" is the seat of the philosophical a priori.[9]

There follows a highly condensed outline of what is basically Husserl's argument for transcendental phenomenological idealism in sections 33–55 of the *Ideen I* (Husserl 1913). To the few cognoscenti, Weyl's orientation to Husserlian phenomenology is readily apparent in the language he uses; to others, it is announced in the book's first footnote. Nor is this a momentary infatuation; despite many changes in the various editions in other parts of the book, these passages in the "Introduction" remain intact from RZM^1 to RZM^5 (1923). Making no apologies for the incursion of philosophy into his technical treatise, at the end of the "Introduction", Weyl stated his belief that philosophical understanding of science was not an idle pursuit but a necessity:

> All beginnings are obscure. [...] from time to time the mathematician, above all, must be reminded that origins [*die Ursprüngen*] lie in depths darker than he is capable of grasping with his methods. Beyond all the knowledge produced by the individual sciences, remains the task of *comprehending* [*zu begreifen*, original emphasis]. Despite philosophy's endless swinging from system to system, back and forth, we must not dispense with it altogether, lest knowledge be transformed into a senseless chaos.[10]

Much of *RZM* and Weyl's related papers in 1918–23 on his geometrical "extension" of general relativity (see immediately below), manifest this striving to attain philosophical comprehension of the new physical knowledge represented in Einstein's theory. On recognition of the many clearly visible intimations of these transcendental-phenomenological leanings, Weyl's texts evince a remarkably sustained attempt to probe the "darker depths" of the "origins" of the knowledge portrayed in classical field theory ca. 1918 through mathematical construction guided by the phenomenological method of "essential analysis". To be sure, Weyl himself judged his treatment of the epistemological questions raised by the new theory as preliminary and tentative, lamenting he had not been able to provide such answers as would salve his scientific conscience (Weyl 1918c, p. 2). Even so, there can be little doubt that his attempt to cast illumination in these dim regions was carried out in agreement with the fundamental thesis of transcendental phenomenological idealism, as stated at the end of section 49 of Husserl's *Ideen I*:

The whole spatiotemporal world, which includes man himself and the human Ego as subordinate single realities is, according to its sense, a merely intentional being, thus one having the merely secondary, relative sense of a being for a consciousness. It is a being which consciousness in its experience posits that, in principle, is only determined and intuited as something identical by motivated manifolds of phenomena: beyond that it is nothing.

Understood as guided by this beacon, Weyl's 1918 re-setting of general relativity within the new framework of a "purely infinitesimal geometry" was principally an attempt to demonstrate that the objects of classical field physics likewise have the "sense of a merely intentional being", of a "being for a consciousness". Broadly speaking, Weyl's case is argued as follows. Because its (pseudo–) Riemannian metric permits direct comparison of vectors and tensors "at a distance", Einstein's general relativity is not in accord with the fundamental principle of *Nahewirkung* (contiguous action) constitutive of field theory. Weyl therefore sought a "world geometry" for field physics in full compliance with this principle, constructed through a phenomenological "essential analysis" of the basic comparison relations of vectors and tensors. The result yielded new mathematical concepts serving as an axiomatic basis for differential (infinitesimal) geometry. These are relations of parallel displacement and congruent displacement that are initially meaningful only "purely infinitesimally", i. e., in the tangent space of a point in a manifold, and so vectors and tensors are immediately comparable only in infinitesimally small regions. As the products of an "essential analysis" into the *meaning* of vector and tensor comparison, these "purely infinitesimal" relations are regarded as *evident*, "given to consciousness" in a localized phenomenological intuition. They also give rise to mathematical structures identical to the vector and tensor fields of gravitation and electromagnetism, the latter enabled by the new mathematical degrees of freedom corresponding to the posit of a local scale ("gauge") arbitrariness. Making these identifications, the electromagnetic field as well as the gravitational field, are incorporated into spacetime geometry. Their construction within a "purely infinitesimal world geometry" comprises a "regional ontology" for classical field physics, displaying it within the bounded space of eidetic possibilities constituting the unity of sense of the objects of field physics. In this way, the sense of these fields as *intentional objects* is exhibited.

11.2.1 "Reine Infinitesimalgeometrie"

Weyl's initial papers, Weyl (1918b,d), inaugurated the program of unified field theory that subsequently occupied the last three decades of Einstein's life. The guiding impetus of unified field theory was geometrical: to account for all fundamental forces of nature (before the discovery of nuclear forces in the 1930s, gravitation and electromagnetism were the only known fundamental forces) within a dynamical spacetime geometry. The cue was taken from Einstein's theory of general relativity (1915–16) where gravitational "force" appears as "spacetime curvature" and so is "geometrized away", but electromagnetism has no organic connection with space-time geometry. Through the above-mentioned "essential analysis", Weyl found a very natural generalization of the pseudo-Riemannian geometry of the Einstein theory in which electromagnetism appeared in the proper geometrical garb of a curvature tensor.[11] As will be seen, the metric in such a "Weyl geometry" has *two* basic terms, the quadratic differential form $g_{\mu\nu}$ corresponding to the gravitational field as in Einstein's theory, as well as a linear differential form φ_μ corresponding to the spacetime four-potential of electromagnetism (here and below, all indices are space-time, running 0–3). Thus Weyl's geometry, which he claimed, improved upon Riemann's, in that it was *"purely infinitesimal"*, led to the first projected geometrical unified field theory, and to an enthusiastic proclamation of the unity of geometry and physics:

Everything real (*Wirkliches*) that transpires in the world is a manifestation of the world-metric. Physical concepts are none other than those of geometry. (Weyl 1918d, p. 385)

As will be seen, Weyl's optimism concerning a purely field theoretic and geometrical basis for physics was remarkably short-lived. For his mathematical construction, Weyl restricted the homogeneous space of phenomenological intuition, the locus of phenomenological *Evidenz*, to what is given at, or neighboring, the experiencing ego:

Only the spatio-temporally coinciding and the immediate spatial-temporal neighborhood have a directly clear meaning exhibited in intuition. [. . .] The philosophers may have been correct that our space of intuition bears a Euclidean structure, regardless of what physical experience says. I only insist, though, that to this space of intuition belongs the ego-center [*Ich Zentrum*] and that the coincidences, the relations of the space of intuition to that of physics, becomes vaguer the further one distances oneself from the ego-center.[12]

Guided by the phenomenological methods of "eidetic insight" and "eidetic analysis", Weyl erected a "purely infinitesimal geometry" upon the epistemologically privileged infinitesimal geometric relations of *parallel transport* of a vector, and *congruent displacement* of a vector magnitude. The task of comprehending "the sense and justification" of the mathematical structures of classical field theory is accordingly to be addressed through a construction or *constitution* of the latter within a "world geometry" (a "regional ontology" in the sense of Husserl) generated from these basic geometrical relations immediately evident within the infinitesimal space of pure intuition. This wholly *epistemological* project coincides with the explicitly *metaphysical* aspirations of Leibniz and Riemann to "understand the world from its behaviour in the infinitesimally small".[13]

By delimiting what Husserl termed "the sharply illuminated circle of perfect givenness", the domain of "eidetic vision", to the infinitely small homogeneous space of intuition surrounding the "ego-center", Weyl could restrict to linear relations, since only these need be considered in passing to the tangent space of a point in a manifold. Linearity, in turn, gave the expectation of "uniform elementary laws".[14] Initially restricting the concept of a coordinate system to the tangent space covering each manifold point P, Weyl essentially assumed a manifold that is Hausdorff, and paracompact; its four dimensionality is to be derived from the "essential laws" of Maxwell's theory (see below). Imposition of a local coordinate system is regarded as the original constitutive act of "a pure, sense-giving ego". Required by the operations of differential calculus on a manifold, a coordinate system always bears an indelible mark of transcendental subjectivity, it is "the unavoidable residue of the ego's annihilation in that geometrico-physical world which reason sifts from the given under the norm of 'objectivity' ".[15] In this necessary presupposition of any differential structure, Weyl recognized an intimation of the phenomenological postulate that "existence is only given and *can* only be given as the intentional content of the conscious experience of a pure, sense-giving ego"(see 11.2.2).

The next steps concern the immediately evident "purely infinitesimal" relations of comparison of direction and magnitude that depend on a specific choice of coordinates and unit of scale. The construction of purely infinitesimal geometry is laid out as taking place in three distinct stages of "connection": topological manifold or "continuous connection" [*stetiger Zusammenhang*], affine connection, and "metric (or, length) connection".[16] The construction itself, "in which each step is executed in full naturalness, visualizability and necessity" [*in voller Natürlichkeit, Anschaulichkeit, und Notwendigkeit*], was "in all essential parts the final result" of the renewed investigation of the mathematical foundations of Riemannian geometry opened up by Levi-Civita's concept of infinitesimal parallel displacement.[17] Taking this as its paradigm relation, Weyl's construction was to satisfy the requirement of phenomenological *Evidenz*, mandating that the consistency and coherence of all integral comparison relations definable between *finitely separated* points are built up by integrating a specified infinitesimal connection along a given path connecting the points. The physical world is then to be distinguished within this "world geometry" through the univocal choice of a gauge invariant action function $W(g_{\mu\nu}, \partial g_{\mu\nu}, \varphi_\mu, \partial\varphi_\mu)$, where $g_{\mu\nu}$ is the (now only conformally invariant) metric tensor and φ_μ is the electromagnetic four potential.[18] However, to Weyl's dismay, it soon became clear that such a function was not uniquely determined and that choice among the various possibilities was essentially arbitrary.[19]

First stage: continuous connection (topology)

Weyl's several discussions of topology in the context of his geometry add nothing new *per se* but take over the modern topological concepts of "point" and "neighborhood", first clarified in his own 1913 book on Riemann surfaces. In fact, there is a clearly identified reason for Weyl's reticence regarding phenomenological constitution of the manifold, and so of the concept of "continuous connection" itself. Referring to the discussion in *Das Kontinuum* of the "deep chasm" separating the intuitive and the mathematical continuum (see 11.2.2), Weyl observed that a "fully satisfactory analysis of the concept of the *n-dimensional manifold* is not possible today" in view of the "difficulty of grasping the intuitive essence [*anschauliche Wesen*] of continuous connection through a purely logical construction".[20] Setting that task aside, Weyl simply assumed that in the tangent space covering each manifold point P, there is an affine linear space of vectors centered on

P in that line elements dx radiating from P are infinitely small vectors. In this way, functions at P and in its neighborhood (in particular, the displacement functions—see below) transform linearly and homogeneously. Weyl's attention then concentrated on the manifold's *Strukturfeld*, its metric, affine (and conformal, and projective) structures, originating the now familiar machinery of connections in a specific epistemological context.

Second stage: affine connected manifold

The concept of parallel displacement of a tangent vector in a Riemannian manifold M was first developed in 1917 by Levi-Civita (and independently by Schouten in 1918). It provided a geometric interpretation—as the parallel displacement of a vector along a path initially connecting a point P to another point P' "immediately nearby" (tangent space $T_P(M)$ of P, where $P' - P = \overrightarrow{PP'} \in T_P(M)$ and $T_{P'}M = T_PM$)—to the hitherto purely analytical Christoffel symbols (of the second kind) of covariant differentiation. Covariant differentiation can then be geometrically understood as a comparison of infinitesimal changes in vector or tensor fields with respect to a parallel transported vector or tensor at the point in question. Parallel transport is "purely infinitesimal" in the sense that directional comparison of vectors at finitely distant points P and Q can be made only by specifying a path of displacement from P to Q and "transporting" to Q a comparison vector defined as "parallel to" the original vector at P. In general, parallel displacement is not integrable, i. e. the new vector arising at Q will depend upon the path taken between the two points.

In Weyl's assessment, Levi-Civita's concept marked a significant advance of "simplicity and visualizability [*Anschaulichkeit*] in the construction of Riemannian infinitesimal geometry." (Weyl 1923b, p. 11) But whereas Levi-Civita had employed an auxiliary construction, embedding M in a Euclidean space where parallel transport was defined, and then projecting it into the tangent space of M, Weyl gave the first intrinsic characterization in terms of bilinear functions $\Gamma(A^\mu, dx)$ that he termed the components of a (symmetrical) *affine connection*.[21] In general, the change δA^μ in a given vector A^μ displaced from P to $P'_{(x^\nu + dx^\nu)}$ is defined

$$\delta A^\mu = -\Gamma^\mu_{\alpha\beta} A^\alpha dx^\beta, \tag{11.1}$$

while the *covariant derivative* of A^μ (a tensor, and so of objective significance) is defined

$$A^\mu_\alpha = \frac{\partial A^\mu}{\partial x^\alpha} + \Gamma^\mu_{\beta\alpha} A^\beta. \tag{11.2}$$

Parallel transport occurs when the components of the affine connection vanish. Next followed the concept of a *manifold with an affine connection*. A point P is affinely connected with its immediate neighborhood just in case it is determined, for every vector at P, the vector at P' to which it gives rise under parallel transport from P to P'. If it is possible to single out a unique affine connection among all the possible ones at each point P, then M is a *manifold with an affine connection*, essentially a conception of space linearly stitched together from infinitely small homogeneous patches. To Weyl, parallel transport was the paradigm comparison relation of infinitesimal geometry for it satisfied the epistemological demand that all integral (and so, not immediately surveyable) relations between finitely separated points cannot be posited but must be constructed from a specified infinitesimal displacement along a given curve connecting them. He also introduced the idea of the curvature of a connection $R\,(\Gamma)$, a (1, 3) tensor analogous to the Riemann-Christoffel tensor of Riemannian geometry, and showed that the calculus of tensors could be developed on the basis of the concept of infinitesimal parallel transport, without any reliance on a metric.[22] However, it was Eddington, not Weyl, who first fully exploited this idea in physics.[23]

In Weyl's characterization, "the essence of parallel displacement" [*das Wesen der Parallelverschiebung*] is expressed in that, in a given coordinate system covering P and its neighborhood, the components of an arbitrary vector A^μ do not change when it is parallel-displaced from P to a neighboring point P'.[24] Unaltered displacement accordingly depends on a particular "geodetic" (at P) coordinate system, a prolepsis to the fact that at P the $g_{\mu\nu}$ have stationary values, $\frac{\partial g_{\mu\nu}}{\partial x^\sigma} = 0$, and so the components of the affine connection vanish. According to Einstein's principle of equivalence, such geodetic coordinates at a point always exist. In this dependence on a particular coordinate system, parallel displacement of a vector or tensor without "absolute change" is not an invariant or "objective" relation. But a specifically epistemological and non-conventional meaning is intended for the statement that some vector at P' is "the same" as a given vector at P. Namely from the original vector at P, a new vector arises at P' that, in the purely local comparison made, as it were, by a particularly situated consciousness, is affirmed to be "without change". Despite the subjectivity of the "experienced" condition $\frac{\partial g_{\mu\nu}}{\partial x^\sigma} = 0$ required by this construction, such comparison is nonetheless the basis for the invariant relation of covariant differentiation. Obviously, the idea is an analogy formed from Einstein's theory in which the non-tensorial gravitational field strengths $\Gamma^\sigma_{\mu\nu}$ (in Weyl's suggestive terminology, the "guiding field" [*Führungsfeld*]) can be locally, but not generally, "transformed away", an observer-dependent "disappearance" of a gravitational field. At the same time, invariant spacetime curvatures are derived from the $\Gamma^\sigma_{\mu\nu}$ that have an objective significance for all observers.

Third stage: metrically connected manifold

Weyl then demanded that

a truly infinitesimal geometry [wahrhafte Nahegeometrie] should know only a principle of displacement [Übertragung] of a length from one point to another infinitely close by.[25]

As the "essence of space" is metric, the fundamental metrical concept, congruence, also must be conceived "purely infinitesimally". (Weyl 1923b, p. 47) Baptized as "the epistemological principle of relativity of magnitude", a postulate is laid down that direct comparison of vector magnitudes can be immediately made only at a given point P or at infinitesimally nearby points P'. Just as an affine connection governs direct infinitesimal comparisons of orientation, or parallelism, so a *length* or *metric connection* is required to determine infinitesimal comparisons of congruence. This also requires a vector to be displaced from P to P' and, in general, the "length" l of the vector is altered. Thus if l is the length of a vector A^μ at $P_{(x)}$, $l_{P_{(x)}}(A^\mu) = ds^2 = g_{\mu\nu} A^\mu A^\nu$, then on being displaced to P', the change of length is defined to be a definite fraction of l,

$$\frac{dl}{dx^\mu} := -l \frac{d\varphi}{dx^\mu}, \qquad (11.3)$$

where $d\varphi = \sum_\mu \varphi_\mu dx^\mu$ is a homogeneous function of the coordinate differentials. The new vector at P', corresponding to A^μ at P, accordingly has the length

$$l_{P'_{(x+dx)}} = (1 - d\varphi)(g_{\mu\nu} + d g_{\mu\nu}) A^\mu A^\nu, \qquad (11.4)$$

where $(1 - d\varphi)$ is a proportionality factor, arbitrarily close to 1. In analogy to (11.1), the change in length of A^μ is defined

$$\delta l := \frac{\partial l_P}{\partial x^\mu} dx^\mu + l\, d\varphi. \qquad (11.5)$$

Then, just as the vanishing of its covariant derivative means that a vector has been parallel-transported from P to P' without "absolute change", so here the vanishing of δl indicates that A^μ has been *congruently displaced* from P to P':

$$\delta l = 0 \Leftrightarrow dl = \frac{\partial l}{\partial x^\mu} dx^\mu = -l\, d\varphi. \qquad (11.6)$$

Up to this point, an arbitrary "gauge" (unit of scale) has been assumed. Recalibrating the unit of length at P through multiplication by λ, an always positive function of the coordinates, multiplies the length $l_{P_{(x)}}$ by λ, $l' = \lambda l$. Then the change in length at P', $d\, l'$, corresponds to a transformation of the "length connection" $d\varphi$,

$$d l' = d(\lambda l) = l\, d\lambda + \lambda\, dl = l\, d\lambda - \lambda l\, d\varphi$$
$$= -\lambda l \left(d\varphi - \tfrac{d\lambda}{\lambda}\right) = -l'\, d\varphi' \qquad (11.7)$$

($d\lambda/\lambda = d \log \lambda$). A *metrically connected manifold* is then one in which each point P is metrically connected to every point P' is its immediate neighborhood through a *metric connection*. In general, length is not integrable for (11.8) follows from (11.5) by integration,

$$\log l \Big|_P^Q = -\int_P^Q \varphi_\mu dx^\mu, \text{ and so } l_Q = l_P\, e^{-\int_P^Q \varphi_\mu dx^\mu}. \qquad (11.8)$$

As Pauli demonstrated, displacement of a vector along different paths between finitely separated points P and Q will lead to arbitrarily different results at Q.[26] But when the linear form φ_μ vanishes, the magnitude of a vector is independent of the path along which it is displaced, which is just the case of Riemannian geometry. The necessary and sufficient condition for this is the disappearance of the "length curvature" [*Streckenkrümmung*] of Weyl's geometry

$$F_{\mu\nu} = \frac{\partial \varphi_\nu}{\partial x^\mu} - \frac{\partial \varphi_\mu}{\partial x^\nu}, \tag{11.9}$$

just as the vanishing of the Riemann tensor is the necessary and sufficient condition for flat space.

Implementation of the local comparison condition means that the fundamental tensor $g_{\mu\nu}$ of Riemannian geometry induces only a local conformal structure on the manifold. Then there is an immediate meaning given to the angle between two vectors at a point, or to the ratio of their lengths there, but not to their absolute lengths. These transform at a point x as $g'_{\mu\nu}(x) = \lambda g_{\mu\nu}(x)$. This weakening of the metrical structure has two important consequences. Such a metric no longer determines a unique linear (affine) connection, but only an equivalence class of connections. Yet Weyl required, as the "fundamental fact" of infinitesimal geometry, that there be unique affine compatibility in the sense that the transport of tangent vectors along curves associated with the connection, i. e. affine geodesics, leave the vectors congruent with themselves with respect to the metric. Weyl showed that a unique connection, coupled to given choice of a metric tensor, is found by incorporating into its definition the linear differential form φ of his length connection. Then, when the components of φ_μ vanish identically at a point, the affine connection of Weyl's geometry becomes identical to the usual "Levi-Civita" connection of Riemannian geometry:

$$\text{(Weyl):} \quad \Gamma^\sigma_{\mu\nu} = \frac{1}{2} g^{\sigma\tau} \left[\frac{\partial g_{\tau\mu}}{\partial x^\nu} + \frac{\partial g_{\nu\tau}}{\partial x^\mu} - \frac{\partial g_{\mu\nu}}{\partial x^\tau} \right] + \frac{1}{2}(g_{\mu\sigma}\varphi^\nu + g_{\nu\sigma}\varphi^\mu + g_{\mu\nu}\varphi^\sigma) \tag{11.10}$$

Given this "Weyl connection", it is possible to speak of a manifold with an affine connection where, as in the Riemannian case, there is unique determination of parallel displacement of a vector at every point. Only in the case of "congruent displacement" [*kongruente Verpflanzung*], or displacement without alteration of length, is parallel displacement possible, and so it is that infinitesimal length or "tract displacement" [*Streckenübertragung*], the "foundational principle of metric geometry", brings along also directional displacement [*Richtungsübertragung*]. This is to say, according to Weyl, that "*according to its nature, a metric space bears an affine connection.*"[27]

Justification of his "essential analysis" of infinitesimal geometry culminated in Weyl's purely mathematical group-theoretical proof of Riemann's posit of an "infinitesimal Pythagorean (Euclidean) metric", the capstone of his efforts to show that the supposition of the purely infinitesimal character of the geometry underlying field physics was not arbitrary (see 11.2.3 below).[28] Writing to Husserl on 26 March 1921, Weyl could report that he had finally captured the "a priori essence of space [*apriorische Wesen des Raumes*], through (using an expression familiar from Hilbert) a notable deepening of its mathematical foundations [*eine merkliche Tieferlegung der Fundamente*]".[29] This philosophical linkage was publicly announced in a newly appended passage in the fourth (1921) edition of *RZM*. There Weyl declared that his "investigations concerning space [...]

appear to me to be a good example of the essential analysis [*Wesensanalyse*] striven for by phenomenological philosophy (Husserl)".[30] Referring to this passage some thirty years later, Weyl observed that he still essentially held to its implicit characterization of the relation between cognition and reflection underlying his method of investigation, one that combined experimentally supported experience, analysis of essence [*Wesensanalyse*] and mathematical construction. (Weyl 1955, p. 161)

The transition to physics[31]

Just as Einstein required the invariance of physical laws under arbitrary continuous transformation of the coordinates (general covariance), Weyl additionally demanded their invariance under the "gauge transformations"

$$g \Rightarrow g'_{\mu\nu} = \lambda g_{\mu\nu}, \text{ and } \varphi \Rightarrow \varphi'_\mu = \varphi_\mu - \frac{1}{\lambda}\frac{\partial \lambda}{\partial x^\mu}.$$ (11.11)

As the first (homogeneous) system of Maxwell's equations

$$\frac{\partial F_{\mu\nu}}{\partial x^\sigma} + \frac{\partial F_{\nu\sigma}}{\partial x^\mu} + \frac{\partial F_{\sigma\mu}}{\partial x^\nu} = 0$$ (11.12)

follows immediately from (11.7) on purely formal grounds, Weyl made the obvious identifications of his length curvature $F_{\mu\nu}$ with the already gauge-invariant electromagnetic field tensor (of "gauge weight 0"), and his metric connection φ_μ with the spacetime four potential. As a formal consequence of the basic relations of Weyl's geometry, equations (11.8) are held to express "the essence of electricity"; they are an "essential law" [*Wesensgesetze*] whose validity is completely independent of the actual laws of nature. (Weyl 1919b, p. 244) In addition, Weyl could show that a vector density, and contravariant second rank tensor density follow from the *general* form of a hypothetical action function invariant under local changes of gauge $\lambda = 1 + \pi$, where π is an arbitrarily specified infinitesimal scalar field. These are respectively identified with the four current density j^μ and the electromagnetic field density $h^{\mu\nu}$, through the relation

$$\frac{\partial h^{\mu\nu}}{\partial x^\nu} = j^\mu,$$ (11.13)

i. e. the second (inhomogeneous) system of Maxwell equations. Thus Weyl claimed that, without having to specify a particular action function, "the entire structure of the Maxwell theory could be read off of gauge invariance". (Weyl 1919b, p. 251) Again, using only the general form of such a function, he demonstrated that conservation of energy-momentum and of charge follow from the field laws in two *distinct* ways.[32] Accordingly he asserted that, just as the Einstein theory had shown that the agreement of inertial and gravitational mass was "essentially necessary" [*wesensnotwendig*], his theory did so in regard to the facts finding expression in the structure of the Maxwell equations, and in the conservation laws. This appeared to him to be "an extraordinarily strong support" for the "hypothesis of the essence of electricity" [*Wesen der Elektrizität*]. (Weyl 1919b, p. 253) Further support for the "world-metrical nature of electricity" (*weltmetrische Natur der Elektrizität*) was seen in the fact that the simplest invariant integral that exists in a four-dimensional metrical manifold

$$\frac{1}{4}\int F_{\mu\nu}F^{\mu\nu}d\omega, \quad (\omega = x^0, x^1, x^2, x^3), \tag{11.14}$$

is precisely the *action integral* of the electromagnetic field (in the absence of charges). (Weyl 1919b, p. 245) The domain of validity of Einstein's theory of gravitation, with its assumption of a global unit of scale, was originally held to correspond to $F_{\mu\nu} = 0$, the vanishing of the electromagnetic field tensor. By 1919, Weyl substituted his own "dynamical" account of the origin of "the natural gauge of the world" noted above. These details of Weyl's theory will suffice for present purposes; further discussion is available elsewhere.[33] We will only reiterate that Weyl gave a sustained and detailed response to Einstein's well-known criticism, that the theory is incompatible with the observed sharp spectral lines of the chemical elements, replying that his theory, but not Einstein's, provided the possibility of accounting for this "natural gauge of the world". In the fifth (1923) edition of *Raum-Zeit-Materie*, the last to appear in his lifetime, the theory was defended not so much as a physical hypothesis, but as "a theoretically very satisfying amalgamation and interpretation of our whole knowledge of field physics."(Weyl 1923a, p. 308) Weyl abandoned his theory of local scale invariance only in 1928, when he convinced himself that the "epistemological principle of relativity of magnitude" was not valid in the realm of atomic physics.[34] Even so, he revived his principle of "gauge invariance" by reinterpreting "gauge" as a factor of the complex phase of the electron's wave function. Carried over into Dirac's relativistic theory of the electron in 1929, Weyl derived the Maxwell equations from the requirement of local phase invariance, originating the modern principle of local gauge invariance that lies at the basis of contemporary geometrical unification programs.[35]

11.2.2 "Das Residuum der Ich-Vernichtung": *index of Weyl's transcendental idealism*

In his first letter (see 11.5), Becker writes of his conviction that Weyl's theory of the continuum and his "pure infinitesimal geometry" exhibit "the same basic idealistic conception" as Becker holds; as noted in 10.4 above, this is the conception of transcendental-phenomenological constitution of objects in agreement with the fundamentals of part IV of Husserl (1913). As evidence of this agreement, Becker points to Weyl's enigmatic designation of coordinate systems as "the residue of ego annihilation". What is the meaning of this phrase and how might Becker have taken it as comprising evidence of Weyl's concordance with his own standpoint of transcendental idealism? For an answer, we must turn to the context of the cited remark, which occurs twice in texts of Weyl referenced in Becker's monograph and so appearing by 1923.

The first occurrence is in section 6 of Weyl (1918a), entitled "Intuitive and Mathematical Continuum".[36] Let us recall that in *Das Kontinuum* the following position is adopted regarding the mathematical continuum. The set theoretic continuum, and so its epistemology of Platonism (or, in Weyl's later terms, "naïve realism") is rejected. In its place, a purely predicative theory of analysis is proposed, founded upon the natural numbers, the basic category of object. Underlying this category is a single primitive relation [*Urbeziehung*] $S(x, y)$ whose meaning, *that y is the immediate successor of x*, is directly exhibited in "pure intuition" (Weyl 1918a, p. 17). In rare agreement with Poincaré ("whose philosophical position I share in so few other respects"), Weyl maintained that *the idea [Vorstellung] of iteration, i.e., of the sequence of the natural numbers, is an ultimate foundation of mathematical thinking*"(Weyl 1918a, p. 39). Then, from the natural numbers, and from certain immediately experienced individual properties and relations of them, predicative principles (prohibiting a *circulus vitiosus*) are laid down for the construction of the rationals, and subsequently for "real numbers", as arithmetically definable lower Dedekind sections in the set of rational numbers. Moreover, as admissible "functions" of these are considered only such as can be arithmetically defined according to the construction principles specified at the outset.[37] Thus a weaker surrogate for the set theoretic real number continuum is constructed on the basis of the intuitively evidenced succession of natural numbers. The resulting mathematical theory of the continuum admittedly involves surrender of treasured parts of classical analysis, such as the principle that every bounded set of reals has a least upper bound. Yet such a drastic step is, according to Weyl, "in very essential part rationally justified" in that it provides mathematics enough for physics, i.e., sufficient for giving an "exact account of what 'motion' means in the world of physical objectivity".[38]

Section 6 of *Das Kontinuum* is a densely philosophical discussion of the problems encountered in attempting such a justification. In a preliminary consideration, Weyl finds the "deep chasm" between the intuitive and the mathematical continuum most obviously exposed in the attempt to link immediately given phenomenal time, the direct experience of temporal continuity, with objectively measurable quantitative time represented as a one-dimensional continuum of real numbers. The fundamental incongruity between them arises in that the former possesses an "essential and undeniable inexactness"; for example, a point of time in it can only be approximately, never exactly, determined. Indeed, Weyl holds that the direct experience of phenomenal time, due to its genuine primitiveness, cannot even be described (Weyl 1918a, p. 70). Despite this, there are "rational motives which impel us across from (the intuitive) to the (mathematical continuum) in our effort to comprehend the world" in natural science, even if "we are not satisfied with replacing the (intuitive) continuum by the exact concept of a real number".

If Becker's surmise is correct, (as we think it is,) the index of Weyl's attachment to the standpoint of transcendental idealism emerges from within the confines of how such a rational justification of his theory of the continuum must proceed. The purported justification enjoins the application of the theorems of analysis (in particular, as may be purely arithmetically developed in Weyl's predicative continuum) to the space and time of physics. For this, a "transfer principle" is required, according to which real numbers and points may be put into one-one correspondence. Points are the objects *sine qua non* of the analytic development of geometrical concepts, including those employed in physical theory ("curve", "surface", "boundary" etc.). But as Weyl noted, "as basic a notion as that of a point in the continuum lacks the required support in intuition" whereas "(the intuitive continuum) of time- or space-points is homogeneous". So it is that the extensionless mathematical point of space or time is itself a conceptual entity that is only "crystallized into full definiteness in the arithmetico-analytic concept of the real number". Hence the application of analysis to the physical world mandates that the space and time points of physics are themselves to be constructed through mandatory reference to a coordinate system. Earlier in this section, Weyl had given a sketch (essentially repeated in *RZM*—see the discussion below) of the construction of "a mathematical theory of time" from the experienced relations of "earlier than" and "equality" of temporal intervals. He now briefly declares that the totality of points in a portion of space is to be arithmetically constructed as three-dimensional sets of real numbers (in his sense). Accordingly, implementation of the "transfer principle" from analysis to geometry and physics requires the introduction of a coordinate system: the correspondence of points and real numbers can be effected only by an arithmetical construction of points upon the skeleton of three-dimensional directed axes. The latter bear the unavoidable impress of subjectivity, for the positing of a coordinate system is the willful act of "a pure, sense-giving ego":

The exhibition of a single point is impossible. Further, points are not individuals and, hence, cannot be characterized by their properties. (Whereas the "continuum" of the real numbers consists of genuine individuals, that of the time- or space-points is homogeneous.) Therefore, points and sets of points can only be defined relative to (i. e., as functions of) a coordinate system, never absolutely. (The coordinate system is the unavoidable residue of the ego's annihilation [*das unvermeidliche Residuum der Ich-Vernichtung*] in that geometrico-physical world which reason sifts from the given under the norm of 'objectivity'—a final scanty token in this objective sphere that existence [*Dasein*] is only given and *can* only be given as the intentional content of the conscious experience of a pure, sense-giving ego.) (Weyl 1918a, p. 72; Engl. trans., p. 94)

The language is telling. Recall that in (Husserl 1913, § 50)), "pure consciousness" is the "phenomenological residuum" ['*phänomenologische Residuum*'] alone surviving the phenomenological reduction that "brackets" or "puts out of action" all posits of "transcendent" objects, i. e., not immanent to consciousness. Just as "pure consciousness" remains in the suspension of the "natural attitude's" posit of the reality of a Nature transcendent to consciousness, so also *within* that posited world (in particular, the "geometrico-physical world" of physical science), necessary reference to a coordinate system (or a designated reference body, the point remains) is a reminder of this "worldly transcendency's" constitution "as the intentional content of the conscious experience of a pure, sense-giving ego". Lest it be thought that the measurable quantities of physics, mathematically represented as field functions of the coordinates of space and time (or, spacetime) are without any taint of subjectivity, Weyl reminds that these concepts invoke necessary reference to a coordinate system. This, indeed, appears an affirmation of transcendental idealism in something like Becker's sense of the "transcendental constitution of all ontological essences in pure consciousness" (see 10.4). The epistemological issue looming here, to which Weyl will presently give the name of "the problem of relativity" (see immediately below) concerns precisely the relation between the subjective and the objective, between intuitive exhibition and precise, discursive mathematical concepts. The general significance of "the problem of relativity" is that objectivity in physics, that is, the purely symbolic world of the tensor fields of relativistic physics, is constituted or constructed *via* subjectivity, neither postulated nor inferred as mind-independent or transcendent to consciousness. This message is elaborated within the second *Ich Vernichtung* passage with which Becker was undoubtedly familiar.

It occurs in a parenthetical sentence amidst remarks appended, in the fourth edition of *RZM* (Weyl 1921b, p. 8), to the discussion of "the essence of measurement" *[das Wesen des Messens]* in the introductory chapter. Almost all citations of *RZM* in Becker's monograph are to this edition.[39] It must be noted that this discussion follows immediately upon a very broad characterization, without thus being designated, of the problem of transcendental-phenomenological constitution of objects. The guiding principle of such constitution (as seen in 11.1 above) is that "the given to consciousness [*Bewußtseins-Gegebene*] is the starting point [*Ausgangspunkt*] in which we must place ourselves in order to comprehend the sense and the justification of the postulate of reality [*der Wirklichkeitsetzung*],"[40] a principle presumably close to Becker's own "philosophical starting point [*philosophischer Ausgangspunkt*]" identified as "the principle of transcendental idealism". In any event, Weyl's scrutiny of measurement is not merely a nicety; rather it is undertaken in order to make understandable "how mathematics comes to play its role in exact natural science". Crucially, this role is only enabled through the introduction of a coordinate system (or equivalently, a designated reference body).

According to the above-mentioned starting point, time is "the primitive form of the stream of consciousness [*die Urform des Bewußtseinstromes*]"; space is "only a form of our intuition". It is the theory of relativity, Weyl claims, that has first made it apparent within physics, that space and time are *only* forms of phenomena, since it demonstrates that nothing of the essence, given in intuition, of space and time enters into the mathematically construed physical world. (Weyl 1921b, pp. 3 and 5). However, as forms of phenomena, both time and space are completely homogeneous. Only through the introduction of real numbers (here taken as unproblematic), can time points and space points be discriminated to the arbitrary precision required by measurable physical magnitudes.[41] Weyl's illustration shows how this is done for the one-dimensional continuum of time, based upon the primitively experienced relations of "earlier than" and "equality" of time intervals. For measurement, a unit for time must be chosen, a choice that presupposes, on conceptual grounds, three distinct time points standing in the phenomenal relation of "earlier than" (or, "later than"). Taking as unit the "distance" between two of them, the third is then obtained as some real number multiple of this unit. This assignment of numbers enables the individuation of points of time generally; it is possible to conceptually fix any point as a time coordinate along an axis based upon this unit. In this way, time is made amenable, as a magnitude, for the construction of the objective world of physics. On the other hand, this arithmetization of time is achieved only upon a directed axis being "exhibited only through an individual act" of the cognizing, experiencing ego.[42] This is taken as indicative: the exact determination or symbolization of the concepts of physics cannot be accomplished without the introduction of a coordinate system. For this reason, it must not be thought that such a objectification, relative to a coordinate system, is absolute:

> But this objectification [*Objektivierung*] through exclusion of the ego and its immediate life of intuition, is not attained without remainder; the coordinate system, exhibited only through an individual act (and only approximately) remains as the necessary residue of this annihilation of the ego [*das notwendige Residuum dieser Ich Vernichtung*]. (Weyl 1921b, p. 8; Engl. trans., p. 8)

Continuing in a more general vein, Weyl makes it clear that it is measurement that mandates this necessary reference to the ego, and that consequently measurement always gives rise to a "theory of relativity":

> *For measurement the distinction is essential between the 'givenness' [dem 'Geben'] of an object through individual exhibition [individuelle Aufweisung] on the one side, in conceptual ways on the other. The latter is only possible relatively to objects which must be immediately exhibited [unmittelbar aufgewiesen]. That is why a theory of relativity is perforce always involved in measurement.*

What must be exhibited is, of course, a coordinate system, with respect to which an individual object O can be singled out and referred to, with arbitrary exactness, from a continuously extended object domain. A "theory of relativity" then establishes what "lawful connection" exists between the coordinates of one and the same arbitrary object O in two different coordinate systems. For Weyl, these presuppositions of measurement pinpoint the subjective ground of physical objectivity according to which, as we have seen, "existence [*Dasein*] is only given and *can* only be given as the intentional content of the conscious experience of a pure, sense-giving ego." Space and time magnitudes (and so all physical magnitudes that are functions of space and time) can not, in their very definition, entirely exclude the cognizing ego. This means that the introduction of a coordinate system, the willful act of a cognizing consciousness, delimits the attempt to exclude all subjectivity in the symbolic construction of the objective world of relativity theory, a world in which "the physically real is conceptually fixed, in all its determinations, through numbers." (Weyl 1921b, p. 9; Engl. trans., p. 10)

We might note that subsequently, in Weyl (1927), Weyl returns several times to "the problem of relativity", in one case repeating the *Ich Vernichtung* passage of *RZM* almost *verbatim*.[43] In yet another, Weyl makes especially vivid his principal claim that the "objective" world of theoretical physics always carries with it a necessary "relative" amendment, whereas the "absolute" is the sole possession of "subjective" immediate experience.

> Immediate experience is *subjective* and *absolute*; even as hazy as it may be, it is given in its very haziness as it is and not otherwise. On the other hand, the objective world [...] which natural science attempts to crystallize out of our practical lives—through methods that are the consistent development of those criteria according to which we construe reality in the natural attitude of daily life—this objective world is necessarily *relative*; it is only representable in a determinate manner (through numbers or other symbols) after a coordinate system is arbitrarily introduced into the world. This oppositional pair: *subjective-absolute* and *objective-relative* seems to me to contain one of the most fundamental epistemological insights that can be extracted from natural science. (Weyl 1927, pp. 82–83; Engl. trans., p. 116)

For Weyl, epistemology must be ever mindful that knowledge in the exact natural sciences is the joint product of the objective, of description in precise mathematical concepts, *and* the subjective, the "immediate life of intuition". In its broadest terms, his epistemology remains centered upon "the problem of relativity", of individual exhibition in intuition vs. conceptual determination, with the result that epistemology must take into account how the objective is dependent upon the subjective, ultimately upon acts of intuitive 'seeing' or of willing. Thus Weyl, as Husserl, is critical of any epistemological realism that lays claim to knowledge of an objective world completely transcendent to consciousness.

11.2.3 Weyl's solution to the "new problem" of space

The wide-reaching world-geometric "physical application" of his pure infinitesimal geometry, attained in 1918, did not at all exhaust its philosophical significance, although showing this took several more years. For Weyl, the Einstein theory of gravitation necessitated a new solution to the "problem of space", previously thought resolved by the treatment of Helmholtz and S. Lie before the turn of the century. This problem concerns, in contemporary language, the justification for the assumption that the metric of a Riemannian manifold is given by a quadratic differential form, as opposed to the other possibilities afforded by the wider class of Finsler metrics.[44] Riemann's own hypothesis, that the distance element can be expressed as the square root of a homogeneous quadratic differential form ("infinitesimal Pythagorean metric") was not fully demonstrated; as a result, it had come to be seen as "an article of faith, essentially". (Coolidge 1940, p. 410). As is well known, Helmholtz and then Lie, who made Helmholtz's results rigorous through the use of continuous groups, showed that the Riemannian assumption was justifiable in virtue of the "fact" of the existence of freely moveable rigid bodies. Thus the Helmholtz-Lie solution is valid in the restricted domain of spaces of constant (including zero) curvature. But this "fact" is called into question with recognition of variably curved spacetimes in which there is no "free mobility" of rigid bodies and thus the "problem of space" is reopened in the new context of Einstein's theory. In the Helmholtz-Lie tradition, Weyl sought to account for the "uniqueness of the Pythagorean (i. e., quadratic) metric determination" in an n-dimensional differentiable manifold M in terms of a group-theoretic treatment of the congruence properties of bodies. The problem occupied him for several years in the early 1920s. His result was proven for the case of 2 and 3 dimensions in (Weyl 1921b, § 18) and Weyl there speculated that the proof probably could be extended to n dimensions. This was carried out in detail in Weyl (1922); the final presentation was made in Spanish lectures given in Barcelona and Madrid early in 1923, appearing in German later that year (Weyl 1924a). In Weyl (1921b), Weyl rather proudly describes his development of infinitesimal geometry, of which this result is the capstone, to be

a good example of essential analysis [*Wesensanalyse*] striven for by phenomenological philosophy (Husserl); an example that is typical for such cases concerning non-immanent essences.[45]

The final lengthy case-by-case proof, which Weyl himself likened to "mathematical tightrope dancing"[46] is cast in very general terms: it must hold for manifolds of arbitrary dimension n and for every value of metric signature (or "inertial index"; relativistic spacetimes have signature $+1$ or -1 depending on sign conventions) and not only for positive-definite metrics. (Weyl 1922)

Weyl's guiding thought, inspired by the aforementioned "fundamental fact" of infinitesimal geometry, was that the class of Riemannian metrics, as well as the class of Weyl metrics, could be distinguished from the larger Finsler class by seeking those metrics which admit a unique affine connection. Under these conditions, he sought to distinguish the "nature of space", located in its "metrical essence" [das metrische Wesen des Raumes], from the mutual "orientation" that the metrics in the different points adopt with respect to one another. The former, he claimed, is a priori and is the same at each point of the manifold, since at every point P the metric is represented by a group of linear transformations of the same kind. The latter "orientation" of the metrics is a posteriori and depends, as required by the field equations of Einstein's theory, on the fortuitous distribution of matter and energy. Hence he would show how it is that space has what appears to be conflicting properties but in fact are different and complementary aspects. As the form of appearances, space is necessarily homogeneous and fully rationally comprehensible; this is the a priori "nature of space". On the other hand, space is also variable and, as requiring reference to empirical data, only approximately knowable; this is the *a posteriori* "orientation" of the metrics at various points. Kant, therefore, was not altogether wrong in claiming an exact a priori character for space; only with infinitesimal geometry the boundary between a priori and a posteriori has shifted. (Weyl 1924c, pp. 44–45) These ideas are implemented in three steps. (Weyl 1924c, pp. 47 ff.)

Weyl begins with the notion of a "metric at a point P", defining the concept of a "congruent mapping" at a single point P via rotations of a "vector body", a parallelpiped spanned by n basis vectors of the vector space centered at P. These motions form a continuous group (infinitesimal orthogonal group) of linear transformations. As the "volume" of the "vector body" is assumed to be preserved under rotation, Weyl showed that this infinitesimal rotation group is Euclidean, a distinguished subgroup of the general linear group comprising those linear transformations that leave invariant a non-degenerate quadratic form. That is to say, it is distinguished from among all linear groups in that it affords a vector body "free mobility" around a fixed point. Secondly, to show that the metric "Nature" or essence [Wesen] of space is everywhere the same, thus that space, as a "form of appearances", is homogeneous, Weyl demonstrated that the infinitesimal rotation group at P is related to the one existing at some *definite* neighboring P' by single linear congruence transformation between P and P'. This is a similarity transformation in Lie's sense, so the groups share the same abstract Lie algebra, differing only as regards their "orientation". Third and finally, Weyl returns to the concept of a "congruent displacement", the "congruent mapping" of a vector from P to P', in order to extend the established group property to all points. This will establish a *metrical connection* between the metric at P and the metric at *any* neighboring point. Here he makes use of the fact that to every coordinate system at P there corresponds a possible system of infinitesimal parallel transport of vectors expressed in terms of a symmetric affine connection (see 11.2.1 above). Now the variable alteration of the metric field required by the spacetimes of general relativity mandates what Weyl calls the "Postulate of Freedom": that the nature of space imposes no restriction on the metrical relationship. So, just as there are many possible concepts of parallel transport of a vector between P and a neighboring P', so the metrical connection between them must allow for a wide system of infinitesimal congruent displacements of a vector. Then, for a given rotation group at P, the metrical connection to P' is expressed via a linear relation tracking the displacement along each coordinate axis (all directions are equivalent). Nonetheless, according to the fundamental postulate of infinitesimal geometry, that the metric uniquely determines the (symmetrical, i. e., torsion-free) affine connection, there is always one and only one parallel displacement from P to *any* neighboring P' that is at the same time a system of infinitesimal congruent displacement. Assuming the existence of such a unique affine connection,[47] Weyl then has the result that the "congruent mapping" or "length connection" is uniquely defined. Using Scheibe's term,[48] Weyl has here introduced the notion of a *group field*, a field of groups of linear transformations in every tangent vector space by which the congruence of tangent vectors at each point may be defined. Thus Weyl has given a "purely infinitesimal" solution to "the problem of space". Unlike in the Helmholz-Lie approach to the homogeneity of space where the rotation groups at each point are necessarily identical, in Weyl's treatment the rotation groups have differing orientations.[49] In particular, "the Helmholtz requirement of homogeneity, with which the old conception of the essence (*Wesen*) of the metric of space stands and falls" (Weyl 1923a, p. 103), is replaced by two completely different postulates of purely infinitesimal geometry: the above-mentioned "Postulate of Freedom" and the requirement that the metric univocally determine the affine connection. This completes our informal sketch of Weyl's group-theoretic solution to the new "problem of space" raised by the general theory of relativity.

As we shall see (11.3.2 below), in his 1923 monograph, Becker's transcendental-phenomenological grounding of geometry ("the laws of space") distinguishes the a priori Euclidean structure of intuitive space through a condition equivalent to the Helmholtz-Lie free mobility characterization of spatial homogeneity. His letters document various responses to Weyl's objections; indeed, Becker subsequently suggests that Weyl's "Pythagorean measure determination" also lends itself to a similar phenomenological grounding. It is apparent that Weyl's objections to Becker's "transcendental-phenomenological grounding" of intuitive space, as they may be inferred from the letter of 27 vi 1923, principally stem from Becker's inconsistent blend of finite ("free mobility") and "purely infinitesimal" notions. Of course, in the latter, the characterization of homogeneity based on the concept of a rigid body and rigid motions is not legitimate.[50]

11.2.4 The limits of field theory and causality: "free decisions"

Many further details of interest, historical, philosophical, and technical, must here be omitted.[51] For our purposes we need mention only one additional part of the Weylean context relevant to Becker's monograph. Already by the autumn of 1919 (and so within a few weeks of his meetings with Brouwer in the Engadin in the summer of 1919[52]), Weyl surrendered his belief that classical field theory, even with a purely infinitesimal geometric basis, was a suitable vehicle for treating quantum phenomena. In the end of a lecture given at Lugano in September, 1919 entitled "The Relation of the Causal to the Statistical Approach in Physics", Weyl announced that, contrary to most physical opinion, he believed that validity of "pure law physics" did not extend to the elementary "quanta of matter". (Weyl 1920a, p. 741) This lecture was not widely known, not appearing in print until August, 1920, and then in an obscure Swiss medical weekly. But in any case, Weyl's reasoning to this conclusion reflected his recent conversion to Brouwer's intuitionism in the previous summer, and is heavily colored by the enthusiasm of the new convert. In rather flamboyant manner, Weyl extended the freedom or lack of constraint implied by Brouwer's conception of the continuum as an uncompleted and uncompletable "medium of free becoming" to limit the realm of validity of the rigorous causal laws of field physics. Under Brouwer's influence, Weyl now holds that is the essence [*Wesen*] of the continuum that it is only conceivable as an infinite "process of becoming" and not as a "rigid being" [*starres Sein*]. The latter, mathematically considered, resolves the continuum into a set of discrete points, and hence is only a conceptual surrogate. But, Weyl maintains, by the same token, there is no adequate conceptual understanding of the continuous field functions of physics that describe the possible course of values of field magnitudes in the continuum that is a physical field. What is given as completed or finished is only a continuum in which this process of becoming has reached a certain point, say the quantitative relations among the field magnitudes in a four-dimensional region of the world S which is intuitively present to me. But these relations are only approximately ascertainable, "not only because of the limited accuracy of my sense organs but also because *they are in themselves afflicted with a sort of vagueness*." For this reason, even under the postulate of the rigorous causal determinism of the field laws of nature, it is not possible to compute values of these field magnitudes in a region of spacetime S' containing S. Instead,

> Only so to speak "at the end of all time" (which is but a limit idea) would the unending process of becoming that is S be completed, and S would bear that degree of definiteness that mathematical physics postulates as its ideal [...]

The result is that,

> without prejudice to the validity of the laws of nature, there is *room for autonomous decisions, causally independent of one another*, whose locus I consider to be the elementary quanta of matter. These 'decisions' are what is *properly real* [*Reale*] in the world.

According to the new view, physics, like the continuum, is divided into the disjoint realms of law and exact determination on the one hand, and of "chance" and "becoming" on the other. But now the status of pure field theory, already seen as a kind of formal scaffolding that omitted the content of reality,[53] has been further devalued to that of the background forceless transmitter of effects, much as the status of background geometry in pre-general relativistic physics. As Weyl stated in a letter to Felix Klein, at the end of December 1920:

> Field physics no longer appears to me as the key to reality[;] just to the contrary the field, the ether, now appears to me only as the medium of transfer of effects, that in itself is only a completely forceless transmitter; matter, however, is the reality beyond the field and *determines its states*.[54]

Weyl's not-inconsiderable change of attitude regarding the fortunes of causality and field physics was incorporated, somewhat incongruously in view of other passages endorsing the world view of field physics, in similar remarks appended at the end of the fourth edition of *RZM* (1921; preface dated November 1920). There, however, Weyl has strategically concealed his intuitionist views regarding the nature of the continuum, addressing himself directly to physicists. After noting that the facts of the quantum theory appears to make the role of statistics in physics unavoidable, Weyl goes on to declare:

> It must again be affirmed quite clearly that physics, at its present stage, no longer lends any support to the belief in a closed causality of material nature resting upon rigorously exact laws. [...] Freedom of action in the world is [...] no more constrained through the rigorous laws of field physics than it is, according to the customary view, through the validity of the laws of Euclidean geometry. [...] Already *statistical physics,* in the form of the quantum theory, has reached a deeper stratum of reality than was accessible to field physics; however the problem of matter still lies entirely in the dark. (Weyl 1921*b*, pp. 283–84; cf. Engl. trans., p. 311)

Becker will make a great deal (as does Husserl in his letter to Weyl of 9 April 1922, see 10.3) of the transcendental-constitutive significance of this distinction between the "structural lawfulness" of the field laws and the specific "causal" laws of nature for whose discovery physics alone is responsible (see 11.3.3). However, in *RZM*[5] (Weyl 1923*a*), the last to appear in Weyl's lifetime, these programmatic metaphysical declarations were excised. Instead, the book concludes with the sober declaration that a solution to the problem of matter, to which perhaps the quantum theory affords the first illumination, is not to be expected from field theory. (Weyl 1923*a*, p. 317) Yet Weyl neither then, nor subsequently, entirely renounced his "agent-theory of matter" in which metaphysical reconciliation of causality and freedom of action was attempted in the characteristic mode of a physical/mathematical speculation. Here, still some years prior to quantum mechanics, Weyl considers charged quantum particles (the electron) as literally lying "outside of space", and so "beyond" the rigorous medium of causality that are the field laws, in "tubes" or "canals" of a topological manifold that is multiply connected in the small.[55] As late as the 1949, Weyl continued to suggest the possibility that

> a more detailed scrutiny of a surface might disclose that, what we had considered an elementary piece, in reality has tiny handles attached to it which change the connectivity of the piece, and that a microscope of ever greater magnification would reveal ever new topological complications of this type *ad infinitum*.[56]

Such speculations, going back to W. K. Clifford, of treating matter in terms of a multiply connected topology, were not original with Weyl, even in the context of general relativity.⁵⁷ In 1935, Einstein and Nathan Rosen unsuccessfully attempted to treat the "particle problem" in general relativity (elementary particles that are everywhere finite and singularity-free) through a quantitative model in which particles are "bridges" across a double-sheeted physical surface. The idea was revived in the 1950s by the noted relativity physicist John A. Wheeler who claimed that such topological complications, baptized by the extravagant name of "wormholes", are a necessary concomitant of the quantization of general relativity.⁵⁸ The idea was revisited yet again in the 1980s in arguing for the possibility of "time travel".⁵⁹ Weyl's own inclination to the idea, however, appears to inextricably tie together both physical motivations (to avoid the unphysicality of the field theoretic conception of the center of a particle as a singularity) and philosophical meditations on the reconciliation of freedom of the will with the causal laws of classical field physics. (Weyl 1924c, pp. 57–59)

Weyl's analogy between the freedom of action of the electron and the individual ego attracted the philosopher Fritz Medicus in his monograph on *The Freedom of Will and its Limits* (1926). Medicus (1876–1956), a colleague of Weyl's at the ETH in Zurich, was the author of works on Fichte and the editor of an edition of Fichte's selected works published just before WW I. According to a later reminiscence, it was in Medicus's seminar at the ETH in Zürich that Weyl was introduced to the ego-centered philosophy of Fichte, whom Weyl would come to regard as "a constructivist of the purest water" (Weyl 1955, p. 163). In the *Vorwort* to his book (dated November 1925; most of the text, Medicus reports, was written in 1923), Weyl is accorded what can only be described as a lavish appreciation. "Without", Medicus wrote, "Hermann Weyl's magnificent interventions for the fruitful constant shaping of relations between physics and philosophy the present work could not have arisen; its content makes this evident." The essay is nothing less than an attempt to resolve the "eternal philosophical problem" of freedom of the will. Its central argument proceeds by showing that the physics of the present day does not support the twin dogmas of fatalism and determinism. Medicus argued, drawing in the main upon Weyl (1921b) and Weyl (1921c), that Weyl's "dynamical agent theory of matter" carried the possibility of transforming Leibniz's monadic metaphysics into physics. (Medicus 1926, pp. 87–88) The activity of atoms, or of individuals, is no longer expressible in a formula holding with mathematical necessity; rather these are only statistical regularities, which, as in the laws of social science, do not speak of individuals or individual events. But there is no conflict with the Kantian postulate that objectivity itself rests upon the exception-less validity of the causal order, for this order pertains to the world in space and time, whereas, as Weyl has speculated, the quantum, like the individual ego, lies 'beyond' physical spacetime. The world in space and time is identical with the world of objects whereas "the new atomism sets matter above spatio-temporally extended being" [*der neue Atomismus die Materie dem räumlich-zeitlich ausgedehnten Dasein überordnet*]. Matter is as little an object as we, who, as personalities, demand freedom for ourselves. This is, of course, in contrast to Kant, whose insights into the non-objectivity of freedom of will rest on questionable speculations regarding a realm of things-in-themselves, and so the question of freedom became an unsolvable riddle. On the contrary, modern physics has given a new meaning to conditions of a reality outside of space and time which is both clear and accessible to physical experiment. (Medicus 1926, p. 93) While attempts to reconcile freedom of the will and causality in the aftermath of the new quantum theory are legion, Medicus's essay is unusual both in that it predates the actual arrival of quantum mechanics, and also in that its "reconciliation" relies but indirectly on the statistical character of quantum laws. Instead it draws directly from Weyl's purely speculative ideas about matter and is in fact cited by Weyl in the list of references to the section on "causality" in his philosophical monograph (Weyl 1927, p. 162). Medicus's little-known appreciation of Weyl's speculative views on matter and causality bears a close resemblance to Becker's own appropriation of these ideas. Almost fifty years later, without the context of the "agent theory of matter", Weyl's views on the limits of causality are revisited in the widely read, and highly controversial, essay of Paul Forman on Weimar culture and quantum physics (Forman 1971).

Forman's essay is well known and accordingly requires little introduction. Its controversial thesis, that the willingness of German scientists in the early Weimar period to "dispense with causality in physics" was "primarily an effort to adapt the content of their science to the values of their intellectual environment" was warmly debated for a number of years.[60] In the fullness of time, we think it fair to say that a consensus has emerged that Forman's thesis is not sustainable. Nonetheless, Forman's essay is relevant here because it treats Hermann Weyl as a central and even representative figure in the adaptation of physics and mathematics to what is described as the "hostile intellectual environment" created by the collapse of the German Empire in 1918 and the subsequent social, economic and political instability. This hostile environment is characterized by Forman as being dominated by a philosophical tendency,

a neo-romantic existentialist "philosophy of life", reveling in crises and characterized by antagonism toward analytical rationality generally and toward the exact sciences and their technical applications particularly. (Forman 1971, p. 4)

It is in this light that Forman reads such passages of Weyl as we have cited above, portraying them as evidence of a "quasi-religious conversion to acausality". (Forman 1971, p. 80) While Forman is well aware of Weyl's phenomenological leanings, he understands phenomenology at this time to be a philosophy based on "intense introspection" that is "degenerating into existentialism". This characterization does not apply at all, we believe, to the philosophical position of Weyl, though it may, crudely, describe latent tendencies in Becker (see especially the letter of 10 x 1924 and the discussion in 11.3). Becker indeed, as does Medicus (whom Forman does not cite), finds his own sympathetic interpretation for Weyl's distinction between the "realm of law" and that of "free choice"(see 11.3.3 below). We believe, however, that there are far more plausible reasons for Weyl's recognition of the limits of classical field theory than "adaptation to a hostile environment". These have to do both with the intractability of quantum phenomena to classical treatment and with the epistemological motivations surveyed above that have led him to challenge both the Riemannian foundations of general relativity and the classical theory of the continuum.

11.3 Becker's 1923 *Habilitationsschrift*: A Transcendental-Phenomenological Grounding of the Geometry of Space

The declared task of Becker's monograph is that of providing a necessary transcendental phenomenological justification of Euclidean geometry that rests upon the distinctive structure of intuitive space. His target is geometric conventionalism, *à la* Poincaré (1902), Kretschmann (1915), and Dingler (1919). In contrast, Becker argues that the Euclidean axioms of the space of intuition are based not on contingently a priori or conventional suppositions, but upon *a rigorously non-contingent and necessary* a priori phenomenological foundation. However, in the context of Einstein's general relativity, a grounding of this kind appeared impossible for two reasons. On the one hand, Einstein's gravitational theory asserted that large, finite regions of spacetime have a non-Euclidean geometry. Moreover, the dynamical dependence and variability of this metrical structure on mass and energy appears to lend an unavoidable contingency to the geometry of space. On the other, already in the special theory of relativity, conventionalism finds support in Einstein's treatment of the relative simultaneity of distant events. On both counts, the result is that philosophical discussions of relativity physics have acquired an empiricist and conventionalist tone, especially skeptical of "pure intuition" and of attributions of any non-analytic a priori character to the axioms of geometry. Of course, Becker recognized that a phenomenological grounding of geometry must confront and surmount these objections. First, it must furnish the necessary phenomenological foundation for the distinguished role of Euclidean geometry in the "lawfulness" of "actual" space (the space of the world given in the pre-scientific or "natural" attitude). Secondly, there must be a phenomenological account of the meaning and the justification of the use of non-Euclidean metric determinations in relativity physics. Here, Becker will make much of the requirement that a (Riemannian) manifold possesses "planeness in the smallest parts".

According to (Becker 1923, pp. 388–91), geometry has a "double character", grounded on the one hand in formal logic, the domain of pure understanding, and in sensuous intuition, on the other. Among the eidetic or pure a priori disciplines, it straddles the Husserlian distinction between a "material-eidetic ontology" and "formal ontology" (inspired by the Leibnizian idea of a *mathesis universalis*) that includes mathematical logic, set theory, and the theory of parts and wholes. In each case, "ontology" signifies, in a sense distinct from traditional metaphysics, the eidetic field of pure a priori possibility and necessity within which all objects are constituted in the "transcendental subjectivity" of "pure consciousness". Whereas formal ontology delineates the general necessary conditions of "objectivity as such", a material-eidetic ontology exhibits the "necessary material form" of all the objects of a specific "region", i.e., domain of a science. In this regard, it bounds the conception of a "possible object" of such a region. Geometry is distinguished from the pure eidetic mathematical sciences (arithmetic, analysis) founded only on formal ontology in that its objects are morphological essences arising from spatial figures (morphology) in acts of idealizing abstraction (categorial intuition). In this sense, geometrical objects are ideal essences in the Kantian sense of an ideal limit. By the same token, geometrical concepts are "ideal" concepts whose contents are essences drawn not from sensory intuition but from intuition *simpliciter*, and so are essentially distinct from the inexact descriptive concepts of the morphological sciences. Geometry is then the eidetic science dealing with a specific "regional ontology", that of "spatial form" and so in its foundations, geometry is material-eidetic. Yet like arithmetic and analysis, it proceeds in the deductive, formal-logical manner characteristic of formal ontology where cognitions are purely analytic. The linkage between these two opposing faces of geometry is found in the Husserlian idea of a "definite manifold", a domain that, with formal-logical necessity, is exhaustively characterized by a finite number of concepts and propositions. (Husserl 1913, § 74) In agreement with the Husserlian requirement that it is possible to ground an inquiry as an exact science only when it comprises a definite manifold, Becker maintains the possibility of an exact geometry must be founded upon the "essential characteristic" (*Wesenseigentümlichkeit*) of space to be a definite manifold. (Becker 1923, p. 390) In this regard, space itself is held to be an ideal essence, in fact, an Idea in the Kantian sense.

Becker's project is thus an ambitious undertaking that joins together fundamentally distinct types of eidetic science. At the same time, Becker holds that the material eidetic laws of space (including the axioms of geometry) are not yet rational in a rigorous sense; they are but *contingently* a priori. For example, it is an "essential law" [*Wesensgesetz*] that space has three dimensions, yet there is apparently no rational ground for this law. His transcendental-phenomenological foundational project for geometry is accordingly charged with two tasks to which the two main parts of the monograph are respectively dedicated. The first (pp. 385–477) is concerned to give a rational basis of the mental process of geometrical idealization from the "simply visualizable" [*das Schlicht-Anschauliche*], invoking the claim that it is an "essential characteristic" of space to be a definite manifold. This argument rests, in large measure, upon Brouwer's theory of the continuum as a "medium of free becoming". Becker regards this to be of overriding significance for phenomenology because Brouwer's theory accords fundamental recognition to the duality between "reason" and "sensibility", where the intuitive (phenomenological) continuum is distinguished from the mathematical, the phenomenological limit from the mathematical, in the sense of Cauchy. (pp. 417 and 425 ff.) Thus Becker's method of the transcendental-phenomenological constitution of space (see below) will follow Brouwer's account of intuition in passing through successive phenomenological limits at lower constitutive levels of spatiality to ascend to higher levels, ultimately to the constitution of a definite manifold. The main objective of the first part is therefore to establish the possibility that the domain of the intuitive continuum can admit of exact laws, and thus be a definite manifold in Husserl's sense.

The second main part of Becker's monograph (pp. 477–560) takes up the subsequent problem of showing how the *contingency* of the a priori character of the laws of space may be overcome. Its objective is to show how systematic phenomenological constitution from the "origin" (i.e., "pure consciousness"), may provide a *necessary* ground for the material eidetic *contingent* a priori character of the laws of space. In particular, Becker argues that specifically Euclidean geometry—as a material eidetic science—is singled out from other possible geometries on this phenomenological basis. In so doing, Becker will claim that it is possible to broaden the idea of a *mathesis universalis* (formal exactness) into an all-encompassing *"mathesis universalissima"* in which also individual objects have their place. (p. 396). Such a discipline includes the ideal of a rigorously rational geometry and physics, freed from the contingencies that these sciences have acquired in the course of their development.

11.3.1 Constitutive method

Referring to Husserl (1913), part four, entitled "Reason and Actuality" [*Wirklichkeit*], the investigation is to be carried out in accordance with the "principle of transcendental idealism". The method of transcendental phenomenology is that of "transcendental constitution of all ontological essences in pure consciousness."[61] "Pure consciousness" is the vehicle for attaining "essential insight" [*Wesensschau*] into each ontological essence. Becker will apply to geometry the general transcendental constitutive method adumbrated by (Husserl 1913, § 149) as the "universal 'constitution' in transcendental consciousness of objectivities" pertaining to particular "regions". As Husserl had shown, the constitutive transition from level to level is possible because every object is phenomenologically presented (in intentional acts of perception, memory, reflection, imaginative fantasy, and so on), together with its "horizon", the manifold of its possible aspects in which are prescribed rules for going beyond it. These rules are "absolutely comprehensible ideal possibilities of 'limitlessness in the progression' of harmonious intuitions, and more particularly, according to typically determined pre-designated directions."[62] In Becker's terms such rules are the "transcendental guiding threads" [*transzendentale Leitfäden*] that are characteristic of the method of transcendental phenomenological constitution. Each ontological essence is only attained through "essential insight" [*Wesensschau*] into the corresponding individual object or system of objects. Hence there is still an initial dependence on the arbitrariness of any empirical presentation. But from any hyletic data, empirical givenness as such, it is possible to extract an ontological essence that is progressively refined through higher and higher levels. Eventually, in the greatest possible generalization of this process, the method leads to the "essence of nature" itself, to the formal idea of '*hyle*', of Nature in general.

This process is employed in Becker's account of how three-dimensional homogeneous space, the space of intersubjective objects, of which the body of the constituting subject is one object among others, constitutively arises from lower levels of spatiality and prespatiality. Primitive perceptual phenomena are "pre-spatial" or quasi-spatial, the phenomena of the various sensory fields, the most important being the immediate visual field of a fixed percipient, as well as kinaesthetic phenomena having a distinctive subjective coloration ("within me" as opposed to the sensation of "outside me"). Through the method of "passing through the limit" the visual field gives rise, in connection with certain kineaesthetic phenomena (eye and head movements), to the still pre-spatial objects of an "oculomotor field". The next higher stage, "oriented space", arises through consideration of additional tactile and kineasthetic phenomena. Oriented space defines an absolute "here" in relation to every "there"; all distance and all points are defined by a bundle of rays proceeding from an oriented "point eye" centered at O. "Visual space" further emerges with the interpretation of certain phenomenal qualities as pertaining to depth-sight from O into a third dimension. Finally, passing through the intermediate level of an oriented tactile space, full homogeneous three dimensional unbounded space, *spatiality* as such, arises with the kineaesthetically characterized motions of a subject "into the horizon" wherein it is possible to identify a thing as remaining the same despite its changing perspectival ("skiagraphic") aspects. Hence from the lowest levels of "pre-spatiality", beginning with the most primitive sensory fields, full three-dimensional homogeneous space is attained through a phenomenological constitution, in preparation for justification of the "properly spatial geometry".

11.3.2 *Transcendental-phenomenological grounding of the Euclidean metric*

Since this is the main bone of contention that surfaced in the correspondence with Weyl, we here state what appears to be at issue. Becker must now show that the three dimensional homogeneous space whose phenomenological constitution has just been sketched, has a necessary Euclidean structure on transcendental-phenomenological grounds. This attempt faces two central challenges: 1) to distinguish on these grounds between Euclidean and other geometries of constant but non-zero curvature; and 2) to provide a transcendental-phenomenological justification for retaining the a priori structural laws of the Euclidean geometry of the phenomenologically constituted homogeneous space in the context of the non-euclidean measure determinations of the physics of variably curved space-times. We discuss these in turn.

The text of Becker's essay relevant to 1) is § 12: "*Attempt at a transcendental-phenomenological grounding of the validity of Euclidean geometry for the space of a straightforwardly intuitive nature*" (pp. 481 ff). As Becker notes, there are three apparently contingent mathematical conditions on the general form of space for which a phenomenological foundation is required: a.) why such a space has constant null curvature, b.) why it is "open" (a topological open set), and c.) why it has three dimensions. Weyl's objections are apparently targeted on a.) and c.) to which we limit our attention. In such phenomenological investigations, Becker considers it appropriate to methodologically distinguish the more general "ontological" considerations from those properly and purely "phenomenological", treating them separately. The "ontological" investigation into the Euclidean metric rests entirely upon the "basic determination" [Grundbestimmung] of a homogeneous space to be a *principium individuationis*. For by a "simple reflection" upon the "essential meaning" of such a space, it is a "medium of repeatability", characterized by the group of rigid motions (congruent displacements) and according to which identical bodies can everywhere exist. It is this defining property of homogeneous space that Riemann referred to as "the existence of bodies independent of place" and Helmholtz by "free mobility". However, such a condition does not single out Euclidean space (constant null curvature) from other homogeneous spaces of constant curvature.

Now a "sharper reflection" upon "repeatability" (a condition independent of the character of the individual space-filling solid figures) yields its "strict sense", according to which none of the points of a homogeneous space should be distinguished. This properly mandates that any geometrical transformation of the space is a translation, motion in a given direction by a given amount, carrying each point A to a *distinct* point A' in agreement with the distinguishing characteristic of translations that there are no fixed points. On this basis, however, results only a geometry of translations, an *affine geometry*. (Weyl 1921b, pp. 13 ff) Such a geometry does not include those rigid motions that are rotations about a fixed point since points lying on the axis of rotation are distinguished. In order to do so, Becker argues that the basic ontological property of space to be a *principium individuationis* is mathematically rendered by the fact that if the group of congruent displacements (rigid motions) contains the group of translations as a (normal) subgroup, the Euclidean metric is singled out.[63] However, as Becker admitted to Weyl in his letter of 27 vi 1923, this argument will not make a very strong impression on the mathematician.

In any case, it would appear to be the transcendental-phenomenological, not the ontological, grounding of the Euclidean metric that Becker is principally concerned to defend in this letter. Thus Becker appeals to the "independence of direction from place" as the univocal Riemannian designation of Euclidean space (Becker 1923, pp. 484, 489 ff.), with the interpretation that it gives recognition to a purely phenomenological distinction. Namely, the group of rigid motions characteristic of homogeneous space can be split into disjoint subgroups of "pure translations" and of "pure rotations" (about an oriented center = "Here"). Becker's "phenomenological investigation of the Euclidean metric" may be summarized in the following way.[64] In concrete spatial intuition, the percipient stands at the origin of a space whose natural coordinates are not Cartesian, but polar. There are therefore three coordinates, r, ψ and ϕ, where r is the radius representing the distance of a point P from the viewing subject at the origin O, ψ and ϕ are angles marking the direction of P from a conventionally chosen null direction "straight ahead" from O. According to Becker, the radial and angular coordinates are phenomenologically very different kinds, i.e., they play completely independent phenomenological-constitutive roles. The angular coordinates ψ and ϕ are as phenomenologically distinct from r as is the auditory quality of pitch from that of intensity. Then the Riemannian characterization of Euclidean space as manifesting "independence of direction from place" is precisely captured in the phenomenological distinction between kinaesthetic apprehension of radial motions of "progression into the depth" and purely oculomotor angular changes in the line of sight.

This fundamental phenomenological difference is taken to mean that the group of motions in homogeneous space is split into displacements without rotation (i.e., pure translations) and pure rotations (without translation), a partitioning that distinguishes Euclidean space from other homogeneous spaces. As the respective coordinate types show, these subgroups are regarded as having completely distinct *constitutive* properties.

Rotations correspond to a pre-spatial stage of pure oculomotor motions, that is, the transformation of the bundle of rays issuing from the oriented center O into itself. On the other hand, a pure displacement corresponds to the higher constitutive stage of "entrance into the 'depth' ", i. e., a radial motion outwards while keeping the middle point of the visual field fixed (thus avoiding oculomotor movement). Each of these is a necessary stage in the transcendental constitution of intuitive space, but it is the latter (as just seen in 11.3.1 above) in which spatiality per se emerges, completing the phenomenological (not ontological) grounding of the Euclidean metric of intuitive space. But this latter appears indistinguishable from the Helmholtz-Lie condition of "free mobility" and so it is perhaps not surprising that Weyl may well have contested its alleged a priori necessity.

Becker's transcendental-phenomenological grounding of Euclidean geometry must also be extended to treat problem 2) above, i. e., to accomodate the existence of spaces of variable, not constant, curvature that play such a prominent role in general relativity. His attempt is found in §17 of his essay, "*Space forms with variable curvature according to time and place*" (pp. 521 ff.) and may be briefly recapitulated. Just as Becker credited Riemann's designation of Euclidean space as recognizing the phenomenological distinction between "pure rotations" and "pure translations", so here he finds in Riemann's imposition of the condition on all space forms, that they possess "planeness in their smallest parts", the sole a priori structural law remaining in a variably curved metrical manifold. Riemann's requirement is sufficient to secure the phenomenologically constitutive step of the direction bundle whose spherical structure is centered at a point O ("here") of space. As seen in 11.3.1, this is the oriented space from which spatiality itself (requiring "entrance into 'depth'), in the sense of Euclidean intuitive space, is constituted as a *principium individuationis*. Accordingly,

if only the condition of planeness in the smallest parts is fulfilled, in every arbitrary metrical manifold with arbitrary variable curvature the normal Euclidean phantom space is necessarily constituted as *principium individuationis* of the material world. (Becker 1923, p. 528)

Something considerably weaker than this condition follows in general relativity from Einstein's principle of equivalence whereby a gravitational field of a rather artificial type (static and homogeneous) can always be locally "transformed away", and may be considered a necessary presupposition of measurement. Note that Becker is not claiming what is untenable in the light of Einstein's theory: that the metrical relations of physical space are necessarily Euclidean. In fact, he accepts general relativity's demonstration that they are not, at least in the large. Rather his concern is that the transcendental constitution of intuitive space as *principium individuationis* of the material world undermines conventionalist claims regarding these metrical relations, and in particular, that Euclidean geometry rests on conventional posits.

We are now in a position to discuss Becker's claim in the letter of 27 vi 23, that ultimately it is possible to proceed in the transcendental constitution of Euclidean phantom space without the Helmholtz-Lie condition of "free mobility" (so prominently deployed in his *Habilitationsschrift*), but with Weyl's "Pythagorean measure determination". Becker refers to pp. 532–33 of his essay where he essentially recapitulates Weyl's distinction (see

11.2.3 above) between the "nature" (or *Wesen*) and the "orientation" of space. There, on p. 532, Becker affirms a close agreement with Weyl's independent solution to the new problem of space; Weyl's result

stands in the best agreement with our constitutive analysis: the Euclidean structure of the rotation group (or the "spherical" structure of the "direction bundle") is that phenomenon upon which the intuitive evidence of Euclidean phantom space constitutively builds [*konstitutiv aufbaut*].

Now, in the letter, citing Weyl's remark, noted above, that Euclidean space is (infinitesimally!) characterized by the property that its rotation group is comprised of those linear transformations leaving a non-degenerate quadratic form invariant, Becker makes the surprising statement that

that the Euclidean space of the phantom world is, so to speak, to be conceived as the homogeneous extension of the infinitesimal bundle of line elements at O *(Here)*.

If this is the case, then, from the phenomenologically constitutive, not ontological, viewpoint, there is a natural continuity between the Euclidean nature of "phantom space" and the variably curved spaces of Riemann, Einstein, and Weyl. Indeed, Becker sees both as deriving from the phenomenologically distinguished, and constitutively prior, spherical bundle of directions (the infinitesimal Euclidean rotation group) at O.

I now maintain: free variability (Riemann, Einstein, Weyl) and independence from place (Euclid) stem from the same root with respect to their transcendental constitution, and phenomenologically. That is, they are two different possible interpretations (and, among the distinct viewpoints of the phantom and physical space, both in their own way necessary) of the same phenomenological connection: viz., the phenomenological independence (or pure separateness) of the "pre-spatial (layer)" and the "radial depth layer" ['*radiale Tiefen-Schicht*'].

But Becker's claim to place no reliance on Helmholtz-Lie (and so, on "free mobility") in the constitution of intuitive space was surely opaque to Weyl. The reasons appear obvious. While asserting their "phenomenological independence", Becker has nonetheless overlooked that the "pre-spatial layer" has a "purely infinitesimal" mathematical characterization (infinitesimal Euclidean rotation group) while the "radial depth layer" has the finite character of "free mobility" ("homogeneous extension of the infinitesimal bundle [...] at O"). These are notions of space that, Weyl could well have argued, are actually inconsistent at the "ontological" (formal, mathematical) level.[65]

After all, it was to remove the remnants of similar *ferngeometrisch* conceptions (i. e., permitting direct comparisons of lengths at finite separations) that spurred Weyl to develop his *reine Infinitesimalgeometrie*. How is it possible to constitute full intuitive visual space, where, to use Becker's own example, "each star is a 'direction in the depths' ['*Richtung in die Tiefe*']," (Becker 1923, p. 527) via an "homogeneous extension of the infinitesimal bundle of line elements at O" without invoking the finite condition of "free mobility"? Becker provides no clue, but for Weyl, the only possible consistent transition from comparison relations between infinitesimally neighboring points to those between finitely separated points must be solely based on infinitesimal notions, and in particular, the concept of the infinitesimal parallel displacement of a vector. As noted in 11.2.1, in this framework, there is a fundamental distinction between what can be ascertained "infinitesimally" at a point P and its "nearby" points P', and what can be affirmed about objects at finite separation, with the requisite reference to a specific path or line element between them. Thus, in his phenomenological constitution of intuitive space, Becker still relies on some equivalent of "free mobility" and so violates both the spirit and the law of Weyl's *Wesensanalyse* of "the problem of space". This accounts for the less than enthusiastic tone of Weyl's final published comment, in 1927, repeated in 1949, on Becker's constitutive analysis of intuitive space.

Perhaps the postulate of the univocal determination of 'straight progression' is also justifiable from requirements which the phenomenological constitution of space places. Becker would, as before, base the significance of the Euclidean rotation group for the space of intuition upon the Helmholtzian postulate of free mobility.[66]

11.3.3 Relativity physics and non-euclidean metrical determinations

The last two sections (§§ 20–21; pp. 547–60) of Becker's monograph are a phenomenological investigation of the meaning and use of non-Euclidean metric determinations in relativity physics. Becker has earlier argued that the space of material nature that, according to classical physics, is Euclidean, is constituted from an intuitive space that is necessarily Euclidean. In the classical conception since Newton, geometry and physics are completely separated. The Euclidean laws form the framework within which the material-causal laws are plotted. "Place" is completely free of physical significance and the structure of the world, i. e., distribution of matter in space, is completely accidental, subject only to laws of probability. According to Becker, the unfounded presupposition of classical physics is a belief in the Euclidean lawfulness of space in which physical events take place; phenomenologically, this amounts to asserting that a physical event can be fully represented in perceptual intuition. Both presuppositions mean that a physical event is solely subject to a specifically "causal" lawfulness leaving the "world structure" intact. These presuppositions are manifest in classical physics' conception of measurement, which makes the most extensive use of the unalterability of a rigid standard. This belief cannot be justified in classical physics by adopting conventions, but is guided by rational motives. In as much as classical physics proclaims the autocracy of causality in nature, it can with the same justification assume the constancy of the standard whenever no altering cause is present. That is, the choice of a standard of measurement in classical physics depends on certain presuppositions of measurement that in turn presuppose that all laws of nature are causal laws and also that the intensity of effects diminishes with distance from their causes.

This situation, of course, has been transformed by Einstein's theory of general relativity, but for Becker it is only Weyl who has shown how to reconcile the necessary Euclidean character of space (intuitive or "phantom" space) with the use, in the Einstein theory, of non-Euclidean metrical determinations. Special relativity had already taken a step going fundamentally beyond classical physics in speaking of "events in space-time"; as a result, the material event no longer has an "imagelike" character, and so cannot be "substructured" by a corresponding phantom event (representable in intuition). On the other hand, "picturability" is characteristic of classical physics, a substructuring of the material world through a phantom world, purified of secondary qualities. But the decisive move has been made with general relativity. There Weyl has shown how the structural component of the laws of nature (the affine and metrical connections determine the structure field of space in its entirety from the infinitely small) is to be distinguished from the causal laws, since both still depend on the contingent distribution of matter in space.

The real content of Einstein's theory lies, in Becker's view, in the form given it by Weyl, for only Weyl has made clear how the transcendental-constitutive structural laws of the world [*Strukturgesetze der Welt*] which are *a priori*, are to be distinguished from the specific causal laws of matter and matter-field interactions, that, of course, are a posteriori.

The former are comprised by the affine connection or "guiding field" [*Führungsfeld*] corresponding to the laws of gravitation and inertia, together with the laws of the metrical connection, corresponding to the electromagnetic field laws.[67] The "guiding field" determines how a definite world direction is carried over from a given point P to the neighboring P', while the metrical connection does the same for an infinitesimal magnitude of extension (a spatial distance or temporal period). These "structure laws" concern vectors and extensions in an (3 + 1-dimensional) metrical "ether"[68] that, as Weyl provocatively put it, "has nothing to do" with the geometry of measuring rods and clocks of physical theory. (Weyl 1919c, p. 113) The relation between the purely structural "world geometry" [*Weltgeometrie*] and the physical geometry of material bodies [*Körpergeometrie*] is complicated and to be determined by the choice of an appropriate action function (i. e., scalar density) entering into an action (Hamiltonian, i. e., variational) principle that encompasses the structure laws, and from which the field laws of Einstein and Maxwell are obtained. (Becker 1923, p. 555; see 11.2.1 above) Only the general form of Weyl's action function is univocally determined, but a unique function is not determinable a priori. Hence, the "structure laws" of the gravitational and inertial fields are a priori until the particular choice of the action function from among the "essential possibilities" bounded by the restricted set of gauge-invariant integral invariants of a Weyl manifold. Of course, what the field magnitudes are at a determinate spatio-temporal position depends on the matter "generating" the field and this is empirically contingent. But in any case, the structural laws of the world have been attained; at least in their general form they are a priori. Their scope comprises all known exact (not statistical) laws of nature. At this a priori level, it seems as if causality has been banished from the world, for everything apparently is resolved in structural lawfulness and physics becomes four-dimensional geometry. (Becker 1923, p. 557)

Of course, this is not quite true: the phenomenologically fundamental character of time must be upheld. There is no "spatialization of time". Constitutionally, this means that a world line of a subject has a tangent at every point that specifies the "temporal direction". At every point of a world line, a definite "cleavage" of the world into time and space is prescribed. This tangent as well as the world line possesses a definite direction: past to future; "my ego" lives "along" a definite world line, the "now-point" is continually shifted on it: this is the symbol of the "original stream of time". (Becker 1923, pp. 558–59) Once again, we can see the resonance in Becker of Weyl's claim that a coordinate system is the "last residuum of Ego-elimination" in the construction of the objective world (see 11.2.2).

As noted in 11.2.4, in Weyl's "agent-theory of matter", the electron (both "positive" and "negative", in the terminology of 1920) has been "taken out" of the field continuum. The world lines of electrons, i. e., of matter, have become channels, or structures in a non-simply connected topology. In this conception, matter literally lies "beyond" the metrical continuum and is not bound by its laws. Exploiting, as did Medicus, Weyl's analogy between the free will of an Ego and an electron lying "outside" the domain of exact field laws, Becker conceives of an electron as "enduring", but now and then sending "actions" into the four-dimensional world. Matter thus causally operates upon the metrical field, which itself as purely structural, is merely a passive transmitter of actions. Summarizing his main contention, Becker states that the general theory of relativity, but only in the "broadened" form given it by Weyl, provides a radical solution to the problem of the respective roles of the a priori and the a posteriori in the determination of the metric of space(time) and that it does so in terms of the phenomenological laws of essence [*Wesensgesetze*]. The solution consists in the aforementioned decisively sharp separation of the structural and causal laws of the world. Thus the problem of metric determination posed by the appearance in physics of a freely variable curved space is solved and, at the same time, a necessary methodological foundation of physics is established in the distinction between a priori laws of structure and a posteriori causal laws. This in fact is the *fundamental* significance of the general theory of relativity, which exists independently of the state of empirical research at any given time. Becker concludes his *Habilitationsschrift* with an unmistakable pride in his accomplishment:

And so we have phenomenologically investigated all known non-Euclidean forms of space and established their possibilities of application in physics. We have thus illuminated the relation in which they stand to Euclidean geometry in the constitutive construction of nature. The first part of this work has shown the basic possibility of a rational treatment of space. Now this second part has given a complete overview of the formal configurations governing the factual [*sachhaltige*] essence of spatiality, arranging them within the great total system of nature and freeing them from all the contingency that initially adhered to them. But thereby the problem that we posed, of overcoming of the contingency of the geometric axioms, is fundamentally solved. (Becker 1923, p. 560)

Concluding Remarks

As is described in some detail in Chapter 10, during this period Weyl saw the fortunes of intuitionism and phenomenology in the foundations of the exact sciences as closely tied together. Apparently failing to discriminate between the resources available to phenomenology and those of intuitionistic mathematics in accounting for a contentual *Anschauung* capable of grounding the meaning of mathematical statements, Weyl saw the failure of the latter, in the face of Hilbert's finitism, as implicating the failure of the former. Revealing, by the mid-1920s, a new appreciation of Hilbert's developing program of proof theory, Weyl declared that the emerging triumph of Hilbert meant nothing less than a "decisive defeat of the philosophical attitude of pure phenomenology":

If Hilbert's view prevails over intuitionism, as appears to be the case, *then I see in this a decisive defeat of the philosophical attitude of pure phenomenology*, which thus proves to be insufficient for the understanding of creative science even in the area of cognition that is most primal and most readily open to evidence—mathematics (original emphasis) (Weyl 1928a, p. 88; Engl. trans., p. 484)

It is widely known that Weyl went on to argue for a position he termed "symbolic construction", regarded as somehow straddling the conflicting tendencies of formal axiomatics and constructivism. However, this was not Weyl's final judgment on phenomenology. Late in life, in a autobiographical lecture at the University of Lausanne, Weyl recalled in some detail the Husserlian influences on his earlier self. Quoting at length the passage appended to the "Introduction" to the fourth edition of *RZM* whose beginning was cited in 1.2.1 above, Weyl admitted that he "still essentially held fast" to the understanding of phenomenological method articulated there, directing epistemological reflection upon the knowledge of the special sciences. (Weyl 1955, p. 639) More generally, his concern with the relations between knowledge and reflection [*des Verhältnisses von Erkenntnis und Besinnung*] in differential geometry and physics in the period 1918–23 may still be regarded as the paramount illustration of the application of phenomenological method to fundamental physical theory. Weyl's neglected example invites and deserves further attention in contemporary philosophy of physics.

12

"DAS ABENTEUER DER VERNUNFT": O. BECKER AND D. MAHNKE ON THE PHENOMENOLOGICAL FOUNDATIONS OF THE EXACT SCIENCES

Introduction

The relationship between phenomenology and the exact sciences in the 1920s is still largely an unexplored area. We still do not have sufficient information on those phenomenologists who, in addition to Husserl, engaged in a dialogue with the exact sciences. Moreover, much should be done by way of studying those scientists who found in phenomenology a source of inspiration in their scientific work. Among the former one should recall Hans Lipps, Oskar Becker, Dietrich Mahnke, Felix Kaufmann, and Moritz Geiger. Among the latter we have the towering figure of Hermann Weyl but the influence of phenomenology on other prominent scientists, such as Fritz London, should also be investigated. In particular there are several, almost untapped, archival resources that would repay further study. In previous works, Thomas Ryckman and I have

---

I would like to thank Dr. U. Bredehorn (Marburg University) for his help in providing me with copies of the Becker-Mahnke correspondence. Dr Bredehorn was also instrumental in writing up notes with a partial list of the contents of the Mahnke *Nachlaß*, which were kindly typed up by Gudrun Hartmann. I am also very grateful to Frau Astrid Becker for permission to publish the letters from Becker to Mahnke. Special thanks to Bernd Peter Aust and Jochen Sattler for the careful transcription and to Richard Zach and Johannes Hafner for help at earlier stages of the work on the transcription and for deciphering some difficult passages. Mark van Atten has, as usual, provided valuable comments. For a transcription of the correspondence see Becker (2005).

analyzed the relationship between Becker and Weyl with particular emphasis on their relationship to phenomenology.[1] In the course of the research work for those essays I became aware of the importance of the correspondence between Dietrich Mahnke and Oskar Becker, which also contains much of interest on Weyl. The correspondence between Becker and Mahnke gives us a lively view of the problems that were central to the phenomenological reflection on the exact sciences in the second half of the twenties. This was a tumultuous period in the foundations of mathematics and physics and the correspondence reflects how these developments forced a reflection on the part of those phenomenologists who worried about how Husserl's philosophy and its modifications, both by the master and by other members of the phenomenological movement, could account for the recent developments that had changed the scientific landscape. It is my hope that a presentation of the correspondence between Becker and Mahnke, transcribed by Bernd Peter Aust and Jochen Sattler (see Becker 2005), will be useful to those who reflect on the still topical problem of how phenomenology relates to the exact science. I will begin by giving a short overview of Mahnke's and Becker's early career. Then I will describe the correspondence and outline some of the major topics treated in it. I hasten to add that my few introductory pages have only the role of giving a quick guide to the interested reader but that a full analysis of this correspondence is not what I am aiming for and would require a much more extensive treatment.

12.1 Dietrich Mahnke (1884–1939)

Dietrich Mahnke was born in Verden in 1884.[2] He studied mathematics and philosophy in Göttingen from 1902 to 1905. He was a student of both Hilbert and Husserl. From early on Mahnke considered himself a phenomenologist and was very close to Husserl, as the long correspondence between the two testifies.[3] Husserl himself considered Mahnke to be personally closer to him than any of his other students (Husserl to Mahnke, 17.x.21). Another important influence on his philosophical thought was that of Dilthey. In 1906 he passed his examination "pro facultate docendi". He then taught in high school at Hameln and Stader until his move to Greifswald in 1923. From early on he had a keen interest in Leibniz and his first book on Leibniz came out in 1912, *Leibniz als Gegner der Gelehrteneinseitigkeit*. He served four years of duty in the war. During this period he managed to publish two books, *Der Wille zur Ewigkeit* (1917) and *Eine neue Monadologie* (1917). He wrote his doctoral dissertation under Husserl's supervision, who was in Freiburg i. Br. at the time, and got his degree in 1922 with a dissertation on Leibniz. In 1923 he published an article in which he explored the connection between Hilbert's metamathematics and Husserl's phenomenology, if only at an introductory level: "Von Hilbert zu Husserl".[4] His thesis on Leibniz appeared in 1925 in the *Jahrbuch für Philosophie und Phänomenologische Forschung* with the title "Leibnizens Synthese von Universalmathematik und Individualmetaphysik". A second installment of this work was planned but never appeared. In 1926 he habilitated in Greifswald and was finally called as Ordinarius of Philosophy in Marburg in 1927. The publications from 1923 to 1927 show the wide variety of interests that Mahnke pursued. In addition to topics in the history of philosophy he also wrote on history and philosophy of mathematics and science (see the bibliography for a representative list of his publications during this period). Extremely valuable are his writings on the history of the Leibnizian differential and integral calculus. Among his last writings special mention should be made of the book *Unendliche Sphäre und Allmittelpunkt. Beiträge zur Genealogie der mathematischen Mystik* (1937). Mahnke served as Dean of the Philosophy Faculty of the University of Greifswald in 1932 but he resigned in 1934. He died of a car accident in 1939. His *Nachlaß* is preserved at the University of Marburg although it is still unclassified. In addition to the correspondence with Becker, the *Nachlaß* contains correspondence with, among others, Husserl, Dingler, Heidegger, Jaensch, Rickert, and H. Scholz.

12.2 Oskar Becker (1889–1964)

Like Husserl himself, Oskar Becker came to phenomenology from the mathematical sciences. He studied at the University of Leipzig with Hölder and Herglotz and wrote a doctoral dissertation in 1914 on axiomatic geometry. After four years of military duty he went to study philosophy with Husserl in Freiburg i. Br. His Habilitation was completed in 1922 and was on the subject of the phenomenological foundations of geometry and relativity theory: "Beiträge zur phänomenologischen Begründung der Geometrie und ihre physikalischen Anwendungen".[5] Husserl was very impressed with the results, which—as he says in a letter to Hermann Weyl—are "nothing less than a synthesis of Einstein's and yours discoveries with my phenomenological investigations on nature." Becker became Husserl's assistant in 1923. During the same period he was attending seminars by Heidegger. Heidegger's influence leaves a strong mark in the next major work he wrote, "Mathematische Existenz", published in 1927 in the same issue of the *Jahrbuch für Philosophie und Phänomenologische Forschung* in which "Sein und Zeit" also appeared. The book is perhaps one of the most ambitious works ever written on the phenomenology of the exact sciences but received severe criticism from many corners.[6] In particular Becker was accused of defending a form of "anthropologism" in the philosophy of mathematics. We will see that Mahnke could also not accept a foundation of mathematics along these lines. Becker replied, also discussing points raised by Mahnke in correspondence.[7] In 1928, Becker became *Ausserordentlicher Professor* in Freiburg i. Br. He left Freiburg in 1931 to take up a chair in Philosophy at the University of Bonn. During the early thirties he wrote a number of important contributions to the history of Greek mathematics and to modal logic. Becker's *Nachlaß* is preserved at the University of Konstanz. Unfortunately, there is precious little concerning the twenties. In personal correspondence, Frau Astrid Becker, daughter of Oskar Becker, wrote that much of her father's correspondence and manuscripts were destroyed during the second world war. Becker kept an extensive correspondence with several scientists and philosophers including, among others, Weyl, Hilbert, Ackermann, Fraenkel, and Reichenbach.[8]

12.3 The Correspondence between Becker and Mahnke

The above short description of Becker's and Mahnke's interests should make it obvious that their range of interests overlapped to a considerable extent. Both approached philosophical issues from a phenomenological perspective, they shared a common background in the exact sciences and, last but not least, the history of science was essential to their philosophical thinking about the sciences. Becker and Mahnke met in Freiburg in July 1922 in Husserl's house on the occasion of Mahnke's promotion to doctor of philosophy.[9] However, contact is resumed only through a letter from Becker to Mahnke, dated 22 August 1926. This was to begin a correspondence that extends until 1933. The correspondence is preserved at the Universitätsbibliothek Marburg. It contains 11 letters from Becker to Mahnke and 2 letters from Mahnke to Becker.[10]

i. Becker to Mahnke, August 22, 1926; 8 pages. [Letter 1]
ii. Becker to Mahnke, September 16, 1926; 6 pages. [Letter 2]
iii. Becker to Mahnke, July 20, 1927; 2 pages. [Letter 3]
iv. Becker to Mahnke, August 21, 1927; 4 pages. [Letter 4]
v. Mahnke to Becker, September 8, 1927; 6 pages. [Reply to 4]
vi. Becker to Mahnke, September, 1927; 14 pages. [Letter 5]
vii. Mahnke to Becker, October 12, 1927; 4 pages. [Reply to 5]
viii. Becker to Mahnke, October 20, 1927; 4 pages. [Letter 6]
ix. Becker to Mahnke, December 12, 1927; 4 pages. [Letter 7]
x. Becker to Mahnke, April 17, 1928; 4 pages. [Letter 8]
xi. Becker to Mahnke, May 8, 1933; 2 pages. [Letter 9]
xii. Becker to Mahnke, May 20, 1933; 2 pages. [Letter 10]
xiii. Becker to Mahnke, July 7, 1933; 2 pages. [Letter 11]

The correspondence shows a great deal of mutual respect and interest in each other's works, notwithstanding a number of disagreements on some central issues. Becker had already been making use of Mahnke's works on Leibniz while preparing "Mathematische Existenz" and indeed in the first letter Becker starts the correspondence by querying Mahnke on the relationship between Leibniz and the recent debates on the foundations of mathematics. This led to a constructive exchange and to reciprocal esteem. For instance, Mahnke refers to Becker's work in his review of volume VII of Dilthey's *Gesammelte Schriften*[11] and Becker refers to Mahnke's work in "Mathematische Existenz" and in "Das Symbolische in der Mathematik". Husserl was also instrumental in building up this mutual appreciation.[12] In the next section I will give an outline of some of the major topics touched upon in the correspondence.

12.4 Leibniz's Philosophy and the Opposition "Formalism-Intuitionism" in the Foundations of Mathematics

Becker's interest in Mahnke's essays on Leibniz[13] is connected to his writing up of the important historical sections that make up §6.c of "Mathematische Existenz", and in particular the section on Leibniz on pp. 714–27. In the first letter to Mahnke, Becker poses the question of how Leibniz's thought fits in the recent debate on the foundations of mathematics and the opposition between formalism (Hilbert) and intuitionism (Brouwer). Becker is after a characterization of Hilbert's and Leibniz's position with respect to Brouwer's intuitionism. Mahnke had touched upon the topic in his dissertation on Leibniz,[14] while discussing the contributions by Hans Pichler. The central point concerns the relationship between the formal aspects of mathematics and intuition. In the discussion of Pichler, Mahnke seems to indicate that Hilbert and Leibniz seem to be in agreement as they both require a radical logicization and on the other hand they attempt to give intuitive content [*Veranschaulichung*] to the formal concepts by means of "sinnliche 'Charaktere'". Becker, in his discussion of the topic, begins by distinguishing three types of mathematical formations:

(1) Formations accessible to the (idealized) sensible intuition; examples include Euclidean geometrical formations in their application to intuitive space.
(2) Formations accessible to categorial ("eidetic") intuition; these include non-Euclidean geometrical formations that can only be intuited arithmetically-analytically; moreover, all the "objectivities" of Brouwer's "intuitive" mathematics.
(3) Formations which are not accessible through any intuition but are nonetheless consistent. Here the example is given by Hilbert's transfinite elements.[15]

According to Becker, the third sort of formation cannot be accounted for by Husserlian "formal ontology". Indeed, whereas in the past one could successfully find intuitive representation of "ideal elements" (like Gauss' interpretation of the complex numbers in the geometrical plane), Hilbert's formal mathematics has brought in a new category of ideal elements that are not accessible, even in principle, to any intuition. But the admission of a realm of formations that is not accessible, even in principle, to any intuition represents a radical departure from Husserl's phenomenological approach. Thus, Becker points out the insufficiency of formal ontology to account for mathematics, for Hilbert's transfinite elements are objectivities that, according to Becker, are not accessible to formal ontology. Becker asks what is Leibniz's position with respect to this problem and whether he had any premonition that there could be consistency proofs that did not go through an "isomorphical" representation of the formal on the intuitive realm (as in Gauss' representation of the complex numbers in contrast to Hilbert's proof-theoretic consistency proofs, which do not require the "intuition" of the domain described by the theory to be shown consistent). The considerations that follow are meant to probe the exact relation between the formal and the intuitive in Leibniz. Becker poses the following dilemma: one cannot claim Leibniz both on the side of phenomenology and on the side of Hilbert (and the recent Weyl) at the same time. Indeed, classical phenomenology must insist that all formations should be accessible to categorial intuition whereas Hilbert's mathematics, by making use of transfinite assertions, which require for their comprehension an appeal to actual infinities, cannot *in principle* express states of affairs that are categorially intuitable. While Mahnke's description of Leibniz in 1925[16] seems to point at a continuous passage between the intuitive and the non-intuitive realms, Becker points out that for Hilbert there is a radical gap between the level of formal mathematics and metamathematics. This leads Becker to raise a number of issues that will be pursued in the later letters. First of all, Hilbert's representation of formal mathematics in metamathematics is not an "isomorphic" one. This indicates that one cannot confuse Husserl's eidetics with God's "infallibilis visio" in the Leibnizian sense (which, for Becker, coincides with Kant's *intellectus archetypus*).[17] In an important citation from a letter that Weyl had sent him, Becker points out also Weyl's skepticism as to the hope that phenomenology can account for the exact sciences. He continues by remarking that this state of affairs constitutes a crisis for classical phenomenology:

We are facing here a certain crisis of the phenomenological method itself. If Weyl is right that Brouwer's intuitionism cannot support theoretical physics, then phenomenology in Husserl's "classical" sense must also be seen as incapable of securing the modern form of knowledge of nature—and to make it completely understandable.

Becker then expresses his hope that perhaps Leibniz's notion of "representation" may help in solving the crisis. However, Leibniz should not be reinterpreted along the lines of classical phenomenology but rather by allowing for a wide notion of "symbolization". In this way, Becker hints at his idea of "mantic phenomenology", which divines nature through symbols.

We see then from this letter how the conflict between formalism and intuitionism can be read, according to Becker, as the conflict between Leibniz and Kant:

But to adjudicate between Leibniz and Kant, or their possible synthesis, is what the battle over the foundations turns on and in this conflict phenomenology will be inevitably drawn.

The second letter brings some clarification to Becker's point of view. Apparently, Mahnke had expressed the hope that the gap between formal mathematics and intuitive mathematics could be bridged. Becker replies that this is in principle impossible since human categorial intuition can never grasp actual infinities. He then proceeds to explain his idea about grounding the transfinite process phenomenologically by generalizing the Husserlian notion of indexing [*Stufencharacteristik*; Ideen, §100] for the iteration of acts in pure consciousness.[18] However, the process in question is a potentially infinite process (of transfinite length) and thus we do not have a foundation of set theory in terms of actually infinite sets. For Becker the essential distinction is not that between the indefinite and the potentially transfinite but rather that between potential infinite and actual infinite. While the potential processes (whether indefinite or transfinite) can be grasped by categorial intuition, this is not the case for the actual infinite. In a note, Becker points out that what can be legitimated by this potentially transfinite process will however not be enough to ground the set theory needed for mathematics and the applications to theoretical physics (a subject which he was also discussing with Weyl at the same time).

Thus, Becker excludes the possibility of a grasp of the actual infinite by a continuous process of human knowledge, and quotes Husserl's *Philosophie der Arithmetik* as evidence that also Husserl excludes this possibility. This leads to a reflection on how to think the relationship between God's intellect and the human intellect, with special reference to Leibniz, Kant, and Husserl. Becker concludes his letter by claiming that the questions he is raising are topical and that the future of phenomenology depends on them. While he does not want to identify his position with Husserl's he also says that Husserl's position deserves the utmost consideration.

Mahnke had sent a typescript of his "Untergang der abendländischen Wissenschaft?" with his reply to Becker's first letter. In the third letter Becker thanks Mahnke for sending an off-print of "Untergang der abendländischen Wissenschaft?" and reciprocates by sending a copy of "Mathematische Existenz".

12.5 Mahnke's Review of Dingler

The fourth letter from Becker to Mahnke takes its start from a review of Dingler's book "Der Zusammenbruch der Wissenschaften und der Primat der Philosophie"(Dingler 1926). In the case of letters 4 and 5 from Becker to Mahnke we are also fortunate to have the two long replies by Mahnke to Becker. Mahnke had reviewed Dingler's book in an article entitled "Untergang der abendländischen Wissenschaft?" and published in *Archiv für Mathematik und Naturwissenschaft* (Mahnke 1927c). Two points of the review have to be mentioned in order to put the discussion in context. Dingler's book emphasized the "chaos of opinions" that the last developments of mathematics and science have, allegedly, brought about. Dingler saw in "inductionism" (the attempt to infer the eternal laws from empirical experience) and "mathematism" (the trust in a pre-established harmony between empty mathematical formulas and the laws of nature) the wrong methodological turns that characterize this chaotic phase of the sciences. In the area of mathematics Dingler shows the dangers of mathematism in the creation of arbitrary consistent systems that has led to the debate between formalism and intuitionism. In the field of physics it is relativity theory that shows how inductionism and mathematism reinforce each other. Both methods fail to characterize the true laws of nature but only give us consistent systems that are incompatible with each other. This leads to a very "pessimistic" appraisal of the actual situation in the sciences but Dingler has hope that this "chaos of opinions" can be overcome by a method, Dingler's "pure synthesis", that will allow to secure the sciences once and for all. This new method, which I will not describe here, cannot emerge from the individual sciences but is the specific contribution of Dingler's philosophy to the overcoming of the scientific crisis facing the early part of the twentieth century. Mahnke's main question is whether Dingler's negative appraisal of the recent developments is justified and if the situation truly calls for a solution by means of Dingler's "system of pure synthesis". Much of the review deals with specific issues in the history of science but the conclusion makes a number of philosophical points. First of all Mahnke shows that Dingler's position is also an extreme form of mathematism and thus criticizes Dingler's understanding of the way science proceeds. A detailed analysis of Dingler's claims leads Mahnke to the final evaluation that Dingler tries to bring together inconsistent and opposite positions: on the one hand modern irrationalism and fictionalism and, on the other hand, classical rationalism and absolutism.

In the fourth letter, Becker expresses his agreement with Mahnke's critique of Dingler but he disagrees with the "optimistic" evaluation given by Mahnke about the prospects of science. Becker's pessimism is interesting and should also be read in conjunction with similar opinions he expresses in letters to Hermann Weyl. In particular, he raises the important, and still topical, issue of the adequacy of intuitionistic and semi-intuitionistic [predicative] mathematics for the physical sciences by claiming that only the latter mathematics can account for the physical sciences. This also reflects discussions that Becker was having with Weyl on the topic.[19] Becker points out that if the "objectivity" of mathematics, i.e. an understanding of mathematics along the lines of Husserl's formal ontology, is given up then physics becomes a symbolic interpretation of nature and that in any such activity there is much arbitrariness. Indeed, physics amounts to an "irrational," "free" and "creative" accomplishment.

In the second part of the letter Becker refers to Mahnke's claim about the "objectivity" of pure mathematics. Indeed, Mahnke held a conception of mathematics as an "eidetic science" in Husserl's sense.[20]

In the final part of the fourth letter Becker points out that if one gives up the "objectivity" [Sachlichkeit] of pure mathematics then it is inevitable that a certain amount of arbitrariness will determine the construction of physics. Whatever knowledge of physics we have is "undeserved luck", but does not stem from necessity of (Husserlian) essences or from evidence. Becker then ties this thought with Weyl's reflections on this topic and concludes by considering the possibility that physics becomes a sort of "magic" (which of course justifies the need for a mantic phenomenology!). While physics might at times arrive at a mastery of the world there is always the possibility that this mastery will "slip away" from us. Our scientific activity remains an "adventure of reason". Becker declares himself to be a pessimist concerning the possibility that any part of science can be grounded with certainty once and for all. If this were possible, he adds, then science would be degraded to pure manual work and "this would be the saddest thing that could ever happen".

12.6 The Discussion on "Mathematische Existenz"

Mahnke's extensive reply begins with a few more remarks on Dingler's book. While he thinks that he is in agreement with Becker on the issues related to the nature of physical knowledge, he stresses that their viewpoint on mathematics is different. Concerning physics, he is of the opinion that Dingler's hope for a system of physics which is unrevisable is a false ideal. However, this is in no way reason for "pessimism" but rather confirms Mahnke's "Optimismus des Schaffens". In the second part of the letter Mahnke continues the discussion by focusing on the main theses of Becker's "Mathematische Existenz" (which he had meanwhile received from Becker). The main point of contention, according to Mahnke, is:

Is it admissible in theoretical physics or at least in pure mathematics to use conceptual structures that are not constructed intuitionistically (in a broad sense that includes categorical intuition) but that can only be shown to be consistent in a "formalistic" way?

Here Mahnke replies without hesitation that his answer, both as a student of Leibniz and of Hilbert, to the question is yes. Mahnke very perceptively remarks that Becker in "Mathematische Existenz" seems to oscillate between two poles. At the beginning of the book his answer to the question is negative but towards the end he concludes, with the later Weyl,[21] by answering yes. This is indeed a tension in Becker's book. Mahnke allows the possibility that nature is perhaps essentially transfinite and that our finite or "weakly-infinite" (i.e. dealing only with potential infinities) conceptual constructions might not be able to exhaust it. But if that is the case there is nothing else left but to rely on symbolic knowledge, which of course does not provide direct access to the intended object. Against Becker's position that this would go against the principle of transcendental idealism, according to which every objectivity should be accessible in consciousness, Mahnke replies that he does not agree with the Heideggerian interpretation of this principle given by Becker. According to Mahnke, one cannot require that the access to the mathematical ideality be given to a human consciousness, as Becker does; it is sufficient to require that the objects be accessible to an ideal consciousness. It is for this reason that Becker, according to Mahnke, is forced to conclude that nature is in principle unknowable. By contrast, Mahnke sides with Leibniz, Hilbert, and the later Weyl and claims that even without an intuitive access to the transfinite, the latter can be represented symbolically. While Mahnke considers symbolic knowledge to be "sufficient", he also grants to Becker that intuitive knowledge is preferable, when obtainable, for symbolic representations are only crutches to help us overcome our limitations as finite beings. For this reason, notwithstanding the previous criticisms, he holds in great esteem Becker's achievement, e.g. the phenomenological grounding of Cantor's transfinite ordinals. In the conclusion of the letter Mahnke reflects on the importance of joining the history of mathematics and physics to the theoretical reflections on it but warns against the insufficiency of Heidegger's hermeneutical phenomenology in accounting for the exact sciences. Heidegger's approach might be appropriate for the human sciences but for an account of mathematics and physics one has to rely on Husserl's phenomenology of consciousness in general [*Husserls Phänomenologie des Bewusstseins überhaupt*].[22]

In letter 5 Becker muses over the analogies between the developments of artistic styles and how they are reflected in the development of mathematics and physics. Once again, he points out that intuitionistic mathematics, or at least "semi-intuitionistic" (predicative) mathematics is a conditio sine qua non for a non-symbolic physics. In the following pages Becker rebuts some of the criticisms made by Mahnke by emphasizing that Hilbert's mathematics is in principle inaccessible not only to Heidegger's human consciousness but also to any Husserlian pure ego. Thus, he claims that his diagnosis of the situation is not dependent on a human notion of consciousness. In particular, drawing on distinctions one already finds in "Mathematische Existenz" (pp. 310–11) he distinguishes between three possible forms of mathematical knowledge, which correspond to human beings, demons, and God. God's mathematics is essentially different from the human one, whereas the demon's mathematics can be seen as an extension of human mathematics, once we lift every limitation of time. Both humans and demons share in the same finite reason (i.e. they are "pure egos" in Husserl's sense) and there is no bridge, even in principle, to God's consciousness. Becker thinks that Mahnke underestimates the "huge gap" that separates finite egos from the "transcendent-transfinite". This leads him to a technical reflection on the notion of "iso-symbolic mapping", which he developed in "Mathematische Existenz" and whose details need not detain us here. The rest of the letter goes back to emphasize that between the mathematics that can be grounded on intuition (even extended to include a certain amount of symbolic reasoning) and Hilbert's transfinite there is a complete ontological discontinuity that cannot be filled. However, Becker adds that this does not exclude the possibility of a completely different type of symbolic knowledge of the transcendent-transfinite. But, he adds, this cannot be obtained along the lines suggested by Mahnke, according to whom the symbolic realm captures "by broad strokes" the intended, but inaccessible domain of posits. In this Becker seems to find support in Weyl's position, with whom he also shares the philosophical problem of how to account for the effectiveness of Hilbert's transfinite mathematics in physics. The discussion of the interpretation of Hilbert's ideal elements does not add much to the exposition in "Mathematische Existenz" but corrects one passage of the book by pointing out that Hilbert's posits are not, as indicated in the book, "Widersinn" but rather "Leersinn".

Mahnke's reply to the last letter is the last item of philosophical interest in the correspondence. In it Mahnke rejects Becker's accusation that he and Leibniz collapse God's knowledge (which can reach to the actual infinite) with that of the demon (potentially infinite). Through a long series of quotations he shows that Leibniz kept the two qualitatively apart, i.e. that God's knowledge is essentially different from that of a demon. This implies, and Mahnke agrees with Becker on this, that God's knowledge is also essentially different from that of a finite human being. But he continues: "However, would it not be possible that we approximate it [God's knowledge] symbolically, although inadequately but still more and more adequately?" In the rest of the letter Mahnke tries to spell out what is implicit in the suggestion. In the previous letter Becker had defended the internal consistency of allowing two types of procedures: the intuitive construction of mathematics and physics, and then also the "symbolic" interpretation of those parts of mathematics and physics that, although not intuitively justified, can be used to interpret nature. Mahnke concludes his letter also agreeing that both forms of procedure have to be allowed but once again he sees in the latter activity not something mysterious but rather a fact about how we can obtain intuitive knowledge of a transcendent reality:

> A transcendent reality (especially if it is also transfinite) is not accessible in the sense of constitutive and hermeneutical phenomenology. We stay outside and we must try, which one of the keys we freely created will work. But when we have opened the door with the key freely created by us then we see also the reality that was previously barred to us [...]

In the reply to Mahnke's letter, Becker (letter 6) says that in his article "The symbolic in mathematics" he will go back to some of the issues raised by Mahnke (Mahnke 1927a). However, on one things he decisively disagrees with Mahnke: there can be no continuous passage between the potential transfinite of Cantor's ordinal numbers and Hilbert's transfinite.

In letter 7 Becker thanks Mahnke for having sent the review of Dilthey[23] and congratulates Mahnke on having shown the connection between Husserl and Dilthey and for seeing Becker's and Heidegger's work in the direction of a synthesis of Husserl and Dilthey.

This marks the end of the philosophical discussion in the letters. The remaining letters in the correspondence deal mainly with practical matters, related to invitations to lecture and publications.

Conclusion

The above short summary of the correspondence will, I hope, serve as a first guide into the topics discussed. Obviously the two main problems discussed are the relationship between intuition and formalism in the phenomenological foundation of the sciences and, connected to it, which phenomenological approach (Husserlian, hermeneutic, mantic) can best do justice to the specificity of mathematical and physical knowledge. These problems give rise to an incredibly rich correspondence which includes, among other things, the role of ideal elements in Hilbert's proof theory, the limitations of Husserl's formal ontology for a phenomenological foundation of the exact sciences, the role of intuitive knowledge in mathematics and physics, the adequacy of intuitionistic and semi-intuitionistic mathematics for contemporary mathematics and physics, Leibniz's understanding of the relationship between the human mind and God, and many other topics. Since many of the issues discussed in the correspondence are still topical, I hope that each reader will find something to profit from and enjoy in the Becker-Mahnke exchange.

PART IV

Tarski and Quine on Nominalism

Introduction

SUMMARY

Part IV discusses nominalism, a metaphysical position that denies the existence of abstract entities. In the analytic literature, an article by Goodman and Quine of 1947 is usually regarded as the *fons et origo* of the contemporary debate over nominalism. While there is some awareness that earlier thinkers, like Lesniewski and Chwistek, also held nominalistic views, the history of nominalism in the twentieth century has been largely neglected. In this context one can also read Chapter 3 as a study of an early approach to nominalism based on a nominalistic reading of the theory of types. But here the focus is on Quine and Tarski. Tarski, Carnap, and Quine spent the academic year 1940–41 together at Harvard. In their autobiographies, both Carnap and Quine highlight the importance of the ensuing conversations that took place among them. Their discussions centered on semantic issues related to the analytic/synthetic distinction and on the project of a finitist/nominalist construction of mathematics and science. Carnap's *Nachlaß* in Pittsburgh contains more than 80 typescript pages based upon Carnap's detailed notes of these conversations. In Chapter 13 I present an analysis of these notes with special emphasis on Tarski's rejection of the analytic/synthetic distinction, the passage from typed languages to first-order languages, Tarski's finitism/nominalism, and the construction of a finitist/nominalist language for mathematics and science. These debates were quite influential in the later development of nominalism, for instance they gave rise to the program pursued by Goodman and Quine in 1947. The study of the nominalistic engagement of Quine and Tarski is also pursued in Chapter 14 where I widen the focus to study the trajectories of Quine and Tarski on nominalism up to 1953 (the date of an important conference on 'Nominalism and Platonism in Contemporary Logic' organized by Beth in Amersfoort). Through the use of the Quine archive and some other important sources I reconstruct the dialectic of caution and engagement that characterized the approach to nominalism of these two giants of analytic philosophy.

BIBLIOGRAPHICAL UPDATE

Given the recent date of publication of Chapters 13 and 14 there are only a few new items of interest. Frost-Arnold (2008) discusses Tarski's nominalism and Frost-Arnold (2009) contains a section on Quine's trajectory in the rejection of the analytic-synthetic distinction which discusses and reconciles the accounts of such development by Creath and myself (see Chapter 13). Sinaceur (2008) discusses Tarski's nominalism in the context of certain formalistic tendencies found in Tarski's work. The comments delivered by Tarski in Chicago in 1965 are now published in Rodriguez-Consuegra (2007).

13

HARVARD 1940–1941: TARSKI, CARNAP, AND QUINE ON A FINITISTIC LANGUAGE OF MATHEMATICS FOR SCIENCE

13.1 Introduction

In his autobiography, Quine writes:

The fall term of 1940 is graven in my memory for more than just the writing of *Elementary Logic*. Russell, Carnap, and Tarski were all at hand. Tarski had a makeshift research appointment at Harvard and was in need of a job. Carnap had a visiting professorship with us. Russell was giving the William James Lectures, backed by a seminar. (Quine 1985, p. 149)[1]

Quine then describes how Carnap proposed discussion of the manuscript of his forthcoming "Introduction to Semantics" (Carnap 1942) to give structure to small meetings which included Carnap, Quine, Tarski, Russell, and sometimes Goodman, Hempel, or John Cooley. Tarski and Quine found themselves disagreeing with Carnap:

I would like to thank Sol Feferman, Greg Frost-Arnold, Marcus Giaquinto, Daniel Isaacson, John MacFarlane, and Thomas Ryckman for useful comments on a previous draft of this chapter. After my research for this chapter had been completed, I became aware that Greg Frost-Arnold (University of Pittsburgh) was working on a dissertation on the Carnap notes. Once we got in touch, he kindly made remarks on a preliminary draft of this chapter, which I have incorporated by acknowledging his suggestions in the appropriate places. All passages quoted from the Carnap *Nachlaß* are quoted with the permission of the University of Pittsburgh (all rights reserved). Thanks also to the Bancroft Library (U. C. Berkeley) for permission to quote from the Tarski Collection.

My misgivings over meaning had by this time issued in explicit doubts about the notion, crucial to Carnap's philosophy, of an *analytic* sentence: a sentence true purely by virtue of the meanings of its words. I voiced these doubts, joined by Tarski, before Carnap had finished reading us his first page. The controversy continued through subsequent sessions without resolution and without progress in the reading of Carnap's manuscript. (Quine 1985, p. 150)

This succinct description of the meetings between Tarski, Carnap, and Quine is bound to pique our curiosity but it leaves us in the dark as to the nature of the discussion. Carnap is fortunately more forthcoming in his autobiography about the events, to which he comes back in different parts of his narration. The first mention of these meetings resembles the description given by Quine:

During the year 1940–1941 I was a visiting professor at Harvard. During the first semester Russell was there too, giving the William James lectures, and I was glad to have an even better opportunity for talks with him on questions of philosophy as well as on social and political issues. Tarski spent the same year at Harvard. We formed a group for the discussion of logical problems; Russell, Tarski, Quine and I were its most active members. I gave several talks on the nature of logic and on the possibility of defining logical truth as a semantical concept. I discovered that in these questions, even though my thinking on semantics had originally started from Tarski's ideas, a clear discrepancy existed between my position and that of Tarski and Quine, who rejected the sharp distinction I wished to make between logical and factual truth. (Carnap 1963, pp. 35–36; see also pp. 64–65)

So far the account resembles Quine's. However more was discussed:

In other problems we came to a closer agreement. I had many private conversations with Tarski and Quine, most of them on the construction of a language of science on a finitistic basis. (Carnap 1963, p. 36)

In this paper I will try to flesh out the above summary by referring to a set of extensive notes Carnap took while these discussions were taking place. I will begin with the issues on semantics and then move on to the second aspect of the discussion, which will be my main focus of interest. However, in the next section I will show that Tarski's doubts about the distinction between "logical and factual truth" had been a long standing concern of Tarski and that Carnap was aware of them since their first meeting in 1930. Finally, a disclaimer before we start. I have chosen to emphasize only two major sets of topics in these extensive discussions. Thus, several topics of discussion have been simply hinted at or passed over in silence. Only an edition of these texts (which is being prepared by Greg Frost-Arnold) will be able to do full justice to the richness of the material they contain. Moreover, my major interest is in Tarski and thus although I will try to keep a balance in the exposition, I will often emphasize issues of interest to Tarskian themes (for a recent account of Tarski's life and work see Feferman and Feferman (2004); for issues related to Tarski's philosophical views see Woleński (1993, 1995b) and Mancosu (2009, 2008)). I am sure other scholars will consult the same materials from other points of view which might shed more light on Carnap and Quine or other aspects of the discussions that I have not chosen to emphasize.

13.2 Tarski on the Carnapian Distinction between Analytic and Synthetic

Let me begin by pointing out that while Quine's doubts about the distinction between analytic and synthetic were probably just coming to the surface, in the case of Tarski this was a longstanding worry. In a note of Carnap's diary, dated February 22, 1930 we read:

8–11 with Tarski at a Café. About monomorphism, tautology, he will not grant that it says nothing about the world; he claims that between tautological and empirical statements there is only a mere gradual and subjective distinction. (quoted in Haller 1992, p. 5)[2]

The objection is raised once again in 1935 in Paris. In Neurath's summary of the discussion at the Paris Congress published in *Erkenntnis*, we read:

In the discussions the "analytic" vs. "synthetic" opposition came to the fore repeatedly. In reply to comments by Tarski to the effect that one could not formulate the distinction sharply, Carnap explained that it was possible to include already in the metalanguage (Syntax) certain signs which are usually seen as "descriptive" (house, temperature and so on) and to use them as "logical signs" so that the sentences in which these terms occur would be determined on the basis of the language rules. He added that it is really a question how—on the basis of the "feeling" which gives rise to the opinion one could draw a boundary between words of the German language such as "und", "oder" etc. on the one side and "Temperatur", "Haus" on the other—one could arrive at a formulation which would allow us to distinguish rigorously "descriptive" and "logic" and then also "synthetic" and "analytic". This is a problem that can be discussed only in rigorous form if we want to avoid misunderstandings of all sorts. (Neurath 1936, pp. 388–89)[3]

Of course, Tarski in his 1936 paper on logical consequence will explicitly mention the fact that no criterion is known for distinguishing logical from non-logical constants. Indeed, the report by Neurath quoted above reproduces some of the discussion which took place after Tarski's presentation (in Paris in 1935) of his paper "On the concept of logical consequence". This we learn from a letter from Tarski to Neurath:

I have questioned there [in my lecture on logical consequence] the absolute character of the partition of concepts into logical and descriptive as well as that of sentences into analytic and synthetic. I have endeavoured to show that the partition of the concepts is quite arbitrary and the partition of sentences should be relativized to that of the concepts. (Tarski to Neurath, Warsaw, 7.ix.36)[4]

Tarski was not to change his mind on these issues. We have Carnap's typescript of a "Gespräch mit Tarski" which took place in Chicago on 6 March 1940. Tarski is making the same point as in 1935:

'L-true'. 'logic-descriptive'. I [Carnap]: my intuition, clearer in the distinction L-true-F-true, than in logic-descriptive. But I can always explain the latter through displaying [Aufweisung] the logical constants in the usual systems and the claim that everything that can be defined from them should be logical. He [Tarski]: he has no such intuition; one could just as well consider "temperature" as a logical sign. I [Carnap]: the truth of a full sentence [Vollsatz] of the temperature function is determined by measurement. He [Tarski]: one can however decide to [unreadable] to a stipulated truth theory despite all observations. I [Carnap]: Then it is a mathematical function and a logical sign and not the physical temperature function. In the case of a full sentence [Vollsatz] of the physical temperature function one cannot find the truth value by mere computation. He [Tarski]: this proves nothing, for also in the case of mathematical functions this is often also not possible, for there are undecidable sentences; no fundamental distinction between mathematical but undecidable sentences and factual sentences. I [Carnap]: it seems to me there is indeed a difference. (RC 090-16-09 Dated 6 March 1940, Chicago, Conversation with Tarski [p. 4])[5]

We see here how Tarski's position on the analytic-synthetic distinction leads him in the direction of the claim that if we treat certain physical constants as logical, then certain statements about temperature might become unrevisable despite all observations. But this is just the other side of the coin of claiming that logical propositions might be just as revisable as the physical ones. This is a position that Tarski defended, as it is clear from the letter to Morton White in 1944 (Tarski 1987). Indeed, Tarski asserted this position already in front of the Vienna Circle in 1935. In a letter from Neider to Neurath dated 29 June 1935, Neider gives a summary of a discussion led by Schlick of Hempel's 1935 article on truth within the Vienna Circle and quotes Tarski as saying that he had never uttered a sentence which he had not considered to be revisable. ("Tarski: Ich habe noch nie einen Satz gesprochen, dessen Korrigierbarkeit ich ausgeschlossen habe").

In the preface to his publication of Tarski's letter, Morton White hints at the fact that Tarski's views had an obvious influence on Quine and himself. He pointed at two aspects contained in Tarski's 1944 letter. The first was the issue of the revisability of logic and mathematics on a par with the revisability of physical theories. The second aspect was the distinction between analytic and synthetic statements. I think the evidence provided above shows more explicitly that Tarski had been questioning the central distinction in Carnap's philosophy already for 10 years before the 1940–41 meetings took place and that his positions on the lack of a sharp distinction between the analytic and the synthetic and the revisability of both logic and physics predate Quine's position on these issues. It is of course the case that the rationale for this rejection differed in the case of Tarski and Quine. This, however, raises the question of when exactly Quine arrived at the criticism of the analytic-synthetic distinction. Creath seems to distinguish in his introduction to Creath (1986) between "doubts" (see p. 33 of Creath 1986) and "explicit rejection" (Creath 1986, p. 35) and dates the former to 1940–41 (in the conversations that are the subject of this paper) and the latter to 1947. Obviously, I cannot enter here into a detailed reconstruction of Quine's trajectory with respect to the analytic-synthetic distinction but I would like to report a passage from a letter to Woodger in 1942, which seems to me to show that Quine at the time saw his paper "Truth by convention" (Quine 1936) as already containing an explicit rejection of the analytic-synthetic distinction. The passage will also be useful to indicate some of the topics that were discussed in 1940–41 and to which we will turn soon. Quine writes:

Last year logic throve. Carnap, Tarski and I had many vigorous sessions together, joined also, in the first semester, by Russell. Mostly it was a matter of Tarski and me against Carnap, to this effect. (a) C's professedly fundamental cleavage between the analytic and the synthetic is an empty phrase (cf. my "Truth by convention"), and (b) consequently the concepts of logic and mathematics are as deserving of an empiricist or positivistic critique as are those of physics. In particular, one cannot admit predicate variables (or class variables) primitively without committing oneself, insofar to the "reality of universals", for better or worse; and meanwhile C.'s disavowal of "Platonism" is an empty phrase (cf. my "Description and Existence"). Other points on which we took C. to task are (c) his attempt to make a general semantics rather than sticking to a convenient canonical form for object languages and studying the semantics thereof more simply and briefly and yet more in detail; (d) his resuscitation of intensional functions. C. argued reasonably and well, as always, and the discussions were good fun. (Quine to Woodger, 2 May 1942, Woodger papers)

Thus, it does seem to me that already in 1940–41 Quine had explicitly rejected the notion of analyticity, and in 1942, he considered that rejection to be already in his 1936 paper "Truth by convention".

13.3 Harvard 1940–1941

We have seen that both Carnap and Quine emphasize the importance of the discussion concerning "Introduction to Semantics". This likely took place in the Fall of 1940. However, except for the already mentioned "Gespräch mit Tarski" in Chicago (which antedates the academic year 1940–41), the folders I was able to locate in the Carnap archive do not report on the particular aspects of the discussion related to Carnap's typescript (the notes Carnap took on Russell during fall 1940 do not deal directly with his typescript; see RC 102-69-01 to RC 102-69-08). Carnap wrote extensive reports on the meeting which took place in 1940–41 and I think we have a rather complete set of reports for the meetings from October 1940 to late Spring 1941. They are contained in two folders found in the Carnap archive. The first set of documents is classified under RC 090-16-02 to 090-16-30 and has the general title "Tarski (and Quine), Finitism, discussion 1941" . Moreover, a different set of documents is relevant and represents material on Quine and Carnap in 1941: RC 102-63 "Quine disc., 1940–1941".

There is however a part of the reports which is obviously related to the discussion on Carnap's "Introduction to semantics". This has to do with the problem of deciding for which category of languages one should provide a semantical treatment. In this part of the exchange, we notice a remarkable insistence on the part of Tarski, and Quine, that one should not aim at too general a class but rather restrict oneself to type-free (first-order) languages of a certain specified form. I believe these documents bear witness to the important shift which took place in logic during this period from type-theoretic languages to first-order languages as paradigm [see also RC 090-16-08 dated 26 May 1941]. Here is a report from 9 December 1940:

Tarski and Quine: general observation on general semantics:

It is probably not at all worth it to relate the definitions and theorems of a system of general semantics to the class K of all languages that can be treated in M. Rather one should preferably relate them only to a subclass K' which is such that:

(1) Each language in K (or each one that we want to take into consideration) is translatable in a language in K';

(2) All languages in K' have a certain usual structure.

The simplest and for all practical goals sufficient: we take into consideration only languages with individual variables and constants, predicate constants and identity; of course, connectives and quantifiers. Thus lower functional calculus (but without predicate variables [...]). Justification for this: with the help of the special relation 'ε' (which however is not here presupposed as occurring in each such language) we can translate set theory in the lower functional calculus. This is the *difference between logic and mathematics: Mathematics=Logic* + 'ε'. By means of 'ε' the system becomes non-finitistic and incomplete. (RC 090-16-03) [6]

Thus, a general semantics can restrict itself to giving semantics for first order languages. The motivation adduced is that all of mathematics can in fact be translated in a first-order theory with the axioms for the membership relation, e.g. set theory. This leads to a distinction between logic and mathematics. All these points deserve careful scrutiny, if we are to understand the importance of the shift here involved.[7]

Type-free languages

Tarski's 1933 article on truth is built around the type theoretic structure of the languages for which a definition of truth is sought. Let us recall that in §3, Tarski deals with a theory of classes containing only variables over individuals; in §4 he extends the treatment to languages of (bounded) finite type; and in §5 he deals with languages which have all finite types in it. As is well known, for the latter class he claims that a definition of truth cannot be given except axiomatically. However, in the Postscript, added in the 1935 version in German, he basically renounces the theory of semantical categories that had been the rationale for the structure of the article on truth.[8] In the Postscript, we already have a description of the two steps involved here. In the main text of the Postscript, Tarski suggests adding to the typed languages "variables of indefinite order which, so to speak 'run through' all possible orders"(p. 271). Then, in note 1, he adds:

From the languages just considered it is but a step to languages of another kind which constitute a much more convenient and actually much more frequently applied apparatus for the development of logic and mathematics. In these new languages all the variables are of indefinite order. From the formal point of view these are languages of a very simple structure; according to the terminology laid down in §4 they must be counted among the languages of the first kind, since all their variables belong to one and the same semantical category. Nevertheless, as is shown by the investigations of E. Zermelo and his successors (cf. Skolem, Th. (66), pp. 1–12), with a suitable choice of axioms it is possible to construct the theory of sets and the whole of classical mathematics on the basis provided by this language. (Tarski 1935a, p. 271)

One might be surprised here and wonder why, given that Zermelo's axiomatization had been around since 1908, it took so long for Tarski to give up type theoretic languages and focus on first-order languages. However, one must keep in mind that Zermelo's intuitive interpretation of Zermelo-Fraenkel set theory using the cumulative hierarchy was only given in Zermelo (1930). Two passages of Carnap's typescript of the conversations are relevant here. In the conversation between Tarski and Carnap in Chicago in 1940, Zermelo's set theory is singled out as the best form of type-free system:

Logic without types. The best form is that developed originally by Zermelo; on its basis we now have the improved systems of Bernays (distinction between classes and sets) and Mostowski (without this difference). Quine in his system has too many special truths (for example, Cantor's theorem does not hold) which scare away the mathematicians and that show that the system is not suitable. (RC 090-16-09)[9]

The continuation of the conversation is a prelude to issues that we will discuss later on finitism:

I [Carnap]: Should we construct the language of science with or without types?
He [Tarski]: Perhaps something else will emerge. One would hope and perhaps conjecture that the whole general set theory, however beautiful it is, will in the future disappear. With the higher types Platonism begins. The tendencies of Chwistek and others ("Nominalism") of speaking only of what can be named are healthy. The problem is only how to find a good implementation.[10]

I believe that what Tarski here calls general set theory is nothing else than the general theory of classes presented in §5 of the truth article (see for instance his note 2 on p. 210 for the 1956 edition where he refers to that system as general theory of sets). Tarski repeats the point about the Platonist commitment involved in dealing with higher types several times during these conversations. For instance:

Tarski: A Platonism underlies the higher functional calculus (thus the use of a predicate variable, especially of higher type) (RC 102-63-09)[11]

Another important detail about the passage from type-theoretic systems to systems without types emerges from a conversation between Tarski and Carnap dated 13 February 1941. Tarski reminisces on the Warsaw group and the attachment to simple type theory. Here is Carnap's summary:

The Warsaw logicians, especially Leśniewski and Kotarbiński saw a system like PM (but with simple type theory) as the obvious system form. This restriction influenced strongly all the disciples; including Tarski until the "Concept of Truth" (where the finiteness of the level is implicitly assumed and neither transfinite types nor systems without types are taken into consideration; they are discussed only in the Postscript added later). Then Tarski realized that in set theory one uses with great success a different system form. So he eventually came to see this type-free system form as more natural and simpler. (RC 090-16-26)[12]

As for the distinction between logic and mathematics we will come back to the issue later.

13.4 Quine's Lecture "Logic, Mathematics, and Science"

Let us go back to the second part of the discussions. Carnap describes these discussions as having to do with his reflections on constructing a system of science (or of a specific particular science) as "a calculus whose axioms represent the fundamental laws of the field in question":

> This calculus is not directly interpreted. It is rather constructed as a "freely floating system", i.e. as a network of primitive theoretical concepts which are connected with one another by the axioms. On the basis of these primitive concepts, further theoretical concepts are defined. Eventually, some of these are closely related to observable properties and can be interpreted by semantical rules which connect them with observables. (Carnap 1963, p. 78)

In developing this project Carnap was led to structure the language into two parts. One part was the observation language, which was presupposed to be completely understood. The other part was the theoretical language. This is where the discussions in 1940–41 become relevant again:

> My thinking on these problems received fruitful stimulation from a series of conversations which I had with Tarski and Quine during the academic year 1940–1941, when I was at Harvard; later Nelson Goodman participated in these talks. We considered especially the question which form the basic language, i.e. the observation language, must have in order to fulfill the requirement of complete understandability. We agreed that the language must be nominalistic, i.e., its terms must not refer to abstract entities but only to observable objects or events. Nevertheless, we wanted this language to contain at least an elementary form of arithmetic. To reconcile arithmetic with the nominalistic requirement, we considered among others the method of representing the natural numbers by the observable objects themselves which were supposed to be ordered in a sequence; thus no abstract entities would be involved. We further agreed that for the basic language the requirements of finitism and constructivism should be fulfilled in some sense. We examined various forms of finitism. Quine preferred a very strict form; the number of objects was assumed to be finite and consequently the numbers occurring in arithmetic could not exceed a certain maximum number. Tarski and I preferred a weaker form of finitism, which left it open whether the number of all objects is finite or infinite. Tarski contributed important ideas on the possible forms of finitistic arithmetic. (Carnap 1963, p. 79)

One has to be careful to take such a passage as a faithful account of what went on in 1940–41. The autobiography was written almost 20 years after the facts, and it does not always match the content of the documents. For instance, Frost-Arnold has pointed out to me that 'observability' does not seem to play the role in the 1940–41 discussions that Carnap attributes to it in the above passage. While I am sympathetic to the point I should however add that in one instance (see section 13.7) Carnap describes his finitism, in opposition to Tarski's, by appealing to observation statements (see also the main citation in section 13.6).

A good point of entry in much of the following discussion is a lecture given by Quine on 20 December 1940. Carnap writes a five-page report on this lecture given to the Logic Group. In Carnap's notes, the talk is entitled "Logic, Mathematics, Science".[13] The talk starts by going back to an issue already discussed in previous sessions: "general semantics must be limited, otherwise it becomes trivial". The suggestion is to investigate languages that contain only the following:

[Constants, predicates], joint denial, universal quantification. Only individual variables; and only formulas with no free variables. (RC 102-63-04)[14]

As we have seen, the restriction to languages with only one type of variables is often discussed by Carnap, Tarski, and Quine during this period. After the long engagement with type-theoretic systems, the obvious question was whether this might not be too restrictive. Quine's answer is no, as in these languages one can translate all of mathematics (in the form of a theory of the membership relationship 'ε'), protosyntax (with a predicate 'M') and general syntax, again with 'ε'. In effect, this work is carried out in Quine's *Mathematical Logic*, which had just appeared.

Quine refers to *Mathematical Logic* (ML) for the idea that constants and functors can be eliminated through contextual definitions. The advantages, as spelled out in ML, are both technical (a simplification of the theory of quantification) and philosophical (questions of meaningfulness are separated from questions of existence [ML, p. vii]). However, one must also address the more general question: can all languages be translated into such format? This is the problem of the "extensionality thesis".

The general suggestion for demarcating logic, mathematics and science is given as follows:

Logic = Theory of joint denial and quantification
Mathematics = Logic + theory of ε.
Physics = Logic + Mathematics + Theory of other predicates.

A side comment indicates that one could characterize the above opposition between logic and mathematics as that between elementary logic and logic but that perhaps using 'logic' in the narrow sense is more in agreement with the spirit of the classical tradition.

Quine then remarks: "the boundary between logic and the other sciences (including mathematics) is important!"

Let me interrupt this summary by pointing out that in *Elementary Logic*, written in 1940, Quine divides logic into three areas: the theory of statement composition (propositional logic), the theory of quantification, and "the theory of membership". Then he adds:

But there is equal justice in an alternative classification, whereby logic proper is taken as comprising just the first two of those three parts, while the theory of membership is placed outside logic and regarded as the basic extra-logical branch of mathematics.[1] Whether we construe "logic" in the tripartite way or in this narrower bipartite way is a question merely of how far we choose to extend the catalogue of "logical locutions" alluded to earlier. According to the wider version, logic comes to include mathematics [2]; according to the narrower version, a boundary survives between logic and mathematics at a place which fits pretty well with traditional usages. (Quine 1941, p. 3)

In note 1, Quine credits Tarski with having urged on him the point he just made. Overall, Quine thinks that it is just a matter of convenience what we decide to treat as logic and I suspect that Tarski would agree. For instance, in 1944, Tarski writes to Morton White that "sometimes it seems to me convenient to include mathematical terms, like the ε-relation, in the class of logical terms, and sometimes I prefer to restrict myself to terms of 'elementary logic'. Is any problem involved here?" (Tarski 1987, p. 29; see also

Tarski 1986b) Moreover, on account of the principle of tolerance, Carnap would also agree with this.

Let's go back to Quine's lecture, which I will summarize by giving a close paraphrase of it as reported in Carnap's notes. Quine points out two important distinctions between logic (now in the sense of quantification theory) and mathematics. First, he claims that there are no logical propositions since there are no logical predicates. The claim is that the investigation of logical procedure is metatheoretical. The variables 'p', 'q', 'F', 'G' do not occur in sentences. In their stead, we have corresponding syntactic signs. By contrast, there are pure mathematical propositions with 'ε', for instance '$(x)x\varepsilon x$'. This is how one obtains content. Thus, mathematics has subject matter, content, whereas logic concerns only form. (This point of view is spelled out on p. 127 of *Mathematical Logic* (1940))

Second, logic does not demand any special objects, not even a determined size of the universe of discourse; if a sentence is logically true from the point of view of an infinite domain, then it must hold also on all finite domains and viceversa. The logical truths, Quine goes on to claim, are valid for almost all philosophies, including nominalism and realism; the exception is intuitionism, which however might be satisfied through extra-logical limitations (for instance by avoiding the use of non-constructive predicates).

Only predicates, according to Quine, make ontological demands but not because they denote. They are actually syncategorematic, for there are no variables which range over them. Rather, a predicate demands specific objects as values for its argument variables (this position stems from Quine's 1939 article "Designation and Existence"); so, for instance, 'ε' demands classes, universals; on this account while mathematics is committed platonistically, logic is not.

We see that going back to the previous exchange where Tarski claims that as soon as one has quantification over predicates, one is committed to Platonism, Quine resolves the issue by not allowing such quantifications in logic proper. But according to Quine, first-order quantification over classes also commits us platonistically. Another advantage of restricting attention to first-order theories, he claims, is that on account of Gödel's completeness theorem for first-order logic, we have a syntactical definition of 'logical truth' and logical consequence.

Let me remark that we are witnessing here a series of arguments which led to the demise of type-theory as the fundamental background logic in favor of first-order theories. The notes we are examining show (and Quine acknowledges this) that Tarski was certainly pushing in this direction. It is however unfortunate that the notes report no comments by Tarski on how the new position fits with the theory of logical consequence given in Tarski (1936d) (see Mancosu 2006). I should also add that a discussion as to the best form for presenting first-order systems also fills three pages of a discussion which took place on 26 May 1941 (RC 090-16-08) with Tarski and Quine exploring several possibilities.

The lecture continues by addressing the problem of what happens when one moves from logic to mathematics and science. Quine says that one introduces suitable predicates (mathematical, physical, biological...) and axioms. The theorems are the logical consequence of axioms. Logic, he claims, then becomes that common part of all (non-trivial) theories. Thus, all of mathematics can be formalized with the relation 'ε', and the theorems of mathematics are the logical consequences of the mathematical axioms. However, because of Gödel's incompleteness theorem, they do not exhaust mathematical truth.

Moving now to the universal language of science, Quine says that here we are confronted with a large number of predicates. Ontologically speaking, the values of their variables have different objectivity, including electrons, atoms, bacteria, tables, sense-qualities and objects which are not things: centimeters, distances, temperatures, electric charges, energy, straight lines, points, classes (or properties).

While some people find some universals more problematic than others and therefore reduce one to the others (for instance, Whitehead reduces points to volumes; Carnap and Jeffreys reduce distances and temperature to pure numbers), Quine remarks that, in his opinion, all universals have ultimately the same nature as points, centimeters and so on, i.e. they are non-things. 'Classes are probably no exception'. He does not insist on eliminating classes or other non-things objects, for they might be necessary for science. This position obviously foreshadows acceptance of certain parts of platonist mathematics based on an "indispensability" argument. Quine agrees with Carnap that one should accept as ineliminable the non-positivistic or non-phenomenalistic language of science.

> I conjecture that Carnap is right: there is only partial clarification, not complete definitional elimination. This clarification takes place by means of the investigation of the relation of confirmation between statements of the more remote and the more immediate type. (RC 102-63-04)[15]

The above obviously refers to the analysis of confirmation given by Carnap in "Testability and Meaning" (Carnap 1936). The picture of science that emerges from Quine's lecture is then summarized as follows:

> Science is replete with Myth and Hypostasis; Goal: to embed the chaotic behaviour of ordinary things in a more understandable superworld [*Überwelt*]; task: prediction with respect to ordinary things; this is pyschologically possible only as a consequence of the greater "perspicuity" [*Übersichtlichkeit*] of the superworld, which is constructed from science as an intermediary device. The trichotomy: phenomena, common sense world, superworld of science holds only roughly; it is a question of degree. Tables are hypostases just as electrons, but in a lower degree. From the superworld we can infer facts about the ordinary world. But not vice versa (underdetermined); just in the same way the ordinary world is underdetermined by experience. (RC 102-63-04)[16]

Quine goes on to conclude that he sees mathematics in the same way, with the theory of real numbers proving itself in the point of contact with the theory of rational numbers. Moreover, the general theory of classes gives common-sense results for finite classes which are parallel to common sense laws about heaps. But by this the general theory of classes is not univocally determined. For this reason, one has to consciously search for myths such as Russell's, Zermelo's and Quine's systems.

Quine's last remark concerns the fact that the above reflections have allowed him to see Gödel's incompleteness theorem as less of an anomaly than it had appeared to him before. I think the reason here is that our formal systems containing arithmetic are underdetermined with respect to their possible completions just as physical theories are underdetermined with respect to experience.

13.5 The Discussion

This lecture of Quine was the starting point for several discussions. A report dated 20 December 1940 states:

Could not we perhaps consider the higher non-finitary parts of logic (mathematics) in such a way that their relation to the finitary parts is analogous to the relation between the higher parts of physics to the observational sentences? In this way non-finitistic logic (mathematics) would become non-metaphysical (just as physics). Perhaps in this way one could shed light on the question whether there are fundamental differences between logic-mathematics and physics.[17]

This is an interesting way to characterize the issue. It brings to mind two different precursors. First of all, in 1928, Hilbert proposes that the non finitistic part of mathematics can be related to the finitistic parts as the theoretical parts of physics are related to the statements that can be checked directly by observation. But we are also reminded of the positions of Russell and Gödel on 'inductivism' in the foundations of mathematics, where the analogy goes as follows. Just as in physics the theoretical part is introduced to 'explain' the observations, so in the foundations of mathematics certain class or set-theoretical assumptions might play an explanatory role with respect to ordinary, not necessarily finitistic, mathematics (see Mancosu (2001) for a treatment of this conception).

Logical consequence

The discussion continues on 10 January 1941 (RC 102-63-12) where the topic is also whether one should not have a system of logic in which a universal sentence follows from all its instances. This is an important document for the problem of the logical validity of the ω-rule but as it is tangential to our main topic I will not treat it here. However, I should mention a different discussion on logical truth and criteria for logical validity.

In this meeting, Alexander Wundheiler, a Polish philosopher who was present at the meeting, recalls that according to a paper by Tarski and Lindenbaum ('über die Beschränkung der Ausdrucksmittel...') the logically true sentences in a domain of individuals are those that are preserved under an arbitrary automorphism of the domain. Quine agrees and adds that the mathematically true sentences are those in which ε is preserved. Carnap adds that this does not say much more than the claim that mathematics (in a certain formulation) is characterized by ε.

Wundheiler then asks whether one could not characterize the distinction between logic, mathematics and physics by means of the notion of group of transformations just as one characterizes projective, [affine?] and metric geometry through transformation groups.

Tarski replies that it is doubtful whether in this connection the group concept helps much. However, we know that this idea of using the basic strategy of Klein's Erlanger program in geometry (that is looking at the group of automorphisms of a domain) was exploited later by Tarski for characterizing logical notions (see Tarski 1986b)[18] Probably, what Tarski saw as unpromising was simply the use of the group concept to distinguish between logic, mathematics and physics. In any case, Wundheiler's proposal seems to be the first time that the suggestion is made that logic be demarcated through transformation groups (Mautner's "An extension of Klein's Erlanger program: logic as invariant theory" dates from 1946).

At the end of the meeting, the following question is raised: "how should one characterize the distinction between mathematics and physics?"[19]

13.6 Tarski's Finitism (RC 090-16-28)

While it is a known fact that Tarski had nominalistic sympathies, we have very little information about which form they took. Summarizing Tarski's engagement with philosophy, Mostowski writes:

Tarski, in oral discussions indicated his sympathies with nominalism. While he never accepted the "reism" of Tadeusz Kotarbiński, he was certainly attracted to it in the early phase of his work. However, the set-theoretical methods that form the basis of his logical and mathematical studies compel him constantly to use the abstract and general notions that a nominalist seeks to avoid. In the absence of more extensive publications by Tarski on philosophical subjects, this conflict appears to have remained unresolved. (Mostowski 1967, p. 81)

Tarski himself joked about this on the occasion of a series of comments he made in an ASL meeting in Chicago in 1965. There talking about his anti-Platonism, he said:

I happen to be, you know, a much more extreme anti-Platonist. [...] However, I represent this very [c]rude, naïve kind of anti-Platonism, one thing which I could describe as materialism, or nominalism with some materialistic taint, and it is very difficult for a man to live his whole life with this philosophical attitude, especially if he is a mathematician, especially if for some reasons he has a hobby which is called set theory (p. 3, Transcript of remarks, ASL meeting, Chicago, Illinois, 29 April 1965, Bancroft Library)

The anti-Platonistic attitude is also central in the discussions with Carnap and Quine and it is one of the major sources motivating the construction of a finitist system for mathematics and science. But the Carnap transcripts we are looking at give us a richer sense of Tarski's nominalism and finitism (at least around 1940). Before proceeding, I ought to point out that no clear distinction is made in the Carnap notes (and thus in the discussions) between nominalism and finitism. Most of the time, the discussion is about finitism, but some of the features of the finitist position should be, at least from our point of view, classified as nominalistic. Moreover, one of the defining features of the finitism discussed in the Carnap notes centres on the notion of 'understanding', and thus it is very different from Hilbert's finitism (or some other varieties of finitism) where this notion plays no role.

One aspect of this Tarskian finitism emerges already in a note to the truth article (note 94 in original German; note 1 p. 253 in the 1956 translation):

In the course of our investigation we have repeatedly encountered similar phenomena: the impossibility of grasping the simultaneous dependence between objects which belong to infinitely many semantical categories; the lack of terms of 'infinite order'; the impossibility of including in *one* process of definition, infinitely many concepts and so on [...]. I do not believe that these phenomena can be viewed as a symptom of the formal incompleteness of the actually existing languages—their cause is to be sought rather in the nature of language itself: language, which is a product of human activity, necessarily possesses a 'finitistic' character, and cannot serve as an adequate tool for the investigation of facts, or for the construction of concepts, of an eminently 'infinitistic' character. (Tarski 1956, p. 253)

Let us then continue with the 1940–1941 discussions in order to capture what is the nature of the 'finitism' involved here. On 10 January 1941, Tarski comments on the issue of finitism as follows:

I understand basically only languages which satisfy the following conditions:

1. Finite number of individuals
2. Realistic (Kotarbiński): the individuals are physical things;
3. Non-platonic: there are only variables for individuals (things) not for universals (classes and so on)

Other languages I "understand" only the way I "understand" [classical] mathematics, namely as a calculus; I know what I can derive from other [sentences] (or have derived; "derivability" in general is already problematic). By any higher, "platonic" sentences [Aussagen] in a discussion I always interpret them as sentences that a determined proposition can be derived (has been derived, resp.) from certain other propositions. (He means it probably thus: the assertion of a certain proposition is interpreted as saying: this proposition holds in the determined given system; and this means: it is derivable from certain basic assumptions).

Why is elementary arithmetic, with a countable domain, already excluded? Because, according to Skolem, all of classical mathematics can be represented through a countable model, that is it can be expressed in elementary arithmetic, by taking 'ε' as a specific relation between natural numbers.[20]

It is possible that condition two, as Greg Frost-Arnold suggested to me, should be corrected to read 'reistic' as opposed to 'realistic'. This makes sense on account of the reference to Kotarbiński. Whatever the case, the reference to Kotarbiński is a useful qualification to Mostowski's assessment of Tarski's attitude vis-à-vis Kotarbiński's reism.

On 31 January 1941 Carnap describes what makes his understanding with Tarski difficult, mainly in three points:

1. Finitism: which languages having what variables do we understand? (this is the hardest point; for me [Carnap] a question of degree; But it is not very clear)
2. Modalities. 'N'; intensional language
3. L-concepts.

(3) is the easiest. If one takes Quine's form of language (or similar ones) then the logical constants are given by enumeration. Then L-true is easily definable.[...] (2) Once we have defined L-true, then N can be easily explained; essentially: 1. 'N(...)' is translated in '...', in case this is L-true; otherwise in '∼ (...)'. 2. '(x)N(...)' is translated in 'N(x)(...)'. (1) is the hardest. In which sense do we "understand" for instance, arithmetic with bounded numerical variables (for natural numbers)(RC 090-16-25)[21]

Tarski then comes back to the nature of finitism:

I understand correctly only a finite language S_1: only individual variables, its values are things; for their number one does not claim infinity (but perhaps also not the opposite). Finitely many descriptive primitive predicates. Numbers: they can be applied, in a finite domain, by thinking the things as ordered and understanding by the numerical signs the corresponding things [*und unter den Zahlzeichen die betreffenden Dinge verstehen*]. Then we can apply arithmetical concepts; but many arithmetical sentences cannot be proved here, because we do not know how many numbers there are. One can also ascribe a cardinal number to a class.[22]

We see that Tarski here modifies slightly his criterion for a finitistic language by allowing the possibility of a non-finite domain but strengthens the requirement on primitive predicates by requiring that they be finite. He then comes back to the issue of what languages we 'understand'.

Tarski: the psychological puzzle is the following: mathematicians seem to understand in a certain sense also infinite arithmetic. Namely, in the case of an undecidable sentence (e.g. that of Gödel). They are able to say, without considering the axioms, that they acknowledge the sentence to be true. And I (Tarski) share this feeling in a certain degree.

I [Carnap]: It seems to me that I actually understand, in a certain degree, infinite arithmetic, let us say a language S_2: only variables for natural numbers, with operators (so that also negated universals), and in addition recursive definitions. To the question of Tarski and Quine, how I interpret this, when the number of things is perhaps finite: I do not know exactly but perhaps through mere positions instead of things (Tarski: this conception (in Syntax) had impressed him very much at the time, but he finds difficulties with it). A position is an ordering possibility for a thing. I do not have the intuitive [instinctive] rejection of the concept of possibility as Tarski and Quine do. To me the possibility of always proceeding [*immer Weiterschreitens*] seems the foundation of number theory. Thus, potential but not actual infinity (Tarski and Quine say: they do not understand this distinction).

I [Carnap]: there is perhaps an intermediate level, similar to Language I, without negated universal sentences. (Tarski: this seems to him to be no essential distinction, for he understands sentences with free variable as short-hand for sentences with operators [quantifiers]). We can understand a universal sentence for natural numbers as a general claim of all instances, since for every natural number there is an expression (Tarski: but not an actual expression as thing, if the number of things is finite). (RC 090-16-25)[23]

We see here that the notion of 'understanding' is rather vague and leaves open alternative positions as to the sort of languages (and theories) that can be said to be understood. As it will become clear, the notion of 'understanding' appealed to in these passages will not receive a more precise clarification in the remaining part of the discussion. Indeed, the only agreement between Carnap, Tarski, and Quine consists in claiming that set theory is not 'understandable' while the basic system (called S_1 and whose precise form will be the object of later discussions (see section 13.7 below)) is. The disagreements over the 'understandability' of infinite arithmetic S_2 were not resolved.

Tarski then suggests that as metalanguage [Methodensprache] M, one obviously needs something stronger than S_1 if one wants to have the predicate 'true' for a language which is not too weak. But the semantics in M need not be conceived as actually understood rather only as a calculus with finite rules that are formulated in S_1 as a part of M.

The central problem is formulated as follows:

WE TOGETHER: now thus the problem: Which part S of M can be taken as a nucleus so that: 1. S can be understood from us in a certain sense and 2) S suffices for the formulation of the syntax of all of M, in as far as it is necessary for science, so as to handle in M the syntax and the semantics of the universal language of science.[24]

Two options are considered here. The first is to see whether the 'poor nucleus' (the finite language S_1) suffices. If so, this would be the most satisfactory solution. The second option splits into two:

2a. How could we justify the rich nucleus (that is, infinite arithmetic S_2)? That is, in which sense can we say that we actually understand it? If this can be done, then the rules of calculus M can be certainly set up.

2b. If S_1 does not suffice to yield classical mathematics, could one not perhaps nonetheless pick S_1 and perhaps show that classical mathematics is not actually necessary for the application of science in life? Could we not perhaps on the basis of S_1 set up a calculus for a fragmentary mathematics which suffices for all practical tasks (that is, not only for everyday tasks but also for the most complicated technical problems)?[25]

Quine objects to 2a with arguments related to the Löwenheim-Skolem theorem. If we understood S_2, then we could be said to understand also all of set theory which has a countable model in the natural numbers. Carnap objects that one would have to add a predicate R to S_2, and this is not the same theory. Carnap claims to understand S_2 but that set theory is certainly on another plane. Concerning 2b, let me point out that the problem of spelling out which parts of classical mathematics are sufficient for physical applications continues to be of great interest. Feferman (1993b) has conjectured that certain systems of predicative mathematics that happen to be conservative over Peano Arithmetic (such as his system W presented in Feferman 1988a) are sufficient for all applicable mathematics, and has supported this with a considerable body of evidence. The philosophical import of the claim is connected to the Quine-Putnam indispensability arguments (see Hellman 2004) and Feferman (2004a) for the latest discussion of this issue). In this connection, let me also quote from a later letter by Tarski to Woodger where "indispensability" is singled out as the central concern:

The problem of constructing nominalistic logic and mathematics has intensively interested me for many-many years. Mathematics—at least the so-called classical mathematics—is at present an indispensable tool for scientific research in empirical science. The main problem for me is whether this tool can be interpreted or constructed nominalistically or replaced by another nominalistic tool which should be adequate for the same purposes. (Tarski to Woodger, November 21, 1948, Woodger Papers)

Before moving on to the remaining parts of the discussion we have to address a problem that has certainly been bothering the reader. Is there any connection between the first part of the discussion on semantics and those on finitism? Quine's letter from 1942 to Woodger, already cited in section 13.3, indicated that the criticism of the analytic-synthetic distinction accounted for the 'empiricist or positivist' treatment of the concepts of mathematics. The finitistic attitude could have been seen as an implementation of this 'empiricist or positivist' critique. We also obtain some further information not from the Carnap notes but from a letter from Quine to Carnap, dated 1 May 1943 (see Creath 1986, p. 295), which, while not completely clarifying the issue, provides a hint that the discussion on analytic-synthetic and the discussion on finitism were not unrelated. After referring back to the finitistic constitution program discussed in 1941 and its importance for a satisfactory epistemology, Quine added:

> However, I think the essential semantic issues between us are readily divorced from the finitistic program and from epistemology altogether. In our discussions, epistemological matters entered the semantical scene in these ways:
>
> (a) Tarski and I questioned the precise nature of your distinction between analytic and synthetic, and in the course of such discussion it began to appear increasingly that the distinguishing feature of analytic truth, for you, was its epistemological immediacy in some sense. (true by "fiat", perhaps; but then a "subconscious fiat", which is to me as much a metaphor as the Kantian sieve.) Then we urged that the only logic to which we could attach any seeming epistemological immediacy would be some sort of finitistic logic. So here we were in epistemology and envisaging a finitistic constitution system.
> (b) I argued, supported by Tarski, that there remains a kernel of technical meaning in the old controversy about reality and irreality of universals, and that in this respect we find ourselves on the side of the Platonists insofar as we hold to the full non-finitistic logic. Such an orientation seems unsatisfactory as an end-point in philosophical analysis, given the hard-headed, anti-mystical temper which all of us share; and, if this were not enough, evidence against the common-sense admission of universals can be adduced also from the logical paradoxes. So here again we find ourselves envisaging a finitistic constitution system. (Creath 1986, p. 295)

However, it is hard to know what to make of the first part of the Quinean account. In fact, Carnap, in his answer to Quine, denied the allegation that for him 'epistemological immediacy' is what characterizes analytic truth and Quine grants the point in the following letter. The second part of the letter to Carnap, however, read in conjunction with the letter from Quine to Woodger, quoted in section 13.3, seems to indicate that one of the consequences of having given up the analytic-synthetic distinction was that Platonism could not be 'disavowed' (letter to Woodger), i.e. that there was a kernel of technical meaning in the controversy about universals (letter to Carnap). Thus, the attempt to provide a finitistic account of mathematics was the desire, shared by Carnap, Tarski, and

Quine, to avoid the Platonism involved in accepting classical mathematics. The notes are, however, silent on what exactly the problem with 'Platonism' is supposed to be.

13.7 Towards a Finitistic Nucleus: RC 090-16-24 (16 February 1941)

The next set of sheets brings to light in crisper form the tension between Carnap and Tarski cum Quine on the foundations of arithmetic. The report from 16 February 1941 (reflections that Carnap writes for himself) begins by opposing empirical vs. logical finitism:

Tarski's finitism is a logical one; he claims that perhaps the number of things in the world is finite; in this case one can only speak of finitely many natural numbers. By contrast I [Carnap] say: we are empiricists. Therefore we say: our knowledge is limited to the finite; that is, each confirmation is based on a finite amount of evidence, that is a finite set of observational statements. But: we can nonetheless speak about finite classes of arbitrarily high cardinality, thus also about the individual natural numbers (for instance, $1000 \neq 1001$), without taking into consideration the number of things in the world. Thus, *logic and arithmetic become independent of the fortuitous number of things in the world*. Nonetheless, logic and arithmetic remain, in a certain other sense, finitistic, if they have to be really understood.

The arithmetic (of the natural numbers) has in fact been developed without our knowing (up to today) with certainty whether the number of things in the world is finite or not. And the demonstrated statements are not doubted by anyone; especially the concrete statements (without variables) appear indubitable. Thus arithmetic can indeed be independent of a factual hypothesis about the world.

Even if the number of things (for instance, electrons etc.) is finite, nonetheless the number of events can be taken as infinite (not only the number of time-points within an interval on account of density, but also the number of time-points which lie one unit from one another, in other words: infinite length of time). Is this a factual hypothesis? Or is it rather not something related to logical possibility?[26]

The next sheet 'On finitistic syntax' pursues the issue by addressing Tarski's suggestion that in developing a metatheory of syntax one should only consider the actually inscribed expressions, sentences, and proofs (this is obviously a nominalist requirement which could in principle be seen independently of the finitism; but, as I already pointed out, no such articulation is present in the Carnap notes). Carnap objects that this is way too narrow. Carnap plays around with the idea of taking as signs only actual things but as

expressions and proofs not only certain actual spatial orderings of these things, but also (non-spatial) sequences of these things, either indicated through the series of names [in the meta-language, PM] of these things separated by commas (elementary sequence expression) or by means of descriptions, for instance as union of two previously given sequences for which we have introduced abbreviations.[27]

Of course, Carnap now asks himself the question whether "speaking of sequences whose length is greater than the number of things in the world is compatible with the principles of finitism. That is, is such a sentence understandable for a finitist?"

The final reflection in this sheet is the idea of working with finitely many signs in the object language but to name [in the metalanguage, PM] expressions whose length is greater than the number of things in the world by means of abbreviations:

$$||a_7, a_1, a_{100}, \ldots |5||$$

This would be a representation of an expressions of the object language which consists of the expression denoted by juxtaposing five occurrences of $|a_7, a_1, a_{100}, \ldots|$

It is really hard to see how Carnap hoped to carry this out. The trick obviously relies on the possibility of interpreting finitistically the numerals (e.g. 5) and thus the solution works only in as far as we have a finitistic meaning for the numerals.

Tarski and Quine were quick to see the limitation of Carnap's proposal, which is the one opening the meeting of 17 February 1941 (RC 090-16-23). Quine objects that the decisive question is whether one has introduced variables for the sequences proposed by Carnap. This seems to be necessary in order to carry out full arithmetic. But then one is committing oneself ontologically and might as well, according to Quine, assume classes, or classes of classes. In this way one obtains a full arithmetic but gives up the realistic [reistic?] finitism.

Tarski proposes to order in some fashion or other the, perhaps finitely many, things in the world: $0, 0', 0'', \ldots$ The name of things serve at the same time as numerals. Axioms analogous to the Peano axioms hold for these things. Of course, now one must transform the Peano axiom system so that no assumption of infinity is built into it. Tarski suggests that one can reformulate the axioms in such a way that the assumption of infinity becomes an axiom and then dropping it. On the basis of this new axiom system (without the axiom of infinity) one should then try to construct a recursive arithmetic. Tarski also added that stroke-expressions for overly large numbers cannot be inscribed because there are not enough signs in the world.

Quine also put forth the possibility of interpreting all numerals that are so high that they cannot denote any thing in the world, as designations for a determined chosen object in the world. The problem is that if this object is 0, then 0 would have a predecessor; if it is another object, than that object would have two distinct predecessors. In both cases, this would call for a revision of the Peano axioms.

The remaining part of the discussion centers on what would be the ultimate constituents of matter out of which things are generated and whether space and time can be taken to be infinite. Quine and Tarski argue that neither space nor time can be taken to be infinite. Carnap points out that if one insists on accepting only actually inscribed, as opposed to possibly inscribed, sequences, then serious limitations to syntax ensue, such as the fact that closure under conjunction for sentences, or closure under juxtaposition for expressions, fails. But Carnap also suspects that a major conceptual confusion might be going on in this realistic interpretation of arithmetic:

It seems to me that the proposal suffers from a mistaken conception of arithmetic: the numbers are reified; arithmetic becomes dependent on contingent facts, while in reality it is about conceptual connections; if you will: about possible not about actual facts.[28]

It might be worth while here to recall how Goodman and Quine addressed some of these issues in 'Steps towards a constructive nominalism' where they declined to assume that there are infinitely many objects (p. 106). Since arithmetic requires infinitely many numbers one cannot "undertake to identify numbers arbitrarily with certain things in the concrete world" since this would drastically curtail classical arithmetic. On the other hand, classical syntax presupposes expressions of arbitrary length obtained by juxtaposition. If expressions are to be concrete objects, the previous assumption must also be given up. At issue here are the 'tokens' of expressions, and Goodman and Quine claim that their enterprise will not assume that given any two (token) expressions, the concatenation is also an expression, as there might not be an inscription long enough to be a concatenation of the two.

13.8 Carnap's Systems and Tarski's System

At this point, Carnap begins developing a system which might be the finitistic arithmetic required for the task. He writes down several versions. The first on 21 February 1941 (4 pages); on 23 February 1941 (10 pages) we have a related version. There are also five sheets from 18 and 19 March 1941 containing two other attempts at formalizing the system of finitistic arithmetic.

I will not reproduce the details for these systems, whose detailed analysis would require another paper. It suffices here to say that Carnap conceives the first system as a modification of Language I in *The Logical Syntax of Language*. He allows a function symbol $'$ for successor and thus the (potential) generation of infinitely many terms (from 0). If the last thing is denoted by 'k' the problem is to give an interpretation for k', k", etc. He looks at three possibilities:

(a) $k' = k'' = \ldots = k$
(b) $k' = k'' = \ldots = 0$
(c) $k' = 0; k'' = 0'$ etc.

In the end he chooses (a) and modifies the axioms of arithmetic so that they are compatible with (a) (for instance 'if $x' = y'$ then $x = y$' needs to be dropped). The second version (23 February 1941) is actually embedded in the construction of the language of science on a finitistic basis (he calls it Basic System [BS]). Here 'k' is short hand for the denotation of the last thing for which a denotation has been built (not of the last thing in the world, adds Carnap). The universe of discourse is assumed to be a finite class of things in the world. The proposal is discussed at the meeting of 1 March 1941 with Tarski, Quine, and Goodman.

Carnap outlines his system emphasizing that the language in questions refers to a finite number of denoted things with a largest element. Tarski is not happy with this. He would like a system for arithmetic that makes no assumption about the number of given numbers, or one that assumes at most the existence of one number (zero). Then he proposes the following idea:

Let A_n be the system of those sentences of ordinary arithmetic that also hold when only the numbers $< n$ exist; thus A_0 without numbers; A_1 only with 0; and so on. Let A_ω be the ordinary infinite arithmetic. In order to simplify things let us exclude A_0 so that we assume at least the existence of one number.[29]

Tarski wants a system which contains all and only those sentences which are true in each system A_n ($n = 1, 2, \ldots \omega$). To this group belong, for instance, all those sentences of the following form: there are no function symbols, all universal quantifiers occur at the front of the sentence unnegated and no existential quantifiers.

Tarski tentatively proposes that one should consider only predicates and no functions, as the latter carry with them existential assumptions. '0' can be accepted as a basic individual constant, but then one should replace the usual successor function with a successor predicate. Recursive definitions can be allowed: 'not only primitive recursion, and the so called general recursion but also those that appear in the definition of the semantical concept "satisfaction" '. This is probably already enough, says Tarski, to give us the full Peano axioms with the exception of the sentence that every number has a successor. But, he adds, as compensation we get the sentence " $x \neq x'$ " in the form "$(x)(y)(Succ(x, y) \rightarrow x \neq y)$".

What distinguishes this system from Carnap's is that it does not contain the sentences having to do with the specific 'k' which appears in Carnap's system. Tarski adds that if 'k' is used as parameter then perhaps the two systems will turn out to be equivalent. He insists that in actuality, one should not assume a specific determinate number k. Moreover, one should allow from the start variables in the language which range over all things in the world; but it remains open how many things there are. Instead of saying '$prod(2, 3) = 6$' he proposes: 'whenever x is a successor of a successor of 0, and $y\ldots$, and $z\ldots$, then $prod(x, y) = z$'. In similar fashion, one should perhaps translate universal sentences with function symbols [*Funktoren*] in implicit sentences. The question then becomes whether one can in general introduce functions so that one has translation rules of the sort given above. Quine remarks at this point that this proposal is in harmony with the old pre-Russellian conception of mathematics (still defended by Bennett, he says) according to which mathematics makes only conditional claims [*Bedingungsaussagen*].

This basically completes the sketch of what a finitistic mathematics is supposed to do. But once the job is carried out, one might still wonder how to go about the construction of the universal language of science W.

13.9 Construction of the Universal Language of Science

Two possible ways of constructing the universal language of science (W), starting from a basic system (B S) which is assumed to be completely understood, are discussed. The first consists in adding to the Basic System (B S) more and more objects [*Dinge*] by means of definitions (for instance, infinitary arithmetic, theory of real numbers; theory of functions, entire physics). If this could be done, this would be the ideal solution. But it seems doubtful that this could happen. The suggestion is that perhaps one will be able to develop a fragmentary mathematics (in comparison with classical mathematics and physics) and physics W'; in fact perhaps enough mathematics and physics as necessary for the practical needs of science. This would still be quite a good solution. The second option, in case option one does not pan out, is to use BS as syntax language and to develop W as a calculus without demanding an interpretation for it (although W is partially interpreted by means of W').

The final dialogue:

Quine: W is then truly only a myth.

I [Carnap]: No, not a myth, just a machine. It would be only a myth when we gave to the parts of the machine (the signs for the calculus) pseudo-interpretations by pointing at entities that in reality do not exist.

Tarski: The second option is unsatisfactory because it remains mysterious just how the machine works, that is how are we to explain that when we input as premisses true sentences from BS then we get as output (as conclusions) again true sentences.

We: Perhaps this is not an unsolvable mystery. We construct the machine with this task in mind and we throw it away if we observe that it does not accomplish this. Perhaps one could even show in BS that when a machine is constructed in such a way then it always delivers true conclusions from true premisses.[30]

On 18 June 1941 we have the last meeting on the nucleus language. Present are Carnap, Tarski, Quine, Goodman, and Hempel. Carnap in his notes gives a final evaluation of where the discussion stands:

Summary of what has been discussed so far. The nucleus language must serve as syntax language for the construction of the universal language of science (including classical mathematics, physics etc.) The language of science receives a partial interpretation through the fact that the nucleus language is assumed to be understood...

Concerning the logico-arithmetical part of the nucleus language. Unbounded quantifiers ... No objections from [the] finitist [point of view] since the values of the variables are physical things. It remains undecided whether their number is finite or infinite. As numbers one takes the things themselves for which we presuppose an ordering on the basis of a successor relation ...

The descriptive part. We have not managed to find agreement on whether it is better to start with thing-predicates or sense data predicates. I [Carnap], and probably Tarski, favor the first solution. Hempel adds Popper. For the second solution: Goodman and Quine.

Finally: the language must be as intelligible as possible. It is however unclear what we mean by this. Should we perhaps ask children (psychologically) what they learn first or most easily?[31]

13.10 Conclusion

The materials just presented can be analyzed from different points of view, but a full development addressing systematically the issues raised in these rich notes would require several papers. This would involve, among other things, a detailed study of the notion of 'understanding' as it occurs in the discussion, a more precise articulation (from our point of view) of how finitism and nominalism are merged in a single position, and an evaluation of the soundness of the solution proposed for a finitist foundation of mathematics and science.

These notes are obviously interesting as a source of information about specific views of Carnap, Quine, and Tarski around 1940–41. I have done that with special emphasis on Tarski, about whose philosophical views we knew much less than about the views of Carnap and Quine. The documents give us a deeper insight into Tarski's nominalism and finitism (even though we would like to know much more about the Polish background of his views (Lesniewski, Kotarbiński, etc.)). Even concerning Tarski, I had to be selective and skip for instance several aspects of the discussion on the best form of first-order languages (RC 090-16-08) or on the topic of states of affairs and models of formal systems (RC 090-16-11).

Moreover, one can try to establish a connection to more general aspects of the history of logic and analytic philosophy. Here I have emphasized, for instance, the importance of these materials as bearing witness to the shift from type theory to first-order languages. But one could as well emphasize the continuity with the search for a universal language of science of neo-positivist descent or relate these discussions to the overall debate on epistemology which was a central concern to Carnap and Quine (I referred to this at the end of section 13.6).

One obvious topic to explore is the connection between these discussions and 'Steps toward a constructive nominalism' by Goodman and Quine. First of all, Goodman and Quine are explicit about singling out the Tarskian roots of the work carried out in their article:

the idea of dealing with the language of classical mathematics in terms of a nuclear syntax language that would meet nominalistic demands was suggested in 1940 by Tarski. In the course of that year the project was discussed among Tarski, Carnap, and the present writers, but solutions were not found at that time for the technical problems involved. (Goodman and Quine 1947, note 12, p. 112)

The technical solution proposed by Goodman and Quine consists in treating the language of platonist mathematics as a language without meaning, that is as merely a string of inscriptions. However, unlike the proposals explored in the 1940–41 discussions, they adopted a nominalist syntactic metalanguage which, by means of the exploitation of mereological concepts, allowed them to get by without assuming even a part of arithmetic in the metalanguage. For this reason, we do not find in Goodman and Quine (1947) the problem of interpreting the numbers as concrete objects that was such an important part of the earlier discussions. In the conclusion of their article they were thus able to claim as achieved what the final part of the discussions in 1940–41 had indicated as the goal to pursue:

We can thus handle much of classical logic and mathematics without in any further sense understanding, or granting the truth of, the formulas we are dealing with. The gains which seem to have accrued to natural science from the use of mathematical formulas do not imply that those formulas are true statements. No one, not even the hardiest pragmatist, is likely to regard the beads of an abacus as true; and our position is that the formulas of platonistic mathematics are, like the beads of an abacus, convenient computational aids which need involve no question of truth. What is meaningful and true in the case of platonistic mathematics is not the apparatus itself, but only the description of it: the rules by which it is constructed and run. These rules we do understand, in the strict sense that we can express them in a purely nominalistic language. The idea that classical mathematics can be regarded as mere apparatus is not a novel one among nominalistically minded thinkers; but it can be maintained only if one can produce, as we have attempted to above, a syntax which is itself free from platonistic commitments. (Goodman and Quine 1947, p. 122)

Thus, the connection between the discussions of 1940–41 and Goodman and Quine (1947) represents an important chapter in the history of instrumentalism in philosophy of mathematics.[32]

The documents

These are classified under RC 090-16-02 to RC 090-16-30. Moreover, a different set of documents is relevant and represents material on Quine and Carnap in 1941: RC 102-63 Quine disc., 1940–41. Here they are reproduced in chronological order.

090-16-07 Dated 29 May 1937, Discussion with Tarski in 1937 in Paris on Wahrheitsbegriff. [3 pp.]

090-16-01 Dated 4 March 1940, Tarski, allg. [1 p.]

090-16-09 Dated 6 March 1940, Chicago, Gespräch mit Tarski [4 pp.]

102-63-09 Dated 18 October 1940, Für Diskussion mit Russell. In Logikgruppe.[3 pp.]

102-63-10 Undated, no title [1 p.]

102-63-11 Dated 27 October 1940, Für Diskussion in Logikgruppe [1 p.]

102-63-10 Dated 3 November 1940, über 'proposition' [2 pp.]

090-16-02 Dated 23 November 1940, Tarski, über allgemeine Semantik [1p.]

102-63-05 Dated 23 November 1940, Quine, über allgemeine Semantik (und Syntax) [1 p.]

090-16-03 Dated 9 December 1940, Tarski und Quine, über allgemeine Semantik [2pp., 2 copies]

102-63-13 Dated 12 December 1940 Für Logikgruppe (Die Kolmogorov-Doob Deutung der Wahrscheinlichkeit) [1 p.]

102-63-13 Dated 20 December 1940, Quine, MS, (ohne Titel; etwaL) "Lg., Math., Sc." Gelesen in Logikgruppe, 20.121940 [5 pp.]

090-16-29 Dated 20 December 1940, Quine wird diskutiert [1 p.]

102-63-06 Dated 10 January 1941, Bemerkungen zu Quines Vortrag in Logikgruppe, 10.121940 [1 p.]

102-63-12, Dated 10 January 1941 Weitere Diskussion über Quines Bemerkungen, Logikgruppe [3 pp.]

090-16-28 Dated 10 January 1941, Tarski, Finitismus [1p.]

102-63-07 Dated 11 January 1941, Logische, mathematische und faktische Wahrheit [2pp.]

102-63-08, Dated 11 January 1941, Nicht-normale Modelle des Peano AS [1 p.]

102-63-14, Dated 11 January [41], No title [1 p.]

102-63-15, Dated 11 January [41], No title [1 p.]

102-63-16 Undated, No title [1 p.]

102-63-03 Dated 20 January 1941, Logikgruppe [2 pp.]

090-16-25 Dated 31 January 1941 Gespräch mit Tarski und Quine über Finitismus, I: 31 January 1941 [5 pp.]

090-16-26 Dated 13 February 1941, Gespräch mit Tarski, über Systeme ohne Typen [3 pp.]

090-16-24 Dated 16 February 1941, Empiristischer vs. logischer Finitismus [1 p.]

090-16-27 Dated 16 February 1941, über finitistischen Syntax (Angeregt durch Gespräch mit Tarski über Finitismus, 31 January 1941) [2 pp.]

090-16-23 Dated 19 February 1941, Gespräch mit Tarski und Quine über Finitismus, II: 17 February 1941 [5 pp.]

090-16-06 Dated 21 February 1941, Finitistische Sprache, durch modifikation von Sprache I [4 pp.]

090-16-12 Dated 23 February 1941, Die Sprache der Wissenschaft auf finitistischer Basis [10 pp.]

090-16-21 Dated 23 February 1941, Andere Deutung der hohen Zahlausdrücke [2 pp.]

090-16-22 Undated, Zahl und Ding [1 p.]

090-16-04 Dated 1 March 1941. Über Finitismus. Gespräch mit Tarski; auch Quine, Goodman, III [3 pp.]

090-16-19 Dated 2 March 1941, Zahl [identische] Formeln [2 pp.]

090-16-20 Undated, Zur Diskussion über Finitismus [1 p.]

090-16-18 Dated 18 March 1941, Finitistische Sprache [2 pp.]

090-16-16 Dated 19 March 1941, Finitistische Arithmetik (Entwurf auf Grund des Gespräches mit Tarski, III, vom 1 March 1941) [3 pp.]

090-16-17 Dated 16 April 1941, No title [1 p.]

090-16-15 Dated 21 May 1941, Tarski, über Funktionskalkül [1 p.]

090-16-08 Dated 26 May 1941, Gespräch mit Tarski und Quine (und Goodman) [3 pp.]

090-16-13 Dated 26 May 1941, Tarski, formalisiertes System mit quotes [2 pp.]

090-16-14 Dated 26 May 1941, Quine's "frames" [1p.]

090-16-30 Dated 2 June 1941, "Gibt es ein Wahrscheinlichkeitsschluss"? [2 pp.]

090-16-11 Dated 10 June 1941, States of aff. und Modelle [2 pp.]

090-16-05 Dated 18 June 1941, Letztes Gespräch über nucleus-Sprache, mit Tarski, Quine, Goodman; dabei auch Hempel [2 pp.]

090-16-10 Dated 18 June 1941, Tarski über "state of affairs" [3 pp.]

Bibliographical Note

Archival Documents

From **Carnap's archive** (Pittsburgh, Konstanz):

The Tarski, Carnap, Quine discussions are found in two sets of documents. The first set of documents is classified under RC 090-16-02 to 090-16-30 and has the general title

"Tarski (and Quine), Finitism, discussion 1941". The second set of documents represents material on Quine and Carnap in 1941: RC 090-102-63 "Quine disc., 1940–1941". See the detailed list above.

From **Neurath's archive** (Haarlem, Konstanz):
  Neurath-Tarski correspondence.
  Neurath-Neider correspondence.

From **Tarski's archive** (Berkeley):
  Tarski, A., Transcript of remarks, ASL meeting, Chicago, Illinois, April 29, 1965, Bancroft Library

From **Woodger's archive** (University College London, Special Collection, GB 0103 WOODGER):
  The Woodger *Nachlaß* is still uncatalogued but the correspondence with Quine and Tarski quoted in this paper is in BOX J-Z.

14

QUINE AND TARSKI ON NOMINALISM

In this chapter I would like to trace the trajectory of two important nominalists in twentieth century analytic philosophy, Quine and Tarski. Each one of them had his own trajectory with regard to nominalism but their paths intersected at two important points: the academic year 1940–41 when both were at Harvard (together with Carnap) and in 1953 at the conference in Amersfoort organized by Beth on "Platonism and Nominalism in Contemporary Logic". These two points of intersection will be important in the exposition but I will also be interested in their individual development with respect to the topic at hand.

The association of Quine and Tarski is not arbitrary. Quine and Tarski were kindred philosophical spirits. Writing to Marja Tarski after Tarski's death, Quine wrote:

Besides being so much my mentor in logic, Alfred was a kindred spirit philosophically. Invariably when issues arose in the philosophy of logic, whether privately or in a group or at a logic convention, we found ourselves in full agreement. One notable case was our joint effort against Carnap on analytic and synthetic judgments, when we were all three together at Harvard in 1941. (Quine to Marja Tarski, January 7, 1984; Quine archive, MS Storage 299, Box 8; By permission of the Houghton Library, Harvard University [henceforth abbreviated as BPHLHU])

He could just as well have mentioned their engagement with nominalism as another example of intellectual kinship. However, there are also important disanalogies. While in Quine's case we have extensive evidence of his nominalist sympathies in the published

---

I would like to thank Arianna Betti, Wim de Jong, Paul van Ulsen, and Henk Visser for their help in tracking down some of the Beth articles and manuscripts. Special thanks to Mark van Atten who generously translated Beth (1953/54) into English for me. For comments and useful information on mereology I am also grateful to Marcus Rossberg. For specific comments on the paper, I am grateful to Lieven Decock, Marcus Giaquinto, Sol Feferman, Chris Pincock, and Mark van Atten. I am also happy to acknowledge the help I received from the librarians at Houghton Library (Harvard) during my stay there and I want to express my gratitude for permission to reproduce the materials from the Quine archive. All quotations from the Quine Archive are by permission of the Houghton Library, Harvard University. Last but not least, many thanks to Douglas B. Quine and Dean Zimmerman for making this special issue of Oxford Studies in Metaphysics possible.

output, Tarski never published on nominalism and everything about his nominalist commitment needs to be gathered from archival sources.

The archival situation differs drastically for Quine and Tarski. Tarski reached the United States in 1939. The invasion of Poland by the Nazis led to the eventual loss of his Warsaw belongings and thus the notes taken by Carnap on the 1940–41 meetings represent the first useful source for Tarski's nominalism.[1] The Tarski archive at the Bancroft Library in Berkeley contains very little from the pre-war period. Since Tarski did not leave a single paper or lecture discussing his nominalist tendencies the reconstruction of his thought on this matter needs to be carried out with the help of sources other than those kept at the Bancroft Library. By contrast, Quine's archive at the Houghton Library in Harvard contains a rich record of Quine's intellectual engagement with nominalism dating back at least to 1935. The archive is still uncatalogued and thus it is not accessible in its entirety. However, what I was able to consult is enough, I believe, to give a rich picture of Quine's reflections on nominalism from 1935 to its eventual abandonment after the Goodman-Quine 1947 paper. It also contains his lecture at the Amersfoort conference organized by Beth in 1953 (Quine 1953b). Section 14.1 describes Quine's engagement with nominalism up to 1940. Then in section 14.2 I will summarize the impact of the 1940–41 discussions on nominalism between Carnap, Quine and Tarski and mention their influence on Goodman. Section 14.3 will be on Quine's allegiance to nominalism and his subsequent reluctant acceptance of Platonism. The last section will then focus on the Amersfoort meeting and will exploit Beth's reports on the meeting to draw some welcome information about Tarski's defense of nominalism in Amersfoort. This story is too long to be recounted in one paper and thus my strategy will be to emphasize points that go beyond what is already known in the literature.

14.1 Quine on Nominalism: 1932–1940

In his autobiography, Quine claims: "Already in 1932 and 1933 in Vienna and Prague ... I felt a nominalist's discontent with classes."(Quine 1988a, p. 14) If this is correct, Quine's discontent antedates his meeting with Leśniewski. His early correspondence with Leśniewski (dating from 1934) does not contain any discussion of nominalism and Quine's first recorded reflections on nominalism do not seem to be directly influenced by Leśniewski. Moreover, Quine in his *Autobiography* points out that in Warsaw, in 1933, he was trying to convince Leśniewski that quantifying over semantical categories carried ontological commitment (Quine 1988a, p. 13, see also p. 26). This raises the question of the sources of Quine's nominalist discontent. One might think that perhaps Whitehead's mereological investigations and Leonard's thesis on the calculus of individuals might have provided the grounds of the discontent. But, as Marcus Rossberg has pointed out to me, Leonard's work was not developed with nominalist goals in mind and I think the same can be said of Whitehead's mereology. In any case, the texts I was able to consult do not seem to reflect the influence of Leśniewski, Leonard, or Whitehead. My sense is that

Quine takes his start more mundanely from the classical debates on the foundations of mathematics (Russell's no-class theory, Poincaré and Weyl on predicativity, etc.).

The first written source I was able to locate of Quine's reflection on nominalism is a three-page entry in a notebook of 300 pages entitled: *Logic Notes. Mostly 1934–1938*. (This notebook, Quine 1934–38, is mentioned in Quine's autobiography Quine 1988a, p. 44.) On p. 134 we find an entry entitled "Philosophical Background of the conceptual calculus" dated 1935. The problem discussed is denotation of the expressions of the calculus of concepts, that is sentences and nouns. First, Quine presents an interpretation of such denotation and then contrasts it with a nominalist interpretation. Whereas "nouns are frequently regarded as denoting some manner of subsistent entities—classes and relations; ... sentences are not ordinarily regarded as denoting at all". Quine goes on to suggest that for unification's sake, it is convenient to consider sentences as denoting truth values, i.e. truth and falsity. However, one should not believe that classes, relations, and truth values are real:

> But all these elements are obviously hypostatized and gratuitous entities—the relations, classes and truth-values alike. The sound way of approaching the calculus is singly to view it as a calculus whose expressions figure as nouns and sentences; not to look for elements. All the traditional abstract-algebraic reference to elements, e.g. discussions of the multiplicity of elements is reducible to such discussion of the behavior of expressions themselves. (Quine 1934–38, p. 135, BPHLHU)

Quine continues discussing another way in which nouns can denote. This time the discussion is about general nouns, such as 'cat'. According to Quine, 'cat' denotes not the class of all cats, but each cat, just as 'Socrates' denotes Socrates and not the unit class containing Socrates. Similarly, relational nouns denote sequences of individuals.

> In the present sense of denotation, the expressions of the conceptual calculus denote nothing but individuals and sequences of such; no entities of the sort which they might be said to denote in the previous sense of denotation. This is what is described in ordinary terms as the confinement of the elements of the calculus and concepts of first type. Now the erection of concepts of higher type upon this basis amounts to showing that objects higher in type than individuals never need to be assumed to exists at all; it is therefore nothing more nor less than a logical validation of nominalism, a solution of the problem of universals. The ontology on which the conceptual calculus may be regarded as ultimately based comprises concrete individuals; better, simply concrete objects, which is all I envisage for an individual. Actual entities, if one likes. These may be parts spatially or temporally one of another, and various of them may be discontinuous in space or in time; neat balls of substance are not essential, heaps are eligible as well. But they are concrete existents. (Quine 1934–38, pp. 135–36, BPHLHU)

Having shown how to give a nominalist interpretation of the calculus of concepts (predicates refer to single objects; relatives to "objects several at a time in serial fashion" and sentences refer to nothing), Quine points out that for simplicity one might introduce the fictional notion of sequence of concrete objects and regard relations as referring to such sequences and predicates as referring to sequences of length one (i.e. the objects themselves). Finally, true sentences can be seen as referring to a sequence of length zero.

The above interpretation of the calculus of concepts contains already many elements of the later Quinian views on nominalism. The universals are identified with objects of higher type in the classical account and the nominalist solution to the problem of universals consists in showing that nothing need be assumed in addition to concrete objects and their 'sums'. The mereological inspiration of the passage is evident but it is not clear whether any or which of the mereologists mentioned earlier—Leśniewski, Whitehead, and Leonard—might have played a role here.

The mentioned passage also spells out how denotation is to be interpreted when one does not hypostatize that common nouns denote classes (or relations). This is a theme that recurs often in the successive lectures on nominalism. Consider for instance the first lecture Quine gave on nominalism and to which we now turn. The lecture is entitled "Nominalism" and was delivered at the Philosophy Club in Harvard on October 25, 1937 (Quine 1937). The lecture, like later ones on nominalism given by Quine, begins with the opposition between realists, who affirm the reality of universals, and nominalists, who deny their reality. Here universals are "thought of as comprising properties, attributes, qualities, classes, relations" (Quine 1937, p. 1). Through a series of quick reductions, Quine claims qualities are not to be distinguished from properties and the same for attributes. Furthermore, there is no need to distinguish between properties and classes: "for any property can be construed as the class of all object having that property and conversely any class can be construed as the property of belonging to that class". Finally, since relations can be reduced to classes (using a trick due to Wiener) the problem of nominalism can be addressed by simply discussing classes.

So we can think of universals from now on simply as classes. Now the claim of nominalism is that there is no such thing. No abstract objects corresponding to abstract words.(Quine 1937, p. 3, BPHLHU)

It would be a mistake, however, to think that Quine is simplifying the ground for a defense of nominalism. On the contrary, he claims that he will be concerned to object to nominalism.

Quine wants to convey "the *feeling* of the nominalist" by recounting "a fictitious history of the class-concept". One begins with concrete objects and nouns to denote such:

Suppose we have settled what things are to be regarded as concrete objects. *These are all there are.* Men use words and phrases to *denote* concrete objects. A noun (substantive or adjective, word or phrase) may denote many concrete objects: M-a-n denotes Jones, also Smith, etc.; each man. (Quine 1937, p. 3, BPHLHU)

Proper nouns denote only one object and this is a very convenient feature, for "manipulating them is almost like manipulating their denotations"(Quine 1937, p. 4). It is this feature of proper nouns that leads us unconsciously, the story continues, "to force all nouns into the pattern of proper nouns". This is done by postulating for each common noun a single entity, i.e. by inventing the class of cats or the property feline to serve as denotation for c-a-t. Men then come to believe in the class (or the property) just as much as in the concrete cats and thus these creations have proved to be Frankenstein monsters—have taken subsequent developments into their own hands. (Quine 1937, p. 6, BPHLHU)

In fact, having postulated the class of cats and having come to believe in them, one can now let a new noun, such as

> s-p-e-c-i-e-s denote this, that, and the other *class*. Then just as propriefication of c-a-t creates a class of objects as a new alleged object, so propriefication of c-l-u-b or s-p-e-c-i-e-s will create a class of classes of objects as a new object. (Quine 1937, p. 6, BPHLHU)

And Quine here points out that this passage is decisive. For, with first level nouns such as c-a-t we can treat the notion of class as a mere manner of speaking; but "once classes of classes have come in, we can't eliminate reference to classes by any obvious rephrasing of our statements"(Quine 1937, p. 8) Whereas first level nouns can be read as always involving distributive predication ("Men are mortal" can be paraphrased as "Every man is mortal") with second level nouns we have collective predication ("Man are a species" cannot be rephrased with distributive quantification over the individual men) and thus "involve an apparently irreducible reference to the class of all men":

> And such second or higher level nouns occur constantly in discourse; in particular all numerical words are of at least 2nd level; and the noun n-u-m-b-e-r itself is at least 3rd level. (Quine 1937, p. 8, BPHLHU)

It thus appears that the nominalist is in trouble for he cannot reject these new objects without giving up most ordinary discourse.

In an attempt to pursue the nominalist line, Quine argues that perhaps the nominalist can find a way to identify such objects with concrete objects. If this could be done the relationship of belonging would turn out to be a relationship between a concrete object and another concrete object. And classes of concrete objects could now be identified with another concrete object and thus classes of classes could also be identified with concrete objects and so on. The natural strategy to pursue is to use the noun itself as the assigned concrete object for the class. Then "belonging to would coincide with *denoted by*." Since there are several nouns that denote the same class of objects, Quine goes on to suggest that to make the class unique we pick a specific noun (the shortest noun of the kind and among the shortest the first in the lexicographic ordering). Thus, we now have the doctrine that universals are 'mere names'. However, there are two difficulties raised against the suggestion. The first is related to Grelling's paradox. A noun is said to be *heterological* if it doesn't denote itself. Is 'heterological' heterological or not? If it is, it is not and if it is not, it is. What the objection shows is that there is no systematic way of assigning to each class of concrete objects a concrete object (using its name). In fact, Quine goes on to generalize the point by claiming that there is no way to associate to each class of concrete objects a concrete object (not necessarily a name). This is done by appealing to Cantor's theorem according to which the cardinality of the class consisting of the classes of concrete objects is strictly greater than the cardinality of all concrete objects. Quine points out that if one is willing to weaken the logic, say to intuitionistic logic, than there are ways to effect such a nominalization. It is not clear from the context whether

Quine is referring here to the intuitionism of Brouwer or the intuitionism (more precisely, predicativism) of Poincaré and Weyl. In any case, the conclusion he draws would apply to both:

> Nominalism, then, in any sense such as has here been considered, is incompatible with ordinary logic and mathematics; possible only if we are prepared for the intuitionist sacrifices. (Quine 1937, p. 13, BPHLHU)

In the last two pages of the lecture, Quine concludes by remarking that Carnap, although often considered a nominalist, simply nominalizes the problem of universals and thus rejects the meaningfulness of the general question what universals are. Thus, Quine concludes, Carnap has very little to say about the problem of reduction of all statements to statements about concrete things. In arguing against the Carnapian take on universals, Quine remarks on what he takes to be the purposes of nominalism:

(1) To avoid metaphysical questions as to the connection between the realm of universals and the realm of particulars; how universals enter into particulars, or particulars into universals.
(2) To provide for reduction to statements ultimately about tangible things, matters of fact. This by way of keeping our feet on the ground—avoiding empty theorizing. (Quine 1937, p. 14, BPHLHU)

Thus, whereas Carnap has achieved something with respect to goal (1) he has not achieved anything concerning goal (2). But Quine concludes that "a nominalism which will gain this end [goal (2)] must pay for it with a good slice of classical logic and mathematics" (Quine 1937, p. 14, BPHLHU)

It is important to point out that there is a certain ambiguity as to what the thesis of nominalism is supposed to be. Sometimes the abstract/concrete distinction seems fundamental; at times the universal/ particular. These are obviously not the same (see Cohnitz and Rossberg 2006). Thus, as of October 25, 1937, Quine seems unwilling to espouse a nominalistic philosophy of logic and mathematics. The argument does not rely on natural science but rather on the sacrifices to classical logic and mathematics that the nominalist program would require.

On May 5, 1938 we find Quine rethinking the whole issue by reflecting on the fact that Cantor's theorem fails in his *New Foundations for Mathematical Logic*. And since it was appeal to Cantor's theorem that seemed to block the possibility of a nominalist reconstruction of classical logic and mathematics, this leads Quine to reconsider the issue of nominalism:

> In view of the ambiguous position assumed by Cantor's theorem in the light of my liberalization of the theory of types (see "On Cantor's Theorem", Journal of Symbolic Logic, 1937) we are perhaps justified in reopening the question of the nominalistic identifiability of classes with terms (expressions). From the classical point of view, this course is blocked by the fact that expressions can be correlated with the natural numbers (lexicographically) whereas classes cannot. But perhaps the alleged class which would be cited as violating any given correlation of terms and classes is actually a spurious class; perhaps the term purporting to express it is unstratified (see "On the theory of types", Journal of Symbolic Logic, 1938) (Quine 1934–38, p. 209, BPHLHU)

The remaining account of what terms denote is given along the lines of the 1937 lecture. Proper nouns denote single objects; common nouns denote many objects. Classes are then identified with the earliest term in the lexicographic ordering denoting the objects in the class. He concludes:

Just as, under my previous procedure, all objects were construed as classes, so now all objects are to be constructed as terms; and terms of the class kind, i.e. terms which are lexicographically earlier than all coextensive terms. (Quine 1934–38, p. 211, BPHLHU)

We see then in these notes from 1938 a revamped interest in the possibilities of nominalism.

We now need to mention two articles and an unpublished lecture from 1939. The unpublished lecture is the original lecture Quine wrote for the Congress of the Unity of Science in Harvard: "A Logistical Approach to the Ontological Problem". This original version (Quine 1939c) is much longer than the short article which had been pre-circulated in 1939 (Quine 1939b) and which was eventually published in *The Ways of Paradox*. The first part of the lecture found its way as "Designation and Existence" (Quine 1939a). Still, some interesting passages occurring in the unpublished long version did not find their way into the two published articles and are of interest. In 'Designation and Existence', Quine arrives at his famous slogan "to be is to be the value of a variable". On the basis of the analysis carried out in the article the following five claims are taken by Quine to be identical except for wording:

a) "there is such a thing as appendicitis"
b) "the word 'appendicitis' designates"
c) "the word 'appendicitis' is a name"
d) "the word 'appendicitis' is a substituend for a variable"
e) "the disease appendicitis is a value of a variable"

For the nominalist, Quine says, 'appendicitis' "is meaningful and useful in context; yet the nominalist can maintain that the word is not a *name* of any *entity* in its own right." The difference between a nominalist and a realist language consists in whether abstract words such as "appendicitis" can be substituted for the variables:

Words of the abstract or general sort, say "appendicitis" or "horse" can turn up in nominalistic as well as realistic languages; but the difference is that in realistic languages such words are substituends for variables... whereas in nominalistic languages this is not the case (Quine 1939c, p.20, BPHLHU); and (Quine 1939a, p. 708)

Nominalism is then characterized as follows:

As a thesis in philosophy of science, nominalism can be formulated thus: it is possible to set up a nominalistic language in which all of natural science can be expressed. The nominalist, so interpreted, claims that a language adequate to all scientific purposes can be framed in such a way that its variables admit only concrete entities, individuals, as values—hence only proper names of concrete objects as substituends. Abstract terms will retain the status of syncategorematic expressions, designating nothing, so long as no corresponding variables are used. (Quine 1939c, p. 21, BPHLHU); and (Quine 1939a, p. 708)

Now various types of contextual definitions will allow the introduction of fictitious entities; the nominalist might even speak as if there were such entities but in doing so he will not renounce his nominalism for quantification over such entities can be shown to be dispensable. However, the contextual definitions must be shown to be eliminable.[2]

An important clarification, which is somehow obfuscated in the short published version, concerns the boundaries between the concrete and the abstract. This is where the unpublished version of the lecture gives something additional to the published version of 1939:

The essential point in the controversy between nominalism and realism can be made independent of any one view as to the boundary between the concrete and the abstract. All ways of specifying the realm of concrete or individual objects will, I suggest, share this common feature: the totality of individuals will be a I[*Ein*]-Ding rather than a II[*Zwei*]-Ding, in von Neumann's terminology; it will be *immanent* rather than *transcendent* in the terminology of my abstract. That is to say, the totality of individuals will be small enough to have a cardinal number; perhaps an infinite cardinal number, but still a cardinal number. Nominalism, then becomes the doctrine that all science can be expressed in a language the total range of whose variables is immanent rather than transcendent. The nominalist has not yet provided such a language and shown it to be adequate to science, but he thinks he can. (Quine 1939c, pp. 24–25, BPHLHU)

The set up described in the above passage will recur in Quine's and Tarski's presentation at the Amersfoort conference. In addition to declaring that alternative positions as to the boundary between the concrete and the abstract are possible, Quine's point of view is extremely liberal concerning how many concrete objects might be available to the nominalist. He seems worried to exclude the possibility of entertaining proper classes of concrete objects. This is in tension with one of the theses that Quine will defend at least until 1947, i.e. that the concrete objects are finite. This is also the first time that the issue of the adequacy of nominalism to natural science is raised. It might not sound like Quine is committing himself to the nominalist position but I think this lecture marks the debut of Quine's hopeful engagement with nominalism. The day before delivering his lecture he commented on a paper by Tarski and concluded his remarks by saying:

[Tomorrow: my paper] Strong argument for nominalism. Probably can't get classical mathematics. But enough mathematics for physical science? If this could be established, good reason then to consider the problem solved. (Sept. 8, 1939, Occasional lectures, 1939, MS Storage 299, vol. 11; Quine Archive, Houghton Library, BPHLHU)

At this point science becomes the benchmark for the possible success of nominalism. Classical mathematics is hardly recoverable nominalistically but if we recover enough mathematics to do science then the nominalist reconstruction of science can be said to be successful. The short version of the paper (Quine 1939b) gives a similar but completely uncommitted analysis. After raising the issue in the form of "How economical an ontology can we achieve and still have a language adequate to all purposes of science?" Quine concludes the published essay as follows:

If, as is likely, it turns out that fragments of classical mathematics must be sacrificed under all such [nominalistic] constructions, still one resort remains to the nominalist: he might undertake to show that those recalcitrant fragments are inessential to science" (Quine 1939b, p. 69)

With this lecture the topic of nominalism moves from the domain of the semantics for abstract terms to concerns more directly related to the problem of whether a nominalist language for science can be successfully constructed. This topic is central in the discussions on nominalism that Quine, Carnap and Tarski carried out while together at Harvard in 1940–41.

14.2 Harvard 1940–1941

In Mancosu (2005), I have described the context and content of the Harvard conversations exploiting the notes taken by Carnap during the discussions. Many topics were touched upon but two stand out: the criticism of the analytic-synthetic distinction which saw Tarski and Quine together opposed to Carnap; and the project for a nominalistic/finitistic construction of mathematics and science. In 1942, Quine wrote to Woodger summarizing the main topics of discussion:

Last year logic throve. Carnap, Tarski and I had many vigorous sessions together, joined also, in the first semester, by Russell. Mostly it was a matter of Tarski and me against Carnap, to this effect. (a) C[arnap]'s professedly fundamental cleavage between the analytic and the synthetic is an empty phrase (cf. my "Truth by convention"), and (b) consequently the concepts of logic and mathematics are as deserving of an empiricist or positivistic critique as are those of physics. In particular, one cannot admit predicate variables (or class variables) primitively without committing oneself, insofar to the "reality of universals", for better or worse; and meanwhile C.'s disavowal of "Platonism" is an empty phrase (cf. my "Description and Existence"). Other points on which we took C. to task are (c) his attempt to make a general semantics rather than sticking to a convenient canonical form for object languages and studying the semantics thereof more simply and briefly and yet more in detail; (d) his resuscitation of intensional functions. C. argued reasonably and well, as always, and the discussions were good fun. (Quine to Woodger, May 2, 1942, Woodger papers, University College London, Special Collection, GB 0103 WOODGER)

Here I will only give a brief survey of the second topic discussed in the meetings and refer to Mancosu (2005) and Frost-Arnold (forthcoming, 2008) for more details and further references.

The topic of eliminating abstract ('unthingly') objects is discussed in a lecture "Logic, Mathematics and Science" read by Quine at Harvard on Dec. 20, 1940 (Quine 1940). Discussing concrete objects (such as electrons, atoms, bacteria, tables, chairs, sense qualities) versus universals ('unthingly' objects such as centimeters, distances, temperatures, electric charges, energy, lines, points, classes (or properties)) Quine says:

I don't insist on eliminating classes or other unthingly objects. It is not clear that the unthingly can be eliminated without losing science. (Quine 1940, p. 6, BPHLHU)

The conversations between Quine, Carnap and Tarski in 1941 had exactly the aim of seeing how far one could pursue the program of elimination of the "unthingly" without sacrificing science. This was structured in two stages. The first consisted in identifying a nominalistic system of mathematics; the second stage was to provide a reconstruction of science on that basis. The discussion for the first part takes its start from Tarski's proposal for what should be taken to be a nominalistic language. For instance, on (January 10, 1941), Tarski describes as follows his nominalistic commitments:

I understand basically only languages which satisfy the following conditions:

(1) Finite number of individuals
(2) Realistic [reistic?, PM] (Kotarbiński): the individuals are physical things;
(3) Non-platonic: there are only variables for individuals (things) not for universals (classes and so on)

Other languages I "understand" only the way I "understand" [classical] mathematics, namely as a calculus; I know what I can infer from other [sentences] (or have inferred; "derivability" in general is already problematic). In the case of any higher "platonic" sentences [Aussagen] in a discussion I always interpret them as sentences that a determined proposition can be inferred (has been inferred, resp.) from certain other propositions. (He means it probably thus: the assertion of a certain proposition is interpreted as saying: this proposition holds in the determined given system; and this means: it is derivable from certain basic assumptions). (RC 090-16-28)

The requirement on a finite number of individuals was later relaxed by Tarski by leaving open the possibility that the individuals could be infinite. This was a step in the directions of distinguishing the finitism expressed in condition 1 from the properly nominalistic requirements of conditions 2 and 3. Throughout the 1940–4 discussion the distinction is however never made systematically. In addition, the notion of understanding, which conditions 1, 2, and 3 are supposed to ground, is never discussed in detail despite the repeated discussions as to which systems of classical mathematics could be properly said to be understood.

These discussions led to quite a tension between Tarski and Quine on the one side and Carnap on the other side. Carnap was in fact reluctant to ground arithmetic on factual matters (such as the cardinality of the existing concrete objects), whereas Tarski and Quine shared a strong commitment to finitism or at least non-infinitism. Moreover, Carnap was sympathetic to uses of modality (possible sequences etc.) whereas Quine and Tarski rejected any appeal to modality as begging the issue.

Nominalism was certainly important to Tarski. In a letter to Woodger, written in 1948, he wrote:

The problem of constructing nominalistic logic and mathematics has intensively interested me for many-many years. Mathematics—at least the so-called classical mathematics—is at present an indispensable tool for scientific research in empirical science. The main problem for me is whether this tool can be interpreted or constructed nominalistically or replaced by another nominalistic tool which should be adequate for the same purposes (Tarski to Woodger, November 21, 1948, Woodger Papers, University College London, Special Collection, GB 0103 WOODGER)

The Carnap transcripts of the 1940–1 meetings are the best source for giving us a more detailed picture of Tarski's position on nominalism. A number of nominalistic tenets appear already in the report of discussions concerning the nature of typed vs untyped languages. Consider the following conversation:

I [Carnap]: Should we construct the language of science with or without types?
He [Tarski]: Perhaps something else will emerge. One would hope and perhaps conjecture that the whole general set theory, however beautiful it is, will in the future disappear. With the higher types Platonism begins. The tendencies of Chwistek and others ("Nominalism") of speaking only of what can be named are healthy. The problem is only how to find a good implementation. (RC 090-16-09)

Tarski comes back to the same claim about the Platonist commitment involved in higher order quantifications several times during the 1940–41 meetings. For instance:

Tarski: A Platonism underlies the higher functional calculus (thus the use of a predicate variable, especially of higher type) (RC 102-63-09)

Since I cannot summarize here the 80 pages of notes taken by Carnap of these discussions let me simply state what the upshot of the discussion was. The starting point was trying to find a part of classical mathematics that could be rendered intelligible according to the finitist/nominalistic criteria enunciated by Tarski. There was disagreement as to which fragments of classical mathematics could be considered intelligible and common agreement was only found for a quantification-free system of elementary arithmetic (formulated with relations as opposed to functions to avoid commitment to infinitely many entities). An interpretation for the system was then sought by ordering the concrete individuals in the world. Since the individuals in the world might be finite a number of complications ensued. But then the discussion turned to the delineation of a nucleus language that would satisfy the nominalistic requirements and be strong enough to formulate the *metatheory* necessary for the mathematics needed in science. The idea was that even a piece of Platonistic mathematics might be rendered (partially) intelligible by a nominalist metatheory. In one of the last meetings, Carnap summarized the upshot of the discussions. On June 18, 1941 we have the last meeting on the nucleus language. Present are Carnap, Tarski, Quine, Goodman, and Hempel. Carnap in his notes gives a final evaluation of where the discussion stands:

Summary of what has been discussed so far. The nucleus language must serve as syntax language for the construction of the universal language of science (including classical mathematics, physics etc.) The language of science receives a partial interpretation through the fact that the nucleus language is assumed to be understood...

Concerning the logico-arithmetical part of the nucleus language. Unbounded quantifiers... No objections from [the] finitist [point of view] since the values of the variables are physical things. It remains undecided whether their number is finite or infinite. As numbers one takes the things themselves for which we presuppose an ordering on the basis of a successor relation...

The descriptive part. We have not managed to find agreement on whether it is better to start with thing-predicates or sense data predicates. I [Carnap], and probably Tarski, favor the first solution. Hempel adds Popper. For the second solution: Goodman and Quine.

Finally: the language must be as intelligible as possible. It is however unclear what we mean by this. Should we perhaps ask children (psychologically) what they learn first or most easily? (RC 090-16-05)

Several issues would require elaboration at this points such as the importance of 'understanding' in these discussions, the systematic confusion between nominalism and finitism and other topics. These are treated in detail in Mancosu (2005) and Frost-Arnold (forthcoming). In addition, the issue of finitude of the world divided Quine from Carnap and Tarski; finally, what kinds of entities could truly be conceived as concrete? I will come back to some of these issues later. We now go back to Quine and we will not hear about Tarski's nominalism until the description of the Amersfoort meeting.

14.3 Caution, Commitment, and Abandonment (Quine 1941–1948)

Caution

The 1940–41 meetings spurred Goodman and Quine into pursuing a nominalist program. Goodman was in charge of writing a report on the discussions that took place in 1941 at Harvard (and which he witnessed in first person) although he found it hard to say what had really been achieved:

Another confession is that I didn't do much more on the outline of the 4-cornered conversations of last semester. Your letter confirmed my suspicion that all we achieved was a somewhat bare skeleton of a program, and so many difficulties and questions occurred to me that I lost interest in trying to set anything down until there is something more solid to set on. I hope that you and I will be able to work together towards tightening and realizing the program, as we had begun to. Perhaps this season, with all the Tarski and Carnap meetings out, you will have time for it again. (Goodman to Quine, September 12, 1941; MS Storage 299, box 4, folder Goodman, BPHLHU)

This 'program' eventually led to the Goodman-Quine article of 1947. But Quine's position on nominalism, even one year before the Goodman-Quine article, was one of caution. We have a lecture, Quine (1946), from March 11, 1946, which in my opinion is the clearest statement of Quine's conception of nominalism. The lecture, now published in Quine (2008), presents nominalism as the thesis that there are only particulars and that there are no universals. From the outset Quine says that the paper will not be either a defense or a refutation of nominalism. But he adds: "I will put my cards on the table now and avow my prejudices: I should like to be able to accept nominalism." (Quine 2008, p. 6) But the sympathy for nominalism does not blind Quine to the problems that the program faces and playing on the double meaning of 'executing' (as pertaining either to the executive or the executioner) he concludes he lecture by saying: "I feel sure that nominalism can be executed, but I don't know in which sense". (Quine 2008, p. 21) But that in itself could be seen as a positive gain. In comparison to the Carnapian rejection of the problem of universals, Quine argued that 'nominalism' is a meaningful philosophical position. Much of the lecture itself goes over a number of topics with which we are by now familiar. In particular, following the doctrine that 'to be is to be the value of a variable', Quine recasts the nominalistic thesis as: "Discourse adequate to the whole of science can be so framed that nothing but particulars need be admitted as values of the variables." Faced with the objection that mathematics quantifies over abstract objects, Quine proposes a way out for the nominalist:

> Now surely classical math. is part of science; and I have said that universals have to be admitted as values of its vbls.; so it follows that the thesis of nominalism is false. What has the nominalist to say to this?
> He need not give up yet; not if he loves his nominalism more than his math. He can make his adjustment by repudiating as philosophically unsound those parts of science which resist his tenets; and his position remains strong so long as he can persuade us that these rejected parts of science are neither intrinsically desirable as ends nor necessary as means to other parts which are intrinsically desirable. (Quine 2008, p. 14)

If the goal of science is efficacy in predicting experience then one could try to argue that some parts of mathematics "are *dispensable* as a *means* to those parts of science which are effective in prediction".

I will only emphasize a few points of interest. While discussing whether Quine's thought is undergoing a shift here is complicated by Quine's attempt to provide a viable argument on behalf of the nominalist, the general question to be kept in mind is: once science is taken to be authoritative for metaphysics, exactly which aspects of science are relevant to metaphysics? The focus on those parts of science which are "effective in prediction" (1946) seems to stand in contrast to passages mentioned previously where the focus was on "all scientific purposes"(1939) or "all purposes" of science (1939). But already in 1940 the goal of science is characterized as "prediction with regard to ordinary things" (Quine 1940, p. 7). Concerning the relation between science and ontology, it might also be useful to point out that whereas Tarski focuses on languages that can be understood, Quine tries to focus on scientific goals and criteria. And this links Quine's position to that of Russell. Quine seems to agree with Russell, as against Carnap, that philosophy requires metaphysics and that metaphysical questions are genuine questions because they can be answered by appeal to our best science.

Another issue concerns Quine's choice of particulars. Nominalism, Quine says, can be constructed in two versions, mental or physical. The first version starts from mental entities. In this case the particulars are simple experiences: concrete, specific mental events. The physical version starts from physical events, i.e. the particulars are spatio-temporally extended physical objects. One important point is Quine's conviction that modern physics warrants the claim that the universe is finite:

> According to current physics, these things are made up of quanta of energy, each of which is an approximation to a point-event. We may for convenience regard every aggregate of such quanta as a physical object—a *particular*, in the present sense—no matter how scattered its parts may be, no matter how intermingled with extraneous quanta. But an aggregate in the sense not of a class but of a heap of stones: a total concrete object of which the constituent quanta and all aggregates of them are *parts*—spatial parts, really spatio-temporal parts.
> This gives us a lot of things, but, according to current physics, only finitely many. Eddington has computed the total number of quanta in the entire extent and duration of the universe: if we call this number k, then the total number of things in the universe—the total number of particulars,

in the adopted sense—can be shown by a familiar mathematical principle to be 2^k. (Quine 2008, p. 7)

This is a position Quine defended also in the conversations with Tarski and Carnap. Moreover, the first version, but not the published version, of the Goodman-Quine article sent for publication also contained the same finitistic commitment. The reason for the change will be described in the next section.

Commitment: Goodman-Quine 1947

This is of course not the place to give an account of the Goodman-Quine paper from 1947, which has been described in detail elsewhere (Decock 2002, Cohnitz and Rossberg 2006, Gosselin 1990). In it, Goodman and Quine managed to provide a nominalist analysis of the predicates 'proof' and 'theorem' and the latter was an impressive result. I would like to point out only two facts related to this paper that emerge from the study of the correspondence between Goodman and Quine. The first concerns infinity. We have seen that in previous talks and articles, Quine defended the idea that the world contains finitely many individuals and appealed to physics for support. This was also the thesis contained in the first version of the Goodman-Quine article. However, some objections by Church led Quine to send a revision of the paper weakening his finitism to a non-infinitism. In two successive letters, Church had pointed out to Quine to be careful in his claims about modern cosmological theories. On August 13, 1947, Church wrote:

I am not familiar with all the latest cosmologies of the physicists. In fact I do not take them seriously because it seems to me obvious that from observation of the visible portion of the universe, which may be relatively very small, the step to extrapolation to the nature of the whole universe is too great to be in the least trustworthy. However, I have not heard of any cosmological theory which makes space-time finite in all four dimensions. I do know of theories which make three dimensions finite and the fourth or time-like dimension infinite; but on the basis of these, your remarks about finiteness of the number of inscriptions seems to me doubtful. At least in a spoken language an inscription or an utterance may be extended in time as well as in space and this quite apart from difficulties about determining simultaneity over long distances.(MS Storage 325, Letters with editors, box 1, BPHLHU)

This was enough to make Quine change his mind. He wrote to the editor of the *Journal of Symbolic Logic*, Max Black, on August 26, 1947:

I am glad Goodman and my paper is accepted. But now, thanks to some correspondence with Church, I've become uneasy over remarks relating to the finitude of the physical world. Accordingly I'd like to ask you to supplant pages 2 and 3 by the revised pages 2 and 3 here enclosed, and to paste the new footnote 4, herewith enclosed, over the old footnote 4. (MS Storage 325, Letters with editors, box 1, BPHLHU)

The new footnote reads:

According to quantum physics, each physical object consists of a finite number of spatio-temporally scattered quanta of action. For there to be infinitely many physical objects, then, the world would have to have infinite extent along at least one of its spatio-temporal dimensions. Whether it has is a question upon which the current speculation of physicists seems to be divided. (Goodman and Quine 1947, p. 106)

In the Amersfoort talk in 1953, Quine summarizes the issue of the size of the universe as follows:

Nominalism in itself guarantees *no* infinite. For, if the only entities, are the concrete (in some sense), then surely they are finite or, at best, not to be *presumed* infinite except by evidence of natural science. It is not for the nominalistic mathematician to declare the size of the universe; his constructions must be compatible with any finite size, but not require finitude. Here, then, seemingly a clear mathematical, even quantitative, reflection of the difference between nominalism, conceptualism and strict Platonism: non-infinitism, denumerable infinitism, and indenumerable infinitism. (Quine 1953b, p. 4, BPHLHU)

Thus we need to correct what Decock (2002) writes on this matter when he claims that "According to Quine nominalists decline the use of infinities. The reason is that we cannot know whether there are infinitely many objects in the universe or not. Nominalists can only accept a finite universe of objects."(Decock 2002, p. 40) While the first two sentences are correct the textual support given for the third (Quine 1953a, p. 129) does not warrant the claim. What the text referred to by Decock explicitly states is: "the nominalist ... is not going to impute infinitude to his universe of particulars unless it happens to be infinite as a matter of objective fact." As it is evident from the passages I mentioned previously, the nominalist must carry out his work without presupposing either finiteness or infinity of the world. In that sense he is non-infinitist because his constructions must be compatible with the possibility that the universe is finite. Cohnitz and Rossberg (2006, p. 84) also make the same erroneous claim as Decock arguing from the premise that "Since inscriptions are physical marks, the number of variables will be constrained by the size of the universe, which is very big, yet finite as current science tells us".

A longer discussion would be required to clarify the loose talk of what characterizes the entities that the nominalist can allow: concrete, particulars, individuals. The three are not the same (just as much as abstract and universal entities need not be the same); but Quine, in his articles and lectures, does not seem to distinguish between them and proceeds as if the concrete/abstract and particular/universal are equivalent ways of capturing the opposition between nominalism and realism. In particular, Goodman was pushing for a version of nominalism in which the notion of a particular had center stage. This led to some discussions about what terminology to use in the Goodman-Quine paper. Just a few months before the publication Quine wrote to Goodman:

Search your heart regarding the word 'particularism'. I am feeling renewed misgivings. Seems a shame to disavow a noble tradition when we are squarely in line with it. Nominalism is negative and so are we. If we cared here to emphasize a positive stand in favor e.g. of physical objects as against phenomena, or vice versa, then indeed I'd favor dropping 'nominalism' in favor of the appropriate more special term. (Quine to Goodman, June 12, 1947, MS Storage 299, box 4, folder Goodman, BPHLHU)[3]

Since this paper is not about Goodman's nominalism, I will simply refer the reader to chapter 4 of Cohnitz and Rossberg (2006) and to Goodman (1988) and Quine (1988b) for a retrospective evaluation of their differences.

Quine's abandonment of nominalism

In his Autobiography for the Schilpp volume Quine rejects what he perceives as a misunderstanding of his position:

> Renewed sections with Goodman led to "Steps towards a constructive nominalism", an effort to get mathematics into an ontology strictly of physical objects. We settled for a formalistic account of mathematics, but still had the problem of making do with an inscriptional proof theory in a presumably finite universe. Our project was good, I think, and well begun. But our paper created a stubborn misconception that I am an ongoing nominalist. Readers try in the friendliest ways to reconcile my writings with nominalism. They try to read nominalism into "On what there is" and find, or should find, incoherence. (1988a, p. 26)

That a gap exists between "On what there is"(1948) and the 1947 paper with Goodman was obvious to Quine already at the time of publication of "On what there is". This shift can be captured in the remarks made by Quine to Woodger in two successive letters. In the first one, dated January 26, 1948, Quine is discussing an analysis of lexicographic ordering put forth by Woodger and tells Woodger: "your approach suggests that you share our [Quine and Goodman's] nominalistic prejudices." But in the following letter dated March 22, 1948, Quine tells Woodger that his thinking in ontology is undergoing rapid transformations:

> A brief reflection now on ontology. I suppose the question what ontology to accept is in principle similar to the question what system of physics or biology to accept: it turns finally on the relative elegance and simplicity with which the theory serves to group and correlate our sense data. We accept a theory of physical objects, ranging from subatomic particles to island universes, because this gives us the neatest and most convenient filing cabinet yet known in which to file away our experiences. Now the positing of abstract entities (as values of variables) is the same kind of thing. As an adjunct to natural science, classical mathematics is probably unnecessary; still it is simpler and more convenient than any fragmentary substitute that could be given meaning in nominalistic terms. Hence the motive—and a good one—for positing abstract entities (which classical mathematics) needs. The platonistic acceptance of classes leads to Russell's paradox et al., and so has to be modified with artificial restrictions. But so does the acceptance of a physical ontology, in latter days, lead to strange results: the wave-corpuscle paradox and the indeterminacy. It seems, more than ever, that the assumption of abstract entities and the assumptions of the external world are assumptions of the same sort. It remains important to study the boundary between that part of discourse which makes the assumption of abstract entities and that part which does not. I have worked on both sides of the boundary myself, and propose to continue to do so; but I tend nowadays to stress the distinction. These very relativistic and tolerant remarks differ in tone from passages in my paper with Goodman and even in my last letter, I expect. My ontological attitude seems to be evolving rather rapidly at the moment. (MS Storage 299, box 9, folder Woodger, BPHLHU)

This is in fact the ontological attitude that will be displayed in "On what there is", which marks the passage from the "nominalistic predjudice" of the Goodman-Quine paper to the reluctant Platonism of later years. Finally, let me point out that the argument presented to Woodger is not quite the classical 'indispensability' argument. Quine says to Woodger that perhaps classical mathematics might turn out to be unnecessary for natural science but it is on account of the theoretical virtues yielded by a systematization using classical mathematics that we should accept it with its Platonistic commitments. Recent developments of indispensability arguments (Colyvan 2001 and Baker 2005) are more in line with this version of the argument than with the classical Putnam-Quine version of it.

14.4 Tarski Again

In this section we can back to Tarski's nominalism. I will exploit Willem Beth's reports of a meeting in Amersfoort in 1953 to complete the picture of Tarski's engagement with nominalism.

Beth on Nominalism (Brussels 1953)

The analysis of Beth's writings on nominalism reveals an interesting shift. Thanks to the documents contained in the Beth archive we are well informed about this shift. On January 24, 1953, Beth gave a lecture in French in Brussels. The title of the lecture was "La reconstruction nominaliste de la logique" (Beth 1953c). In this lecture Beth surveys the developments in set theory and logic that saw as protagonists Cantor, Zermelo, Frege and Russell. He points out that both Zermelo's system and Russell's type theory emerged as a way to give a solution to the paradoxes. He then emphasizes that both systems contain elements that derive from a "Platonist" conception. In Zermelo's case this is revealed first of all by the fact that each individual results from the compression of a multitude into a unity, "which reminds one of the methods of the theory of forms." Moreover, Beth adds, in Zermelo's theory this Platonism is even more evident on account of the fact that each individual is the result of such a compression. By contrast, Russell's system allows for true individuals that are not the result of a compression. When we compress a multitude we end up with an entity of higher order. But Beth points out that Quine has insisted on the presence of Platonist elements also in the theory of types. The reason given is that the higher-order entities are treated just like the individuals. This is especially evident by the fact that one can quantify not only over individuals but we can quantify also over higher-order entities. In this way the higher-order entities take on a concrete or substantial character so that the "universals solidify themselves". This leads Beth to reflect on whether one could give an alternative, "nominalist", interpretation of the systems in question so as to avoid the Platonist commitments. What seemed clear to him is that the analysis of the paradoxes has shown that one cannot have at the same time a uniform quantification and an unlimited compression. A requirement for nominalism, given without argument, is that nominalism must require a uniform quantification. He grants that one can construct systems of logic that satisfy nominalist requirements; first-order logic is a prominent example, but the problem is that first-order logic is not sufficient to reconstruct mathematics. And even allowing a certain amount of compressibility (I think Beth is here thinking of "predicative theories") one ends up with non-standard models and thus with an unsatisfactory reconstruction of mathematics. One thing that is common to both Zermelo's system and Russell's system is the notion of "membership" (either in the form of "a universal inheres in an individual" or "an individual is a member of a class"). Perhaps this is the notion at the root of both systems that one should try to eliminate. Beth mentions that such an attempt could be made replacing "membership" by the relation of inclusion (i.e. part and whole). However this solution also presents various problems and Beth concludes his talk by stating that the nominalist reconstruction of logic and mathematics faces considerable difficulties and that the nominalist requirements are not in harmony with the needs of classical logic and mathematics. However, he sees more of a future for nominalism at the level of metalogic. Although he does not mention here explicitly the work by Goodman and Quine, it is clear that he was aware of it as he mentioned it in the article published on February 14, 1953 in *Folia Civitatis* (Beth 1953a).

Summarizing, in Beth's first contribution on nominalism we find: (a) a sympathetic description of the program while not renouncing a Platonist attitude; (b) a critical attitude with respect to the prospects of success for a nominalist reconstruction of mathematics;(see also the letter to Scholz quoted in van Ulsen 2000, p. 62) only mention of Quine and Goodman as prominent nominalists.

The Amersfoort summer conference of 1953

An important development in Beth's account of nominalism takes place as a consequence of a summer meeting he organized in Amersfoort on the topic of "Nominalism and Platonism in Contemporary Logic". The change can be traced most clearly in a series of publications. First of all, the report "Summer conference 1953" and the article "Reason and Intuition" also published in 1953 (both in Dutch). More elaborate accounts are then to be found in the volumes *L'Existence en Mathématiques* (1955) and *The Foundations of Mathematics* (1959). The two prominent speakers at the conference were Quine and Tarski. I have already mentioned that Quine's lecture for the summer meeting is still extant in the Houghton Library and I have already quoted a passage from it.[4]

However, most of Quine's lecture is centered on the opposition between ontology and ideology, and this does not add much to the published Quine (1951). My interest in this material rests especially in the amount of information we can extract from these articles concerning Tarski's nominalism. This provides a welcome complement to the information we were able to extract from the Carnap notes of the Harvard meetings in 1940–41.

Let's begin with the report on the summer conference (Beth 1953/54). Beth presented Quine and Tarski as "the most authoritative spokesmen" of the nominalist efforts in logic. After describing the Quinian distinction between ontology and ideology, and the nominalist rejection of a gap between the two, he mentioned Quine's taxonomy [this is not in Quine's written notes for the lecture] as to the possible views that a nominalist can take vis-à-vis platonistic mathematics:

(1) he may try to build a new, nominalist mathematics;
(2) he may, after the example of Hilbert's formalism, conceive of classical mathematics as a formal system, the structure of which can be described nominalistically, but which does not require a nominalist interpretation;
(3) he may try to find a nominalist interpretation of classical mathematics.

Quine's sympathies lie with the third option. As an aside, let me point out here that the 1940–41 discussions contain elements of all three strategies and that Goodman and Quine (1947) carries out a version of the second strategy. The difficulties of the enterprise are well known. In what sense could *Principia Mathematica* be given, despite its Platonist commitments, a nominalist interpretation? Already when quantifying over the natural numbers, construed as classes in *Principia*, we can find no obvious nominalist interpretation. There is the alternative possibility of identifying the natural numbers with concrete objects "but then one has to assume that there are infinitely many concrete objects, and such an assumption a nominalist will hardly be prepared to make."

The difficulties become even worse when we move to analysis, given the ubiquitous use of impredicative definitions: they not only give rise to universals but also quantify over universals. Impredicative definitions are objected to also by non-nominalists. Thus, in addition to nominalism and Platonism there is a third option: conceptualism. This position accepts quantification over universals but rejects impredicative definitions. However, on account of Löwenheim-Skolem's theorem and Gödel's results on the construction of an inner model for ZF, through the constructible sets L, the opposition between Platonism and conceptualism is not as stark ("this contrast begins to wane", says Quine in the lecture) as it may appear at first sight.

Then Beth proceeds to summarize Tarski's lecture:

Prof. Tarski agreed with Quine's exposition of the nominalist view, and elaborated on various points. He distinguished a 'basic ontology' B and an 'extended ontology' E. The sciences may serve as an illustration: there, B consists of the objects accessible to macroscopic observation, while atoms, electrons and so on belong to E. The line between B and E cannot be drawn sharply, and transferring certain elements from B to E or vice versa does not, from the nominalist view, yield any gain. We may however try to reduce B and E simultaneously by assuming an empiricist stance with respect to B and a nominalist stance with respect to E. B then provides us with a minimal supply of concrete objects, and E should provide an ontology of universals acceptable to the nominalist. The problem here lies, as we have seen, in finding a nominalist interpretation of the quantifiers whose range consists in universals (classes). (Beth 1953/54, p. 43)

Thus, we see that Tarski agrees with Quine on the third strategy for nominalism, i.e. a reinterpretation of formal systems acceptable by nominalistic standards. It is at this point that we find an important and rare reference to Leśniewski, which shows that Leśniewski's point of view had left a mark on Tarski's approach to nominalism:

Tarski reminded the audience that the nominalistically acceptable universals are those for which a specific expression is available. Lesniewski has pointed to the possibility of restricting the range of the quantifiers in question to universals that are acceptable in this sense. The question arises whether on accepting such an ontology the usual laws of logic remain valid, as these laws are taken over from a system that is founded on platonist assumptions. According to Tarski, the results of Gödel's mentioned suggest that there will be no difficulties at this point. But at other points there are difficulties. (Beth 1953/54, p. 43)

The idea would thus be that of using substitutional quantification instead of objectual quantification. Tarski saw the work by Gödel as providing an ontology of abstract universals (classes) that could be named. Perhaps he saw in this result the vindication of the hope he had expressed in 1941:

The tendencies of Chwistek and others ("Nominalism") of speaking only of what can be named are healthy. The problem is only how to find a good implementation. (RC 090-16-09)

But other difficulties for the nominalist project were pointed out by Tarski. First of all, the domain B must yield infinitely many objects:

(i). E has to supply an infinite number of objects. But this is only justified if B covers an infinite number of objects, and this is an assumption that sits ill with an empiricist stance toward B. (Beth 1953/54, p. 43)

We have discussed at length the issue of the cardinality of the concrete objects and I will not insist more on it (but see below for how the connection between B's infinity and E's infinity was likely established by Tarski). A second issue concerns the kind of predicates that can be applied in B. For instance, a use internal to B of the truth predicate seems to be unavailable:

(ii) The empiricist stance toward B leads to more problems. What kind of logic can be applied in B? Predicates like 'red' are admissible here; the introduction of the predicate 'true' presupposes a transition to a higher-order logic, but that becomes possible only in E. (Beth 1953/54, p. 43)

But E should at least contain the natural numbers. But how can this be justified?

(iii) The problems mentioned under (i) and (ii) entail that even in E the construction of elementary arithmetic gives rise to objections. Namely, one of Peano's axioms implies the existence of an infinite number of objects and thus crosses the limits of E. (Beth 1953/54, pp. 43–4)

Tarski's proposed solution clarifies the passage in the 1940–4 notes by Carnap when he said:

Other languages I "understand" only the way I "understand" [classical] mathematics, namely as a calculus; I know what I can infer from other [sentences] (or have inferred; "derivability" in general is already problematic). In the case of any higher "platonic" sentences [Aussagen] in a discussion I always interpret them as sentences that a determined proposition can be inferred (has been inferred, resp.) from certain other propositions. (He means it probably thus: the assertion of a certain proposition is interpreted as saying: this proposition holds in the determined given system; and this means: it is derivable from certain basic assumptions). (RC 090-16-28)

The if-thenism suggested by the previous quotation is stated quite explicitly in Amersfoort:

> That this objection is not insurmountable was shown by Tarski in the following way. Let X be the axiom in question and let A be an arithmetical theorem whose proof involves axiom X. We now consider the statement: if X then A. This statement too is an arithmetical theorem, and to prove it we do not have to make an appeal to axiom X; by now casting all arithmetical theorems in the hypothetical form meant here, we evade any appeal to axiom X and thereby the necessity to cross the limits of E. In the application of arithmetical theorems to concrete numerical examples however, the hypothetical form: if X then A, is just as useful as the original theorem A. (Beth 1953/54, p. 44)

This concludes Beth's report. What is contained in section 13 of "Reason and Intuition" (Beth 1954) only confirms the if-thenism attributed to Tarski. There are also elements in Beth (1954) to think that Tarski went into more details on how to construct out of B the nominalistically acceptable universals contained in E. In 1954 Beth presents the construction by stages for both the Russell's type system and Zermelo's set theory and claims in note 16 to be following Tarski's exposition in Amersfoort. The idea is simple: one starts with a countable domain of individuals S1. Then one constructs out of those only classes which are definable. The union of the two sorts is the species S. The material bodies are those obtained in this way. Beth names this the 'cosmological hypothesis':

> The objects of the species S we can now identify with the material bodies, appealing to a 'cosmological hypothesis', according to which the universe contains countably many material bodies.(Beth 1954, Engl. transl. p. 95)

The latter assumption relies on the fact that B (called S1 in Beth 1954), the starting point of the construction, already gives us countably many objects. This leads us back to the difficulties already mentioned in Beth 1953/54. But Beth goes on to indicate that now the solution to the difficulties for set theory and type theory is to be found in the same if-thenism proposed for arithmetic:

> Naturally one can object to the appeal to a cosmological hypothesis as mentioned above; its necessity resulted from the acceptance of the axiom of infinity. A similar difficulty already arises with regard to elementary arithmetic, and we can adopt the solution given for this case. Let X be the axiom of infinity and A a theorem of Zermelo's theory (respectively Russell's) in the proof of which X plays a role. According to the deduction theorem, the theorem 'if X then A' can then be proved in the theory in question without having to make an appeal to axiom X. We can therefore leave the axiom of infinity out provided we give all the theorems in the proof of which it plays a role the hypothetical form just indicated. For all practical purposes 'if X then A' is just as useful to us as A. The interpretation given above remains tenable now, but we can leave open the question of how many material objects the universe contains.(Beth 1954, Engl. transl. p. 96)

Thus, it seems that Tarski was more sanguine than Quine in holding on to the nominalist project. He thought that if-thenism could provide acceptable nominalistic interpretations of classical calculi whereas Quine seems to have opted for conceptualism, a position he did not identify with nominalism.

14.5 Conclusion

I have attempted to provide new elements for a better understanding of Quine's and Tarski's nominalist engagement. Before concluding I would like to compare, very briefly, the nominalist strategies followed by Quine and Tarski (reinterpretation of classical axiomatic systems) with those present in the contemporary literature on nominalism.

In *A Subject with No Object*, Burgess and Rosen distinguish between revolutionary and hermeneutic nominalism. In the first approach, the revolutionary conception, the goal is reconstruction or revision: "the production of novel mathematical and scientific theories to replace current theories." (Burgess and Rosen 1997, p. 6) Most of the reconstructed theories sound exactly like the classical theories but they are reconstructed or reinterpreted according to nominalist standards. In the case of hermeneutic nominalism, the nominalist claims that his preferred reconstruction or reinterpretation is what the classical theories have meant all along. I see no trace of hermeneutic nominalism in Quine and Tarski; their approach lies squarely in the revolutionary tradition. This is especially evident if we think of the position they defended in Amersfoort using Gödel's results on L as the reinterpretation of classical set theory and type theory. No claim is made there as to the fact that that's what set theory meant all along; rather they propose a reinterpretation according to which the nominalist standards are satisfied.

In itself, this does not mark a major difference with contemporary nominalist programs, which are usually revolutionary rater than hermeneutic. But nominalist programs differ greatly among each other. Some, following Goodman and Quine (1947), rely on mereological ideas. Tarski himself does not seem to emphasize mereology in his nominalist reconstruction; as for Quine, mereology shows up in some of the lectures from the 1930s and 1940s and in the article with Goodman. But in Amersfoort his approach does not rely on mereology. Other contemporary programs rely on modality. Here Tarski and Quine reject appeal to modality explicitly; this is especially clear in the discussions with Carnap. However, it is interesting to point out that Tarski's (partial) if-thenism corresponds to the first step in Hellman's modal approach to nominalism. Tarski's if-thenism is partial as it seems limited to the axiom of infinity (or the successor axiom in arithmetic). Hellman's favorite reconstruction of classical mathematics begins by constructing for each theorem of, say, classical arithmetic an 'if-then' version and then by prefixing the 'if-then' statement with a possibility operator. Finally, Field's version of nominalism appeals to points and regions of physical space. While Tarski and Quine would not be unsympathetic, I am not sure how they would have reacted to the proposal but certainly the issue of infinity of such points and regions would have been an issue. In any case, they would have found Field's proposal of interest in addressing an issue that had been central to their concern: can nominalism be sufficient to account for the mathematics used in the natural sciences?

What is striking about Tarski and Quine in comparison to contemporary nominalism is the fact that the motivation for nominalism is not argued on epistemological grounds. Contemporary nominalism has been, by and large, an attempt to reply to Benacerraf's dilemma on how we can have access to abstract entities. Tarski and Quine seem to proceed to nominalism without the mediation of the epistemological problems. This might not be surprising as causal theories of knowledge were not dominant in the thirties and forties as they became after the sixties. Their anti-platonism originates from metaphysical qualms and from methodological commitments favoring paucity of postulated entities. Perhaps this should be qualified by recalling that the issue of "understanding", obviously an epistemic notion, was central to Tarski's characterization of nominalism; nonetheless, this concern is different from those explicitly related to causal theories of knowledge.

PART V

Tarski and the Vienna Circle on Truth and Logical Consequence

Introduction

SUMMARY

Part V contains three contributions to issues in the history and the philosophy of logic stemming from the groundbreaking work of the Polish logician Alfred Tarski. The development of formal semantics in the 1930s was a truly momentous event for the disciplines of logic, linguistics, and philosophy. Semantics is the discipline that studies the relationship between expressions of a language and their meaning or their denotation. Typical concepts of semantics are denotation, meaning, definability, and truth. Since at least Megarian times, certain semantical notions were considered suspect by a variety of philosophers due to the fact that their unrestricted use had given rise to paradoxes, of which the most famous is Epimenides' 'Liar Paradox'. For this reason, a widespread attitude in the early part of the twentieth century, especially among anti-metaphysical oriented philosophers, was to avoid the use of "truth" and other related notions. The group of philosophers known as the Vienna Circle, in particular Neurath and Carnap, had developed a minimalist conception of philosophy in the late twenties and early thirties that identified philosophy with the study of the logical syntax of language. However, Tarski's work on definability and truth in the early 1930s established semantics as a rigorous logical discipline. Moreover, Tarski was also able to use the techniques of semantics to give a precise account of the central notion of logic, namely that of logical consequence.

Yet Tarski's achievement was not unanimously recognized within the Vienna Circle. While Carnap embraced wholeheartedly the new discipline and saw no disharmony between semantics and the anti-metaphysical stance of the Vienna Circle, Neurath strongly disagreed and thought that accepting semantics was a dangerous step backwards in the direction of metaphysics.

Chapter 15 investigates the exact nature of the debate about the admissibility of semantics. This is more involved than it might seem at first sight because an agreement was made at the time within the Vienna Circle not to bring these differences of opinion out in print, lest the image of a splintered group be given to the outside scholarly community. As a consequence of this decision, the story of the debate concerning the acceptability of semantics within the Vienna Circle can only be reconstructed from the archives of Neurath, Carnap, and Tarski. From these resources we find that much of the debate focused on whether Tarski's theory of truth was indeed metaphysically loaded, as Neurath seemed to fear, and on Neurath's own proposal as to the legitimate use of the word 'true'.

In addition to truth, an essential notion in semantics is that of logical consequence which is used in the explication of the notion of a logically valid argument. An argument is logically valid if and only if it is not possible for the premises to be true and the conclusion false. This is basically the definition (although given in the more abstract terminology of models) that Tarski gave of logical consequence in 1936. In the last twenty years there has been a heated controversy on the exact nature of Tarski's theory of logical consequence (see for instance Etchemendy (2008)). This debate has important consequences for the philosophical problem of giving an exact characterization of this central notion of logic. However, the intense focus on the original publication by Tarski has led to several disagreements with respect to important interpretative issues related to Tarski's contribution. In Chapter 16, I argue that up to at least 1940 Tarski held a fixed domain conception of logical consequence. The paper uses Tarski's publications and some unpublished work to argue that in 1936 and 1940 Tarski defended a notion of logical consequence at odds with the model-theoretic one common today. Parts of my interpretation have been recently criticized by Gomez-Torrente (2009). I address some of his criticisms in the following addendum to Chapter 16.

One of the central pieces of evidence used in Chapter 16 is an unpublished lecture by Tarski entitled "On the completeness and categoricity of deductive theories"(1940). Chapter 17 provides a step-by-step commentary of this lecture that in addition to providing insights into Tarski's notion of logical consequence also contains important developments concerning the notions of categoricity and semantical completeness in higher order logic that have recently been the object of renewed attention by mathematical logicians. Tarski's original lecture is here published for the first time as an archival appendix in Chapter 18.

BIBLIOGRAPHICAL UPDATE

There is obviously a great deal of research that directly or indirectly refers to Tarski's theory of truth and logical consequence. Here I will limit myself to contributions that are more historically oriented and/or touch upon some of the theoretical issues that I address in my Chapters. The reader should consult the recent volume edited by D. Patterson, *New Essays on Tarski and Philosophy* (OUP, 2008). It contains several essays that touch upon Tarski's production during the period investigated in this part such as Murawski and Woleński (2008), Betti (2008), Feferman (2008b), Hodges (2008). Recent publications touching on the fixed domain interpretation of logical consequence include Etchemendy (2008), Gomez-Torrente (2008, 2009).

15

TARSKI, NEURATH, AND KOKOSZYŃSKA ON THE SEMANTIC CONCEPTION OF TRUTH

CARNAP'S *Autobiography* reports that Tarski's presentation of the semantic conception of truth at the Paris congress in 1935 gave rise to conflicting positions. While Carnap and others hailed Tarski's definition as a major success in conceptual analysis others, such as Neurath, expressed serious concerns about Tarski's project.[1] Tarski (1944) contains only indirect references to these debates and it avoids explicit mention of those whose objections were not formulated in print (such as Neurath).[2] My goal in this Chapter is to review the debate that accompanied the international recognition of Tarskian semantics by using not only the published sources but also the extended correspondence between Neurath, Tarski, Lutman-Kokoszyńska (hencejorth Kokoszyńska), and Hempel.

It is well known that Tarski's theory of truth had a lasting impact on some members of the Vienna Circle, such as Carnap. It is less known, at least outside the community of

---

I would like to thank Johannes Hafner, Greg Frost-Arnold, and Thomas Uebel for their useful comments on a previous version of this chapter. I would also like to thank Dr Brigitte Parakenings for her wonderful kindness during my stay in Konstanz (October 2004) while working on the Carnap and Neurath Nachlaß. I am also grateful to Solomon Feferman for helping me track down the Popper-Tarski correspondence. I would like to acknowledge the comments of the audiences of the Colloque International "Le Rayonnement de la philosophie polonaise au XX siècle"(École Normale Supérieure, Paris, 2/7/05), of the Séminaire de Philosophie des Sciences of the IHPST, Paris (February 2005), of the Department of Philosophy at the Catholic University in Milan (April 2005), of the group 'Logic in the Humanities' at Stanford University (March 2006), and of the Escuela Latinoamericana de Logica Matematica in Oaxaca (August 2006), where I presented parts of this chapter. All passages quoted from the Carnap Nachlaß are quoted with the permission of the University of Pittsburgh (all rights reserved). All passages quoted from the Neurath Nachlaß are quoted with the permission of the Wiener Kreis Stichting (all rights reserved). Finally, I am very grateful to Sandra Lapointe for improving the style of the translations from German into English.

historians of the logical positivist movement, that Tarski's work appeared in the midst of a debate on the nature of truth which involved several members of the Vienna Circle.[3] The most important interventions in this debate, prior to Tarski's work, were Schlick's article "On the Foundations of Knowledge" (1934), Neurath's reply "Radical Physicalism and the 'Real World'"(1934), and Hempel's "On the Logical Positivists' Theory of Truth" (1935). The influence of Tarski's work is evident in successive articles related to this debate such as Carnap's "Truth and Confirmation" (1936) and Kokoszyńska's "On the Absolute Concept of Truth and Some Other Semantical Concepts" (1936b). Of course, as it will become clearer in what follows, there were substantial differences between the disputants as to the issues they addressed concerning truth and thus when I say that the debate concerned "the nature of truth"(as if there was a single notion being explicated), this should be taken with a grain of salt.

The key date here is the Paris meeting (the First International Congress of the Unity of Science) of 1935, where Tarski was invited to present his work on semantics and the theory of truth. Ayer says in his autobiography that "philosophically the highlight of the Congress was the presentation by Tarski of a paper summarizing his theory of truth"(Ayer 1977, p. 116).

Tarski's theory of truth seemed to many to give new life to the idea of truth as correspondence between language and reality. The discussion following Tarski's presentation was summarized by Neurath in his long overview of the Paris congress published in *Erkenntnis*. Already at the Paris meeting, Neurath had suggested that

From the point of view of terminology he [Neurath] thinks that one should reserve the use of the term "true" for that Encyclopedia, among the many consistent ones which are controlled by protocol sentences, that has been chosen, so that each consequence of this Encyclopedia and each new sentence accepted into it would be called "true" and any one contradicting it would be called "false". (Neurath 1936, p. 400)

It was this proposal for the use of "truth"(given in prior but similar formulations) that had led Schlick to attack, in 1934, Neurath's position as a "coherence" theory of truth. It is important to point out from the outset that from Schlick's point of view Neurath's proposal amounts to a philosophical position on the nature of truth (a coherence theory), and that some passages in Neurath support this reading. However, from Neurath's point of view the proposal is more radical and perhaps he would even reject the idea that he was defending a conception, or a theory, of truth. Indeed, as will become clearer below, Neurath was calling for a replacement of a methodology of science that thinks of itself as methodology of truth attainment by a scientific metatheory that systematically explores how warrant obtains and spreads both across systems and across communities of investigators. Whenever I use "conception of truth" in connection to Neurath, the reader should keep in mind the remarks just made.

Neurath's concerns about the Tarskian definition of truth were already obvious from the above mentioned report but their full articulation can only be grasped from the correpondence he had with, among others, Tarski, Carnap, Lutman-Kokoszyńska, and Hempel on the subject.[4]

15.1 The Correspondence Between Tarski and Neurath

The correspondence between Neurath and Tarski contains forty-one letters from Tarski and forty-two from Neurath spanning the period 1930–39. Three of them were published in 1992 by Rudolf Haller (Haller 1992). In particular, the letters published by Haller are important for historical matters concerning the mutual influences between the Polish logicians and the Vienna Circle (see also Woleński (1989b) and Woleński and Köhler (1998)). However, I will focus on another aspect of the correspondence having to do with Neurath's objections to Tarskian semantics. The letters of interest for us come after the 1935 Paris meeting and they are hitherto unpublished.

However, I would like to point out that reading the letters devoted to the discussion of the issues of the mutual influence between the Vienna Circle and the Polish logicians, one is already struck by Tarski's extensive familiarity with the philosophical tenets of the Vienna Circle. For instance, in his letter to Neurath dated 7.9.36 the discussion centers around the following four claims by Neurath (made in Neurath 1935), each of which is disputed by Tarski:

(a) The admissibility of sentences about sentences, the possibility to speak unobjectionably about a language, was accepted broadly by the Vienna Circle before Tarski's lectures in Vienna in 1930.

(b) The claim that the Vienna Circle and the Polish logicians arrived at the same time and independently at the claims contained in (a).

(c) The claim according to which sentences, parts of sentences, etc. are physical entities [Gebilde] was discussed by the Vienna Circle before Tarski's 1930 lectures in Vienna. Tarski does not dispute this but points out that in Warsaw this position was held since 1918.

(d) The claim according to which for the goals of the real sciences one can get by with a universal language.

Some of these issues we will have to come back to. In the same letter Tarski added:

I cannot understand why you also continue to regard semantics as "objectionable" although you have nothing to object to Carnap's discussions of "tautology," "analytic,"which run parallel and are closely related to it. I have looked rather carefully at your correspondence with Ms Lutman but it has not helped me at all.[5]

We learn here that Neurath had been corresponding with Maria Kokoszyńska, who in the second half of the 1930s was considered to be one of the representatives of the Lvov-Warsaw school on philosophical issues related to semantics.[6]

What were then Neurath's objections to Tarski's theory of truth?[7] Let us follow the correspondence between Neurath and Tarski. Neurath and Tarski had first met in Vienna during Tarski's visit in 1930. Then Neurath visited Warsaw, twice in 1934, and invited Tarski to be part of the so called Prag-Vorkonferenz, which was planned as a preliminary meeting for the Paris conference of 1935. Thus, they had had several occasions to discuss all kinds of issues related to matters of philosophical and logical interest. In any case, apart from an interesting letter (published in Haller 1992) from Tarski in 1930 concerning the Polish scholars involved in philosophy of the exact sciences, most of the correspondence between them until 1935 is taken up by more mundane things. Meanwhile Tarski had arrived in Vienna in January 1935 with a Rockfeller Foundation fellowship. Most of the correspondence during this period (Neurath was already established in Holland) was concerned with the publication of an article by Tarski in the proceedings of the Prag-Vorkonferenz. It is however interesting that Neurath in a letter dated May 2, 1935 concerning the Paris congress writes to Tarski:

I hope you will contribute something that will be of service to EMPIRICISM. I am constantly worrying that some fine day a book 'METAPHYSICA MODO LOGISTICA DEMONSTRATA' will appear. And then we will be blamed even for that.[8]

This already points to a recurrent theme in Neurath, i.e. the fear that the logical formalism might seduce people into metaphysical positions. In a letter from Neurath to Carnap from 1943 (written in English) in which the danger of semantics is at issue, a full genealogy is also provided:

I am really depressed to see here all the Aristotelian metaphysics in full glint and glamour, bewitching my dear friend Carnap through and through. As often, a formalistic drapery and hangings seduce logically-minded people, as you are very much... It is really stimulating to see how the Roman Catholic Scholasticism finds its way into our logical studies, which have been devoted to empiricism. The Scholasticism created Brentanoism, Brentano begot Twardowski, Twardowski begot Kotarbiński, Łukasiewicz (you know his direct relations to the Neo-Scholasticism in Poland), both together begot now Tarski etc., and now they are Godfathers of OUR Carnap too; in this way Thomas Aquinas enters from another door Chicago ... (January 15, 1943, RC 102-55-02).[9]

We see then that Neurath's comment in 1935 already contained the seeds of a worry which would not go away.

15.2 1935: The Paris Congress and Its Aftermath

The Paris congress of 1935 (from 15 to 23 September) represents a turning point in the history of the Vienna Circle and in Tarski's career. During his stay in Vienna in 1935, Tarski had had the opportunity to explain to Carnap and Popper his theory of truth (Polish 1933, German 1935) and upon Carnap's insistence he decided to lecture on it in Paris. In addition, other scholars, such as Arne Naess, came to know Tarski's theory reading the galley proofs during the same period.[10] In Paris, Tarski also gave a second lecture on the concept of logical consequence.[11] At the Congress there were important lectures by Carnap and Kokoszyńska that already built on Tarski's theory of truth. While Carnap and Popper became immediate converts to Tarski's theory of truth, the Paris congress revealed a wide variety of reactions to Tarski's work. In the *Autobiography*, Carnap says:

> To my surprise, there was vehement opposition even on the side of our philosophical friends ... Neurath believed that the semantical concept of truth could not be reconciled with a strictly empiricist and anti-metaphysical point of view ... I showed that these objections were based on a misunderstanding of the semantical concept of truth, the failure to distinguish between this concept and concepts like certainty, knowledge of truth, complete verification and the like (1963, p. 61)

Neurath touches on the topic in correspondence with Tarski (although there were surely discussions in Paris) already on November 26, 1935 discussing the issue of terminology:

> As far as I can see, the terminology you and Dr Lutman propose seems to give rise to all kinds of confusions. You could maybe emphasize what bearing it has. I still think that my suggestion to call true any Encyclopedia singled out at any point, and accordingly to call "true" all the acknowledged sentences we acknowledge as following from it or contained in it and "false" all those we reject, is terminologically less hazardous. But this is, so to speak, more of a pedagogical problem.
> I think that your expositions are in general very important for the questions of logical empiricism. Especially the question as to how "propositions" occur among other "things" etc., and also the problem as to how analytic propositions are to be delimited etc. Unfortunately, I will hardly be able to study these questions more closely in the immediate future. But hopefully not too long from now. The proceedings of the congress in which your paper is to appear, will certainly be very useful to me.[12]

And after having read Tarski's technical article on truth, Neurath wrote:

> I have read the work you were so kind to send me. Though I do not mean to criticize it in the least, I nonetheless want to say it will certainly give rise to confusion. The restrictions you impose on the concept of truth will not be observed and your formulations will be used for all kinds of metaphysical speculations. But this is a sociological comment which as such is not unimportant. (Neurath to Tarski, March 24, 1936)[13]

We thus see that Neurath feared a metaphysical usage of Tarski's theory, due to an inappropriate extension of its field of validity (from formal languages to ordinary languages) and was also opposed to Tarski's use of the notion of "truth". At the same time, he recommended using the term "true" in connection to talk of acceptability and rejection from the Encyclopedia. In his reply, dated 21.4.36, Tarski tried to diffuse the issue by claiming that between him and Neurath on the issue of "truth" there were only terminological differences. However, Neurath thought much more was at stake and from his next letter we glimpse at the constellation of elements that were fueling his resistance:

> I thank you very much for your reflections on our "truth definitions." Of course there are to begin with only terminological differences but I have the strong impression that in the discussion concerning the domain of the real sciences your intuition slips very easily into metaphysics. One should fully speak one's mind on this issue. I wrote something to Dr Lutman Kokoszyńska about this. When you hold that it is trivial to say that one speaks with the language about the language then I can only rejoin that an essential part of science consists in defending trivialities against errors. From the beginning of the Vienna Circle, for instance, I have fought against Wittgenstein's

attempt to introduce a sort of "elucidations" and thus "illegitimate," almost non- or pre-linguistic considerations in order to then speak of the opposition between "the" language and "the" reality, and hence to speak outside the language [...] And insofar as your terminological choice suggests objectionable consequences, it has perhaps not come about independently of these consequences. On the one hand one emphasizes that this concept of truth holds only for formalized languages. On the other hand the concept of truth is of practical interest precisely in non formalized domains. For this reason, if one is not simply to get rid of the term, I am in favor of my terminology, for the latter remains applicable also in non formalized domains. By contrast the terminology you and Lutman use leads to bad things when it is applied to non formalized domains. (Neurath to Tarski, March 24, 1936) [14]

The quotation clearly shows that Neurath envisaged a radical reinterpretation of the term "true", perhaps one so extreme that it could not even be classified as an "explication" of what its meaning had been all along.

In a subsequent letter Neurath gives more details about the Viennese roots (see Frank (1997)) of his objection:

Long before we made contact with Warsaw, there was a disagreement within the Vienna Circle concerning the question as to whether it makes any sense to compare language with "reality"(for instance whether the language is more complex or less complex than the reality or just as complex and so on) from a position, so to speak, outside of both. The rejection of propositions about "the" reality originated with Frank and within the "Circle" in Vienna chiefly from me. The discussion was connected with a second one which concerned the question, whether "propositions about propositions" are meaningful or not. Wittgenstein, Schlick and others—who however defended their viewpoint less rigorously—and Waismann strictly rejected propositions about propositions, so that the discussion about propositions and reality had to be carried out so to speak *outside* of language, in terms of "clarifications" as "ladder" so to speak that one would afterwards throw away. (Neurath to Tarski, May 7, 1936)[15]

In order to understand what this amounts to we have to step back and look at some of Neurath's previous work and the debate on the nature of truth which had divided the Vienna Circle.

15.3 Neurath Against the Right Wing of the Circle

Wittgenstein's *Tractatus* played an important role in the development of the Vienna Circle. When asked by Tarski in October 1935 (see letter from Tarski to Neurath dated 7.ix.36), Carnap (as reported by Tarski) characterized the Wittgensteinian influence as both stimulating and confining. It was stimulating in that Wittgenstein brought attention to the importance of the problems that relate to language, i.e. the reducibility of philosophical problems to linguistic problems. On the other hand, Wittgenstein disputed and rejected the possibility of speaking about language in a legitimate way. In a preceding quote by Neurath we have already seen how this second aspect of Wittgenstein's position was central to the Vienna Circle discussions as well as the related problem about the relation between language and reality. Wittgenstein had essentially espoused a correspondence theory of truth[16] whereby the truth of a non-tautological statement consists in its being a picture of a fact. Wittgenstein recognized as acceptable only the logical sentences (which are *sinnlos* but not *unsinnig*) and the sentences of science. In this way he was forced to declare even the propositions contained in the *Tractatus* as "explanations," or "elucidations," which have to be thrown away after one has arrived at an understanding of the *Tractatus* just as one can throw away the ladder after one has climbed upon it (*Tractatus* 6.54). Neurath was a relentless opponent of these Wittgensteinian theses. A constant refrain in Neurath is his rejection of anything that smacks of the "absolute": the "World", the "Truth" etc. In his 1931 article "Physicalism," Neurath attacks central tenets of Wittgenstein's conception which were also shared by other members of the Vienna Circle, such as Schlick and Waismann:

> Wittgenstein and others, who admit only scientific statements as 'legitimate', nevertheless also acknowledge 'non-legitimate' formulations as preparatory 'explanations' which later should no longer be used within pure science. Within the framework of these explanations the attempt is also made to construct the scientific language with the help, so to speak, of pre-linguistic means. Here we also find the attempt to confront the language with reality; to use reality to verify whether the language is serviceable. Some of this can be translated into the legitimate language of science, for example, as far as reality is replaced by the totality of other statements with which a new statement is confronted ... But much of what Wittgenstein and others say about elucidations and the confrontation of language and reality cannot be maintained if unified science is built on the basis of scientific language from the beginning; scientific language itself is a physical formation whose structure, as physical arrangement (ornament), can be discussed by means of the very same language without contradictions. (Neurath 1931b, pp. 52–53)

This dense passage contains many characteristic themes of Neurath. The conception of language as a physical formation; the rejection of Wittgensteinian "elucidations"; the possibility of speaking about (parts of) the language within a (part of the) language; the rejection of the comparison between language and reality; the replacement of such a comparison by means of a confrontation of a group of statements with an other statement.

What science is about, according to Neurath, is making predictions. At the beginning of this process are observation statements (what later came to be called protocols) by means of which one formulates laws, which are instructions for finding predictions that can then be tested by further observation statements. What is peculiar to Neurath's position is the claim that even at the level of observation statements we do not compare the statement with reality. Rather it is always a matter of agreement or disagreement between a body of sentences and the sentence being considered:

> Thus *statements are always compared with statements*, certainly not with some 'reality', nor with 'things', as the Vienna Circle also thought up till now. This preliminary stage had some idealistic and some realistic elements; these can be completely eliminated if the transition is made to pure unified science ... If a statement is made, it is to be confronted with the totality of existing statements. If it agrees with them, it is joined to them; if it does not agree, it is called 'untrue' and

rejected; or the existing complex of statements of science is modified so that the new statement can be incorporated; the latter decision is mostly taken with hesitation. *There can be no other concept of 'truth' for science.* (Neurath 1931b, p. 53)

Thus, in Neurath's account of truth there is no issue of comparing language to reality. Everything is intralinguistic:

Language is essential for science; within language all transformations of science take place, not by confrontation of language with a 'world', a totality of 'things' whose variety language is supposed to reflect. An attempt like that would be metaphysics. *The one scientific language can speak about itself, one part of the language can speak about the other;* it is impossible to turn back behind or before language. (Neurath 1931b, p. 54)

According to his anti-absolutism, Neurath denies that alongside existing science there exist a "true" science:

Unified science formulates statements, changes them, makes predictions; however, it cannot itself anticipate its future condition. Alongside the present system of statements there is no further "true" system of statements. To speak of such, even as a conceptual boundary, does not make any sense. (Neurath 1931a, p. 61)

The last quote comes from "Sociology and Physicalism," where Neurath expounded on the same claims as his previously cited article on physicalism. In the same vein as in the quotes previously given, Neurath remarks:

Science is at times discussed as a system of statements. *Statements are compared with statements,* not with "experiences", not with a "world" nor with anything else. All these meaningless *duplications* belong to more or less refined metaphysics and are therefore to be rejected. Each new statement is confronted with the totality of existing statements that have already been harmonized with each other. *A statement is called correct if it can be incorporated* in this totality. What cannot be incorporated is rejected as incorrect. (Neurath 1931a, p. 66)

It was in "Protocol Sentences" (Neurath 1932/3) that some of the implicit consequences of Neurath's claim came fully to the fore. In particular, Neurath defends an antifoundationalist theory of science. Sentences are checked against bodies of sentences for agreement or disagreement. When a conflict is detected a decision is made as to what to alter. Nothing is sacrosanct. Even observation statements, or protocols, can be given up and thus every statement of science is revisable:

There is no way to establish fully secured, neat protocol statements as starting points of the sciences. There is no tabula rasa. We are like sailors who have to rebuild their ship on the open sea, without even being able to dismantle it in dry-rock and reconstruct it from the best components. (Neurath 1932/3, p. 92)

The fate of being discarded may befall even a protocol statement. There is no 'noli me tangere' for any statement. (Neurath 1932/3, p. 95)

It was against this picture of science and the radical proposal for the use of "truth" that went along with it that Schlick attacked Neurath in 1934.

15.4 Schlick, Neurath, Hempel, and the Debate on Truth in Neo-Positivism

Schlick found the fallibilist position defended by Neurath, and as of 1932 also by Carnap, unacceptable and published a sharp attack against it in 1934. This led to replies by Neurath and Hempel. In Paris in 1935, Carnap tried to reconcile the two camps but no unity was to be achieved. As of 1935, Schlick retained his foundationalist outlook, Carnap had moved to his semantic stage, and Neurath kept defending his views in the form of an "encyclopedism"(see Übel (1992)).

Schlick's "On the Foundation of Knowledge" (1934) is a rebuttal to what the author saw as the relativism of Neurath and Carnap. Against the conception of protocols as descriptions of special empirical facts, which can always be revised if need be, Schlick introduced the notion of an affirmation [*Konstatierung*]. It is through this notion that Schlick aimed at recovering what he saw as the rationale for introducing protocol sentences in the first place and in doing so he spelled out the connection to the problem of truth:

> The purpose [of introducing protocols] can only be that of science itself, namely to provide a true account of the facts. We think it self-evident that the problem of the foundations of all knowledge is nothing else but the question of the criterion of truth. The term 'protocol propositions' was undoubtedly first introduced so that by means of it certain propositions might be singled out, by whose truth it should then be possible to measure, as if by a yardstick, the truth of all other statements. According to the view described, this yardstick has now turned out to be just as relative as, say, all the standards of measurement in physics. And that view with its consequences has been commended, also, as an eviction of the last remnant of 'absolutism' from philosophy.
>
> But then what do we have left as a criterion of truth? Since we are not to have it that all statements of science are to accord with a specific set of protocol propositions, but rather that all propositions are to accord with all others, where each is regarded as in principle corrigible, truth can consist only in the *mutual agreement of the propositions with one another*. (Schlick 1934, p. 374)

Thus, Schlick proceeded to characterize Neurath's position as a "coherence theory of truth" in contrast to the older "correspondence" theory of truth. Against Neurath, Schlick argued that the only plausible meaning that "agreement" between propositions can have in such a truth theory is "absence from contradiction". But then any fictional story which is coherent in itself, would have as much a right as scientific knowledge:

> Anyone who takes coherence seriously as the sole criterion of truth must consider any fabricated tale to be no less true than a historical report or the propositions in a chemistry textbook, so long as the tale is well enough fashioned to harbour no contradiction anywhere. (Schlick 1934, p. 376)

According to Schlick it is not consistency with any sort of statements that can provide the criterion of truth but rather lack of contradiction with quite specific statements. These are the "affirmations" ("Here now so and so") and for this kind of consistency, Schlick concludes, "there is nothing to prevent... our employment of the good old phrase 'agreement with reality'." My interest here is not in explicating Schlick's foundationalist viewpoint but only to point out his disagreement with Neurath on the issue of the criterion of truth and his dubbing of Neurath's position as a coherentist position.

Neurath replied to Schlick's article in "Radical Physicalism and the 'Real' World" (1934). Several points of disagreement with Schlick were addressed. Two of them are particularly important for our understanding of Neurath's reaction to Tarskian semantics. The first concerned the accusation that physicalism did not have an unambiguous criterion of truth; the second, that it did not address the relationship between language and reality. On both these points Neurath clarified and reiterated his previous position. Concerning truth he held once again that

> We shall call a statement 'false' if we cannot establish conformity between it and the whole structure of science; we can also reject a protocol sentence unless we prefer to alter the structure of science and make it into a 'true' statement. (Neurath 1934, p. 102)

The second point concerned the comparison of language and reality:

> The verification of certain content statements consists in examining whether they conform to certain protocol statements; therefore we reject the expression that a statement is compared with 'reality', and the more so, since for us 'reality' is replaced by several totalities of statements that are consistent in themselves but not with each other. (Neurath 1934, p. 102)

While much more would need to be said both about Schlick's and Neurath's positions what we have covered does at least give the sense of the nature of the opposition between these two members of the Vienna Circle. Finally, it should be mentioned that this part of the debate also included an article by Hempel and a few more items by Schlick. Hempel sided with Neurath and Carnap against Schlick and he also characterized Neurath's position as a "restrained coherence" theory and defined this conception of truth as "a sufficient agreement between the system of acknowledged protocol-statements and the logical consequences which may be deduced from the statement and other statements which are already adopted."(Hempel 1935, p. 54)

Schlick responded with "Facts and Propositions" (1935) and defended his approach to truth as a comparison between facts and propositions by discussing the example of checking a statement in a travel guide against a fact:

> What on earth could statements express but facts?... saying that certain black marks in my Baedecker express the fact that a certain cathedral has two spires is a perfectly legitimate empirical assertion. (Schlick 1935, p. 402)

He admitted that at times one compares a sentence with another sentence but that there are also the cases where "a sentence is compared with the thing of which it speaks"(401) By contrast, Hempel's reply reasserted that Schlick's talk of comparing a statement from the travel guide with reality simply amounted to the comparison of two statements, i.e. the statement in the travel guide and the statement expressing "the result (not the act!) of counting the spires"(Hempel 1935, p. 94). Neurath strongly objected, as we will see, to be classified as a coherence theorist both in print and in correspondence.

In order to clarify that Neurath was not simply contraposing a "coherence" theory of truth to a "correspondentist" one, it might be useful to say a few things about the reasons that led him to deny that he held such a theory. This is also topical, for some of the secondary literature still claims that Neurath held a coherence theory of truth. Despite the fact that it was Schlick who dubbed Neurath as a coherentist, Schlick himself knew that Neurath defended no such theory. In reply to a letter by Carnap, where Carnap pointed out that Neurath does not accept a coherence theory of truth, Schlick wrote (June 5, 1934):

> I have never doubted that he would refuse to count as a follower of the usual coherence theory. However, I just meant to say that the coherence theory follows from his statements, if one is to take them seriously. I assumed that this was not even clear to himself for his thoughts are too unclear. (RC 029-28-10)[17]

Neurath was incensed at being dubbed a coherentist. He touched upon the topic with several correspondents including Carnap (15.xi.35 (remarks on a preliminary version of Carnap's *Wahrheit und Bewährung* [RC 110-02-01]), 23.xii.35, 27.i.36), Hempel (11.iii.35, 18.ii.35, 25.iii.35, 29.xi.35, 12.xii.35), Kokoszyńska (8.iv.36, 23.iv.36, 3.vi.36), Nagel (26.ii.35), Neider (2.iv.35), and Stebbing (9.iii.35). Let me quote from the letter to Stebbing (written in tentative English), as Neurath is voicing his thoughts in preparation for the Paris congress of 1935:

> Mr Schlick and also Mr Hempel use the name "coherence-theory"... All right—*but I fear*, that for English readers this terminus produces psychological associations which make a connection with the modern Idealism in England... The terminus "coherence theory" seems more used by Metaphysicians than by Scientists. Is it not so? If I enough know about Bradley, a.s.f. Joachim a.s.f. is the basis: the "coherence" of the total system (a subject adapted to the well known spirit of Laplace) And if I understand is every statement more or less right in proportion to the quantity of the total coherence, which is inherent in the single statement. That means: the coherence theory of the modern English Idealism seems to be founded in the *absolutism* of the total-world-coherence. But my thesis is directly *against* such a absolutism and for a relativism. The science is a parcel of statements without contradiction and founded on Protocol-Statements. It is possible to make variations of all statements, to bring new statements, to reduce the statements. An[d] we have not an approach to an *absolute system of coherence*/ the quasi *one and right world* / as the highest judge. And if we see, that we cannot make an confrontation of our parcel of statements with *this total and ideal system of coherence* must the idealistic philosopher, as Joachim use the single statements, and the harbour of refuge for the man of totality-coherence is—that seems so—correspondence. That means. For the idealistic philosophers of such type is the "correspondence" theory very relative and the "coherence" theory the ideal-type of an absolute theory. But for us is the correspondence theory/ with Atom-Statements and so further/ a form of absolutism and the Theory of "Radical Physicalism" a form of relativism... Excuse please this discurs [sic] about terminology. But I wish to collect terms for Paris (Neurath to Stebbing, March 9, 1935)

In addition, Neurath objects to the fact that "coherence theory" is too strongly associated in the literature with Neo-idealism and thus with an absolutism he has always rejected. Neurath was later to write to Carnap:

I have never claimed... that truth consists in the agreements between propositions but only in the agreement with a preferred collection of propositions. This "preference" contains all those elements that are essential for a "realistic" conception. (RC 102-50-01, December 23, 1935)[18]

However, Neurath would also reject a coherence theory in the sense that the mere consistency of the set of sentences would be enough to consider a set of sentences as true. His emphasis on the "preferred" class of statements points at an extra condition determined by pragmatic factors. Notice moreover how Neurath's claims on 'truth' can be at times stated in such a way that he appears to be giving a theory of truth rather than a proposal for an altogether different usage of the term.

Let us now move to Carnap's use of Tarski's theory as a possible mean to bring peace between Neurath and Schlick.

15.5 Back to the Paris Congress

Of course, Tarski did not provide all the details of his theory of truth at the Paris congress but he emphasized the most general aspects of his strategy. Central to Tarski's informal characterization were formulations of the project that were bound to bother Neurath. Consider, for instance, the definition of semantics:

The word 'semantics' is used here in a narrower sense than usual. We shall understand by semantics the totality of considerations concerning those concepts which, roughly speaking, express certain connexions between the expressions of a language and the objects and states of affairs referred to by these expressions. As typical examples of semantical concepts we might mention the concept of denotation, satisfaction, and definition... (Tarski 1936a, p. 401)

As for truth:

The concept of truth also—and this is not commonly recognized—is to be included here, at least in its classical interpretation, according to which 'true' signifies the same as 'corresponding with reality'. (Tarski 1936a, p. 401)

After all the background work we have done we can have a better sense of what's at stake (at least for Neurath) in this sentence. The concept of truth Tarski is after is certainly not the one that was proposed by Neurath but rather the one corresponding to the classical conception. It is thus not surprising that Tarski's work could be interpreted, among other things, as a vindication of Schlick's position in the protocol debate.[19] In addition to Tarski, Kokoszyńska also gave a paper, "Syntax, Semantik und Wissenschaftslogik," which certainly disturbed Neurath's anti-absolutist tendencies. Arguing for the need to extend Carnapian Syntax to Tarskian semantics in the analysis of science, Kokoszyńska used as example "the absolute concept of truth":

Of late, the classical concept of truth, according to which—as one usually says—the truth of a proposition consists in its agreement with reality, has been labeled as the absolute conception of truth. This conception of truth, as is well known, is called the correspondence theory. This theory is to be contrasted with the coherence theory of truth, according to which the truth of a proposition consists in a certain agreement of this proposition with other propositions. Some logical positivists have in the last few years made a transition from a correspondence theory of truth to a coherence theory of truth. In this transition have found expression both the conviction that the absolute conception of truth is an unscientific concept which should be excluded from philosophical investigation and—as it appears—the opinion that it can be replaced by a syntactic one with the same extension and thus be defined in the syntax language.[20]

She went on to claim that Tarski's investigations had shown how to treat scientifically the absolute conception of truth. Thus, the notion of "absolute truth" which had been previously taken to be metaphysical could now be seen as part of the logic of science in its post-syntax phase. As a consequence, problems like "How is the real world?" can be shown not to be pseudo-problems but rather to be suitable for scientific analysis (p. 13). We can see why Neurath felt that Tarskian semantics might end up reviving metaphysical issues that he had tried to dispose of once and for all as pseudo-problems. Moreover, he found himself classed as a coherence theorist, something that, as we have already seen, annoyed him very much.[21] However, in his long review of the Paris meeting for the readers of *Erkenntnis*, Neurath reports on the discussion that followed the talks by Tarski and Kokoszyńska and although reporting on several objections, many of which due to him, he fairly claimed that the talks had found most people in agreement.

An important development during this meeting stemmed from Carnap's application of Tarski's theory of truth to the protocol debate. Carnap began his lecture "Truth and Confirmation" by sharply distinguishing two notions:

The difference between the two concepts 'true' and 'confirmed' ('verified', 'scientifically accepted') is important and yet frequently not sufficiently recognized. 'True' in its customary meaning is a time-independent term; i.e. it is employed without a temporal specification. For example, one cannot say that "such and such a statement is true today (was true yesterday; will be true tomorrow)", but only "the statement is true". 'Confirmed', however, is time-dependent. When we say "such and such a statement is confirmed to a high degree by observations" then we must add: "at such and such a time." (Carnap 1936, p. 18, translation Übel 1992, p. 198)

Carnap diagnosed the source of the equivocation between the two terms in the misgivings logicians had about the concept of truth, due to the antinomies that had emerged from its unrestricted use, which led to an avoidance of the concept. In an interesting letter to Kokoszyńska, Carnap reflects on the situation before the appearance of Tarski's results:

After partly reading the proofs of Tarski's essay and seeing that he gives a fully correct definition of the concept of truth, I agree with you thoroughly that "true" and the other concepts related to it are to be seen as scientifically sound. My earlier scepticism, and that of other people, concerning this concept was in fact historically justified, inasmuch as no definition was known which was on the one hand formally correct and on the other hand avoided the antinomies. And the theory that employs these concepts, "semantics" in Tarski's sense, seems to me to be an important scientific domain. I consider it very deserving on Tarski's part that he opened up this new domain. (Carnap to Lutman-Kokoszyńska, July 19, 1935)[22]

This was thus, according to Carnap, the historical reason for the misuse of the term "true" for "confirmed". This usage was of course in conflict with ordinary usage according to which any declarative sentence is either true or false, something that is not the case for the concept of confirmation. Carnap, in his lecture, then points out the new situation created by Tarski's definition of truth, which allows, under certain restrictions, the consistent use of the adjective "true". As a consequence

> The term 'true' should no longer be used in the sense of 'confirmed'. We must not expect the definition of truth to furnish a criterion of confirmation such as is thought in epistemological analyses. (Carnap 1936, p. 19, trans. Übel 1992, p. 198)

Using these distinctions Carnap outlined the essential tenets of a theory of confirmation distinguishing between direct confirmation, obtained by confronting the statement with observations, and indirect confirmation, obtained by confronting sentences with sentences. And although he pointed out the danger involved in the talk of "comparison" between sentences and facts he also allowed as unobjectionable the idea that sentences can be confronted with observations thereby striking a middle ground between Neurath and Schlick (see Übel (1992) for more details).

The published version of Carnap's talk was actually the subject of correspondence between Neurath and Carnap. Neurath asked Carnap to present his conception of truth as a "proposal" but Carnap refused. Moreover, Carnap decided only to present his point of view and not to try to characterize the previous debate for he was convinced he could not do this "without upsetting both of you [Neurath and Schlick]" (letter of December 4, 1935). Neurath, in a final desperate attempt, replied by using "scare tactics":

> You'll soon see how questionable it is 1. that one has pinned us with the tag of coherence theory ... and 2. that Tarski's and Lutman's indeed valuable considerations circulate with the label "true". If you still can, you should choose a different name for it. I cannot conceive of this term ever contributing to clarification but on the contrary that it will constantly create confusion ... I just want to say it again really clearly and sternly because I find painful what, for instance, Rougier said in the conclusion about the shift of the demarcation line in support of metaphysics. (Neurath to Carnap, December 8, 1935)[23]

Carnap was not to be deterred from his terminological choices concerning "true" although he agreed with Neurath's criticism of Kokoszyńska's terminology of "absolute truth."(Carnap to Neurath, 27.i.36)

We are finally back to Tarski and Neurath.

15.6 Tarski's Reply to Neurath

With a better understanding of Neurath's background we can now return to Tarski's reply to Neurath. In his letter dated 28.iv.36, Tarski replied as follows to Neurath's comments:

I completely agree that "to defend trivialities against errors" is an important task of science. I have indeed for this very reason stressed many times that one must always speak in a language about another language—and not outside the language (from the reductive standpoint, just about my entire 'semantics' should be seen as a triviality; this does not upset me in the least). It seems to me that it is a big mistake, when Wittgenstein, Schlick etc. speak of "the" language instead of (a number of) languages; that might be the true source of the Wittgensteinian "metaphysics". Incidentally, all those who speak about the unified language of science with the slogan "Unity of Science" [Einheitswissenschaft] seem to commit the same mistake. We all know—because of arguments from semantics and syntax—that there is strictly speaking no unified language [Einheitssprache] in which science as a whole could be expressed. It is not enough to say that this is just a temporary, imprecise formulation. For, what should then the final, precise formulation be? Kokoszyńska recently held a lecture on the problem of a Unified Science for the local philosophical society and subjected this point to her criticism; an article from her on this subject is forthcoming in Polish. (Tarski to Neurath, April 28, 1936)[24]

Here it is interesting to point out that Tarski had drawn attention to the danger of speaking of a single language for science already in Paris in 1935 (see Neurath (1936), p. 401). Tarski's point here is quite simple. Since the universal language of science would have to be semantically closed it would end up being inconsistent. Thus, one ought to speak about languages (in the plural) and not about a single language. The argument was later developed at length by Kokoszyńska in her "Bemerkungen über die Einheitswissenschaft" (1937) which in all likelihood is the printed version of the lecture referred to by Tarski in the previous quote. Tarski continues:

Now, as far as my "terminological choice" is concerned, I can assure you, firstly, that it came about completely independently of Wittgenstein's metaphysics and, secondly, that it was in no way a "choice". The problem of truth came up very often, especially in the Polish philosophical literature. One was constantly asking (see for instance Kotarbiński's "Elements"), whether it was possible to define and apply the concept of truth unobjectionably, using such and such properties (which I spelled out in my later work). I simply provided a positive solution to this problem and noted that this solution can be extended to other semantic concepts. Like you, I am certain that this will be misused, that a number of philosophers will "overinterpret" this purely logical result in an unacceptable manner. Such is the common destiny of both small and great discoveries in the domain of the exact sciences (at times, one compares the philosophers to the "hyenas of the battle field"). (Tarski to Neurath, April 28, 1936)[25]

Concerning his relationship to metaphysics, here is what he had to say:

But I must confess to you that even if I do not underestimate your battle against metaphysics (still more from a social than from a scientific point of view), I personally do not live in a constant and panic fear of metaphysics. As I recall, Menger once wrote something witty on the fear of antinomies; it seems to me that one could apply it—mutatis mutandis—to the fear of metaphysics. It is a hopeless task to caution oneself constantly against metaphysics. This becomes all the clearer to me when I hear, here at home, various attacks on the very metaphysics of the Vienna Circle (going, namely, in your direction and in that of Carnap), when, for instance, Łukasiewicz talks, with respect to the *"Logical Syntax"*, about Carnap's philosophy, philosophizing etc. (in his mouth this has roughly the same sense as 'metaphysics' in yours). What you blame me for on account of the concept of truth, one blames Carnap for on account of the introduction of the terms 'analytic', 'synthetic', etc. ("Regression to the Kantian metaphysics"); and it seems to me that I was even more justified than Carnap to designate as truth the concept that I discuss. In general it is a valuable task to fill old bottles with new wine. (Tarski to Neurath, April 28, 1936)[26]

Finally, Tarski points out that to be coherent Neurath should also criticize all the formally defined concepts that are central to syntax and semantics (thus most of those found in Carnap's work in the *Logical Syntax of Language*):

Another point in this connection: my concept of truth, you claim, holds only in formalized languages. But on the contrary, the concept of truth is of practical significance precisely in non formalized domains. One can extend this literally to all precise concepts of syntax and semantics (consequence, content, logical and descriptive concepts, etc.): all these concepts can only be related approximately to the non formalized languages (thus to the actual languages of all non formal sciences [*Realwissenschaften*]): truth here is no exception. (Tarski to Neurath, April 28, 1936)[27]

The remaining part of the exchange did not add much to this picture and I have already quoted in section two some passages from the later discussion. Neurath was however to pursue the discussion with Kokoszyńska and we now turn to that part of the exchange.

15.7 Neurath and Kokoszyńska

Let me recall that at the Paris congress Kokoszyńska had presented a paper on issues concerning semantics entitled "Syntax, Semantik und Wissenschaftslogik." During this meeting she was part of the lively discussions on the concept of truth (basically on Tarski's side) and this led to an extensive correspondence with Neurath. The correspondence between Neurath and Kokoszyńska contains nineteen letters from Kokoszyńska and fourteen letters from Neurath. As a consequence of the Paris discussions, Kokoszyńska had promised to send Neurath some reflections on the viability of a 'sociological' definition of truth (her own label), by which name she meant Neurath's distinctive position as opposed to a coherence theory of truth. She had come to make this distinction under pressure from Neurath who, as we have seen, refused to be classified as a coherentist. She apparently sent her comments in the form of a short essay (which I was not able to locate) which accompanied a letter dated 22.iii.1936. We can gather the contents of these essay both from Neurath's reply and from later letters by Kokoszyńska. One central argument against the sociological theory of truth was the following. If we consider as a requirement of any theory of truth that it allows a derivation of the instances of the schema " 'p' is true iff p" then the sociological theory should give rise to " 'p' is acknowledged iff p." But herein lies the absurdity of the proposal, for from the fact that a statement 'p' is acknowledged we then would be able to conclude that p and from p that 'p' is acknowledged. In both directions we can come up with innumerable counterexamples.

Neurath replied with a letter (dated 23.IV.36) containing three dense pages of comments. Neurath's letter is a point by point commentary to Kokoszyńska's essay divided into four parts: (1) linguistic use; (2) coherence theory; (3) "acknowledgement" [*Anerkennung*] theory of truth (sociological definition), and (4) dangers of the Tarksi–Lutman conception of truth. The first part of the letter points out the variety of uses of the word 'truth' in natural language and refers to the empirical work by Arne Naess of the issue.[28] This was meant to undermine the idea that the "semantic" conception had any better right to claim to capture some sort of ordinary concept of truth than the sociological definition proposed by Neurath. Neurath points out that in different circles, with different linguistic practices, what decides the partition between "true" and "false" depends on a criterion [*Instanz*] against which the partition is decided. In most cases this criterion turns out to be metaphysical and not in harmony with empiricism. In the case of his proposal, the criterion is empirically given as it consists of the sentences accepted by a specific group of human beings at a certain moment in time. The second part of the letter questions whether anyone at all defends a theory of coherence as defined by Schlick and discussed also by Kokoszyńska. In the same section Neurath gives an overview of how the problem of truth originated in the Vienna Circle and was pursued in connection to the protocol debate. In the third part, Neurath reiterated his position that a statement should be called true if acknowledged at a certain time by a determined group of people under certain circumstances. The objections to Tarski and Kokoszyńska reassert the generic claim about the metaphysical dangers of the conception. In particular Neurath objected to the fact that the starting point of the Tarskian conception is an appeal to the ordinary usage but at the same time the realm of validity of the theory is limited to formal languages and thus it cannot be applied to natural language; however, these restrictions will not be observed, or so Neurath conjectures, and this will lead to metaphysical abuses of Tarski's theory. Finally, Neurath objected to certain formulations by Tarski and Kokoszyńska such as, e.g., "a proposition can be acknowledged without its holding" or "there can be life on Venus without man experiencing it." Against this type of talk Neurath states that he does "not think to be able to include them in the total body of science." The "holding" [*zutreffen*] of a statement according to Neurath can only be a question of being a recognition by someone. Not accepting this is tantamount to slip into metaphysics. In his summary of the major points of the letter, Neurath wrote:

> The "sociological" definition of truth can be upheld, and certain propositions can thus be characterized as true "now" in its sense. The "sociological" definition of truth corresponds to certain elements of the traditional conception. The Tarski–Lutman definition does not correspond to the ordinary usage in any privileged way (historical question). The Tarski–Lutman definition of truth is only applicable within formalized languages. The Tarski–Lutman terminology lures one into applying it to non-formalized languages and to interpret it in an absolute way. The justificatory explanations by Tarski and Lutman on "acknowledged but not holding" immediately seem to entail absolutist elements and seem not to be applicable within whole science neither according to Neurath's conception (Carnap, Hempel, and so on) nor even according to the very conception expressed elsewhere by Tarski and Lutman.[29]

In her reply Kokoszyńska explicitly stated that she could not accept as a theory of truth any theory which would not prove (all the instances of) the T-schema. That is the reason why she rejects the "sociological" theory as a theory of truth. Concerning the limited domain of applicability of Tarski's theory, Kokoszyńska pointed out that natural science can be formalized (say as in Carnap's language II) and thus Tarski's definition could immediately be applied. However, on this point Kokoszyńska underestimated the roots of Neurath's objection which rested on the idea that natural science is expressed in a great part through natural language and presents vague concepts [Ballungen] which make its full formalization hopeless. This aspect of Neurath's thoughts can be traced back to his anti-Cartesianism (see Mormann (1999)).

I will not pursue in detail the remaining letters except to point out a constant tendency on Neurath's part to push Kokoszyńska into claiming (especially by means of suggesting revisions to her forthcoming article in *Erkenntnis*) that there was no contradiction between his views and those defended by Tarski and Kokoszyńska. Eventually, Kokoszyńska reacted firmly against this attempt and wrote the following (6.9.36):

> As far as I understand, you want me to describe the situation as if there were no contradiction between the position you have defended so far concerning the classical concept of truth and the thoughts contained in my comments in *Erkenntnis*. But such a contradiction seems to exist after all. The issue is whether one can reliably use a concept which, so to speak, involves talk of an "agreement with reality". You have to some extent completely rejected this concept for you thought that the determination of such an "agreement" would require one to go beyond the framework of language—which is impossible—and you have tried, to a certain extent, to replace this concept by a sociological-syntactic one. It appears now from Tarski's investigations that one can speak of an "agreement between sentences and reality"—and therefore consider it within language—in positing propositions in which not only names of propositions occur but also names of other things. You have nothing to object to positing such propositions except—what affects mainly you—that they are not necessary in empirical sciences. It thus transpires that one can deal adequately precisely with the concepts which you had rejected so far. The contradiction mentioned above seems to lie therein.[30]

Kokoszyńska concluded by saying that Neurath had only made skeptical remarks in print about the classical conception of truth but that he had never treated the topic exhaustively and publicly. Her intention in corresponding with Neurath was to set limits to such skepticism.

The correspondence with Kokoszyńska is quite lengthy and often repetitive on Neurath's part. However, it does provide a detailed glimpse of the set of issues that were motivating Neurath while at the same time increasing the reader's frustration for the lack of a clear articulation of Neurath's rationale in his criticism of the theory of truth. He did not object formally to the theory nor to its application within formalized languages. He saw the danger of a possible misapplication of the theory of truth by overextending its limits of application and giving rise to metaphysical pseudo-talk of comparison of language and reality. But while Neurath was focusing on these possible dangers he did not focus on the opposite danger, which consisted in using the word "truth" for talking about "acknowledgement," certainly a quite unintuitive move from the point of view of the ordinary usage of the expression "true." The final chapter in this story I want to consider is the private section on semantics which saw Carnap and Neurath on opposite camps at the 1937 "Congrès Descartes" in Paris.

15.8 Neurath vs. Carnap: Paris 1937

The archives on Neurath and Carnap contain two documents which, taken together, mark a culminating point of the debate on semantics within the Vienna Circle.[31] On occasion of the Congrès Descartes (Paris, 1937) Carnap, Neurath, and others met for a private discussion on semantics. Among the invited people were Tarski and Kokoszyńska.[32] Both Carnap and Neurath presented written contributions. Neurath's paper was entitled "The Concept of Truth and Empiricism" (Neurath 1937d) and Carnap's "The Semantical Concept of Truth" (Carnap 1937).

Neurath's contribution is ten pages long and it is dated July 12, 1937. He begins by acknowledging that he should have made clear, already from his 1931 *Scientia* article on physicalism, that he had only intended to make proposals as opposed to presenting dogmas. On the other hand he claims to have individuated clearly, unlike the other participants in the truth debate, the real opponent, i.e. Wittgenstein and those close to him. His proposal then is to delimit a subject of investigation "where we constantly compare sentences with sentences, investigate their logical extent and their systematic position etc. If one analyzes science in this way then one is engaged in what Carnap called the logic of science." Neurath's proposal is to see "how much can be handled *within* the logic of science"(Neurath 1937d, p. 1).

Neurath then proceeds to rehearse the origin of the debate on truth with which we are by now familiar, including Wittgenstein's theses on the comparison between language and "the" reality and the idea that verification consists in a reference to the given. Against this type of talk, Neurath had suggested that both sentences and facts (or states of affairs) were types of objects, objects of the sentence type and objects of the non-sentence type. With the help of new sentences one could now talk about these sentences and non-sentences and thus confront sentences about sentences and sentences about non-sentences.

It is the goal of the logic of science to investigate the logical relationship between, among other things, real sentences. Suppose one wants to study the relationship between theory and experiment in behavioral terms. This normally refers to the activity of scientists first in relation to the experimental apparatus and then in their theoretical formulations. The logic of science, he adds, uses the following trick: it expresses by an observation sentence the outcome of the experimental work, say "At location A ice melts at (temperature) -3 degree"; then it compares it with a theoretical statement, say, "Ice melts at (temperatures) greater than 0 degree." It then investigates how much of an incoherence with a given class of statements it would be to use both sentences simultaneously. This is the way to move away from talk of comparison between 'language' and 'reality' or between 'thought' and 'being'.

Neurath then suggests to apply the 'trick' to semantics. This, he claims, he had already suggested in 1935 at the previous meeting in Paris but he had found no adherents. Carnap and Hempel went along with the formulations of Tarski and Kokoszyńska which, he adds, "can become dangerous for empiricism." He then goes on to make his proposal in terms of the "acknowledgement theory": suppose we are given a sentence of the Encyclopedia which describes (structurally) a sentence, say "it snows", by describing the letters composing it. Then this expression is called a "true sentence" if and only if I am given a sentence of the encyclopedia: it snows. All of this is done within the logic of science and there is no need to use expressions such as "relations between expressions of language and designated objects." Thus, he proposes to investigate how far one can proceed this way in the framework of logical empiricism.

Moving now to more criticism of the Tarski-Kokoszyńska line, he first points out that it might simply be better to use "accepted (in the Encyclopedia)" and "rejected (in the Encyclopedia)" instead of "true" and "false", which are too loaded with meaning. Against Kokoszyńska he objects that she takes for granted that what she calls "the absolute concept of truth" agrees with the ordinary concept of truth. Against this he adduces the investigations by Arne Naess which "show that there are many common concepts of truth." He proposes the same argument against Tarski who is described as the defender of the traditional philosophical concept of truth, as evidenced by his references to Kotarbiński who, adds Neurath, despite his general sympathies with the logical empiricists, on the issues of truth displays the absolutist tendencies of the Brentano school. Neurath objects that it is not the role of a defender of logical empiricism to discuss more closely a plea [Plädoyer] for the traditional concept of truth, until one shows to him the need to apply this concept in his analysis of science. Later in his paper Neurath added that "one had already seen in 1935 in Paris, how Tarski and Lutman were actually interpreted and probably not without justice, given that both show a certain 'connivence' vis-à-vis the traditional conception"(Neurath 1937d, p. 9). In conclusion, Neurath asked Carnap, Hempel, and the "Polish friends" to discuss whether and how far his "proposal" could be carried out and whether they thought that in this way "semantical and related problems could be brought within the logic of science."

Carnap's typescript is entitled "The Semantic Conception of Truth." It is dated 18.7.37 and it is twelve pages long. Carnap begins by listing four theses he would like to propose for discussion:

(1) The semantical conception of truth is correct and unobjectionable;
(2) It cannot be replaced by merely syntactical method;
(3) It is useful and important;
(4) It is in agreement with the concept of truth used in ordinary language.

Under (1) Carnap gives an informal description of the legitimacy of introducing a binary relation Bez (x, y) which captures the notion of denotation. Then he claims that in terms of denotation one can define truth. Both denotation and truth are examples of semantical concepts.

In section (2), Carnap addresses directly Neurath's proposal which, in a way, was an attempt to show the eliminability of talk of truth in terms of syntax. Here Carnap shows that this is not possible. Carnap grants that there are cases where sentences which contain semantical concepts (denotation, truth, etc.) can be transformed in purely syntactical sentences (in the technical sense of the *Logical Syntax*). For instance, "the expression '3 + 4' denotes (the number) 7" can be translated in the syntactical sentence " '3 + 4' is logically-synonymous with '7'." The semantical sentences which are translatable in syntactical sentences are called unessentially semantical sentences. The others are called essentially semantical sentences. There are also cases in which semantical concepts are eliminated by translating the sentence into a sentence of the object-language. For instance " 'Paris is a city' is true" can be translated into a sentence of the object-language "Paris is a city." Carnap points out that he had given examples of both strategies in the *Logical Syntax*. He then restates Neurath's proposal as: can one always eliminate the semantical concepts? He answers negatively. Carnap explains that what Neurath calls the 'trick' of science is nothing else than the elimination of a non-essential semantical sentence in the syntactical language. However, he disagrees with Neurath when the latter proposes to translate "truth" by "sentence of the Encyclopedia" or "acknowledged." Carnap argues for the difference between "true" and "acknowledged" by remarking that in the case of 'true' one does not need to give any temporal or pragmatic parameters, which are however necessary in the second case. This is the solution he had already proposed for distinguishing the two concepts in Paris 1935. Consider the sentence A: "the moon has in its dark side a crater which is even greater than the one it has in the visible side." While it can certainly be agreed that "A does not belong in 1937 to the Encyclopedia"(or "A is not scientifically acknowledged"), this is not the case for "A is not true" or its translation "the moon does not have in its dark side a crater which is even greater than the one it has in the visible side." Thus, Carnap concludes that "A does not belong in 1937 to the Encyclopedia" and "A is not true" do not have the same meaning; thus "true" and "(scientifically) acknowledged" (or "sentence of the Encyclopedia," "scientifically accepted", or "scientifically believed") are different concepts.

Moving on to point (3), Carnap expresses his belief that semantical concepts will turn out to be useful and important for epistemological work. As an example he gives a possible analysis of "x knows y"as "x believes y and y is true." Thus, Carnap concludes, "knows" is a semantical concept. A discussion of some examples with "knows" leads Carnap to observe that while in some cases the semantical notions can be eliminated by moving to sentences of the object language (as in the cases we have looked at before), this cannot be done when the sentence of the object language is not referred to by name. Examples would be: "Each sentence…" or "There exists a sentence…" Similar to semantical concepts are "seeing", "hearing", "perceiving". By contrast, Carnap adds, "believing", "thinking", "dreaming", "meaning", "imagining", are not semantical concepts.

In section (4) on truth and ordinary language, Carnap claims that he is not interested in the concept of truth of the metaphysicians but only that used in ordinary language.

Setting aside the iteration of semantical concepts, which leads to antinomies, one can arrive at an unobjectionable concept of truth for ordinary language which has the same degree of clarity as other concepts used in natural language. The argument proceeds by comparing two sentences: B: "It is true that Goethe died in Weimar in 1832" and C: "Goethe died in Weimar in 1832." Carnap holds that the word "true" is used in ordinary language in such a way that B and C are accepted as synonymous:

> A proposition of the form [B] which contains the word 'true' is more rarely used than [C] namely only when it was preceded by questions, doubts or disputes or when, for some other reason, one wishes to express a stronger emotional emphasis... But that is just a psychological and not a logical difference. And this is shown by the fact that no one who would be asked to decide between two propositions such as [B] and [C] would accept the first but reject the other or even leave the latter undecided.[33]

Carnap concluded that since the two sentences are recognized by ordinary speakers as logically equivalent and that the semantical theory of truth also treats them as logically equivalent that there is thus agreement between the ordinary usage and the semantic conception.

The concluding section of the paper gave some practical advise about how to proceed concerning the disagreements that were obviously present in the circle containing the notion of truth. On the side of the semantic concept of truth Carnap mentions the "Chicagoans"(Carnap, Hempel, Helmer), the Polish (Kotarbiński, Tarski, Lutman[-Kokoszyńska]), and in the opposite camp "Neurath and maybe Ness [sic] and others." Given that the debate cannot be immediately resolved Carnap expresses his conviction that the differences are due to a lack of clarity and misunderstandings that would disappear within a few years. As for the two groups, he gave the following suggestions:

1. Suggestions for the group of those who want to pursue semantics while their approach is empiristic and antimetaphysical. The latter will set up their terminology and formulations so that the delimitation of metaphysical problems always remains as clear as possible. They will do so not only with consideration for themselves but also for their readers. They will also keep in mind the question of the extent to which the semantical propositions are translatable into non-semantical ones; this especially in favor of those in our circles who, for whatever reason, strive to avoid semantical concepts.

2. Suggestions for the group of those who have reservations about the semantical concepts. They will at first temporize and they will not carry out public polemics against semantics as a whole until the further development let transpire, first, whether or not the work in the domain of semantics is fruitful for science and especially for the general task we have set ourselves of an analysis of science and, second, whether or not the feared danger of slipping back into metaphysics is real. Therefore, they will not characterize semantical concepts as a whole as metaphysical but will only criticize single specific formulations that they might find objectionable especially if they indeed give rise to pseudo-problems.[34]

15.9 Coda

While the 1937 Paris congress marks a culminating point in the debate on semantics among members of the Vienna Circle it was not the end of the story. Indeed, the conflict flared up anew with the publication of Carnap's "Introduction to Semantics" (1942) which led to renewed expressions of skepticism and outright dismissal on the part of Neurath. For instance, on 22.12.42 Neurath wrote: "Of Tarski's metaphysics I do no longer say anything. It is trivial sad. Aristotle redivivus, nothing more." Eventually, Carnap became exasperated with Neurath:

> As you can imagine, I am very sorry about the bad impression you got of my book, and that you even think it is a revival of Aristotelian metaphysics. I try to remember the many and sometimes long conversations we had in the past on Semantics. The first was in the train to Paris 1935. Then there was the public discussion at the Pre-Conference at Paris, with you and Ness [sic] on the one side, and Tarski and me on the other side. After these two discussions I remember I had the definite impression that there were no rational arguments left on your side. When Tarski and I showed that your arguments were based on misconceptions concerning the semantical concept of truth you had nothing to reply. What was left, as far as we saw it, were merely your emotional reactions, namely your dislike of the term "truth" and your vague fear that this would finally lead us back to old metaphysics. Later we sometimes had discussions on the same topic in America; but I did not have the impression that we came any step forward towards a mutual understanding, still less to an agreement... In any case, in spite of the disappointing experiences in the past, I am very willing to continue the discussion with you. (Carnap to Neurath, May 11, 1943, Neurath *Nachlaß*, English in original)

Needless to say, there was no reconciliation on this issue and the discussion on whether semantics was loaded with metaphysics continued in the correspondence between Carnap and Martin Strauss in the early 1940s.

As we know, Tarski addressed many of the issues we have discussed in his 1944 paper on truth (Tarski 1944). That paper is well known and I need not enter into Tarski's reply to the criticisms that had been raised against semantics. Many of those criticisms go back to Neurath. In particular, section 14 ("Is the semantic conception of truth the 'right' one?"), section 16 ("Redundancy of semantic terms—their possible elimination"), section 19 ("Alleged metaphysical elements in semantics"), section 20 ("Applicability of semantics to special empirical sciences"), and section 22 ("Applicability of semantics to the methodology of empirical science") of Tarski's 1944 article directly address, without mentioning him, issues that Neurath had been raising since 1935.

In conclusion, there were two parts to Neurath's criticism of semantics. On the one hand a background set of strongly held beliefs that led Neurath to his own proposal for using "truth" as "acknowledgement". On the other, the more specific criticisms to the semantic conception of truth that Neurath raised in consonance with those deeply held beliefs. The aim of this chapter has been to show how the criticisms to the semantic conception of truth emerge from those background beliefs and to spell out the discussion which emerged as a consequence with Tarski, Kokoszyńska, and Carnap.[35]

One could now ask how coherent are those deeply held beliefs. More specifically, is Neurath's proposal a defensible one? Ideas going back to Neurath's position are often discussed and criticized in the epistemological literature on coherentism (see Pollock and Cruz 1999, ch. 3, and BonJour 1985) where however the discussion is on justification/confirmation rather than truth. Indeed, BonJour (1985) defends a coherentist position of justification but a correspondentist account of truth. Hoffmann-Grüneberg (1988) for one endeavors to defend a conception of truth inspired by Neurath's position; in addition, there might be other "vindications" of Neurath that build on the rejection of philosophy of science as the methodology of truth attainment and emphasize the pragmatic component of how warrants are obtained and transmitted within scientific practice. Were any of these positions to mark an interesting and coherent approach to the problem of truth then we would have to recognize that behind the, at times frustratingly vague and unclear, objections by Neurath to semantics there was not just a negative drive but an idea which could be turned into a workable alternative.

Archival Documents

From Neurath's Archive (Haarlem, Konstanz)

Carnap–Neurath correspondence.
Hempel–Neurath correspondence.
Kokoszyńska–Neurath correspondence.
Morris–Neurath correspondence.
Naess–Neurath correspondence.
Neider–Neurath correspondence.
Stebbing–Neurath correspondence.
M. Strauss–Neurath correspondence.
Tarski–Neurath correspondence.
Neurath (1937d), Wahrheitsbegriff und Empirismus (Verbemerkungen zu einer Privatdiskussion mit Carnap im Kreis der Pariser Konferenz), Call number: K.30
Neurath (1937c), "Fuer Die Privatsitzung, 30 Juli 1937"; Call number: K.31
Neurath (1937b), "Diskussion Paris 1937 Neurath-Carnap"; Call number: K.32
Neurath (1937a), "Bemerkungen zur Privatdiskussion", Call number: K.33

From Carnap's Archive (Pittsburgh, Konstanz)

Carnap (1937), Ueber den semantischen Wahrheitsbegriff, RC 080-32-01
The correspondence between Carnap and Lutman-Kokoszyńska is classified under RC 088-57. It contains sixteen items.
The correspondence between Carnap and Neurath is found in different parts of the *Nachlaß*. Refer to the notes for specific call numbers of the correspondence mentioned in this chapter. While quoting from the Carnap–Neurath exchange if I use a source from

the Carnap *Nachlaß* I give the call number beginning with RC. Otherwise, the item comes from the Neurath *Nachlaß*.

M. Strauss–Carnap correspondence, RC 102-74 and 102-75
Schlick–Carnap correspondence. RC 029-27 and RC 029-28.

From Popper's Archive (Hoover Institution, Stanford)

Popper–Tarski correspondence. Box number 354. Folder ID: 8.

16

TARSKI ON MODELS AND LOGICAL CONSEQUENCE

IN the last two decades there has been a heated debate on the exact nature of Tarski's theory of logical consequence (Tarski 1936d,c).[1] Since the publication of Etchemendy's papers and book in the 1980s, contributions by, among others, Sher, Ray, Gomez-Torrente, and Bays have provided a much more detailed picture of Tarski's seminal 1936 paper. However, this intense focus on the original publication has led to several disagreements with respect to important interpretative issues related to Tarski's contribution. One of the bones of contention, and the only one I will treat in this chapter, concerns Tarski's notion of model, the key element of Tarski's famous definition of logical consequence. In this chapter, I shall offer new arguments (see sections 16.2 and 16.3) to show that Tarski upheld a fixed domain conception of model in his 1936 paper and that he was still propounding in 1940.

Let us begin with the definition of logical consequence:

The sentence X follows logically from the sentences of the class K if and only if every model of the class K is also a model of the sentence X. (Tarski 1936d, p. 417)

What then is a model? First of all note that X is a sentence and K a class of sentences. In order to define the notion of model for a sentence, Tarski uses the notion of satisfaction of a sentential function. He says:

One of the concepts which can be defined in terms of the concept of satisfaction is the concept of model. Let us assume that in the language we are considering certain variables correspond to every extra-logical constant, and in such a way that every sentence becomes a sentential function if the constants in it are replaced by the corresponding variables. (Tarski 1936d, p. 416)

I would like to thank John MacFarlane, Aldo Antonelli, Daniel Isaacson, Marcus Giaquinto, Ignacio Jané, José Ferreirós, Johannes Hafner, and Sol Feferman for their invaluable feedback on previous versions of this paper. I am also grateful to the Bancroft Library at U.C. Berkeley for having granted permission to quote from Tarski's unpublished correspondence to Corcoran and from the 1940 lecture.

Thus for L a class of sentences we obtain, by replacing non-logical constants by variables, a class L' of sentential functions. The notion of model is introduced next:[2]

> An arbitrary sequence of objects which satisfies every sentential function of the class L' will be called a model or a realization of the class L of sentences (in just this sense one usually speaks of models of an axiom system of a deductive theory). (Tarski 1936d, p. 417)

The definition of logical consequence provided by Tarski looks prima facie just like the model-theoretic definition of logical consequence we are used to from our courses in logic and model theory. However, whether this is so depends on the notion of model used by Tarski. This is where interpretations differ. Etchemendy in his 1988 paper claims that Tarski in 1936 is not working with our notion of model, in that there is no variability of domain of quantification in Tarski's approach:

> For the standard account, besides requiring that we canvass all reinterpretations of the nonlogical constants, also requires that we vary the domain of quantification. However, as long as the quantifiers are treated as logical constants, Tarski's analysis always leaves the domain of quantification fixed. Because of this, sentences like (15) will come out logically true on Tarski's account:
>
> (15) $(\exists x)(\exists y)(x \neq y)$
>
> This simply because on the present selection of logical constants, there are no nonlogical constants in the sentence to replace with variables. Thus, such sentences are logically true just in case they happen to be true; true of course in the intended interpretation. (Etchemendy 1988, p. 69)

He then later says: "Not only is Tarski's 1936 account of logical consequence not equivalent to the model-theoretic definition, he clearly avoided such an account with open eyes" (p. 72)

By the fixed domain conception of model one usually refers to an account of logical consequence of the sort described by Etchemendy, i.e. one in which one does not allow domain variation. We will see in sections 16.3 and 16.4 that the fixed conception of model comes in at least two main different versions.

Many scholars are unpersuaded by Etchemendy's interpretation. For instance, Sher (1991), Ray (1996) and Gomez-Torrente (1996) have argued against Etchemendy's account of Tarski's 1936 paper. In particular, Gomez-Torrente (1996) is considered by some as having solved all the historical issues. Discussing Etchemendy's historical claims (including the one on non-variability of domains), Scott Soames says:

> In my view, all of these essentially historical criticisms have been refuted by a variety of scholars, with the most thorough and penetrating refutation I am aware of being given by Mario Gomez-Torrente. (Soames 1999, p. 18)

What is Gomez-Torrente's view on the issue of models? According to him the reason why a sentence like $(\exists x)(\exists y)(x \neq y)$ is not a logical consequence of an arbitrary consistent theory rests on the following consideration:

> Tarski had in mind mathematical theories in whose canonical formulation the domain of objects of the intended interpretation or interpretations is the extension of a primitive predicate of the

language of the theory. Assuming that to specify a domain for an interpretation is nothing but to give an interpretation for such a predicate, Tarski's definition would allow for domain variation in the test for logical consequence. Under this assumption, sentences like '$(\exists x)(\exists y)(x \neq y)$' would not be declared logical truths, because they would be mere unofficial abbreviations for other sentences with relativized quantifiers ('$(\exists x)(\exists y)(Nx \& Ny \& (x \neq y))$' for example). It is natural to picture Tarski as having in mind the idea that the domain of the intended model or models is denoted by an extra-logical predicate, but without even thinking of formulating or caring to formulate this as an explicit requirement for the application of his definition." (Gomez-Torrente 1996, p. 143)

Despite Soames' claim, Gomez-Torrente's solution leaves many questions unanswered. For instance, Bays (2001) marshals several arguments against interpreting Tarski as defending a variable domain conception of model in the 1936 paper.[3]

In this chapter I will provide new evidence and considerations which lead me to the conclusion that Tarski was working with a fixed domain conception of model in 1936. The chapter proceeds as follows. The first section provides background on the notions of interpretation and model up to Tarski (1936d). In section 16.2, I will criticize Gomez-Torrente's position on the issue of the domain variability of models by showing that it does not account for a wide variety of theories considered by Tarski. Section 16.3 will provide new evidence from an unpublished lecture by Tarski which in my opinion tilts the balance in favor of the fixed-domain conception of model. Finally, section 16.4 will discuss a number of open problems.

16.1 Axiomatic Systems as Sets of Propositional Functions and Their Interpretations

The study of axiom systems for various disciplines goes back to the third part of the nineteenth century (Pasch, Peano).[4] It is the group of scholars centered around Peano that spent most care in trying to specify what is involved in the axiomatic method. In particular, Padoa and Pieri wrote important articles on the axiomatic method. Pieri (1901) asserted that the primitive notions of any deductive system whatsoever "must be capable of arbitrary interpretations within certain limits assigned by the primitive propositions", subject only to the restriction that the primitive propositions must be satisfied by the particular interpretation. He applied this notion of interpretation to discuss independence between propositions and referred to Padoa for a more extensive treatment. Alessandro Padoa was another member of the group around Peano. Just like Pieri, Padoa (1901, 1902) also speaks of systems of postulates as a pure formal system on which one can reason without being anchored to a specific interpretation, "for what is necessary to the logical development of a deductive theory is not *the empirical knowledge of the properties of things, but the formal knowledge of relations between symbols*" (Padoa 1901, p. 319; transl. van Heijenoort, 1967a, p. 121). It is possible, Padoa continues, that there are several, possibly infinitely many, interpretations of the system of undefined symbols which verify the system of basic propositions and thus all the theorems of a theory. He then adds:

> The system of undefined symbols can then be regarded as the abstraction obtained from all these interpretations, and the generic theory can then be regarded as the abstraction obtained from the specialized theories that result when in the generic theory the system of undefined symbols is successively replaced by each of the interpretations of this theory. Thus, by means of just one argument that proves a proposition of the generic theory we prove implicitly a proposition in each of the specialized theories. (Padoa 1901, pp. 319–20; transl. van Heijenoort, 1967a, p. 121)

The most natural reading of Pieri and Padoa on the issue of interpretations is that once the new specification of meaning for the primitive terms is given the newly interpreted system gives rise to a new set of propositions, which can be either true or false in the new interpretation. This interpretation foreshadows the conception of axiomatic systems as propositional functions, to be discussed shortly.

Throughout the 1910s the terminology for interpretations of axiomatic systems remains rather stable. Interpretations are given by reinterpreting the meaning of the original constants so that they refer to systems of objects with certain relationships defined on them. Bôcher suggests the expression 'mathematical system' to "designate a class of objects associated with a class of relations between these objects". (Bôcher 1904, p. 128)

During the first 15 years of the twentieth century we encounter a flurry of publications by a group of mathematicians collectively known as postulate theorists (see Scanlan 1991). Inspired by Hilbert and Peano's approaches to axiomatic theories their goal was to investigate systems of objects satisfying certain laws. The laws are expressed in terms of certain undefined primitives and taken as postulates. In 1906–07 Huntington describes the approach as follows:

> The only way to avoid this danger [using more than is stated in the axioms, PM] is to think of our fundamental laws, not as axiomatic propositions about numbers, but as blank forms in which the letters a, b, c, etc. may denote any objects we please and the symbols + and × any rules of combination; such a blank form will become a proposition only when a definite interpretation is given to the letters and symbols—indeed a true proposition for some interpretations and a false proposition for others. ... From this point of view our work becomes, in reality, much more general than a study of the system of numbers; it is a study of any system which satisfies the conditions laid down in the general laws of §1. (Huntington 1906–07, pp. 2–3)

In Huntington (1913) the approach is defined explicitly in terms of the concept of propositional function:[5]

> We agree to consider a certain set of *postulates* (namely, the postulates stated in chapter II), involving, besides the symbols which are necessary for all logical reasoning, only the following two *variables*:
>
> (1) The symbol K, which may mean any *class of elements* $A, B, C, \ldots$; and
> (2) the symbol R, which may mean any *relation* $A R B$, between two elements

These postulates are not definite propositions—that is they are not in themselves true or false. Their truth or falsity is a function of the logical interpretation given to the variables in such an equation. They might therefore be called 'propositional functions' (to use a term of Russell's) since they become definite propositions (true or false) only when definite 'values' are given to the variables K and R. (Huntington 1913, pp. 525–26)

Notice that this approach differs from the one we found in Padoa, as the basic postulates of the theory are not considered to be propositions.

For the postulates under consideration (defining geometry by means of the relationship of inclusion between spheres), Huntington (1913) gives immediately two interpretations. In the first interpretation K is the class of ordinary spheres including the null sphere; R is interpreted as the relation of inclusion. This interpretation satisfies all the postulates. The second interpretation has $K = \{2, 3, 5, 7, 10, 14, 15, 21, 210\}$ and $R =$ 'factor of'. This does not satisfy all of the postulates as postulate 4 is false in this interpretation.[6]

There are two points which are relevant here. First, the conception of an axiomatic system in terms of propositional functions. Second, the notion of a mathematical system as an interpretation of an axiomatic theory conceived as a set of propositional functions, to be discussed below.[7]

The conception of axiomatic systems in terms of propositional functions is quite widespread in the 1910s and 1920s. It is found, among others, in Whitehead (1907), Huntington (1911, 1913), Korselt (1913), Keyser (1918b,a, 1922), Langford (1927a), Lewis and Langford (1932), Carnap (1927, 2000, 1930).[8] Tarski probably encountered it for the first time in Ajdukiewicz (1921), who seems to have proposed it independently of the above sources (Tarski was also familiar with Huntington (1913), which he quotes in Tarski (1929b)). Talking about Hilbert's system for geometry (with interpreted logical symbols) Ajdukiewicz says:

Let $A(X)$ denote the logical product of the axioms of geometry whose consistency is to be shown. These axioms are not unambiguous sentences but are susceptible to various "interpretations" i.e. they are sentential functions defined for a system of variables such as "point", "straight" etc. This whole system of variables is represented by the letter X in the symbol $A(X)$. The totality of objects represented by it forms the 'domain' of geometry. The domain of geometry is thus a set of variables whose values are again sets, relations, etc. The axioms are, therefore, neither true nor false but turn into true or false if values are substituted for all variables. (Ajdukiewicz 1921, pp. 23–24)

Thus, for instance, when Hilbert replaces 'point', 'straight' and 'point a lies on straight line b' with a set of arbitrary pairs of real numbers $(x; y)$, a ternary relation of real numbers $(u; v; z)$ and 'the equation $ux + vy + w = 0$', respectively, then he has transformed the axiom system $A(X)$ into $A(\Omega)$.

This widespread conception of axiomatic system in terms of propositional functions is at the source of Tarski's notion of model. While in the 1936 article it is in the background (through the process of elimination of the non-logical constants by means of variables) it is explicitly stated in various publications of the period. At the 9th Congress of International Philosophy (1936), and thus at the same time of the publication of Tarski (1936d), Tarski presented a paper entitled "Sur la méthode deductive". In it he describes the conception as follows:

Let us imagine that in the axioms and theorems of the constructed science, we have replaced everywhere the primitive terms with corresponding variables (in order not to complicate the discussion let us ignore the theorems containing defined terms). The laws of the science have ceased to be propositions and have become what in contemporary logic are called propositional functions. These are expressions having the grammatical form of propositions and which become propositions when one replaces the variables occurring in them by appropriate constant terms. Considering arbitrary objects, one can examine whether they satisfy the axiom system transformed in the way described, that is if the names of these objects, once put in place of the variables, turn these propositions into true propositions; if this turns out to be the case we say that these objects form a model of the axiom system under consideration. For instance, the objects designated by the primitive terms constitute such a model. This model does not play any privileged role in the construction of the science; in deducing this or that theorem from the axioms we do not think at all of the specific properties of this model; on the contrary, from the way in which we reason it follows that not only this special model but every other model of the system of axioms under consideration must satisfy the theorems which we prove. (Tarski 1937b, pp. 331–32)

One finds, almost verbatim, the same characterization in Tarski's "Einführung in die mathematische Logik" (Tarski 1937a, pp. 81–82). There is thus no question that this is the conception of model that Tarski in the 1936 paper says comes from the methodology of the deductive sciences. Let me simply remark, to avoid confusion on the part of the reader, that when Tarski speaks of 'arbitrary objects' this is taken to include individuals, classes, relations etc. Another source of confusion is that Tarski at times speaks of 'concepts' instead of 'objects' (see below). To avoid confusion, I will talk of objects throughout.

16.2 Tarski on Models and Gomez-Torrente's Interpretation

We still need to get a more precise idea about what a model is. In Tarski (1935b), Tarski thinks of concrete deductive theories as 'models' ('realizations') of a general deductive theory that contains four primitive predicates (Tarski call them 'concepts') and specifies that any quadruple of 'concepts' (his terminology) that satisfies the axioms is called a model of the theory (see Tarski 1986a, vol. 2, note 1, p. 28). Thus, when dealing with a system of axioms that have only finitely many primitive (non-logical) constants, a model for that system is a finite sequence of objects (of the appropriate kind) that satisfies the propositional functions corresponding to the postulates (see also below, section 16.3).

Going back now to logical consequence it is easy to see how the notion of model of an axiomatic system is generalized and put to service in the explication of the relation 'X is a logical consequence of K'. We take every sentence in $K \cup X$ and we replace the non-logical constants by variables of the appropriate kind. Thus $K' \cup X'$ is the resulting set of propositional functions. A model of K' is a sequence of objects (of the appropriate kind) satisfying every propositional function in K' (notice that if a sentence L does not have any extra-logical constants then $L' = L$). Then, X is a logical consequence of K iff every model of K' is a model of X'.

This notion of model is obviously different from the one we are accustomed to. For instance, the domain of the model (the set over which the individual variables range) is not specified explicitly. Moreover, since K and X might contain constants for higher-order objects and quantifications over higher-order objects we must assume that the background logic the models in question have to "support" is a higher-order logic. The range of these higher-order variables is also not mentioned explicitly. Finally, note that in contemporary model theory we reinterpret the constants over different domains. Here the non-logical constants are not reinterpreted; rather we replace the non-logical constants by variables and consider the satisfaction of certain propositional functions by certain objects (see also note 9 in Bays 2001).

Usually, during the 1930s, Tarski uses as a background logic for his theories a simple theory of types (with or without axiom of infinity).[9] Examples are the theory of real numbers developed in 1931 and the theory of classes developed in Tarski (1933b). This usage is however not without exception. Often, he focuses on first-order logic (which can be seen as the first-order fragment of the simple theory of types). This is the case for instance when he is studying elementary theories (see the elementary theory of dense orders given in Tarski (1936b); Tarski 1986a, vol. II, pp. 232–34)[10] Quite often, Tarski is silent about the background system of logic. For instance, the axiomatic theories of Boolean algebra presented in Tarski (1935c, 1938b) or the theory of Abelian groups studied in Tarski (1938a) are given by Tarski without specifying the background logic.

It is worthwhile, for later discussion, to single out two important logical systems: $STT + I$ and STT. 'STT' stands for the simple theory of types; 'I' stands for the axiom of infinity. Of course, *Principia* has a ramified theory of types but Tarski, starting in 1931, always prefers to use the theory of types in its simple form and for this reason I have singled out the above systems. In any case, whenever a deductive theory is formalized within the background of $STT + I$, or a fragment thereof, the objects constituting the 'universe of discourse' of the theory are taken to be part of the individuals that form the class of lower complexity in STT. Thus, in the 1931 article on the theory of real numbers the real numbers are taken to be individuals.

So far we have established that models are in general finite sequences of 'objects' that satisfy the propositional functions corresponding to a theory in which the non-logical constants have been replaced by variables of the appropriate kind. The conflict of interpretations begins when we ask what is the range of the individual variables. We have seen that in 1988 Etchemendy claimed that Tarski's 1936 account of logical consequence is not equivalent to the usual model-theoretic definition. The reason, according to him, is that there is no mention of domain variability in Tarski's definition of logical consequence and that models share a fixed domain of individuals. In my opinion, the great virtue of Etchemendy's position here is the natural account it yields of Tarski's claim, let us call it (LM), in the 1936 paper (p. 419) according to which if one treats all non-logical constants as logical constants, the notion of logical consequence coincides with that of material consequence. In a variable domain conception of model it is hard, if not impossible, to make sense of the claim (see also Bays 2001, pp. 1078–79).[11] We have seen that Gomez-Torrente (1996) disagrees with Etchemendy but remarkably he says nothing at all about Tarski's claim (LM). He might be probably tempted to take on this issue the line taken by Ray (1996):

Tarski's 1933 work has, in effect, already shown us how truth-theoretic semantics will fit together with his semantic account of the logical properties for which domain-relativization is crucial. I think it gives us strong (though not historically decisive), additional reason *not* to suppose that Tarski just missed the need for domain relativization when it came to logical consequence, i.e. we have further reason to attribute to Tarski only the lesser of the two errors [the mistake of thinking that logical and material consequence coincide when all terms of a language are treated as logical constants, PM]. This judgment, in turn, tends to undermine Etchemendy's divergence argument, because that argument presupposes that Tarski made the greater error [the mistake of missing the need for domain variation, PM]. (Ray 1996, p. 630)

In short, those who claim that there is domain variability in Tarski's account of logical consequence either do not address Tarski's (LM) claim or simply dismiss it as an obvious mistake. Obviously, this is quite unsatisfactory.

Let us now move to the positive part of Gomez-Torrente's 1996 account. While I agree with much of Gomez-Torrente (1996), I find myself disagreeing with his treatment of the issue of domain variation in the definition of model contained in section 4 of his paper. I will repeat it for the reader's convenience:

Tarski had in mind mathematical theories in whose canonical formulation the domain of objects of the intended interpretation or interpretations is the extension of a primitive predicate of the language of the theory. Assuming that to specify a domain for an interpretation is nothing but to give an interpretation for such a predicate, Tarski's definition would allow for domain variation in the test for logical consequence. Under this assumption, sentences like '$\exists x \exists y \neg (x = y)$' would not be declared logical truths, because they would be mere unofficial abbreviations for other sentences with relativized quantifiers ('$\exists x \exists y (Nx \& Ny \& \neg(x = y))$' for example). It is natural to picture Tarski as having in mind the idea that the domain of the intended model or models is denoted by an extra-logical predicate, but without even thinking of formulating or caring to formulate this as an explicit requirement for the application of his definition. (Gomez-Torrente 1996, p. 143)

The claim derives its prima facie plausibility from the fact that many mathematical theories considered by Tarski at the time of writing his 1936 paper contain an extra-logical predicate that is meant to characterize the 'domain of discourse' of the mathematical theory. Gomez-Torrente refers to the theories presented by Tarski in his 1937 introductory book. Consider for instance the theory T on the congruency of segments presented by means of a language containing S and $\cong$ as primitives and two non-logical axioms:

1. for any element x in S, $x \cong x$
2. for any element x, y, z in S, if $x \cong z$ and $y \cong z$ then $x \cong y$

A model for T is given by any pair (K, R) satisfying the axioms. Variations of K would correspond to the variation of domains for individual variables in the standard model-theoretic account. Since $\exists x \exists y(\neg(x = y))$ would here be a short hand for $\exists x \exists y(Sx \& Sy \& \neg(x = y))$, this immediately yields that $\exists x \exists y(\neg(x = y))$ is not a logical consequence of T (as it is easy to display a one element model of the theory).

Despite the prima facie plausibility of such a claim, the examples of mathematical theories provided by Gomez-Torrente as illustration of his point do not settle the issue concerning the fixed domain vs variable domain issue, because those examples (taken from Tarski (1937a) and other articles of the period) can be accounted for straightforwardly in a fixed-domain interpretation (see also Bays 2001, pp. 1711–12). We could think of S as taking different interpretations within the same 'universal' class of individuals provided by the background theory of types. The problem then is: can these two interpretations be shown to differ substantially so as to adjudicate the issue of where Tarski stands in 1936? I claim they can.

First of all notice that the metatheoretical constraint posed by Gomez-Torrente's interpretation has the following two consequences: (1) The constraint would force the predicate S to have a non empty extension. In fact $\exists x(x = x)$ is a theorem of the background logic and thus the metatheoretical constraint would always force the theory T to have as an axiom $\exists x(S(x) \& x = x)$; (2) The constraint is such that every theory T formulated with a non-logical constant S characterizing the domain must be inconsistent with the claim $\exists x \neg S(x)$. Since the variable x must be restricted to objects in S the claim is equivalent to $\exists x(S(x) \& \neg S(x))$. But that's obviously unsatisfiable.

It is not hard to show that Tarski would not accept (1). On p. 145 of his *Einführung in die mathematische Logik*, Tarski is discussing a theory A formulated by means of the predicate Zl (for Zahl) and he remarks that the theory does not prove or disprove the statement $\exists x Zl(x)$. Since, as I pointed out, under Gomez-Torrente's constraint this is equivalent to $\exists x(Zl(x) \& x = x)$ (and thus $\exists x Zl(x)$) would have to be an axiom (resp. a theorem) of the theory), it follows that the constraint is incorrect when judged against Tarski's practice.

Concerning 2, in what follows I will show that Tarski would reject that every mathematical theory he is considering must be inconsistent with the statement that there are individuals which fall outside the 'domain of discourse'. While the argument given for (1) already proves the point (since in the case Zl is empty the theory A still proves the theorem $\exists x(x = x)$ forcing an element to be in the complement of Zl), I will use a different set of examples. I will provide evidence that Tarski and other logicians at the time made a distinction between 'range of the quantifiers' (or range of significance of the individual variables) and 'domain of discourse', that is they entertained theories that, while presenting a predicate S for the 'domain of discourse', either prove $\exists x \neg S(x)$ or simply do not decide the issue either way.

In order to simplify the discussion, from now on I will restrict myself to axiomatic systems with two undefined primitive symbols, i.e. two non-logical constants, and treat them in terms of propositional functions $f(x)$ (for a class of objects) and $g(x, y)$ for a relation.[12] The first thing to observe is that, as we have seen, most axiomatic systems studied at the time were intended to characterize 'mathematical systems' (K, R) where K is the value of $f(x)$ and R (generally, but not always, a relation on K) the value of $g(x, y)$, so that all the postulates are true on (K, R). This fact has led many interpreters (including Gomez-Torrente) to forget that the range of significance of the individual variables occurring in the propositional functions is wider than the 'universe of discourse' K. The distinction was obviously salient to many people involved in axiomatic studies. Langford, one of the foremost postulate theorists, says:

The set [of postulates] (a)–(d) [for dense linear orders, PM] places no restrictions upon things not satisfying the function f, that is, not belonging to the class determined by this function. In connection with any interpretation of the set, we are interested merely in relationships among those things belonging to the class determined by the interpretation put upon f; this class constitutes, as it were, the 'universe of discourse'. If, for instance, we allow $f(x)$ to mean "x is a planet" and $g(x, y)$ to mean "x is larger than y" we are concerned solely with size among planets. Nevertheless, the propositions resulting from this interpretation will be significant for other things, and will always be satisfied by them. (Lewis and Langford 1932, p. 353)

Some interesting metatheoretical applications required keeping the distinction between 'universe of discourse' and 'range of significance' of the individual variables clearly in mind. In Langford (1927a), which Tarski studied carefully and used in his seminars in Warsaw in the 1920s, we find an axiomatization of dense linear orders without end-elements that explicitly requires two axioms determining how objects not contained in the 'universe of discourse' behave. The last two axioms of Langford's axiomatization are very important for our goals.

While the first axiom states that no object x (no restriction to K) is such that xRx, axioms 2 to 8 state properties relativized to objects in K. For instance axiom 2 says that for all x, y, if x, y are in K and $x \neq y$, then Rxy or Ryx. Langford then adds:

These properties with some modifications, are the ones usually assigned for this type of order. They are, however, with the exception of (1), confined to assertions relevant to elements in the class K, and it is customary to omit any mention of properties belonging to elements not in K. But it is necessary in the present case to consider such properties—otherwise some important theorems break down. (Langford 1927a, p. 21)[13]

He then adds postulates 9 and 10 to the effect that

(9) If x and y do not both belong to K, then Rxy fails.

(10) there are at least n elements not in K.

This fact is highly relevant for the debate that has surrounded the interpretation of Tarski's notion of model in the article on logical consequence.

I will first show that Tarski entertains theories which, like the one given above by Langford, state facts about objects falling outside the 'universe of discourse', i.e. K in the previous example.

Consider a given axiomatic theory with only two primitives. I will denote the axiomatic theory by $A(f(x), g(x, y))$ to point out that the two non-logical constants have been replaced by variables which stand for classes of individuals and binary relations between such individuals. A model of such a theory, if there is any, is given by a pair (K, R). It is quite possible that this model also satisfies a sentential function such as $\exists x \neg f(x)$. Thus, if $\exists x \neg f(x)$ can be satisfied in a model (K, R) we have to remark that the range of the quantified variables is not something that is given explicitly in the presentation of the model (K, R). We have seen that Langford discussed theories of this sort. Did Tarski?

There are two articles by Tarski written around the period of his work on logical consequence that show that Tarski was in fact careful to keep the above distinction between 'universe of discourse' and 'range of the quantifiers' carefully in mind. The articles are Tarski (1934/35) and Tarski and Lindenbaum (1936). The context in which the issue emerges is the following. In (1934/35) Tarski is discussing the problem of the completeness of concepts. One of his main results consists in showing how this notion of completeness of concepts relates to the notion of categoricity. He says: "As is well known a set of sentences is said to be categorical when two arbitrary 'interpretations' ('realizations') of this set are isomorphic". In an appended note he refers to Veblen (1904) as the source of the notion. Let us recall that in this article, Tarski is working within the background of a simple theory of types. In particular there is a 'universal' class of individuals V.

For the theorem he is after, Tarski needs, however, a stronger notion of categoricity than the one given by Veblen. In note 15 he says:

We use the word 'categorical' in a different, somewhat stronger sense than is customary: usually it is required of the relation R [...] only that it maps $x', y', z', \ldots$ onto $x'', y'', z'', \ldots$ respectively, but not that it maps the class of all individuals onto itself. The sets of sentences which are categorical in the usual (Veblen's) sense can be called intrinsically categorical, those in the new sense absolutely categorical. The axiom systems of various deductive theories are for the most part intrinsically but not absolutely categorical. It is, however, easy to make them absolutely categorical. It suffices, for example, to add a single sentence to the axiom system of geometry which asserts that every individual is a point (or more generally one which determines the number of individuals which are not points). (Tarski 1983, pp. 310–11, note 1)[14]

Theorem 4 in Tarski's paper is false, as he warns us in note 19, if categoricity is taken in Veblen's sense.[15]

A similar warning is given in Tarski and Lindenbaum (1936) concerning the applicability of a certain theorem proved in the text. This "must be restricted to such axiom systems from which it follows that there are no individuals outside of the domain of discourse of the theory discussed" (Tarski 1983, p. 392). From the above quotes the following facts are evident:

(1) Most theories which are intrinsically categorical (such as geometry and arithmetic) do not prove that every individual is a point (in geometry) or every individual is a number (in arithmetic). One can add such axioms to the original theory and then absolute categoricity and relative categoricity coincide.

(2) In general, the 'domain of discourse' of a theory does not coincide with the range of significance of the individual variables (i.e. the range of the individual quantifiers). This only happens when a special axiom is added to the theories in question.

The upshot of the above for the Gomez-Torrente interpretation is that there is no evidence that Tarski thought of most of the theories he treated as presupposing a restriction of the range of the quantifiers to the 'universe of discourse' of the theory as he claims that most examples of theories which are intrinsically categorical fail to be absolutely categorical because they do not have as an axiom or as a theorem the statement $\neg \exists x \neg S(x)$. Thus, $\exists x \neg S(x)$ is at least consistent with them or, in some cases, even provable in them. But then one cannot relativize the existential quantifier to that same predicate S.

Now it might seem paradoxical that I appeal to the articles Tarski (1934/35) and Tarski and Lindenbaum (1936) as evidence against Gomez-Torrente as the results on categoricity stated in those papers are used even by (Bays 2001, p. 1710, note 14) as evidence that Tarski despite presupposing the fixed domain conception of model in the 1936 paper on logical consequence, also worked comfortably with a variable-domain conception of model:

> In [1934/35], for instance, Tarski proves two theorems concerning the categoricity of several (second-order) systems of axioms. First, he proves that the axioms for second-order arithmetic are categorical *on the assumption that these axioms include an axiom stating that every individual is a number.* Second, he proves that there is a categorical set of axioms which characterize the real numbers. Clearly, these two theorems cannot be jointly accepted on a fixed-domain conception of model. For, on such a conception, the first result would show that the number of objects in the world is merely countable (since we can find some model of the natural numbers with the whole world as its domain), while the second result would show that the world is uncountable (since it contains enough individuals to construct a model for second-order analysis). (Bays 2001, p. 1710)

I will come back in the conclusion to these issues raised by Bays.

In my opinion, Tarski (1934/35) and Tarski and Lindenbaum (1936) show that Tarski in publications around the logical consequence paper is dealing, among other things, with theories expressed within the background of simple type theory and that the range of significance of the individual variables is determined by the class of individuals corresponding to the interpretation of the individual variables assumed for the simple theory of types (that is, the universe V). For this reason, when specifying a model for a theory, there is no need to specify what the range of the individual variables is; as for the 'universe of discourse' of the mathematical theory in question this will be given by a class taken as the value of the variable corresponding to one of the primitives of the theory (K as the value of $f(x)$ in the examples above).[16]

Further evidence can be adduced against Gomez-Torrente's claim. Tarski did in fact come back to explicating what was involved in his notion of categoricity and explicitly denied that his mathematical theories are to be thought as having a predicate characterizing the domain. This is found in correspondence Tarski had with Corcoran concerning changes for the second edition of *Logic, Semantics and Metamathematics*. In 1979, John Corcoran had raised some objections to Tarski's claim on Veblen vs absolute categoricity to which Tarski replied in 1980. (This material is found in the Tarski papers at the Bancroft Library at U.C. Berkeley).

Tarski first instructs Corcoran to replace on p. 311 (line 14) 'usual (Veblen)' by 'customary'. Then he adds a clarification in reply to two objections by Corcoran (contained in folder 9.13, sheets dated 8–22–79) where Corcoran objects first, that Veblen categoricity is not the same, as Tarski implies, as intrinsic categoricity and second, that intrinsic categoricity and absolute categoricity are the same. Tarski replies:

Re your objection to this line and the following ones. I do not claim that my definition of intrinsic categoricity is exactly equivalent to the one of Veblen. To decide this question I would have to know what Veblen understands by a model of a theory. What would be a model for him if the theory discussed were provided, say, with 2 or 100 of unary predicates and (for simplicity) with no n-ary predicate for $n \geq 2$? Or a contrary case: the theory provided with no unary predicates but with some predicate of rank $n \geq 2$? From what you find on the same p. 311 you will see e.g. that I am interested in a theory provided with a ternary predicate as the *only* [underlined by Tarski, PM] primitive predicate—this is not Veblen's case. At any rate I have removed, as you see, the reference to Veblen, so that the objection is not applicable. As regard your second objection—the alleged equivalence of intrinsical and absolute categoricity, your argument is wrong. Consider an intrinsically categorical system with two primitive predicates a unary U and a ternary R; to simplify matters assume that one of the axioms of this system is

$$(x, y, z) : R(x, y, z) . \supset .U(x).U(y).U(z)$$

Certainly there may be two models of this system $\langle U', R' \rangle$ and $\langle U'', R'' \rangle$ such that there are just two elements of the universal class V which do not belong to U', and just three such elements which [[do not]] belong to $U''(\star)$. Then obviously there is no function which maps in a one-one way both V onto V and U' onto U''; thus our system is not absolutely categorical.

($\star$) I assume that there are no axioms of our system which ascertain anything about elements of V not belonging to U, since this is irrelevant for intrinsical categoricity. I also assume that the axioms [[secure]] the infinity of the set U. (Tarski Papers, Box 9. Folder 9.11, p. 43a)

Two points are essential for our discussion:

(1) Tarski emphasizes that in the case discussed he is "interested in a theory provided with a ternary predicate as the only primitive predicate". Thus, one cannot claim that Tarski always presupposes that there will be a unary predicate characterizing the domain of the theory. By the way, in (Tarski and Lindenbaum 1936; see Tarski 1986a, vol. 2, 208–09) the theory of geometry appealed to has only a four-placed predicate of congruency between point pairs.

(2) The simplification adopted in the discussion is revealing. Tarski says: "Consider an intrinsically categorical system with two primitive predicates a unary U and a ternary R; to simplify matters assume that one of the axioms of this system is $(x, y, z) : R(x, y, z) . \supset .U(x).U(y).U(z)$". This means that in general one does not assume that the objects satisfying the relationship $R(x, y, z)$ have to be in the extension characterized by a unary predicate U. In other words no assumption of cross-binding is given in general between U and R as would be necessary in the Gomez-Torrente interpretation (since the interpretation of U would have to be the domain of discourse of the model over which $R(x, y, z)$ must take its values). Of course, specific mathematical theories might force the cross-binding through the axioms, but that's another issue.

There is one more important issue related to Gomez-Torrente's reconstruction that cannot be left unmentioned. In order for his proposed reconstruction to go through, it is essential that the background logic contains no existential assumptions. He says:

> Tarski naturally intended his definition to be applicable not only to purely logical theories, but also to mathematical theories with special mathematical primitives and postulates. In the works of this period Tarski considers several mathematical theories formalized using a logical apparatus, or 'logical basis', to use Tarskian terminology, without any cardinality assumption. Generally, this logical basis for formalization is again the calculus of levels, but without the axiom of infinity. (Gomez-Torrente 1996, p. 141)

But that is a puzzling comment as Tarski did in fact consider many theories as formalized within a logic with strong existential assumptions. I will come back to this in the next section but the relevance of the point is related to one aspect of the notion of model that Tarski emphasizes (see Tarski 1937a, p. 83). If we have a model of an axiomatic system T then that model also satisfies all the theorems of the axiomatic system. But notice that all the axioms and theorems of the background logic will also be theorems of the axiomatic system T. In particular if the axiom of infinity is part of the background logic the model would have to satisfy the axiom of infinity. If we restrict the range of the quantifiers to the 'universe of discourse' this would come into conflict immediately with any interpretation in which the 'universe of discourse' is finite. But if we make the distinction between 'universe of discourse' and 'range of the individual variables' then there is no conflict.

Where does this leave us? I think that the evidence adduced so far shows that the original strategy by Gomez-Torrente does not work because it forces in every theory the identity between the intended domain of the theory and the range of the quantifiers. But Tarski thinks that most theories (arithmetic, analysis, geometry) are such that that identity does not hold. This suggests that the range of the quantifiers is always to be taken as the 'universal' domain of individuals. But against this speaks the fact that we have strong categoricity results for arithmetic, analysis and geometry which would seem to force, on a fixed domain conception of model, the (fixed) universe of individuals to be at times countable and at times uncountable. I will discuss in the conclusion whether these absolute categoricity results can be accounted for in the fixed conception of model.

I thus take the above considerations to show the weakness of the Gomez-Torrente strategy but not as decisive as evidence for the fixed-domain interpretation of model, or a variation thereof.

In the next section I will provide a further piece of evidence which speaks in favor of the fixed-domain interpretation.

16.3 Tarski's 1940 Lecture on Completeness and Categoricity

In this section I will give new archival evidence that speaks in favor of thinking that Tarski had in mind a fixed conception of model in his theory of logical consequence. In 1940 Tarski gave a lecture at the Logic Club at Harvard entitled "On the completeness and the categoricity of deductive systems" (Tarski 1940).[17] It contains some important (and I claim decisive) evidence concerning the issues I have been discussing.

Tarski's general aim in the lecture is to develop semantical analogues of the notion of syntactic completeness for a deductive theory by means of the notion of semantic completeness and to carry out related investigations on the notion of categoricity.[18]

The motivation for developing these semantical notions of completeness is given by Tarski through some historical reflections on attempts to capture syntactically the notion of logical validity. He begins with a familiar distinction between logical and non-logical sentences as applied to an arbitrary deductive theory:

In what follows we assume that the concepts (constant terms) of a deductive theory are divided into two classes, the logical and the non-logical, to the first of which belong in any case the constants of the calculus of sentences and the quantifiers. Correspondingly, we divide the sentences of our theory into two classes, the logical and the non-logical, depending on whether they contain exclusively logical constants or not. (Tarski 1940, p. 3)

The logically valid sentences are a special subset of the logical sentences:

Among the logical sentences we single out the logically valid sentences. This is usually done in an axiomatic way: the logically valid sentences are defined as those which can be obtained by applying the determined rules of inference to the given logical axioms.

The class of logically valid sentences forms the logical basis of the given deductive theory. (Tarski 1940, p. 4)

I hasten to point out that the axiomatic notion of logically valid sentence cannot coincide, exactly for the reasons given below by Tarski, with the class of sentences true in all models.[19] Tarski specifies that he will consider deductive theories for which a concept of derivability has been specified in such a way that "every logically valid sentence is derivable from any system of logical or non-logical sentences" (p. 4). In short, any theory T will have as consequences all the logical axioms and the theorems derivable from the logical axioms. If the background logic has an axiom of infinity, the theory will (trivially) have that axiom as part of its theorems.

As an example of the sort of deductive theories Tarski has in mind he mentions the logic of *Principia Mathematica*, or a fragment theoreof, with additional non-logical constants for geometry through which a system of axioms for Euclidean geometry is formulated:

Taking any deductive theory let us consider an arbitrary system of non-logical sentences of this theory. As the theory let us think e.g. of the system of *Principia Mathematica* or of a fragment of it, but in either case enriched by certain non-logical, viz. geometrical constants and a system of sentences, a system of axioms for Euclidean geometry. (Tarski 1940, p. 4)

Thus, notice that Tarski is thinking of the background logic for such theories as being a theory of types and allows the possibility that it be a fragment of *Principia* (in particular, no assumption has been made that the axiom of infinity be among the logically valid sentences). In general, if the logical basis (the class of logically valid sentences) is strong enough, one encounters the phenomenon of syntactic incompleteness:

If the logical basis of our theory is rich enough, we can formalize within its boundaries the arithmetic of natural numbers and for this reason its logical basis is incomplete: i.e. there are logical sentences which are not logically valid and whose negations are likewise not logically valid. In other words, there are problems belonging entirely to the logical part of our theory which cannot be solved either affirmatively or negatively with the purely logical devices at our disposal. (Tarski 1940, p. 4)

Now, it is a well known fact that one needs the axiom of infinity to develop arithmetic in a theory of types (see *Introduction to Logic*, Tarski 1941, pp. 81, 130; and Tarski 1937a, pp. 51, 80 (note 1), 87). This shows that Tarski does not insist on a logic which carries no existential assumptions and thus considers freely mathematical theories which have as a basis a theory of types with the axiom of infinity (contrary to claims by Gomez-Torrente mentioned in the previous section).

It is on account of the incompleteness phenomenon that Tarski introduced the notion of semantical completeness:[20]

Now let us introduce the concept of semantic completeness. Because incompleteness is such a general phenomenon, the problem arises as to whether this is due to our conception of derivability. The concept of derivability developed in modern logic and reduced there to the concept of constructive rules of inference was intended to be a formal analogue of the intuitive concept of logical consequence. Thus, a doubt arises as to whether this intention has been realized. I discussed this question in some of my papers published some years ago and in this lecture I can only avail myself of the final result of this discussion. It turned out that between the intuitive concept of logical consequence and the formal concept of derivability there was a big gap. If we want to formulate an exact definition of the concept of logical consequence we must apply quite different methods and concepts. (Tarski 1940, pp. 4–5)

Here I will simply remark that Tarski is not distancing himself from any part of the analysis of logical consequence given in the 1936 paper, and thus this lecture can be considered to be in line with the positions put forward in that paper. Now we finally arrive at Tarski's summary of his approach to logical consequence:

The most important role is played here by the concept of model or realization. Let us consider a system of non-logical sentences and let, for instance "C_1", "C_2"..."C_n" be all the non-logical constants which occur. If we replace these constants by variables "X_1", "X_2"..."X_n" our sentences are transformed into sentential functions with n free variables and we can say that these functions express certain relations between n objects or certain relations to be fulfilled by n objects. Now we call a system of n objects O_1, O_2... O_n a model of the considered system of sentences if these objects really fulfill all conditions expressed in the obtained sentential functions. It is of course possible that the whole system reduces to one sentence; in this case we speak simply of the [sic] model of this sentence. We now say that a given sentence is a logical consequence of the system of sentences if every model of the system is likewise a model of this sentence. (Tarski 1940, p. 5)

This is very much in line with Tarski's 1936 definition. However, notice that Tarski is here more explicit about the fact that models of axiomatic theories will in general be finite sequences, since there are usually finitely many non-logical constants used in the formulation of the theory. This, I also pointed out above by referring to Tarski (1935b).

After remarking on the semantical nature of the concepts involved ('model', 'fulfillment') Tarski goes on to formulate the notion of semantical completeness, which is obtained by replacing the concept of derivability with the semantical concept of logical consequence in the definition of completeness:

> Thus a system of sentences of a given deductive theory is called semantically complete if every sentence which can be formulated in the given theory is such that either it or its negation is a logical consequence of the considered set of sentences. (Tarski 1940, p. 5)

Then Tarski claims the following:

> It should be noted that the condition just mentioned is satisfied by any logical sentence: hence we can deduce without difficulty that the concept of semantical completeness is a generalization of the concept of relative completeness: every system that is relatively complete is likewise semantically complete (but it can be shown by an example that the converse is not true). (Tarski 1940, p. 5)

Tarski makes here an important claim, i.e. "that the condition just mentioned is satisfied by any logical sentence", and then proceeds to draw a mathematical result from it. The fact that he draws a mathematical result from the first claim shows that the claim cannot be dismissed lightly as an oversight. Thus, I will first analyze what Tarski's claim implies and then give the details of the mathematical result he draws from it.

Tarski's claim

I will refer to Tarski's claim as 'C'. Tarski premisses his discussion of alternative notions of completeness by saying that he will be interested in theories that do have non-logical constants in them. On page 4 he says:

> We shall be interested here only in such deductive theories in which non-logical sentences actually do occur, and with respect to their completeness we shall consider exclusively systems consisting of non-logical sentences. To simplify our discussion let us assume moreover that there are no non-logical constants of our theory which do not occur in sentences of the considered system. (Tarski 1940, p. 4)

Let $L(T)$ be the language (logical and non-logical) in which the theory is formulated. To clarify what C amounts to, let us simplify the situation and consider as the system of sentences in question the entire theory T, identified here for convenience with all the statements containing non-logical expressions of $L(T)$. T is semantically complete iff for every P expressed in $L(T)$, either P is a logical consequence of T or $\neg P$ is a logical consequence of T. If L is a sentence in $L(T)$ that contains only logical symbols then, according to C, it is automatically the case that either L is a logical consequence of T or $\neg L$ is a logical consequence of T.[21] Thus, either all models of T make L true or all models of T make $\neg L$ true. Hence, each model of T gives us complete information about which logical sentences are true in it and all models of T agree on the truth values of the logical sentences. In the face of it this claim commits one to a conception of logical consequence which cannot allow for full domain variation. Let, for instance,

T be the axioms for the theory of groups. Consider the logical sentence '$\exists x \exists y \neg(x = y)$'. According to Tarski's claim either '$\exists x \exists y \neg(x = y)$' is a logical consequence of T or its negation is. But this cannot be accounted for in a conception of logical consequence which allows for full domain variation. For in that case we could easily find a one element domain satisfying the axioms and a two element domain satisfying the axioms thereby showing that neither '$\exists x \exists y \neg(x = y)$' not its negation is a logical consequence of T.

Semantical completeness implies relative completeness

Lest the reader thinks that claim C is just an off hand remark by Tarski, or worse a slip of the pen, I hasten to point out that Tarski relates his claim to the notion of relative completeness. Before introducing the notion of semantical completeness, Tarski considers a different concept 'completeness with respect to the logical basis' or simply 'relative completeness'. Consider a deductive theory and an arbitrary system [set] of non-logical sentences of the theory. If the logic is strong enough we are faced with incompleteness. Requiring the system of non-logical sentence to be complete would mean that the non-logical sentences can decide the problems which are left undecided by the logic. This is in general too strong of a requirement. However, "we can at least require that the considered system of sentences should not extend the incompleteness of the logical part of our theory" (p. 4). The condition of relative completeness is then stated as follows:

In order to formulate this requirement in an exact way, let us denote two given sentences as equivalent with respect to the considered system of sentences, when if this system is enriched by the addition to it of the first of these sentences, the second becomes derivable and vice versa. Our requirement can now be stated as follows: For any sentence of our theory, there must be a logical sentence which is equivalent to it with respect to the given system of sentences. If this condition is satisfied, we say that the considered system is *complete with respect to its logical basis, or simply, that it is relatively complete*. It is clear that in case the logical basis is itself complete, relative completeness reduces to absolute completeness. (Tarski 1940, p. 4)

Let us now show how claim C is used by Tarski to show that "every system that is relatively complete is likewise semantically complete" (p. 5) The argument can be spelled out as follows. Suppose S is relatively complete. As Tarski does let us assume that from S all logically valid sentences can be derived. Let ϕ be an arbitrary sentence in $L(S)$ ($L(S)$ includes also all the logical symbols). We want to show that either ϕ or $\neg \phi$ is a logical consequence of S. By definition of relative completeness for any sentence ϕ there is a logical sentence ϕ^* such that $S \cup \{\phi\} \vdash \phi *$ and $S \cup \{\phi *\} \vdash \phi$. If ϕ is a logical sentence then, by claim C, it or its negation is a logical consequence of S, so there is nothing to prove. If ϕ is not logical then let ϕ^* be a logical sentence satisfying the condition given in the definition of relative completeness for S. We consider two cases. Because ϕ^* is a logical sentence either all models of S are models of ϕ^* or all models of S are models of $\neg\phi^*$. First assume all models of S are models of ϕ^*. Then any model M of S is also a model of $S \cup \{\phi^*\}$. Since $S \cup \{\phi *\} \vdash \phi$ and the logical system is assumed sound, M is a model of ϕ. So ϕ is a logical consequence of S. Now assume that every model of S is a model of $\neg\phi *$. Furthermore, by way of contradiction, assume there is a model M' of

S such that M' is not a model of $\neg\phi$. Thus M' is a model of ϕ. Because $S \cup \{\phi\} \vdash \phi^*$ hence by the soundness of the logical system, M' is a model of ϕ^*. This contradicts all models of S being models of $\neg\phi^*$. So $\neg\phi$ is a logical consequence of S. Consequently, S is semantically complete.

Notice that this claim by Tarski gives a general theorem about conceptual relationships between relative completeness and semantical completeness and it is stated in such a way that no qualification on the logic is made. Tarski meant this result to hold, as he explicitly says in a quote given in the main text, at least for the system of *Principia Mathematica* and all fragments of it, including the first-order fragment without the axiom of infinity.

'Strong' fixed conception of model vs 'Absolute relative' fixed conception of model

The formalization of any theory would have two parts. First, the logical basis expressed in a type-theoretic framework, let us call it STT. Then there would be a set of axioms for the theory formulated by means of non-logical constants.

It is the type-theoretic framework that in the first place decides what the class of individuals (V) is. We can think of this as follows. The type-theoretic framework comes interpreted. In particular, the quantifiers play the role of logical constants and thus, as Etchemendy pointed out, there is no reinterpretation of the quantifiers. For this reason every model of the mathematical theory T in question will either make true '$\exists x \exists y \neg (x = y)$' or its negation. However, it is not all obvious that '$\exists x \exists y \neg (x = y)$' will in fact be the sentence which turn out to be a logical truth. That will depend of the class of individuals over which the variables of lowest order are ranging.

Here we have to clarify what has been an ambiguity in the fixed-domain interpretation all along. It seems to me that up to this point the fixed domain conception has been understood in the literature as follows. The theory of types comes already with the meaning of the quantifiers fixed and that determines the range of the individual variables in all possible theories that can be formulated over the theory of types. The range of the individual variables (thus the class V) is taken to be the 'real' universe of individuals. Obviously on such a conception '$\exists x \exists y \neg (x = y)$' would have to be a logical consequence of all theories T formulated within the background of the theory of types. I would like to dub this the 'absolute' (or 'strong') fixed domain conception of model.

But if we look at the practice of using the theory of types as background for a mathematical theory we notice that something else is going on. Every theory T comes equipped with its background theory of types and with its own interpretation of the theory of types. Thus, for instance if one has decided to study an axiomatic system for Peano arithmetic over a theory of types (as Gödel does in 1931) one can take the meaning of the individual quantifiers to be such that they range over a superclass of the natural numbers or, if one prefers, over just the natural numbers. The same for a theory of real numbers (see Tarski 1931) where the class of individuals might be larger than the class of real numbers or be exactly identical with it. Thus, there is a certain flexibility in choosing what the class of individuals is that will be assumed in the background. I would like to call this the 'relative' (or 'weak') fixed domain conception of model. It is this kind of flexibility,

I claim, that accounts for the categoricity results that have given so much trouble to the original interpretation of the fixed domain conception of logical consequence.

I thus think the above argument makes it plausible that in Tarski's definition of logical consequence all models come equipped with a fixed domain in the background and that logical validity is not a matter of truth in all models with variable domains.

16.4 Conclusion

The contemporary debate on Tarski's notion of logical consequence has ranged widely from historical issues about interpreting Tarski's 1936 text to philosophical debates about what the correct notion of logical consequence should be. Obviously, in this chapter my intention was only to discuss the historical claims that have been made concerning the notion of model in the 1936 article. In particular, I do not touch at all the theoretical issue of whether Tarski's notion of logical consequence is what we want as an account of logical consequence and whether it undergenerates or overgenerates with respect to such a desideratum. Once again, I am aiming here at clarifying what the historical Tarski, rightly or wrongly, thought. However, the evidence provided heightens the problem of finding a coherent interpretation of Tarski's position. This is not an easy task. The problem is that some of the claims by Tarski point in the direction of a fixed-domain conception of model and others point in the direction of a variable conception of model. Is this an unresolvable tension or can one find a point of view which will manage to accommodate the evidence in a unified framework without doing violence to the Tarskian texts? The following reflections are proposed as a tentative suggestion for a possible solution.

Let me briefly summarize the situation. In favor of the fixed-domain conception of model one can adduce a) Tarski's claim that if every constant of a language is taken as logical then logical consequence reduces to material consequence; b) Tarski's claim contained in the 1940 lecture that for every logical sentence L and for any mathematical theory T formulated in a given background logical theory, either L or $\neg L$ is a logical consequence of T (see claim C in section 16.3).

In favor of the variable domain conception we can mention: c) the claim that intrinsic categoricity and absolute categoricity coincide when the theory in question has as an axiom or consequence a sentence to the effect that the domain of individuals coincide with the intended domain. Second-order arithmetic and analysis are both absolutely categorical in this sense but in the first case the class of individuals, V, turns out to be countable and in the second case uncountable (see quote by Bays in section 16.3). Thus there seems to be variability of V after all; d) Tarski's use in his metamathematical work of upwards and downwards forms of Löwenheim-Skolem. Consider the upward Löwenheim-Skolem theorem. On the fixed-domain conception of model, if the class of individuals has a specified infinite cardinality, then any theory which has an infinite model will also have models of cardinality higher than the cardinality of the class of individuals. But that's obviously not possible, since no model could have more elements than the class of all individuals.

I will sketch in broad outline here what I think is a plausible solution for accounting for all these claims at once. The most important thing, which is often forgotten, is that the notion of logical consequence for Tarski is always tied to specific interpreted languages. Consider an axiom system for geometry with simple type theory as the logical basis. On the reconstruction I have given the axiom system can be represented as superposing to the simple theory of types a set of axioms, which for convenience we can abbreviate as AS(**R**) where **R** stands for a vector of predicates and relations symbols needed for the specific axioms. The type theory is already interpreted, e.g. the quantifiers range over a fixed class of individuals. Under general circumstances we let this interpretation be the natural one, e.g. 'all' individuals. However, there are (as pointed out in Tarski and Lindenbaum (1936)) special axiomatic investigations in which it is convenient to assume that the class of individuals coincides with that of points of with that of numbers. This is exactly what happens, with minor differences in Gödel's 1931 paper and in Tarski's 1931 paper on definability of real numbers. Let us go back to the case of geometry. Let L' stand for the interpreted language in which the range of the quantifiers consist of 'all' individuals. Let L'' stand for the interpreted language (same syntax as L') in which the range of the quantifiers is limited to points. Because the two languages are different (despite their having the same syntax) so will in general also the respective extensions of the consequence relationship. But notice the following. Once the interpreted language is given, the logical consequence relationship is reduced to truth in that interpretation of a universal sentence. To judge whether a certain statement $Th(\mathbf{R}^*)$ ($\mathbf{R}^*$ a subset, possibly empty, of **R**) is a logical consequence of $AS(\mathbf{R})$ we ask whether $\forall \mathbf{X}(AS(\mathbf{X}) \rightarrow Th(\mathbf{X}^*))$ is true in the intended interpretation. How to account then for Löwenheim-Skolem type theorems in this context? Simply, we look at them as metatheorems about the various interpretations that can be given to an axiomatic system based on the first-order fragment of the simple theory of types (obviously, we need to limit the type theory otherwise the theorems in question do not hold). Consider the upward Löwenheim-Skolem theorem. Consider a first-order axiomatization of geometry. In the metatheory then we prove that if we have an interpretation of a type theory L' which is such that there is an appropriate sequence of objects $\mathbf{O}'$ satisfying the propositional function corresponding to the axioms and which are sets, then we we can find an interpreted language L'' of type theory and a sequence $\mathbf{O}''$ (which are sets) such that the cardinality of the domain from which they are taken is higher than that corresponding to the objects $\mathbf{O}'$. The use of set theory in the metatheory is absolutely essential. Type theory alone would not be sufficient to generate the required cardinalities.

What I just said, can be read in two ways. Either as a claim that for any chosen universe of individuals, which is a set, we can always go on and come up with a new universe of individuals or as claiming that there is a universal class V from which we can manage to carve set interpretations of higher and higher cardinality. If we go the second way then it is essential to insist that the domain of all individuals V underlying every single one of these interpretations be a proper class in the set-theoretic sense, i.e. it can have no cardinality. Here the picture would be the following. We assume that we are dealing with a unique interpretation of the type theory which is common to all the possible axiomatic systems that can be superimposed on it. The domain of individuals then is a proper class and every interpreted language L' will have all its variables ranging over the same class V. In order to account, in this picture for the categoricity results, we would have to assume that we are investigating the consequence of a contrary to fact assumption (this way of reading the situation has been suggested by Marcus Giaquinto). And there is evidence that points in this way. For instance, Tarski says explicitly that "the axiom systems of various deductive theories are for the most part intrinsically but not absolutely categorical". Examples would include geometry and arithmetic. So, one could argue that the ordinary systems of arithmetic and geometry are only intrinsically categorical because the domain of individuals is wider than the domains of natural numbers (in the case of arithmetic) or the domain of points (in the case of geometry). That is, the identity of all individuals with the numbers or with the points which is forced by adding to the systems of arithmetic or geometry sentences to that effect would result in theories which are consistent but contrary to fact. If this were the case, then the apparent relativization of the universe of individuals V to a countable or uncountable set would just be a contrary to fact assumption.

How about the first alternative? In that case there is no single, given once and for all, class of individuals V that is fixed in advance. Every interpretation (i.e. arbitrary choices for what counts as the class of individuals) of the type-theoretic apparatus is fine. Once fixed, that interpretation gives rise to an interpreted language L' with its notions of truth in L' and logical consequence for L'. But notice that despite the variability of V for different interpretations of the type theory, once the interpretation of V is fixed questions of logical consequence for that specific interpreted language are to be answered only by looking at models (i.e. sequences of appropriate objects) coming out from that specific interpretation for V. Thus all models for that interpretation share the same domain V (e.g. the domain of the interpretation). Obviously, in this set up there is no problem in accounting for the categoricity results. In conclusion, according to this second reading, what seems to have been a source of confusion in understanding the notion of a fixed domain for deciding what logical consequence consists in is not having kept clearly in mind that the notion is always relativized to a specific interpreted language. It is with respect to that interpretation that the domain for all the models become fixed.

Whichever of the two options outlined above one accepts it will not affect the fact that to determine what logical consequence consists in for a specific language L' we need to focus on that interpretation of L' and all the models in questions will then share the same domain.

Let me conclude by pointing to one further historical problem that would deserve more detailed study.

In his 1988, Etchemendy, addressing the issue of why Tarski moved from his 1936 account of logical consequence to that of 1953 (*Undecidable Theories*, written with Mostowski and Robinson), where he embraces a standard model-theoretic account, says:

How about our second question? Tarski was surely aware of the above flaw in his 1936 definition, and it was no doubt partly responsible for his later giving them up. But this in itself does not explain why he subsequently came to endorse the standard account, in spite of the considerations raised in the earlier article. Unfortunately there is little to go on here, since Tarski never addressed the philosophical issues raised in the 1936 article. Indeed, in his introductory logic text [41m], although he discusses similar issues in some detail, there is no mention at all of analyzing logical truth or logical consequence semantically. Instead he sems to offer a syntactic gloss of the consequence relation (pp. 118–119), perhaps a sign of his dissatisfaction with his 1936 definition. Then later, when he gives the standard model-theoretic definitions in [Undecidable Theories, PM] he says nothing about the divergence of these definitions from the earlier account. (Etchemendy 1988, p. 73)

I agree that it would be interesting to know more in detail how Tarski moved from one account to another. Indeed, the 1940 lecture discussed in section 16.3 provides us with the specific information that until 1940 Tarski stood by his 1936 analysis. This shows that Etchemendy's tentative conjecture about what is going on in the 1937 textbook (from which the 1941 English translation is derived) is not warranted; moreover, one should recall that the 1937 book (but not the 1941 English version) is a literal translation from a book written in 1936 in Polish. It is still possible, as Etchemendy speculates, that Tarski's change was influenced by discussions with Carnap and Quine in 1941. And while I have began to make a few forays in this direction (see Mancosu 2005) the results will have to wait for another occasion.

Addendum to Chapter 16
(Added March 24, 2009)

Chapter 16 contains a sustained criticism of Gomez-Torrente's interpretation of Tarski's notion of logical consequence given in Gomez-Torrente (1996). Gomez-Torrente has recently replied to my criticisms (Gomez-Torrente 2009) and I would like to offer a few considerations to clarify where we now stand. However, I do not intend this to be a full rejoinder.

Let me begin with some preliminary remarks. The bone of contention concerns the issue of whether Tarski, during the thirties, allowed for domain variation in his application of the notion of logical consequence, as Gomez-Torrente claimed, or whether, as I claimed, no domain variation was needed to account for the application of logical consequence to all the formal theories discussed by Tarski. While there are other scholars who have defended various positions on this issue, and they are quoted at length in Chapter 16 and in Gomez-Torrente (2009), I will consider here only my interpretation and Gomez-Torrente's alternative in order not to add to the complexities of the discussion other positions that differ, either on major or minor points, from our views of the matter.

Gomez-Torrente defends a 'pluralism' in Tarski's practice at the time, namely "a disposition to work within a variety of different frameworks, each with its own assumptions and conventions." (2009, pp. 250–51) By contrast, while allowing for the pluralism of practices, I argued in 2006 that a single fixed-domain conception of logical consequence could account for all the different applications contemplated by Tarski of his notion of logical consequence. Lest misunderstandings be generated by this claim, I hasten to point out that I also have no doubt that Tarski could comfortably work with varying domains. For me the issue is whether such contexts can be comfortably recast in the mold of his definition of logical consequence with fixed domain.

Gomez-Torrente (1996) posited a stark contrast between Tarski's different treatments of theories that contained only logical vocabulary and theories containing also extra-logical mathematical predicates. In the latter case he makes the following unqualified remark:

However, I note, first, that Tarski clearly required domain variation in the test for logical consequence within languages containing extra-logical mathematical primitives; and second, I offer textual evidence that in these cases he assumed that the domain of the standard model or models was denoted by an extra-logical predicate, hence subject to reinterpretation in the test for logical consequence. (p. 127)

Other passages, such as the following, strengthened the impression that whether or not the extra-logical predicate denoting the intended model was present (and Gomez-

Torrente did discuss theories in which such a predicate was absent), Tarski was implicitly assuming that such a predicate was part of the background conventions:

It is natural to picture Tarski as having in mind the idea that the domain of the intended model or models is denoted by an extra-logical predicate, but without even thinking of formulating or caring to formulate this as an explicit requirement of the application of his definition.(p. 143)

For this reason I assumed that Gomez-Torrente was here suggesting a general strategy that would apply to all theories containing non-logical constants. The general strategy, as I read the paper, consisted of dealing with theories containing non-logical predicates as presupposing the convention described in the following quote:

Tarski had in mind mathematical theories in whose canonical formulation the domain of objects of the intended interpretation or interpretations is the extension of a primitive predicate of the language of the theory. Assuming that to specify a domain for an interpretation is nothing but to give an interpretation for such a predicate, Tarski's definition would allow for domain variation in the test for logical consequence. Under this assumption, sentences like '$\exists x \exists y \neg (x = y)$' would not be declared logical truths, because they would be mere unofficial abbreviations for other sentences with relativized quantifiers ('$\exists x \exists y (Nx \& Ny \& \neg(x = y))$' for example). It is natural to picture Tarski as having in mind the idea that the domain of the intended model or models is denoted by an extra-logical predicate, but without even thinking of formulating or caring to formulate this as an explicit requirement for the application of his definition. (p. 143)

Chapter 16 was then a reaction to what I saw as an unwarranted claim in Gomez-Torrente (1996), namely the claim that formalized theories containing non-logical mathematical predicates were considered by Tarski as being subject to the condition that the domain of discourse of the theory was usually expressed by a non-logical predicate in such a way that the domain of quantification of the quantifiers varied along with the interpretation of the domain predicate. I took Gomez-Torrente to hold that:

(1) No domain variation was applied by Tarski in the case of theories with merely logical vocabulary, such as the simple theory of types, in the determination of relationships of logical consequence.

(2) In the case of theories with mathematical vocabulary, whether or not containing a non-logical predicate meant to characterize the domain of discourse of theory, Tarski had appealed to domain variation in the determination of relationships of logical consequence by making use of appropriate conventions in the formulation of the theory in question.

Interpreting Gomez-Torrente in this way, I then provided evidence to show that claim 2) was open to exceptions for the following types of theories: a) first-order theories with predicates characterizing the domain of discourse of the theory; b) theories with non-logical predicates characterizing the domain of discourse of the theory in higher order logic, including the full simple theory of types; c) ordinary varieties of first order theories such as those presented by Tarski in his 1937 textbook.

This move was instrumental in order to show that a single conception of logical consequence was sufficient to unify what at first sight might have looked like a set of heterogeneous irreconcilable practices on Tarski's part.

Gomez-Torrente and I agree on the theories described in 1) so those will not be discussed further. In his 2009, he disputes my characterization of 2) since he declares that already in his 1996 paper the scope of his claim had been much weaker than I allege. In particular, he now asserts that he had explicitly allowed for the possibility that the conventions mentioned did not apply to all theories, not even to all first-order theories with non-logical mathematical constants. I reread carefully his 1996 article and I must confess that while he is clear about the fact that not all mathematical first-order theories considered by Tarski have a non-logical constant characterizing the domain of discourse of the theory, the article does not qualify sufficiently his claims concerning the extent of application of the conventions that he claimed were implicitly presupposed by Tarski in 1936. As I mentioned, many passages in his text invited a construal according to which the strategy applied generally to theories with non-logical mathematical constants and was not limited to a rather restricted subset of them. In 2009, Gomez-Torrente remarks that in 1996 he had qualified his claim when stating that "it can be amply documented that *many* of the theories containing mathematical primitives that Tarski formalizes (with the help of a logic without cardinality assumptions) explicitly contain extra-logical predicates which are true of all objects in the domain of the intended interpretation (or interpretations, in algebraic theories) of the theory". In 2009 he underlines 'many' but in the context of his 1996 paper, which did not have the emphasis on 'many', the statement seemed, at least to me, to carry an emphasis on 'explicitly'. Indeed, the aforementioned passage occurred in a discussion which was addressed to counter Etchemendy's interpretation with the general strategy according to which Tarski had "in mind the idea that the domain of the intended model or models is denoted by an extra-logical predicate, but without even thinking of formulating or caring to formulate this as an explicit requirement for the application of his definition."

Thus, while I understand Gomez-Torrente's desire to minimize the impact of my criticism, I do think that the exposition in 1996 invited a different reading of the paper from the more qualified reinterpretation given in 2009. This qualification is made in 2009 by stating that:

In TLC [Gomez-Torrente (1996)] I claimed that, for languages like a typical first order language for elementary arithmetic, Tarski frequently (though not always) used at the time the conventions (C1) of naming the domain of quantification of the intended interpretation with a non-logical predicate to which quantifiers must be relativized, and (C2) requiring the domain of quantification of alternative interpretations to be named by that same predicate. (p. 251)

What we are told in 2009 is that such conventions were compatible with claiming

as TLC did also claim, that Tarski did not require (C1) or (C2) in other contexts of investigation, even for typical first-order mathematical languages (p. 252)

To repeat, I do find in Gomez-Torrente (1996) the mention of first-order theories that do not satisfy (C1) or (C2) but that is consistent with the implication running throughout the paper that conventions (C1) and (C2) could be deployed even when Tarski did not explicitly invoke them. But perhaps that is enough regarding my reading of the 1996 article.

Gomez-Torrent in 2009 accepts one part of my criticism and at the same time limits the scope of his claim more explicitly. The part of my criticism he allows concerns theories of the form b):

> However, TLC obviously failed to make clear the exact scope of my claims, which I will now attempt to make precise. This will also allow me to correct a secondary suggestion made in TLC, on Tarski's views about the relation of logical consequence in languages where the vocabulary of the simple theory of types is supplemented with mathematical non-logical constants; here an unpublished lecture, Tarski (1940), unearthed by Mancosu, will be the main piece of evidence. But this correction of a tangential claim of TLC will not affect the variable-domains thesis.(p. 260)

While I do not agree, for the reason given above, that the correction is 'tangential' to the way TLC was set up, I welcome our agreement on what is in fact a large class of mathematical theories. With the scope of his claim restricted to theories that explicitly satisfy conditions (C1) and (C2) our disagreement gains in focus and this is already progress. Indeed, the discussion can now focus on the theories discussed by Tarski in his *Introduction to Logic*, for those are the primary examples of theories adduced by Gomez-Torrente as clearly satisfying conventions (C1) and (C2).

When writing my paper in 2006, I thought I could definitely establish that a close look at the theories in Tarski's *Introduction to Logic* showed that at least one of them failed to satisfy (C2). By implication, this undermined the plausibility that (C2) was assumed in the remaining theories in Tarski's book. My argument was textual but rested on the realization that Gomez-Torrente's conventions (C1) and (C2) had the following effect on the mathematical theories in question, namely 1) that they forced the domain of the theory denoted by the non-logical predicate characterizing such domain to be non-empty and 2) that the complement of such domain should always be empty.

I thought I had found the smoking gun in the following quote concerning a theory of real numbers expressed by means of a non-logical predicate $Zl(x)$ described by Tarski in his *Introduction to Logic*. Tarski at one point says:

> Der Bestand von Sätzen, die sich aus dem Axiomensystem A ableiten lassen, ist nämlich recht klein: auf Grund dieses Systems ist man z.B. nicht imstande, die elementare Frage, ob es überhaupt Zahlen gibt, bejahend oder verneinend zu beantworten. (Tarski 1937a, p. 145)

The English translation reads:

> One reason for this is the fact that the variety of theorems which can be derived from the axiom system A is very small indeed; it is, for instance, not possible to give, on its basis, an answer to the very elementary question as to whether any numbers exist at all. (Tarski 1941, p. 205)

I read these passages as claiming that the theory A could not prove or disprove the statement '$(\exists x)Zl(x)$' and concluded that this theory could not therefore satisfy condition (C2). Gomez-Torrente has objected to my reading by rightly pointing out that my reading does not square with an exercise given by Tarski in *Introduction to Logic* and that reads as follows:

24. In section 48 we have given the definition of the symbol '0' by way of example. If one wishes to have the certainty that this definition does not lead to a contradiction, one must have first the following theorem: There is exactly one number z such that, for any number x, the formula $x + z = x$ holds. Prove this sentence on the basis of axioms 6–9 [of the theory A] alone. (Tarski 1937a, p. 131; Tarski 1941 (with minor variations), p. 189).

The axioms 6–9 are the following:

6. $\forall x \forall y (Zl(x) \& Zl(y) \supset \exists z (Zl(z) \& x + y = z));$
7. $\forall x \forall y (Zl(x) \& Zl(y) \supset x + y = y + x);$
8. $\forall x \forall y \forall z (Zl(x) \& Zl(y) \& Zl(z) \supset x + (y + z) = (x + y) + z);$
9. $\forall x \forall y (Zl(x) \& Zl(y) \supset \exists z (Zl(z) \& x = y + z));$

Gomez-Torrente argues as follows. If from axioms 6–9 one can prove the statement of exercise 24 then obviously the theory proves that there exists at least an x such that $Zl(x)$. But this contradicts my claim that the theory does not decide $(\exists x) Zl(x)$ either way. Hence, Gomez-Torrente concludes,

Since clearly Tarski thought that $(\exists x) Zl(x)$ was a theorem of A, it follows that he was using the convention that Zl names the domain of quantification in every interpretation of the language of A (p. 255)

I think that Gomez-Torrente's observation weakens the force of my counterexample but, contrary to the consequence he draws, the failure of my attempt provides no positive evidence for his claim (i.e. that A satisfies convention (C2)). Here is why. First of all, it should be noticed that rigorously speaking the statement of exercise 24 does not follow using axioms 6–9 alone. Indeed, it does not follow even if all the axioms of A are used (Gomez-Torrente agrees here; hence the need for postulating an unstated convention). This is because we can provide models of the system A in which $Zl(x)$ has an empty extension. How do we handle the problem? One possibility is to claim that on account of an oversight Tarski demands something in exercise 24 that in fact cannot be done. When I posed the question to Steven Givant, a former collaborator of Tarski, he pointed out to me that such oversights are not unusual when writing textbooks. Opting for an 'oversight' would allow me to insist on my original reading of the passage from p. 145 and thereby resist Gomez-Torrente's preferred interpretation of those passages according to which what Tarski is saying is that "on the basis of the system [A] it is not possible, for example, to answer positively or negatively the question whether there are numbers at all" where the emphasis is on the plural "numbers". What Tarski is claiming, according to Gomez-Torrente, is that it is not possible to show that there are at least two numbers.

While the possibility of an oversight is quite real, I will not take this route. I feel that in the literature generally, too many mistakes are attributed to Tarski and I would prefer not to have my argument rest on such an assumption. It would seem then that I ought to adopt Gomez-Torrente's conclusion to the effect that accepting the derivability of $(\exists x) Zl(x)$ from axioms 6–9 forces an acceptance that Tarski "was using the convention that Zl names the domain of quantification in every interpretation of the language of A".

But I take this to be a non-sequitur. Indeed, a very reasonable way to interpret Tarski here would be to claim that the convention deployed in this context is to consider only interpretations in which the non-logical predicate characterizing the domain of discourse of the mathematical theory is non-empty. This could be argued for instance by appealing to the fact that otherwise the interpretations are uninteresting. In short, whereas Gomez-Torrente is reading Tarski as presupposing, among other things, $(\forall x)Zl(x)$ and from there infers $(\exists x)Zl(x)$, I notice that to explain the presence of exercise 24 it is sufficient to postulate a background convention according to which $(\exists x)Zl(x)$ is part of the unstated assumptions. While my original claim is undermined by my accepting the derivability of exercise 24 from the axioms of A, Gomez-Torrente's interpretation receives no positive support from his rebuttal of my criticism.

While the possible failure of my original counterexample to theory A as a theory satisfying (C2) leaves Gomez-Torrente's interpretation as a viable one, nothing said so far shows that a fixed domain interpretation in the determination of logical consequence is unviable, even for the theories presented by Tarski in his *Introduction to Logic*. For instance, my suggestion that the convention adopted by Tarski is that of excluding from consideration non-empty domains of the theory still leaves open the possibility that condition (C2) might be defeated in case the theory in question does not refute $(\exists x)\neg Zl(x)$.

Gomez-Torrente (2009) not only defends his interpretation against my criticisms but also points out difficulties for the fixed-domain interpretation. I will not address those here, otherwise this 'addendum' would quickly become a lengthy essay (Gomez-Torrente's discussion of my criticism runs to 17 pages). The important point is to have gained a clearer understanding of where we stand. Gomez-Torrente has qualified his position to be the claim that the definition of logical consequence given by Tarski in 1936d can be straightforwardly applied to theories interpreted with variable domains if the theories in question obey conventions (C1) and (C2). He points out that this is a relatively weak claim as it does allow that many theories with non-logical mathematical predicates that do not satisfy the conventions are indeed treated by Tarski within the fixed domain conception. As examples of theories satisfying conventions (C1) and (C2) he adduces the theories from Tarski's *Introduction to Logic*. I welcome the acceptance on the part of Gomez-Torrente of the evidence I adduced in favor of showing that Tarski treated a large class of mathematical theories within a fixed domain conception of logical consequence. I also agree that in the qualified version of his claim, given in 2009, such theories do not constitute a counterexample to his relatively weak claim. Further, I agree that my original counterexample to his claim that the theory A satisfies condition (C2) is problematic on account of exercise 24.

However, I believe that all of this does not settle all the central issue concerning variable domain versus fixed-domain interpretations of logical consequence, at least in the versions that Gomez-Torrente and I defend. Needless to say, any further attempt to resolve the dispute will have to be left for another occasion.

17

TARSKI ON CATEGORICITY AND COMPLETENESS: AN UNPUBLISHED LECTURE FROM 1940

In the Tarski archive at Berkeley there is a lecture entitled "On the Completeness and Categoricity of Deductive Systems" which Tarski never published.[1] The lecture is an important historical document for the development of abstract notions of semantics, such as semantical completeness and categoricity. Versions of these notions had played an important role in the logical investigations of the twenties and thirties (most notably in Fraenkel, Carnap, and Gödel) but in this lecture Tarski has at his disposal the tools of semantics he had recently developed. In addition, the lecture provides welcome information on a topic of central interest to Tarski and to contemporary philosophy of logic, e.g. the Tarskian notion of logical consequence. Finally, some of the technical problems raised in the lecture are still open and have very recently been the object of renewed attention.

The lecture itself can be divided ideally into three parts. The first part introduces the notion of deductive completeness and surveys some of the elementary theories for which deductive completeness has been established. However, on account of Gödel's incompleteness theorems, deductive completeness is a rare phenomenon. Most theories, including logic (understood as containing the theory of simple types), are incomplete. This leads to the second part of the lecture, where Tarski presents the notion of seman-

---

Many thanks to Steve Awodey, John Burgess, Dana Scott, and Stewart Shapiro for many useful conversations that have helped me see straight on many historical and technical issues raised by Tarski's paper. In particular, I thank them for their insight into the equivalence of relative completeness and non-forkability. In addition, I thank John Burgess for coming up with the nice example of a theory that is semantically complete but not relatively complete given in the main text. I am also grateful to Johannes Hafner and José Sagüillo for their valuable comments on a previous version of this chapter.

tical completeness as a way to regain the completeness that was lost at the syntactic level. While presenting the notion of semantical completeness, Tarski makes a claim that has important ramifications for an adequate understanding of his conception of logical consequence. Finally, the third part of the lecture deals with the problem of how to establish semantical completeness and brings into the picture categoricity. A number of results concerning semantical completeness and categoricity and some open questions are stated at the end of the presentation.

My goal in this chapter is to spell out the key elements in the lecture and to provide extensive commentary on the topics discussed in the lecture with the aim of clarifying the conceptual background of the lecture and to point to the relevance of some of the issues discussed by Tarski to contemporary discussions.

17.1 Dating the Lecture

A summary of the lecture was published as an appendix to Wolénski and Jan Tarski's publication of Tarski's lecture "Some current problems in metamathematics" (Tarski 1995). The editors conjectured that the lecture was probably delivered in 1939. We can be much more precise using two sources. The first is a letter from Tarski to Quine replying to a request of information concerning the relationship between completeness and categoricity:

Dear Van,

The precise relationship between the completeness and the categoricity (both relative to logic) is the following one. Every categorical axiom-system is complete; the problem whether the converse holds remains open; if, however, a system which is complete possesses an interpretation in logic, it is categorical. I wonder when and where my paper concerned ("On completeness and categoricity of deductive theories") will be published. You can quote the paper by Lindenbaum and myself "Über die Beschränkheit der Ausdrucksmitteln..." in Mengers "Ergebnisse Math. Coll." (7 or 8?) since it contains essentially the same results. In this case you would have to add that the concept of "Nicht-gabelbarkeit" which is discussed there is equivalent to the relative completeness. But I would be glad if you could mention the forthcoming paper on completeness and categoricity or refer to my lecture at Harvard. (Tarski to Quine, July 1, 1940. Quine archive, MS Storage 299, box 8, folder Tarski)

Quine was happy to oblige with Tarski's request and through Quine's acknowledgment we can date the lecture with great precision. In the following footnote found in Quine and Goodman (1940) we read:

The latter notion (synthetically complete), under the name 'completeness relative to logic', is due to Tarski. It is easier to formulate than the older concept of categoricity, and is related to the latter as follows: systems which are categorical (with respect to a given logic) are synthetically complete, and synthetically complete systems possessed of logical models are categorical. These matters were set forth by Tarski at the Harvard Logic Club in January, 1940 and will appear in a paper "On completeness and categoricity of deductive theories". See also Lindenbaum and

Tarski, Über die Beschränkheit der Ausdrucksmitteln deduktiver Theorien, Ergebnissen eines mathematischen Kolloquium, Heft 7 (1936), pp. 15–22 where 'Nichtgabelbarkeit' answers to 'synthetic completeness'. (Quine and Goodman 1940, footnote 3, p. 109)

I also found no evidence to suggest that this lecture was conceived, as conjectured by J. Wolénski and J. Tarski, as the second lecture in a series of which "Some current problems in metamathematics" was supposed to be the first installment.

17.2 The First Part of the Lecture

I will not spend much time summarizing this part of the lecture. Tarski uses 'absolute completeness' to discuss what we now call syntactic (or deductive) completeness. The notion is defined as follows:

A system of sentences of a deductive theory is called absolutely complete or simply complete if every sentence which can be formulated in the language of this theory is decidable, that is, either derivable or refutable in this system. (Tarski 1940, p. 1)

Modulo some trivial facts about the theory, the condition is equivalent to stating that a theory is complete "if for every sentence either it or its negation is derivable" (p. 1). Among the systems satisfying the condition, Tarski quotes "certain systems of the calculus of statements (Post and others), of Boolean algebra (Löwenheim and others), of the theory of linear order (Langford and others), and finally the theory of addition of the natural numbers (Presburger)." (pp. 1–2) Tarski had been heavily involved in these investigations. One of the techniques investigated in Tarski's seminar in Warsaw was what he called the elimination of quantifiers. The method was originally developed in connection to decidability problems in Löwenheim (1915) and Skolem (1920). It basically consists in showing that one can add to the theory certain formulas, perhaps containing new symbols, so that in the extended theory it is possible to demonstrate that every sentence of the original theory is equivalent to one of the new theory that is quantifier-free (or that belongs to a special class of formulas that can be easily decided). This idea was cleverly exploited by Langford to obtain, for instance, decision procedures for the first-order theories of linear dense orders without end elements, with first but no last element and with first and last element (Langford 1927a) and for the first-order theory of linear discrete orders with a first but no last element (Langford 1927b). As Langford emphasizes at the beginning of Langford (1927a), he is concerned with "categoricalness", i.e. with the property that the theories in question determine the truth value of all their sentences (something he obtains by showing that the theory is syntactically complete). Many such results were obtained afterwards, such as Presburger's elimination of quantifiers for the additive theory of numbers and Skolem's decision procedure (published in 1931) for the theory of order and multiplication (but not addition!) on the natural numbers. Moreover, Tarski extended the results by Langford to the first-order theory of discrete order without a first or last element and for the first-order theory of discrete order with first and last element.

Going back to the lecture, what characterizes all the above theories is their "elementary logical structure" and the "meager mathematical content" they can capture. By elementary logical structure, Tarski means that the systems are formalized either in propositional or first-order logic ("the restricted functional calculus"). In addition, these systems can only capture very basic parts of mathematics in which "deep" mathematical facts cannot even be expressed. This is not the case for the systems of classical algebra and elementary Euclidean geometry. While still formalizable within an elementary logic their mathematical power is quite remarkable. Tarski is here reporting on results he had obtained in Warsaw in the late twenties, and had announced publicly in 1931, on the absolute (i.e. syntactic) completeness of both elementary algebra and elementary geometry. The "effectiveness" of the proof given for completeness yields also decision procedures for the theories in question. These results will eventually be published in Tarski (1948).

Despite these positive results, most interesting mathematical theories are not syntactically complete. This is a consequence of Gödel's incompleteness results. As soon as an elementary part of the arithmetic of the natural numbers is formalizable in a consistent and effectively presented system, we have incompleteness. This has also consequences for systems of logic:

> In general the domain of application of this result is very extensive, and is not essentially limited by the premises which condition the incompleteness of the system. For it is well known that the arithmetic of whole numbers can be formalized within any deductive theory with a sufficiently rich logical structure, even if the concepts of arithmetic themselves do not occur explicitly in this theory. (p. 3)

Such passages remind us that in 1940 it was still common practice to refer to simple type theory as logic (see Ferreiros 2001, Mancosu 2005) and thus to see arithmetic as being derivable within one such logic. The conclusion of this part of the discussion is that absolute completeness is rather the exception than the rule in the domain of the deductive sciences. In order to remedy the situation, Tarski proposes two weaker notions of completeness, relative completeness and semantical completeness. This leads to the second part of the lecture. But before getting there I will devote the next section to provide the historical background on studies of completeness in the twenties and thirties.

17.3 Fraenkel and Carnap on Completeness

Early occurrences of various concepts of completeness and categoricity in authors such as Hilbert, Huntington and Veblen, have been extensively studied in the literature (see Awodey and Reck 2002a; Scanlan 2003, and Chapter 1 of this volume). For instance, Huntington (1902) defines a 'complete' set of postulates for a theory as one that satisfies the following properties:

(1) The postulates are consistent;

(2) They are sufficient;

(3) They are independent (or irreducible).

Condition 1 says that there exists an interpretation satisfying the postulates and condition 2 asserts that there is essentially only one such interpretation possible. Condition 3 says that none of the postulates is a "consequence" of the others.

A system satisfying the above properties one and two we would nowadays call categorical rather than complete. Indeed, the word "categoricity" was introduced in this context by Veblen in a paper on the axiomatization of geometry in 1904. Veblen credits Hungtington with the idea and Dewey for having suggested the word 'categoricity'. When one looks carefully at Veblen's definition, which I will not state here, one immediately notices a certain ambiguity between defining categoricity as the property of admitting only one model (up to isomorphism) and defining it by means of a notion which is a consequence of the first definition, namely what we would call semantical completeness (see Awodey and Reck 2002a).[2] According to contemporary terminology, a system of axioms is categorical if all its interpretations (or models) are isomorphic. In the early part of the twentieth century it was usually mentioned, for instance, that Dedekind had shown that every two interpretations of the axiom system for arithmetic are isomorphic. One thing on which there was already clarity at the time is that two isomorphic interpretations make the same set of sentences true. But, as I remarked, this notion of categoricity was not clearly distinguished from what we call semantical completeness.

An early attempt to provide a terminological clarification concerning different meanings of completeness is found in the second edition (1923) of the *Einleitung in die Mengenlehre*, where Fraenkel distinguishes between completeness in the sense of categoricity and completeness as decidability [*Entscheidungsdefinitheit*]. In the third edition of *Einleitung in die Mengenlehre* (1928), Fraenkel adds a third notion of completeness, the notion of *Nichtgabelbarkeit* ('non-forkability'), meaning essentially that every two interpretations satisfy the same sentences. Fraenkel attributes this third notion to Carnap and says that he has seen his work on this. In 1927 Carnap claimed to have proved the equivalence of all three notions (which he calls monomorphism, decidability and non-forkability; see Carnap (1927) for the *Gabelbarkeitssatz*). The proofs were supposed to be contained in his manuscript "Untersuchungen zur allgemeinen Axiomatik" (Carnap 2000) but his approach there is marred by the inability to distinguish between object language and metalanguage, and between syntax and semantics, and thus to specify exactly to which logical systems the proofs are supposed to apply (for an analysis of these issues see: Awodey and Reck 2002a,b and Reck 2008); Carnap's unpublished investigations on general axiomatics are now edited in Carnap (2000)). Gödel, however, had access to the manuscript and in fact Gödel's 1929 dissertation acknowledges the influence of Carnap's investigations (as does Kaufmann 1930). Awodey and Carus (2001, p. 23) also point out that Gödel's first official mention of the incompleteness theorem in 1930 in Königsberg (see Gödel 1995, p. 29 and the introduction by Goldfarb 1995) was aimed specifically at Carnap's claim. Indeed, speaking of the meaning of the completeness theorem for axioms systems he pointed out that in first-order logic monomorphicity (Carnap's terminology) implies (syntactic) completeness [*Entscheidungsdefinitheit*]. If completeness also held of higher-order logic then (second-order) Peano arithmetic, which by Dedekind's classical result is categorical, would also turn out to be syntactically complete. But, and here is the first announcement of the incompleteness theorem, Peano's arithmetic is incomplete (Gödel 1930c, pp. 28–30; a similar point is stressed in Tarski 1934–35, p. 319).

Carnap had discussed his investigations with Tarski in 1930 when the latter visited Vienna. At that time, Tarski had recognized the defects in Carnap's work and called them to his attention, after which Carnap abandoned the project. However, it should be stressed that Carnap had formulated the main concepts and questions in this area and proved correctly that if a system T is categorical then it is also semantically complete. He also conjectured the converse and provided a faulty proof of the claim, as he could not properly state the relevant condition of having a definable model. In light of these facts it is quite surprising, to say the least, that Tarski does not give Carnap proper credit either in the paper with Lindenbaum or in this lecture.

17.4 Relative Completeness

In order to introduce the new notions of completeness, Tarski stipulates that we have a way to distinguish between logical and non-logical constants of the systems under consideration and that such systems are to be thought of as built upon a logical framework through specific axioms involving the non-logical constants. Logical sentences are those in which non-logical constants do not appear. Tarski uses the terminology of 'logically valid sentence' to denote theorems of logic. As example of such a system Tarski mentions a system of geometry built upon the logic of *Principia Mathematica* (or a fragment thereof). If the logical basis is 'rich' (i.e. contains quantification over higher-order entities and axioms such as the axiom of infinity) we can formalize the theory of natural numbers in it. By Gödel's incompleteness theorem it is therefore incomplete. Tarski summarizes the situation thus:

If the logical basis of our theory is rich enough, we can formalize within its boundaries the arithmetic of natural numbers and for this reason its logical basis is incomplete: i.e. there are logical sentences which are not logically valid and whose negations are likewise not logically valid. In other words, there are problems belonging entirely to the logical part of our theory which cannot be solved either affirmatively or negatively with the purely logical devices at our disposal. (p. 4)

Thus a requirement of absolute completeness for the theory of geometry mentioned above would end up in requesting that the addition to the logic of the geometrical axioms decides the problems that were left undecided by the logic. By Gödel's incompleteness that's too much to ask. But a weaker request consists in the demand that the geometrical axioms do not "extend the incompleteness of the logical part of our theory"(p. 4). The following two definitions introduce the required notion.

Definition 1. *Two sentences A and B are equivalent with respect to a theory T iff $T \cup \{A\} \vdash B$ and $T \cup \{B\} \vdash A$.*

Definition 2. *A theory T is relatively complete (or complete with respect to its logical basis) iff for every B in L(T) there is an A expressed in the logical vocabulary of T such that A and B are equivalent with respect to T.*

Tarski mentions two results concerning the notion just introduced. First of all, if the logical part of T is absolutely complete and T is relatively complete then T is absolutely complete. Second, it is not always the case that if a system is relatively complete then it is absolutely complete.

The first result is clear. If the background logic is absolutely complete then by relative completeness for any sentence B in $L(T)$ we can find an A, which contains only logical vocabulary, equivalent to it with respect to T. Thus, all we need to do is to exploit the absolute completeness of the logic to decide whether A or $\neg A$ is derivable in the logic and that automatically will tell us whether B is derivable from T or whether its negation is.

As an example of a system that is relatively complete but not absolutely complete one might think of second-order Peano arithmetic formalized with the simple theory of types (and the axiom of infinity) as background logic. Because every sentence of second-order Peano arithmetic can be translated into a sentence of the simple theory of types and shown to be equivalent to its translation inside the simple theory of types we have a case of relative completeness. However, second-order Peano arithmetic is not absolutely complete by Gödel's incompleteness.

In the letter to Quine mentioned at the outset, Tarski points out that relative completeness in this paper is equivalent to non-forkability (or non-ramifiability) in his 1936 paper with Lindenbaum. For a finitely axiomatized theory expressed by the conjunction of its axioms—say $a(a, b, c, \ldots)$, where $a, b, c \ldots$ stand for the non-logical constants appearing in the axioms—the definition reads as follows:

The axiom system $[a(a, b, c, \ldots)]$ is said to be non-ramifiable if, for every sentential function '$\sigma(x, y, z, \ldots)$', the disjunction

$$(x, y, z, \ldots) : a(x, y, z, \ldots). \to .\sigma(x, y, z, \ldots) : \vee :$$
$$(x, y, z, \ldots) : a(x, y, z, \ldots). \to .\neg \sigma(x, y, z, \ldots)$$

is logically provable. (Tarski 1983, p. 390)

Quine and Goodman, following Tarski's remarks in the letter to Quine, repeat the claim of equivalence between relative completeness and non-ramifiable. The claim is indeed correct.

Let us prove first that relative completeness implies non-ramifiability. Since $a(a, b, c, \ldots)$ is relatively complete we can prove in the logic, for any $\sigma(a, b, c, \ldots)$ and equivalent logical τ modulo $a(a, b, c, \ldots)$:

$$\vdash a(a, b, c, \ldots) \rightarrow (\sigma(a, b, c, \ldots) \leftrightarrow \tau)$$

But since this is provable in pure (higher-order) logic we can also prove the universally quantified version of it:

$$\vdash (x, y, z, \ldots)(a(x, y, z, \ldots) \rightarrow (\sigma(x, y, z, \ldots) \leftrightarrow \tau))$$

To prove non-ramifiability of $a(a, b, c, \ldots)$ is now enough to reason by *reductio ad absurdum* within the logic. Assume the negation of non-ramifiability. Then we can infer that there are x, y, z satifying $a(x, y, z, \ldots)$ & $\sigma(x, y, z, \ldots)$ and $x, y,$ and z satisfying $a(x, y, z, \ldots) \& \neg \sigma(x, y, z, \ldots)$. By relative completeness we can replace whatever instances of σ by τ and whatever instance of $\neg \sigma$ by $\neg \tau$. In this way we see that the assumption of the negation of non-ramifiability leads to $\tau \& \neg \tau$. Thus, non-ramifiability holds under the assumption that the theory $a(a, b, c, \ldots)$ is relatively complete.

In order to show that non-ramifiability implies relative completeness, let $a(a, b, c, \ldots)$ be non-ramifiable. Let $\sigma(a, b, c, \ldots)$ be any sentence in the language of a. Consider now the following two sentences expressed with only logical vocabulary:

$$\tau = (x, y, z, \ldots)(a(x, y, z, \ldots) \rightarrow \sigma(x, y, z, \ldots))$$

$$\tau' = (x, y, z, \ldots)(a(x, y, z, \ldots) \rightarrow \neg\sigma(x, y, z, \ldots))$$

We have $a(a, b, c, \ldots) \cup \{\tau\} \vdash \sigma(a, b, c, \ldots)$ and $a(a, b, c, \ldots) \cup \{\tau'\} \vdash \neg\sigma(a, b, c, \ldots)$.

Since $a(a, b, c, \ldots)$ is non-ramifiable we have

$$[(x, y, z, \ldots) : a(x, y, z, \ldots). \rightarrow .\sigma(x, y, z, \ldots) : \vee :$$

$$(x, y, z, \ldots) : a(x, y, z, \ldots). \rightarrow .\neg\sigma(x, y, z, \ldots)]$$

$$(i.e. \, a(a, b, c, \ldots) \vdash \tau \vee \tau')$$

and thus

$$a(a, b, c, \ldots) \cup \{\sigma(a, b, c, \ldots)\} \vdash (x, y, z, \ldots)$$

$$(a(x, y, z, \ldots). \rightarrow .\sigma(x, y, z, \ldots)).$$

We have thus shown that $\sigma(a, b, c, \ldots)$ is equivalent to τ with respect to $a(a, b, c, \ldots)$.

17.5 Semantical Completeness

Defining the notion of semantical completeness requires a few semantical preliminaries. Building on his previous work on models and logical consequence, Tarski first defines these notions through the notions of realization (or model) which is ultimately based on the notion of 'fulfillment' (i.e. satisfaction). The preliminary move is syntactical. Whenever we are considering a sentence (resp. a set of sentences) containing non-logical constants for satisfaction (resp. logical consequence) we replace all the non-logical constants by variables of the appropriate type.

The most important role is played here by the concept of model or realization. Let us consider a system of non-logical sentences and let, for instance "C_1", "C_2"... "C_n" be all the non-logical constants which occur. If we replace these constants by variables "X_1", "X_2"... "X_n" our sentences are transformed into sentential functions with n free variables and we can say that these functions express certain relations between n objects or certain conditions to be fulfilled by n objects. Now we call a system of n objects $O_1, O_2... O_n$ a model of the considered system of sentences if these objects really fulfill all conditions expressed in the obtained sentential functions. It is of course possible that the whole system reduces to one sentence; in this case we speak simply of the model of this sentence. We now say that a given sentence is a logical consequence of the system of sentences if every model of the system is likewise a model of this sentence. (p. 5)

The definition of semantical completeness appears next:

Thus a system of sentences of a given deductive theory is called *semantically complete* if every sentence which can be formulated in the given theory is such that either it or its negation is a logical consequence of the considered set of sentences. (p. 5)

The confusion found in the previous literature (Fraenkel, Carnap) between a syntactic notion and a semantical notion of completeness is here completed removed. But then Tarski makes a claim immediately after giving us his definition of semantical completeness which might make us think that some confusions have crept in. In fact he goes on to say:

It should be noted that the condition just mentioned is satisfied by any logical sentence: hence we can deduce without difficulty that the concept of semantical completeness is a generalization of the concept of relative completeness: every system that is relatively complete is likewise semantically complete (but it can be shown by an example that the converse is not true). (p. 5)

Let us call the claim that "the condition just mentioned is satisfied by any logical sentence" claim C. The reason why this might seem hopelessly confused is the following. Consider the logical sentence $(\exists x)(\exists y)(x \neq y)$. In first-order logic, just as well as in higher-order logic, we have models for it and models for its negation. In what sense then does the sentence in question automatically satisfy the condition that "either it or its negation is a logical consequence of the considered set of sentences"? The answer to this question has important ramifications for a proper understanding of Tarski's notion of logical consequence. But before discussing those ramifications I would like to spell out the proof of the claim that "every system that is relatively complete is likewise semantically complete." In addition to clarifying the notions involved the proof will also serve to pre-empt the possible criticism that Tarski's claim about logical sentences should be dismissed as an oversight or a mistake.

17.6 Every System That Is Relatively Complete Is Semantically Complete

The argument can be spelled out as follows. Suppose S is relatively complete. As Tarski does, and following his terminology, let us assume that from S all logically valid sentences can be derived. Let ϕ be an arbitrary sentence in L(S) (L(S) includes also all the logical symbols). We want to show that either ϕ or $\neg\phi$ is a logical consequence of S. By definition of relative completeness for any sentence ϕ there is a logical sentence ϕ^* such that $S \cup \{\phi\} \vdash \phi^*$ and $S \cup \{\phi^*\} \vdash \phi$.

If ϕ is a logical sentence then, by claim C, it or its negation is a logical consequence of S, so there is nothing to prove. If ϕ is not logical then let ϕ^* be a logical sentence satisfying the condition given in the definition of relative completeness for S. We consider two cases. Because ϕ^* is a logical sentence either all models of S are models of ϕ^* or all models of S are models of $\neg\phi^*$. First assume all models of S are models of ϕ^*. Then any model M of S is also a model of $S \cup \{\phi^*\}$. Since $S \cup \{\phi^*\} \vdash \phi$ and the logical system is assumed sound, M is a model of ϕ. So ϕ is a logical consequence of S. Now assume that every model of S is a model of $\neg\phi^*$. Furthermore, by way of contradiction, assume there is a model M' of S such that M' is not a model of $\neg\phi$. Thus M' is a model of ϕ. Because $S \cup \{\phi\} \vdash \phi^*$ hence by the soundness of the logical system, M' is a model of ϕ^*. This contradicts all model of S being models of $\neg\phi^*$. So $\neg\phi$ is a logical consequence of S. Consequently, S is semantically complete.

Notice that this claim by Tarski gives a general theorem about conceptual relationships between relative completeness and semantical completeness and it is stated in such a way that no qualification on the logic is made. Tarski meant this result to hold, as he explicitly says, at least for the system of *Principia Mathematica* and for fragments of it.

Tarski also mentions that the converse does not hold but provides no counterexample. The following counterexample is due to John Burgess and uses only resources that Tarski could have had at his disposal. Let $D(<)$ express that $<$ orders the universe in order type ω. By the incompleteness theorems there is an $E(<)$ such that

$$C = \forall X(D(X) \rightarrow E(X))$$

is semantically valid but not syntactically demonstrable. Let T be $C \rightarrow D(<)$.

Then T is semantically equivalent to $D(<)$, which is categorical, hence semantically complete.

Claim: T is not relatively complete. Specifically, if $B = D(<)$ then there is no purely logical A such that $T \rightarrow (B \leftrightarrow A)$ is syntactically demonstrable, which is to say, such that

(1) $(C \rightarrow D(<)) \rightarrow (D(<) \leftrightarrow A)$

is syntactically demonstrable.

Proof of Claim: For suppose there were such an A. First note that the following are trivially syntactically demonstrable:

(2) $\neg \forall X D(X)$

(3) $\forall X \neg D(X) \to C$

Then the following are also syntactically demonstrable, being syntactically deducible from 1–3:

(4) $\neg C \,\&\, A \to D(<)$ 1
(5) $\neg C \,\&\, \neg A \to D(<)$ 1
(6) $\neg C \,\&\, A \to \forall X D(X)$ 4
(7) $\neg C \,\&\, \neg A \to \forall X \neg D(X)$ 5
(8) $\neg C \to \forall X D(X) \lor \forall X \neg D(X)$ 6, 7
(9) $\neg C \to \forall X \neg D(X)$ 8, 2
(10) $\neg C \to C$ 9, 3
(11) C 10

But C was chosen to be not syntactically demonstrable.

17.7 Consequences for (Logical) Consequence

The importance of the above for a proper interpretation of Tarski's notion of logical consequence cannot be overemphasized. Since the publication of Etchemendy's book *The Concept of Logical Consequence* much attention has been devoted to Tarski's notion of logical consequence. A heated debate has opposed those like Etchemendy, who claimed that Tarski does not allow for variable domains, and those like Gomez-Torrente (1996) and Ray (1996), who favor reading Tarski in terms of our contemporary notion of logical consequence with variable domains. To fix ideas, the major difference between the two interpretations is this. According to Etchemendy, Tarski works with 'the' intended domain of the theory of types in the background and the quantifiers are already interpreted on this domain (see also Corcoran 2003). In the test for logical consequence the quantifiers do not change interpretation (they are logical constants and the domain is fixed); only the non-logical constants do. Consequently, when the sentence being tested for logical validity is purely logical (as in "there are two distinct objects") the sentence is valid if and only if it is true on the intended domain and thus "there are two distinct objects" turns out to be valid (since the universe of individuals is usually assumed to have more than one object). For Gomez-Torrente and Ray this is a *reductio ad absurdum* of the interpretation. Tarski, they argue, could not have missed the importance of letting the quantifiers vary in the test for logical consequence; and indeed, if quantifiers are allowed to vary over different domains the sentence in question does not end up being classified as a validity. In Mancosu (2006) (see Chapter 16 in this volume with Addendum), I argued that the evidence speaks in favor of Etchemendy's interpretation. This is not the place to present the detailed and complex evidence to support my case, but let me single out two major points.

First, I take seriously Tarski's claim (made in the 1936 article) that if we treat the non-logical constants as if they were logical constants then logical consequence and material consequence would coincide. Etchemendy gives a perfectly coherent account of this claim whereas Gomez-Torrente (1996) simply ignores the claim (but see now Gomez-Torrente 2009) and Ray (1996) attributes it to a mistake on Tarski's part. Second, Tarski's claim, detailed above, that every relatively complete theory is semantically complete is a significant mathematical result which depends essentially on the claim that every logical sentence automatically satisfy the condition that "either it or its negation is a logical consequence of the considered set of sentences." On a fixed domain interpretation for the quantifiers, the claim makes perfect sense as questions of logical validity for logical sentences reduce to questions of truth in the intended interpretation. However, on a variable domain conception the claim makes no sense. In which sense could "there are at least two objects" be such that automatically "either it or its negation is a logical consequence of the considered set of sentences"? If one considers a theory that does not imply facts about the cardinality of its models, it would seem clear that neither the sentence nor its negation ought to be a logical consequence of the theory. Thus, this claim is evidence for the fact that in 1936 and also in the 1940 lecture Tarski is offering a fixed domain conception of logical consequence.[3]

17.8 Categoricity

The last part of the lecture is devoted to the issue of how to determine whether theories are semantically complete. The importance of categoricity for Tarski consists in the fact that it is closely related to relative and semantical completeness. Tarski attributes the concept of categoricity to Veblen but immediately goes on to distinguish two meanings of categoricity:

We shall distinguish here two variants of the concept of categoricity: semantical categoricity, and categoricity with respect to the logical basis or relative categoricity, which parallel respectively semantical and relative completeness. But we do not know and therefore do not introduce any concept of categoricity which would parallel the absolute completeness.

We shall call a system of sentences semantically categorical if any two models of this system are isomorphic. The concept of isomorphism is a well known logical concept, not a methodological or semantical one, and thus I could take it that it is understood; but in any case I should like to have it noted that the precise definition of this concept depends on the logical foundation of our methodological investigations. Roughly speaking [and adapting ourselves to the language of *Principia Mathematica*, but disregarding the theory of types] we can say that two system of objects [classes, relations, etc.] $O_1, O_2, \ldots, O_n$ and $P_1, P_2, \ldots, P_n$ are isomorphic if there is a one-one correspondence which maps the class of all individuals onto itself and simultaneously the objects $O_1, O_2, \ldots, O_n$ onto $P_1, P_2, \ldots, P_n$ respectively. (pp. 5–6)

First of all, one should notice that this is a stronger notion than the one defined originally by Veblen because of the insistence that the correspondence be an automorphism of the domain which was not present in Veblen's 1904 definition of categoricity. One should also keep in mind that the notion of categoricity just presented does not require that the logic be strong enough to establish the facts about the isomorphism as a theorem. By contrast, the following definition, relative categoricity, brings the deductive resources of the logic to bear. The concept of relative categoricity, or categoricity relative to a logical basis, is defined after a few preliminary considerations:

In order to obtain the second variant of the concept of categoricity we shall confine ourselves for the sake of simplicity, to such a case in which the considered system of sentences is finite and contains only one non-logical constant, say "C". Let "$P(C)$" represent the logical product of all these sentences. "C" can denote, for instance, a class of individuals or a relation between individuals, or a class of such classes or relations etc. We assume further that the logical structure of the deductive theory is rich enough to express the fact that two objects X and Y of the same logical type as C are isomorphic and we assume that this fact is expressed by the formula "$X \sim Y$". We can now correlate with the semantical sentence stating that our system is semantically categorical an equivalent sentence formulated in the language of the deductive theory itself. This is the following sentence:

> For every X and Y, if $P(X)$ and $P(Y)$ then $X \sim Y$
> Or in symbols
> $(X)(Y)[P(X) \& P(Y) \to X \sim Y]$

Now we say that the considered system of sentences is *categorical with respect to its logical basis*, or *relatively categorical*, if the sentence formulated above is logically valid. (p. 6)

This second notion of categoricity corresponds exactly to the one presented in the article "On the definability of concepts" (Tarski 1983, p. 310). There it is called "provable categoricity" (see also the article Tarski and Lindenbaum 1936, p. 390 of Tarski 1983 where theories satisfying the property are called "categorical or monomorphic"). It is interesting that the notion of semantic categoricity is not given a prominent place in either Tarski (1934–35) or Tarski and Lindenbaum (1936), although the possibility for the investigation of the notion is pointed out by reference to the Carnapian distinction between a-concepts (roughly: syntactic concepts) and f-concepts (roughly: semantical concepts) and the possibility of redefining a-concepts (such as provable category) by means of f-concepts. Tarski points out that moving to f-concepts (semantical concepts) gives different conceptual relationships, for instance f-completeness coincides with f-non-ramifiability (Tarski 1983, p. 391), which is not the case for a-completeness and a-non-ramifiability.

In the 1940 lecture, Tarski states two theorems about categoricity and completeness:

Theorem I Every system of sentences which is categorical with respect to its logical basis is also complete with respect to this basis.

Theorem II Every semantically categorical system of sentences is also semantically complete. (p. 6)

Since the proof of theorem I is clear from the text of the lecture, let us sketch the proof of theorem II. Assume $T(C_1, C_2, \ldots, C_n)$ is semantically categorical. Thus, any two models of the system $(O_1, O_2, \ldots, O_n)$ and $(P_1, P_2, \ldots, P_n)$ are isomorphic, i.e. there is an isomorphism R which maps all individuals one-one onto themselves and simultaneously $O_1, O_2, \ldots, O_n$ onto $P_1, P_2, \ldots, P_n$. Let $S(C_1, C_2, \ldots, C_n)$ be any sentence. Claim: either every model of $T(C_1, C_2, \ldots, C_n)$ is a model of $S(C_1, C_2, \ldots, C_n)$ or every model of $T(C_1, C_2, \ldots, C_n)$ is a model of $\neg S(C_1, C_2, \ldots, C_n)$. Because every two models of $T(C_1, C_2, \ldots, C_n)$ are isomorphic a proof by induction on the complexity of formulas easily shows that the same sentences will be true in both models and this proves the result.

It is important to remark that the theorem would not hold without the strong condition that the isomorphism R is an automorphism of the domain. Consider the theory of a single predicate S, which asserts that the extension of S does not coincide with any finite subset of V, where V is the domain of individuals (assumed infinite). Consider now two models for the theory, S^* and S^{**} which are such that V-S^* contains exactly two elements and V-S^{**} contains exactly three elements. There is certainly a one to one mapping sending S^* onto S^{**} but it cannot be an automorphism of V. And in this case semantical completeness would not hold since the first model makes "there are exactly two objects satisfying $\neg S$" true while the second model makes it false.[4]

Tarski emphasized the importance of the strengthened notion of categoricity. In a note to Tarski (1934–35) he says:

we use the word 'categorical' in a different, somewhat stronger sense than is customary: usually it is required of the relation $R[\ldots]$ only that it maps $x', y', z', \ldots$ onto $x'', y'', z'', \ldots$ respectively, but not that it maps the class of all individuals onto itself. The sets of sentences which are categorical in the usual (Veblen's) sense can be called *intrinsically categorical*, those in the new sense *absolutely categorical*. The axiom systems of various deductive theories are for the most part intrinsically but not absolutely categorical. It is, however, easy to make them absolutely categorical. It suffices, for example, to add a single sentence to the axiom system of geometry which asserts that every individual is a point (or more generally one which determines the number of individuals which are not points). (Tarski 1983, pp. 310–11, note 1)

This helps clarify the final comments of the lecture. In concluding the talk Tarski drew attention to unsolved problems and to the pervasiveness of the notion of categoricity:

There are some problems concerning the mutual relations of the concepts of completeness and categoricity which still remain open; for example, the problem as to whether the converses of theorems I and II are true.

We know many systems of sentences that are categorical; we know, for instance, categorical systems of axioms for the arithmetic of natural, integral, rational, real, and complex numbers, for the metric, affine, projective geometry of any number of dimensions etc. All these systems are semantically categorical, but if we base them on a sufficiently rich logic they become also categorical with respect to the logical basis. From theorems I and II we see that all mentioned systems are at the same time semantically or relatively complete. Thus, in opposition to absolute completeness, relative or semantical completeness occurs as a common phenomenon. (p. 7)

The above comments have to be read in the light of what Tarski had pointed out in 1934/35. The systems mentioned are only intrinsically categorical. Thus, in order for the claim that they are sematically categorical to hold we need to add the assumption, explicitly stated by Tarski in 1934/35, that additional axioms are added to the theory specifying that every individual is a point or a number. Under that condition they become semantically complete.

17.9 Conclusion: Some Open Problems

Far from being just of historical interest the problems raised by Tarski have recently resurfaced in the literature on higher order logic. In the article with Lindenbaum, Tarski had already shown that "Every categorical axiom system [with respect to its logical basis, PM] is non-ramifiable" (Tarski 1983, p. 390). Since relative completeness is equivalent to non-ramifiability this yields Theorem I as presented in the talk we are discussing. Tarski points out that the converse of Theorem I is an open question. However, he does prove a partial result: "Every non-ramifiable system which is effectively interpretable in logic is categorical"(Tarski 1983, p. 391). As mentioned earlier, this was actually one of Carnap's main results, and Tarski's proof essentially follows Carnap's although in the more rigorous setting developed by Tarski.

So, what can be said about the converse implication of theorem II? In first-order logic it obviously fails. Consider a syntactically complete theory T admitting infinite models. It is also semantically complete. But by Löwenheim-Skolem it has models of different infinite cardinalities and thus it is not categorical. What about higher-order logic? If the theorem is stated by allowing theories in the sense of arbitrary sets of sentences in some given language it fails also in higher-order logic (Awodey and Reck 2002b, p. 83).

Tarski and Lindenbaum (Tarski 1983, pp. 391–2) mention that semantical completeness and semantical non-ramifiability have the same extension. They then state that if a theory (implicitly assumed to be finitely axiomatizable) is semantically complete and it has a definable model in the background logic then it is categorical (in some semantical sense which is not more precisely specified in the article). In the paper we are commenting on, the notion of semantical categoricity is spelled out by requiring an automorphism of the domain of individuals. This is a rather strong condition and it is not quite the same notion of categoricity as used in contemporary model theory (where we have non-fixed domains and thus appealing to an automorphism of 'the' domain of individuals makes little sense).

In the remaining considerations, I will mention some results which have been formulated within the current model theoretic framework for higher-order logic (in particular, second-order logic) and which bear an immediate connection to the converse of theorem II as stated by Tarski.

First of all, there are some other conditions that also allow one to infer categoricity from semantical completeness, provided the theory is finitely axiomatizable, such as "having a model with no proper submodel" and "being categorical in some power" (see Awodey and Reck 2002b, p. 84).

The restriction to finitely axiomatized theories is in line with the context of the early work on these issues (see Awodey and Reck 2002a, p. 25) since researchers were interested in theories with finitely many axioms. In the 1940 lecture by Tarski, restrictions to finiteness show up implicitly in the restriction to finitely many non-logical constants and the discussion of categoricity is carried out (for simplicity, says Tarski) with respect to axiom systems with finitely many axioms and a single non-logical constant. So, there are different possible questions to be raised depending on how strictly we take the various restrictions on the theories in question. Let me mention two of them but many more variations could be added. I will for simplicity state them, following some of the literature, in terms of second-order logic.

First: is there a theory in second-order logic with finitely many axioms and no non-logical constants that is semantically complete but not categorical?

Second: is there a theory in second-order logic with finitely many axioms (and finitely many non-logical constants) that is semantically complete but not categorical?

A negative answer to the first questions has been given by Dana Scott (see Awodey and Reck 2002b, p. 82). Some other new results in this area are also given in Weaver and George (2003; 2005).

With respect to the second of the questions mentioned above, let me mention an important result obtained by Solovay and posted on FOM [foundations of mathematics] in May 2006:

(Solovay, see FOM, 5/16/06): Let A be a second-order theory. $ZFC + V = L$ proves "if A is finitely axiomatizable and semantically complete then A is categorical".

(Solovay, see FOM, 5/16/06): there is a model of $ZFC + V \neq L$ in which there is a finitely axiomatizable and semantically complete second-order theory that is not categorical.

18

APPENDIX: "ON THE COMPLETENESS AND CATEGORICITY OF DEDUCTIVE SYSTEMS" (1940) BY ALFRED TARSKI

In this lecture I should like to discuss two very important and closely related concepts of contemporary methodology of the deductive sciences, namely completeness and categoricity. We start with the concept of completeness; we shall distinguish here three variants of this concept. But a considerable part of my lecture will be devoted to the first, and from the intuitive point of view the most important type of completeness, viz. absolute completeness, which I shall refer to also simply as completeness.

The definition of absolute completeness is as follows: A system of sentences [statements] of a deductive theory is called absolutely complete or simply complete if every sentence which can be formulated in the language of this theory is decidable, that is, either derivable or refutable in this system. We assume that it is determined for the given deductive theory which expressions are treated as sentences, and under what conditions one sentence is said to be derivable from other sentences, or refutable by them. In order to explain what we mean by derivability we must formulate so-called rules of inference and we shall say that a sentence is derivable from other sentences if it can be obtained from them by applying the rules of inference any finite number of times. We assume that these rules of inference have a constructive character, in other words that they concern only the form of sentences (not their meaning) and that we are able to decide in any case whether a given sentence is obtainable from other sentences by one application of any of these rules.

Concerning the concept of refutability, in most cases it is sufficient to assume that a sentence is refutable if and only if its negation is derivable. Thus we can say that a system is complete if for every sentence either it or its negation is derivable. It is worth recalling at this point that a system of sentences is called consistent if for any sentence, the sentence itself and its negation are not both derivable.

It is easy to understand the importance of the concept of completeness. The development of every deductive science consists in formulating in the language of the theory problems of the form "Is so and so the case?" and then in attempting to solve them on the basis of the assumed sentences, called axioms. It is clear that every problem of this form can be decided in one of two ways: either in the affirmative or in the negative. On the first alternative, the answer runs: "so and so is the case", on the second it runs: "so and so is not the case". Now the completeness of a given system of axioms guarantees that every problem of this kind formulated entirely in the terms of our theory can be solved in at least one way on the basis of the theory itself. It may be added that we are exclusively interested in those deductive theories which, apart from whether they are complete or not, are consistent. Now the consistency of a theory guarantees that no problem of the kind mentioned above can be solved in two ways, that is, both affirmatively and negatively. Thus if a theory is both complete and consistent, we are certain that any such problem has one and only one solution.

Various deductive theories have been tested with regard to their completeness. Hitherto, in the literature of this problem, few results of a positive kind are to be found. Among the systems which have proved to be complete on the basis of investigations so far published may be mentioned, among others, certain systems of the calculus of statements (*Post* and others), of Boolean algebra (*Loewenheim* and others), of the theory of linear order (*Langford* and others), and finally of the theory of the addition of natural <<p. 2>> numbers (*Presburger*). The deductive theories in which all these systems are contained have an elementary logical structure and a meager mathematical content. The simplicity of structure finds its expression in the fact that in each of these systems all variables belong to one logical type, and all logical concepts are drawn from the most elementary parts of mathematical logic—the calculus of statements and the restricted functional calculus; no very profound, factually or historically important, problem can be formulated within the boundaries of the systems in question, and all problems which can be formulated can be decided with the simplest means and in a uniform manner. Also, during the period when no exact proof of their completeness was yet forthcoming, presumably no one would seriously have considered the possibility of finding a sentence in one of these systems which was neither derivable nor refutable. Nor is this contradicted by the circumstance that the proofs of completeness which were found for these systems are in no way trivial, but are interesting and important from the methodological point of view.

I have been interested in the problem of completeness in relation to other mathematical disciplines, namely, classical algebra and elementary Euclidean geometry. Many years ago, I succeeded in giving proofs of completeness for certain systems of algebra and geometry, but they have not yet been published. The logical structure of the theories which I investigated is not less elementary that that of previously investigated theories. All these theories can be formulated within the bounds of the restricted functional calculus. However, it seems to me that the mathematical content of the theories with which I dealt is considerably richer. Thus the formalized algebra which I considered contains such concepts as the relations of equality and inequality, the four arithmetical operations of addition, subtraction, multiplication and division, and further, any individual natural number such as 0, 1, 2, etc. (although the general concept of natural number finds no place in this theory). With the help of these, algebraical polynomials and equations of any desired degree can be constructed, various problems regarding the divisibility of polynomials, the solubility of equations, etc., can be formalized [~~within the considered theory~~] and solved, and thus comprehensive parts of classical algebra can be carried over into the considered theory.

As regards the system of geometry investigated by me it is important to establish the fact that the whole of elementary geometry, as it occurs for example in *Euclid's Elements* (in particular the theory of congruences, parallels, proportions, etc.), can be carried over into it almost without remainder. The same holds for certain branches of higher geometry, e.g. the theory of conic sections or, more generally, of algebraic curves of determined degree. Outside the system there remain problems and theorems in which the general concept of natural number explicitly or implicitly appears, e.g. theorems regarding polygons of any number of sides, as well as various parts of modern geometry which are essentially influenced by set theory, e.g. topology or the theory of convex bodies.

In connection with the relatively rich mathematical content of the considered system of algebra and geometry it would be possible to mention numerous historically and materially important problems which have arisen in the course of the history of mathematics on the basis of these theories and which can be decided with their means of proof. In the construction of these theories, and especially in the solution of the problems mentioned, various and by no means trivial modes of inference are applied. The positive solution of the problem of completeness for the system of algebra and geometry here treated will scarcely seem obvious or even very plausible to anyone.

It should be emphasized that all proofs of completeness hitherto found have an "effective" character in the following meaning: it is not merely theoretically shown that every sentence of a given theory is provable or refutable, but at the same time a procedure is given which permits every such sentence actually to be proved or refuted <<p. 3>> by the rules of inference of this theory. By the end of such a proof not only the problem of completeness, but also the decision-problem is solved for the considered system. In other words, every such proof shows that it is possible to construct a machine which provides an automatic solution for every problem which can arise in the given theory. In particular it follows from my results that such a machine can be constructed for elementary algebra and geometry.

In contradiction to the positive results discussed so far an important negative result concerning the problem of completeness is known at the present day. It was shown by *Gödel* that any system of [~~sentences~~] axioms of a given deductive science cannot be complete if the theory of the addition and multiplication of whole numbers can be formulated within the language of this science and can be founded upon the considered system of [sentences] axioms; and if, moreover, some additional conditions are fulfilled which concern mainly the consistency of the system in question.

Among other systems which come within the scope of this result, are also those which can be obtained by a natural extension of algebra and geometry of which I have spoken before. In general the domain of application of this result is very extensive, and is not essentially limited by the premisses which condition the incompleteness of the system. For it is well known that the arithmetic of whole numbers can be formalized within any deductive theory with a sufficiently rich logical structure, even if the concepts of arithmetic themselves do not occur explicitly in this theory. And, as to the condition of consistency, it can be said that we are not interested in systems that do not satisfy this condition because the failure of this condition in any system gives rise to the well-founded suspicion that the system in question contains sentences which are false from the intuitive point of view; and moreover, we are usually able to show that the systems with which we deal satisfy this condition provided that we are willing to avail ourselves of sufficiently strong methods of proof.

It should be noted that the systems which are shown to be incomplete by Gödel's method are, in a certain sense, essentially incomplete, that is to say, they cannot be completed by devices of a constructive character; in particular by the addition of any finite number of [sentences] axioms or of a finite number of constructive rules of inference.

On the basis of the foregoing we see that absolute completeness occurs rather as an exception in the domain of the deductive sciences, and by no means can it be treated as a universal methodological demand.

In this connection, I want to call your attention to certain concepts very closely related to the concept of absolute completeness, which are the result of a weakening of this concept and whose occurrence is not such an exceptional phenomenon. I have in mind here, firstly, completeness with respect to the logical basis, denoted briefly as relative completeness, and secondly, semantic completeness. Both of these concepts are of a somewhat more complicated nature and in order to make them more precise we must enlarge the domain of concepts with which we have so far dealt.

In what follows we assume that the concepts (constant terms) of a deductive theory are divided into two classes, the logical and the non-logical, to the first of which belong in any case the constants of the calculus of sentences and the quantifiers. Correspondingly, we divide the sentences of our theory into two classes, the logical and the non-logical, depending on whether they contain exclusively logical constants or not. <<p. 4>>

Among the logical sentences we single out the logically valid sentences. This is usually done in an axiomatic way: the logically valid sentences are defined as those which can be obtained by applying the determined rules of inference to the given logical axioms.

The class of logically valid sentences forms the logical basis of the given deductive theory, and we assume that the concept of derivability has been determined for the considered deductive theory in such a way that every logically valid sentence is derivable from any system of logical or non-logical sentences. We shall be interested here only in such deductive theories in which non-logical sentences actually do occur, and with respect to their completeness we shall consider exclusively systems consisting of non-logical sentences. To simplify our discussion let us assume moreover that there are no non-logical constants of our theory which do not occur in sentences of the considered system.

Taking any deductive theory let us consider an arbitrary system of non-logical sentences of this theory. As the theory let us think e.g. of the system of Principia Mathematica or of a fragment of it, but in either case enriched by certain non-logical, viz. geometrical constants and as a system of sentences, a system of axioms for Euclidean geometry; the non-logical concepts which occur in these axioms are usually called the "primitive" or [~~undefined~~] fundamental concepts of the given axiom system. If the logical basis of our theory is rich enough, we can formalize within its boundaries the arithmetic of natural numbers and for this reason its logical basis is incomplete: i.e. there are logical sentences which are not logically valid and whose negations are likewise not logically valid. In other words, there are problems belonging entirely to the logical part of our theory which cannot be solved either affirmatively or negatively with the purely logical devices at our disposal.

If we required our system of non-logical sentences to be complete we should have to take the logical problems, which were found to be undecidable by purely logical means, and to solve them on the basis of the non-logical axioms. It seems to be clear thus that such a requirement [and it follows from the] is excessive, but we can at least require that the considered system of sentences should not extend the incompleteness of the logical part of our theory. In order to formulate this requirement in an exact way, let us denote two given sentences as equivalent with respect to the considered system of sentences, when if this system is enriched by the addition to it of the first of these sentences, the second becomes derivable and vice versa. Our requirement can now be stated as follows: For any sentence of our theory, there must be a logical sentence which is equivalent to it with respect to the given system of sentences. If this condition is satisfied, we say that the considered system is *complete with respect to its logical basis*, or simply, that it is *relatively complete*. It is clear that in case the logical basis is itself complete, relative completeness reduces to absolute completeness. In general, these two concepts are not identical: although every complete system is a fortiori relatively complete, the converse does not always hold. But we are at least sure that on the basis of a relatively complete system sentences are decidable to the same extent, so to speak, as logical sentences.

Now let us introduce the concept of semantic completeness. Because incompleteness is such a general phenomenon, the problem arises as to whether this is due to our conception of derivability. The concept of derivability developed in modern logic and reduced there to the concept of constructive rules of inference was intended to be a formal analogue of the intuitive concept of logical consequence. Thus, a doubt arises as to whether this intention has been realized. I discussed this question in some of my papers published some years ago and in this lecture I can only avail myself of <<p. 5>> the final result of this discussion. It turned out that between the intuitive concept of logical consequence and the formal concept of derivability there was a big gap. If we want to

[~~make~~] formulate an exact definition of the concept of logical consequence [~~precise, it is necessary to~~] we must apply quite different methods and concepts.

The most important role is played here by the concept of model or realization. Let us consider a system of non-logical sentences and let, for instance "C_1", "C_2"..."C_n" be all the non-logical constants which occur. If we replace these constants by variables "X_1", "X_2"..."X_n" our sentences are transformed into sentential functions with n free variables and we can say that these functions express certain relations between n objects or certain conditions to be fulfilled by n objects. Now we call a system of n objects $O_1, O_2...O_n$ a model of the considered system of sentences if these objects really fulfill all conditions expressed in the obtained sentential functions. It is of course possible that the whole system reduces to one sentence; in this case we speak simply of the model of this sentence. We now say that a given sentence is a logical consequence of the system of sentences if every model of the system is likewise a model of this sentence.

The concepts which we just used, such as "fulfillment" and "model", have a semantical character, because they express some relation between expressions of a language and objects which are "talked about" in these expressions. Since the concept of logical consequence is based on semantical concepts it also belongs to the domain of semantics. And if in the definition of completeness, we replace the concept of derivability by the semantical concept of logical consequence, we arrive at the concept of semantical completeness. Thus a system of sentences of a given deductive theory is called *semantically complete* if every sentence which can be formulated in the given theory is such that either it or its negation is a logical consequence of the considered set of sentences. It should be noted that the condition just mentioned is satisfied by any logical sentence: hence we can deduce without difficulty that the concept of semantical completeness is a generalization of the concept of relative completeness: every system that is relatively complete is likewise semantically complete (but it can be shown by an example that the converse is not true).

We ask now the question as to by which means it can be shown that a given system is relatively or semantically complete. At first blush it might seem that the methods that have to be applied here are not essentially different or simpler than those which are used in the investigation of absolute completeness. Here as well as there, we have to establish that every sentence of the given theory has a certain property and it might seem that this cannot be established without an exhaustive methodological investigation concerning all possible forms of sentences. But it turns out that this is not the case. We shall see, namely, that the concept of relative and semantical completeness are closely related to the concept of categoricity (due to *Veblen*) and the investigation of the latter concept does not require in general any special and subtle methodological investigations.

We shall distinguish here two variants of the concept of categoricity: semantical categoricity, and categoricity with respect to the logical basis or relative categoricity, which parallel respectively semantical and relative completeness. But we do not know and therefore do not introduce any concept of categoricity which would parallel the absolute completeness.

We shall call a system of sentences *semantically categorical* if any two models of this system are isomorphic. The concept of isomorphism is a well known logical concept, not a methodological or semantical one, and thus I could take it that it is understood; but in any case I should like <<p. 6>> to have it noted that the precise definition of this concept depends on the logical foundation of our methodological investigations. Roughly speaking [and adapting ourselves to the language of *Principia Mathematica*, but disregarding the theory of types] we can say that two system of objects [classes, relations, etc.] $O_1, O_2, \ldots, O_n$ and $P_1, P_2, \ldots, P_n$ are isomorphic if there is a one-one correspondence which maps the class of all individuals onto itself and simultaneously the objects $O_1, O_2, \ldots, O_n$ onto $P_1, P_2, \ldots, P_n$ respectively.

In order to obtain the second variant of the concept of categoricity, we shall confine ourselves for the sake of simplicity, to such a case in which the considered system of sentences is finite and contains only one non-logical constant, say "C". Let "P(C)" represent the logical product of all these sentences. "C" can denote, for instance, a class of individuals, or a relation between individuals, or a class of such classes or relations etc. We assume further that the logical structure of the deductive theory is rich enough to express the fact that two objects X and Y of the same logical type as C are isomorphic; and we assume that this fact is expressed by the formula "$X \sim Y$". We can now correlate with the semantical sentence stating that our system of sentences is semantically categorical an equivalent sentence formulated in the language of the deductive theory itself. This is the following sentence:

For every X and Y, if P(X) and P(Y) then $X \sim Y$

or in symbols :

$(X)(Y)[P(X) \& P(Y) \rightarrow X \sim Y]$

Now we say that the considered system of sentences is *categorical with respect to its logical basis*, or *relatively categorical*, if the sentence formulated above is logically valid.

It is not quite easy to bring out the intuitive importance of the concept of categoricity. We could say that it is important for us to know that a given system of axioms is categorical because in this case, from the deductive point of view, the model of this system is uniquely determined. But saying it, we merely repeat the definition of categoricity in a less precise way. It seems that the best way to exhibit the significance of the concept of categoricity is to connect it with that of completeness. These connections find their expression in the following two recently proved theorems.

> Theorem I. Every system of sentences which is categorical with respect to its logical basis is also complete with respect to this basis.
> Theorem II. Every semantically categorical system of sentences is also semantically complete.

As the proofs of these theorems are analogous, we shall outline only the proof of the first. Let us assume, for simplicity, that in the sentences of the given theory only one non-logical constant, say, "C", occurs. Let, further, "P(C)" be the logical product of all sentences of the considered categorical system. Thus, according to the definition of relative categoricity, the following sentence is logically valid:

(1) $(X)(Y)[P(X) \ \& \ P(Y) \rightarrow X \sim Y]$

Let us consider now, an arbitrary sentence of our theory, say "$S(C)$". It was proved with respect to all known logical systems that the following sentence is also logically valid:

(2) $(X)(Y)[X \sim Y \ \& \ S(X) \rightarrow S(Y)]$

From 1 and 2 we see that the following sentence is logically valid:

(3) $(X)(Y)[P(X) \ \& \ P(Y) \ \& \ S(X) \rightarrow S(Y)]$

If we substitute "C" for "X" in 3 we see by a slight transformation that the following sentence is derivable from the considered system of sentences:

$$S(C) \rightarrow (Y)[P(Y) \rightarrow S(Y)]$$

This means that if we enrich our system of sentences by adding the sentence $S(C)$, the following sentence

$$(Y)[P(Y) \rightarrow S(Y)]$$

which is obviously logical, becomes derivable. But it is also obvious that if we enrich our system by this logical sentence, then the sentence $S(C)$ becomes derivable. Thus, both sentences are equivalent with respect to our system of sentences. We have therefore proved that to any sentence of the considered theory, an equivalent logical sentence can be constructed. In other words, our system of sentences is complete with respect to the logical basis, and out theorem is proved.

There are some problems concerning the mutual relations of the concepts of completeness and categoricity which still remain open; for example, the problem as to whether the converses of theorems I and II are true.

We know many systems of sentences that are categorical; we know, for instance, categorical systems of axioms for the arithmetic of natural, integral, rational, real, and complex numbers, for the metric, affine, projective geometry of any number of dimensions etc. All these systems are semantically categorical, but if we base them on a sufficiently rich logic they become also categorical with respect to the logical basis. From theorems I and II, we see that all mentioned systems are at the same time semantically or relatively complete. Thus, in opposition to absolute completeness, relative or semantical completeness occurs as a common phenomenon.

NOTES

Chapter 1

1. Each author has been responsible for specific sections of the essay: PM for I–III, VII, and VIII; RZ for itineraries V and VI; and CB for itinerary IV. While responsibility for the content of each section rests with its author, for the sake of uniformity of style we use "we" rather than "I" throughout. A book length treatment of the topics covered in itinerary IV is Badesa (2004). Itinerary V contains passages from Richard Zach, "Completeness before Post: Bernays, Hilbert, and the development of propositional logic", *The Bulletin of Symbolic Logic* 5 (1999) 331–66, ©1999, Association for Symbolic Logic, which appear here with the kind permission of the Association for Symbolic Logic. Itinerary VI contains passages from Richard Zach, "The practice of finitism: Epsilon calculus and consistency proofs in Hilbert's program", *Synthese* 137 (2003), 79–94, ©2003, Kluwer Academic Publishers, which appear here with the kind permission of Kluwer Academic Publishers.
2. On Zermelo's contribution to mathematical logic during this period see Peckhaus (1990, chapter 4); see also Peckhaus (1992).
3. In 1914, Philip Jourdain drew the same distinction but related it to two different conceptions of logic:

   > We can shortly but very accurately characterize the dual development of the theory of symbolic logic during the last sixty years as follows: The calculus ratiocinator aspect of symbolic logic was developed by Boole, De Morgan, Jevons, Venn, C. S. Peirce, Schröder, Mrs Ladd Franklin and others; the lingua characteristica aspect was developed by Frege, Peano and Russell. (Jourdain 1914, p. viii)

   Couched in the Leibnizian terminology, we thus find the distinction of logic as calculus versus logic as language, which van Heijenoort (1967b) made topical in the historiography of logic.
4. On Peano's contributions to logic and the foundations of mathematics and that of his school the best source is Borga *et al.* (1985), which also contains a rich bibliography. For Peano's contributions to logic and the axiomatic method, see especially Borga (1985), Grattan-Guinness (2000), and Rodriguez-Consuegra (1991). See also Quine (1987).
5. This idea of Padoa is at the root of a widespread interpretation of axiomatic systems as propositional functions, which yield specific interpreted theories when the variables are replaced by constants with a definite meaning. This view is defended in Whitehead (1907), Huntington (1913), Korselt (1913), Keyser (1918b, 1922), and Ajdukiewicz (1921). Such an interpretation also influences the development of the notion of model in Carnap and Tarski.
6. A similar result is stated which shows that the set of basic propositions of a system is irreducible, that is, that no one of them follows for the others: "To prove that the system of unproved propositions [P] is irreducible it is necessary and sufficient to find, for each of these propositions, an interpretation of the system of undefined symbols that verifies the other unproved propositions but not that one." (1901, 123)
7. See also Hilbert's lectures on geometry (2004).

8. On the various meanings of completeness in Hilbert, see Awodey and Reck (2002a, pp. 8–15) and Zach (1999).
9. On the debate that opposed Hilbert and Frege on this and related issues, see Demopoulos (1994).
10. Padoa later criticizes Hilbert for claiming that there might be other ways of proving the consistency of an axiom system. After Hilbert's talk in 1900, Peano claimed that Padoa's lecture would give a solution to Hilbert's second problem. Hilbert was not present at the lecture, but the only proof of consistency given by Padoa for his system of integers was by interpreting the formal system in its natural way on the domain of positive and negative integers. It is hard to believe that this led to an acrimonious article in which Padoa (1903) attacked Hilbert for not acknowledging that his second problem was only a "trifle". After a refusal to buy into the hierarchical conception of mathematics displayed by the reduction of the consistency of geometry to arithmetic, Padoa stated that Hilbert could modify at will all the methods used in the theory of irrational numbers but that this would never give him a consistency proof. Indeed, only statements of inconsistency and dependence could be solved by means of deductive reasonings, not issues of consistency or independence. According to Padoa, a consistency proof could only be obtained by displaying a specific interpretation satisfying the statements of the theory. Hilbert never replied to Padoa; in a way, the problem Padoa had raised was also a result of the vague way in which Hilbert had conjectured how it could be solved. It should be pointed out that Pieri (1904) takes position against Padoa on this issue remarking that perhaps one could find a direct proof of consistency for arithmetic by means of pure logic.
11. On the relationship between the axiom of completeness and the metalogical notion of completeness, see section 1.5.3.
12. We will follow, for consistency, Awodey and Reck (2002a) when providing the technical definitions required in the discussion. An axiomatic theory T is called categorical (relative to a given semantics) iff all models of T are isomorphic.
13. An axiomatic theory is called semantically complete (relative to a given semantics) if any of the following four equivalent conditions hold:
    1. For all formulas φ and all models M, N of T, if $M \models \varphi$, then $N \models \varphi$.
    2. For all formulas φ, either $T \models \varphi$ or $T \models \neg\varphi$.
    3. For all formulas φ, either $T \models \varphi$ or $T \cup \{\varphi\}$ is not satisfiable.
    4. There is no formula φ such that both $T \cup \{\varphi\}$ and $T \cup \{\neg\varphi\}$ are satisfiable.
14. This idea is expressed quite clearly in Bôcher (1904, 128).
15. "Suppose we express a law by a formal sentence S, and A is a structure. Different writers have different ways of saying that the structure A obeys the law. Some say that A satisfies S, or that A is a model of S. Many writers say that the sentence S is true in the structure A. This is the notion in the title of my talk. This use of the word *true* seems to be a little over fifty years old. The earliest occurrence I find is 'wahr in N^*' in a paper of Skolem (1933) on non-standard models of arithmetic (Padoa in (1901) has 'vérifie' [p. 136])" (Hodges 1986, p. 136).
16. A few more examples. "The assignment of an admissible meaning, or value, to each of the undefined elements of a postulate system will be spoken of as an interpretation of the system. By 'admissible' meanings are meant meanings that satisfy the postulates or that, in other words, render them true propositions." (Keyser 1918a, p. 391)

    "The logical structure of axiomatic geometry in Hilbert's sense—analogously to that of group theory—is a purely hypothetical one. If there are anywhere in reality three systems of objects, as well as determined relationships between these objects, such that the axioms hold of them (this means that by an appropriate assignment of names to the objects and relations the axioms turn into true statements [die Axiome in wahre Behauptungen übergehen]), then all theorems of geometry hold of these objects and relationships as well." (Bernays 1922a, p. 192)

17. For Russell's abandonment of idealism, see Hylton (1990).
18. For recent work on reconstructing Frege's system without Basic Law V, see Demopoulos (1995) and Hale and Wright (2001).
19. For an overview of the role of paradoxes in the history of logic see Cantini (2009). See the references given in section 1.2.1 for extensive analyses of the paradoxes.
20. For a survey of the history of predicativity, see Feferman (2004b).
21. For Poincaré on predicativity, see Heinzmann (1985).
22. See Chihara (1973), de Rouilhan (1996), and Thiel (1972) for detailed analyses of the various versions of the vicious circle principle.
23. There is even disagreement as to whether the types are linguistic or ontological entities and on the issue of whether the type distinction is superimposed on the orders or vice versa; see Landini (1998) and Linsky (1999).
24. On Russell's reasons for ramification, see also Goldfarb (1989).
25. See the extensive treatment in Grattan-Guinness (2000), and also Potter (2000) and Giaquinto (2002). Recent work has also been directed at studying the differences between the first and second edition of *Principia*; see Linsky (2004) and Hazen and Davoren (2000). The reader is also referred to the classic treatment by Gödel (1944). Hazen (2004) has pursued Gödel's suggestion that there is a new theory of types in the second edition.
26. We disagree with those who claim that metatheoretical questions could not be posed by Russell on account of his "universalistic" conception of logic. However, a detailed discussion of this issue cannot be carried out here. For this debate, see van Heijenoort (1967b), Dreben and van Heijenoort (1986), Hintikka (1988), Goldfarb (1979, 2001), de Rouilhan (1991), Tappenden (1997), Rivenc (1993).
27. On the development of set theory see, among others, Dauben (1971), Ferreiros (1999), Garciadiego (1992), Grattan-Guinness (2000), Kanamori (2003), Hallett (1984), and Moore (1982).
28. On Zermelo's role in the development of set theory and logic, see also Peckhaus (1990).
29. It should be pointed out that Russell had independently formulated a version of the axiom of choice in 1904.
30. The best treatment of the debate about the axiom of choice and related debates is Moore (1982).
31. On the antinomy see Garciadiego (1992). The antinomy is a transformation of an argument of Burali-Forti, made by Russell. If there were a set Ω of all ordinals, then it can be well ordered. Thus it is itself an ordinal, that is, it belongs and it does not belong to itself.
32. On the connection between Weyl (1910a) and (1918a), see Feferman (1988a).
33. On Zermelo's reaction to Skolem's paradox, see van Dalen and Ebbinghaus (2000).
34. Studies on the independence of the remaining axioms of set theory were actively pursued. See for instance Fraenkel (1922a).
35. On Mirimanoff, see the extended treatment in Hallett (1984).
36. On replacement, see Hallett (1984).
37. On von Neumann's system and its extensions, see Hallett (1984) and Ferreiros (1999).

38. Zermelo investigated the metatheoretical properties of his system, especially issues of categoricity (see Hallett 1996a).
39. In (1870), Peirce used the word "relative" in place of "relation" employed by De Morgan. In 1903, p. 367, n. 3, Peirce called De Morgan his "master", and regretted his change of terminology.
40. To our knowledge, van Heijenoort was the first to grasp the real historical interest of Löwenheim's paper. In "Logic as Calculus and Logic as Language" (1967b) he noted the elements in Löwenheim's paper that made it a pioneering work, deserving a place in the history of logic alongside Frege's *Begriffsschrift* and Herbrand's thesis. For the history of model theory, see Mostowski (1966), Vaught (1974), Chang (1974); the historical sections of Hodges (1993), and Lascar (1998).
41. For a detailed exposition and defense of the thesis presented in this contribution, see Badesa (2004).
42. On Tarski's suggestion, McKinsey (1940) had given an axiomatization of the theory of atomic algebras of relations. The 45 years that Tarski mentions is the time elapsed between the publication of the third volume of *Vorlesungen* and McKinsey's paper. A brief historical summary of the subsequent developments can be found in Jónsson (1986) and Maddux (1991).
43. It cannot be said to be totally algebraic, given the absence of an algebraic foundation of the summands and productands that range over an infinite domain.
44. Traditionally, "logic of relatives" is used to refer to the calculus or, depending on the context, to the theory of relatives. Our use of this expression is not standard.
45. Schröder showed how to develop the logic of predicates within the logic of binary relatives in his *Vorlesungen* 1895, §27. The proof that every relative equation is logically equivalent to a relative equation in which only binary relatives occur is due to Löwenheim (1915, Theorem 6).
46. Quantifiers were introduced in the algebraic approach to logic by Peirce (1883, p. 464). The word *quantifier* was also introduced by him in (1885, p. 183).
47. Expressions of the form $A \in B$ (called *subsumptions*) are also used as formulas, but the canonical statements are the equations. Depending on the context, the subsumption symbol ($\in$) denotes the inclusion relation, the usual ordering on $\{0, 1\}$, or the conditional. Löwenheim does not consider this symbol to belong to the basic language of the logic of relatives; this explains why he does not take it into account in the proof of his theorem.
48. In 1920, Skolem used *Zählaussage* instead of Löwenheim's *Zählausdruck*. Gödel erroneously attributes the term *Zählaussage* to Löwenheim (Gödel 1929, pp. 61–62).
49. In fact, Skolem (1922, p. 294) used the term *Lösung* (solution) to refer to the assignments of truth values to the relative coefficients that satisfy a given formula in a domain.
50. He probably intended not only to simplify the proof but also to make it more rigorous, but he did not doubt its correctness. See, for example, Skolem (1920, p. 254; 1922, p. 293; and 1938, pp. 455–456).
51. Löwenheim also generalized (1.1) to the case of formulas with multiple quantifiers, but this generalization is trivial. For typographical reasons, we use Σ in place of Löwenheim's double sigma.
52. See van Heijenoort (1967a, pp. 230), Wang (1970, p. 27), Vaught (1974, p. 156), Goldfarb (1979, p. 357), and Moore (1988, p. 122).
53. See van Heijenoort (1967a, pp. 229–230) and Moore (1988, p. 121).
54. Which the possible systems are depends on whether the fleeing indices are functional terms. More exactly, certain alternatives are only possible when fleeing indices are not functional

terms. For example, a system of equalities in which $1 = 2$ and $3 \neq 4$ is not compatible with a functional interpretation of the fleeing indices, because $3 = k_1$ and $4 = k_2$. Löwenheim repeatedly insists that two different numerals can denote the same element without placing restrictions on this, but he does not explicitly clarify which systems of equalities are admissible.

55. Skolem (1929a), proved again the weak version of the theorem. In this paper, Skolem corrects some deficiencies of his previous proof in (1922) (Wang 1974, pp. 20ff.) and introduces the functional form. As it is well known, the functional form of a formula such as $\forall x \exists y \forall z \exists u\, A(x, y, z, u)$ is $\forall x \forall z\, A(x, f(x), z, g(x, z))$. Skolem (1929a) states explicitly the informal procedure to which Gödel refers, but some of his assertions reveal that he lacks a clear understanding of the completeness problem.

56. The use of substitution is indicated at the beginning of *2. A substitution rule was explicitly included in the system of Russell (1906b), and Russell also acknowledged its necessity later (e.g., in the introduction to the second edition of *Principia*). For a discussion of the origin of the propositional calculus of *Principia* and the tacit inference rules used there, see O'Leary (1988).

57. This becomes clear from Bernays (1918), who makes a point of distinguishing between correct and provable formulas, "to avoid a circle". In (Hilbert 1920a, p. 8), we read: "It is now the first task of logic to find those combinations of propositions, which are always, i.e., without regard for the content of the basic propositions, *correct*."

58. This connection between the completeness theorem and the completeness axiom is tenuous: Hilbert's completeness axioms do not in general guarantee the categoricity of the axiom systems, nor its completeness in the sense that the system proves or disproves every statement. See Baldus (1928) for a counterexample and Awodey and Reck (2002a) for more detailed discussion.

59. Note that here, as indeed in Post (1921), syntactic completeness only holds if the rule of substitution is present.

60. Post (1921) gives the same definition and establishes similar results; see section 1.8.3.

61. The interested reader may consult Kneale and Kneale (1962, pp. 689–694) and, of course, Bernays (1926). The method was discovered independently by Łukasiewicz (1924), who announced results similar to those of Bernays. Bernays' first system defines Łukasiewicz's 3-valued implication.

62. Gödel (1932b) quotes the independence proofs given by Hilbert (1928a).

63. These results extend the method of the previous sections insofar as the independence of rules is also proved. To do this, it is shown that an instance of the premise(s) of a rule always takes designated values, but the corresponding instance of the conclusion does not. This extension of the matrix method for proving independence was later rediscovered by Huntington (1935).

64. This is not stated explicitly, but is evident from the derivation on p. 11.

65. Paul Bernays, notes to "Mathematische Logik", lecture course held winter semester 1929–30, Universität Göttingen. Unpublished shorthand manuscript. Bernays Nachlaß, ETH Zurich Archive, Hs 973.212. The signs '&' and '∨' were first used as signs for conjunction and disjunction in Hilbert and Bernays (1923b). The third axiom of group I and the second axiom of group V are missing from the system given in Hilbert and Bernays (1934). The first (*Simp*), third (*Comm*), and fourth axiom (*Syll*) of group I are investigated in the published version of the *Habilitationsschrift* (Bernays 1926), but not in the original version (1918).

66. Hilbert (1905a, p. 249); see Zach (1999, pp. 335–36) for discussion.

67. See Mancosu (1999a) for a discussion of this talk.

68. For extensive historical data as well as an annotated bibliography on the decision problem, both for classes of logical formulas as well as mathematical theories, see Börger et al. (1997).
69. On Curry's work, see Seldin (1980).
70. For more details on the work of Hertz and Gentzen, see Abrusci (1983) and Schröder-Heister (2002).
71. On the ε-calculus, see Hilbert and Bernays (1939) and Avigad and Zach (2002).
72. Hilbert (1920b, pp. 39–40). Almost the same passage is found in Hilbert (1922c, pp. 1127–28).
73. In a letter to Hilbert dated June 27, 1905, Zermelo mentions that he is still working on a "theory of proofs" which, he writes, he is trying to extend to "'indirect proofs', 'contradictions' and 'consistency'" (Hilbert Papers, NSUB Göttingen, Cod Ms Hilbert 447:2). Unfortunately, no further details on Zermelo's theory are available, but it seems possible that Zermelo was working on a direct consistency proof for Hilbert's axiomatic system for the arithmetic of the reals as discussed by Hilbert (1905a).
74. Hilbert developed a second approach to eliminating ε-operators from proofs around the same time, but the prospects of applying this method to arithmetic were less promising. The approach was eventually developed by Bernays and Ackermann and was the basis for the proof of the first ε-theorem in Hilbert and Bernays (1939). On this, see Zach (2003).
75. See Zach (2003) for an analysis of this proof and a discussion of its importance.
76. Von Neumann (1927) is remarkable for a few other reasons. Not only is the consistency proof carried out with more precision than those of Ackermann, but so is the formulation of the underlying logical system. For instance, the set of well-formed formulas is given a clear inductive definition, application of a function to an argument is treated as an operation, and substitution is precisely defined. The notion of axiom system is defined in very general terms, by a rule that generates axioms (additionally, von Neumann remarks that the rules used in practice are such that it is decidable whether a given formula is an axiom). Some of these features von Neumann owes to König (1914).
77. This is problem IV in Hilbert (1929).
78. See Gödel's recollections reported by Wang (1996, pp. 82–84).
79. On the reception of Gödel's incompleteness theorems more generally, see Dawson (1989), and Mancosu (1999b, 2004).
80. On Brouwer's life and accomplishments see van Atten (2003), van Dalen (1999), and van Stigt (1990). For an account of the foundational debate between Brouwer and Hilbert see Mancosu (1998a) and the references contained therein.
81. A good account of the French intuitionists is found in Largeault (1993b,a).
82. On the Kantian themes in Brouwer's philosophy see Posy (1974) and van Atten (2003, ch. 6).
83. Troelstra (1982) gives a detailed account of the origin of the idea of choice sequences.
84. On Brouwer's intuitionistic mathematics see van Atten (2003), van Dalen (1999), Dummett (1977), Franchella (1994), van Stigt (1990), and Troelstra and van Dalen (1988).
85. Indeed, in intuitionistic mathematics one can actually prove the negation of certain valid classical principles. For instance, one can prove in intuitionistic analysis that "it is not the case that every real number is either rational or irrational". These counterexamples are called strong counterexamples, and they are consequences of mathematical principles, such as the continuity principle, that are proper to intuitionism (as opposed to other forms of constructive mathematics or classical mathematics). Brouwer gave the above-mentioned counterexample in his (1928). On the continuity principle in intuitionistic analysis, see van Atten (2003, ch. 3), and on the difference between weak and strong counterexamples see van Atten (2003, chs. 2, 4, 5).

86. The best historical account of the debates surrounding intuitionism in the 1920s is Hesseling (2003).
87. We refer the reader to Thiel (1988), Mancosu and van Stigt (1998), and Hesseling (2003) for a more detailed treatment.
88. In Mancosu (1998a, p. 280) it was stated by mistake that Church had committed a faux pas at this juncture.
89. We should remark that Kolmogorov (1925) rejects the principle "ex falso sequitur quolibet", which he however accepts in 1932. There is some contemporary discussion on whether the principle is intuitionistically valid. For a first introduction see van Atten (2003, pp. 24–25).
90. Gentzen (1933a) (in collaboration with Bernays) had arrived at the same result, but Gentzen withdrew the article from publication after Gödel's paper appeared in print. The similarity between Gödel's and Gentzen's articles is striking. This parallelism can be explained by noting that both of them relied on the formalization of intuitionistic logic given by Heyting (1930a) and the axiomatization of arithmetic given by Herbrand (1931a).
91. See Mancosu (1998c) on finitism and intuitionism in the 1920s.
92. On all these contributions, see the useful introductions by Troelstra in (Gödel 1986).
93. On Łukasiewicz's logical accomplishments and the context in which he worked see Woleński (1989a).
94. Słupecki, like Łukasiewicz, used the Polish notation; for the reader's benefit, we have used the *Principia* notation in this section.
95. Among the few variations one can mention "concrete representation" (Veblen and Young 1910, p. 3; Young 1917, p. 43). It should be pointed out here that although the word "model" was widespread in physics (see, e.g., "dynamical models" in Hertz 1894) it is not as common in the literature on non-Euclidean geometry, where the terminology of choice remains "interpretation" (as in Beltrami's 1868 interpretation of non-Euclidean geometry). However, "*Modelle*", i.e., desktop physical models, of particular geometrical surfaces adorned the German mathematics departments of the time. Many thanks to Jamie Tappenden for useful information on this issue.
96. Following Russell, structure-theoretic terminology is found all over the epistemological landscape. See, for instance, Carnap's *Der logische Aufbau der Welt* (1928).
97. A similar approach is found in Lewis (1918, p. 355).
98. See Dreben and van Heijenoort (1986, pp. 47–48) for a clarification of some delicate points in Hilbert and Ackermann's statement of the completeness problem.
99. In the 1929 dissertation, the result for countable sentences is obtained directly and not as a corollary to compactness. For the history of compactness, see Dawson (1993).
100. The notion of *allgemeingültig* can be relativized to specific types of domains. So, for instance, $(Ex)F(x) \vee (x)F(x)$ is *allgemeingültig* for those domain consisting of only one element. See Bernays and Schönfinkel (1928, p. 344).
101. Gödel did not provide the foregoing explanations in the published version of the thesis (1930a), but the same definition occurs in later published works (Gödel 1933b, p. 307), where the same idea is used to define the notion of a model over I (a domain of individuals).
102. An early case is Weyl (1910a) and concerns the continuum-problem. Weyl says (p. 304) that the continuum problem will not admit a solution until one adds to the system of set theory an analog of the opposite of Hilbert's completeness axiom: From the domain of Zermelo's axioms one cannot cut out a subdomain which already makes all the axioms true.
103. Nowadays we call the second notion "syntactic completeness". As the notion of categoricity as isomorphism is already found, among other places, in Bôcher (1904), Huntington

(1906–07), and Weyl (1910a) (also Weyl 1927), we cannot agree with Howard (1996, p. 157), when he claims that Carnap (1927) is "the first place where the modern concept of categoricity, or monomorphism in Carnap's terminology, is clearly defined and its relation to issues of completeness and decidability clearly expounded. Moreover, it was through Carnap's relations with Kurt Gödel and Alfred Tarski that the concept of categoricity later made its way into formal semantics". The first conjunct is made false by the references just given, the second by the fact that Carnap's claims as to the equivalence of categoricity and decidability turned out to be unwarranted. As for Carnap's influence, it is certainly the case that Tarski was familiar with the concept of categoricity before he knew of Carnap's investigations (see Tarski 1930b, p. 33). Howard's article is to be recommended for exploring the relevance of the issue of categoricity for the natural sciences. On completeness and categoricity see Awodey and Carus (2001), Awodey and Reck (2002a), and also Read (1997).

104. Weyl's reflection on *Entscheidungsdefinitheit* are related to the great attention given to this notion in the phenomenological literature, including Husserl, Becker, Geiger, London, and Kaufmann.
105. See the review by Rosser (1937).
106. Scanlan (2003) deals with the influence of Langford's work on Tarski. See Zygmunt (1990) on Presburger's life and work. Tarski's early results are discussed by Feferman (2004c), who uses them to reply to some points by Hodges (1986). On Tarski's quantifier elimination result for elementary algebra and geometry, see the extensive study by Sinaceur (2007). For a treatment of the main concepts of the methodology of deductive sciences according to Tarski, see Czelakowski and Malinowski (1985) and Granger (1998).
107. One should also not forget the possible influence of Łukasiewicz; see Woleński (1994). On the Polish school see Woleński (1989a, 1995a).
108. For the interpretation of the differences between the original article (1933b) and the claims made in the postscript in (1935a) see de Rouilhan (1998).
109. Gödel was aware of the result before Tarski published it; see the discussion in Murawski (1998). However, the author makes heavy weather of Gödel's use of the word *richtig* as opposed to *wahr*. To this it must be remarked that *richtig* is used in opposition to *falsch* throughout the writings of the Hilbert school. Moreover, Gödel himself speaks of *wahr* in his dissertation (Gödel 1929, pp. 68–69). See also Feferman (1984).
110. On the issue of whether Tarski defines truth in a structure, see Hodges (1986) and Feferman (2004c). On logical consequence, see, among the many contributions, Etchemendy (1988, 1990), Ray (1996), Gomez-Torrente (1996), Bays (2001), and Mancosu (2006).

Chapter 2

1. See Feferman (1988b) for an overview of work on Hilbert's problems related to the foundations and further references. For biographical information on Hilbert and Bernays see Blumenthal (1935), Dieudonné (1948), Fang (1970), Lauener (1971, 1978), Müller (1978), Reid (1970), Specker (1979), Weyl (1944). Hilbert's programme continues to generate work also from the philosophical side. Although I have kept these contributions in mind (e.g. Detlefsen, Kitcher, Kreisel, Parsons, Prawitz, Rolf, Tait; see also Moriconi 1987) it would be impossible to discuss at length these works within the limited extent of this chapter.
2. For extensive references to the literature on the *Foundations of Geometry*, see Peckhaus (1990), Schüler (1983) and Toepell (1986).
3. See Bernays (1928a, p. 200).

4. Bernays (1922a) offers an appraisal of the significance of Hilbert's work on the foundations of geometry that emphasizes its close relation to the development of mathematics in the nineteenth century and its broad epistemological significance. On the foundational significance of Hilbert's work in the foundations of geometry see also Webb (1980) and Hallett (1996b).

5. In a series of lectures given in 1919, Hilbert emphasized the importance of the notion of intuition for mathematics: "Wäre die dargelegte Ansicht zutreffend, so müßte die Mathematik nichts anderes als eine Anhäufung von übereinander getürmten logischen Schlüssen sein. Es müßte ein wahlloses Aneinanderreihen von Folgerungen stattfinden, bei welchem das logische Schließen allein die treibende Kraft wäre. Von einer solchen Willkür ist aber tatsächlich keine Rede; vielmehr zeigt sich, daß die Begriffsbildungen in der Mathematik beständig durch Anschauung und Erfahrung geleitet werden, so daß im großen und ganzen die Mathematik ein willkürfreies, geschlossenes Gebilde darstellt." (Hilbert 1992, p. 5) Compare Bernays: "Diese methodische Loslösung von der Raumanschauung ist nicht gleichzusetzen mit einem Ignorieren des raumanschaulichen Ausgangspunktes der Geometrie." (Bernays 1928a, p. 203)

6. However, there is also the meaning of completeness given by the axiom of completeness; that is, the domain of the interpretation must be non-extendible.

7. (See Hallett 1995a) and references to Hilbert's passages contained therein.

8. Hilbert identifies consistency and "existence" in mathematics for most of his career. In 1928, however, the identification is considered problematic. The identification of consistency and existence is stated explicitly in a letter to Frege dated 27.xii.1899: "You write: '[I call axioms propositions]... From the truth of the axioms it follows that they do not contradict one another'. I was very interested to read exactly this, since for as long I have been thinking, writing, and lecturing on these things, I have asserted exactly the opposite: If the arbitrarily chosen axioms do not contradict each other with all their consequences, then they are true, and the things defined by the axioms exist. That for me is the criterion of truth and existence." Frege (1980) And in 1905: "A concept exists or does not exist in the mathematical sense according to whether or not the axioms by which the concept is fixed form a consistent system." (Hilbert 1905b)

The following passage by Bernays is important for clarifying the relationship between consistency, existence, and completeness: "Und jedenfalls ist hier, wo es sich um die Aufstellung von Postulaten handelt, die Deutung der Existenz im Sinne der Widerspruchslosigkeit mit den Postulaten nicht angängig.

"Die Gleichsetzung von Existenz und Widerspruchslosigkeit ist in zweierlei Sinn berechtigt: erstens mit Bezug auf den geometrischen Raum, dessen Existenz in der Tat nur in der Widerspruchslosigkeit der ihn definierenden Postulate besteht, zweitens auch mit Bezug auf die geometrischen Gebilde, jedoch nur unter der Bedingung der *Vollständigkeit des Systems der Postulate*.

"Wenn das System der Postulate vollständig ist, d.h. wenn durch die Postulate bereits für jede Kombination (jede Synthese) von Elementarbegriffen entschieden wird, ob sie zugelassen oder ausgeschlossen ist, dann fällt allerdings die Möglichkeit (Widerspruchslosigkeit) eines Gebildes mit seiner Existenz zusammen.

"Solange man sich aber noch auf dem Wege zur Gewinnung eines Postulatensystems, d.h. zur schrittweisen Bestimmung des geometrischen Raumes befindet, muß man zwischen Existenz und Widerspruchslosigkeit unterscheiden. Aus dem Nachweis der Widerspruchslosigkeit einer Synthese folgt ja dann nur, daß diese mit den *bisher eingeführten* Postulaten im Einklang steht; es konnte uns dennoch freistehen, diese Synthese durch ein weiteres Postulat auszuschließen. Ein *Existenzbeweis* dagegen besagt, daß wir durch die bisherigen Postulate bereits logisch *genötigt* sind, die Betreffende Synthese zuzulassen......Um also, wie es Strohal will, die Existenz der geometrischen Gebilde durch die Widerspruchslosigkeit mit den Postulaten charakterisieren zu können, muß man ein *vollständiges* System von Postulaten haben." (Bernays 1928a, p. 202) For more on the issue of consistency and existence see Hallett (1995b, pp. 33–52).

9. "Die Zahlen sind für uns nur ein Fachwerk von Begriffen geworden, auf das wir allerdings nur durch die Anschauung geführt werden, mit dem wir aber desungeachtet ohne jede Zuhilfenahme der Anschauung operieren können. Damit aber dies Begriffssystem auf die uns umgebenden Dinge anwendbar wird, ist es von Grund aus so construiert worden, dass es überall eine volle Analogie mit den trivialsten Anschauungen und damit den Thatsachen der Erfahrung bildet." (quoted in Peckhaus 1990, p. 60)

10. The possible reduction of arithmetic to logic is considered inadequate: "If we observe attentively, however, we realize that in the traditional exposition of the laws of logic certain fundamental arithmetical notions are already used, for example, the notion of set and, to some extent, also that of number. Thus we find ourselves turning in a circle, and that is why a partly simultaneous development of the laws of logic and of arithmetic is required if paradoxes are to be avoided." (Hilbert 1905b, p. 131)

11. For more details on Kronecker see Edwards (1988, 1989), Gauthier (1993), Sieg (1990a), (Webb 1980, pp. 71–74), and Marion (1995). Of great interest for Hilbert's attitude to Kronecker are the passages from Hilbert (1920b), translated in Ewald (1996, vol. II, pp. 943–46).

12. "Kronecker, der als erster die Anforderungen des finiten Standpunktes geltend gemacht hat, ging darauf aus, die nichtfiniten Schlußweisen allenthalben aus der Mathematik auszuschalten. In der Theorie der algebraischen Zahlen und Zahlenkörper ist er damit zum Ziel gekommen. Hier gelingt auch die Einhaltung des finiten Standpunktes noch in solcher Weise, daß man von den Sätzen und Beweismethoden nichts Wesentliches aufzugeben braucht." (Hilbert and Bernays 1934, p. 42) There is an important difference between this quote and the previous mentioning of Kronecker in Hilbert's writing. In 1934 Hilbert and Bernays distinguish clearly between Kronecker's finitism and intuitionism, since the latter is seeing as more encompassing than finitism. In the 1931 article *On the foundation of the elementary theory of numbers* Hilbert also identifies Kronecker's position with the finitistic position of proof theory but he does not seem to make a distinction between intuitionism and Kronecker's finitism. See the section on finitism and intuitionism (see also Bernays 1930b, pp. 251, 253).

13. On Poincaré's philosophy of mathematics see Heinzmann (1985), Mooij (1966), and Schmid (1978).

14. See Poincaré (1906) and (1908, pp. 179–91) [1952]. For the relationship between Hilbert and Poincaré see Sinaceur (1995).

15. See, for instance, Hilbert (1928a, p. 473).

16. For an interesting overview see Heyting (1955, pp. 42–60) and Abrusci (1978).

17. An analysis of Hilbert's lecture notes 1920a, pp. 46–62 and 1922b, pp. 52–60 and 1a–9a reveals the existence of different reconstructions of elementary arithmetic. A thorough analysis of these lectures is necessary for an accurate study of Hilbert's developing conception of finitism from the early 1920s to *Grundlagen der Mathematik* (1934). For obvious reasons this cannot be done here.

18. Note that Hilbert is even more radical than Brouwer. In 1908 Brouwer asserts that the unlimited use of excluded middle, although not justified, will never lead to contradictions. For the mistakes engendered by the use of infinite sums and products, see Hilbert's 1919 lectures published in Hilbert (1992).

19. This quote by Hilbert also invites a reflection on the relationship between the debates in the foundations of mathematics in the 1920s and the issue of verificationism. In addition to Hilbert's statements, one should look at Löwy (1926) for a verificationist reading of intuitionistic mathematics. Löwy was a close affiliate of the Vienna Circle.
20. For the epistemological importance of the atomic debate in physics see Nye (1976).
21. Smoryński (1989, p. 39) sees here a more complicated distinction between real propositions, finitary general propositions, and ideal propositions. See also the objections to Smorynski's position presented in Detlefsen (1990).
22. The word *formalism* is actually not used there and seems to appear for the first time, in Brouwer's writings, in the review of Mannoury's *Methodologisches und Philosophisches zur Elementarmathematik* (1909), which he published in 1911. From that review it is clear that the terminology was used by Mannoury and covers a wide spectrum of positions: "Finally the book expresses the author's conviction about the origin of the mathematical truths, and there he defends the 'formalist' conception, which has also been advocated by Dedekind, Peano, Russell, Hilbert and Zermelo, against 'intuitionists' like, for instance, Poincaré and Borel. This formalist conception recognizes no other mathematics than the mathematical language, and it considers it essential to draw up definitions and axioms and to deduce from these other propositions by means of logical principles which are also explicitly formulated beforehand." (Brouwer 1975, p. 121)
23. Detlefsen proceeds to give a more precise characterization of what this amounts to. See Detlefsen (1993b,a). Detlefsen's position on the two different phases of formalism is nicely corroborated by Hilbert (1922b, pp. 31–32).
24. Skolem (1922, p. 300) and Weyl (1928a) criticize this aspect of Hilbert's programme. See the enlightening introduction by van Heijenoort to Weyl (1928a). Fraenkel sees in Hilbert's solution a victory for Poincaré: "Letztlich ist . . . .die in Hilberts Anschauungsgrundlage vorkommende Rekursion bzw. Induktion nichts wesentlich Anderes als die von Poincaré an die Spitze der Mathematik gestellte (und ja gerade auch von ihm zu den synthetischen Urteilen a priori im Sinne Kants gerechnete!) vollständige Induktion, ja selbst als der von Brouwer als Quelle aller mathematischen Konstruktionen reklamierte Begriff der natürlichen Zahl." (Fraenkel 1928, p. 379)
25. Note that Hilbert has dropped the qualification expressed in 1923 that induction be used on finite totalities. Compare Hilbert (1929, p. 232).
26. This is a complicated problem. In the appendix to *Mathematical Knowledge* (1975) Steiner attempts to show how one can defend Hilbert's approach on this count. However the treatment is only tentative. See also Detlefsen's defence of Hilbert in Detlefsen (1986). On the issue of the role of the axiom of induction in the metatheory, I find myself in agreement with van Heijenoort (1967a, pp. 480–82). Around 1930, Hilbert and Becker corresponded on the issue of the role of induction in metamathematics. The correspondence is preserved as Cod MS Hilbert 457 in the archives of the Mathematisches Institut in Göttingen. It contains an incomplete draft of a letter from Hilbert to Becker and the reply by Becker (dated October 4, 1930).
27. "Es ist nicht vermeidbar, den Begriff der positiven ganzen Zahl auch inhaltlich einzuführen. Es ist nicht möglich, die Beweistheorie aufzubauen, ohne daß die positive ganze Zahl und alle ihre intuitionistisch, d.h. inhaltlich, ableitbaren Eigenschaften schon a priori zur Verfügung stehen." (von Neumann 1927, p. 4) Von Neumann asserts this point of view also in von Neumann (1931, pp. 61–62). This position is also defended by Herbrand in 1930.

28. "Der methodische Standpunkt des 'Intuitionismus', den Brouwer zugrunde legt, bildet eine gewisse *Erweiterung der finiten Einstellung* insofern, als Brouwer zuläßt, daß eine Annahme über das Vorliegen einer Folgerung bzw. eines Beweises eingeführt wird, ohne daß die Folgerung bzw. der Beweis nach anschaulicher Beschaffenheit bestimmt ist". (Hilbert and Bernays 1934, p. 43) The reason behind this change of view is the double negation interpretation, discovered independently by Gödel and Gentzen in 1933, which showed the consistency of classical arithmetic relative to intuitionistic arithmetic. For a substantiation of this point, see Sieg (1984, pp. 173–74).

29. "Nun haben aber Philosophen—und Kant ist der hervorragendste, der klassische Vertreter dieses Standpunktes—behauptet, daß wir außer der Logik und der Erfahrung noch a priori gewisse Erkenntnisse über die Wirklichkeit haben. Daß die mathematische Erkenntnis letzen Ende auf einer Art anschaulicher Einsicht beruht, daß wir sogar zum Aufbau der Zahlentheorie eine gewisse anschauliche Einstellung und wenn man will apriorische Einsicht nötig haben, daß die Anwendbarkeit der mathematischen Betrachtungsweise auf die Gegenstände der Wahrnehmung eine wesentliche Bedingung für die Möglichkeit exakter Naturerkenntnis ist, das erscheint mir sicher." (Hilbert 1988, pp. 87–88) For Hilbert's thought on the role of intuition in physics see Majer (1993b).

30. "Zur *philosophischen Ergänzung* dieser Beweistheorie ist eine *methodische Erörterung* erforderlich, durch die jene in der Beweistheorie systematisierten Prinzipien eine Art von Deduktion erhalten im Sinne einer Klärung ihrer erkenntnis-methodischen Bedeutung. Diese Erörterung müßte zugleich die Methoden der mathematischen Idealisierung klarstellen und damit eine befriedigende Antwort geben auf die Frage Nelsons, worin denn die Norm für eine Idealisierung bestehen könne, wenn sie nicht in der reinen Anschauung liege." (Bernays 1930a, p. 111)

31. Apart from the virulent attacks of the intuitionists, there are various objections raised from other viewpoints. For example, the Italian mathematician Cipolla wrote a rather critical survey of Hilbert's new approach to the foundations of mathematics (Cipolla 1924). He criticized in particular Hilbert's postulation of the transfinite axiom $A(\tau A) \to A(a)$ to be read as: If the object τA satisfies the predicate A, so does any object a. Cipolla criticized both the intuitive evidence for such an axiom and the alleged justification which would follow from a consistency proof à la Hilbert. Cipolla sees the admission of the transfinite τ operator on a par with the admission of Zermelo's principle of choice. His general criticism can be seen to be a variation on those criticisms that refused to accept a formal proof of consistency as guaranteeing anything until a concrete interpretation of the axioms is given. Fraenkel (1928) is also rather critical of Hilbert's approach.

In this section I will only treat the philosophical debates that saw Hilbert and Bernays as protagonists. Interesting confrontations with other philosophical movements could be made. They would include, just to name a few cases, the relationship to the philosophy of the Als Ob (see Hilbert's appeal to "fictions" in 1926, Betsch 1926, and Bernays 1930b, note 20), Cassirer's interpretation of the foundational debates (Cassirer 1929, vol. III) and the relationship to phenomenology (see Mahnke 1923).

32. On Müller see Kluge (1935). Bernays (1923b) is not mentioned by either Müller or Bernays. It might be the case that this negative review is what prompted Müller to attack Hilbert.

33. See the remark by Bernays in Hilbert (1935, p. 163, note 2).

34. "Zur Lösung dieser Aufgabe ist es offenbar nötig, innerhalb jeder mathematischen Disziplin eine scharfe Trennung vorzunehmen zwischen dem, was logisch beweisbar ist, und dem, was als Voraussetzung solcher Beweise aus der Anschauung hinzukommt" (Nelson 1928, p. 2).
35. Nelson (1928, p. 8); Nelson's summary of Bernays seems accurate: see Bernays (1928c, pp. 144–45) and 1930a.
36. A complete account of the technical work originating from Hilbert's programme would have to cover most of proof theory. For a technical survey see Feferman (1988b), Prawitz (1981), Sieg (1984, 1988, 1996), and Simpson (1988).

Addendum

1. "Es haben in der Tat philosophen—und Kant ist der klassische Vetreter dieses Standpunktes—behauptet, dass wir ausser der Logik und der Erfahrung noch *a priori* gewisse Erkenntnisse über die Wirklichkeit haben. Nun gebe ich zu, dass schon zum Aufbau der theoretischen Fachwerke gewisse *apriorische Einsichten* nötig sind und dass stets dem Zustandekommen unserer Erkenntnisse solche zu Grunde liegen. Ich glaube, dass die math. Erkenntnis letzten Endes auf einer Art anschaulicher Einsicht beruht und dass wir sogar zum Aufbau der Zahlentheorie eine gewisse anschauliche Einstellung *a priori* nötig haben. Damit behält also der allgemeinste Grundgedanke der Kantschen Erkenntnistheorie seine Bedeutung: nämlich das philosophiche Problem, jene anschauliche Einstellung a priori festzustellen und damit die Bedingungen der Möglichkeit jeder begrifflichen Erkenntnis und zugleich jeder Erfahrung zu untersuchen. Ich meine, dass dies im Wesentlichen in meinen Untersuchungen über der Prinzipien der Math. geschehen ist." (Hilbert 1923b, pp. 424–25 of lecture 3) I would like to thank Ulrich Majer and Tilman Sauer for having provided me with copies of the transcription of these lectures long before their publication in Hilbert (2009).

Chapter 3

1. The talk was delivered on December 7, 1920, and was entitled "Elemente der Logik." The results were published in Schönfinkel (1924).
2. The list of corrections was sent to Russell by Behmann. The first mention of Boskovitz's corrections is found in a letter from Behmann to Russell dated August 23, 1923 (Behmann Archive, Erlangen). Behmann sent a list of corrections, including those of Boskovitz with a letter dated September 19, 1923 (Russell Archive, McMaster University, 110205e). Indeed, a letter from Boskovitz to Russell—dated July 3, 1923—containing the mentioned corrections is preserved in the Russell archive at McMaster University, call number 110240a. The latter letter contains also side remarks by Behmann concerning the validity of Boskovitz's list of errata for PM.

   After his stay in Göttingen Schönfinkel moved to Moscow. As for Boskovitz he ended up in Budapest (see Behmann to Scholz, November 3, 1927, Behmann Archive, Erlangen). Behmann praised Boskovitch in the letter just quoted as one of the very few people with a deep mastery of PM.
3. I do not mean to imply that Hilbert had lost all interest in mathematical logic. Indeed, according to Behmann, this was not the case: "So weit ich mich erinnere, habe ich das Verfahren der Normalform ... durch Hilbert, der sich ja vor dem Kriege selbständig mit dem Problem der symbolischen Darstellung der Logik, und zwar insbesondere der Aussagenlogik,

beschäftigte, kennen gelernt" (Behmann to Scholz, December 29, 1927, Behmann Archive, Erlangen). The normal form procedure is found in Hilbert (1905a). Sieg (1999, p. 8) gives additional evidence from Hilbert's lecture course given in 1910 "The quadrature of the circle and related problems." Peckhaus (1990) also provides very relevant information for the period in question. However, Behmann does not appear in Peckhaus' account.

4. Sieg (1999, pp. 37–38) has edited the postcard exchange between Russell and Hilbert concerning a possible invitation for Russell to visit Göttingen. In particular Hilbert wrote to Russell on April 12, 1916 that "we have been discussing in the Mathematical Society your theory of knowledge already for a long time and ... we had intended, just before the outbreak of the war, to invite you to Göttingen, so that you could give a sequence of lectures on your solution to the problem of the paradoxes" (quoted in Sieg 1999, p. 12). The visit, however, never materialized. I have found further unpublished details on this invitation for Russell in the acts of the Universitätskuratorium Göttingen. See, in particular, document 5a of Hilbert XVI, IV. Mention of the discussion of Russell's work in Göttingen is also found in a letter from Hugo Dingler to Hilbert dated January 2, 1915 (Dingler *Nachlaß*, Aschaffenburg) where Dingler, in reply to a previous letter by Hilbert, expresses his regret for having missed the discussion on Russell's work (Hilbert *Nachlaß*, Staats- und Universitätsbibliothek, Göttingen, Cod. Ms. D. Hilbert 74). In Hilbert's original letter to Dingler (December 26, 1914, Dingler *Nachlaß*), Hilbert mentions Paul Hertz, Bernstein, and Grelling as his closest associates on foundational issues.

5. The information is taken from the brief biography appended to the dissertation and from Haas and Stemmler (1981). Further details about Behmann's academic career are also found in the acts of the Universitätskuratorium of the University of Göttingen.

6. "Trotz alledem bietet die richtige Auflösung all dieser geometrischen Widersprüche, sobald sie einmal gegeben ist, dem Verständnis keine erhebliche Schwierigkeit dar. Sie besagt nämlich einfach, daß die idealen Elemente nicht im eigentlichen Sinne Gegenstände der Geometrie sind, sondern zunächst einmal nur Worte und als solche Teile von Ausdrucksweisen, deren man sich in der Geometrie bedient, um eine gewisse Klasse von Sätzen auf eine möglichst einfache Form zu bringen." (p. 10)

7. "Das Axiom sagt also aus, daß in dieser einen Stufe schon genügend viele Funktionen vorhanden sind, um alle überhaupt möglichen Klassen zu definieren; es läst sich daher auch als eine Art Vollständigkeitsaxiom für die prädikativen Funktionen auffassen." (p. 14)

8. In 1922 he writes: "Mag man zu der auf den ersten Anblick gewiss befrendenden Russell-Whiteheadschen 'Theorie der logischen Typen' stehen, wie man will, man wird nicht bestreiten können, daß es ihr in der Tat gelungen ist, nicht nur die Mengenlehre [...] sondern die gesamte Logik, von der hier die Arithmetik im weitesten Sinne nur als ein Ausschnitt erscheint, auf eine Grundlage zu stellen, mit der sich jedenfalls vollkommen sicher arbeiten läßt und die, wie der Erfolgt lehrt, trotz der in ihr liegenden unvermeidlichen Beschränkungen das stolze Werk der Begründung der Arithmetik aus bloßer Logik in einer—grundsätzlich gesprochen—völlig strengen und bewundernswert geschlossenen Form tatsächlich geleistet hat—was selbstverständlich nicht ausschließt, daß dieses Werk sogar in wesentlichen Punkten der Vervollkommnung fähig und bedürftig ist." (Behmann 1922b, p. 56) Behmann, however, expressed doubts whether the axiom of reducibility is to be considered as an axiom and spoke of it as a postulate (1918, p. 196) showing that it fails in certain restricted domains. For Hilbert's similar position on the axiom of reducibility see Sieg (1999, p. 17), which also contains additional information on the development of Hilbert's views regarding Russell's logicism.

9. "It was, in fact, that work of yours [PM] that first gave me a view of that wonderful province of human knowledge which ancient Aristotelian Logic has nowadays become by the use of an adequate symbolism. But, I daresay, it might be said of your work just as well what H. Weyl said of his own book, that 'it offers the fruit of knowledge in a hard shell,' requiring indeed a considerable amount of labour in order to be accustomed to that particular manner of thinking, equally different from that of common life and that of college logic and philosophy, which is absolutely necessary for a rigorous treatment of the topic.

"Several years ago, I had therefore resolved to write something like an introduction or commentary to that work, providing a way by which the unavoidable difficulties of understanding are separately treated and, in consequence of it, may be clearly grasped and overcome soon by the unacquainted reader in order that the Principia Mathematica might become as well known as both the work and the topic deserve." Behmann to Russell, August 8, 1922 (English in original) Although I originally used the letter preserved in the Behmann archive in Erlangen, there is also a copy in the Russell archive at McMaster University (1368).

Some of the ideas of the dissertation can also be found in the text for a talk delivered in 1927 in Kiel entitled "Die Russelsche Theorie der logischen Typen." In this connection see also Hilbert's correspondence with Russell (quoted in Sieg 1999, p. 12 and appendix) which shows that Russell's theory of knowledge was being widely discussed in Göttingen earlier than 1916. Sieg (1999, p. 32) asks what was being read by Russell at the time. Behmann quotes the following works in his dissertation: *Principia Mathematica*, *Mathematical Logic as Based on the Theory of Types* (1908), *Principles of Mathematics* (1903), *The Problems of Philosophy* (1912). I have found no additional texts by Russell explicitly quoted during the period between 1914 and 1921 in the writings (published or unpublished) of Hilbert, Behmann, or Bernays. Bernays, who only returned to Göttingen in 1918, wrote to Russell that he began studying Russell's *Principia* only in 1917/18 after Hilbert's 1917/18 lecture course. See Bernays (1920).

10. Russell to the librarian of the Göttingen University Library, July 2, 1924, Behmann Archive, Erlangen.

11. "Es wäre infolgedessen von vornherein verfehlt, wenn wir uns anschicken wollten, die abstrakten Begriffe und Relationen der Arithmetik für sich allein zu untersuchen, ohne dabei des Umstandes zu gedenken, dass alle derartigen Begriffe immer erst im Zusammenhange des *Satzes* etwas bedeuten und dass auch die bestgebildeten Begriffe uns nichts nützen und uns auch nicht vor Widersprüchen bewahren, solange wir nicht wissen, wie wir sie richtig zu *verwenden* haben." (p. 34)

12. "so müssen wir nun auch, wie uns die letzte Ueberlegung zeigte, den Begriff der *Aussage* von vornherein in den Mittelpunkt unserer gesamten Untersuchungen stellen. Das bedeutet, wir dürfen die in Aussagen auftretenden Begriffe—wie "Menge," "Zahl," "Aussage," "Raum," "Kraft" u.s.f.—niemals für sich allein—als selbstständige logische Gegenstände—betrachten, sondern immer nur im Zusammenhang derjenigen Aussagen, in denen sie sinnvoll auftreten können." (p. 34)

13. "Wir bauen nach und nach ein gewisses festes *System von Aussagen* auf, das wir so einrichten, dass keine zwei in ihm enthaltenen oder—was übrigens auf dasselbe hinauskommt—durch einwandfreie Schlüsse aus ihm ableitbaren Aussagen, sofern sie als richtig erkannt sind, jemals einander widersprechen, dass andererseits aber gewiss all arithmetisch brauchbaren (richtigen und falschen) Aussagen in ihm vorkommen." (p. 35)

14. "Da wir zufolge der Natur unseres Gegenstandes nicht einmal die einfachsten logischen Begriffe in ihrem überkommenen Gebrauch voraussetzen dürfen, müssen wir infolgedessen

zunächst *allem* abstrakten Denken gegenüber misstrauisch sein. Weil wir aber doch irgend einen Ausgangspunkt haben müssen, so darf dies nur ein solcher sein, bei dem unser abstraktes Denken noch nicht beteiligt ist, dessen Existenz also nicht, wie z.B. die der Zahlen, erst auf der Möglichkeit des Denkens beruht und der darum auch nicht durch es verfälscht sein kann. Die einzige Dinge, die dieser Forderung genügen, sind nun aber diejenigen, die ohne Zuhilfenahme des Denkvermögens, d.h. also durch die sinnliche Wahrnehmung allein, unmittelbar erkannt werden, die somit dem Denken erst den zu seiner Möglichkeit unbedingt notwendigen Stoff geben. (Fast alle falsche Metaphysik krankte bisher an dem Bestreben, Gegenstände des Denkens selbst denkend erzeugen zu wollen. Wie überall, muss aber auch hier der Archimedische Punkt notwendig ausserhalb liegen.) Infolgedessen is unsere *erste Annahme* die, dass es gestattet sei, von *Gegenständen der Erfahrungswirklichkeit* (individuals) *als dem vor allem Denken Existierenden auszugehen* (Damit halten wir uns natürlich für berechtigt, von einer etwa möglichen Verfälschung durch die Wahrnehmung vollständig abzusehen.) Diese Individuen betrachten wir somit als fertig vorliegend und für unser Denken unmittelbar verfügbar." (pp. 44–45)

15. "Denn da die objektive Welt, d.h. die Gesamtheit der Individuen mit all ihren Eigenschaften und Beziehungen, wie wir hier unbedenklich voraussetzen dürfen, im letzen Grunde gewiss einen widerspruchslosen Bereich bildet, so muss offenbar *jede* Aussage, die nur Individuen zu Gegenständen hat, die also keine anderen Dinge als existierend voraussetzt, dem wirklichen Tatbestand entweder entsprechen oder widersprechen und somit den Sätzen vom Widerspruch und vom ausgeschlossenen Dritten genügen." (p. 107)

16. A few quotes from PM on individuals. "For this purpose, we will use such letters as a, b, c, x, y, z, w, to denote objects which are neither propositions nor functions. Such objects we shall call *individuals*. Such objects will be constituents of propositions or functions, and will be *genuine* constituents, in the sense that they do not disappear on analysis, as (for example) classes do, or phrases of the form 'the so-and-so.' " (Whitehead and Russell 1910, p. 51)

"The symbols for classes, like those for descriptions, are, in our system, incomplete symbols: their *uses* are defined, but they themselves are not assumed to mean anything at all. That is to say, the uses of such symbols are so defined that, when the definiens is substituted for the definiendum, there no longer remains any symbol which could be supposed to represent a class. Thus classes, so far as we introduce them, are merely symbolic or linguistic conveniences, not genuine objects as their members are if they are individuals." (Whitehead and Russell 1910, p. 72)

"We may explain an individual as something which exists on its own account" (Whitehead and Russell 1910, p. 162)

17. "Wir wollen nun eine Aussage, in der eine unmittelbare Wahrnehmung zum Ausdruck kommt oder doch wenigstens vorgestellt wird, als '*einfache Aussage*' bezeichnen." (p. 47)

18. Explaining why certain propositions, such as "A propositional function fx cannot be considered as an object in the system of propositions constructed," do not imply that propositional functions are objects, Behmann explains that: "Es steht uns jedoch frei, den obigen Satz folgendermassen zu deuten 'Das Zeichen $f\hat{x}$—also ein konkretes Individuum—darf nicht so betrachtet werden, als ob ihm [...] ein eigentlicher Gegenstand entspräche, der mithin als Gegenstand der Aussage gelten dürfte, d.h. es hat nur als ein künstlich abspaltbarer Bestandteil eines umfassenderen Zeichens zu gelten.' " (p. 64)

In chapter 3 Behmann goes on to the construction of propositional functions and of propositions of second order. These are obtained by variation of the propositional and functional components occurring in proposition of zero and first order (p. 95). Generalizations to higher order can be obtained in the obvious way. In PM we read concerning propositions: "Owing to the plurality of objects of a single judgment, it follows that what we call a "proposition" (in the sense in which this is distinguished from the phrase expressing it) is not a single entity at all. That is to say, the phrase which expresses a proposition is what we call an "incomplete" symbol." (Whitehead and Russell 1910, p. 44)

19. "In gleichem Sinne [as the differential dx] sind nun aber auch die Aussagefunktionen genau wie alle Abstrakta keine echten Gegenstände, sondern sie erwecken nur, indem sie dem in grösserer Entfernung von seinen eigentlichen konkreten Gegenständen sich bewegenden abstrakten Denken gewisse Ruhepunkte bieten, den trügerischen Schein einer Gegenständlichkeit." (p. 106)

20. Behmann appeals explicitly to Russell's "no-class" theory and then says: "Das bedeutet, dass wir die Klassen nur als formale Bestandteile der sie betreffenden Aussagen anerkennen, deren eigentliche Aussagegegenstände hingegen einzig die in den Klassen enthaltenen—oder auch nicht enthaltenen—Individuen sind." (p. 165)

In PM the idea is expressed as follows:

"The symbols for classes, like those for descriptions, are, in our system, incomplete symbols: their *uses* are defined, but they themselves are not assumed to mean anything at all. That is to say, the uses of such symbols are so defined that, when the definiens is substituted for the definiendum, there no longer remains any symbol which could be supposed to represent a class. Thus classes, so far as we introduce them, are merely symbolic or linguistic conveniences, not genuine objects as their members are if they are individuals." (Whitehead and Russell 1910, p. 72)

21. "Die Klassen—und gleicherweise übrigens die Zahlen—sind dann, wie wir schon früher andeuteten, gar nichts weiter als Redensarten, die für die bequeme und übersichtliche Darstellung der Arithmetik von äusserstem Nutzen sind, aber gleichwohl recht bedenklich werden können, sobald man sie ernst nimmt und ihrer Natur zuwider für Namen von Gegenständen hält." (p. 159)

22. "Es liegt nun recht nahe, zu vermuten—und lässt sich, wie in den Principia Mathematica gezeigt wird, auch streng beweisen, sobald man alle für die mathematische Logik benötigten Axiome wirklich heranzieht—dass man mit den Individuenklassen beim praktischen Schliessen genau so verfahren kann, als ob sie Individuen wären, solange man sich nur hütet, sie mit den wirklichen Individuen geradezu in eine Reihe zu stellen. Denn die wesentliche Eigenschaft, die der Bereich der Individuenklassen—aber nicht der Bereich aller Klassen überhaupt—mit dem Individuenbereich gemeinsam hat, ist ja die der Konsistenz. Und wir dürfen die Elemente eines konsistenten abstrakten Bereiches, solange wir ihn gegen alle anderen Bereiche scharf abgrenzen, d.h. die Typenfestsetzungen streng innehalten,—so z.B. auch die des Bereiches der natürlichen oder der reellen Zahlen—innerhalb gewisser Grenzen so betrachten, *als ob* sie dem Denken ursprünglich gegebene Dinge wären." (pp. 187–88)

23. "Gegenüber jenen Theorien der (endlichen oder unendlichen) Arithmetik, die die blosse Möglichkeit des symbolischen Formalismus—wir könnten auch sagen: des schriftlichen Rechnens als Grundlage haben, liegt der Hauptvorzug der hier vertretenen Theorie offensichtlich darin, dass unsere Arithmetik ihrer ganzen Entstehung nach keinesfalls bloss ein Spiel mit Zeichen ist, sondern von vornherein einen objektiven Inhalt hat." (pp. 313–14)

24. "Die letzte Wirklichkeit, auf die sich die der Menge noch zurückführen liess, konnte natürlich einzig die der zu zählenden *Dinge* sein. Aber von welcher Art werden nun diese Dinge

sein müssen, um dem Denken als sicherer, nicht schon von Widersprüchen bedrohter Ausgangspunkt dienen zu können?" (p. 335)

The individuals provide arithmetic with its objective content: "Bereits an früherer Stelle hatten wir es ja ausgesprochen, dass die Arithmetik nur dadurch als Wissenschaft—d.h. als ein System objektiv gültiger Erkenntnisse—Berechtigung hat, dass sie einen von aussen her gegebenen Stoff vorfindet, den sie in ihren Untersuchungen nach einer gewissen Seite zu ergründen hat. Ohne einen solchen Stoff—als welcher bekanntlich die Individuen dienen—wäre sie in der Tat gerade so unmöglich, wie die Astronomie, wenn es keine Gestirne gäbe." (p. 153)

25. "Da somit alle abstrakten Dinge—wenigstens soweit sie überhaupt zur Arithmetik in Beziehung treten können—bereits ausgeschlossen sind, bleiben einzig noch die konkreten Gegenstände der Erfahrungswirklichkeit." (p. 335)

26. "Die Anschauung, zu der man von hier aus gelangt, ist nun die in dieser Abhandlung zugrunde gelegte und ausführlich erörtete, dass die Arithmetik in Wahrheit gar nicht ihre eigenen Gegenstände hat, sondern keine andere Gegenständlichkeit voraussetzt als andere Wissenschaften auch, nämlich die der Erfahrung (die zwar nicht unbedingt als wirklich, aber doch wenigstens als möglich vorgestellt werden muss). Da wir jede vernünftige Aussage über diese Wirklichkeit mit gutem Recht als dem Satz vom Widerspruch gemäss voraussetzen dürfen, ist damit auch die Widerspruchslosigkeit der Arithmetik grundsätzlich gewärleistet." (p. 336)

27. "Die Zahlen fassen wir als blosse formale Bestandteile von Aussagen...In diesem Sinne sind auch die Zahlen nichts weiter als Mittel, um gewisse Erkenntnisse bequem abzuleiten und auszusprechen, aber abgesehen von dieser Bestimmung und für sich allein sind sie gar nichts." (p. 337)

28. "Da nun, wie gezeigt worden, die Begriffe ihren Stoff von der anschauenden Erkenntnis entlehnen, und daher das ganze Gebäude unserer Gedankenwelt auf der Welt der Anschauungen ruht; so müssen wir von jedem Begriff, wenn auch durch Mittelstufen, zurückgehen können auf die Anschauungen, aus denen er unmittelbar selbst, oder aus denen die Begriffe, deren Abstraktion er wiederist, abgezogen worden: d.h. wir müssen ihn mit Anschauungen, die zu den Abstraktionen im Verhältnis des Beispiels stehen, belegen können. Diese Anschauungen also liefern den realen Gehalt alles unseres Denkens, und überall, wo sie fehlen, haben wir nicht Begriffe, sondern blosse Worte im Kopfe gehabt." (quoted in Behmann 1918, pp. 341–42)

29. "Gegenüber dieser letzten Feststellung bleibt allerdings die Tatsache bemerkenswert, dass wir uns beim wirklichen Aufbau der Arithmetik—wir meinen hier nicht nur die Zahlentheorie, sondern auch die Analysis im allgemeinen—um diese ihr wesensnotwendige Beziehung zur Erfahrung und damit um den wahren Inhalt der Zahlformeln und Schlussweisen kaum in höherem Masse zu kümmern pflegen als etwa um den als zuverlässig erprobten inneren Bau einer Rechenmaschine, uns nichtsdestoweniger aber innerhalb dieser Gebiete vor Widersprüchen mit Recht für völlig geschützt halten." (p. 343)

30. "Wohlgemerkt: nicht die Zahl ist eine Fiktion—sie ist vielmehr in dem früher erklärten Sinne ein bloss formaler Bestandteil arithmetischer Aussagen—sondern wir *bedienen* uns einer Fiktion, wenn wir, was an sich ja keineswegs nötig ist, ihr—nicht aus Rücksichten der Denknotwendigkeit, sondern der Denkbequemlichkeit—eine ihr in Wahrheit nicht zukommende Gegenständlichkeit im tatsächlichen Denken dennoch zuerkennen, d.h. die für echte Gegenstände geltenden logischen Gesetze unbedenklich, bez. nachdem wir uns der

Zulässigkeit dieses Verfahrens innerhalb des fraglichen Bereiches versichert haben, auch auf sie anwenden." (p. 346)
31. "Göttingen, den 1. Februar 1918,
"Die eigentliche Aufgabe und Absicht der vorliegenden Abhandlung von Behmann ist es, in die Gedankenwelt derjenigen Disziplin einzuführen, die man als symbolische Logik bezeichnet, die unter den Händen einer Reihe bedeutender Mathematiker und Logiker ein wichtiger Bestandteil der Erkenntnistheorie geworden ist und schliesslich in den letzten Jahren in dem gross-angelegten Werke 'Principia Mathematica' von Russell und Whitehead die reifste Bearbeitung und Darstellung erfahren hat. Während lange Zeit hindurch die symbolische Logik nichts anderes als eine formalistische äusserliche Weiterbildung der Aristotelischen Schlussfigurentheorie zu sein schien, gelang es Russell, mit der Anwendung der symbolischen Logik auf die schwierigsten erkenntnistheoretischen Fragen zum ersten Male sichere Erfolge zu erziehen. Obenan in der Russelschen Theorie steht—als oberstes Denkaxiom—das sogennante Reduzierbarkeitsaxiom. Dieses Axiom, einschliesslich der damit verbundenen Typentheorie von Russell bietet dem Verständnis ausserodetliche Schwierigkeiten. Um diese zu beseitigen zieht Behmann das von mir in die Arithmetik eingeführte Vollständigkeitsaxiom heran, welches nicht nur seinem logischen Charakter nach mit dem Reduzierbarkeitsaxiom gleichartig ist, sondern auch sachliche innere Zusammenhänge mit jenem zeigt. Indem Behmann diese Zusammenhänge herausarbeitet, gelingt es ihm nicht nur jenes Reduzierbarkeits-axiom klar zu fassen, sondern auch die Anwendung der Russelsche[n] Theorie auf ein spezielles tiefliegendes Problem—die Auflösung der Antinomie der transfiniten Zahl—über Russell hinaus durchzuführen. Sein Resultat bleibt im Wesentlichen: alle transfinite Axiomatik ist ihrer Natur nach etwas unabgeschlossenes, aber die mengenth[eoretische] Begriffe von Cantor sind streng zulässig. Jedes an sich vernünftlich gestellte mengentheoretische Problem behält Bedeutung und ist daher einer Lösung fähig.
"Bei der Darstellung ist besonderer Wert darauf gelegt, dass dabei keinerlei spezifisch mathematische Kenntnisse vorausetzt werden; dieselbe ist daher auch dem Nicht-Mathematiker verständlich. Ich hoffe für die Behemmanische Schrift einen Verleger zu finden, da sie—wie mir scheint—einem mathematisch-philosophischen Bedürfnis in der Gegenwart Rechnung trägt. Ich beantrage, bereits 2 Druckbogen als Dissertation gelten zu lassen.
"Praedikat
"Hilbert"
UAG, Phil. Fak. Promotionen: B vol. VI (1917–), Universitätsarchiv Göttingen
32. For an introduction to Hilbert's program see Mancosu (1998c) and the references given there.
33. Hilbert (1922c), p. 134 of the translation. This position is repeated verbatim in many later publications by Hilbert.
34. See note 18 for Behmann's interpretation of signs as concrete individuals.
35. Consider, for instance, the emphasis on relying on primitive forms of cognition as a starting point for the construction of arithmetic. "An appeal to an intuitive grasp of the number series as well as to the multiplicity of magnitudes is certainly to be considered. But this could certainly not be a question of an intuition in the primitive sense; for, certainly no infinite multiplicities are given to us in the primitive intuitive mode of representation. And even though it might be quite rash to contest any farther-reaching kind of intuitive evidence from the outset, we will nevertheless make allowance for that tendency of exact science which aims as far as possible to eliminate the finer organs of cognition [*Organe der Erkenntnis*] and to rely only on the most primitive means of cognition." Bernays (1922b, trans., p. 215) Compare

Bernays (1923a, p. 226): "Hilbert's theory does not exclude the possibility of a philosophical attitude that conceives the numbers as existing, nonsensible objects ... Nevertheless the aim of Hilbert's theory is to make such an attitude dispensable for the foundations of the exact sciences." In this connection it might be worthwhile to investigate the possible influence of Russell's theory of knowledge on Hilbert and Bernays.

36. The terminology is undoubtedly Behmann's. He writes to Russell in 1922: "It was what I call the Problem of Decision, formulated in the said paragraph that induced me to study the logical work of Schröder. And I soon recognized that, in order to solve my particular problem, it was necessary first to settle the main problem of Schröder's Calculus of Regions, his so-called Problem of Elimination ... Indeed, the chief merit of the said problem [The Decision Problem] is, I daresay, due to the fact that it is a problem of fundamental importance on its own account, and, unlike the applications of earlier Algebra of Logic, not at all imagined for the purpose of symbolic treatment, whereas, on the other hand, the only means of any account for its solution are exactly those of symbolic logic." (Behmann 1922c) In this connection see also Behmann to Scholz (December 27, 1927, Behmann Archive, Erlangen) where, in addition to reiterating his originality in the formulation of the Entscheidungsproblem, Behmann also provides a historical overview of the early history of the Entscheidungsproblem. One should not confuse Behmann's introduction of the Entscheidungsproblem either with the Entscheidungsproblem formulated, in connection to Grelling's paradox, in Hessenberg (1906) or with the Kroneckerian requirement of decidability for the introduction of concepts in mathematics (for the latter see Pasch 1918). Sieg (1999, p. 19) tentatively claims that the first occurrence of the word Entscheidungsproblem is found in lectures of Hilbert from 1922–23. The assertion should be corrected in light of the above.

37. The metaphor of the chess game to speak about formalized mathematics had great fortune in the twenties. However, it is usually attributed to Weyl who formulated it in print in Weyl (1924b, 1925).

38. "Bekanntlich läßt sich die symbolische Logik *axiomatisieren*, d. h. auf ein System verhältnismäßig weniger Grundformeln und Grundregeln zurückführen, sodaß auch das *Beweisen von Sätzen* nunmehr als ein *bloßes Rechenverfahren* erscheint. Man braucht nur noch zu gegebenen Formeln neue hinzuschreiben, wobei durch Regeln bereits festgelegt ist, was man jeweils hinschreiben darf. Das Beweisen hat sozusagen den *Charakter eines Spieles* angenommen. Es ist etwa wie beim Schachspiel, wo man durch Verschieben eines der eigenen Steine, gegebenenfalls mit Wegnahme einer gegnerischen, die *jeweils vorliegende Stellung in eine neue* verwandelt, wobei nun das Verschieben und das Wegnehmen durch die Regeln des Spieles erlaubt sein muß.

"Aber gerade dieser Vergleich zeigt uns auch in krasser Weise, daß uns der eben geschilderte *Standpunkt der symbolischen Logik* für unser Problem *noch keineswegs genügen* kann. Denn dieser sagt uns wie die Regeln des Schachspiels *nur, was man tun darf,* und *nicht, was man tun soll.* Dies bleibt in dem einen wie in der anderen Falle eine Sache des erfinderischen *Nachdenkens,* der glücklichen *Kombination.* Wir verlangen aber weit mehr: daß nicht etwa nur die erlaubten Operationen im einzelnen, sondern auch der *Gang der Rechnung selbst* durch Regeln festgelegt sein soll, m.a.W. eine *Ausschaltung des Nachdenkens zugunsten des mechanischen Rechnens.* Ist irgend eine logische mathematische Aussage vorgelegt, so soll das verlangte Verfahren eine vollständige Anweisung geben, wie man *durch eine ganz zwangläufige Rechnung nach endlich vielen Schritten* ermitteln kann, ob die gegebene Aussage *richtig oder falsch* ist. Das oben formulierte Problem möchte ich *das allgemeine Entscheidungsproblem* nennen.

"Für das Wesen des Problems ist von grundsätzlicher Bedeutung, daß als Hilfsmittel des Beweises *nur das ganz mechanische Rechnen* nach einer gegebenen Vorschrift, ohne irgendwelche Denktätigkeit im engeren Sinne, zugelassen wird. Man könnte hier, wenn man will, von *mechanischem* oder *machinenmässigem* Denken reden. (Vielleicht kann man es später sogar durch eine Machine ausführen lassen)." (Behmann 1921, pp. 5–6)

Chapter 4

1. Felix Bernstein did his Habilitation in 1901 under Hilbert. He taught in Göttingen from 1907 to 1933. See Frewer (1981) and Peckhaus (1990) for more biographical details.
2. Kurt Grelling was in Göttingen from 1906 to 1910 and for a short period in 1914. See Peckhaus (1990, 1993, 1994b) for more biographical details.
3. Leonard Nelson was in Göttingen from 1903 to 1928. For more biographical details and on Nelson's role in foundational thinking, especially as it relates to Hilbert, see Peckhaus (1990).
4. Hermann Weyl obtained his doctorate in 1908 under Hilbert, his Habilitation in 1910 and taught in Göttingen until 1913 when he took up a professorship at the ETH in Zurich. For more details on Weyl's foundational thinking during this period see Mancosu (1998b).
5. Ernst Zermelo was in Göttingen from 1897 to 1910 when he obtained a position at the ETH in Zurich. On Zermelo and his role in the development of mathematical logic during this period see Peckhaus (1990).
6. On Leon Chwistek's correspondence with Russell during this period see Jadacki (1986). The article contains a transcription of the correspondence and a useful bibliography.
7. Peckhaus (1990, p. 3) writes: "Ausgelöst wurde diese Umorientierung durch die Veröffentlichung der logischen und mengentheoretischen Antinomien durch Russell (1903) and Frege (1903)".

    I should point out that interest in the paradoxes remained alive also among the phenomenologists. Husserl seems to have been thinking about Russell's paradox in the early 1910s (see Zambelli 1999, p. 306) and Koyré, at the time a student of Husserl, even wrote an essay on the topic of the paradoxes, which he submitted to Husserl for a doctoral degree. The essay was however rejected by Husserl. For the whole episode see Schuhmann (1987) and Zambelli (1999). Zambelli also contains a full trascription of Koyré's proposed dissertation "Insolubilia. Eine logische Studie über die Grundlagen der Mengenlehre". Although the work was probably written in late 1911 or early 1912, it contains no discussion of *Principia Mathematica* (it does however discuss Russell's paradox).
8. Indeed, Hilbert and Zermelo claimed that Zermelo had found the Russellian antinomy independently (see Zermelo 1908a and Rang and Thomas 1981).
9. It should also be added that many neo-Friesians were critical of Russell's logicism (see the correspondence between Hessenberg and Nelson, quoted in Peckhaus 1990, p. 178).
10. "Nachdem ich schon seit längerer Zeit durch das Buch von Couturat und durch verschiedene Ihrer Aufsätze Ihre Ansichten über die Grundlagen der Mathematik kennen gelernt hatte, habe ich mich vor kurzem an die Lektüre Ihrer "Principles" gemacht und bin sehr bald zu der überzeugung gelangt, daß es sehr gut wäre, wenn man Ihr Buch dem deutschen Publikum zugänglich machen könnte.

    "Um Ihnen nicht als Schmeichler zu erscheinen, will ich mein eigenes Urteil über Ihr Buch beiseite lassen und Ihnen höchstens sagen, was eigentlich selbstverständlich ist, daß ich in einer Reihe von Punkten Ihnen vorläufig nicht zustimmen kann. Soviel läßt sich aber ganz objektiv feststellen, daß die ganze moderne Literatur über die Grundlagen der Mathematik kein Werk aufzuweisen hat, das diesen Gegenstand in ebenso umfassender und gleichermaßen für Philosophen und Mathematiker verständlicher und genießbarer Weise behandelte, geschweige denn, daß Ihr Buch von einem anderen in diesen Eigenschaften übertroffen würde" (Grelling to Russell, Göttingen 9.10.1909, Russell Archive, McMaster).

11. Chwistek was trying to show that induction could be proved without appealing to impredicative procedures. At the end of the essay, sent with a letter from Göttingen dated 18.11.1909, Chwistek wrote: "The essay above is based upon the paper: 'Mathematical Logic as Based on the Theory of Types' (Amer. Journ. of Mathem. XXX) of Bertrand Russell". The letter and the essay are transcribed in Jadacki (1986). For Russell's general influence on Chwistek see the letter Chwistek sent Russell from Krakow on January 30, 1921: "In 1909 I have read for the first time your paper concerning the theory of types and from this time I was obsessed by the idea of building up a system of formal logic based on your theory, but not containing between its primitive propositions the axiom of reducibility. Now, I have finished the first part of my work, including inductive cardinals" (Russell archive, McMaster; Jadacki 1986).

12. See the list of lectures given in logic and foundations of mathematics at the Colloquium between 1914 and 1921, as reported yearly in the *Jahresbericht der Deutschen Mathematiker-Vereinigung*. In addition to the lectures from 1914 to 1917 quoted in the main text, one should also consider the ones given from 1917 to 1921:

> July 31, 1917, D. Hilbert, *Referat über seine Vorlesungen über Mengenlehre* [see Hilbert 1918b]
>
> November 20, 1917, P. Bernays, *Weyl über die Grundlagen der Analysis* [see Bernays 1917]
>
> November 27, 1917, D. Hilbert, *Über axiomatisches Denken* [see Hilbert 1918a]
>
> December 7, 1920, M. Schönfinkel, *Elemente der Logik* [see Schönfinkel 1924]
>
> February 1 and 8, 1921, R. Courant and P. Bernays, *Über die neue arithmetischen Theorien von Weyl und Brouwer*
>
> February 21, 22, 1921, D. Hilbert, *Eine neue Grundlegung des Zahlbegriffes* [see Hilbert 1922c]
>
> May 10, 1921, H. Behmann, *Das Entscheidungsproblem der mathematischen Logik* [see Behmann 1922b]
>
> December 6, 1921, P. Bernays and M. Schönfinkel, *Das Entscheidungsproblem im Logikkalkül* [see Bernays and Schönfinkel 1928].

13. "... haben wir dort [Mathematisches Kolloquium] soeben vor den Ferien eine ausführliche Diskussion über das grosse dreibändige Werk von Russell gehabt, die äusserst lebhaft und belehrend verlief. Es sind besonders hier die Herren Grelling, Hertz u. Bernstein, die sich für die erkenntnistheoretischen Seiten der Mathematik hier interessieren" (Hilbert to Dingler, December 26, 1914, Dingler archive, Aschaffenburg).

14. On Grelling see Peckhaus (1990, 1993, 1994b, 1995a). Grelling met Russell in 1908 in Rome at the International Congress of Mathematicians. On August 3, 1910 Grelling wrote to Russell (from Berlin): "Ich bin dabei, über Ihre Typentheorie einen Bericht zu schreiben und bin auf eine, wie mir scheint, nicht unwesentliche Schwierigkeit gestoßen" (Russell archive, McMaster University).

15. Paul Hertz (1881–1940) was, in the period under analysis, mainly concerned with the foundations of statistical mechanics. His work in logic dates from the 1920s. On Hertz's life and work see Rudolf Hertz, 'Biography of Paul Hertz', undated typescript, Paul Hertz *Nachlass*, Pittsburgh, [PH01-24-02]. On Hertz's work in logic see Abrusci (1983).

16. For Bernstein's foundational work during this period see Bernstein (1919). For a short account of his career see Frewer (1981) and Peckhaus (1990).
17. On Behmann, see Chapter 3. For the Behmann *Nachlaß* see Haas and Stemmler (1981).
18. "Trotz alledem bietet die richtige Auflösung all dieser geometrischen Widersprüche, sobald sie einmal gegeben ist, dem Verständnis keine erhebliche Schwierigkeit dar. Sie besagt nämlich einfach, daß die idealen Elemente nicht im eigentlichen Sinne Gegenstände der Geometrie sind, sondern zunächst einmal nur Worte und als solche Teile von Ausdrucksweisen, deren man sich in der Geometrie bedient, um eine gewisse Klasse von Sätzen auf eine möglichst einfache Form zu bringen" (Behmann 1914, p. 10).

    We find traces of this idea in a review of Müller's work *Der Gegenstand der Arithmetik* written by Bernays in 1923: "In fact, the mathematician knows that in his science an especially fruitful and continually applied procedure consists in the introduction of "ideal elements" that are introduced purely formally as subjects of judgments, and that, however, when detached from the statements in which they occur formally, are nothing at all" (Bernays 1923b, p. 521).
19. "Das Axiom sagt also aus, daß in dieser einen Stufe schon genügend viele Funktionen vorhanden sind, um alle überhaupt möglichen Klassen zu definieren; es läst sich daher auch als eine Art Vollständigkeitsaxiom für die prädikativen Funktionen auffassen" (Behmann 1914, p. 14).
20. Behmann seems to have accepted the logicist program as viable at least until 1922. In 1922 he writes: "Mag man zu der auf den ersten Anblick gewiss befrendenden Russell-Whiteheadschen "Theorie der logischen Typen" stehen, wie man will, man wird nicht bestreiten können, daß es ihr in der Tat gelungen ist, nicht nur die Mengenlehre [...] sondern die gesamte Logik, von der hier die Arithmetik im weitesten Sinne nur als ein Ausschnitt erscheint, auf eine Grundlage zu stellen, mit der sich jedenfalls vollkommen sicher arbeiten läßt und die, wie der Erfolgt lehrt, trotz der in ihr liegenden unvermeidlichen Beschränkungen das stolze Werk der Begründung der Arithmetik aus bloßer Logik in einer - grundsätzlich gesprochen—völlig strengen und bewundernswert geschlossenen Form tatsächlich geleistet hat—was selbstverständlich nicht ausschließt, daß dieses Werk sogar in wesentlichen Punkten der Vervollkommnung fähig und bedürftig ist" (Behmann 1922b, p. 56). See however Section 4.7.3.
21. "Das Ziel, die Mengenlehre und damit die gebräuchlichen Methoden der Analysis auf die Logik zurückzuführen, ist heute nicht erreicht und ist vielleicht nicht erreichbar" (Hilbert 1920b, p. 33). I completely agree with Sieg: "Clearly, as documented here, logicism had been given up as a viable option in the summer of 1920 explicitly; implicitly, that recognition is already in the background for the lectures in the winter of 1920" (Sieg 1999, p. 23). The lectures WS 1920 were given in the early part of 1920 before the SS 1920 lectures.
22. "Wie Sie wohl wissen, beschäftigt sich Herr Professor Hilbert—dessen Assistent zu sein ich die Ehre habe—seit einigen Jahren besonders eifrig mit den Problemen der mathematischen Logik. Durch eine Vorlesung, welche Prof. Hilbert im Winter 1917/18 über diesen Gegenstand gehalten hat, bin ich zum genaueren Studium dieser Fragen und insbesonder auch Ihrer Theorie der mathematischen Logik angeregt worden" (Bernays to Russell, April 8, 1920, Russell archive, McMaster University, 110208b).
23. "Wir wollen nicht nur imstande sein, einzelne Theorien für sich von ihren Prinzipien aus rein formal zu entwickeln, sondern wollen die Grundlagen der mathematischen Theorien selbst auch zum Gegenstand der Untersuchung machen und sie darauf hin prüfen, in welcher Beziehung sie zu der Logik stehen und inwieweit sie aus rein logischen Operationen und Begriffsbildungen gewonnen werden können; und hierzu soll uns der logische Kalkül als Hilfsmittel dienen" (Hilbert 1918c, p. 188; translation from Sieg 1999, p. 15).

24. "Wenn wir nur in diesem Sinne von dem logischen Kalkül Gebrauch machen, so werden wir dazu gedrängt, die Regeln des formalen Operierens in einer gewissen Richtung zu erweitern. Während wir nämlich bisher die Aussagen und Funktionen von den Gegenständen völlig trennten und demgemäss auch die unbestimmten Aussage-und Funktionszeichen von den Variablen, welche Argumente bilden, streng gesondert hielten, werden wir nunmehr zulassen, dass unbestimmte Aussagen und Funktionen in gleicher Weise wie eigentliche Gegenstände als Werte von logischen Variablen genommen werden und dass unbestimmte Aussagezeichen und Funktionszeichen als Argumente von symbolischen Ausdrücken auftreten" (Hilbert 1918c, p. 188).

25. "Die Notwendigkeit, Funktionen von Funktionen einzuführen, stellt sich insbesondere bei der Erörterung des *Anzahlbegriffs* heraus. Eine Anzahl ist kein Gegenstand im eigentlichen Sinne, sondern eine Eigenschaft (ein Prädikat). Welche sind aber die Gegenstände, denen eine Anzahl zukommt? Den gezählten Dingen kann die Anzahl nicht als Eigenschaft zugeschrieben werden, da jedes von den Dingen nur eines ist, sodass eine von Eins verschiedene Anzahl danach gar nicht vorkommen könnte. Dagegen lässt sich die Zahl als eine Eigenschaft desjenigen Begriffes auffassen, unter welchem die gezählten Gegenstände vereinigt werden. So kann z.B. die Tatsache, dass die Anzahl der Erdteile Fünf ist, nicht so ausgedrückt werden, dass jedem Erdteil die Anzahl Fünf zukommt; wohl aber ist es eine Eigenschaft des Prädikates "Erdteil-sein", dass er auf genau fünf Gegenstände zutrifft.

"Die Zahlen erscheinen hiernach als Eigenschaften von Prädikaten, und für unseren Kalkül stellt sich also eine Zahl als eine bestimmte Funktion eines veränderlichen Prädikates dar. Die Wichtigkeit dieser Darstellung der Zahlen besteht darauf, dass die Prädikaten-Funktionen, welche die Zahlen bilden, sich vollständig mit Hilfe der logischen Symbole ausdrücken lassen, wodurch es möglich wird, die Zahlenlehre in die Logik einzubeziehen" (Hilbert 1918c, pp. 192–93).

26. For a certain predicate P, let $m(P)$ be the set corresponding to P (i.e., $m(P) = \{x : Px\}$). Two sets defined by two predicates P and Q are identical iff P and Q are logically equivalent, i.e., $m(P) = m(Q)$ iff $P \leftrightarrow Q$. The possibility of interpreting the predicate-function F(P) as a predicate of sets now rests on whether, as we would say, F is extensional, i.e.,

$$(P)(Q)\{Aeq(P, Q) \to (F(P) \leftrightarrow F(Q))\}.$$

This property is for the numbers as predicate-functions always satisfied. Thus: "Aus der angestellten Ueberlegung geht hervor, dass die Beziehung der Logik zur Zahlentheorie einen Sonderfall einer allgemeineren Beziehung zwischen Logik und Mengenlehre bildet" (Hilbert 1918c, p. 195). The relationship between set theory and logic is then treated with a set of examples including the correspondence between sets of sets, such as the power set, and the corresponding predicate of predicates defining them.

27. "Wir müssen also aus den gefundenen Widersprüchen schliessen, dass unsere Methode des formalen Operierens in irgend einer Beziehung fehlerhaft ist. Und zwar kann dieser Fehler nur darauf beruhen, dass wir bei der Erweiterung des Ursprünglich für den Funktionen-Kalkül aufgestellten Axiomen-Systems nicht vorsichtig genug verfahren sind; denn das ursprüngliche Axiomen-System liess sich ja widerspruchslos nachweisen" (Hilbert 1918c, pp. 218–19).

28. "Bei der anfänglichen Methode des Funktionen-Kalküls haben wir ein System oder mehrere Systeme (Gattungen) von Gegenständen als von vornherein gegeben angenommen, und durch die Beziehung auf solche Gesamtheiten von Gegenständen erhielt das Operieren mit den Variablen (insbesondere mit den Klammerzeichen) seine logische Bedeutung. Die

Erweiterung des Kalküls bestand nur darin, dass wir die Aussagen, Prädikate und Relationen als Arten von Gegenständen betrachteten und demnach symbolische Ausdrücke zuliessen, deren logische Deutung eine Bezugnahme auf die Gesamtheit der Aussagen, bezw. der Funktionen erforderte. Dies Vorgehen ist nun in der Tat bedenklich, insofern nämlich dabei jene Ausdrücke, welche erst durch die Bezugnahme auf die Gesamtheit der Aussagen bezw. der Funktionen ihren Inhalt gewinnen, ihrerseits wieder zu den Aussagen und Funktionen hinzugerechnet werden, während wir doch andererseits, um uns auf die Gesamtheit der Aussagen oder Funktionen beziehen zu können, die Aussagen bezw. die Funktionen als von vornherein bestimmt ansehen müssen. Hier liegt also eine Art von logischem Zirkel vor, und wir haben Grund zu der Annahme, dass dieser Zirkel die Ursache für das Zustandekommen der Paradoxien bildet" (Hilbert 1918c, pp. 219–20). He also says: "Eine solche Bezugnahme auf die Gesamtheit aller Aussagen (Sätze, Behauptungen, Urteile) findet bei allen Formulierungen dieser Paradoxie statt" (Hilbert 1918c, p. 223).

29. "Jetzt fragt es sich aber, ob auf diese Weise nicht der Kalkül zu sehr eingeengt wird, sodass uns manche von denjenigen Schlussweisen abgeschnitten werden, welche für die Grundlegung der Mathematik eine wesentliche Rolle spielen und bei denen wir überzeugt sind, dass sie keinen Anlass zu Widersprüchen geben" (Hilbert 1918c, p. 226).

30. "Ganz ähnlich wie bei diesem Beispiel tritt noch an mehreren entscheidenden Stellen der Fall ein, dass wir durch die Sonderung der Stufen daran verhindert werden, gewisse mathematische Schlussweisen durch den logischen Kalkül nachzubilden. Es zeigt sich somit, dass unsere Methode des Stufen-Kalküls die Möglichkeiten des Schliessen über das zulässige Mass hinaus beschränkt, und wir werden daher suchen eine gewisse Modifikation, und zwar auf möglichst ungezwungene Art, anzubringen, durch der Kalkül eine grössere Bewegungsfreiheit erhält" (Hilbert 1918c, p. 231).

31. "Zu jedem im Stufen-Kalkül vorkommenden Funktions-Ausdruck gibt es einen äquivalenten prädikativen Ausdruck" (Hilbert 1918c, p. 231).

32. Conclusion: "So zeigt sich, dass die Einführung des Axioms der Reduzierbarkeit das geeignete Mittel ist, um den Stufen-Kalkül zu einem System zu gestalten, aus welchem die Grundlagen der höheren Mathematik entwickelt werden könne" (Hilbert 1918c, p. 246).

33. "Die methodische Bedeutung der Aufstellung dieses Postulates besteht vor allem darin, dass wir im Gegensatz zu der bisher anerkannten Forderung, wonach jede Funktion entweder als Grundfunktion durch ein individuelles Zeichen eingeführt werden oder als abgeleitete Funktion durch einen symbolischen Ausdruck darstellbar sein muss, nunmehre gewisse Prädikate und Relationen als etwas an sich Bestehendes betrachten, sodass ihre Mannigfaltigkeit weder von den tatsächlich gegebenen Definitionen noch auch überhaupt von unseren Möglichkeiten des Definierens abhängt. Ein solches Vorgehen erscheint zunächst sehr befremdlich, es ist aber unvermeidlich, sofern wir darauf ausgehen, die Grundlagen der Mengenlehre und der Analysis aus dem System der Funktionen unseres Kalküls zu entwickeln. Denn solange wir daran festhalten, dass jede Funktion sich explizit angeben lassen muss, können die Funktionen unseres Kalküls nur eine abzählbare Gesamtheit bilden und können daher nicht ausreichen, um die überabzählbaren Mannigfaltigkeiten zu repräsentieren, mit denen es in jenen mathematischen Disziplinen zu tun haben" (Hilbert 1918c, pp. 232–33).

34. "Es hängt das Resultat einer solchen Definition davon ab, was es für reelle Zahlen gibt, andrerseits ist das Ergebnis der Definition mitbestimmend dafür, was es für reelle Zahlen gibt" (Bernays 1917, pp. 1–2).

35. "Vielmehr herrscht eine vollkommene Sicherheit und Präzision des Schliessen und Einhelligkeit aller Ergebnisse (Von einem "Chaos" ist da keine Rede)" (Bernays 1917, p. 7). See Hilbert (1922c) for similar claims. Indeed it is obvious that Bernays (1917) served in part to write the first part of Hilbert (1922c).
36. "Dieselbe axiomatische Begründungsweise in umfassender Anwendung liegt vor, wenn die Zahlentheorie, Analysis und Mengenlehre aus einem einheitlichen System von Dingen entwickelt werden. Ein solcher Aufbau findet sich einerseits in der Zermeloschen Mengentheorie, in welcher die Grundbeziehung die Zugöherigkeit eines Dinges zu einer Menge bildet, andrerseits in der Russellschen Logik, welche auch ein Axiomen-System darstellt und zwar mit der prädikativen Beziehung als Grundbeziehung. (Von den gewohnten inhaltlichen Auffassung dieser Beziehungen müssen wir hier abstrahieren.) (Allerdings hat Russell nicht die rein axiomatische Tendenz)" (Bernays 1917, pp. 11–12).
37. See note 21.
38. "Es zeigt sich, dass auch hier das Ziel einer Zurückführung der Arithmetik auf die Logik nicht erreicht wird" (Hilbert 1922b, p. 76).
39. "Wie aber eine solche Ergänzung im Rahmen rein logischer Begriffsbildung zu erreichen ist, ersieht man nicht, und es besteht keine Aussicht auf die Entdeckung eines solchen Verfahrens. Somit bleibt nur die Möglichkeit, das System der Prädikate und Relationen erster Stufe als einen an sich bestehenden Inbegriff vorauszusetzen, auf welchen das Axiom der Reduzierbarkeit zutrifft" (Hilbert 1922b, p. 99).
40. "Hiermit kehren wir aber zu dem axiomatischen Standpunkt zurück, und das Ziel der Begründung von Arithmetik und Analysis durch die Logik wird preisgegeben. Denn eine Zurückführung auf die Logik besteht nun bloss noch dem Namen nach" (Hilbert 1922b, p. 99).
41. Whitehead and Russell are explicitly mentioned in the 1921/22 lectures (pp. 75–76).
42. The general exposition of the theory starts with a set of individuals and basic predicates, which could be taken to be of an intuitive nature ("Mann kann sich diese Grundfunktionen etwa als anschaulicher Natur vorstellen" (Hilbert 1918c, p. 100)). This idea about the intuitive nature of the basic predicates is found in the 1921/22 lectures as follows: "Um nun hier zu einer scharfen Bestimmung zu gelangen, machen wir uns klar dass alle Aussagen über Individuen sich aus gewissen elementaren Aussagen mit Hülfe der logischen Operationen zusammensetzen; diese elementaren Aussagen bestehen darin, dass eine anschauliche Eigenschaft einem Gegenstande zugeschrieben oder eine anschauliche Beziehung von Gegenständen behauptet wird, oder aber sie betreffen die individuelle Unterscheidung der Gegenstände." (Hilbert 1922b, pp. 81–82) The emphasis on the intuitive nature of the basic properties is not accidental and it is related to Behmann's reconstruction of the *Principia* system (see below).
43. "Ich möchte mich nur gestatten, Ihnen eine Bemerkung mitzuteilen, welche ich bezüglich des Kalküls Ihrer *Principia Mathematica* gemacht habe und von der ich annehme, dass sie Ihnen ein gewisses Interesse bietet, wenngleich sie nichts mit den grossen philosophischen Fragen zu tun hat, um derentwillen Sie Ihren logischen Kalkül ausgebildet haben. Es handelt sich dabei um den ersten Abschnitt der mathematischen Logik, in welchem die "elementary propositions" betrachtet werden. Als Grundlagen für die Beweise nehmen Sie hier die fünf Sätze, welche Sie abkürzungsweise mit "Taut", "Add", "Perm", "Assoc", "Sum" bezeichnen, ferner die Definition der Implikation (*1.01) und ausserdem die Regel "Anything implied by a true elementary proposition is true". Indem ich diese Grundlagen vom axiomatischen Standpunkt untersuchte, fand ich, dass die Formel "Assoc" entbehrlich ist, indem sie aus den übrigen "primitive propositions" abgeleitet werden kann. Diese Ableitung habe ich

Ihnen auf dem beiliegenden Blatt aufgeschrieben, wobei ich mich ganz an Ihre Symbolik halte (Eine gewisse Abkürzung, die ich mir zur einfacheren Darstellung erlaube, wird in den "Vorbemerkungen" erklärt, die ich dem Beweis vorausschicke). Sie werden gewiss den Beweisgang—sofern er Ihnen überhaupt neu ist—mit einem Blicke übersehen. Die nunmehr sich erhebende Frage, ob etwa auch einer der übrigen Sätze (Taut, Add, Perm, Sum) als Axiom entbehrt werden kann, habe ich verfolgt (durch Betrachtung von endlichen Systemen, für deren Elemente die Negation und das logische Produkt in geeigneter Weise definiert wird). Es zeigt sich, dass keiner der Sätze Taut, Add, Perm, Sum sich auf die vier übrigen primitive propositions (einschliesslich Assoc) zurückführen lässt, insbesondere ergibt sich auch, dass Taut nicht ersetzt werden kann durch: $p \supset p$. Die Beweise für diese Behauptungen habe ich in einer (im Sommer 1918 eingereichten) Habilitationsschrift ausgeführt. Diese wird aber wohl nicht im Druck erscheinen. Sollten Sie sich für das Nähere dieser Beweismethode interessieren, so würde ich Ihnen natürlich gern genaueres mitteilen.

"Ihr ergebener Hochachtung empfiehlt sich Ihnen
"Paul Bernays
"Privatdozent an d. Universität
"Göttingen, Nikolausbergenweg 43"
(Bernays to Russell, 8.20, Russell archive, McMaster University, 110208b, Bernays 1920).

44. Here is the summary given by Bernays to Russell in a letter he sent in 1921: "Was nun die Unabhängigkeits-Beweise anlangt, welche Sie auf den beiliegenden Blättern finden, so handelt es sich darum, zu zeigen, dass von Ihnen primitive propositions: "Taut", "Add", "Perm", "Sum" mit Ausnahme von "Assoc", keiner auf die übrigen 4 zurückzuführen ist, und zwar auch dann nicht, wenn jedesmal die Formel $p \supset p$ hinzugenommen wird. Die Beweismethode besteht in dem bekannten Verfahren der Ausweisung von endlichen Systemen. Die Buchstaben $p, q, \ldots$ werden als Variable genommen, welche jeweils nur 3, beziehungsweise 4 mögliche Werte besitzen. Jedem Wert von p wird ein Wert von "$\sim p$" durch Festsetzung zugeordnet, ebenso jedem Wertepaar von p, q ein Wert von "$p \vee q$". Unter den 3 (bezw. 4) Werten wird einer, (oder auch mehrere) ausgezeichnet, und die aus den Variablen $p, q, \ldots$ mit den Zeichen "$\sim$", "$\vee$" gebildeten Ausdrücke (elementary propositions) werden nun daraufhin betrachtet, ob sie für *jedes* Wertsystem der Variablen jenen ausgezeichneten Wert, (bezw. einen der ausgezeichneten Werte), ergeben. Wenn dies der Fall ist, so soll der betreffenden Ausdruck (mit Bezug auf die getroffenen Festsetzungen) eine "richtige Formel" heissen. Dabei werden die Festsetzungen so eingerichtet, dass immer, wenn A und $(\sim A) \vee B$ richtige Formeln sind, auch B eine richtige Formel ist,—entsprechend Ihrer proposition *1.1: "Anything implied by a true elementary proposition is true". Hierdurch wird erreicht, dass jede Formel, die aus einem System S von Ausgangsformel, aufgrund jener primitive proposition *1.1, ableitbar ist, eine "richtige" Formel sein muss, sofern alle Formeln aus S richtige Formeln sind.
"Um daher die Unabhängigkeit einer Formel F (elementary proposition) von einem System (S) von Formeln zu beweisen, hat man nur nötig, die willkürlichen Festsetzungen so zu treffen, dass zunächst jene eben genannte Bedingung erfüllt ist un dass ferner die Formeln aus S sämtlich "richtige Formeln" werden, die betrachtete Formel dagegen nicht." (Bernays to Russell, 19.iii.1921, Russell archive, McMaster University, Bernays 1921).

45. "Was Sie mir in Ihrem Briefe von dem Schefferschen Zeichen erzählten, war mir damals ganz neu und natürlich sehr interessant. Ich habe über diese Reduktion der logischen Zeichen den Göttinger Mathematikern berichtet, und sie hat dort Anregung zu weiteren Überlegungen in

dieser Richtung gegeben. Insbesondere hat Herr Schönfinkel gefunden, dass auch im Gebiete des Kalküls mit Variablen alle logischen Zeichen auf ein einziges: $\varphi(x)/^x\psi(x)$ zurückführbar sind, welchem man die Bedeutung geben kann: "für kein x besteht $\varphi(x)$ mit $\psi(x)$ zusammen", symbolisch $(x)..\neg\varphi(x). \vee .\neg\psi(x)$. Herr Professor Hilbert (dem ich damals Ihren Gruss ausgerichtet habe) beschäftigt sich weiter sehr lebhaft mit den Fragen der mathematischen Logik, insbesondere mit der Aufgabe des Nachweises der Widerspruchslosigkeit für die Arithmetik (Analysis). Seit dem Sommer hat er seinen Ansatz zur Behandlung dieses Problems wesentlich weiter geführt. Gleichwohl befinden sich diese Untersuchungen einstweilen noch in den Anfängen. Bestens empfiehlt sich Ihnen Ihr ergebener, Paul Bernays" Bernays to Russell, 19.iii.1921 (Russell archive, McMaster University).

46. "The theory of Dr Scheffer, to which you kindly called my attention, is well known here at Göttingen. I think you take interest in learning that some Russian named M. Schönfinkel, who is living in Göttingen, has taken great pains in pursuing Scheffer's plan to its utmost consequences. I shall try to give you a short outline of the theory. Firstly, by a device of Prof. Bernays, instead of Sheffer's dash the sign $/^x$ (where $fx/^x gx$ means $(x).\neg f(x) \vee \neg g(x)$) is taken as the only undefined symbol." Behmann to Russell, May 10, 1923, (Russell archive, McMaster University, 110205d, Behmann 1923).

47. "Ausserdem es ist vielleicht von Interesse, dass ich im Auftrag von Herrn Prof. Bernays versucht habe, die in den *Principia Mathematica* erwähnte Begründung der Mengenlehre unter Elimination des Reduzierbarkeitsaxioms (durch Axiome) tatsächlich durchzuführen. Ich nehme allerdings an dass dies bei der Verfassung der Principia schon geschehen ist (ich hätte übrigens sehr gerne Sicherheit darüber). Doch für den Fall dass es noch nicht geschehen wäre, möchte ich bemerken dass es auf verschiedene Weisen geht, immer unter Aufstellung von 2 neuen Axiomen. Der Anschluss an *13 (ev. auch *14) und *20, sowie *21 gelingt in jedem Falle (Für die Relationen brauchen wir zwei weitere Axiome). (Zu etwa erwünschten nähere Ausführungen bin ich selbstverständlich immer gerne bereit)" Boskovitz to Russell, July 3, 1923 (Russell archive, McMaster University, 110240a).

48. "It was, in fact, that work of yours [PM] that first gave me a view of that wonderful province of human knowledge which ancient Aristotelian Logic has nowadays become by the use of an adequate symbolism. But, I daresay, it might be said of your work just as well what H. Weyl said of his own book, that "it offers the fruit of knowledge in a hard shell", requiring indeed a considerable amount of labour in order to be accustomed to that particular manner of thinking, equally different from that of common life and that of college logic and philosophy, which is absolutely necessary for a rigorous treatment of the topic. Several years ago, I had therefore resolved to write something like an introduction or commentary to that work, providing a way by which the unavoidable difficulties of understanding are separately treated and, in consequence of it, may be clearly grasped and overcome soon by the unacquainted reader in order that the *Principia Mathematica* might become as well known as both the work and the topic deserve." Behmann to Russell, August 8, 1922, Behmann (1922c) (English in original). Although I originally used the letter preserved in the Behmann archive in Erlangen, there is also a copy in the Russell archive at McMaster (1368). Some of the ideas of the dissertation can also be found in the text for a talk delivered in 1927 in Kiel entitled "Die Russellsche Theorie der logischen Typen". In this connection see also Hilbert's correspondence with Russell (quoted in Sieg 1999, pp. 37–38) which shows that Russell's theory of knowledge was being widely discussed in Göttingen earlier than 1916. Sieg (1999, p. 32) asks what works by Russell were being read at the time. Behmann quotes the following works in his dissertation: *Principia*

Mathematica, Mathematical Logic as Based on the Theory of Types (1908), *Principles of Mathematics* (1903), *The Problems of Philosophy* (1912). I have found no additional texts by Russell explicitly quoted during the period between 1914 and 1921 in the writings (published or unpublished) of Hilbert, Behmann, or Bernays.

49. Russell to the librarian of the Göttingen University Library, July 2, 1924. Behmann Archive, Erlangen.
50. "Es wäre infolgedessen von vornherein verfehlt, wenn wir uns anschicken wollten, die abstrakten Begriffe und Relationen der Arithmetik für sich allein zu untersuchen, ohne dabei des Umstandes zu gedenken, dass alle derartigen Begriffe immer erst im Zusammenhange des *Satzes* etwas bedeuten und dass auch die bestgebildeten Begriffe uns nichts nützen und uns auch nicht vor Widersprüchen bewahren, solange wir nicht wissen, wie wir sie richtig *zu verwenden* haben" (Behmann 1918, p. 34).
51. "... so müssen wir nun auch, wie uns die letzte Ueberlegung zeigte, den Begriff der *Aussage* von vornherein in den Mittelpunkt unserer gesamten Untersuchungen stellen. Das bedeutet, wir dürfen die in Aussagen auftretenden Begriffe–wie "Menge", "Zahl", "Aussage", "Raum", "Kraft" u.s.f.—niemals für sich allein—als selbstständige logische Gegenstände—betrachten, sondern immer nur im Zusammenhang derjenigen Aussagen, in denen sie sinnvoll auftreten können" (Behmann 1918, p. 34).
52. "Wir bauen nach und nach ein gewisses festes *System von Aussagen* auf, das wir so einrichten, dass keine zwei in ihm enthaltenen oder—was übrigens auf dasselbe hinauskommt—durch einwandfreie Schlüsse aus ihm ableitbaren Aussagen, sofern sie als richtig erkannt sind, jemals einander widersprechen, dass andererseits aber gewiss all arithmetisch brauchbaren (richtigen und falschen) Aussagen in ihm vorkommen" (Behmann 1918, p. 35).
53. "Da wir zufolge der Natur unseres Gegenstandes nicht einmal die einfachsten logischen Begriffe in ihrem überkommenen Gebrauch voraussetzen dürfen, müssen wir infolgedessen zunächst *allem* abstrakten Denken gegenüber misstrauisch sein. Weil wir aber doch irgend einen Ausgangspunkt haben müssen, so darf dies nur ein solcher sein, bei dem unser abstraktes Denken noch nicht beteiligt ist, dessen Existenz also nicht, wie z.B. die der Zahlen, erst auf der Möglichkeit des Denkens beruht und der darum auch nicht durch es verfälscht sein kann. Die einzige Dinge, die dieser Forderung genügen, sind nun aber diejenigen, die ohne Zuhilfenahme des Denkvermögens, d.h. also durch die sinnliche Wahrnehmung allein, unmittelbar erkannt werden, die somit dem Denken erst den zu seiner Möglichkeit unbedingt notwendigen Stoff geben. (Fast alle falsche Metaphysik krankte bisher an dem Bestreben, Gegenstände des Denkens selbst denkend erzeugen zu wollen. Wie überall, muss aber auch hier der Archimedische Punkt notwendig ausserhalb liegen.) Infolgedessen ist unsere *erste Annahme* die, dass es gestattet sei,*von Gegenständen der Erfahrungswirklichkeit* (individuals) *als dem vor allem Denken Existierenden auszugehen* (Damit halten wir uns natürlich für berechtigt, von einer etwa möglichen Verfälschung durch die Wahrnehmung vollständig abzusehen.) Diese Individuen betrachten wir somit als fertig vorliegend und für unser Denken unmittelbar verfügbar" (Behmann 1918, pp. 44–45).
54. A few quotes from PM on individuals.

"For this purpose, we will use such letters as a, b, c, x, y, z, w, to denote objects which are neither propositions nor functions. Such objects we shall call *individuals*. Such objects will be constituents of propositions or functions, and will be *genuine* constituents, in the sense that they do not disappear on analysis, as (for example) classes do, or phrases of the form "the so-and-so"". (PM, p. 51)

"The symbols for classes, like those for descriptions, are, in our system, incomplete symbols: their *uses* are defined, but they themselves are not assumed to mean anything at all. That is to say, the uses of such symbols are so defined that, when the definiens is substituted for the definiendum, there no longer remains any symbol which could be supposed to represent a class. Thus classes, so far as we introduce them, are merely symbolic or linguistic conveniences, not genuine objects as their members are if they are individuals." (PM, p. 72)

"We may explain an individual as something which exists on its own account." (PM, p. 162)

I should add that the reconstruction of arithmetic will also force the assumption of the existence of infinitely many individuals (see pp. 276–77). Behmann recognizes that these are not given to us in any empirical experience and thus the assumption is only a postulate.

55. "Wir wollen nun eine Aussage, in der eine unmittelbare Wahrnehmung zum Ausdruck kommt oder doch wenigstens vorgestellt wird, als *"einfache Aussage"* bezeichnen" (Behmann 1918, p. 47).

56. "Denn da die objektive Welt, d.h. die Gesamtheit der Individuen mit all ihren Eigenschaften und Beziehungen, wie wir hier unbedenklich voraussetzen dürfen, im letzen Grunde gewiss einen widerspruchslosen Bereich bildet, so muss offenbar jede Aussage, die nur Individuen zu Gegenständen hat, die also keine anderen Dinge als existierend voraussetzt, dem wirklichen Tatbestand entweder entsprechen oder widersprechen und somit den Sätzen vom Widerspruch und vom ausgeschlossenen Dritten genügen"(Behmann 1918, p. 107)

57. Explaining why certain propositions, such as "A propositional function fx cannot be considered as an object in the system of propositions constructed", do not imply that propositional functions are objects Behmann explains that: "Es steht uns jedoch frei, den obigen Satz folgendermassen zu deuten 'Das Zeichen $f\hat{x}$—also ein konkretes Individuum—darf nicht so betrachtet werden, als ob ihm [...] ein eigentlicher Gegenstand entspräche, der mithin als Gegenstand der Aussage gelten dürfte, d.h. es hat nur als ein künstlich abspaltbarer Bestandteil eines umfassenderen Zeichens zu gelten.'" (Behmann 1918, p. 64)

58. "In gleichem Sinne [*as the differential dx*] sind nun aber auch die Aussagefunktionen genau wie alle Abstrakta keine echten Gegenstände, sondern sie erwecken nur, indem sie dem in grösserer Entfernung von seinen eigentlichen konkreten Gegenständen sich bewegenden abstrakten Denken gewisse Ruhepunkte bieten, den trügerischen Schein einer Gegenständlichkeit" (Behmann 1918, p. 106).

59. Behmann appeals explicitly to Russell's "no-class" theory and then says: "Das bedeutet, dass wir die Klassen nur als formale Bestandteile der sie betreffenden Aussagen anerkennen, deren eigentliche Aussagegegenstände hingegen einzig die in den Klassen enthaltenen—oder auch nicht enthaltenen—Individuen sind." (Behmann 1918, p. 165) In PM the idea is expressed as follows: "The symbols for classes, like those for descriptions, are, in our system, incomplete symbols: their uses are defined, but they themselves are not assumed to mean anything at all. That is to say, the uses of such symbols are so defined that, when the definiens is substituted for the definiendum, there no longer remains any symbol which could be supposed to represent a class. Thus classes, so far as we introduce them, are merely symbolic or linguistic conveniences, not genuine objects as their members are if they are individuals." (PM, p. 72)

60. "Die Klassen—und gleicherweise übrigens die Zahlen—sind dann, wie wir schon früher andeuteten, gar nichts weiter als Redensarten, die für die bequeme und übersichtliche Darstellung der Arithmetik von äusserstem Nutzen sind, aber gleichwohl recht bedenklich werden

können, sobald man sie ernst nimmt und ihrer Natur zuwider für Namen von Gegenständen hält" (Behmann 1918, p. 159). How the strategy is supposed to work is made explicit by comparison with the elimination of ideal elements in projective geometry (p. 165) and with an example from the vector calculus (p. 160).

61. "Es liegt nun recht nahe, zu vermuten—und lässt sich, wie in den *Principia Mathematica* gezeigt wird, auch streng beweisen, sobald man alle für die mathematische Logik benötigten Axiome wirklich heranzieht—dass man mit den Individuenklassen beim praktischen Schliessen genau so verfahren kann, als ob sie Individuen wären, solange man sich nur hütet, sie mit den wirklichen Individuen geradezu in eine Reihe zu stellen. Denn die wesentliche Eigenschaft, die der Bereich der Individuenklassen—aber nicht der Bereich aller Klassen überhaupt—mit dem Individuenbereich gemeinsam hat, ist ja die der Konsistenz. Und wir dürfen die Elemente eines konsistenten abstrakten Bereiches, solange wir ihn gegen alle anderen Bereiche scharf abgrenzen, d.h. die Typenfestsetzungen streng innehalten,—so z.B. auch die des Bereiches der natürlichen oder der reellen Zahlen—innerhalb gewisser Grenzen so betrachten, *als ob* sie dem Denken ursprünglich gegebene Dinge wären" (Behmann 1918, pp. 187–88).

62. "Ganz von selbst gelangen wir von hier aus auf die Frage, ob das zugrunde gelegte Axiomsystem die Eigenschaft der Vollständigkeit hat, d.h. ob jede vernünftig gestellte Frage einer Beantwortung in dem gemeinten Sinne fähig ist" (Behmann 1918, p. 324).

63. "Vielleicht ist es sogar möglich, die Unentscheidbarkeit ihrerseits zu beweisen oder gar die Unentscheidbarkeit der Entscheidbarkeitsfrage" (Behmann 1918, p. 325).

64. I will give the full text (UAG, Phil. Fak. Promotionen: B vol. VI, (1917–), Universitätsarchiv, Göttingen): "Göttingen, den 1. Februar 1918, Die eigentliche Aufgabe und Absicht der vorliegenden Abhandlung von Behmann ist es, in die Gedankenwelt derjenigen Disziplin einzuführen, die man als symbolische Logik bezeichnet, die unter den Händen einer Reihe bedeutender Mathematiker und Logiker ein wichtiger Bestandteil der Erkenntnistheorie geworden ist und schliesslich in den letzten Jahren in dem gross-angelegten Werke "*Principia Mathematica*" von Russell und Whitehead die reifste Bearbeitung und Darstellung erfahren hat. Während lange Zeit hindurch die symbolische Logik nichts anderes als eine formalistische äusserliche Weiterbildung der Aristotelischen Schlussfigurentheorie zu sein schien, gelang es Russell, mit der Anwendung der symbolischen Logik auf die schwierigsten erkenntnistheoretischen Fragen zum ersten Male sichere Erfolge zu erziehen. Obenan in der Russelschen Theorie steht—als oberstes Denkaxiom—das sogenannte Reduzierbarkeitsaxiom. Dieses Axiom, einschliesslich der damit verbundenen Typentheorie von Russell bietet dem Verständnis ausserordentliche Schwierigkeiten. Um diese zu beseitigen zieht Behmann das von mir in die Arithmetik eingeführte Vollständigkeitsaxiom heran, welches nicht nur seinem logischen Charakter nach mit dem Reduzierbarkeitsaxiom gleichartig ist, sondern auch sachliche innere Zusammenhänge mit jenem zeigt. Indem Behmann diese Zusammenhänge herausarbeitet, gelingt es ihm nicht nur jenes Reduzierbarkeits-axiom klar zu fassen, sondern auch die Anwendung der Russelsche[n] Theorie auf ein spezielles tiefliegendes Problem—die Auflösung der Antinomie der transfiniten Zahl—über Russell hinaus durchzuführen. Sein Resultat bleibt im Wesentlichen: alle transfinite Axiomatik ist ihrer Natur nach etwas unabgeschlossenes, aber die mengenth.[eoretische] Begriffe von Cantor sind streng zulässig. Jedes an sich vernünftlich gestellte mengentheoretische Problem behält Bedeutung und ist daher einer Lösung fähig. Bei der Darstellung ist besonderer Wert darauf gelegt, dass - dabei keinerlei spezifisch mathematische Kenntnisse vorausgesetzt werden; diesselbe ist daher

auch dem Nicht-Mathematiker verständlich. Ich hoffe für die Behemmanische Schrift einen Verleger zu finden, da sie—wie mir scheint—einem mathematisch-philosophischen Bedürfnis in der Gegenwart Rechnung trägt. Ich beantrage, bereits 2 Druckbogen als Dissertation gelten zu lassen.

"Praedikat

"Hilbert"

65. "Da ein Individuenbereich, der dieses Axiom erfüllen würde, und naturlich in keiner wirklichen Erfahrung gegeben werden kann und wir überdies die Vermutung, dass die Anzahl aller wirklichen Dinge—so z.B. insbesondere der Atome und Elektronen -unendlich sei, durch keinerlei triftige Gründe zu stützen vermögen [...]" (Behmann 1918, p. 277).

66. On Hilbert's program in the 1920s see Chapter 2 and Sieg (1999).

67. This position is repeated verbatim in many later publications by Hilbert.

68. "An appeal to an intuitive grasp of the number series as well as to the multiplicity of magnitudes is certainly to be considered. But this could certainly not be a question of an intuition in the primitive sense; for, certainly no infinite multiplicities are given to us in the primitive intuitive mode of representation. And even though it might be quite rash to contest any farther-reaching kind of intuitive evidence from the outset, we will nevertheless make allowance for that tendency of exact science which aims as far as possible to eliminate the finer organs of cognition [*Organe der Erkenntnis*] and to rely only on the most primitive means of cognition" (Bernays 1922b, E.t., pp. 215–16). Compare Bernays (1923a, p. 226): "Hilbert's theory does not exclude the possibility of a philosophical attitude that conceives the numbers as existing, nonsensible objects ... Nevertheless the aim of Hilbert's theory is to make such an attitude dispensable for the foundations of the exact sciences". In this connection it might be worthwhile to investigate the possible influence of Russell's theory of knowledge on Hilbert and Bernays.

Chapter 5

1. In recent years I have been occupied, among other things, with the development of the foundations of mathematics in the twenties (Mancosu, 1998a, 1999a,b). My decision to work in this area, and in particular on the development of Hilbert's program and Weyl's predicative approach to mathematics, is very much a consequence of Sol's influence on my interests. In addition to his technical engagement in proof theoretical work Sol has also displayed a refined interest in the historical development of the programs to which he so successfully contributed (see, for instance, the essays contained in Feferman (1998b) and his general editorship of Gödel's *Collected Works*). My paper for this Festschrift is also a contribution to this part of my research. The debate I discuss centers around the topic of how "constructive" classical mathematics is—a theme dear to Sol—and contains, among other things, some hitherto unknown information concerning Gödel's contributions to the foundations of mathematics. Sol's influence on my work has been a decisive one and it extends to all of my research since my days as a graduate student (including my work on mathematical logic, on seventeenth century mathematics, and on mathematical explanation). And that this influence reaches far beyond Sol's specific expertise points to the fact that—in addition to what might be specific transmission of knowledge—Sol's greatest legacy is that of his style of thought. I dedicate this paper to him in gratitude.

2. On the literature on proofs by contradiction see Mancosu (1996, chapter 4).

3. Recent work on Behmann include Mancosu (1999a), Thiel (2002). For a catalogue of the Behmann Nachlaß see Haas and Stemmler (1981).
4. "Es handelt sich um die Durchführung einer Aufgabe, zu der mich Herr Kaufmann während meines Aufenthalts in Wien im September angeregt hat. Obgleich ich ja nicht selbst die Auffassung des Konstruktivismus teile, halte ich derartige Untersuchungen doch für wichtig, damit die praktische Tragweite der konstruktivistischen Prinzipien nicht überschätzt und infolgedessen unberechtigterweise große Teile der Mathematik in Frage gestellt werden." [Bernays Nachlaß, ETH Zurich, Hs 975: 275] For Behmann's attitude towards constructivism see also [Behmann to Kaufmann, October 22, 1930, Kaufmann archive, Konstanz, 006313].
5. "Herr Carnap hat den Aufsatz bereits geprüft und ist in jeder Hinsicht einverstanden." [Bernays Nachlaß, ETH Zurich, Hs 975: 275]
6. See [Kaufmann to Hempel, November 30, 1930] and [Behmann to Kaufmann, Sept. 5, 1930]; both letters are in the Kaufmann archive in Konstanz.
7. See [Behmann to Kaufmann, July 9, 1930, Kaufmann archive, Konstanz].
8. "Ich möchte Ihnen gern noch erzählen, daß ich, als ich am Montag Abend in Dresden wegen des im Speisewagen genossenen Kaffees—koffeinfreien gab es leider nicht—nicht sogleich einschlafen konnte, über Ihr Problem der Konstruktivität der Existenzbeweise nachgedacht habe und dabei zu dem Ergebnis gekommen bin, daß Sie mit Ihrer Vermutung nach meiner Beurteilung Recht haben." [Behmann to Kaufmann, September 26, 1930, Kaufmann archive, Konstanz, 006271]
9. Copies of the essay by Behmann are found in the Bernays Nachlaß (ETH, Zurich, Hs 974:18) in the Rudolf Carnap Nachlaß (Pittsburgh/Konstanz), in the Behmann Nachlaß (Erlangen) and in the Felix Kaufmann Nachlaß (Konstanz). The correspondence between Behmann and Kaufmann can be reconstructed by using jointly the Behmann and the Kaufmann archives.
10. In the original to Kaufmann it is "Der Satz von Kaufmann."
11. See Moore (1982).
12. "Die Anforderung (2) wird nur in dem Fall eine positive Beschränkung bedeuten, daß es Existenzbehauptungen gibt, die kein nicht-konstruktives Axiom voraussetzen und nur auf die zweite, indirekte Art bewiesen werden können." (Behmann 1930) As Bernays will be quick to point out the opposition between constructive by instance vs proof by contradiction is misleading. See below.
13. "Hierzu hat neuerdings Herr Kaufmann eine interessante Vermutung ausgesprochen durch welche die angedeutete Frage verneinend beantwortet würde. Er ist der Ansicht, daß auch ein indirekter Existenzbeweis, sofern er der Anforderung (1) genügt, seiner Natur nach konstruktiv sei, genauer gesagt, auf einer versteckten Aufweisung beruhe, die sich durch eine geeignete Analyse des Beweises zwangsläufig zur Evidenz bringen lassen müsse." (Behmann 1930)
14. On the elimination of second-order quantification see Kaufmann archive, 006243.
15. "Das gleiche Schema gilt für jeden einzelnen Schluß des Beweises, wenn im ersten Fall, d.h. falls der Schluß einer direkt Argumentation angehört, a, b, c, (usw.) als die (auf Grund der vorausgehenden direkten Beweisschritte durchweg wahren) Vordersätze und d als der Schlußsatz des Einzelschlusses [...] betrachtet werden." (Behmann 1930, p. 3)
16. "Ganz unten steht (mit abwärts gerichtetem Schlußpfeil) die Sternfigur des letzten Schlusses des Beweises. Soweit die an den Bedingungspfeilen stehenden Aussagen nicht Voraussetzungen des Gesamtbeweises sind, schließt sich an diese je eine weitere Sternfigur mit ihrem Schlußpfeil (der ungeachtet etwaiger Überdeckungen der Sternfiguren wiederum abwärts

gerichtet vorgestellt werden möge), an jeden Bedingungspfeil dieser Sternfiguren, der nicht durch eine Voraussetzung des Gesamtbeweises besetzt ist, in genau der gleichen Weise je eine weitere Sternfigur, usf. Mit Ausnahme der Vordersätze und des Schlußsatzes des Gesamtbeweises steht also jede auftretende Aussage am Anfang und am Ende genau eines Pfeiles." (Behmann 1930, p. 3)

17. "Es ist nun anschaulich evident, daß innerhalb der beschriebenen Struktur als solcher, wenn man vom Richtungssinn der Pfeile absieht, die Stelle des Schlußsatzes mit denen der Vordersätze topologisch gleichberechtigt ist, daß man also das Schema in genau der gleichen Weise von irgend einem 'Zweigende' her aufgebaut werden kann. Drehen wir nämlich die zugehörige Sternfigur so, daß das fragliche Zweigende abwärts (d.h. der Pfeil aufwärts) gerichtet ist, so gilt wiederum, daß an diejenigen Sternzacken, an welchen weder Vordersätze noch der Schlußsatz des Beweises stehen, sich genau eine weitere Sternfigur anschließt, deren berührenden Pfeil wir wiederum senkrecht gestellt denken können, usf." (Behmann 1930, pp. 3–4)

18. For the history of graph theory see (Biggs 1976).

19. I use the version due to Bernays, who had shown the redundancy of one of the original axioms in the propositional system of *Principia* and explicitly states the need for a substitution rule. For Bernays' work on propositional logic see (Zach 1999).

20. "Herzlichen Dank für Ihren Brief und den erwarteten Beweis. Er erscheint mir als absolut schlüssig und von mustergültiger Klarheit." Kaufmann archive, Konstanz, 006231

21. "Die von Herrn Behmann bewiesene, in meinem genannten Buche [*Das Unendliche in der Mathematik und seine Ausschaltung*] S. 66 f. aufgestellte Behauptung bringt (wie dortselbst S. 59 ff näher ausgeführt wurde) m.E. die Entscheidung in dem Methodenstreit zwischen Formalismus und Neo-Intuitionismus (Brouwer). Dies ergibt sich zunächst im Hinblick auf diejenigen Sätze, deren Beweis ohne Hilfe des Auswahlaxioms durchgeführt wird. Denn der mathematische Kern der Brouwerschen Lehre liegt in der Aufstellung des folgenden Trilemmas:

"(1) Eine mathematische Existentialbehauptung E z.B. das Auftreten einer Sequenz 0, 1, 2, 3, ... 9, in der Dezimalbruchentwicklung von π ist durch direkte Aufweisung verifiziert.

"(2) der Beweis der Absurdität von E ist geliefert, d.h. $\neg E$ ist für alle Zahlen bewiesen.

"(3) Die Absurdität von (2) also die Absurdität der Absurdität von (1) wird bewiesen, aber ohne dass zugleich eine Konstruktion (1) angebbar wäre.

"Die Möglichkeit von (3) würde bedeuten, dass aus der Absurdität der Absurdität einer Existenzialbehauptung nicht die Gültigkeit dieser Existenzialbehauptung gefolgert werden könnte und die Annahme dieser Möglichkeit ist es, die Brouwers These, dass in unendlichen Bereichen der Satz vom ausgeschlossenen Dritten keine Geltung habe, zugrundeliegt. Diese These deckt sich, wie Brouwer selbst mit Nachdruck hervorgehoben hat (z.B. Intuitionistische Mengenlehre, Jahresber. d. Deutsch. Math.-Ver., Bd 28, S 203–208, 1919) mit seiner Negierung der Entscheidungsdefinitheit der Arithmetik und Analysis.

"Die Basis für diese Behauptung fällt also zunächst für alle Sätze, die ohne Hilfe des Auswahlaxioms bewiesen werden, weg.

"Hier erhält man das folgende Ergebnis:

"Die intuitionistische Forderung nach durchgängiger Konstruktivität ist berechtigt; aber sie erscheint auch in den sogenannten Existenzbeweisen der klassischen Mathematik implizit erfüllt, daher bringt ihre Verwirklichung keine Einbusse des Besitzstandes der Mathematik mit sich." [Kaufmann archive, Konstanz, 006228–006230]

22. "Heute habe ich eine längere, sehr schöne Diskussion mit Herrn Carnap gehabt; er ist äusserst erfreut über den gelungenen Beweis, den er—wie ich es nicht anders erwartet habe—für vollkommen greifbar hält und dessen gewaltige Tragweite er erkennt. Das Gleiche gilt von Herrn Waismann, mit dem ich heute telephonisch gesprochen habe." [Kaufmann archive, Konstanz, 006287]
23. "Carnap möchte auch Gödel gerne Ihren Beweis zeigen, wofür Ihre Genehmigung erbitte." [Kaufmann to Behmann, Vienna, October 9, 1930, Kaufmann archive, Konstanz, 006231]
24. Kaufmann archive, Konstanz, 006218.
25. According to Dawson (1997) the diary entry is to be found under RC 028-06-19.
26. [Kaufmann archive, Konstanz, 006304] and [Behmann archive, Erlangen]. The transcription is from the copy in the Behmann archive. ⟦The latter is now published in Gödel 2003a, pp. 565–68.⟧

    "Wien, am 19. Oktober 1930.

    "Sehr geehrter Herr Behmann!

    "Eben ist Herr Dr. Gödel bei mir, der gegenüber meiner Konstruktivitätsbehauptung bzw. Ihrem Beweise derselben folgende Einwände erhebt:

    "Er behauptet sehr wohl Beispiele konstruieren zu können, wo offenkundig eine Existenzialbehauptung bewiesen ist, ohne daß doch eine Konstruktion angegeben werden könne. Das einfachste von Herrn Gödel genannte Beispiel, welches aber auch prinzipiell als Repräsentant für die übrigen Beispiele gelten darf, ist folgendes:

    "Gegeben sei eine eineindeutige Zuordnung zwischen den natürlichen Zahlen und gewissen Rationalzahlen des Intervalles (0, 1). Dann läßt sich auf die bekannte Art zeigen, daß die angegebene Folge von Rationalzahlen einen Häufungspunkt hat, ohne daß es jedoch allgemein möglich wäre, einen solche anzugeben. Betrachten wir ein Beispiel, wo man einen Häufungspunkt sicher nicht angeben kann! Eine Zahl heiße Goldbach-Zahl, wenn alle kleinern geraden Zahlen Summe zweier Primzahlen sind. Die Folge rationaler Zahlen werde nun so definiert: Der natürlichen Zahl n sei die Zahl $\frac{1}{n}$ zugeordnet, wenn n eine Goldbach-Zahl ist und die Zahl $1 - \frac{1}{n}$, wenn n keine Goldbach-Zahl ist. Dann kann man beweisen, daß diese Folge einen Häufungspunkt hat (und zwar einen rationalen) nämlich entweder 0 oder 1. Es läßt sich aber ohne Lösung des Goldbach'schen Problems kein Häufungspunkt angeben.

    "Herr Gödel hat auch mit Herrn Carnap darüber gesprochen und diesem erscheint der gemachte Einwand sehr einleuchtend.

    "Ich wäre Ihnen sehr dankbar, wenn Sie mir mitteilen wollten, wie Sie sich zu diesem Einwand stellen, und will mir inzwischen auch selbst darüber der Kopf zerbrechen.

    "Herr Gödel fügt noch hinzu, daß es sich hier um eine Disjunktion zwischen *zwei* Möglichkeiten handle, daß dies aber nicht prinzipiell sei; es könne auch die Disjunktion zwischen unendlich vielen Fällen unentschieden bleiben.

    "Mit den herzlichsten Grüßen bin ich Ihr,

    "Felix Kaufmann"
27. "Zunächst ist zu sagen, daß der zweifellos sehr interessante Einwand Herrn Gödels jedenfalls meinen Beweis als solchen nicht berührt; er betrifft ja nur die Voraussetzung meines Beweises, daß jeder direkte Existenzbeweise auf einer Aufweisung beruht. Auch Ihre Vermutung wird genau genommen nicht betroffen, wenn man sie dahin versteht, daß 'in gleicher Weise wie der direkte Existenzbeweis auch der indirekte eine Aufweisung implizit enthält.'" [Kaufmann archive, Konstanz, 006306]

28. Löwenheim presented the paper in Berlin in 1917. The German version was never published but a translation into English was published by Quine in 1946 (see Löwenheim 1946).
29. "Die Schwierigkeit des Nachweises liegt in der Mannigfaltigkeit der Formen, in denen das imprädikative Moment auftreten kann." [Bernays to Behmann, October 31, 1930, Bernays Nachlaß, ETH Zurich, Hs 975: 5768]
30. Indeed, it should be pointed out that in intuitionistic logic it holds that if $\exists x\, Ax$ is provable then $A(t)$ (for some t) is provable, which is exactly the property Behmann is after.
31. This result appears in volume I of (Hilbert and Bernays 1934).
32. "Es muss gezeigt werden, dass im Falle der Ableitbarkeit einer Formel E mit vorausstehenden unnegierten All- und Seinszeichen auch eine Formel E^* ableitbar ist, welche aus E hervorgeht, indem jedem Seinszeichen in E eine entsprechende "Aufweisung" zugeordnet ist; und zwar ist zu zeigen, dass die Ableitung von E^* so geführt werden kann, dass darin keine Seinszeichen und keine negierten Allzeichen vorkommen." [Bernays to Behmann, November 11, 1930, Bernays Nachlaß, ETH Zurich, Hs 975: 5769]
33. "Hiernach erscheint es doch fraglich, ob die Veröffentlichung meines Aufsatzes, insbesondere meines Beweises für die Ersetzbarkeit jedes indirekten Beweises durch einen direkten sich im Augenblick lohnt. Zum mindesten dürfte nicht einfach vorausgesetzt werden, daß jeder direkte Existenzbeweis eine Aufweisung enthält, sondern dies müßte ausdrücklich als bloße Vermutung eingeführt werden." (November 2, 1930) [Kaufmann archive, Konstanz, 006329]
34. "Einen Aufsatz 'Zur Frage der Konstruktivität von Beweisen' habe ich vor einigen Jahren ausgearbeitet, aber nicht veröffentlicht. Ich wollte hauptsächlich darauf hinaus, daß der indirekte Existenzbeweis wie der direkte, sofern er nicht vom Auswahlaxiom Gebrauch macht, 'konstruktiv' ist, d.h. die explizite Aufweisung eines Dinges der fraglichen Art enthält. Ein wesentlicher Schritt hierfür war der Nachweis, daß ein indirekter Beweis nichts weiter als eine Umwandlung eines direkten ist.

"Prof. Bernays, dem ich die Ausarbeitung zur Beurteilung geschickt hatte, schrieb mir darauf, daß dieser letzte Nachweis schon vor längerer Zeit von Löwenheim gegeben worden sei, der ihn den Berliner Mathematikern vorgetragen, aber nicht veröffentlicht habe. Er gab weiterhin eine ausführliche Kritik des Aufsatzes im ganzen, die ich freilich in ihren Einzelheiten nicht klar durchschaut habe und von der ich daher nicht weiß, inwieweit sie berechtigt war; es mag vielleicht sein, daß er sich durch einen Schreibfehler von mir ('nicht-prädikativ' statt 'nicht-konstruktiv') etwas irreführen ließ. Demzufolge war ich jedenfalls im Zweifel, ob die Veröffentlichung noch angebracht war." [Behmann to Hirano, December 8, 1934, Behmann archive, Erlangen]

Chapter 6

1. For details about Behmann's work in the foundations of mathematics, see Mancosu (1999a).
2. See Mancosu (2002, §1), for a discussion of the meaning of that passage. The paper also reconstructs the debate surrounding the discussion of Behmann's 'proof'.
3. "1. Ausschluß nicht-konstruktiver Existenzbeweise (insbesondere des Auswahlaxioms). 2. Ausschluß nicht-prädikativer Existenzbeweise (soweit nicht bereits durch (1) betroffen). 3. Ausschluß von Begriffen höherer Stufe (Beschränkung auf den sogenannten 'engeren Funktionenkalkül).'" (Behmann 1930, p. 1)
4. "Hierzu hat neuerdings Herr Kaufmann eine interessante Vermutung ausgesprochen [...] Er ist der Ansicht, daß *auch ein indirekter Existenzbeweis*, sofern er der Anforderung (1) genügt, *seiner Natur nach konstruktiv sei*, genauer gesagt, auf einer verstecken Aufweisung beruhe,

die sich durch eine geeignete Analyse des Beweises zwangläufig zur Evidenz bringen lassen müsse." (Behmann 1930, p. 1)
5. On Gödel's objection to Behmann's paper, see Mancosu (2002, §5).
6. This aspect of Behmann's paper has been studied by Mancosu (2002).
7. For a list of various proofs of the infinity of prime numbers, see Dickson (1951, pp. 413–15).
8. Euclid's proof, a typical case of *reductio ad absurdum*, is as follows: suppose that there is a greatest prime number, p_n. One defines from the primes $p_1, \ldots, p_n$ the number $N = (p_1 \times p_2 \times p_3 \times \ldots \times p_n) + 1$, which must be either prime or composite. If it is prime, then we have a contradiction, since it would be bigger than all the prime numbers smaller or equal to p_n. If it is composite, it must be divisible exactly by a prime number. But this prime divisor cannot be any prime smaller or equal to p_n because they would all leave a remainder of 1. Therefore there must be another prime number, bigger than p_n.
9. Notice that this is a special form of Riemann's zeta function $\zeta(z) = \sum_{n=1}^{\infty} \frac{1}{n^z}$. The zeta function for real z appears in a work by Euler dated 1744, where he introduced $\zeta(z) = \sum_{n=1}^{\infty} \frac{1}{n^z} = \prod_{n=1}^{\infty} (1 - \frac{1}{p_n^z})^{-1}$, where the p_n's are prime numbers.
10. The previous equation involves the summation formula for the geometric series, i.e. for any prime p we have $\sum_{n=0}^{\infty} \frac{1}{p^n} = \frac{1}{1-\frac{1}{p}} = \frac{p}{p-1}$. Note that on account of the unique factorization theorem (which is independent of the infinitude of the prime numbers) every natural number appears only once as denominator in the expanded product. Moreover, the series obtained by multiplying absolutely convergent series is also absolutely convergent and its sum is the product of the sums in the series' factors.
11. "Auch diesem Beweise kann man ebenso wie dem Euklidischen eine solche Form geben, dass sich aus ihm ein Intervall $(m \ldots n)$ ergiebt, innerhalb dessen sicher eine neue Primzahl $p > m$ sich befindet, wie groß m auch angenommen worden ist; erst dann ist ja auch der letzten und höchsten Anforderung genügt, die man an einen strengen mathematischen Beweis zu stellen hat." (Kronecker 1901, p. 270)
12. As pointed out in Hasse (1950, p. 183), while proofs of specific cases rely on a generalization of Euclid's proof, Dirichlet's general result relies on Euler's proof. Kronecker had already pointed this out in his *Vorlesungen* (1901, p. 273).
13. "Eine Überführung dieses Beweises in die konstruktive Form durch Herrn Wittgenstein ist mir gleichzeitig mit der Kaufmann'schen Vermutung bekannt geworden" (Kaufmann archive, Konstanz, 006239).
14. "Da ich die Ausführungen Herrn Waismanns wegen der Wittgensteinschen Ergänzung des Eulerschen Beweises in den Einzelheiten nicht mehr in Erinnerung habe, würde es mich sehr interessieren, wie weit die von mir ziemlich unabhängig auf Grund des von mir entwickelten allgemeinen Schemas durchgeführte Beweisumwandlung mit der Wittgensteinschen übereinstimmt. Jedenfalls wäre es auch wohl notwendig, sich mit Herrn Wittgenstein wegen der Erwähnung dieses Punktes, die aus Billigkeitsgründen nicht gut zu umgehen ist, zu verständigen, was wohl am besten durch Herrn Waismann geschehen kann" (Kaufmann archive, Konstanz, 006273).
15. "Was den Zusammenhang Ihres als Beispiel für den allgemeinen Satz gegebenen Beweises mit demjenigen von Wittgenstein betrifft, so meint Herr Waismann, daß wohl eine Verwandtschaft bestehe, daß aber Ihre Anmerkung diesem Sachverhalte hinreichend Rechnung trage, umsomehr als das Ergebnis von zentraler Wichtigkeit in Ihren allgemeinen Ueberlegungen liege. Er maßt sich aber eine endgültige Entscheidung hierüber nicht an und wird daher, ganz in Einklang mit Ihrem Wunsche, Wittgenstein Ihren Beweis schicken und ihn über die Begleitumstände informieren. Für keinen Fall liegt irgendein Grund zur

Vorzögerung der Publikation vor. Vielleicht darf ich Sie noch bitten, die in Frage kommende Anmerkung in der Weise zu ändern, daß Sie schreiben, Sie hätten die Ueberführung des Euler'schen Primzahlbeweises in die konstruktive Form durch Wittgenstein vor Ihrer Ausarbeitung dieses Beweises erfahren, ohne in diesem Zusammenhange die Kenntnisnahme von meiner Vermutung zu erwähnen, da ja diese schon in meinem Buche (S. 66 f.) enthalten ist" (Kaufmann archive, Konstanz, 006287).

16. "Eine mir kürzlich bekannt gewordene Überführung dieses Beweises in die konstruktive Form durch Herrn Wittgenstein hat die Aufstellung meines obigen allgemeinen Verfahrens mit veranlaßt. Eine konstruktive Vervollständigung (nicht Umformung) des Eulerschen Beweises is bereits von Kronecker (Vorlesungen über Zahlentheorie, S. 270ff.) angegeben worden" (Behmann 1930, p. 6, note 1).
17. For details about Wittgenstein's *Nachlaß*, see von Wright (1993).
18. The notes of conversations with Wittgenstein and Schlick taken by Waismann also contain a reference to Euler's proof during the meeting of September 25, 1930: "Euler's proof is immediately in error, as soon as prime numbers are written down in the form $p_1, p_2 \ldots, p_n$. For if the next index n is to mean an *arbitrary number*, then this already presupposes a law of progression, and this law can be given only in terms of an induction. Thus the proof presupposes what it is supposed to prove" (Wittgenstein 1979, pp. 108–09). This remark is rather unhelpful but at least it indicates that Euler's proof was at the forefront of Wittgenstein's preoccupations in late September 1930, just a few days after Behmann had left Vienna.
19. For the remarks on Euler's proof, see pp. 321–25.
20. Behmann wrote to Kaufmann on October 11: "Außerdem will ich u.a. noch ergänzend bemerken, daß eine Vervollständigung des Eulerschen Beweises im Sinne der Konstruktivität bereits von Kronecker (*Vorlesungen Über Zahlentheorie*, S. 270–273) angegeben worden ist. Doch betrachtet Kronecker dies durchaus als Zusatz zu dem gegebenen nicht-konstruktiven, nicht als Umwandlung in die konstruktive Form" (Kaufmann archive, Konstanz, 006286).
21. One should notice that if one does not use real but only natural numbers, this simply means that for any n in $\mathcal{N}$ it is enough to sum up 2^{2n-1} elements of the harmonic series to go beyond n.
22. For a discussion of the numerous passages where Wittgenstein says things that are in basic agreement with this reasoning, see Marion (1998, chapters 4 and 6).
23. On these matters, see Marion (1998, chapter 1).
24. The full letter reads: "Sehr geehrter Herr Kaufmann! Besten Dank für die Zusendung Ihres Buches. Ich bin leider zu sehr beschäftigt, es lesen zu können. Ich höre von Herrn Waismann, daß unsere Ansichten in manchen Hinsicht übereinstimmen. Hochachtungsvoll, Ihr ergebener L. Wittgenstein" (Kaufmann archive, Konstanz, 006096).
25. This story is not to be found in Ray Monk biography of Wittgenstein (Monk 1990a) but he told it in a newspaper article that appeared at the same time as his biography (Monk 1990b).
26. See Marion (1998).
27. These papers were published in *Erkenntnis*, vol. 2 (1931), pp. 91–121. For English translations, see Benacerraf and Putnam (1983, pp. 41–65).

Chapter 7

1. "Was ist mit Herrn Goedel los? Ich höre zwar allerhand aufregende Dinge, kann aber nicht herausbekommen, worum es sich handelt." Carnap *Nachlaß* : RC 102-72-11. Gödel's original

article on incompleteness is found in Gödel (1986). All the citations from the Carnap and Reichenbach archives are quoted with permission of the University of Pittsburg. All rights reserved.

2. It is interesting to note that the possibility of incompleteness was discussed by Bernays and Tarski as early as 1928. In a letter from Bernays to Scholz dated April 9, 1947 we read: "Die Möglichkeit der Unableitbarkeit der Glieder einer ableitbaren Alternative von der Form (A)

(A) $(X)(\phi(X) \to \gamma(X)) \vee (X)(\phi(X) \to \neg\gamma(X))$

wurde übrigens schon etliche Zeit vor dem Erscheinen des Gödelschen Theorems erwogen. Ich sprach über derartiges seiner Zeit mit Herrn Tarski auf dem Kongress in Bologna." The letter is preserved in the Bernays *Nachlaß* at the Wissenschaftshistorisches Archiv, Zurich, Hs 975: 4123. I thank Bernd Buldt for having called my attention to the above passage.

3. The proceedings of the conference were published in *Erkenntnis*, 2 (1931), pp. 91–131. The English translation of the main talks by Carnap, Heyting, and von Neumann is in Benacerraf and Putnam (1983). Dawson (1984) contains a translation of the discussion following the three talks.

4. During the period in which Hempel was corresponding with Kaufmann on Gödel's theorems, the French mathematician Jacques Herbrand was visiting Berlin. Sieg (1994) provides, also by means of unpublished correspondence, an account of Herbrand's activities and thinking during this period. After learning about Gödel's incompleteness from von Neumann, Herbrand also received the galleys of Gödel's article from Bernays (see Dawson 1997, p. 74). On April 7, 1931 he wrote to Gödel a long letter which contains several reflections on the importance of the results. The letter will be published with commentary in Gödel's *Collected Works*, vol. 5 [see Gödel 2003a, pp. 14–31]. On the same day Herbrand sent Bernays a carbon copy of the letter to Gödel with a cover letter. Dawson located a copy of the letter in the Gödel *Nachlaß* and I found another copy in the Wissenschaftshistorisches Archiv of the ETH in Zürich (Bernays Hs 975: 2124). Given the scarcity of information on Herbrand's activity during this period I will add two reports on Herbrand. They are found in letters to Kaufmann written by Hempel and Behmann. On February 3, 1931 Behmann wrote to Kaufmann from Halle:

"Am Freitag habe ich zufällig Herrn Herbrand kennen gelernt, der von Berlin zu unserem algebraischen Kolloquium gekommen war. Am Sonntag haben wir beide eine längere Besprechung gehabt, aber ohne rechte Klärung, da wir immer wieder dahin kamen, an einander vorbei zu reden. Er hat mir versprochen, mir ein Exemplar seiner Schrift über die Theorie des Beweises zu schicken. Er kommt voraussichtlich in einer oder zwei Wochen wieder nach Halle, um über ein algebraisches Thema vorzutragen" (Kaufmann *Nachlaß* 006034).

On May 2, 1931 Hempel wrote to Kaufmann from Berlin:

Vielleicht interessiert es Sie, wenn ich Ihnen noch erzähle, dass ich kürzlich im mathematischen Kolloquium an der hiesigen Universität den französischer Mathematiker Herbrand über die Hilbertsche Beweistheorie und die Metamathematik vorgetragen hörte. Leider sprach Herbrand ein so schwer verständliches Deutsch (auch rein akustisch), dass ich nicht einmal sagen könnte, worin das Neue und eigentlich Wesentliche in seinem Vortrage bestanden hat. Immerhin wurde ich lebhaft an Ihre einschlägigen Überlegungen erinnert, als Herbrand den Löwenheim-Skolemschen Satz für seine Überlegungen über die Widerspruchsfreiheit heranzog. Hierüber bemerkete er, wenn ich recht verstanden habe, folgendes: Der übliche Beweis dieses Satzes operiert mit den inhaltlichen Mitteln einer Mathematik, die als widerspruchsfrei erst erwiesen werden müsste, insbesondere wird skrupellos mit transfiniten Begriffsbildungen gearbeitet. Ob diese Überlegungen finitisierbar sind, darüber äusserte sich Herbrand nicht, dagegen führte er aus, dass er jedenfalls für seine Zwecke mit einer Formulierung der Gedanken auskomme, die nicht mit unendlichen Bereichen, sondern mit endlichen, aber immer nach Wunsch verlängbaren Individuenbereichen operiere. (immer bis zu einer 'Stufe' n, wie er sagte, wobei aber die Stufen hier nichts mit den verschiedenen Typen Russels zu tun haben dürften.) Ich bedaure es ausserordentlich, Ihnen nichts Genaueres über diese Dinge mitteilen zu können, möglicherweise findet sich aber Näheres in Herbrands Publikationen (Kaufmann *Nachlaß* 006016–006017)

5. Hempel was at that time a graduate student in Berlin. "Kurz vor Beginn des letzten Kolloquium wurden wir durch die von Herrn v. Neumann übermittelte Nachricht stark betroffen, dass es Herrn Gödl gelungen sei, die Unentscheidbarkeit der Widerspruchslosigkeit der Mathematik zu beweisen. Selbstverständlich findet diese Mitteilung unser brennendstes Interesse, und ich habe gleich gestern Herrn Carnap gebeten, Herrn Gödl um die Übersendung eines Exemplars seiner Arbeit nach Berlin anzugehen. Wenn es rechtzeitig eintrifft, soll noch vor Weihnachten eine besondere Sitzung zur Besprechung der Arbeit eingelegt werden. Es wäre sehr freundlich von Ihnen, wenn auch Sie noch Herrn Gödl einen kleinen moralischen Rippenstoss geben würden—wofern eine Ausarbeitung seines Beweises überhaupt schon vorliegt". (Kaufmann *Nachlaß* 006052)

6. It might be worth mentioning that in a letter (Kaufmann *Nachlaß* 006023) to Kaufmann dated March 4, 1931, Hempel specifies that von Neumann's lectures on Hilbert's proof theory were delivered in the Winter Semester 1930–31 (which had just ended as Hempel was writing his letter). In this same letter Hempel mentions that he is including passages from these lectures where von Neumann gives the formal structure of an essential type of non-constructive existential proof. Unfortunately, I have been unable to find this enclosure in Kaufmann's papers, but I have not had the time to do a thorough search.

7. "Herrn Gödel habe ich sogleich nach Erhalt Ihres Briefes angerufen und er hat mir versprochen, an Herrn Neumann Druckbogen seines Aufsatzes über die Kritik der formalistischen Widerspruchsfreiheit-Untersuchungen, der vermutlich Ende Jänner in den mathematischen Monatsheften erscheinen wird, zu senden. Meine Auffassung in dieser Frage kennen Sie ja. Ich bin der Meinung, daß man die Widerspruchsfreiheit der Mathematik unmittelbar mit aller anstrebbaren und erzielbaren Evidenz unmittelbar aus den Peano'schen Axiomen entnehmen kann, wenn man das fünfte Axiom in der von mir angegebenen Weise umformt, daß aber jeder Versuch, mit Hilfe von Widerspruchslosigkeitsbeweisen das Operieren im überabzählbar Unendlichen zu rechtfertigen, aussichtslos ist." (Kaufmann *Nachlaß* 006040) For Kaufmann's position on completeness see (Kaufmann, 1930, chapter 6).

8. Rudolf Carnap *Nachlaß*: RC 081-07-07. [Also in Stadler (1997, pp. 278–80)]

Schlick-Zirkel Protokoll am 15.1.1931
Ueber Widerspruchsfreiheit und Entscheidbarkeit in Axiomensystemen. (Wechselrede z. Referat Herrn Gödels)
Kaufmann fragt, wie es mit der Entscheidbarkeit der Sätze eines Teilsystems steht. *Gödel* erwidert, dass soweit sie sich beweisen lässt, dieser Beweis Mittel in Anspruch nehmen muss, die sich innerhalb des Teilsystemes selbst nicht formalisieren lassen. Das sei auch in Uebereinstimmung mit seinen Beweisführungen.
Auf eine Frage von *Hahn* ruft
Gödel noch einmal den Leitgedanken seines Beweises für die Unmöglichkeit des Widerspruchsfreiheitbeweises in Erinnerung. Adjungiert man die Widerspruchsfreiheit eines Systems dem System selbst— und diese Adjunktion lässt sich formal durchführen—dann wird in diesem erweiterten System ein im

Ursprünglichen unentscheidbarer Satz entscheidbar, folglich kann die Widerspruchsfreiheit eines systems im System selbst nicht gezeigt werden.

Auf eine Frage *Schlicks* formuliert Gödel die Vermutung Herrn v. Neumanns: Wenn es einen finiten Widerspruchsbeweis überhaupt gibt, dann lässt er sich auch formalisieren. Also involviert [involviert in the typescript] der Gödelsche Beweis die Unmöglichkeit eines Widerspruchsbeweises überhaupt.

Hahn fragt nach der Anwendung auf das Axiomensystem von Heyting:

Gödel Das System Heyting ist enger als das von Russell. Ist es o-widerspruchsfrei dann lassen sich in ihm unentscheidbare Sätze angeben.

Hahn verweist darauf, dass einer der ["den" in the typescript] Grundgedanken des Beweises "Es gibt keine sinvolle Gesammtheit des Konstruierbaren schlechthin" seit dem [seitdem in the typescript] Cantorschen Diagonalverfahren in der Mengenlehre eine entscheidende Rolle spielt.

Gödel bemerkt, dass die Anwendung eben dieses Gedankens auch fraglich erscheinen lässt, ob die Gesammtheit aller intuitionistisch einwandfreien Beweise in einem formalen System Platz findet. Das sei die schwache Stelle in der Neumannschen Argumentation.

Kaufmann fragt wie es etwa um die Widerspruchsfreiheit von Sätzen steht, die kein Begriffspaar gemeinsam haben oder etwa um die Peano-Axiome. Es gibt eine erste, es gibt eine letzte Zahl.

Gödel entgegnet, dass es bei einem Widerspruchsfreiheitbeweis auf die Begriffe als solche nicht ankomme.

Es handelt sich hier überhaupt nicht um Widerspruchsfreiheit im Sinne inhaltlichen Denkens.

Auf den Einwurf Kaufmanns, dass inhaltliche Widerspruchsfreiheitbeweise nicht ausgeschlossen seien, stellt Gödel klar:

dass es sich bei solchen "Einsichten" überhaupt nicht um Beweise im Sinne einer formalistischen Theorie handelt.

Neumann fragt an, ob es so einfache Systeme gibt, dass sie die konkrete Form des unentscheidbaren Satzes in durchsichtiger Weise angeben lässt.

Gödel erwidert, dass es dabei auf das System ankommt in dem man ihn darstellt. Er erinnert an den entscheidenden Kunstgriff seines Verfahrens. Die isomorphe Abbildung der Schlussfiguren auf Folgen f_2 von Zahlenfolgen f_1, der es überhaupt erst ermöglicht, die Beweisbarkeit intern zu formulieren. Bezeichnet dann z.B. $S(f_2)$ eine Schlussfigur $l(f_2)$ die "Länge" der zugehörigen Kette, dann schreibt sich die Beweisbarkeit von f_1

$$Bew f_1 \equiv (\exists f_2)\{S(f_2) \& f_2[l(f_2)] = f_1\}$$

Damit kann man sich begnügen oder das Symbol S weiter auflösen.

Hahn macht auf das Buch Lusin aufmerksam "Sur les ensembles analytiques"

Lusin unterscheidet bei den Existenzbeweisen für die Borelschen Mengen höherer Klasse sorgfältig, ob das Diagonalverfahren eingeht oder nicht.

Anschliessend fragt *Hahn*: ob sich auch aus der Gödelschen Beweisführung das Diagonalverfahren ausschliessen lasse.

Gödel stellt dem gegenüber fest, dass die von ihm angegebene unentscheidbare Formel wirklich konstruierbar ist. Ihr Inhalt ist finit wie der des Goldbachschen oder Fermatschen Satzes. Auf eine Bemerkung *Kaufmanns* meint *Gödel* schliesslich:

dass der Intuitionismus nach der Auffassung Brouwers durch seine Arbeit nicht berührt werde, weil er eben in keinem formalen System enthalten sein will.

9. The reader should not identify this Neumann (Robert) with Johann von Neumann (see Stadler 1997, p. 271).

10. Rudolf Carnap *Nachlaß*: RC 029-08-01.

Berlin, den 7.6.1931
Lieber Herr Carnap!
Vielen Dank für Ihren Brief und seine Anregungen.

Ich habe in meinem Beitrag auf die ausführlicheren Literatur-Angaben aus Weyls Handbuch-Artikel verwiesen, und selbst keine Literatur angegeben. Ich tat dies hauptsächlich, weil ich durch meine Auswahl keine Wertschätzungen geben wollte. Es gibt manche programmatische Veröffentlichungen Hilberts, in denen das bewiesen-Sein oder beinahe-bewiesen-Sein von Dingen behauptet wird, für die das auch nicht approximativ der Fall ist (Kontinuum-Problem, u. s. w.). Ich möchte dieselben daher weder citieren noch richtigstellen, noch verschweigen, und glaube daher, dass ich durch einen generellen Verweis auf den Handbuch-Artikel noch am besten handle.

Bezüglich Ihrer Anregung über das gemeinsame Versenden der Separata möchte ich folgendes sagen: Ich habe in Königsberg und nacher im Briefwechsel, von K. Gödel (den Sie auch kennen?) über seine Resultate erfahren. Diese (seither veröffentlichten) Sätze zeigen, dass gewisse logische und mathematische Sätze, so z. B. die Widerspruchfreiheit der Analysis, in gewissen formal-logischen Systemen unbeweisbar sind. M. E. zeigen sie aber auch, dass dies inhaltlich unmöglich ist, denn ich bin der Ansicht, dass alle inhaltlichen Schlussweisen in *einem* gewissen formalen System (das ich hier nicht näher beschreiben will, aber für welches Gödels Sätze gelten) wiederholbar sind. Ich bin daher heute der Ansicht dass:

1. / Gödel die Undurchführbarkeit von Hilberts Programm erwiesen hat
2. / kein Anlass zur Ablehnung des Intuitionismus mehr vorliegt (wenn man vom Ästhetischen absieht, der freilich im praxi auch für mich der ausschlaggebende sein wird.

Ich halte daher den Königsberger Stand der Grundlagendiskussion für überholt, da Gödels fundamentale Entdeckungen die Frage auf eine ganz veränderte Platform gebracht haben. (Ich weiss, Gödel ist in der Wertung seiner Resultate viel vorsichtiger, aber m.E. übersieht er die Verhältnisse an diesem Punkte nicht richtig.)

Ich habe mit Reichenbach mehrfach besprochen, ob es unter diesen Umständen überhaupt Sinn hat, mein Referat zu publicieren—hätte ich es 4 Wochen später zu halten gehabt, so hätte es ja wesentlich anders gelautet. Wir kamen schliesslich überein, es als Beschreibung eines gewissen, wenn auch überholten Standes der Dinge doch niederzuschreiben.*)

[* Ich möchte betonen: *Nichts* an Hilberts Absichten ist *falsch*. Wären sie durchführbar, so würde aus ihnen durchaus das von ihm Behauptete folgen. Aber sie sind eben undurchführbar, das weiss ich erst seit Sept. 1930.]

Wie Sie sich auf Grund dieser Zeilen denken können, würde ich von mir aus kein einziges Separatum meines Referats verschicken. Ich will aber eine gemeinsame Aktion nicht stören, und schliesse mich daher Ihrem Vorschlag, sowie der von Ihnen vorgeschagenen Liste an.

Mit den besten Grüssen, Ihr Johann von Neumann
Berlin W. 10, Hohenzollernstrasse 23.
⟦The text and the translation has been improved following von Neumann (2005, pp. 85–86). I would also like to thank Miklós Redéi for his improved reading of one important passage in the letter.⟧

11. Hermann Weyl, Philosophie der Mathematik und Naturwissenschaften, 1926.
12. Rudolf Carnap *Nachlaß*: RC 029-08-02.

[Prof. Dr. Rudolf Carnap
Wien XIII/5
Stauffergasse 4]
Wien, den 11. Juli 1931
Lieber Herr von Neumann!
Die Bedenken, die Sie gegen die Publikation Ihres Königsberger Vortrages hatten, verstehe ich sehr wohl. Aenliche Bedenken hatte ich nicht nur gegen die Veröffentlichung, sondern auch schon gegen die Abhaltung meines Vortrages. Ich habe mich dann aber auf den Standprunkt gestellt, dass es sich hier nicht darum handelt, eine endgültige Lösung darzustellen, oder meine persönliche Auffassung zu den verschiedenen Punkten darzulegen, sondern über die Grundgedanken zu berichten, von denen der Logizismus ausgeht und über die Möglichkeiten und Schwierigkeiten für seine Durchführung, soweit

sie sich heute überblicken lassen. Meine eigene Auffassung ist nicht die Russellsche, sondern beruht auf einer Verbindung Russelscher und Hilbertscher Gedanken: sie wird daher durch die Gödelschen Resultate auch sehr stark betroffen. Für die Veröffentlichung hat die Schilderung der Problemsituation im Königsberger Stadium sicherlich doch einen Wert, wenigstens solange die Konsequenzen, die sich aus den neuen Resultaten für die verschiedenen Auffassungen ergeben, noch nicht klar zu überblicken sind.

Die Literaturangaben bei Weyl sind doch ziemlich spärlich. Wäre es nicht vielleicht doch zweckmässig, statt dessen oder ausserdem auf Fraenkels Mengenlehre (3.A.) zu verweisen, der ja sehr reichhaltige Literaturangaben bis 1928 macht.

Mit besten Grüssen Ihr, R. C.

13. Adolf Fraenkel, *Einleitung in die Mengenlehre* (Springer, Berlin, 1928, 3rd edition).
14. Reichenbach Nachlaß: HR 015-25-02

Lieber Herr Professor Reichenbach!
Anbei schicke ich den gewünschten Bericht. Da das Manuskript beim Eintreffen Ihres Briefes schon fertig war, hat es nicht den gewünschten Titel, bitte ändern Sie den Titel derart ab, wie es in Ihre Zusammenstellung passt. Übrigens habe ich mich entschlossen, Gödel nicht zu erwähnen, da die Ansicht, dass noch eine gewisse Hoffnung für die Beweistheorie existiert, Vertreter gefunden hat: u.a. Bernays und Gödel selbst. Zwar ist m.E. diese Ansicht irrig, aber eine Diskussion dieser Frage würde aus dem vorliegenden Ramen [sic] hinausführen, ich möchte daher bei einer anderen Gelegenheit darüber sprechen
Mit den besten Grüssen
Ihr, J. v. Neumann
P.S. Entschuldigen Sie bitte, dass ich Ihnen die längst versprochenen Separata jetzt nicht schicken kann da sie in Berlin liegen. Ich schäme mich sehr und schicke dieselben sobald ich wieder in Berlin bin.

Von Neumann's talk was advertised as "Die axiomatische Begründung der Mathematik" (*Erkenntnis*, 1 (1930), p. 80) and published as "Die formalistische Grundlegung der Mathematik" (*Erkenntnis*, 2 (1931), pp. 116–21).

15. "Es wird hier viel über die Arbeit Herrn Gödels gesprochen. Es wäre mir sehr erwünscht diese Arbeit zu haben. Ich wäre Ihnen sehr dankbar, wenn Sie Herrn Gödel dazu bewegen möchten, mir diese Arbeit zu schicken.[...] Das Resultat Gödels scheint mir für den Nominalistischen Standpunkt sehr erwünscht zu sein. Zunächst ersieht man, dass die Hilbertsche Methode keineswegs wesentlich ist, denn aber, dass die einfache Typenlehre keineswegs zur reinen Logik, vielmehr aber zu einem Ontologischen System führt, was ich seit langer Zeit unterstrichen habe. Mir scheint aber das Resultat Gödels in der reinen Typenlehre also auch in der klassichen Mathematik ausgeschlossen zu sein.

"Die Einführung des Begriffs 'aller Eigenschaften von x' in die Metamathematik, wo man ja praktisch mit Eigenschaften höheren Ordnung zu tun hat ist ein offenbares Missverständnis. Führt sie zu keinen Antinomien, so ist nichtsdestoweniger klar, dass sie zu Resultaten führen muss, deren Wert problematisch ist". (Kaufmann archive 006008–0060011)
16. Kaufmann *Nachlaß* 005983–005986.

Zakopane 11. VIII. 1931
Sehr geehrter Herr Kollege,
Ich danke Ihnen bestens für die Liebenswürdigkeit mir die Arbeit Dr Gödels besorgt zu haben. Ich bitte Sie meinen besten Dank Herrn Dr Gödel mitzuteilen.

Die Methode Dr Gödels hat in mir einige Zweifel erregt, zunächst weswegen, dass sie streng symbolisch nicht durchführbar zu sein scheint. Es handelt sich um die theoretische Semantik, die ich treibe [?], und welche zu der Typenhierarchie von Ausdrücken führt. Ich habe noch Bedenken, ob es nicht möglich wäre eine Antinomie [?] System Dr Gödels zu konstruiren. Man kann ja statt von Klassenzeichen von den *sinnvollen Klassenzeichen* sprechen, was ja Dr Gödel selbst zugibt (S.174).

Dann können wir statt der Formel (1) [S.175] die Formel:

$$n \in K \equiv \overline{[R(n); n]}$$

benutzen und das Klassenzeichen $R(q)$ ins Auge fassen, so dass $[R(q); n]$ inhaltlich gedeutet besagt, dass n zu K gehört. Es ist dann leicht zu zeigen, dass der Satz $[R(q); q]$ widerspruchsvoll ist.

Ich wäre Ihnen sehr dankbar mir mitzuteilen, wie es eigentlich mit dieser Sache steht.[...] Ich bin jetzt mit der Vorbereitung einer Streng Symbolischen Systems der theoretischen Semantik und Metamathematik beschäftigt in welchem das Klassensymbol $f\{x\}$ aud das Substitutionsschema (abcd) zurückgeführt wird.

Mit besten Grüßen, ergebenst Ihr L. Chwistek

17. Kaufmann *Nachlaß* 005979–005980.

Lwow 3.10.1931
Sehr geehrter Herr Kollege,
In meinem letzten Briefe habe ich einige Bedenken über die Arbeit Dr Gödels ausgesprochen, die nach einem tieferen Eindringen in das Problem gänglich verschwunden sind. Ich habe zunächst gedacht, dass es sich um eine schweigende Einführung der nichtprädikativen Urteilfunktionen handelt, welche die Benutzung des streng symbolischen Verfahrens unmöglich macht, und sogar habe ich eine Antinomie befürchtet. Nun sehe ich dass es davon keine Rede sein kann. Die Methode ist wirklich wunderbar und der reinen Typenlehre gänglich angepasst. Ich bitte sie diese Berichtigung Herrn Prof Carnap mitzuteilen. Den Aufsatz für die "Erkenntnis" werde ich in nächster Zeit bearbeiten. Ich hoffe, dass der Aufsatz "Neue Grundlagen" 2te Mitteilung bald erscheinen wird.

Mit besten Grüssen und ergebenst, Ihr L. Chwistek

18. In a letter to Chwistek dated February 16, 1932, (Kaufmann Nachlaß 005946) Kaufmann writes that he has informed Gödel, as per Chwistek's wishes, that Chwistek had retracted his objections. Kaufmann had in fact informed Carnap and Gödel in letters dated November 12, 1930 (Carnap Nachlaß : RC 028-24-08) and February 16, 1932 (Kaufmann Nachlaß 005947), respectively.
19. *Erkenntnis*, 3, 1932/3, pp. 367–88.
20. "Neue Grundlagen... Zweite Mitteilung", *Mathematische Zeitschrift* 34 (1932).

Chapter 8

1. See also Kreisel (1980), Köhler (2002), and Feferman's "Gödel's life and work" in Gödel (1986, pp. 1–36).
2. Extracts from quite a few of Gödel's letters to his mother are published in Köhler (2002, Vol. 1, pp. 185–207) in a section entitled "Gödels Briefe an seine Mutter".
3. Perhaps the most extreme case here is Wilson Follett's outburst concerning the publication of "Gödel's Proof" by Nagel and Newman. See Vol. IV, p. 419.
4. More precisely, Zermelo shared with Gödel the criticism of the "finitistisches Vorurteil" mathematically but he did not appreciate the metamathematical point of view. The Gödel-Zermelo correspondence had already been studied by Dawson (1985) and Grattan-Guinness (1979).
5. This letter to Balas was already discussed in Feferman (1984), Wang (1981, p. 654), and Wang (1987, pp. 84–85).
6. It could be argued that Gödel did at times fall prey to what he called the prejudices of the time. See in particular what Gödel says about Platonism in Gödel (*1933) and Feferman's introduction to that lecture in CW, Vol. III.

7. See also Gödel (1934), in CW, Vol. I, pp. 262–63, where he already states the undefinability of truth in the language and explains the relation of his proof of incompleteness to the Liar paradox. The relations to Tarski's work were footnoted later, for its reproduction in Davis (1965).
8. The correspondence between Gödel and Herbrand had already been discussed in Dawson (1993) and Sieg (1994).
9. Interesting passages from the Gödel-von Neumann correspondence were already used in Köhler (1991).
10. Sieg points out that Gödel's change of mind can be dated to 1933; see CW, Vol. V, p. 8.
11. Von Neumann repeats his position in a letter to Carnap dated June 6, 1931:
    I am today of the opinion that
    1. Gödel has shown the unrealizability of Hilbert's program
    2. There is no more reason to reject intuitionism (if one disregards the aesthetic issue, which in practice will also for me be the decisive factor). Therefore I consider the state of the foundational discussion in Königsberg to be outdated, for Gödel's fundamental discoveries have brought the question to a completely different level. (I know that Gödel is much more careful in the evaluation of his results, but in my opinion on this point he does not see the connections correctly).

    (For the full original text see Mancosu 1999b).
12. Sieg (CW, Vol. V, p. 332) points out that what Gödel found problematic here was "the claim that the totality of all intuitionistically correct proofs is contained in one formal system" and refers to the minutes kept by Rose Rand of the discussion following Gödel's presentation of his results to the Schlick Circle (see also Mancosu 1999b for the full text of the discussion).
13. This is not the place for providing all the evidence required to make my case. I simply want to point out that more work needs to be done to settle the issue.

Chapter 9

1. For biographical information see Chevalley and Weil (1957), Newman (1957), Reid (1970), and Dieudonné (1976). This Chapter has greatly benefited from the works Feferman (1988a) and van Dalen (1995). Useful commentary on Weyl can also be found in Largeault's extended introduction to H. Weyl, *Le Continu et Autres Écrits*, (Paris:Urin), 1994.
2. I will use the term "impredicative definition" throughout. At first such definitions were referred to as "non-predicative". (Thiel 1972, note 31, p. 139) mentions as one of the earliest occurrences of the term "impredicative" the work by Julius König, *Neue Grundlagen der Logik, Arithmetik und Mengenlehre*, Leipzig, 1914.
3. "With the help of a tradition bound up with that complex of notions which even today enjoys absolute primacy in mathematics and which is connected above all with the names of Dedekind and Cantor, I have discovered, traversed, and here set forth my own way out of this circle." (Weyl 1994, p. 47)
4. Of course, I used the above notation only for ease of exposition. Frege does not talk about sets but about extensions of concepts.
5. Zermelo (1909) and Poincaré's objections to Zermelo in Poincaré (1909) are also of the greatest interest in this connection. Poincaré showed that the use of impredicative definitions in Cauchy's proof of the fundamental theorem of algebra was only apparent by giving a proof that avoided recourse to impredicative definitions.

6. "Die Definitionsprinzipien gewinnen in der Mengenlehre eine besondere Bedeutung, wie ich glaube, dadurch, daß der Begriff "definierbar" in die Axiome dieser Disziplin selbst hineinspielt. Eines der von Zermelo aufgestellten axiome [Axiom III] bahauptet nämlich, daß diejenigen Elemente x einer vorgelegten Menge M, welche irgend eine definite Aussage erfüllen, stets wieder eine Menge bilden, und dabei ist nach Zermelos Erklärung eine definite Aussage eine solche, deren Zutreffen oder Nichtzutreffen eindeutig und ohne Willkür auf Grund der zwischen den Dingen der Mengenlehre bestehenden Grundbeziehungen $\in$ entschieden werden kann. Hier ist meinem Empfinden nach eine noch größere Präzision vonnöten, insofern mir die Redeweise von der 'eindeutigen und ohne Willkür zu treffenden Entscheidung' etwas zu vage erscheint." (Weyl 1910a, p. 304)
7. Weyl's conception of the axiomatic method is, however, more in line with the classical conception than with Hilbert's. Weyl emphasizes that the axioms must be true judgments.
8. The reader should be careful not to be misled by Weyl's use of "iteration" in a narrow and generalized form, treated in §5 and §7, respectively. Weyl refers to the former as the principle of "definition by complete induction" (Weyl 1994, p. 26).
9. I am very grateful to Michael Hallett for having drawn my attention to this manuscript.
10. "Bei diesem Standpunkt gibt der Begriff der reellen Zahl nur einen Rahmen, in welchen all das eingefügt wird, was man jeweils durch die Methode der Schnitte (bzw. der Zahlfolgen) erhält. Aber es wird nicht ein bestimmter Bereich abgesteckt; der Begriff der reellen Zahl wird nicht bestimmt umgrenzt." (Hilbert, n.d., Cod. Ms. 685, p. 4)
11. "Nun stellt man sich aber gar nicht immer auf diesen Standpunkt, vielmehr geht man zumeist von der Auffassung aus, dass der Begriff einer Einteilung der rationalen Zahlen (bzw. einer Folge von rationalen Zahlen) ein seinem Inhalt nach ganz scharfer und seinem Umfang nach genau begrenzter ist." (Hilbert, n.d., Cod. Ms. 685, p. 5)
12. On Kaufmann, see Kaufmann (1930) and Gillies (1980).
13. For details on the personal relationship between Brouwer and Weyl, see van Dalen (1995) and van Stigt (1990, pp. 75–77 and 80–82).
14. For example, Skolem in 1929 wrote: "Die Arbeiten Brouwers sind schwer lesbar, und der Intuitionismus ist deshalb mehr durch einige Arbeiten Weyls bekannt geworden." (Skolem 1929a, p. 217)
15. On Weyl's conception of the continuum, see also Breger (1986) and (1988).
16. "In seiner ersten Mitteilung zur "Neubegründung der Mathematik" hat sich Hilbert in heftiger Polemik gegen die von Brouwer und mir vertretene Auffassung gewendet. Mir scheint, selbst von seinem Standpunkt mit geringem Recht; denn soviel ich sehe, stimmen wir in dem entscheidensten Punkte miteinander überein. Auch für Hilbert reicht die Kraft des inhaltlichen Denkens nicht weiter als für Brouwer; es ist für ihn ganz selbstverständlich, daß die 'transfiniten' Schlußweisen der Mathematik nicht trägt, daß es keine Rechtfertigung für alle die transfiniten Aussagen der Mathematik als *inhaltlichen Wahrheiten* gibt." (Weyl 1924b, p. 448, 1925, p. 136; 1928a, p. 483) On the relationship between Hilbert and Weyl see also Jahnke (1990).
17. See also the discussion on Mach's phenomenalism in Chapter 2.
18. Commenting on *The Continuum*, Husserl wrote to Weyl: "Endlich ein Mathematiker, der Verständnis zeigt für die Notwendigkeit phänomenologischer Betrachtungsweisen in allen Fragen der Klärung der Grundbegriffe, und der sich also zurückfindet auf den Urboden logisch-mathematischer Intuition, auf dem allein eine wirklich quellenmässige Begründung der Mathematik und eine Einsicht in den Sinn mathematischer Leistung möglich ist!" (van Dalen 1984, p. 3)

19. See Lorenzen (1965), Feferman (1964, 1988a), and Heinzmann (1985) for a start and further references.

Chapter 10

1. ⟦Since the original letters are not reprinted here, I will give page references to the original German whenever quoting from those letters.⟧
2. For biographical details on Weyl see Chevalley and Weil (1957), Newman (1957), and Dieudonné (1976).
3. See Mancosu (1998b) for a more extensive introduction to Weyl's changing approaches to the foundations between 1910 and 1930.
4. Commenting on *Das Kontinuum* Husserl wrote to Weyl: 'Endlich ein Mathematiker, der Verständnis zeigt für die Notwendigkeit phänomenologischer Betrachtungsweisen in allen Fragen der Klärung der Grundbegriffe, und der sich also zurückfindet auf den Urboden logisch-mathematischer Intuition, auf dem allein eine wirklich quellenmässige Begründung der Mathematik und eine Einsicht in den Sinn mathematischer Leistung möglich ist!' Husserl to Weyl, April 10, 1918 (van Dalen 1984, p. 3; Husserl 1994a, p. 287)
5. Weyl to Husserl, March 26–27, 1921 (Husserl 1994a, p. 290) In the same letter Helene Weyl also showed her low opinion of Schlick's work:

    Mein Mann hat den sachlichen Teil auf sich genommen, aber ich möchte Ihnen auch gern noch selbst sagen, wie dankbar, und ehrfürchtig wir Ihre Gabe angenommen haben: ich habe mich aus der Schlick'schen Erkenntnislehre, die ich gezwungener Weise sehr eingehend las und die mich mit ihrer abscheulich unsauber gedachten Erkenntnis—und Wirklichkeitstheorie ganz unglücklich machte, wie erlöst in die harte und helle Atmosphäre der Logischen Untersuchungen geflüchtet. Helene Weyl to Husserl, March 26–27, 1921 (Husserl 1994a, p. 292).

    Translations are ours unless otherwise stated. We only give the German original when citing in note or when the original text is a manuscript. When a standard translation is available we also give the page number of the translation.
6. "Ich habe seinerzeit 12 Semester Mathematik (bei Hölder und Herglotz in Leipzig) studiert und dann in diesem Fach promoviert, mit einer axiomatisch-geometrischen Arbeit (1914). Nach 4 Jahren Kriegsdienst habe ich mich dann der Philosophie zugewandt und mich 1922 bei Husserl hier habilitiert. Auch noch in den letzten Jahren habe ich mit Mathematikern brieflich und mündlich viel verkehrt, auch mit Herren aus Ihrem engeren Kreise (W. Ackermann und J. v. Neumann), ferner mit H. Weyl, und hier natürlich oft mit E. Zermelo, in dessen Seminar ich sogar über Probleme aus der Theorie der transfiniten Ordnungszahlen mehrfach Vorträge gehalten habe." Becker to Hilbert, October 4, 1930 (Cod Ms Hilbert 457, sheet 6). See also note 9.
7. Mr Richard Wilhelm Ackermann, son of Wilhelm Ackermann, has kindly informed us that there are no letters from Becker extant in his father's *Nachlaß*.
8. The correspondence Oskar Becker-Arend Heyting is preserved at the Archief A. Heyting in the Rijksarchief in Haarlem. It contains 6 items. Five letters from Becker to Heyting and 1 (draft) letter from Heyting to Becker:

    i. Becker to Heyting, January 5, 1931; 4 pages
    ii. Becker to Heyting, February 16, 1931; 6 pages
    iii. Becker to Heyting, September 19, 1933; 4 pages

iv. Heyting to Becker, September 23, 1933; 6 pages
v. Becker to Heyting, September, 1933; 6 pages
vi. Becker to Heyting, September, 1934; 2 pages

We would like to thank Professor A. Troelstra for having kindly provided us with a copy of the correspondence. Part of the correspondence is published in Troelstra (1990) [[see now van Atten (2005a)]]. In addition to the latter, see Mancosu and van Stigt (1998) and Sundholm (1994) for Becker's influence on the interpretation of intuitionistic logic.

9. The correspondence David Hilbert-Oskar Becker contains one letter from Becker to Hilbert and a partial draft of a reply by Hilbert. It is preserved in the Universitätsbibliothek Göttingen under the call number Cod Ms Hilbert 457. The letter from Becker to Hilbert is dated October 4, 1930 (6 pages). On Hilbert and Becker see Roetti (1996) and Chapter 2.

10. The correspondence Felix Kaufmann-Oskar Becker is to be found in a folder of the Felix Kaufmann *Nachlaß* (Konstanz) entitled 'Mathematische Korrespondenz'. It contains 6 letters, 5 from Becker to Kaufmann and one from Kaufmann to Becker:

   i. Becker to Kaufmann, December 31, 1928; 2 pages; sheets 006198-006199.
   ii. Becker to Kaufmann, March 25, 1929; 2 pages; sheets 006191-006192.
   iii. Becker to Kaufmann, July 12, 1929; 2 pages; sheets 006178-006179.
   iv. Kaufmann to Becker, July 22, 1929; 2 pages; sheets 006174-006175.
   v. Becker to Kaufmann, July 30, 1929; 1 page; sheet 006173.
   vi. Becker to Kaufmann, March 8, 1930; 6 pages; sheets 006127-006132.

We would like to thank Dr Martin Endress for having helped us in obtaining a copy of the Kaufmann's folders containing the above correspondence.

11. We have found no trace of the J. von Neumann-Oskar Becker correspondence. The von Neumann archive in Washington contains very little material for the period before von Neumann's move to the States in 1930.

12. The Hans Reichenbach-Oskar Becker correspondence, spanning the period 1928–31, is being edited by A. Kamlah for the complete works of Reichenbach. It contains 11 items, 7 letters from Becker to Reichenbach and 4 from Reichenbach to Becker. The correspondence is kept in the Reichenbach *Nachlaß* in the Special Collections of the University of Pittsburgh Library (and available in microfilm at the University of Konstanz). We would like to thank Dr Brigitte Uhlemann (Universität Konstanz) for her help in reproducing the Becker-Reichenbach correspondence. On the debate between Becker and Reichenbach on the foundations of geometry see Volkert (1994).

13. The correspondence Dietrich Mahnke-Oskar Becker [[now published in Becker (2005)]] is preserved at the Universitätsbibliothek Marburg. It contains 11 letters from Becker to Mahnke and 2 letters from Mahnke to Becker, (*Nachlaß* D. Mahnke, Ms. 862).

   i. Becker to Mahnke, August 22, 1926; 8 pages.
   ii. Becker to Mahnke, September 16, 1926; 6 pages.
   iii. Becker to Mahnke, July 20, 1927; 2 pages.
   iv. Becker to Mahnke, August 21, 1927; 4 pages.
   v. Mahnke to Becker, September 8, 1927; 6 pages.
   vi. Becker to Mahnke, September, 1927; 14 pages.
   vii. Mahnke to Becker, October 12, 1927; 4 pages.

viii. Becker to Mahnke, October 20, 1927; 4 pages.
ix. Becker to Mahnke, December 12, 1927; 4 pages.
x. Becker to Mahnke, April 17, 1928; 4 pages.
xi. Becker to Mahnke, May 8, 1933; 2 pages.
xii. Becker to Mahnke, May 20, 1933; 2 pages.
xiii. Becker to Mahnke, July 7, 1933; 2 pages.

We would like to thank Dr U. Bredehorn for his help in providing us with the above correspondence.

14. Oskar Becker mentions his correspondence with Abraham Fraenkel in a letter to Weyl. Much of the Fraenkel's correspondence is in the possession of his son. Unfortunately, all our attempts to get in contact have met with no success.
15. The Weyl *Nachlaß* is kept at the Wissenschaftshistorische Sammlungen at the ETH in Zurich. We would like to thank Frau Dr Yvonne Vögeli for her generous help in reproducing the correspondence and granting permission to publish on behalf of the ETH. See section 10.4 for details of the correspondence Becker-Weyl.
16. As Oskar Becker was a colleague of Ernst Zermelo in Freiburg they did not need to correspond. There is however one letter from Becker to Zermelo in the Zermelo *Nachlaß* in Freiburg. The letter, seven pages long, is written from Freiburg and it is dated December 31, 1930. We would like to thank Dr Winfried Hagenmaier, Abteilung Handschriften, Universitätsbibliothek Freiburg im Breisgau, for his help in sending us a copy of Becker's letter.
17. Husserl to Weyl, April 9, 1922 (Husserl 1994a, pp. 293–94)
18. Husserl wrote to Weyl: 'Bitte senden Sie uns doch nach Freiburg alle Ihre phänom-<enologisch> irgend relevanten Arbeiten und seien Sie, ohne Antwort abzuwarten, überzeugt, daß Sie bei uns wirken!' Husserl to Weyl, April 9, 1922 (Husserl 1994a, pp. 294–95)
19. In the introductory part of Becker (1923) we read: 'Unsere Dankbarkeit bei der Abfassung dieser Arbeit gebührt also in erster Linie Edmund Husserl dessen Forschung das Fundament ist, auf dem sie sich erhebt, und in zweiten Linie Hermann Weyl, dessen Darstellung der mathematisch-physikalischen Probleme uns ein für die phänomenologische Analyse um so geeigneteres Material bot, als er selbst der Phänomenologie nahe steht.' (Becker 1923, p. 388)
20. 'Es ist meine Überzeugung, dass gerade die von Ihnen vertretene Auffassung sowohl des Kontinuumsproblems als auch der Struktur von Raum und Zeit, es ermöglicht, eine geschlossene phänomenologische Begründung der Geometrie (im Sinne von "Weltgeometrie") zu geben und ich habe in meiner Arbeit die Umrisse einer solchen Begründung zu skizzieren versucht.' Becker to Weyl, April 12, 1923.
21. On Mahnke see Mahnke (1977) and Wohltmann (1957). ⟦See also Chapter 12.⟧
22. Weyl (1925) is published in *Symposion*. Benary was also the publisher of Carnap's *Der Logische Aufbau der Welt*. There is no correspondence between Benary and Weyl in the Weyl *Nachlaß*.
23. 'Ich sprach neulich mit Weyl und dieser meinte das auch, er sagte sogar, es wäre die eigentlich *neue* Errungenschaft der *modernen* Mathematik gegenüber der antiken, für jedes Erscheinungsgebiet das spezifisch angemessene (ganz "autarkische", d.h. ganz losgelöst, durch Axiome definierte) Zahlen System (den Körper oder die Gruppe *etc.*) finden zu können, während man früher nur das einzige System der reellen Zahlen für alle Probleme zugrunde gelegt habe' (Becker to Mahnke, August 21, 1927, Mahnke *Nachlaß*, Ms 862). Friedman (1995, 1999) proposes the following conjecture on the relationship between Becker

and Weyl: 'Now this phenomenological analysis of Becker's was of course published after Weyl's own group-theoretic work on the "space problem." Nevertheless, it appears likely that Weyl and Becker communicated about these questions much earlier, through their common participation in the phenomenological circle around Husserl in Göttingen.' (Friedman 1999, p. 54) In light of our reconstruction it should be evident that the conjecture is not warranted. In particular, Becker was never part of the Göttingen circle having associated himself with Husserl in 1918, that is after the latter's move to Freiburg in 1916. Concerning Friedman's more substantial claim that 'Weyl conceives his group-theoretic analysis of the space problem as a generalization and refinement of Becker's' (Friedman 1999, p. 55) see Chapter 11.

24. 'Dabei ist von Wichtigkeit, dass mein philosophischer Ausgangspunkt (in Übereinstimmung mit der Freiburger Richtung der Phänomenologie im Gegensatz zu der Münchner-Kölner Richtung) das Prinzip des transzendentalen Idealismus ist, woraus sich das grundlegende Problem der phänomenologischen Konstitution der Natur ergibt.' Becker to Weyl, April 12, 1923.

25. The reader is referred to the informative and clear exposition of Spiegelberg for more details. Talking about the group of students faithful to *Logische Untersuchungen* Spiegelberg says: 'But to this lively group and to its varying membership and fringe, phenomenology meant something different from what it meant to Husserl at this stage, *i.e.*, not the turn toward subjectivity as the basic phenomenological stratum, but towards the 'Sachen', understood in the sense of the total range of phenomena, and mostly towards the objective, not the subjective ones [...] phenomenology meant to the circle primarily a universal philosophy of essences [*Wesensphänomenologie*], not merely a study of the "essence of consciousness".' (Spiegelberg and Schuhmann 1982, p. 168).

26. Becker constantly identifies Brouwer's and Weyl's theories of the continuum. However, the two are not identical. See note 36 for further references.

27. Weyl (1918a, p. 72); Weyl (1994, p. 94). The passage is also found in the fourth edition of *Raum, Zeit, Materie* 1921b. See Chapter 11 for discussion.

28. See the instructive criticisms in Geiger (1928) on this issue.

29. See also Becker (1927, p. 808).

30. Weyl (1925, pp. 23–24); Mancosu (1998a, p. 136).

31. For an overview of the first three phases see Mancosu (1998b), to which the reader is referred for further references.

32. For an account of *Das Kontinuum* see Feferman (1988a, 1988b), and Mancosu (1998a).

33. In 1923 Weyl wrote 'Wir [Brouwer und Weyl] arbeiten zusammen an einer Revolutionierung und Neubegründung der Analysis' (Frei and Stammbach 1992, p. 49). For details on the personal relationship between Brouwer and Weyl see van Dalen (1995), van Stigt (1990, pp. 75–77 and 80–82), and van Dalen (1999, pp. 317–25). On the relationship between Brouwer and Weyl see also Majer (1988) and van Atten (1999). For the immediate reception of intuitionism see Hesseling (2003).

34. 'Daneben arbeite ich wieder über die Grundlagen der Analysis. Eine Zusammenkunft mit Brouwer im Sommer hat der Sache neuen Impuls gegeben; ich modifiziere meinen Standpunkt wesentlich. Brouwer ist ein Mordskerl und ein wunderbar intuitiver Mensch. Ich war durch ein paar Stunden Zusammensein mit ihm ganz beglückt. Wenn sie den an Heckes

Stelle nach Göttingen bekommen, beneide ich Sie.' (Weyl to Bernays, Zurich, January 9, 1920; *Nachlaß* Weyl, ETH, Hs 91:10)

35. For example, Skolem in 1929 wrote: 'Die Arbeiten Brouwers sind schwer lesbar, und der Intuitionismus ist deshalb mehr durch einige Arbeiten Weyls bekannt geworden' (Skolem 1929a, p. 217).

36. Brouwer himself was quite aware of that. In a short handwritten note containing comments on Weyl (1921c) (preserved in the Brouwer Archief) we read:

> *Limiting Arithmetic and Analysis to general statements about numbers and free-becoming sequences.* This restriction of mathematics to mathematical entities and species of the lowest order is totally unjustified. This clearly refers to p. 70, where he dismisses my theory of species as meaningless, and it shows that in the end Weyl only half understands what intuitionism is about. (Quoted from Mancosu 1998a, p. 122; p. 70 corresponds to p. 109 in the translation of Weyl 1921c)

Dirk van Dalen and Mark van Atten have recently brought to our attention a manuscript 'Alter und neuer Intuitionismus' in which Brouwer is quite negative in his appreciation of Weyl's contribution to intuitionism:

> Als schaffender Neo-Intuitionist kommt neben Brouwer ausschliesslich dessen Schüler Heyting in Frage, der die geometrische Axiomatik auf der Grundlage des neuen Intuitionismus neu gestaltet hat. Weyl aber kann bis heute in diesem Zusammenhange kaum erwähnt werden: zwar hat er als erster, halb-verstehender Nachfolger Brouwers einige Popularisierungsversuche publiziert, aber die in denselben enthaltenen einigen Zutaten sind sämtlich falsch und führen das Publikum irre; nur seine Richtigstellung des Fundamentalsatzes der Algebra ist stichhaltig, aber auch diese wirkt infolge als notwendig vorgestellter, aber in Wirklichkeit überflüssiger Komplikationen verwirrend. (Brouwer Archief, Utrecht; transcription by M. van Atten)

The fragment is hard to date but from internal evidence it must have been written after 1926, probably at the turn of the decade.

37. See van Dalen (1995), Majer (1988), and Mancosu (1998b) for details.

38. On the issue of the constructivization of the fundamental theorem of algebra (Weyl 1924b) see also the correspondence between Weyl and Skolem preserved in Weyl's *Nachlaß* at the ETH (Hs 91: 750).

39. Weyl (1925, pp. 23–24); Mancosu (1998a, p. 136); cf. Weyl (1927, p. 44); Weyl (1949, p. 54).

40. 'Für mich hat der Kampf Brouwer-Hilbert freilich eine ganz prinzipielle Bedeutung. Ich bin zu sehr Mathematiker, um mich dem Eindruck verschliessen zu können, dass praktisch die Brouwersche Mathematik nicht das ist, was wir brauchen, und sich nicht durchsetzen wird. Ich bin in dieser Hinsicht geschichtsgläubig und ein frommes Weltkind, dass ich mit Hilbert finde, der Erfolg ist das Entscheidende. *Siegt Hilbert über Brouwer ob, so bedeutet das für mich zugleich, dass damit die Phänomenologie als philosophische Grundwissenschaft gerichtet ist.* ... Für mich beginnt sich als erkenntnistheoretisches Hauptproblem herauszuschälen *das theoretische Bauen*, das sich letztlich an keine anschaulichen Fundamente als absoluten unüberschreitbaren Rahmen kehrt. Ich bin skeptisch geworden gegen den Versuch, die metaphysischen Stütze, auf die es sich bei Leibniz gründet, das anschaulich Gegebene, durch Phänomenologie zu ersetzen' (Weyl as quoted by Becker in a letter to Mahnke, August 22, 1926; Mahnke *Nachlaß*, Ms 862).

41. 'Es handelt sich also hier um eine gewisse Krise der phänomenologischen Methode selbst. Wenn Weyl damit Recht hat, dass der Intuitionismus Brouwers die theoretische Physik nicht

tragen kann, dann muss auch die Phänomenologie im 'klassischen' Sinne Husserls als unfähig angesehen werden, die moderne Form der Naturerkenntnis zu sichern—und bis ins Letzte verständlich zu machen' (Becker to Mahnke, August 22, 1926; Mahnke *Nachlaß*, Ms 862).

42. See Frank (2000) and Billinge (2000) for overviews.
43. See Tieszen (2000). For a recent study relating Husserl and Brouwer see van Atten (1999).
44. This is an assumption that Becker seems to share; *cf.* Lohmar (1989, p. 195).
45. Although not directly relevant to our claim here, it is interesting to point out that Weyl claimed that intuitionism did not go beyond Hilbertian 'finitistic' reasoning: 'Vielleicht nimmt Hilbert in dieser Frage [the extent of 'contentual thought'] einen noch radikaleren Standpunkt ein als Brouwer; ich glaube aber, daß solche durchgeführten Beweise wie der eben angegebene für den funktionentheoretischen Fundamentalsatz der Algebra erkennen lassen, daß die Grenzen des finiten Denkens in der Brouwerschen Analysis—wenigstens in der von mir gegebenen Interpretation—nicht überschritten werden' (Weyl 1924b, p. 449). We should add that the conflation of intuitionism and finitism was rather typical in this period (see Mancosu 1998c, pp. 167–68).
46. It is important to point out that the notion of belief used by Weyl in this context must be understood within the background of Weyl's appeal to the metaphysical notions of 'potentiality' or 'possibility', which, according to him, play an essential role in 'symbolic construction' (*cf.* Weyl 1940, p. 708).

Weyl also claims that theoretical physics provides a type of knowledge that cannot be completely realized in intuition. Here the impossibility of a phenomenological grounding of physics can be obtained in two ways. The first is to point out that physics must avail itself of classical mathematics, which however—as we have seen—cannot be phenomenologically grounded (this is the inference Becker draws, in the letter to Mahnke, from Weyl's claim that Brouwer's intuitionism cannot support theoretical physics). Alternatively, there are physical statements and laws for whose meaning no isolable sensory (or categorial) intuition can be provided (examples include relativity theory, etc.). Weyl says:

In theoretical physics we have before us the great example of a [kind] of knowledge of completely different character from the common intuitive [intuitive] or phenomenal knowledge that expresses purely what is given in intuition.(*) While in this case every judgment has its own sense that is completely realizable within intuition, this is by no means the case for statements of theoretical physics. (Weyl 1924b, p. 451)

The note (*) clarifies that ' "Intuition" [*Anschauung*] will not be restricted here to sensible intuition, but rather denotes any presenting act [*jeden gebenden Akt*]'. If for Weyl, as is likely, the argument for the insufficiency of phenomenology in the area of physics rests on such a consideration then one should raise to it the same objections we have raised to the cogency of the argument in the case of mathematics. In other words, one would need to show that Husserlian phenomenology does not have the resources to account for physics. However, no such argument is found in Weyl. Weyl seems to rest his case with general observations such as the one reported by Becker:

Your thought that in the course of its (recent) history physics is moving more and more away from the intuitively-graspable and that the time has come, where the last rope is cut that keeps it at the 'phenomenal shore', the recognition of formal ontology, is certainly momentous. (Original German in Mancosu and Ryckman 2002, pp. 191–92)

In his first letter to Mahnke, Becker seems to agree with the Weylean analysis of the situation in physics:

Man könnte diese Bezeichnung 'amphibium inter ens et non-ens' [the reference is to Leibniz's characterization of complex numbers] viel besser auf Hilberts Transfinites (seine 'idealen' oder wenn man will 'imaginären' Aussagen) anwenden, während man von den imaginären Zahlen sagen müsste, dass Gauss sie aus Amphibien zu Landtieren des Seins gemacht habe, sodass sie keine 'schwankende Haltung' mehr haben (nach Gauss eigenem Ausdruck). Wir stehen heute den Hilbert'schen transfiniten Aussagen genau so gegenüber wie die Mathematiker vor Gauss dem Imaginären: man gelangt von verständlichen ('reellen') Aussagen über unverständliche transfinite ('imaginäre') zu verständlichen Resultaten. *Weyl* hat dies neuerdings so interpretiert, dass die moderne theoretische Physik auch nur als ganzes *System* verifizierbar sei und ihre einzelnen Aussagen u. Formeln keinen verständlichen Sinn hätten. (In der Tat haben ja, eben in der relativitäts-theoretischen Physik, die weitgehend mit Tensoren arbeitet, die Formeln der Zwischenrechnung, deren Seiten ⟨?⟩ nicht 'invariant' sind gegen affine Transformationen, die also keine Tensoren darstellen, *keinen* physikalisch interpretierbaren Sinn.). Daher mache es nichts aus, ob manche einzelne Aussagen bloss 'ideal' seien, wenn nur am Schluss die Verbindung mit der 'Erfahrung', die übrigens sehr locker ist, hergestellt werden könne. (Becker to Mahnke, August 22, 1926; Mahnke *Nachlaß*, Ms 862).

47. For a first introduction to Husserl's philosophy of mathematics, which also emphasizes the constructivist-platonist debate in the interpretation of Husserl's philosophy, and further references see Tieszen (1989, 1995).

48. The literature on Husserl's notions of categorial intuition and formal ontology is too vast to be given here. For a first orientation see Smith and Smith (1995).

49. 'Hier ist mir nun ein wesentlicher Punkt unklar: Man kann (rein sachlich, von unserer heutigen Problemlage aus) unterscheiden zwischen den ohne Weiters (einigermassen) anschaulichen Figuren der Euklidischen Geometrie, in ihrer Anwendung auf den Anschauungsraum (die *materiale*, wiewohl unselbständige (abstrakte) Region Raum nach Husserl) und den nicht-euklidischen Gebilden, die (wenn man vorsichtig sein will) nur im 'Zahlenraum', also arithmetisch-analytisch gegeben werden können, also als Wesenheiten der 'formalen Region' im Sinne *Husserls*, auch als solche aber noch der *kategorialen Anschauung* zugänglich. Man muss aber *weiterhin* unterscheiden (und dieser zweite Unterschied darf mit dem ersten nicht verwechselt werden!!) zwischen den der *Brouwer*'schen 'intuitiven' Mathematik zugänglich[en] arithmetisch-analytischen Gegenständlichkeiten, zu denen auch unendliche Mengen in Form von *Prozessen* gehören und sogar die ersten Zahlen der Cantor'schen II. Zahlklasse, zum wenigsten bis zur ersten 'Epsilon-Zahl' ($\varepsilon = \omega^{\omega^{\omega^{\cdot^{\cdot^{\cdot}}}}}$),—und den *Hilbert*'schen *transfiniten* Gebilden, über die nicht mehr eigentliche Aussagen, die wahr oder falsch sein können, möglich sind, sondern nur noch uneigentliche, 'ideale Aussagen', die den eigentlichen analog wie die komplexen Zahlen den rellen [sic], die idealen Punkte der projektiven Geometrie u. die idealen Zahlen in der Theorie der algebraischen Zahlen den gewöhnlichen adjungiert werden. Diese im *Hilbert*'schen Sinn 'transfiniten' Gebilde sind von Husserls 'kategorialer Anschauung' nicht mehr fassbar. Sie unterscheiden sich dadurch fundamental von den nicht-euklidischen Gebilden (soweit diese, was aber auch bei 'euklidischen' eintreten kann, nicht 'transfinit' (Hilbert) sind.). Denn diese sind natürlich kategorial anschaubar. Man muss also (mindestens) 3 Arten von mathematischen 'Gebilden' unterscheiden: 1. der (idealisierten) sinnlichen Anschauung zugängliche Gebilde. 2. der kategorialen Anschauung zugängliche Gebilde. 3. keiner Anschauung zugängliche, aber widerspruchsfreie Gebilde. (Die Gebilde 3. Art gehören nicht mehr in die Husserl'sche 'formale Ontologie'; sondern in die Sphäre der 'widerspruchsfreien Gesetztheiten', in der es keine 'Wahrheit', sondern nur 'Konsequenz' gibt.)' (Becker to Mahnke, August 22, 1926; Mahnke *Nachlaß*, Ms 862).

50. More precisely, Becker (1927, p. 483), claims that Hilbert's transfinite sets with all their properties are:

    1. consistently thinkable, that is 'existent objectivities' in the sense of logic of meaning [*Bedeutungslogik*];
    2. not 'countersensical' [*widersinnig*] also in the sense of formal ontology;
    3. in principle not accessible to 'categorial intuition' and therefore not positively existing.

51. Weyl (1925, pp. 30–31); Mancosu (1998a, p. 140).
52. Thus, empirical meaning, as furnished in sensory intuition, can be attributed to physical theories only as a whole. One should point out that Weyl's route to semantical holism regarding theories appears to stem from his growing recognition of the insufficiency of intuition to provide meaning for isolated statements of theory. Thus, Weyl's argument for holism has been arrived at independently of that of Duhem (whom Weyl does cite in Weyl (1927), at the end of §21 on theory formation).
53. Weyl (1927, p. 49); Weyl (1949, pp. 61–62).
54. Weyl (1927, p. 53); Weyl (1949, pp. 65–66). Cf. Weyl (1925).
55. Fichte's influence on Weyl is emphasized in Scholz (1994b).
56. 'Ich sprach im August mit *Weyl*, der mir sagte, er sei mit meiner *Hilbert*-Kritik heute noch durchaus einverstanden. Hilbert selbst habe er öfters nach dem Sinn seiner (H.'s) "idealen Aussagen" gefragt und H. habe "regelmäßig den Kopf weggewendet und von etwas anderem gesprochen." Aber Weyl hält trotzdem, trotz der ontologischen völligen Unbegreiflichkeit der H.schen "Mathematik", eine symbolische physikalische Erkenntnis für möglich, die er aber für frei schöpferisch, künstlerischer Tätigkeit verwandt hält.' (Becker to Mahnke, September 1927; Mahnke *Nachlaß*, Ms 862). For discussion and further references on Hilbert's position on ideal elements see Mancosu (1998c, pp. 159–61).
57. See Becker (1927, p. 543). Of course, this is only one aspect, and not the most radical, of Becker's criticism of Husserl. In the second part of Becker (1927) the Heideggerian influence, and especially the idea of mathematics as a means of fleeing death, shows how far Becker had moved away from Husserl. See van Atten (1999, p. 71, n. 23):

    Husserl's letter to Weyl quoted above, p. 70 [Husserl to Weyl, April 9, 1922], tells that Husserl studied Becker's *Habilitationsschrift* intensively. That appears not to have been the case with 'Mathematische Existenz'. In the period between the two works, Becker shifted his position from Husserlian to Heideggerian phenomenology. (Letter from Husserl to Heidegger, May 24, 1927: 'Haben Sie Beckers Arbeit mitgelesen? Direkte Anwendung der Heid⟨eggersche⟩n Ontologie' (Briefwechsel IV: p. 143)) In a letter to Ludwig Landgrebe of October 1, 1931 (Briefwechsel IV: p. 269), Husserl expresses his regret about this shift on his former assistant's part. Was this why Husserl read the second half of *Mathematische Existenz* only 10 years after it was published (Husserl-Chronik, p. 484)?

58. For the distinction between 'historical' and 'natural' time, which appears at this point in the correspondence and which Becker takes from Heidegger, see Becker (1927, pp. 660–68). For a discussion of the 'set' W, see Becker (1927, pp. 550–59).
59. Becker's criticism of Hilbert's consistency proofs is discussed at length in Mancosu (1998c) and Roetti (1996). For reflections in Becker (1927), which closely parallel this section of the letter to Weyl, see pp. 772–75. For a detailed discussion of Weyl's interpretation of quantification see Majer (1988) and Marion (1998).
60. For closely related remarks see Becker (1927, pp. 780–81).
61. Weyl's conjecture turned out to be correct in the following sense. The ε numbers can be reached from below (the predicative ordinals) and the construction breaks down at the

countable ordinal Γ_0 (the first non-predicative ordinal). We thank Giuseppe Longo for having emphasized this point to us.
62. Related material is discussed in Becker (1927, pp. 772–75).
63. See Becker (1927, pp. 747–68) for a fuller account of mantic phenomenology and, in particular, for a more explanatory account of the notion of Ratio, Trieb, and Geist mentioned in the letter to Weyl. See also Poeggeler (1969) for an overview of Becker's philosophical position.

Chapter 11

1. The equation of motion of these particles is the geodetic equation $\frac{d^2 x_\mu}{ds^2} + \Gamma^\mu_{\alpha\beta} \frac{dx_\alpha}{ds} \frac{dx_\beta}{ds} = 0$, where ds^2 is the infinitesimal spacetime interval, and $\Gamma^\mu_{\alpha\beta}$ is the (symmetrical) affine connection.
2. Weyl's term "*Führungsfeld*" for the combined gravitational-inertial field of general relativity is advanced as promoting conceptual clarification of the implicit claim of the relativity of all motions within Einstein's theory. In contrast to Einstein, Weyl pioneered the general relativistic conception of the gravitational field as a four-dimensional affine-connected manifold (the affine connection is, however, uniquely determined by the metric), considered responsible for the observable tendency of "guidance" or "persistence of world direction" of physical bodies. In this way, Weyl maintained, the Einstein theory's resolution of the dilemma between the "evident" principle of relativity of motion and the existence of inertial forces is to be understood. In particular, from the standpoint of the general theory of relativity, according to which all bodies may be momentarily transformed to rest, the concept of the relative motion of two bodies with respect to one another makes no more sense than does that of a body's absolute motion (e.g. Weyl 1923a, pp. 219 ff.; cf.; 1927, p. 74, Engl. trans., p. 105). The motion of a body is properly regarded as a conflict ("Kampf") between such "guidance" and "force", as so all "forced motions" are motions with respect to the "persistence tendency" of the *Führungsfeld*. The lack of clarity on the status of relative motion in Einstein's theory became apparent to Weyl after Einstein's confrontation with the critics of relativity, in particular, the Nobel prize winner and future doyen of Nazi science, Philip Lenard, during the 1920 annual meeting of German natural scientists at Bad Nauheim on 19–25 September (see the reports in *Physikalische Zeitschrift*, 21 (1920), pp. 649–99). Weyl's *kennzeichnende* designation of the gravitational field as *Führungsfeld* appears shortly thereafter in his writings (Weyl e.g. 1920b, p. 610). The term does not appear in *RZM* until the 4th edition (e.g. Weyl 1921b, p. 200).
3. Andrade (1898); Reech (1852); see the discussion in Dugas (1955).
4. Since there is no reference to Andrade in Weyl's work on relativity prior to this letter, it may well be that it is due to Becker that Weyl includes a footnote to Andrade (1898) in the dialogue "Massenträgheit und Kosmos" (Weyl 1924c, p. 65, note 1). The reference occurs in Weyl's discussion of his central conception that in general relativity, the dynamical opposition between "guiding field" and force, replaces the older geometrical distinction (still also in special relativity) between uniform and accelerated motions. Noting that the new conception is reminiscent of the Aristotelian opposition between "natural" and "violent" motions, Weyl observes that Andrade did give a new interpretation of classical mechanics in this manner. However, it is clear that Weyl has not completely agreed with Becker's appraisal, in particular, that Andrade's conception of "the natural course of things" is "identical" with Weyl's conception of "guiding field" (see the discussion of the letter of 12 April 1923 in

11.1 above): "(Andrade) distinguished the natural course of things (inertial motions including gravitation [...] and so on), for which no mechanical forces were invoked and to be treated purely descriptively, from the complusion which bodies exercise on one another. Of course, one certainly seeks in vain in him the exact theoretical comprehension of the guiding field and the laws of its interaction with matter."

5. "Massenträgheit und Kosmos", in (Weyl 1924c, p. 86); these passages appears in English in (Weyl 1949, pp. 115–16).
6. Blaschke (1923), esp. §§ 76–79 on "Affinsphären" and § 89 on "W-Flächen". "Wurf" surfaces are surfaces on which every curve permits a complete group of projective transformations such that any given point is sent into any other point; they were first recognized by Lie on the basis of his theory of continuous groups.
7. See "Massenträgheit und Kosmos", in (Weyl 1924c, pp. 71–74); here Weyl notes that the Einstein "cylinder universe" allows the possibility that every star has infinitely many "ghosts" while the most natural interpretation of the De Sitter hyperboloid universe is that of an expanding universe. Weyl's comments, of course, are five years before Hubble published his velocity-distance "law".
8. Letter of Albert Einstein to Hermann Weyl of 8 March 1918; item #476 in *The Collected Papers of Albert Einstein*, v. 8 (Princeton: Princeton University Press, 1998), English translation, p. 491.
9. RZM 1st–5th, pp. 3–4.
10. RZM 1st–5th, p. 9, cf. Engl. trans., p. 10.
11. The most comprehensive account is (Vizgin 1994, chapter 3); see also Scholz (1994a, 1999, 2001), and Ryckman (2005, chapters 4 and 5).
12. Weyl (1931b, pp. 49, 52). Weyl here alludes to the fact that that the tangent space at a point P of a Riemannian manifold M is a vector space associated with P, not part of M itself. However, there is always a neighborbood of any vector in the tangent space to P that can be mapped diffeomorphically onto an open neighborhood of P in M by the so-called exponential mapping, traceable back to Riemann. Thus the quotation continues: "That is mirrored in theoretical construction in the relation between the curved surface and its tangent plane at the point P: both cover the immediate surroundings of the center P, but the further one proceeds from P, the more arbitrary becomes the continuation of an unambiguous correspondence of the covering relation between surface and plane." Cf. Weyl (1927, p. 98; Eng. trans., p. 135): "a space of intuition whose metrical structure on essential grounds [*aus Wesensgründen*] fulfills the Euclidean laws does not contradict physics in so far as it clings to the Euclidean character of the infinitely small region of a point O (at which I momentarily find myself)."
13. (Weyl 1918c, p. 82) and 1923a, p. 86. Also Weyl (1927, p. 61; Eng. trans., p. 86).
14. Weyl (1927, p. 61); Eng. trans., p. 86.
15. Weyl (1918a, p. 72; Eng. trans., p. 93)): "The coordinate system is the unavoidable residuum of the ego's annihilation [*das unvermeidliche Residuum der Ich-Vernichtung*] in that geometrico-physical world which reason sifts from the given under the norm of 'objectivity'—a final scanty token in this objective sphere that existence [*Dasein*] is only given and *can* only be given as the intentional content of the conscious experience of a pure, sense-giving ego." See also (Weyl 1921b, p. 8; Eng. trans., p. 8).
16. Weyl (1918d, 1919b, 1921c, 1923b).
17. "Vorwort zur dritten Auflage", in (Weyl 1919c, p. vi).
18. (Weyl 1918d, p. 385); (1968, Vol. 2, p. 2).
19. See Weitzenbock (1920).

20. (Weyl 1918d, p. 386); (1968, Vol. 2, p. 3).
21. Following Cartan (1923), such a connection is called "without torsion".
22. Weyl (1923b, p. 17). A metric tensor is needed only to raise or lower indices.
23. Eddington (1921); for discussion, see (Ryckman 2005, chapter 7).
24. (Weyl 1923a, p. 113); (1921b, p. 542), reprinted in (1968, Vol. 2, p. 238). See also (Scholz 1994a, 2001).
25. (Weyl 1918b, p. 466); (1968, Vol. 2, p. 30). Emphasis in original.
26. Pauli (1921); Eng. trans., pp. 195–6.
27. (Weyl 1923a, p. 124): "ein metrischer Raum trägt von Natur einen affinen Zusammenhang". Laugwitz (1958) proved Weyl's conjecture, showing that this condition singles out infinitesimal Euclidean metrics from the wider class of Finsler metrics.
28. (Weyl 1921a, p. 497); (1968, Vol. 2, p. 235); the full treatment appears in Weyl (1923b).
29. (Husserl 1994a, p. 291).
30. Weyl (1921c, p. 133); cf. Eng. trans., p. 148: "The investigations made concerning space in chapter two appear to me to be a good example of the essential analysis [*Wesensanalyse*] striven for by phenomenological philosophy (Husserl), an example that is typical for such cases where a non-immanent essence is dealt with. We see in the historical development of the problem of space, how difficult it is for us actuality-prejudiced humans to hit upon what is decisive. A long mathematical development, the great unfolding of geometrical studies from Euclid to Riemann, the physical exploration of nature and its laws since Galileo, together with all its incessant boosts from empirical data, finally, the genius of singularly great minds—Newton, Gauss, Riemann, Einstein—all were required to tear us loose from the accidental, non-essential characteristics to which we at first remain captive. Certainly, once the true standpoint has been attained, Reason [*Vernunft*] is flooded with light, recognizing and accepting what is understandable out-of-itself [*das ihr aus-sich-selbst Verständliche*]". "The example of space", Weyl continued, "is most instructive for that question of phenomenology that seems to me particularly decisive: to what extent the delimitation of the essentialities [*Wesenheiten*] rising up to consciousness express a characteristic structure of the domain of the given itself and to what extent mere convention participates in it."
31. Weyl (1918b, 1919b)
32. (Weyl 1919b, pp. 251–2) and (Weyl 1923a, pp. 314–15); for discussion, see Brading (2002).
33. See Scholz (2001) and (Ryckman 2005, chapter 5).
34. This is because of the existence of an "absolute length", the so-called "Compton length" in the quantum theory of the electron. See (Weyl 1931b, p. 55). For discussion, see Ryckman (1994), and (Ryckman 2005, chapters 4 and 6).
35. For example Yang (1981) and the comprehensive discussion in O'Raifeartaigh (1997).
36. (Weyl 1918a, pp. 65–74); "Anschauliches und mathematisches Kontinuum".
37. Feferman (1988a); (Mancosu 1998b, pp. 70–74).
38. (Weyl 1918a, p. 71); Engl. trans. (1994, p. 94); For a defense of this perspective, see Feferman (1998b).
39. (Weyl 1921b, p. 8; Engl. trans., p. 8). Becker several times calls attention to the fact that his citations are to the 4th edition of *RZM* by appending an exclamation point to the citation, e.g., (1923, p. 387 fn. 1): "'Raum, Zeit, Materie'. 4. Aufl. (!)"; see also p. 536, fn. 2.
40. (Weyl 1921b, p. 4); cf. Engl. trans., p. 5.
41. Weyl writes some ten years before Heisenberg's Uncertainty Principle (1927).
42. Of course, to attempt to locate "the direction of time" in a physical process would be question-begging from the standpoint of phenomenology.

43. (Weyl 1927, p. 57; Engl. trans., p. 75): "On the basis of objective geometrical relations, with which the axioms are concerned, it is not possible to determine a point absolutely, but conceptually only *relative* to a coordinate system, through numbers. For understanding the application of mathematics to reality the distinction between the 'givenness' of an object through individual exhibition on the one side and, on the other in conceptual ways, is fundamental. The objectification through exclusion of the ego and its immediate life of intuition [*Objektivierung durch Ausschaltung des Ich und seines unmittelbaren Lebens der Anschauung*] is not attained without remainder, the coordinate system, exhibited only through an individual act (and only approximately) remains as the necessary residue of this eradication of the ego [*das notwendige Residuum dieser Ich-Vernichtung*]."
44. See Chern (1996).
45. Weyl (1921b, p. 133; cf. Engl. trans., p. 147). See also the discussion of this passage in Friedman (1995, p. 254). Friedman correctly draws attention to Weyl's contrast here between his own method of *Wesensanalyse*, involving step-by-step mathematical construction from the "purely infinitesimal" basis of geometry, and the employ of a quasi-perceptual *Wesensanalyse* by Carnap in his early work on space (Carnap 1922), where intuitive space is constituted in a "single exemplary act of making-present" [*Vergegenwärtigung*].
46. Borel (1986, p. 54).
47. For a proof, see Laugwitz (1965, pp. 185 ff.).
48. See Scheibe (1988, pp. 68 ff.).
49. See the discussion in Hawkins (2000, pp. 435–36).
50. Weyl (1923a, p. 85): "The Euclidean distant-geometry [*Fern-Geometrie*] is made for the investigation of straight lines and planes, to these problems it has oriented itself. But as soon as one goes over to infinitesimal geometry, the most natural and reasonable course is to place at its foundation the infinitesimal *Ansatz* of Riemann: no complication thereby results and one is protected from inappropriate distant-geometrical considerations."
51. See the works of Vizgin, Scholz and Ryckman cited in the Bibliography.
52. See van Dalen (1999, p. 308).
53. Witness the concluding passage from Weyl (1918c) and (1919c): "The further physics develops, the more understandable it becomes that the relations between the reality that each one of us lives, and these objective essences [*Wesenheiten*], treated by physics in mathematical symbols, are not at all as simple as it appears to the naïve conception, and that nothing of the contentual [*Inhaltliches*] of that immediately experienced reality fundamentally enters into the physical world. [...] (Physics) has for reality [*Wirklichkeit*] no further reaching meaning than formal logic has for the domain of truth. What formal logic teaches is certainly based on the essence [*Wesen*] of truth, and no truth violates its laws. But it plainly teaches nothing at all of whether a concrete affirmation is true or not, the contentual [*Inhaltliche*] of truth it entirely leaves aside. The ground of the truth of a judgment lies in the judged facts and not in logic. I maintain that physics has to do only with what, in an entirely analogous sense, would be designated as the formal composition [*Verfassung*] of reality. Its laws are just as little violated in reality as there are truths which are not in agreement with logic. However, regarding the contentual essentials [*inhaltlichen Wesenheiten*] of this reality, it settles nothing, the ground of reality is not grasped in it. If the delusion of the scholastic method is to wish to deduce essentials [*Wesenhaften*] out of mere formulas, then the world view that one designates by 'materialism' is only a variety of scholasticism."
54. Weyl to Felix Klein, 28 December 1920, as quoted and translated in Sigurdsson (1991, p. 204).

55. Weyl (1921a, pp. 556 ff.); and Weyl (1924c, pp. 56–59).
56. Weyl (1927, p. 64; Engl. transl., p. 91).
57. See Flamm (1916); for a recent discussion, see Visser (1996, p. 45).
58. Wheeler, "Geons", *Physical Review* 97 (1955), pp. 511–36; C. W. Misner and J. A. Wheeler, "Classical Physics as Geometry", *Annals of Physics* 2 (1957), pp. 525–603; both reprinted in Wheeler (1962). For a brief review of the idea of electric charge as "electric lines of force trapped in the topology of a multiply connected space", see Wheeler (1968, pp. 300–01, fn. 36).
59. Morris and Thorne (1988); for discussion see Thorne (1994), and more technically, Visser (1996).
60. See Hendry (1980).
61. Becker (1923, p. 394); referring to Husserl (1913), §§ 136–53 and especially to §§ 148–50.
62. Husserl (1913), § 149, quoted in Becker (1923, p. 395).
63. Becker refers, *inter alia*, to Hilbert (1903), where this result is proved.
64. With reference to his 1923 monograph, Becker himself does so in a later work (Becker 1959, pp. 50–51).
65. See also Friedman (1995, p. 136).
66. Weyl (1949, p. 137; translation modified to accord with (1927, p. 100)). Just before this passage on p. 99, Weyl also expresses a rather damning assessment of Becker's transcendental justification of the three-dimensionality of space. Friedman (1995, p. 136), cites the quoted passage as evidence that "Weyl conceives of his own analysis of the 'space problem' as a generalization and refinement of Becker's". But this claim contravenes the chronicle of contacts between Weyl and Becker, a record that definitively shows that Weyl had already outlined and given a first version of his group-theoretic result in (Weyl 1921b, § 18); (cf. Engl. transl., pp. 138–48), i.e., *before* learning of Becker's *Habilitationsschrift* in Husserl's letter of 9 iv 1922 (see 10.3). But even disregarding the documented contacts between Weyl and Becker, we find Friedman's claim implausible for two substantive reasons. First, Weyl apparently regards his solution to the "problem of space" as the crowning arch of his elaboration of "pure infinitesimal geometry", as clearly appears from his remark appearing at the end of (Weyl 1921b, § 18), cited in § 11.2.1 above, claiming that, as thus group-theoretically justified, his construction of purely infinitesimal geometry is a "good example" of Husserlian *Wesensanalyse*. Secondly, Weyl's couples *Wesensanalyse* with, as Friedman noted, mathematical construction; Becker, on the other hand, attempts to constitute intuitive space from pre-spatial sensory fields, a project that has more in common with Poincaré and Helmholtz, the conventionalists both he and Weyl abhor. We also recall that for Weyl, "relativity theory has made it entirely evident that of the essence of space and time given in intuition nothing enters into the mathematically construed physical world." (Weyl 1921b, p. 3; cf. Engl. trans., p. 3)
67. Becker (1923, pp. 554 ff.). On Becker's claim there that the laws of electromagnetism are a priori "laws of structure", see the end of 11.2.1 above. Becker also alludes to Weyl claim discussed there that since the general form of the Maxwell action function is the simplest integral invariant that can exist in a Weyl manifold, Weyl's theory "explains why the world is four dimensional". Weyl (1923a, p. 301; cf. Engl. trans., pp. 284, 292).
68. Cf. Weyl (1923a, p. 302): "If our hypothesis is correct, then we have nothing more to do with two fields subsisting beside one another in no inner connection, rather *the ether, which binds the different material individuals into a causal whole, is a $(3 + 1)$-dimensional extensive medium of metrical structure*" (original emphasis).

Chapter 12

1. Mancosu and Ryckman (2002).
2. On Mahnke see Wohltmann (1957), from which much of the bibliographical information is taken.
3. Published in Husserl (1994b).
4. That Hilbert and Husserl were the two teachers who most influenced him is nicely expressed in a letter to Husserl: "Ueber Gebiete zu reden oder zu schreiben, von denen ich keine 'Sach'-kenntniss habe, das habe ich bei Ihnen und auch bei Hilbert nicht gelernt" (Mahnke to Husserl, 30.xii.1927 [*Durchschlag*], in Husserl 1994b, p. 463).
5. Becker (1923).
6. See Geiger (1928); Scholz (1928); Cassirer (1929).
7. Becker (1927/28, 1928/29, 1930a).
8. For more information see Mancosu and Ryckman (2002).
9. Becker to Mahnke, 22.viii.1926. On this occasion Mahnke also met Heidegger, with whom he will be a colleague later in Marburg (see Mahnke to Husserl, 30.xii.1927, in Husserl 1994b, p. 465).
10. *Nachlaß* D. Mahnke, Ms. 862. [[Published in Becker (2005)]].
11. Mahnke (1927b).
12. "Inzwischen werden Sie sich gefreut haben zu sehen, wie sehr Dr. Becker Ihre 'Doktor'arbeit schätzt, was ja in unserem phänomenologischen Kreis selbstverständlich ist. Ich hoffe, daß Sie nun umgekehrt für diese ausgezeichnete Leistung B[ecker]s Interesse haben werden" (Husserl to Mahnke, 3.viii.1927, in Husserl 1994b, p. 455).
13. Mahnke (1925a,b, 1926, 1927a).
14. Mahnke (1925b).
15. For appropriate background information on Hilbert's program and Becker's position on it see Mancosu (1998c) and Mancosu and Ryckman (2002).
16. Mahnke (1925a).
17. On why Husserl could not accept Kant's *intellectus archetypus* see Kern (1964).
18. See for details Mancosu and Ryckman (2002).
19. See Mancosu and Ryckman (2002).
20. The reference seems to be to Mahnke's lost letter but we can glance at Mahnke's position on the foundations of pure mathematics from a letter he sent to Husserl a few months afterwards. I would like to quote parts of it because it sheds light on the rest of the exchange between Becker and Mahnke: "Ich [...] werde mich immer an die Sachen selbst hingeben, mich in ihr eigenes Wesen und ihre wahre Wirklichkeit vertiefen und so zu objektiver Erkenntnis zu gelangen suchen, so weit das für ein subjektives Individuum möglich ist. Das scheint mir das Wichtigste zu sein, was ich bei Ihnen gelernt habe. Freilich ist dabei streng zu scheiden zwischen Tatsachen- und Wesenswissenschaften, wie Sie es ja tun, während ich fürchte, dass Heidegger in dem Versuche, apriorische Erkenntnisse über das menschliche 'Dasein' zu gewinnen, zu weit geht. Die Mathematik kann und muss rein eidetisch begründet werden. Zum 'Verständnis' der Natur und Geschichte dagegen reicht, wie ja auch Sie in Ihrem Briefe hervorheben, die Phänomenologie allein nicht aus: neben die constitutive Phänomenologie, die den apriorischen Wesenskern erkennt, muss die empirische Realitätserkenntnis treten, und es können nicht etwa beide durch ein Mittleres, eine hermeneutische Phänomenologie ersetzt

werden" (Mahnke to Husserl, 30.xii.1927). Although the letter is concerned with Heidegger it is obvious that the position defended by Mahnke is also opposed to Becker's approach.
21. See Weyl (1927, 1928a).
22. This is a theme touched upon also in Mahnke (1927b) (see letter 8 from Becker to Mahnke).
23. Mahnke (1927b).

Chapter 13

1. Harvard in 1940–41 was at the center of much activity in philosophy of science, some of which involved Tarski, Carnap, and Quine, such as The Harvard Science of Science Group studied in Hardcastle (2003).
2. "8–11h mit Tarski im Café. über Monomorphie, über Tautologie, er will nicht zugeben, daß sie nichts über die Welt sagt; er meint zwischen tautologischen und empirischen Sätzen sei ein bloß gradueller und subjektiver Unterschied." (quoted in Haller 1992, p. 5)
3. "In den Diskussionen wurde die Gegenüberstellung "analytisch" und "synthetisch" mehrfach behandelt. Auf äusserungen TARSKIS hin, daß man diese Gegenüberstellung nicht scharf durchführen könne, erklärte CARNAP, daß es denkbar wäre, gewisse gewöhnlich als "deskriptiv" angesehene Zeichen (Haus, Temperatur usw.) bereits in der Metasprache (Syntax) aufzunehmen und sie als "logische Zeichen" zu behandeln, wodurch die Sätze, in denen diese Termini auftreten, auf Grund der Sprachregeln determiniert wären. Es sei wirklich eine Frage, wie man [auf Grund des "Gefühls", das uns zu der Meinung veranlaßt, man könne zwischen Worten der deutschen Sprache, wie "und", "oder" usw. einerseits, "Temperatur", "Haus" andererseits, eine Grenze ziehen,] zur einer Formulierung gelangen könnte, die uns erlaubt, "deskriptiv" und "logisch" und dann auch "synthetisch" und "analytisch" streng zu scheiden—eine Problemstellung, die nur in strenger Form diskutiert werden kann, sollen nicht Mißverständnisse aller Art entstehen." (Neurath 1936, pp. 388–89)
4. The full text should be read in the context of a polemic between Tarski and Neurath occasioned by Tarski's claim that Neurath was not giving proper credit to the influential role of the Warsaw group on the Vienna Circle. The full passage gives a better sense of Tarski's annoyance for how Neurath had portrayed the exchange with Carnap: "Ich habe Ihnen einmal ein paar Worte über die "Enstehung der Legenden" geschrieben. Ich kann nun Ihnen auf das Beispiel einer "Legende" sozusagen "in statu nascendi" hinweisen; einige polnische Bekannten, die im Pariser Kongress teilgenommen haben, haben mich darauf vor kurzer Zeit aufmerksam gemacht. Ich habe in Paris einen Vortrag über den Begriff der logischen Folgerung gehalten; ich habe dort u.a. den absoluten Charakter der Einteilung der Begriffe in logische und deskriptive sowie der Sätze in analytische u. synthetische bestritten: und ich bestrebte mich zu zeigen, daß die Einteilung der Begriffe ziemlich willkürlich ist und die Einteilung der Sätze auf die Einteilung der Begriffe relativiert werden muß. In der Diskussion hat Carnap erklärt, daß er meine diesbezügliche Bemerkungen als sehr tiefe ansieht, und hat in einer klarer und populärer Form meine Grundgedanken noch einmal dargelegt; ich war ihm dafür freilich sehr dankbar. Nun soll man sehen, wie der Bericht über diesen Vortrag und die ihm nachfolgende Diskussion in Erkenntnis 5, Heft 6, S. 388–389, gefaßt ist! Meinem Vortrag is nicht einmal eine ganze Zeile gewidmet (es ist sogar nicht erwähnt, dass ich einen Vortrag über dieses Thema gehalten habe), die Rede Carnaps ist dagegen sehr genau und ausführlich in 13–14 Zeile wiedergegeben. Der Leser muß den Eindruck haben, daß Tarski eine Frage

gestellt hat, daß Carnap aber diese Frage sehr ausführlich und zutreffend beantwortet hat; es ist absolut unmöglich, aus dem Bericht die wirkliche Sachlage zu erraten!" (Tarski to Neurath, Warsaw, 7.ix.36) The issue about the arbitrariness of the distinction between logical and descriptive signs was also discussed by Carnap and Tarski in Cambridge (MA) in 1939 as it transpires from a set of notes written by Carnap for himself (see RC 089-06-03:1). I wish to thank Dr Brigitte Parakenings for having transcribed the relevant passages from Stolze-Schrey into German.

5. "1. 'L-wahr'. 'logisch-deskri[ptiv]'. Ich: meine Intuition. Klarer in der Unterscheidung L-wahr—F-wahr, als in logisch-descri[ptiv]. Die Letztere kann ich aber immerhin erklären durch Aufweisung der einfachsten logischen Konstanten in den üblichen Systemen und Angabe, dass Alles daraus Definierbaren auch logisch sein soll. Er: Er hat gar keine solche Intuition; man könnte ebenso gut 'Temperatur' auch als logisch rechnen. Ich: Die Wahrheit eines Vollsatzes des Temperaturfunktoren bestimmt man durch Messung. Er: Man kann aber beschliessen, an einem festgesetzten Wahrheitslehre [unleserlich] trotz aller Beobachtungen. Ich: Dann ist es eine mathematische Funktion, und ein logisches Zeichen, und nicht der physikalische Temperaturbegriff. Bei einem Vollsatz des physikalischen Temperaturfunktores können wir nicht durch blosses rechnen den Wahrheitswert finden. Er: Das beweist nichts, denn auch für eine mathematische Funktion ist das oft nicht möglich, weil es unentscheidbare Sätze gibt; kein fundamentaler Unterschied zwischen den mathematischen, aber unentscheidbaren Sätzen und den faktischen Sätzen. Ich: Das scheint mir doch." [RC 090-16-09; Dated 6 March 1940, Chicago, Conversation with Tarski, 4 pp.]

6. "*Tarski und Quine*: Allgemeine Bemerkung zur allgemeinen Semantik: Es ist wohl kaum lohnend, die Definitionen und Theoreme eines Systems der allgemeinen Semantik auf die Klasse K aller Sprachen zu beziehen, die in M behandelt werden können, sondern lieber nur auf eine Teilklasse K' die so ist, dass: 1.) Jede Sprache von K (oder jede solche, die wir mit in Erwägung ziehen wollen) ist übersetzbar in eine Sprache von K', 2.) Alle Sprachen von K' haben gewisse üblichen Strukturen. Am Einfachsten, und für alle praktischen Zwecke [hinreichend] weit: wir beziehen uns nur in Sprachen, die Individualvariable und Konstanten, Prädikatenkonstanten, und Identität haben; doch Verknüpfungen und Operatoren. Also niederer Funktionskalkül (aber ohne Prädikatenvariable [...]) Rechtfertigung hierfür: mit Hilfe der besonderen Relation 'ε' (die aber hierbei nicht als in jeder Sprache vorkommend vorausgesetzt wird) können wir die Mengenlehre und Mathematik in den niederen Funktionskalkülen übersetzen. Dies ist die *Unterscheidung zwischen Logik un Mathematik: Mathematik=Logik+ 'ε'*. Durch das 'ε' wird das System non-finitistisch, und unvollständig." (RC 090-16-03)

7. In the historiography of logic, this shift has been perceptively pointed out by Ferreiros. See Ferreiros (1997, 1999, 2001).

8. See de Rouilhan (1998).

9. "Logik ohne Typen. Die beste Form ist die, die ursprünglich von Zermelo gemacht worden ist; auf Grund davon jetzt verbesserte Systeme von Bernays (Unterscheidung von Klassen und Mengen) und [Mostowski] (ohne diese Unterscheidung). Quine macht in seinen Systemen zu viele Sonderwahrheiten (z.B. so, dass Cantors Theorem nicht stimmt), die die Mathematiker abschrecken, und die anzeigen, dass die Systemform nicht zweckmässig ist." (RC 090-16-09) For related material see RC 090-16-26.

10. "Ich: Sollen wir vielleicht die Sprache der Wissenschaften mit oder ohne Typen machen? Er: Vielleicht wird sich etwas ganz Anderes entwickeln. Es wäre zu wünschen und vielleicht zu

vermuten, dass die ganze allgemeine Mengenlehre, so schön sie auch ist, in der Zukunft verschwinden wird. Mit den höheren Stufen fängt der Platonismus an. Die Tendenzen von Chwistek und anderen ("Nominalismus"), nur über Bezeichenbarem zu sprechen, sind gesund. Problem nur, wir gute Durchführung zu finden." (RC 090-16-09)

11. "Tarski: Ein *Platonismus* unterliegt dem höheren Funktionskalkul (also den Gebrauch einer Prädikatenvariable, besonders höherer Stufe)." (RC102-63-09)

12. "T: *Die Warschauer Logiker*, besonders Leśniewski und Kotarbiński, sahen ein system wie PM (aber mit einfacher Typentheorie) ganz selbstverständlich als die Systemform an. Diese Beschränkung wirkte stark suggestiv auf alle Schüler; auf T. selbst noch bis zu "Wahrheitsbegriff" (wo weder transfinite Stufen noch stufenloses System betrachtet wird, und Endlichkeit der Stufen stillschweigend vorausgesetzt wird, erst im später hinzugefügten Anhang werden sie besprochen). Dann aber sah T., dass in der *Mengenlehre* mit grossem Erfolg eine ganz andere Systemform verwendet wird. So kam er schliesslich dazu, diese stufenlose Systemform als natürlicher und einfacher zu sehen." (RC 090-16-26). For background on the Lvov-Warsaw school see Woleński (1989a).

13. The detailed summary is in German; it is unclear whether Quine presented his ideas in German; however, some words appear in the Carnap summary in English and this might point to the fact that Carnap was translating the talk into German using English occasionally for convenience. [[See now 13.4]]

14. "Vorschlag: wir wollen Sprache untersuchen, die nur folgendes enthält: *[Konstante, Prädikate]*, *joint denial, univ. quantification. Mit nur einer Art von Variable; nur geschlossene Sätze.*" (RC 102-63-04)

15. "Ich vermute, C. hat recht: es gibt nur teilweise Klärung, nicht vollständige definitionale Eliminierung. Diese Klärung geschieht durch Untersuchung der Relationen der Konfirmation zwischen Sätze der entfernteren und solchen der mehr unmittelbaren Art." (RC 102-63-04)

16. "*Wissenschaft ist voll von Myth und hypostasis;* Zweck: das chaotische Verhalten der gewöhnlichen Dingen einzubetten in eine mehr verständliche überwelt; Endaufgabe: prediction inbezug auf die gewöhnliche Dinge; das ist psychologisch möglich nur infolge der grösseren "Ubersichtlichkeit" [sic] der überwelt, die von der Wissenschaft als intermediary devise construiert wird. Die Trichotomie: Phänomene, common sense Welt, überwelt der Wissenschaft, gilt nur grob; es handelt sich um Grade. Von der überwelt können wir auf die gewöhnliche Welt schliessen. Nicht umgekehrt (unterdeterminiert); ebenso ist die gewöhnliche Welt underdeterminiert durch experience." (RC 102-63-04)

17. 20.12.1940. "*Quine wird diskutiert*: Können wir vielleicht die höheren, nicht-finitistischen Teile der Logik (Mathematik) so auffassen, dass ihre Beziehung zu den finitischen Teilen analog ist mit der Beziehung der höheren Teile der Physik zu den Beobachtunssätzen? Dadurch wurde die nicht-finitistische Logik (Mathematik) nicht-metaphysich (wie die Physik). Vielleicht wird dadurch auch leicht [sic] geworfen auf die Frage, ob fundamentaler Unterschied zwischen Logik-Mathematik und Physik besteht." (RC 090-16-29)

18. More on the Wundheiler suggestion on RC 102-63-03 (meeting of 20 January 1941) where Quine suggests that a sentence is logically true in the narrow sense (as against mathematically true) when its truth is preserved under an arbitrary automorphism of all objects and not just the individuals. There is also more on L-wahr.

19. Quine's first reply to the question: it is a distinction in the type of evidence; in mathematics one does not need experiments as in physics; thus, a priori in a behavioral sense. Carnap remarks that he would rather not characterize the *a priori* behavioristically.

20. "*Tarski, Finitismus.* Bemerkung in Diskussion in der Logikgruppe, *10 January 1941.* Tarski: Ich verstehe im Grunde nur eine Sprache die folgende Bedingungen erfüllt: [1] *Finite* Anzahl der Individuen; [2] *Realistisch* (Kotarbiński): Die Individuen sind physikalische Dinge; [3] *Nicht-platonisch*: Es kommen nur Variable für Individuen (Dinge) vor, nicht für Universalien (Klassen usw.). Eine andere Sprache "verstehe" ich nur so, wie ich die klassische Mathematik "verstehe", nämlich als Kalkül; ich weiss, was ich aus anderem Ableiten kann (oder abgeleitet habe; "Ableitbarkeit" im Allgemeinen schon problematisch). Bei irgendwelchen höheren, "platonischen" Aussagen in einer Diskussion deute ich sie mir als Aussagen, dass ein bestimmter Satz aus gewissen anderen Sätzen ableitbar (bzw. abgeleitet) ist. (Er meint wohl so: Die Behauptung eines gewissen Satzes wird gedeutet als besagend: dieser Satz gilt in dem bestimmten, vorausgesetzten System; und das heisst: er ist ableitbar aus gewissen Grundannahmen). Warum wird auch schon die elementare Arithmetik, mit abzählbarem Bereich, ausgeschlossen? Weil, nach Skolem, die ganze klassische Mathematik sich durch ein abzählbares Modell darstellen lässt, also in der elementare Arithmetik ausdrücken lässt, z.B. indem man ε als eine gewisse Beziehung zwischen natürlichen Zahlen nimmt." (RC 090-16-28)

21. "*Gespräch mit Tarski und Quine über Finitismus, I: 31.1.42; Ergebnis: p. 4.* Ich: Schwierigkeiten der Verständigung zwischen Tarski und mir, hauptsächlich in drei Punkten: *[1] Finitismus.* D.h.: Sprachen mit *was für Variablen verstehen* wir? (Das ist der *schwierigste* Punkt; für mich Frage des Grades. Aber nicht ganz klar.) *[2] Modalitäten.* 'N'; intensionale Sprache. *[3] L-Begriffe. [3] ist am Leichtesten.* Nehmen wir Quines Sprachform (oder andere, ähnliche). Wir geben die logischen Konstanten durch Aufzählung an. Dann ist 'L-wahr' leicht definierbar. [...] [2] Wenn 'L-wahr' definiert, so kann 'N' leicht erklärt werden; im Wesentlichen: 1, 'N(...)' wird übersetzt in '...' falls dies L-wahr; anderenfalls in '∼(...)'. (Wir nehmen hierbei an: nur geschlossene Sätze, wie bei Quine.) 2, '(x)N(...)' wird übersetzt in 'N(x)(...)'. [1] ist am Schwierigsten. In welchem Sinn "verstehen" wir z.B. Arithmetik mit gebundene Zahlvariablen (für natürlichen Zahlen)." (RC 090-16-25)

22. "*Finitismus. Tarski:* Ich verstehe richtig nur eine endliche Sprache S_1: nur Individuumsvariable, ihre Werte sind Dinge; für deren Anzahl wird nicht Unendlichkeit behauptet (aber vielleicht auch nicht das Gegenteil). Endlich viele deskr[iptive] primitiven Prädikate. Zahlen: sie können verwendet werden, in endlichem Bereich, indem wir die Dinge geordnet denken, und unter den Zahlzeichen die betreffenden Dinge verstehen. Wir können dann [arithmetische] Begriffe verwenden; aber viele arithmetische Sätze können hier nicht bewiesen werden, weil wir nicht wissen, wie viele Zahlen vorhanden sind. Man kann auch einer Klasse eine Kardinalzahl zuschreiben." (RC 090-16-25)

23. "*Tarski:* Das psychologisches Rätsel ist folgendes: Die Mathematiker scheinen in einem gewissen Sinn auch die unendliche Arithmetik zu verstehen. Nämlich bei einem unentscheidbaren Satz (z.B. dem von Gödel), sind sie imstande, ohne Rüchsicht auf die Axiome, zu sagen, dass sie ein Satz als wahr anerkennen. Und ich (Tarski) teile dieses Gefühl in einem gewissen Grade. Ich: Mir scheint, in einem gewissen Sinn verstehe ich wirklich die *unendliche Arithmetik*, sagen wir etwa Sprache S_2: nur Variable für natürlichen Zahlen, mit Operatoren (sodass auch negierten Allsätzen), dazu rekursive Definitionen. Auf Tarskis und Quines Frage, wie ich das deute, wenn die Anzahl der Dinge doch vielleicht endlich ist: ich weiss nicht genau, aber vielleicht durch blosse Stellen anstatt Dinge (Tarski: diese Auffassung in (Syntax) hat ihm damals grossen Eindruck gemacht, er findet doch Schwierigkeiten dabei). Eine stelle ist eine Anordnungsmöglichkeit für ein Ding. Ich habe nicht die gefühlmässige [Ablehnung]

gegen die Möglichkeitsbegriff wie Tarski und Quine. Mir scheint die Möglichkeit des immer Weiterschreitens die Grundlagen der Zahlenreihe. Also potentiales, nicht aktuales unendlich (Tarski und Quine sagen: sie verstehen diesen Unterschied nicht.) Ich: Vielleicht gibt es auch noch Zwischenstufe, ähnlich Sprache I, ohne negierte Allsätze. (Tarski: Dies scheint ihm kein wesentlicher Unterschied, da er Satz mit freier Variable als Abkürzung für Satz mit Operatoren auffasst.) Ein allstaz für natürliche Zahlen können wir auffassen als gemeinsame Behauptung aller Instanzen, da ja für jede natürliche Zahl ein Ausdruck vorhanden ist (Tarski: aber nicht ein wirklicher Ausdruck als Ding, falls die Anzahl der Dinge endlich ist.)" [RC, 090-16-25]

24. "*Wir zusammen*: also jetzt Problem: Was für einen Teil S von M können wir nucleus nehmen derart, dass 1.) S in einem gewissen Sinn von uns verstanden wird, und 2.) S hinreicht zur Formulierung der Syntax von ganz M, soweit sie nötig ist für die Wissenschaft, um in M die Syntax und Semantik der Gesamtwissenschaftssprache zu behandeln." (RC 090-16-25)

25. "2a. Wie können wir den *rich nucleus* (d.h. unendliche Arithmetik S_2) rechtfertigen? D.h. in welchem Sinn können wir vielleicht doch sagen, dass wir ihn wirklich verstehen? Wenn ja, so können wir damit sicherlich die Regeln des Kalküls M aufstellen.

"2b. Wenn S_1 nicht ausreicht, um die klassische Mathematik zu erreichen, könnte man dann nicht vielleicht trotzdem S_1 nehmen und vielecht zeigen, dass die klassische Mathematik nicht wirklich nötig ist für die Anwendung der Wissenschaft im Leben. Können wir vielleicht auf Grund von S_1 einen Kalkül für eine fragmentare Mathematik aufstellen, die für alle praktischen Zwecke genügt (d.h. nicht etwa nur für Alltagszwecke, sondern auch für die kompliziertesten Aufgaben der Technik.)." (RC 090-16-25)

26. "*Empiristischer vs. logischer Finitismus*. Tarskis Finitismus ist ein logischer. Er meint: vielleicht die Anzahl der Dinge in der Welt endlich; in diesem Fall kann man auch nur von endlich vielen natürlichen Zahlen sprechen, Ich dagegen: Wir sind Empiristen. Daher sagen wir: unser Wissen ist auf Endliches beschränkt; d.h. jede Konfirmation ist basiert auf eine endliche Menge von Evidenz, d.h. endliche Menge von Beobachtungsausdrücke. *Aber*: Wir können trotzdem über endliche Klassen von beliebig hoher Kardinalzahl sprechen, also auch über die einzelnen natürlichen Zahlen (z.B. 1000 ≠ 1001), ohne die Anzahl der Dinge in der Welt inbetracht zu ziehen. So werden *Logik und Arithmetik unabhängig von der zufälligen Anzahl der Dinge in der Welt*. Trotzdem bleiben auch Logik und Arithmetik in einem gewissen anderen Sinn finitistisch, wenn sie wirklich verstanden werden soll. Die Arithmetik (der natürlichen Zahlen) ist ja tatsächlich entwickelt worden, ohne dass wir [bis heute] mit Sicherheit wissen, ob die Anzahl der Dinge in der Welt endlich ist oder nicht. Und die bewiesenen Sätze werden von niemandem bezweifelt; besonders die konkreten Sätze (d.h. ohne Variable) scheinen doch unzweifelhaft. Also kann die Arithmetik doch wohl abhängig sein von einer faktischen Hypothese über die Welt. Auch wenn die Anzahl der Dinge (z.B. Elektronen usw) endlich ist, so kann trotzdem *die Anzahl der Ereignisse als unendlich* [angenommen] werden (nicht nur die Anzahl der Zeitpunkte innerhalb eines Intervalls infolge der Dichte, sondern auch die Anzahl der Zeitpunkte im Einheitsabstand von einander, mit anderen Worten: unendliche Länge der Zeit). Ist dies eine faktische Hypothese? Oder hängt es nicht auch wieder mit *logischer Möglichkeit* zusammen?" (RC 090-16-24)

27. "Aber das is viel zu eng. Dann enthält PM nicht einen einzigen Beweis für ein Theorem. Aber wir können es *doch finitistisch* machen: wir nehmen als Zeichen nur wirkliche Dinge, aber als Ausdrücke und Beweise nicht nur gewisse wirkliche räumliche Anordnungen von diesen Dingen, sondern (nicht-räumliche) *Sequenzen dieser Dinge*, entweder bezeichnet durch die

Reihe der Namen dieser Dinge, getrennth durch Kommata (elementarer Sequenzausdruck), oder durch Deskriptionen, z.B. als Verbindung zweier früher angegebener Sequenzen, für die wir Abkürzungen eingeführt haben." (RC 090-16-27)

28. "Mir scheint der ganze Vorschlag an einer Fehlauffassung der Arithmetik zu kranken: Die Zahlen werden reifiziert; die Arithmetik wird von kontingenten Fakten abhängig gemacht, während sie in Wirklichkeit von begrifflichen Zusammenhängen handelt; wenn man so will: von möglichen, nicht von wirklichen Fakten." (RC 090-16-23)

29. "*Tarski*: Ich möchte ein Arithmetiksystem haben, dass keine Annahmen über die Anzahl der vorhandenen Zahlen macht, oder höchstens eine Zahl (0) annimmt. A_n sei das System derjenigen Sätze der gewöhnlichen Arithmetik, die auch gelten wenn es nur die Zahlen < n gibt; also A_0 ohne Zahlen; A_1 nur mit 0; usw. A_ω sei die ganze unendliche gewöhnliche Arithmetik. Wir wollen zur Vereinfachung A_0 ausschliessen, also wenigstens die Existenz *einer* Zahl annehmen. Mein (d.h. Tarskis) System soll alle die und nur die Sätze enthalten, die in jedem der Systeme A_n ($n = 1, 2, \ldots \omega$) gelten. Hierher gehören zum Beispiel alle Sätze von folgender Form: keine Funktoren kommen vor, alle universellen Operatoren stehen unnegiert am Anfang, keine Existenzoperatoren." (RC 09-16-04)

30. "*Quine*: W is dann eigentlich nur ein Mythos. *Ich*: Nein, kein Mythos, einfach eine Maschine. Es wäre nur ein Mythos, wenn wir den Maschinenteil (Kalkülzeichen) Pseudointerpretation beilegen würde, durch Hinweise auf Entitäten, die es in Wirklichkeit nicht gibt. *Tarski*: Weg (2) hatte aber dies unbefriedigende, dass es eigentlich mysteriös bleibe, wie so die Maschine richtig wirkt, d.h. wie es zu erklären ist, dass wenn wir wahre Sätze von $B S$ in die Maschine stecken (als Prämissen), dann auch wahre Sätze (als Konklusionen) wieder herauskommen. *Wir*: Das ist vielleicht kein unlösbares Geheimnis. Wir bauen ja die Maschine zu diesem Zweck, und verwerfen sie, wenn wir merken, dass sie dies nicht leistet. Vielleicht kann man sogar in $B S$ zeigen: wenn eine Maschine so und so konstruiert ist, so liefert sie zu wahren Prämissen stets wahren Konklusionen." (RC 090-16-04)

31. "*Zusammenfassung der Bishergesprochenen*. Nucleus lang., soll dienen als Syntaxsprache für Aufbau der Gesamtwissenschaftssprache (einschliesslich klassische Mathematik, Physik, usw.). Die Wissenschaftssprache bekommt eine teilweise Interpretation dadurch, dass die n.l. als verstanden vorausgesetzt wird . . . 1, Der logisch-arithmetische Teil der n.l.: unbeschränkte Operatoren, . . . Hiergegen keine Bedenken vom finitistischen [unleserlich], weil die Werte der Variablen nur physikalische Dinge sind. Dabei bleibt es unbestimmt, ob deren Anzahl endlich oder unendlich ist. Als Zahlen werden die Dinge selbst genommen, für die eine Ordnung vorangesetzt wird, auf Grund einer Nachfolgerrelation.... 2. Der deskriptive Teil. Wir haben uns nicht geeinigt, ob man besser mit Ding-Prädikaten oder sense data Prädikaten anfängt, Für das erstere: Ich, wohl auch Tarski; Hempel führt Popper an. Für das zweite: Goodman und Quine. [Zuletzt]: die Sprache soll möglichst gut intelligible sein. Es ist aber nicht klar, was wir eigentlich damit meinen. Sollen wir vielleicht die Kinder psychologisch fragen, was das Kind zuerst oder am leichtesten lernt?" (RC 090-16-05)

32. Some of the questions raised at the beginning of the conclusion are being pursued by Greg Frost-Arnold in his dissertation (University of Pittsburgh).

Chapter 14

1. The notes will be published in their entirety in Frost-Arnold (forthcoming).
2. The theory of virtual classes offered first in Quine (1944) was also a tool that the nominalist could exploit. On the philosophical import of such a theory see Martin (1964).

3. On the same topic it is interesting to read what Goodman writes to Quine on June 28, 1948: "I have finally been forced to give up the term 'nominalism' for the purpose for which it had been used in the thesis because the difficulty of keeping this use distinct from the other was too great. As a result, I use 'particularism' for what I earlier called nominalism. Had it not been that your articles and our joint one used 'nominalism' as they did, I probably would have kept the term for the use made of it in the thesis and used 'particularism' for the other purpose; as I think a fairly good case could be made out for the thesis that what I now call 'Particularism'— the refusal to countenance any other individuals than concrete ones—is closer to the rather amorphous traditional nominalism than is what you and we have called nominalism—the refusal to countenance any other entities than individuals." (MS Storage 299, box 4, folder Goodman, BPHLHU)

4. It is quite likely that Tarski never wrote a finished text for the lecture. Before the conference, on June 20, 1953, he wrote to Quine: "Dear Van, Some time ago Beth asked me whether I would be willing to take part in the conference on nominalism and Platonism and to give a talk there. I told him that I was interested in the topic and would be glad to attend the conference (should my plans of going this summer to Europe eventually materialize), but I also pointed out to him that it would be too late for me to prepare any formal talk and that I could only promise to make some contribution to the discussion. Now I have received a formal invitation and I am planning to answer in the same style. On the other hand it occurs to me that, if we are both to spend in Belgium a part of August and find time to refresh Washington talks in our memory, we might be able to concoct something by joining our forces. There are some points which interest me very much—e.g. the possibility of a semantic interpretation of quantifiers with variables of higher orders." (MS Storage 299, box 8, folder Tarski)

Chapter 15

1. See Carnap's *Autobiography* (1963, p. 61) and *Introduction to Semantics* (1942, p. x). While my attention here will be exclusively on Neurath, I should point out that among the early objectors to the semantic theory of truth one finds Jørgensen, Juhos, Nagel, and Naess.
2. Neurath (1936) is referred to in Tarski (1944) but only as reference for a survey of the discussions which took place in Paris in 1935. For a biography of Tarski see Feferman and Feferman (2004). For other aspects of Tarski's philosophical engagement see Mancosu (2004, 2009), Woleński (1993, 1995b).
3. See Hoffmann-Grüneberg (1988), Grundmann (1996), Hempel (1982), Rutte (1991), and Übel (1992). For an earlier account see Tugendhat (1960). For a general account of the protocol debate see Cirera (1994), Oberdan (1993), Übel (1992).
4. For the location of the unpublished correspondence quoted in the paper see the details given in the notes. Unpublished materials from the Carnap archive have a call number that always begin with "RC" followed by a string of numbers, i.e. "RC 102-55-05." All original sources quoted without a call number are from the Neurath *Nachlaß*.
5. "Ich kann nicht begreifen, warum Sie auch weiterhin die Semantik als 'bedenklich' halten obgleich sie den parallel laufenden und sehr nahe verwandten Erörterungen Carnaps, die die Begriffe 'Tautologie,' 'analytisch' u.s.w. betreffen nichts vorzuwerfen haben. Ich habe Ihre Korrespondenz mit Fr. Lutman ziemlich sorgfältig durchgeschaut, das hat mir aber nichts geholfen." (Tarski to Neurath, 7.ix.36, Neurath *Nachlaß*)
6. Maria Lutman-Kokoszyńska was born in 1905. She earned a Ph.D. under Twardowski after having studied philosophy and mathematics in Lvov. Her Ph.D. thesis was finished in 1928 and

was on the topic of "General and ambiguous names." A curriculum vitae dated 1961 is found in Carnap's *Nachlaß* under RC 088-57-07. For a bibliography of Kokoszyńska's works see Zygmunt (2004).

7. The only contributions I am aware of that treat to a certain extent Neurath's objections to Tarski are Mormann (1999) and Hoffmann-Grüneberg (1988). There is also a useful discussion in Übel (1992). While all of them consulted the Neurath-Tarski correpondence, they made very limited use of it and did not refer to the other archival sources I am using.

8. "Ich hoffe Sie bringen etwas, das auch dem EMPIRISMUS zugute kommt. Ich habe ja stets die Sorge, dass eines schönen Morgens ein Buch erscheinen wird: 'METAPHYSICA MODO LOGISTICA DEMONSTRATA'. Und dann werden wir noch daran schuld sein sollen." (Neurath to Tarski, 2.v.35, Neurath *Nachlaß*) See also the letter from Neurath to Carnap dated 19.iv.35 (RC 029-09-60) where Neurath gives voice to the same concern (the correspondence Carnap-Neurath was first studied in Hegselmann (1985)). A concern similar to the one expressed in the above quote is expressed as late as 1944: "I have the feeling to continue your *Logical Syntax* period, before you became Tarskisized with Aristotelian flavour, which I detest. I always fear, that you, a calculatory genius, supports [sic] a kind of possible scholasticism who [sic] leads away from scientific empiricism." (Neurath to Carnap, 1.4.44, RC 102-55-05) See also Neurath to Morris dated 18.xi.44.

By the way, to Neurath's request quoted above (letter dated 19.iv.35), Tarski replied through Kokoszyńska. In a letter from Kokoszyńska to Neurath (dated: Paris, 25.vii.35) the discussion is about Tarski's lecture: "Ich bin von Tarski beauftragt, Ihnen mitzuteilen, dass der endgültige Titel seines referates lauten wird 'Die Grundlegung der wissenschaftlichen Semantik.' Er hat, Ihrem Wünsche gemäss, sein ursprüngliches Thema geändert, um auf dem Kongresse solche Fragen zu behandeln, die von ziemlich prinzipieller Bedeutung für das Wissenschaftsganze sind." (Kokoszyńska to Neurath, 19.iv.35, Neurath *Nachlaß*) In my opinion Tarski's famous comment on physicalism and semantics in Tarski (1936a) is to be seen in light of Tarski's eagerness to please Neurath. I will not however be dealing with this topic. For a recent discussion see Frost-Arnold (2004).

9. In general, Neurath uses similar formulations with different correspondents. For instance, the lineage Brentano-Twardowski-Łukasiewicz-Tarski is also given in a letter to Hempel dated 20.ii.1943: "It is a sad situation that one has now to object to the Aristotelian metaphysics well formalized by Tarski and Carnap. I shall touch this point only but I think another day I shall explain this point in detail. The Scholasticism via Brentano-Twardowski-Łukasiewicz-Tarski appears now within a calculus but I think the calculus may be useful even within empiricism with a different interpretation, but hardly as it stands."(Neurath to Hempel, 20.ii.43, Neurath *Nachlaß*) See also Neurath to Martin Strauss dated 16.i.43 and Neurath to Carnap dated 27.viii.38. For an exposition of theories of truth from Brentano to Tarski see Woleński and Simons (1989). For a survey of the Lvov-Warsaw school see Woleński (1989a).

10. See Naess to Neurath, dated 8.vii.36.

11. On Tarski's theory of logical consequence and further references see Mancosu (2006).

12. "So viel ich sehe, scheint die von Ihnen und Dr Lutman vorgeschlagene Terminologie allerlei Verwirrung anzurichten. Vielleicht können sie betonen, welches die Tragweite ist. Ich denke nach wie vor, dass der von mir gemachte Vorschlag, die jeweils ausgezeichnete Enzyklopädie wahr zu nennen und demgemäss alle aus ihr folgenden oder in sie hinzukommenden anerkannten Sätze 'wahr' alle abgelehnten 'falsch', bezüglich des Terminus weniger gefährlich ist. Aber das ist sozusagen mehr ein pädagogisches Problem.

"Ich glaube, dass Ihre Darlegungen für die Fragen des logischen Empirismus im allgemeinen sehr wichtig sind. Insbesondere die Frage, wie die 'Sätze' neben anderen 'Dingen' auftreten usw., auch das Problem, wie analytische Sätze abzugrenzen wären, usw. Leider komme ich in der nächsten Zeit kaum dazu diese Fragen genauer zu studieren. Aber hoffentlich nicht in zu ferner Zeit. Da wird mir der Kongressbericht, in dem Ihre Arbeit erscheint sicher viel nützen." (Neurath to Tarski, 26.xi.35, Neurath *Nachlaß*)

13. "Ihre so liebenswürdig übersandte Arbeit habe ich gelesen. Ohne damit die geringste interne Kritik üben zu wollen, möchte ich doch sagen, dass sie sicher Verwirrung stiften wird. Die Einschränkungen, die Sie für den Wahrheitsbegriff vorbringen wird man nicht beachten, wohl aber Ihre Formulierungen als [Beweis] für allerlei metaphysische Spekulationen verwenden. Aber das ist eine soziologische Bemerkung, die deshalb nicht unwichtig sein muss." (Neurath to Tarski, 24.iii.36, Neurath *Nachlaß*)

14. "Ich danke Ihnen für die Mitteilungen über unsere 'Wahrheitsdefinitionen'. Natürlich liegen zunächst nur terminologische Unterschiede vor, aber ich habe sehr den Eindruck, dass bei der Diskussion auf realwissenschaftlichem Gebiet Ihre Anschauung sehr leicht ins Metaphysische abgleitet. Darüber müsste man sich ausführlich aussprechen. Ich habe einiges darüber an Dr Lutman-Kokoszyńska geschrieben.

"Wenn sie meinen, dass es eine Trivialität ist zu sagen, man spreche mit der Sprache über die Sprache, so kann ich darauf nur sagen, dass die Wissenschaft zu einem wesentlichen Teil darin besteht Trivialitäten gegen Irrtümer zu vertreten. Ich habe z.B. vom Beginn des Wiener Kreises an mich gegen die Versuche von WITTGENSTEIN gewehrt eine Art 'Erläuterungen' also 'nichtlegitime', quasi nicht- oder vorsprachliche Betrachtungen einzuführen, um dann über die Gegenüberstellung von 'der' Sprache mit 'der' Wirklichkeit zu reden, also *ausserhalb* der Sprache. Ich glaube, dass die 'Konstatierungen' von Schlick, die Sätze und doch wieder nicht Sätze sind aus dieser WITTGENSTEINSCHEN Metaphysik herzuleiten sind.

"Und sofern Ihre terminologische Wahl bedenkliche Konsequenzen nahelegt, ist sie vielleicht nicht ganz unabhängig von diesen Konsequenzen zustandegekommen. Auf der einen Seite wird betont, dass dieser Wahrheitsbegriff nur für formalisierten Sprachen gelte, andererseits ist der Wahrheitsbegriff gerade in nicht formalisierten Bereich von praktischer Bedeutung. Deshalb bin ich, wenn man den Terminus nicht überhaupt fallen lässt mehr für meine Terminologie, die im nicht formalisierten Bereich verwendbar bleibt. Während die von Ihnen und Lutman verwendete Terminologie im nicht formalisierten Bereich verwendet zu schlimmen Dingen führt. [Ich wüsste sehr gerne, ob Sie mit der Darstellung übereinstimmen, die Rougier in Paris von der Grenzverschiebung gegeben hat, die durch Ihre Thesen zwischen Metaphysik und Wissenschaft erfolgt sei.]" (Neurath to Tarski, 24.iv.36, Neurath *Nachlaß*)

15. "Innerhalb des Wiener Kreises war lange bevor der Kontakt mit Warschau aufgenommen wurde ein Gegensatz da, der sich auf die Frage bezog, ob es einen Sinn hat einen Vergleich der Sprache mit der 'Wirklichkeit' (z.B. die Sprache ist komplexer oder weniger komplex als die Wirklichkeit oder ebenso komplex usw.) sozusagen von einem Punkt *ausserhalb beider*, anzustellen.

"Die Ablehnung der Sätze über 'die' Wirklichkeit ging von Frank und innerhalb des 'Zirkels' in Wien vor allem von mir aus.

"Die Diskussion verknüpfte sich mit einer zweiten, die mit der Frage zusammenhing, ob 'Sätze über Sätze' sinnvoll seien oder nicht. Wittgenstein, Schlick und andere—die aber ihren Standpunkt weniger scharf vertraten—[aus] Waismann lehnten Sätzen über Sätzen strikt ab, so dass die Diskussionen über Sätze und Wirklichkeit sozusagen *ausserhalb* der Sprache vor sich gehen mussten, als 'Erläuterungen', sozusagen als 'Leiter', die man später wegwerfen müsse usw." (Neurath to Tarski, 7.v.36, Neurath *Nachlaß*)

16. For an account of Wittgenstein's theory of truth in the *Tractatus* see Newman (2002) and Glock (2006). See also Mulligan (1984).

17. "Ich habe nie daran gezweifelt, dass er es ablehnen würde, als Anhänger der üblichen Kohärenztheorie zu gelten; aber ich wollte auch nur behaupten, dass aus seinen Aeusserungen, wenn man sie erst nimmt, die Kohärenztheorie folge. Ich nahm an, dass ihm das selbst nicht klar sei, weil seine Gedanken zu undeutlich sind." (Schlick to Carnap, 5.vi.34, RC 029-28-10)

18. "Niemals habe ich das behauptet ... dass nämlich die Wahrheit in der Uebereinstimmung der Sätze besteht, sondern nur, in der Uebereinstimmung mit einer *bevorzugten* Satzmasse. Diese 'Bevorzugung' enthält alle jene Elemente, die für eine 'realistische' Auffassung wesentlich sind." (Neurath to Carnap, RC 102-50-01, 23.xii.1935)

19. For instance, Rougier in his unsigned introduction to Neurath (1935) contraposes in a note the positions of Carnap and Neurath (he also adds Popper, Poznanski and Wundheiler [1934]) to those of Schlick, Tarski, and Lutman (see Neurath (1935), p. 5). See also the comment reported by Neurath in Neurath (1936), p. 400 where Rougier sees Tarski's theory as a vindication of Schlick's position that sentences and reality can be compared for agreement.

20. "Als absoluter Wahrheitsbegriff wird letztens der klassische Wahrheitsbegriff bezeichnet, nach dem—wie man zu sagen pflegt—die Wahrheit eines Satzes in seiner Übereinstimmung mit der Wirklichkeit besteht. Man bezeichnet diese Auffassung der Wahrheit bekanntlich als Korrespondenztheorie. Dieser Theorie steht die Kohärenztheorie der Wahrheit gegenüber, nach der die Wahrheit eines Satzes in gewisser Übereinstimmung dieses Satzes mit andern Sätzen besteht. Bei einigen logischen Positivisten lässt sich in den letzten Jahren ein Übergang von einer Korrespondenztheorie der Wahrheit zu einer Kohärenztheorie nachweisen. In diesem Übergange hat teilweise die Überzeugung Ausdruck gefunden, der absolute Wahrheitsbegriff sei ein unwissenschaftlicher Begriff, der aus den philosophischen Untersuchungen ausgeschaltet werden soll, teilweise aber—wie es scheint—auch die Meinung, er liesse sich durch einen umfangsgleichen syntaktischen ersetzen und auf diese Weise in der Syntaxsprache definieren." (Kokoszyńska 1936a, p. 11)

21. Neurath was unsuccessful in changing this widespread perception. Indeed, Russell (1940) quipped that according to this view of Neurath "empirical truth can be determined by the police." BonJour (1985, p. 213), ascribes to Neurath a notion of coherence as mere consistency.

22. "Nachdem ich einige Korrekturbogen von Tarskis Aufsatz gelesen und gesehen habe, dass er für den *Wahrheitsbegriff* eine vollkommen korrekte Definition aufstellt, stimme ich Ihnen durchaus zu, dass 'wahr' und die andern mit ihm zusammenhängenden Begriffe als wissenschaftlich einwandfrei anzusehen sind. Meine und anderer Leute frühere Skepsis gegen diese Begriffe war ja insofern historisch berechtigt, als keine Definition bekannt war, die einerseits formal korrekt war und andrerseits die Antinomien vermied. Und die diese Begriffe verwendende Theorie, die 'Semantik' im Sinn von Tarski scheint mir ein wichtiges Wissenschaftsgebiet zu sein. Ich halte es für sehr verdienstvoll von Tarski, dass er dieses neue Gebiet erschlossen hat." (Carnap to Lutman-Kokoszyńska, July 19, 1935, RC 088-57-16)

23. "Du wirst bald sehn, wie bedenklich es ist 1. dass man uns den Titel Kohärenztheorie angehängt hat ... und 2. dass die wirklich wertvollen Betrachtungen von Tarski und Lutman mit den Terminus 'wahr' herumlaufen. Wenn du noch kannst, solltest Du dafür einen anderen Namen verwenden. Ich kann mir nicht denken, dass dieser Terminus je zur Klärung dient,

wohl aber, dass er ständig Verwirrung stiften wird ... Ich wills nur noch einmal recht nett und ernst Dir gesagt haben, weil ich nur schmerzlich es empfinde, was z.B. Rougier im Schlusswort über die Verschiebung der Demarkationslinie zugunsten der Mataphysik sagte." (Neurath to Carnap, December 8, 1935, RC 102-50-04) On Rougier's comment see Neurath (1936, p. 401).

24. "Ich bin völlig Ihrer Meinung, daß es eine wichtige Aufgabe der Wissenschaft ist 'Trivialitäten gegen Irrtümer zu vertreten.' Eben deshalb habe ich ja selbst vielmals betont, daß man stets in einer Sprache über eine andere Sprache sprechen muß—und nicht außerhalb der Sprache (vom rein deduktiven Standpunkte aus ist übrigens meine ganze 'Semantik' fast als eine Trivialität anzusehen; das ärgert mich nicht im wenigsten). Es ist—wie mir scheint—ein großer Fehler, wenn Wittgenstein, Schlick usw. von 'der' Sprache anstatt von Sprachen (in Mehrzahl) sprechen; das ist vielleicht die echte Quelle der Wittgensteinschen 'Metaphysik'. Nebenbei gesagt, denselben Fehler scheinen auch alle diejenigen zu begehen, die im Zusammenhang mit dem Stichwort 'Einheitswissenschaft' über die Einheitssprache der Wissenschaft reden. Wir wissen ja alle—auf Grund der Erörterungen aus der Semantik und Syntax—, daß es streng genommen keine Einheitssprache gibt, in der die ganze Wissenschaft ausdrückbar wäre. Es genügt nicht zu sagen, daß das nur eine vorläufige, unpräzise Formulierung ist; denn wie soll die endgültige, präzise Formulierung lauten? (Kokoszyńska hatte vor kurzer Zeit einen Vortrag über das Problem der Einheitswissenschaft in der hiesigen Phil. Gesell. und hat u.a. diesen Punkt einer Kritik unterworfen; es soll ein Aufsatz von ihr in der polnischen Sprache darüber erscheinen)." (Tarski to Neurath, 28.IV.36, Neurath *Nachlaß*)

25. "Was nun meine 'terminologische Wahl' betrifft, so kann ich Ihnen versichern, daß sie erstens ganz unabhängig von der Wittgensteinschen Metaphysik zustandegekommen ist und daß es zweitens überhaupt keine 'Wahl' war. Das Problem der Wahrheit kam speziell in der polnischen Philosophischen Litteratur sehr oft vor, man hat immer gefragt, ob man den Wahrheitsbegriff mit den und den Eigenschaften (die ich später in meiner Arbeit genau präzisiert habe) in einwandfreier Weise definieren und verwenden kann (vgl. Z.B. die 'Elemente' von Kotarbiński). Ich habe einfach dieses Problem positiv gelöst und habe bemerkt daß sich diese Lösung auf andere semantische Begriffe ausdehnen läßt. Ebenso wie Sie bin ich sicher, daß man daraus verschieden Mißbräuche machen wird, daß verschiedene Philosophen dieses Ergebnis rein logischer Natur in unerläßlicher Weise 'hinausinterpretieren' werden—das ist das gemeinsame Schicksal aller kleineren und größeren Entdeckungen aus dem Bereiche der exakten Wissenschaften (man vergleicht ja manchmal die Philosophen mit den 'Hyänen des Schlachtfeldes')." (Tarski to Neurath, 28.iv.36, Neurath *Nachlaß*)

26. "Aber ich muß Ihnen offen gestehen: wenn ich auch Ihren Kampf gegen die Metaphysik keinewegs unterschätze (noch mehr unter sozialem, als unter wissenschaftlichem Gesichtspunkt), so lebe ich persönlich nicht in einer ständigen, panischen Angst vor der Metaphysik. Wie ich erinnere, hat einmal Menger etwas geistreiches über die Furcht vor Antinomien geschrieben; es scheint mir, daß man das alles—mutatis mutandis—auch auf die Angst vor der Metaphysik übertragen könnte. Es ist eine hoffungslose Aufgabe, sich stets vor dem Vorwurf einer Metaphysik zu warnen. Das wird mir besonders klar, wenn ich hier bei uns verschiedenen Angriffe eben auf die Metaphysik des Wiener Kreises (und zwar Ihrer und Carnapschen Richtung) höre, wenn z. B. Łukasiewicz a propos der 'Logischen Syntax' über Carnaps Philosophie, Philosophieren usw. spricht (das hat in seinem Mund ungefähr denselben Sinn wie in Ihrem "Metaphysik"). Dasselbe, was Sie mir wegen des Wahrheitsbegriff vorwerfen,

wirft man Carnap wegen der Einführung der Termini 'analytisch', 'synthetisch' u.s.w. vor ('Rückkehr zu der Kantschen Metaphysik'); und es scheint mir, daß ich im Grunde noch mehr als Carnap berechtigt war den von mir erörterten Begriff als Wahrheit zu bezeichnen. Im allgemeinen ist es eine wertvolle Aufgabe alte Gefässe mit neuem Trunk zu füllen." (Tarski to Neurath, 28.iv.36, Neurath *Nachlaß*)

27. "Noch ein Punkt in diesem Zusammenhang: mein Wahrheitsbegriff gelte nur für die formalisierten Sprachen, andrerseits ist der Wahrheitsbegriff gerade im nicht formalisierten Bereich von praktischer Bedeutung. Das kann man wörtlich auf alle präzisen Begriffe der Syntax und Semantik (Konsequenz, Gehalt, logischer und deskriptiver Begriff u.s.w.) übertragen: alle diese Begriffe können nur annährungsweise auf die nicht-formalisierten Sprachen (also auf die aktuellen Sprachen aller Realwissenschaften) bezogen werden; Wahrheit ist hier keine Ausnahme." (Tarski to Neurath, 28.iv.36, Neurath *Nachlaß*)

28. Appeal to Naess' work is also found in a letter from Neurath to Tarski dated 27 May, 1937: "Dass meine Bedenken gegen Semantik sich nur auf die Interpretationen beziehen, die im Empirismus in Frage kommen erwähnte ich schon. Ich glaube, wenn Sie an den Wahrheitsbegriff bei Kotarbiński anknüpfen, Lutman von einen üblichen Wahreitsbegriff spricht, so ist etwas zu wenig Vorsorge getroffen die lebendige Diskussion damit erreicht zu haben, denn es gibt sehr viele Auffassungen von Wahrheit, wie NESS [*sic*] festgestellt hat." For Naess' work see Naess (1936) and (1938).

29. "Die 'soziologische' Wahrheitsdefinition lässt sich aufrechterhalten, so lassen sich in ihrem Sinne gewisse Sätze als 'jetzt' wahr kennzeichnen.

"Die 'soziologische' Wahrheitsdefinition entspricht gewissen Elementen traditioneller Auffassung. Die T.L. Definition entspricht dem Sprachgebrauch nicht in bevorzugten Weise (historische Frage).

"Die T.L. Definition ist nur innerhalb formalisierter Sprachen verwendbar.

"Die T.L. Terminologie verlockt dazu sie für nicht-formalisierte Sprachen zu verwenden und absolutistisch zu deuten.

"Die begründenden Erörterungen von L.K. über 'zwar anerkannt, aber nicht zutreffend' scheinen unmittelbar absolutistische Elemente zu enthalten und innerhalb der Gesamtwissenschaft nicht verwendbar zu sein, weder nach Auffasung von Neurath (Carnap, Hempel, usw.) noch auch nach sonst geäusserten Auffassung von T. and L." (Neurath to Kokoszyńska, 23.iv.36, Neurath *Nachlaß*)

30. "Sie wollten—soweit ich verstehe—dass ich die Sachlage so schildere, als ob kein Widerspruch zwischen Ihrer bisherigen Haltung gegen über dem klassischen Wahrheitsbegriff un[d] den in meinen *Erkenntnis*-Bemerkungen enthaltenen Gedanken bestehe. Ein solcher Widerspruch scheint aber doch zu bestehen. Es handelte sich ja darum, ob man sich in verlässlicher Weise mit einem Begriffe bedienen kann, in dem sozusagen von einer 'Übereinstimmung mit der Wirklichkeit' die Rede war. Sie haben teilweise diesen Begriff ganz abgelehnt, da Sie meinten, die Feststellung einer solcher 'Übereinstimmung' müsse ein Ausgehen aus den Rahmen der Sprache erfordern/was unmöglich ist/und teilweise haben sie versucht, diesen Begriff durch einen soziologisch-syntaktischen zu ersetzten. Nun zeigt es sich aus den Untersuchungen von Tarski, dass man von jener 'Übereinstimmung zwischen Sätzen und Wirklichkeit' sprechen/also sie innerhalb der Sprache betrachten/kann, indem man nämlich Sätze aufstellt, in denen nicht nur Namen von Sätzen aber auch von anderen Dingen auftreten. Gegen Aufstellung von solchen Sätzen haben Sie nichts einzuwenden ausser—was speziell Sie

betrifft—dass sie in empirischen Wissenschaften nicht nötig sind. Es hat sich also gezeigt, dass man eben diese von Ihnen bisher abgelehnten Begriffe in korrekter Weise behandeln kann. Darin scheint mir der vorher erwähnte Widerspruch zu stecken." (Kokoszyńska to Neurath, 6.ix.36, Neurath *Nachlaß*)

31. In addition to Carnap (1937) and Neurath (1937d) there are three additional pieces by Neurath entitled respectively "Fuer Die Privatsitzung, 30 Juli 1937" (1937c); "Diskussion Paris 1937 Neurath-Carnap" (1937b); "Bemerkungen zur Privatdiskussion" (1937a), classified as K.31, K.32 and K.33 in Neurath's *Nachlaß*.

32. Tarski reports positively about the discussion in a letter to Popper dated 4.x.37 (Popper-Tarski correspondence, Hoover Institution, Stanford, Box number 354. Folder ID: 8).

33. "Einen Satz von der Form A 1b [B], der das Wort 'wahr' enthält, wird man seltener verwenden als A 1c [C], nämlich gewöhnlich nur dann, wenn Frage, Zweifel oder Bestreitung vorausgegangen sind, oder wenn aus sonstigen Gründen eine stärkere emotionelle Betonung der Behauptung zum Ausdruck kommen soll... Aber das ist nur ein psychologischer, kein logischer Unterschied. Das zeigt sich darin, das niemand in der Umgangssprache, dem zwei solche Sätze wie A 1b [B] und A 1c [C] zur Entscheidung vorgelegt werden, den einen akzeptieren wird, den andern aber ablehnen oder auch nur unentschieden lassen wird." (Carnap 1937, p. 9, RC 080-52-01)

34. "1. Anregung für die Gruppe derer, die Semantik betreiben wollen und dabei empiristisch und antimetaphysisch eingestellt sind. Diese werden sich—nicht nur mit Rücksicht auf sich selbst, sondern vor allem auch auf ihre Leser—ihre Terminologie und Formulierungen möglichst so einrichten, dass die Abgrenzung gegen metaphysische Probleme immer deutlich bleibt. Sie werden ferner auch die Frage im Auge behalten, in welchem Ausmaß semantische Sätze in nicht-semantische übersetzbar sind; dies vor allem zugunsten derjenigen in unsern Kreisen, die Q aus was für Gründen immer Q sich bemühen, semantische Begriffe zu vermeiden. 2. Anregung für die Gruppe derer, die Bedenken gegen die semantischen Begriffe haben. Diese werden zunächst eine abwartende Haltung einnehmen und nicht in der Oeffentlichkeit schon gegen die Semantik im ganzen polemisieren, bis die weitere Entwicklung erstens erkennen läßt, ob die Arbeit auf dem Gebiet der Semantik für die Wissenschaft und besonders für unsere Gesamtaufgabe der Wissenschaftsanalyse fruchtbar ist oder nicht, und zweitens, ob die gefürchtete Gefahr des Abgleitens in die Metaphysik wirklich eintritt oder nicht. Sie werden also nicht die semantischen Begriffe im ganzen als metaphysisch bezeichnen, sondern nur die einzelnen etwa auftretenden ihnen bedenklich erscheinenden Formulierungen kritisieren, insbesondere, soweit sie etwa tatsächlich zu Scheinproblemen Anlaß geben." (Carnap, 1937, pp. 11–12, RC 080-52-01)

35. This task can be seen as complementary to that carried out in Mormann (1999) which is not as detailed on the reconstruction of the debate but takes a broader view of the philosophical positions held by Neurath and traces his opposition to semantics to his anti-Cartesianism. However, I believe that that is only one of the sources of Neurath's objections. We have seen that his anti-Wittgensteinianism was a powerful factor.

Chapter 16

1. For Tarski (1936d) (originally in German) I will use the English translation from Tarski (1983). For Tarski (1936c) (originally in Polish) I will use the English translation from Tarski (2002).

Although the two versions are different at times (see Introductory notes to Tarski (2002)) I will still refer to 'the' 1936 paper. Page numbers will refer to the translation, when available.

2. When did the word 'model' become common currency in axiomatics? 'Model', as an alternative terminology for interpretation, seems to make its appearance in the mathernatical foundational literature with von Neumann (1925), where he talks of models of set theory. However, the new terminology owes its influence and success to Weyl's 'Philosophy of Mathematics and Natural Science' (1927). In introducing techniques for proving independence, Weyl describes the techniques of 'construction of a model [*Modell*]' (p. 18) and described both Klein's construction of a Euclidean model for non-Euclidean geometry and the construction of arithmetical models for Euclidean geometry (or subsystems thereof) given by Hilbert. Once introduced in the axiomatical literalure by Weyl, the word 'model' finds a favorable reception. It occurs in Carnap (1927, 2000 [1927–29], 1930), Kaufmann (1930), and in articles by Gödel (1930b), Zermelo (1929, 1930) and Tarski (1935b, 1936b). The usage is, however, not universal. The word 'model' is not used in Hilbert and Ackermann (1928) (but it is found in Bernays 1930b). Fraenkel (1928) speaks about realizations or models (p. 353) as does Tarski 1936d. They do not follow Carnap in making a distinction between realizations (concrete, spatio-temporal interpretations) and models (abstract interpretations). 'Realisation' is also used by Baldus (1924) and Gödel (1929). Among the few variations one can mention 'concrete representation' (Veblen and Young (1910, p. 3) and Young (1917, p. 43)). It should be pointed out here that while the word 'model' was common currency in physics (see 'dynamical models' in Hertz (1894) and Bolzmann (1974)), it is not as common in the literature on non-Euclidean geometry, where the terminology of choice remains 'interpretation' (as in Beltrami's 1868 interpretation of non-Euclidean geometry). However, 'Modellen', i.e. desktop physical models, of particular geometrical surfaces adorned the German mathematics departments of the time. Many thanks to Jamie Tappenden for useful information on this issue.

The reader should see Sinaceur (1997), Webb (1995), Guillaume (1994) for the notion of interpretation in mathematics in the nineteenth century and earlier periods.

As for 'structure' it is not used in the 1920s as an equivalent of 'mathematical system'. Rather, mathematical systems have structure. In *Principia Mathematica* (vol. 2 (1912), part iv, pp. 150 ff.) and then in Russell (1919, ch. 6) we find the notion of two relations having the same structure, In (Weyl 1927, p. 21), two isomorphic systems of objects are said to have the same structure. This process will eventually lead to the idea that a 'structure' is what is captured by an axiom system: 'An axiom system is said to be monomorphic when exactly one structure belongs to it [up to isomorphism]' (Carnap 2000 [1927–29], p. 127). Here it should be pointed out that the use of the word 'structure' in the algebraic literature was not yet widespread, although the structural approach was. It seems that 'structure' was introduced in the algebraic literature in the early 1930s by Oystein Ore to denote what we nowadays call a lattice (see Vercelloni 1988, Corry 2004b).

For general information about Tarski's metatheoretical conceptions during the period in question see, among others, Bellotti (2002), Czelakowski and Malinowski (1985), de Rouilhan (1998), Sinaceur (2000, 2001), Blok and Pigozzi (1988), Woleński (1989a), Feferman (2004c).

3. Bays (2001) is an extended attempt to show that a fixed-domain conception of model creates fewer difficulties than Sher, Gomez-Torrente and others claim. In consonance with the spirit of this paper, whose main aim is to determine which notion of model is defended in the 1936 paper, I will not enter into a detailed discussion of how to account for the Löwenheim-Skolem

theorem in this context (but see the comments in the conclusion) or for the logical validity of inferences involving the natural numbers. This is left for another paper. However, Bays' article is the obvious starting point for any such discussion. John Corcoran has since the 1970s claimed as evident that Tarski upheld the fixed domain conception of model. An overview of his position is now given in the unpublished nine page typescript "The Absence of Multiple Universes of Discourse in the 1936 Tarski Consequence-definition Paper". Unfortunately, given the debates to which the 1936 paper has given rise to, one cannot be left satisfied by claiming either the fixed domain conception or the relative domain conception of model as evident. A paper which defends the fixed domain conception and which comes close, as far as I can tell, to my final concluding remarks is Sagüillo (1997).

4. See Chapter 1 of this volume for an introduction to the study of metatheoretical properties of axiomatic systems and semantics from 1900 to 1935.

5. See (Huntington 1911, p. 172) for an earlier statement by Huntington. Whitehead (1907) is the earliest clear statement of the conception I have been able to find: "Par conséquent (la classe des points étant indéterminée) les axiomes ne sont pas du tout des propositions: ils sont des fonctions propositionelles. Un axiom, en ce sens, n'étant pas une proposition, ne peut être ni vrai ni faux." (Whitehead 1907, p. 35)

6. Postulate 4 says: 'If X is a point of the segment [AB], then [AB] is the simple sum of the two segments [AX] and [BX]'. (Huntington 1913, p. 537)

7. The quotes are also important for the issue of 'truth in a structure'. I will not deal with the problem in this paper but see Chapter 1 of this volume.

8. For reasons of space I cannot get into Carnap's conception of model, which is very similar to Tarski's. See Carnap (1927, 1930, 1934a,b) and Carnap and Bachmann (1936). For important aspects of Carnap's work during this period see the works by Awodey cited in the bibliography. In general, from the point of view of Tarski's 1936 paper, one should devote a whole paper to the relationship between Carnap and Tarski on the notion of model and logical consequence.

9. In 1939 Tarski also entertains the possibility of having, in alternative to a type theory, a type-free system resembling Zermelo's set theory as background logical theory.

10. Concerning the theories Tarski treated in the 1920s, here is a brief attempt to survey the background logic. In the early articles on set theory, the background theory is usually that given by Zermelo (1908c) (see (Tarski (1924b), vol. I of 1986, p. 41); (Tarski (1924a), vol. I of 1986, p. 67)) but no specific background logical system is mentioned. Tarski and Lindenbaum (1926) points out that most of the set-theoretical results of the paper can just as well be obtained in the systems of *Ontology* of Leśniewski or in *Principia Mathematica*. Exceptions are noted on p. 186 of vol. I of Tarski (1986a). Of interest in this article is the description of the theory of arithmetic of ordinal numbers (p. 196, vol. I of Tarski (1986a)). This is given in terms of four axioms involving the predicate '<' with *Principia Mathematica* as the background system. In this part of the article Tarski shows his mastery of independence proofs. Remarkable is also the fact that the system of *Principia* and the *Ontology* of Leśniewski are seen as systems of set theory (see p. 200; the same in (Tarski 1929a, p. 241), where Tarski remarks that the system of *Principia*, unlike Zermelo-Fraenkel set theory, cannot even prove the existence of a single infinite cardinal number). The next item of interest is "Les fondements de la géométrie des corps". As the English translation is quite different from the original 1929 article, it is quite important to refer to the original. The background system here is Lesniewski's ontology. However, in note 1 on p. 229 Tarski specifies that he is using 'class' as it is used in *Principia* and that he is treating spheres as objects of the lowest rank in the system of *Principia*, that

is as individuals, and points (classes of spheres) as objects of second rank. He states that the axiom system considered is categorical but the reader should note that the notion of categoricity appealed to is the Veblen notion and not the stronger one which Tarski considers in 1935–36.

In 1930a Tarski begins emphasizing that a theory can become object of metatheoretical study only if it based on a determinate logic (see (Tarski (1930a), pp. 347–38 of vol. I of Tarski (1986a)). Moreover, the metatheoretical enterprise requires a general logical basis, which Tarski says can be taken to be that offered by *Principia* without assuming choice or infinity (he uses infinity but claims that its use is not essential). The (1931) article "On definable sets of real numbers" (vol. I of Tarski 1986a, pp. 519–48) is, as far as I know, the first place where Tarski explicitly uses the simple theory of types as the background logical theory for the theory of real numbers (and in the metatheory for studying the system of real numbers). For the importance of simple type theories as a standard logical framework in the twenties and thirties see Ferreiros (1999).

11. I would like to point out that the Polish text makes an 'iff' claim while the German text has an 'if' claim: "The extreme would be the case in which we treated all terms of the language as logical: the concept of following formally would then coincide with the concept of following materially: the sentence X would follow from the sentences of the class K if and only if either the sentence X were true or at least one sentence of the class K were false." (Tarski 2002, pp. 188–89) For a pointed answer by Etchemendy to Ray see (Etchemendy 2008, note 12).

12. Generalizing to systems with n non-logical constants for classes and m non-logical relations is easy. Much of the work on postulate theory at the time involve the study of mathematical systems formed by a class with operations defined on them, say (M, +). My discussion in the text cover these cases too, as we can think of n-ary operations in terms of $n+1$-ary relationships. This is already pointed out in Bôcher (1904).

13. The reason why Langford thinks he needs to state things that way need not detain us here. On Langford's work and its influence on Tarski see Scanlan (2003).

14. A possible example to clarify the above distinction is the following. Suppose V the class of individuals is countable. Take the axioms for the first-order theory of a dense linear ordering without endpoints. Consider an ordering $<$ on V which satisfies the axioms for the theory. Consider now a countable subset A of V such that $A \neq V$ and define a relationship $<_A$ on it satisfying the axioms of the theory. Since $(V, <)$ and $(A, <_A)$ are both countable they are isomorphic, by a classical theorem of set theory. However, no isomorphism between $(V, <)$ and $(A, <_A)$ can send V one-one onto V, for by assumption A is strictly included in V. Thus the theory is Veblen categorical but not absolutely categorical.

15. There are several passages in the article where Tarski keeps reminding the reader that certain results only works if the theory has an axiom that forces the identity of the universe V with, say, the points of geometry or the natural numbers. See for instance (Tarski 1983, p. 313, note 2), where Tarski says: "This is exact only if the axiom system of arithmetic contains a sentence to the effect that every individual is a number".

16. Note moreover, that not in all cases are we going to have a model given with a 'universe of discourse' over which the primitive relations and/or function symbols are defined. In Tarski (1935b), for instance, Tarski gives an axiomatization of dense linear orders without endelements in which the only primitive is a binary relationship R. A model for such a theory is given by a finite sequence containing only the relation R (the 'universe of discourse' is implicitly defined by the field of R).

17. A brief summary of this text has been published by Jan Tarski and Jan Woleński in History and Philosophy of Logic without an exact specification of its location (see Tarski (1995)). I was recently able to locate this text in Carton 15 of Tarski's archive in the Bancroft Library at U.C. Berkeley. J. Tarski and J. Woleński conjecture that this might be the text of the second of a series of lectures given by Tarski at Harvard in 1939. However, this does not seem right. I base my dating of the lecture on the following footnote found in Quine and Goodman (1940): The latter notion (synthetically complete), under the name 'completeness relative to logic', is due to Tarski. It is easier to formulate than the older concept of categoricity, and is related to the latter as follows: systems which are categorical (with respect to a given logic) are synthetically complete, and synthetically complete systems possessed of logical models are categorical. These matters were set forth by Tarski at the Harvard Logic Club in January, 1940 and will appear in a paper "On completeness and categoricity of deductive theories" (Quine and Goodman 1940, footnote 3, p. 109).

18. Let us recall that a theory is syntactically complete (Tarski says 'absolutely complete or simply complete') "if every sentence which can be formulated in the language of this theory is decidable, that is, either derivable or refutable in this system." (Tarski 1940, p. 1) Modulo some trivial facts about the theory, the condition is equivalent to stating that a theory is complete "if for every sentence either it or its negation is derivable." (p. 1) Tarski points out that due to Gödel's incompleteness theorems the propery of "absolute completeness occurs rather as an exception in the domain of the deductive sciences, and by no means can it be treated as a universal methodological demand." (Tarski 1940, p. 3)

19. Since the term 'logically valid sentence' has already been used by Tarski in this lecture to indicate axioms of logic and the theorems derivable from it, we need a new term to indicate the sentences which are true in all models. The Tarski (1936d) paper suggests, following Carnap, to call a sentence 'analytical' if every sequence of objects is a model of it. And modulo a few assumptions on the languages considered, Tarski claims in 1936 "that those and only those sentences are analytical which follow from every class of sentences (in particular from the empty class)."(Tarski 1936d, p. 418)

20. Before introducing the notion of semantical completeness, Tarski considers a different concept: 'completeness with respect to the logical basis' or simply 'relative completeness'. This will be treated below.

21. That every logical sentence is determinate is one of the central claims of Carnap's account of analyticity. I plan to carry out a comparative analysis of Tarski and Carnap in a different paper.

Chapter 17

1. The text is translated in Chapter 18. I should point out that the first three pages of Tarski's unpublished typescript overlap significantly with another text Tarski prepared for publication in 1940 but which saw the light of day only in 1967 (Tarski 1967). For an introduction to Tarski's life and work see Feferman and Feferman (2004). Some parts of this chapter follow the exposition found in Mancosu (2006) and Chapter 1 of this volume.

2. An axiomatic theory is called semantically complete (relative to a given semantics) if any of the following four equivalent conditions holds:

   (1) For all formulas φ and all models M, N of T, if $M \vDash \varphi$ then $N \vDash \varphi$.

   (2) For all formulas φ, either $T \vDash \varphi$ or $T \vDash \neg\varphi$.

(3) For all formulas φ, either $T \vDash \varphi$ or $T \cup \{\neg\varphi\}$ is not satisfiable.

(4) There is no formula φ such that both $T \cup \{\varphi\}$ and $T \cup \{\neg\varphi\}$ are satisfiable.

The formulations are those given in Awodey and Reck (2002a).

3. Among defenders of the fixed domain conception of logical consequence one finds Quine. On Quine's conception of truth and logical consequence see the enlightening paper Sagüillo (2002).

4. The example stems from one given by Tarski to Corcoran to show the difference between intrinsic and absolute categoricity. For details see Mancosu (2006, pp. 224–25).

BIBLIOGRAPHY

ABRUSCI, MICHELE (1978), 'Autofondazione della matematica. Le ricerche di Hilbert sui fondamenti della matematica', in V. Michele Abrusci (ed.), *David Hilbert. Ricerche sui Fondamenti della Matematica*, 14–131 (Naples: Bibliopolis).
—— (1983), 'Paul Hertz's logical works: Contents and relevance', in *Atti del Convegno Internazionale di Storia della Logica, Le teorie delle modalità. 4–8 December 1982, San Gimignano*, 369–74 (Bologna: CLUEB).
ACKERMANN, WILHELM (1924), 'Begründung des "tertium non datur" mittels der Hilbertschen Theorie der Widerspruchsfreiheit', *Mathematische Annalen*, 93, 1–36.
—— (1928a), 'Über die Erfüllbarkeit gewisser Zählausdrücke', *Mathematische Annalen*, 100, 638–49.
—— (1928b), 'Zum Hilbertschen Aufbau der reellen Zahlen', *Mathematische Annalen*, 99, 118–33.
—— (1940), 'Zur Widerspruchsfreiheit der Zahlentheorie', *Mathematische Annalen*, 117, 162–94.
AJDUKIEWICZ, KAZIMIERZ (1921), *Z metodologii nauk dedukcyjnych* (Lwów). English translation by J. Giedymin, *From the Methodology of the Deductive Sciences*, Studia Logica, 19, (1966), 9–46.
ANDRADE, JULES F. C. (1898), *Leçons de Méchanique Physique* (Paris: Gauthier-Villars et fils).
ANDREWS, PETER B. (2003), 'Herbrand Award Acceptance Speech', *Journal of Automated Reasoning*, 31, 169–87.
AVIGAD, JEREMY and ZACH, RICHARD (2002), 'The epsilon calculus', *Stanford Encyclopedia of Philosophy*, http://plato.stanford.edu/entries/epsilon-calculus/.
AWODEY, STEVE and CARUS, ANDRE W. (2001), 'Carnap, Completeness, and Categoricity: the Gabelbarkeitssatz of 1928', *Erkenntnis*, 54, 145–72.
AWODEY, STEVE and RECK, ERICH (2002a), 'Completeness and Categoricity. Part I: Nineteenth-century axiomatics to twentieth-century metalogic', *History and Philosophy of Logic*, 23, 1–30.
—— (2002b), 'Completeness and Categoricity. Part II: 20th century metalogic to 21st century semantics', *History and Philosophy of Logic*, 23, 77–94.
AYER, ALFRED J. (1977), *A Part of my Life* (London: Collins).
BADESA, CALIXTO (2004), *The Birth of Model Theory. Löwenheim's Theorem in the Frame of the Theory of Relatives* (Princeton: Princeton University Press).
BAIRE, RENÉ-LOUIS, BOREL, ÉMILE, HADAMARD, JACQUES, and LEBESGUE, HENRI (1905), 'Cinq lettres sur la théorie des ensembles', *Bulletin de la Société Mathématique de France*, 33, 261–73. English translation in Appendix 1 of Moore (1982).
BAKER, ALAN (2005), 'Are There Genuine Mathematical Explanations of Physical Phenomena?', *Mind*, 114, 223–38.
BAKER, GORDON and HACKER, PETER M. S. (1985), *Wittgenstein. Rules, Grammar and Necessity* (Oxford: Blackwell).
BALDUS, RICHARD (1924), *Formalismus und Intuitionismus in der Mathematik* (Karlsruhe: Braun).
—— (1928), 'Zur Axiomatik der Geometrie I: Über Hilberts Vollständigkeitsaxiom', *Mathematische Annalen*, 100, 321–33.

BARZIN, MARCEL and ERRERA, ALFRED (1927), 'Sur la logique de M. Brouwer', *Académie Royale de Belgique, Bulletin*, 13, 56–71.

BAYS, TIMOTHY (2001), 'On Tarski on models', *The Journal of Symbolic Logic*, 66, 1701–26.

BECKER, OSKAR (1923), 'Beiträge zur phänomenologischen Begründung der Geometrie und ihrer physikalischen Anwendungen', *Jahrbuch für Philosophie und phänomenologische Forschung*, 6, 385–560. Partially translated in T. Kisiel and J. Kockelmans, *Phenomenology and Natural Science* (Evanston, Ill.: Northwestern University Press, 1975).

—— (1927), 'Mathematische Existenz', *Jahrbuch für Philosophie und phänomenologische Forschung*, 8, 439–809.

—— (1927/28), 'Das Symbolische in der Mathematik', *Blätter für Deutsche Philosophie*, 1, 329–48.

—— (1928/29), 'Über den sogenannten "Anthropologismus" in der Philosophie der Mathematik', *Philosophischer Anzeiger*, 3, 369–87.

—— (1930a), 'Die apriorische Struktur des Anschauungsraumes', *Philosophischer Anzeiger*, 4, 129–62.

—— (1930b), 'Zur Logik der Modalitäten', *Jahrbuch für Philosophie und phänomenologische Forschung*, 11, 497–548.

—— (1959), *Grösse und Grenze der mathematischen Denkweise* (Freiburg and Munich: Verlag Karl Alber).

—— (2005), 'Briefwechsel mit Dietrich Mahnke', in Volker Peckhaus (ed.), *Oskar Becker und die Philosophie der Mathematik*, 245–78 (Munich: Wilhelm Fink Verlag). Edited by Bernd Peter Aust and Jochen Sattler.

BEHMANN, HEINRICH (1914), 'Über mathematische Logik', Unpublished typescript dated December 1, 1914. Behmann Archive, Erlangen.

—— (1918), *Die Antinomie der transfiniten Zahl und ihre Auflösung durch die Theorie von Russell und Whitehead*, Dissertation, Universität Göttingen. 352 pp.

—— (1921), 'Entscheidungsproblem und Algebra der Logik', Unpublished typescript dated May 10, 1921. Behmann Archive, Erlangen.

—— (1922a), 'Beiträge zur Algebra der Logik, insbesondere zum Entscheidungsproblem', *Mathematische Annalen*, 86, 163–229.

—— (1922b), 'Die Antinomie der transfiniten Zahl und ihre Auflösung durch die Theorie von Russell und Whitehead', *Jahrbuch der Mathem.-Naturwiss. Fakultät in Göttingen*, 23, 55–64.

—— (1922c), 'Letter to Bertrand Russell, August 8, 1922', Unpublished letter. Behmann Archive, Erlangen. Also found in the Russell Archive, McMaster, Hamilton, 1638.

—— (1923), 'Letter to Bertrand Russell, May 10, 1923', Unpublished letter. Behmann Archive. Also found in the Russell archive, McMaster, Hamilton, 110205d.

—— (1930), 'Zur Frage der Konstruktivität von Beweisen', Unpublished typescript, Wissenschaftshistorische Sammlung, ETH, Zurich, Hs 974: 18.

BELL, JOHN (2000), 'Hermann Weyl on intuition and the continuum', *Philosophia Mathematica*, 8, 259–73.

—— (2004), 'Hermann Weyl's later philosophical views: his divergence from Husserl', in Richard Feist (ed.), *Husserl and the Sciences*, 173–85 (Ottawa: Ottawa University Press).

BELLOTTI, LUCA (2002), 'Tarski on Logical Notions', *Synthese*, 135, 401–13.

BELTRAMI, EUGENIO (1868), 'Saggio di interpretazione della geometria non-euclidea', *Giornale di Matematiche*, 6, 284–312.

BENACERRAF, PAUL and PUTNAM, HILARY (eds.) (1983), *Philosophy of Mathematics* (Cambridge: Cambridge University Press), 2nd edn.
BERNAYS, PAUL (1910), 'Das Moralprinzip bei Sidgwick und bei Kant', *Abhandlungen der Fries'schen Schule*, III(3), 501–82.
—— (1913a), 'Über den transzendentalen Idealismus', *Abhandlungen der Fries'schen Schule*, IV(2), 365–94.
—— (1913b), 'Über die Bedenklichkeiten der neueren Relativitätstheorie', *Abhandlungen der Fries'schen Schule*, IV(3), 459–82.
—— (1917), 'Bernays über Weyls Kritik der Analysis', Hilbert *Nachlaß*, Cod. Ms. Hilbert 685, Nr. 3, Blätter 13–20. (Staats- und Universitätsbibliothek, Göttingen).
—— (1918), *Beiträge zur axiomatischen Behandlung des Logik-Kalküls, Habilitationsschrift*, Universität Göttingen. Bernays *Nachlaß*, WHS, ETH Zurich Archive, Hs 973.192.
—— (1920), 'Letter to Russell', Russell archive, McMaster, Hamilton, 110208b.
—— (1921), 'Letter to Russell dated 19.III. 1921' (with 11 page attachment)', Russell archive, McMaster, Hamilton.
—— (1922a), 'Die Bedeutung Hilberts für die Philosophie der Mathematik', *Die Naturwissenschaften*, 10, 93–99. English translation in Mancosu (1998a), pp. 189–97.
—— (1922b), 'Über Hilberts Gedanken zur Grundlegung der Arithmetik', *Jahresbericht der Deutschen Mathematiker-Vereinigung*, 31, 10–19. English translation in Mancosu (1998a), pp. 215–22.
—— (1923a), 'Erwiderung auf die Note von Herrn Aloys Müller: Über Zahlen als Zeichen', *Mathematische Annalen*, 90, 159–63. English translation in Mancosu (1998a), pp. 223–26.
—— (1923b), 'Review of Aloys Müller's "Der Gegenstand der Mathematik"', *Die Naturwissenschaften*, 11(26), 520–22.
—— (1926), 'Axiomatische Untersuchungen des Aussagen-Kalkuls der "Principia Mathematica"', *Mathematische Zeitschrift*, 25, 305–20.
—— (1927), 'Probleme der theoretischen Logik', *Unterrichtsblätter für Mathematik und Naturwissenschaften*, 33, 369–77.
—— (1928a), 'Die Grundbegriffe der reinen Geometrie in ihrem Verhältnis zur Anschauung', *Die Naturwissenschaften*, 16(12), 197–203.
—— (1928b), 'Review of: Heinrich Behmann, "Beiträge zur Algebra der Logik, insbesondere zum Entscheidungsproblem"', *Jahrbuch für die Fortschritte der Mathematik*, 48(8), 1119–20.
—— (1928c), 'Über Nelsons Stellungnahme in der Philosophie der Mathematik', *Die Naturwissenschaften*, 16, 142–45.
—— (1928d), 'Zusatz zu Hilberts Vortrag über "Die Grundlagen der Mathematik"', *Abhandlungen aus dem Mathematischen Seminar der Universität Hamburg*, 6, 88–92. English translation in van Heijenoort (1967a), pp. 485–89.
—— (1930a), 'Die Grundgedanken der Fries'schen Philosophie in ihrem Verhältnis zum heutigen Stand der Wissenschaft', *Abhandlungen der Fries'schen Schule*, 5, 99–113.
—— (1930b), 'Die Philosophie der Mathematik und die Hilbertsche Beweistheorie', *Blätter für deutsche Philosophie*, 4, 326–67. Reprinted in Bernays (1976), pp 17–61. English translation in Mancosu (1998a), pp. 234–65.
—— (1976), *Abhandlungen zur Philosophie der Mathematik* (Darmstadt: Wissenschaftliche Buchgesellschaft).
—— (2004), *Philosophie des Mathématique* (Paris: Vrin).
BERNAYS, PAUL and SCHÖNFINKEL, MOSES (1928), 'Zum Entscheidungsproblem der mathematischen Logik', *Mathematische Annalen*, 99, 342–72.

BERNSTEIN, FELIX (1919), 'Die Mengenlehre Georg Cantors und der Finitismus', *Jahresberich der Deutschen Mathematiker-Vereinigung*, 28, 50–63.
BETH, EVERT W. (1953a), 'Nominalisme in de hedendaagse logica', *Folia Civitatis*.
—— (1953b), 'On Padoa's method in the theory of definition', *Indagationes Mathematicae*, 15, 330–39.
—— (1953c), 'La Reconstruction nominaliste de la logique', Lecture delivered in Bruxelles on January 24, 1953. Archief E. W. Beth, Amsterdam.
—— (1953/54), 'Zomerconferentie 1953/4a', *Algemeen Nederlands Tijdschrift voor Wijsbegeerte en Psychologie*, 46, 41–5. English translation by Mark van Atten, unpublished.
—— (1954), 'Verstand en Intuïtie', *Algemeen Nederlands Tijdschrift voor Wijsbegeerte en Psychologie*, 46, 213–24. English translation in E. Beth, *Aspects of Modern Logic* (Dordrecht: Reidel), 1971, pp. 86–101.
BETSCH, CHRISTIAN (1926), *Fiktionen in der Mathematik* (Stuttgart: Frommanns Verlag).
BETTI, ARIANNA (2008), 'Polish axiomatics and its truth: On Tarski's Lesniewskian background and the Ajdukiewicz connection', in Douglas Patterson (ed.), *New Essays on Tarski and Philosophy*, 44–71 (Oxford: Oxford University Press).
BIGGS, NORMAN L. (1976), *Graph Theory, 1736-1936* (Oxford: Oxford University Press).
BILLINGE, HELENE (2000), 'Applied constructive mathematics: On Hellman's 'Mathematical Constructivism in Spacetime", *British Journal for the Philosophy of Science*, 51, 299–318.
BIMBÓ, KATALIN (2008), 'Combinatory Logic', Stanford Encyclopedia of Philosophy.
BLASCHKE, WILHELM (1923), *Vorlesungen über Differentialgeometrie und Geometrische Grundlagen von Einsteins Relativitätstheorie, Bd. II Affine Differentialgeometrie*, vol. 2 (Berlin: Springer).
BLOK, W. J. and PIGOZZI, DON (1988), 'Alfred Tarski's work on general metamathematics', *The Journal of Symbolic Logic*, 53, 36–50.
BLUMENTHAL, OTTO (1935), 'Lebensgeschichte', in David Hilbert (ed.), *David Hilbert. Gesammelte Abhandlungen. Dritter Band: Analysis, Grundlagen der Mathematik, Verschiedenes*, vol. 3, 388–429 (Berlin: Springer).
BÔCHER, MAXIME (1904), 'The fundamental conceptions and methods of mathematics', *Bulletin of the American Mathematical Society*, 10, 115–35.
BOI, L., PATRAS, F., KERZBERG, P. (2007), *Rediscovering Phenomenology. Phenomenological essays on mathematical beings, physical reality, perception and consciousness* (Dordrecht: Springer).
BOLZMANN, LUDWIG (1974), 'Model', in B. McGuinness (ed.), *Theoretical Physics and Philosophical Problems. Selected Writings*, 213–20 (Dordrecht: Reidel).
BONDONI, DAVIDE (2007), *La Teoria delle Relazioni nell'Algebra della Logica Schroederiana* (Milan: Led).
BONIFACE, JACQUELINE (2004), *Hilbert et la Notion d'Existence en Mathématiques* (Paris: Vrin).
BONJOUR, LAURENCE (1985), *The Structure of Empirical Knowledge* (Cambridge, Mass.: Harvard University Press).
BOREL, ARMAND (1986), 'Hermann Weyl and Lie Groups', in K. Chandrasekharan (ed.), *Hermann Weyl 1885–1985, Centenary Lectures delivered at the ETH Zurich*, 53–82 (Berlin and New York: Springer Verlag).
BOREL, ÉMILE (1905), 'Quelques remarques sur les principes de la théorie des ensembles', *Mathematische Annalen*, 60, 194–95.
BORGA, MARCO (1985), 'La logica, il metodo assiomatico e la problematica metateorica', in Borga et al. (1985), 11–75.

BORGA, MARCO, FREGUGLIA, PAOLO, and PALLADINO, DARIO (eds.) (1985), *I Contributi Fondazionali della Scuola di Peano* (Milan: Franco Angeli).

BÖRGER, EGON, GRÄDEL, ERICH, and GUREVICH, YURI (1997), *The Classical Decision Problem* (Berlin: Springer).

BRADING, KATHERINE (2002), 'Which symmetry? Noether, Weyl, and the Conservation of Charge', *Studies in History and the Philosophy of Modern Physics*, 23, 3–22.

BRADING, KATHERINE and RYCKMAN, THOMAS A. (2008), 'Hilbert's "Foundations of Physics": Gravitation and electromagnetism within the axiomatic method', *Studies in the History and Philosophy of Modern Physics*, 39, 102–53.

BREGER, HERBERT (1986), 'Leibniz, Weyl und das Kontinuum', *Studia Leibnitiana, Supplementa*, 26, 316–30.

—— (1988), 'Möglichkeit, Konstruktion, Geschichte: Bemerkungen zur Erkenntnistheorie Hermann Weyls', in W. Deppert (ed.), *Exact Sciences and their philosophical foundations*, 325–41 (Frankfurt am Main: Lang).

BROUWER, LUITZEN EGBERTUS JAN (1907), *Over de Grondslagen der Wiskunde* (Amsterdam: Maas & van Suchtelen). English translation in Brouwer (1975), pp. 13–101.

—— (1912a), 'Intuitionism and Formalism', *Bulletin of the American Mathematical Society*, 20, 81–96. Reprinted in Benacerraf (1983), and Putnam pp. 77–89.

—— (1912b), *Intuitionisme en Formalisme* (Amsterdam: Clausen). English translation in Brouwer (1912a).

—— (1918), 'Begründung der Mengenlehre unabhängig vom logischen Satz vom ausgeschlossenen Dritten. Erster Teil: Allgemeine Mengenlehre', *KNAW Verhandelingen 1e Sectie*, XII(5), 1–43.

—— (1921), 'Intuitionistische Verzamelingsleer', *KNAW Verslagen*, 29, 797–802. English translation in Mancosu (1998a), 23–27.

—— (1923a), 'Intuitionistische splitsing van mathematische grondbegrippen', *KNAW Verslagen*, 32, 877–80. English translation in Mancosu (1998a), 286–89.

—— (1923b), 'Über die Bedeutung des Satzes vom ausgeschlossene Dritten in der Mathematik, insbesondere in der Funktionentheorie', *Journal für die reine und angewandte Mathematik*, 154, 1–7. English translation in van Heijenoort (1967a), pp. 334–341.

—— (1928), 'Intuitionistische Betrachtungen über den Formalismus', *KNAW Proceedings*, 31, 374–79. English translation in Mancosu (1998a), 40–44.

—— (1975), *Collected Works*, vol. 1 (Amsterdam: North-Holland).

BURGESS, JOHN and ROSEN, GIDEON (1997), *A Subject with no Object* (Oxford: Oxford University Press).

CANTINI, ANDREA (2009), 'Paradoxes, self-reference and truth in the 20th century', in Dov Gabbay and John Woods (eds.), *Handbook of the History of Logic*, vol. 5. *Logic from Russell to Church*, 875–1013 (Amsterdam: Elsevier).

CANTOR, GEORG (1883), *Grundlagen einer allgemeinen Mannigfaltigkeitslehre* (Leipzig: B. G. Teubner).

CARNAP, RUDOLF (1922), *Der Raum. Ein Beitrag zur Wissenschaftslehre*, vol. 56 of *Kant-Studien Ergänzungsheft* (Berlin: Reuther & Reichard).

—— (1927), 'Eigentliche und uneigentliche Begriffe', *Symposion*, 1, 355–74.

—— (1928), *Der Logische Aufbau der Welt* (Berlin: Weltkreis-Verlag).

—— (1930), 'Bericht über Untersuchungen zur allgemeinen Axiomatik', *Erkenntnis*, 1, 303–10.

—— (1931), 'Die Logizistische Grundlegund der Mathematik', *Erkenntnis*, 1, 91–105. English translation in Benacerraf and Putnam (1983), pp. 41–52.

—— (1934a), 'Die Antinomien und die Unvollständigkeit der Mathematik', *Monatshefte für Mathematik und Physik*, 41, 263–84.

—— (1934b), *Logische Syntax der Sprache* (Vienna: Springer). English translation, *Logical Syntax of Language* (Routledge and Kegan Paul, London, 1937).

—— (1936), 'Wahrheit und Bewährung', in *Actes du Congrès International de Philosophie Scientifique*, vol. 4, 18–23 (Paris: Hermann).

—— (1937), 'Ueber den semantischen Wahrheitsbegriff', Carnap *Nachlaß*, RC 080-32-01.

—— (1942), *Introduction to Semantics* (Chicago: University of Chicago Press).

—— (1963), 'Intellectual autobiography', in P. A. Schilpp (ed.), *The Philosophy of Rudolf Carnap*, volume 11 of *Library of Living Philosophers* (La Salle, Ill.: Open Court).

—— (2000) [1927–29]. *Untersuchungen zur allgemeinen Axiomatik* (Darmstadt: Wissenschaftliche Buchgesellschaft).

CARNAP, RUDOLF and BACHMANN, FRIEDRICH (1936), 'Über Extremalaxiome', *Erkenntnis*, 6, 166–88.

CARTAN, HENRI (1923), 'Sur les variétés à connexion affine et la théorie de la relativité généralisée', *Annales de l'École Normale Supérieure*, 40, 325–412. Translation by Anne Magnon and Abhay Astekar, *On Manifolds with an Affine Connection and the Theory of General Relativity* (Naples: Bibliopolis, 1986).

CASSIRER, ERNST (1929), *Die Philosophie der symbolischen Formen*, vol. 3 (Bruno Cassirer). English translation, *The Philosophy of Symbolic Forms*, vol. III (Yale University Press, New Haven, 1957).

CHANG, CHEN CHUNG (1974), 'Model Theory 1945–1971', in Henkin et al. (1974), 173–86.

CHERN, SHIING-SHEN (1996), 'Finsler Geometry is Just Riemannian Geometry without the Quadratic Restriction', *Notices of the AMS*, 43, 959–63.

CHEVALLEY, CLAUDE and WEIL, ANDRÉ (1957), 'Hermann Weyl (1885–1955)', *Enseignement mathématique*, 3, 157–87. Reprinted in Weyl (1968a), pp. 655–85.

CHIHARA, CHARLES S. (1973), *Ontology and the Vicious-Circle Principle* (Ithaca, NY, and London: Cornell University Press).

CHURCH, ALONZO (1928), 'On the law of the excluded middle', *Bulletin of the American Mathematical Society*, 34, 75–78.

—— (1932), 'A set of postulates for the foundation of logic', *Annals of Mathematics*, 33, 346–66.

—— (1933), 'A set of postulates for the foundation of logic (Second Paper)', *Annals of Mathematics*, 34, 839–64.

—— (1936a), 'A note on the Entscheidungsproblem', *Journal of Symbolic Logic*, 1, 40–41.

—— (1936b), 'An Unsolvable Problem of Elementary Number Theory', *American Journal of Mathematics*, 58, 345–63.

CIPOLLA, MICHELE (1924), 'Sui fondamenti logici della matematica secondo le recenti vedute di Hilbert', *Annali di Matematica*, 4, 19–29.

CIRERA, RAMON (1994), *Carnap and the Vienna Circle* (Amsterdam: Rodopi).

COHEN, PAUL J. (1966), *Set Theory and the Continuum Hypothesis* (New York: W. A. Benjamin).

COHNITZ, DANIEL and ROSSBERG, MARCUS (2006), *Nelson Goodman* (Montreal: McGill-Queen's University Press).

COLYVAN, MARK (2001), *The Indispensability of Mathematics* (Oxford: Oxford University Press).

COOLIDGE, JULIAN L. (1940), *A History of Geometrical Methods* (Oxford: Clarendon Press).

CORCORAN, JOHN (2003), 'The Absence of Multiple Universes of Discourse in the 1936 Tarski Consequence-definition Paper', Unpublished typescript, 9 pages.

CORRY, LEO (1997), 'David Hilbert and the axiomatization of physics (1894–1905)', *Archive for the History of the Exact Sciences*, 51, 83–198.

—— (2004a), *David Hilbert and the Axiomatization of Physics (1898–1918): From Grundlagen der Geometrie to Grundlagen der Physik* (Dordrecht: Springer).

—— (2004b), *Modern Algebra and the Rise of Mathematical Structures* (Basel: Birkhäuser), 2nd edn.

CREATH, RICHARD (ed.) (1986), *Dear Carnap, Dear Van. The Quine–Carnap Correspondence and Related Work* (Berkeley and Los Angeles: University of California Press).

CURRY, HASKELL B. (1929), 'An analysis of logical substitution', *American Journal of Mathematics*, 51, 363–84.

—— (1930), 'Grundlagen der kombinatorischen Logik', *American Journal of Mathematics*, 52, 509–36, 789–834.

CZELAKOWSKI, JANUSZ and MALINOWSKI, GRZEGORZ (1985), 'Key notions of Tarski's methodology of deductive systems', *Studia Logica*, 44, 321–51.

DA SILVA, JAIRO (1997), 'Husserl's phenomenology and Weyl's predicativism', *Synthese*, 110, 277–96.

DAUBEN, JOSEPH W. (1971), *Georg Cantor: His Mathematics and Philosophy of the Infinite* (Boston: Harvard University Press).

DAVIS, MARTIN (ed.) (1965), *The Undecidable: Basic Papers on Undecidable Propositions, Unsolvable Problems and Computable Functions* (Hewlett, NY: Raven Press).

DAWSON, JOHN W. (1984), 'Discussion on the foundations of mathematics', *History and Philosophy of Logic*, 5, 111–129.

—— (1985), 'Completing the Gödel-Zermelo correspondence', *Historia Mathematica*, 66–70.

—— (1989), 'The reception of Gödel's Incompleteness Theorems', in Stuart G. Shanker (ed.), *Gödel's Theorem in Focus*, 74–95 (London: Routledge).

—— (1990), 'The Reception of Gödel's Incompleteness Theorems', in Thomas Drucker (ed.), *Perspectives on the History of Mathematical Logic*, 84–100 (Basel: Birkäuser). Originally in Philosophy of Science Association 1984, vol. 2.

—— (1993), 'The compactness of first-order logic: from Gödel to Lindström', *History and Philosophy of Logic*, 14, 15–38.

—— (1997), *Logical Dilemmas* (Wellesley, Mass.: A.K. Peters).

DE RISI, VINCENZO (2009), 'Mahnke and Reichenbach', in *La Réception de Leibniz en sciences et philosophie des sciences aux 19e et 20e siècles, Archives Henri Poincaré, Nancy, 3-5 aprile 2008*, (Basel: Birkhäuser).

DE ROUILHAN, PHILIPPE (1991), 'De l'universalité de la logique', in Jacques Bouveresse (ed.), *L'Âge de la science*, vol. 4, pt. I, 93–119 (Paris: Jacob).

—— (1996), *Russell et le Cercle des Paradoxes* (Paris: Presses Universitaires de France).

—— (1998), 'Tarski et l'universalité de la logique', in F. Nef and D. Vernant (eds.), *Le Formalisme en Question. Le Tournant des Années 30*, 85–102 (Paris: Vrin).

DECOCK, LIEVEN (2002), *Trading Ontology for Ideology* (Dordrecht: Kluwer).

DEDEKIND, RICHARD (1872), *Stetigkeit und Irrationale Zahlen* (Braunschweig: Vieweg). English translation in Ewald (1996), pp. 756—79.

—— (1888), *Was sind und was sollen die Zahlen?* (Braunschweig: Vieweg). English translation in Ewald (1996), pp. 787–833.

DEMOPOULOS, WILLIAM (1994), 'Frege, Hilbert, and the Conceptual Structure of Model Theory', *History and Philosophy of Logic*, 15, 211–225.

DEMOPOULOS, WILLIAM (ed.) (1995), *Frege's Philosophy of Mathematics* (Cambridge: Cambridge University Press).

DETLEFSEN, MICHAEL (1986), *Hilbert's Program* (Dordrecht: Reidel).

—— (1990), 'On an alleged refutation of Hilbert's program using Gödel's first incompleteness theorem', *Journal of Philosophial Logic*, 19, 343–77.

—— (1993a), 'The formal character of Hilbert's Formalism', in J. Czermak (ed.), *Philosophie der Mathematik. Akten des 15. Internationalen Wittgenstein-Symposiums*, 195–205 (Vienna: Verlag Hölder-Pichler-Tempsky).

—— (1993b), 'Hilbert's formalism', *Revue Internationale de Philosophie*, 47(4), 285–304.

DICKSON, LEONARD E. (1951), *History of the Theory of Numbers*, vol. 1 (New York: Chelsea).

DIEUDONNÉ, JEAN (1948), 'David Hilbert (1862–1943)', in F. Le Lionnais (ed.), *Les Grands Courants de la Pensée Mathématique*, 291–97 (Marseilles: Cahiers du Sud).

—— (1976), 'Hermann Weyl', in Charles C. Gillispie (ed.), *Dictionary of Scientific Biography*, vol. 14 (New York: Charles Scribner's Sons).

DINGLER, HUGO (1919), *Die Grundlagen der Physik. Synthetische Prinzipien der mathematischen Naturphilosohie* (Berlin and Leipzig: Walter de Gruyter).

—— (1926), *Der Zusammenbruch der Wissenschaft und der Primat der Philosophie* (Munich: Reinhardt).

DREBEN, BURTON, ANDREWS, PETER, and AANDERAA, STÅL (1963), 'False lemmas in Herbrand', *Bulletin of the American Methematical Society*, 69, 699–706.

DREBEN, BURTON and VAN HEIJENOORT, JEAN (1986), Introductory note to Gödel 1929, 1930 and 1930a, 44–59, vol. 1 of Gödel (1986).

DUGAS, RENÉ (1955), *Historie de la Méchanique* (Neuchâtel: Éditions du Griffon). English translation by J. R. Maddox, *A History of Mechanics* (New York, Dover, 1988).

DUMITRIU, ANTON (1977), *History of Logic* (Tunbridge Wells: Abacus Press).

DUMMETT, MICHAEL (1977), *Elements of Intuitionism* (Oxford: Oxford University Press), 2nd edn. 2000.

EBBINGHAUS, HEINZ DIETER (2007), *Ernst Zermelo. An Approach to his Life and Work* (Berlin: Springer).

EDDINGTON, ARTHUR S. (1921), 'A Generalization of Weyl's Theory of the Electromagnetic and Gravitational Fields', *Proceedings of the Royal Society of London*, A99, 104–22.

EDWARDS, HAROLD (1988), 'Kronecker's place in history', in W. Aspray and P. Kitcher (eds.), *History and Philosophy of Modern Mathematics*, 139–44 (Minneapolis: University of Minnesota Press).

—— (1989), 'Kronecker's view on the foundations of mathematics', in D. E. Rowe and J. McCleary (eds.), *The History of Modern Mathematics*, vol. I, 67–77 (San Diego: Academic Press).

—— (2005), *Essays in Constructive Mathematics* (Dordrecht: Springer).

EINSTEIN, ALBERT (1922), *Vier Vorlesungen über Relativitätstheorie* (Braunschweig: Vieweg).

EMRICH, JOHANNES (2005), 'Beckers Anwendung der Denkfigur des offenen Horizonts auf mathematische Objekte', in Volker Peckhaus (ed.), *Oskar Becker und die Philosophie der Mathematik*, 143–52 (Munich: Wilhelm Fink Verlag).

ETCHEMENDY, JOHN (1988), 'Tarski on Truth and Logical Consequence', *The Journal of Symbolic Logic*, 53, 51–79.

ETCHEMENDY, JOHN (1990), *The Concept of Logical Consequence* (Cambridge, Mass.: Harvard University Press).

—— (2008), 'Reflections on consequence', in D. Patterson (ed.), *New Essays on Tarski and Philosophy*, 263–99 (Oxford: Oxford University Press).

EWALD, WILLIAM BRAGG (ed.) (1996), *From Kant to Hilbert. A Source Book in the Foundations of Mathematics*, vol. 2 (Oxford: Oxford University Press).

FANG, J. (1970), *Hilbert* (New York: Paideia).

FEFERMAN, ANITA and FEFERMAN, SOLOMON (2004), *Alfred Tarski: Life and Logic* (Cambridge: Cambridge University Press).

FEFERMAN, SOLOMON (1964), 'Foundations of predicative analysis', *The Journal of Symbolic Logic*, 29, 1–30.

—— (1984), 'Kurt Gödel: conviction and caution', *Philosophia naturalis*, 21, 546–62. Reprinted in Feferman (1998), pp. 150–64.

—— (1986), 'Gödel's Life and Work', in Gödel (1986), pp. 1–36.

—— (1988a), 'Weyl vindicated: "Das Kontinuum" 70 years later', in *Atti del Congresso Temi e prospettive della logica e della filosofia della scienza contemporanee. Cesena, 7–10 gennaio 1987*, Vol. I, 59–93 (Bologna: CLUEB). Reprinted in Feferman 91998), 249–83.

—— (1988b), 'Hilbert's program relativized: Proof-theoretical and foundational reductions', *The Journal of Symbolic Logic*, 53(2), 364–384.

—— (1993a), 'What rests on what? The proof-theoretic analysis of mathematics', in Johannes Czermak (ed.), *Philosophy of Mathematics. Proceedings of the Fifteenth International Wittgenstein-Symposium, Part 1*, 147–171 (Vienna: Hölder-Pichler-Tempsky).

—— (1993b), 'Why a little bit goes a long way: Logical foundations of scientifically applicable mathematics', *PSA 1992*, 2, 442–55. Reprinted in Ch. 14, Feferman (1998b), pp. 284–298.

—— (1998a), 'Deciding the Undecidable: Wrestling with Hilbert's Problem', in Feferman (1998b), pp. 3–27.

—— (1998b), *In the Light of Logic* (New York and Oxford: Oxford University Press).

—— (2000), 'Does reductive proof theory have a viable rationale?', *Erkenntnis*, 53, 63–96.

—— (2004a), 'Comments on "Predicativity as a philosophical position" by G. Hellman', *Revue Internationale de Philosophie*, 229, 313–23.

—— (2004b), 'Predicativity', in Stewart Shapiro (ed.), *The Oxford Handbook of the Philosophy of Mathematics and Logic*, 590–624 (New York: Oxford University Press).

—— (2004c), 'Tarski's conceptual analysis of semantical notions', in Ali Benmakhlouf (ed.), *Sémantique et épistémologie*, 79–108 (Casablanca and Paris: Éditions Le Fennec and J. Vrin).

—— (2008a), 'Lieber Herr Bernays! Lieber Herr Gödel! Gödel on finitism, constructivity and Hilbert's program', *Dialectica*, 62(2), 179–203.

—— (2008b), 'Tarski's conceptual analysis of semantic notions', in D. Patterson (ed.), *New Essays on Tarski and Philosophy*, 72–93 (Oxford: Oxford University Press).

FEIST, RICHARD (2002), 'Weyl's appropriation of Husserl's and Poincaré's thought', *Synthese*, 132, 273–301.

—— (2004), 'Husserl and Weyl: Phenomenology, mathematics, and physics', in *Husserl and the Sciences*, 153–72 (Ottawa: Ottawa University Press).

FERREIROS, JOSÉ (1997), 'Notes on Types, Sets and Logicism', *Theoria*, 28, 91–124.

—— (1999), *Labyrinth of Thought. A History of Set Theory and its Role in Modern Mathematics* (Basel: Birkhäuser).

—— (2001), 'The road to modern logic—an interpretation', *Bulletin of Symbolic Logic*, 7, 441–84.

FLAMM, LUDWIG (1916), 'Beiträge zur Einsteinschen Gravitationstheorie', *Physikalische Zeitschrift*, 17, 448–54.

FOLINA, JANET (1992), *Poincaré and the Philosophy of Mathematics* (New York: St Martin's Press).

—— (2008), 'Intuition between the Analytic-Continental divide: Hermann Weyl's philosophy of the continuum', *Philosophia Mathematica*, 16, 25–55.

FORMAN, PAUL (1971), 'Weimar Culture, Causality, and Quantum Theory, 1918–1927: Adaptation by German Physicists and Mathematicians to a Hostile Intellectual Environment', *Historical Studies in the Physical Sciences*, 3, 1–115.

FRAENKEL, ADOLF ABRAHAM (1922a), 'Axiomatische Begründung der transfiniten Kardinalzahlen. I', *Mathematische Zeitschrift*, 13, 153–88.

—— (1922b), 'Der Begriff "definit" und die Unabhängigkeit des Auswahlaxioms', *Sitzungsberichte der Preussischen Akademie der Wissenschaften, physikalische-mathematische Klasse*, 253–57. English translation in van Heijenoort (1967a), 253–57.

—— (1922c), 'Zu den Grundlagen der Cantor-Zermeloschen Mengenlehre', *Mathematische Annalen*, 86, 230–37.

—— (1923), *Einleitung in die Mengenlehre* (Berlin: Springer), 2nd edn.

—— (1928), *Einleitung in die Mengenlehre* (Berlin: Springer), 3rd edn.

FRANCHELLA, MIRIAM (1994), *L. E. J. Brouwer Pensatore Eterodosso* (Milan: Guerini).

FRANK, MATTHEW (2000), 'Constructive Mathematics and Mathematical Physics: A Program and Progress Report', Unpublished typescript.

FRANK, PHILIP (1997), *The Law of Causality and its Limits* (Dordrecht: Kluwer).

FREGE, GOTTLOB (1893), *Grundgesetze der Arithmetik, begriffsschriftlich abgeleitet*, vol. 1 (Jena: Pohle). Partial translation in Frege (1964).

—— (1903), *Grundgesetze der Arithmetik, begriffsschriftlich abgeleitet*, volume 2 (Jena: Pohle). Partial translation in Frege (1964).

—— (1964), *Basic Laws of Arithmetic* (Berkeley and Los Angeles: University of California Press). Translated and with an introduction by Montgomery Furth.

—— (1980), *Philosophical and Mathematical Correspondence* (Oxford: Basil Blackwell).

FREGUGLIA, PAOLO (1985), 'Il calcolo geometrico ed i fondamenti della geometria', in Borga et al. (1985), 174–236.

FREI, GÜNTHER and STAMMBACH, URS (eds.) (1992), *Hermann Weyl und die Mathematik an der ETH Zürich, 1913-1930* (Basel: Birkhäuser Verlag).

FREWER, M. (1981), 'Felix Bernstein', *Jahresberich der Deutschen Mathematiker-Vereinigung*, 83, 84–95.

FREYTAG, WILLY (1902), *Der Realismus und das Transcendenzproblem* (Halle: Niemayer).

—— (1904), *Die Erkenntnis der Aussenwelt* (Halle: Niemayer).

FRIEDMAN, MICHAEL (1995), 'Carnap and Weyl on the Foundations of Geometry and Relativity Theory', *Erkenntnis*, 42, 247–60. Reprinted in Friedman (1999), pp. 44–58.

—— (1999), *Reconsidering Logical Positivism* (New York: Cambridge University Press).

FROST-ARNOLD, GREG (2004), 'Was Tarski's theory of truth motivated by physicalism?', *History and Philosophy of Logic*, 25, 265–80.

—— (2008), 'Tarski's Nominalism', in Douglas Patterson (ed.), *New Essays on Tarski and Philosophy*, 225–46 (Oxford: Oxford University Press).

—— (2009), 'Quine's evolution from Carnap's disciple to the author of "Two Dogmas"', unpublished.

—— (Forthcoming), *Carnap, Tarski, and Quine in Conversation: Logic, Science, and Mathematics* (La Salle, Ill.: Open Court).

GABBAY, DOV and WOODS, JOHN (eds.) (2009), *Handbook of the History of Logic*, vol. 5, *Logic from Russell to Church* (Amsterdam: Elsevier).

GARCIADIEGO, ALEJANDRO R. (1992), *Bertrand Russell and the Origins of Set-theoretic "Paradoxes"* (Basel: Birkhäuser).

GAUTHIER, YVON (1993), 'Hilbert et la logique interne des mathématiques', *Revue Internationale de Philosophie*, 47, 305–18.

—— (2002), *Internal Logic. Foundations of Mathematics from Kronecker to Hilbert*, vol. 310 of *Synthese Library* (Dordrecht: Kluwer).

GEIGER, MORITZ (1928), 'Review of Becker's "Mathematische Existenz"', *Göttingische Gelehrte Anzeigen*, 47, 401–19.

GENTZEN, GERHARD (1933a), 'Über das Verhältnis zwischen intuitionistischer und klassischer Logik', First published in *Archiv für mathematische Logik und Grundlagenforschung* 16 (1974), 119–32. English translation in Gentzen (1969), pp. 53–67.

—— (1933b), 'Über die Existenz unabhängiger Axiomensysteme zu unendlichen Satzsystemen', *Mathematische Annalen*, 107, 329–50. English translation in Gentzen (1969), pp. 29–52.

—— (1934), 'Untersuchungen über das logische Schließen I–II', *Mathematische Zeitschrift*, 39, 176–210, 405–31. English translation in Gentzen (1969), pp. 68–131.

—— (1935), 'Die Widerspruchsfreiheit der reinen Zahlentheorie', Published as "Der erste Widerspruchsfreiheitsbeweis für die klassische Zahlentheorie" in *Archiv für mathematische Logik und Grundlagenforschung* 16 (1974), 97–118. English translation in Gentzen (1969), pp. 132–213.

—— (1936), 'Die Widerspruchsfreiheit der reinen Zahlentheorie', *Mathematische Annalen*, 112, 493–565. English translation in Gentzen (1969), pp. 132–213.

—— (1969), *The Collected Papers of Gerhard Gentzen* (Amsterdam: North-Holland).

GETHMANN, CARL FRIEDRICH (2002), 'Hermeneutische Phänomenologie und Logischer Intuitionismus. Zu O. Beckers Mathematische Existenz', in Annemarie Gethmann-Siefert and Jürgen Mittelstrass (eds.), *Die Philosophie und die Wissenschaften Zum Werk Oskar Beckers*, 87–108 (Munich: Wilhelm Fink Verlag).

GIAQUINTO, MARCUS (1983), 'Hilbert's philosophy of mathematics', *British Journal for Philosophy of Science*, 34, 119–32.

—— (2002), *The Search for Certainty* (Oxford: Oxford University Press).

GILLIES, DONALD (1980), 'Phenomenology and the Infinite in Mathematics', *British Journal for the Philosophy of Science*, 31, 289–98.

GIUGLIANO, ANTONELLO (2005), 'Zahl und Zeit: Becker zwischen Nietzche und Heidegger', in Volker Peckhaus (ed.), *Oskar Becker und die Philosophie der Mathematik*, 47–58 (Munich: Wilhelm Fink Verlag).

GLIVENKO, VALERII (1928), 'Sur la logique de M. Brouwer', *Académie Royale de Belgique, Bulletin*, 14, 225–28.

—— (1929), 'Sur quelques points de la logique de M. Brouwer', *Académie Royale de Belgique, Bulletin*, 15, 183–88. English translation in Mancosu (1998a), pp. 301–05.

GLOCK, HANS-JOHANN (2006), 'Truth in the Tractatus', *Synthese*, 148, 345–68.

GÖDEL, KURT (1929), *Über die Vollständigkeit des Logikkalküls*, Dissertation, Universität Wien. Reprinted and translated in Gödel (1986), pp. 60–101.

—— (1930a), 'Die Vollständigkeit der Axiome des logischen Funktionenkalküls', *Monatshefte für Mathematik und Physik*, 37, 349–360. Reprinted and translated in Gödel (1986), pp. 102–23.

—— (1930b), 'Einige metamathematische Resultate über Entscheidungsdefinitheit und Widerspruchsfreiheit', *Anzeiger der Akademie der Wissenschaften in Wien*, 67, 214–15. Reprinted and translated in Gödel (1986), pp. 140–43.

—— (1930c), 'Vortrag über die Vollständigkeit des Loggikkalkuls', in Gödel (1995), pp. 6–28.

—— (1931), 'Über formal unentscheidbare Sätze der *Principia Mathematica* und verwandter Systeme I', *Monatshefte für Mathematik und Physik*, 38, 173–98. Reprinted and translated in Gödel (1986), pp. 144–95.

—— (1932a), 'Ein Spezialfall des Entscheidungsproblem der theoretischen Logik', *Ergebnisse eines mathematischen Kolloquiums*, 2, 27–28. Reprinted and translated in Gödel (1986), pp. 130–235.

—— (1932b), 'Zum intuitionistischen Aussagenkalkül', *Anzeiger der Akademie der Wissenschaften in Wien, Mathematisch-Naturwissenschaftliche Klasse*, 69, 65–66. Reprinted and translated in Gödel (1986), pp. 222–25.

—— (1933a), 'Eine Interpretation des intuitionistischen Aussagenkalküls', *Ergebnisse eines mathematisches Kolloquiums*, 4, 39–40. Reprinted and translated in Gödel (1986), pp. 300–03.

—— (1933b), 'Zum Entscheidungsproblem des logischen Funktionenkalküls', *Monatshefte für Mathematik und Physik*, 40, 433–43. Reprinted and translated in Gödel (1986), pp. 306–27.

—— (1933c), 'Zur intuitionistischen Arithmetik und Zahlentheorie', *Ergebnisse eines mathematisches Kolloquiums*, 4, 34–38. Reprinted and translated in Gödel (1986), pp. 286–95.

—— (1940), *The Consistency of the Axiom of Choice and of the Generalized Continuum Hypothesis*, Annals of Mathematics Studies, vol. 3 (Princeton: Princeton University Press). Reprinted in Gödel (1990), pp. 1–101.

—— (1944), 'Russell's mathematical logic', in Paul Arthur Schilpp (ed.), *The Philosophy of Bertrand Russell*, 125–53 (La Salle, Ill.: Open Court). Reprinted in Gödel (1990), pp. 102–41.

—— (1986), *Collected Works*, vol. 1 (Oxford: Oxford University Press).

—— (1990), *Collected Works*, vol. 2 (Oxford: Oxford University Press).

—— (1995), *Collected Works*, vol. 3 (Oxford: Oxford University Press).

—— (2003a), *Collected Works*, vol. 5 (Oxford: Oxford University Press).

—— (2003b), *Collected Works*, vol. 4 (Oxford: Oxford University Press).

GOLDFARB, WARREN D. (1979), 'Logic in the twenties: the nature of the quantifier', *Journal of Symbolic Logic*, 44, 351–68.

—— (1989), 'Russell's reasons for ramification', in C. W. Savage and C. A. Anderson (eds.), *Rereading Russell* (Minneapolis: University of Minnesota Press).

—— (1993), 'Herbrand's Error and Gödel's Correction', *Modern Logic*, 3, 103–18.

—— (1995), *Introductory note to Gödel *1930c*, in Gödel (1995), pp. 13–15.

—— (2001), 'Frege's conception of logic', in Juliet Floyd and Sanford Shieh (eds.), *Future Pasts: the Analytic Tradition in Twentieth-Century Philosophy*, 25–41 (New York: Oxford University Press).

GOLDSTEIN, CATHERINE, GRAY, JEREMY, and RITTER, JIM (2007), *The Shaping of Arithmetic: After C. F. Gauss's Disquisitiones Arithmeticae* (Berlin: Springer).

GOMEZ-TORRENTE, MARIO (1996), 'Tarski on logical consequence', *Notre Dame Journal of Formal Logic*, 37, 125–51.

—— (2008), 'Are There Model-Theoretic Logical Truths that Are not Logically True?', in Douglas Patterson (ed.), *New Essays on Tarski and Philosophy*, 340–68 (Oxford: Oxford University Press).

—— (2009), 'Rereading Tarski on Logical Consequence', *The Review of Symbolic Logic*, 2(2), 249–95.

GOODMAN, NELSON (1988), 'Nominalisms', in L. E. Hahn and P. A. Schilpp (eds.), *The Philosophy of W.V. Quine*, 159–61 (La Salle, Ill.: Open Court).

GOODMAN, NELSON and QUINE, WILLARD VAN ORMAN (1947), 'Steps towards a constructive nominalism', *The Journal of Symbolic Logic*, 12, 105–22.

GOSSELIN, MIA (1990), *Nominalism and Contemporary Nominalism* (Dordrecht: Kluwer).

GRANGER, GILLES GASTON (1998), 'Le problème du fondement selon Tarski', in F. Nef and D. Vernant (eds.), *Le Formalisme en Question. Le Tournant des Années 30*, 37–47 (Paris: Vrin).

GRATTAN-GUINNESS, IVOR (1977), *Dear Russell—Dear Jourdain* (London: Duckworth).

—— (1979), 'In memoriam Kurt Gödel: His 1931 correspondence with Zermelo on his incompletability theorem', *Historia Mathematica*, 6, 294–304.

—— (2000), *The Search For Mathematical Roots 1870–1940. Logics, Set Theories and the Foundations of Mathematics from Cantor through Russell to Gödel* (Princeton: Princeton University Press).

GRAY, JEREMY (2000), *The Hilbert Challenge* (Oxford: Oxford University Press).

GRELLING, KURT and NELSON, LEONARD (1908), 'Bemerkungen zu den Paradoxieen von Russell und Burali-Forti', *Abhandlungen der Fries'schen Schule, Neue Folge 2*, 3, 301–34.

GRUNDMANN, THOMAS (1996), 'Can science be likened to a well-written fairy tale? A contemporary reply to Schlick's objection to Neurath's coherence theory', in E. Nemeth and F. Stadler (eds.), *Encyclopedia and Utopia. The life and work of Otto Neurath (1882-1945)*, 127–33 (Dordrecht: Kluwer).

GUILLAUME, MARCEL (1994), 'La Logique mathématique en sa jeunesse', in J.-P. Pier (ed.), *Development of Mathematics 1900–1950*, 185–367 (Basel, Boston, and Berlin: Birkhäuser).

HAAS, GERRIT and STEMMLER, ELKE (1981), 'Der Nachlaß Heinrich Behmanns (1891–1970) - Gesamtverzeichnis', *Aachener Schriften zur Wissenschaftstheorie, Logik und Logikgeschichte*, 1, 1–39.

HAHN, H., CARNAP, R., GÖDEL, K., HEYTING, A., REIDEMEISTER, K., SCHOLZ, H., and VON NEUMANN, J. (1931), 'Diskussion zur Grundlegung der Mathematik', *Erkenntnis*, 2, 135–51. English translation in Dawson (1984).

HALE, BOB and WRIGHT, CRISPIN (2001), *The Reason's Proper Study. Essays Towards a Neo-Fregean Philosophy of Mathematics* (Oxford: Oxford University Press).

HALLER, RUDOLF (1992), 'Alfred Tarski: Drei Briefe an Otto Neurath', *Grazer Philosophische Studien*, 43, 1–32.

HALLETT, MICHAEL (1984), *Cantorian Set Theory and Limitation of Size* (Oxford: Oxford University Press).

—— (1990a), 'Physicalism, reductionism and Hilbert', in Andrew D. Irvine (ed.), *Physicalism in Mathematics*, 183–257 (Dordrecht: Reidel).

—— (1995a), 'Hilbert and Logic', in M. Marion and R.S. Cohen (eds.), *Québec Studies in the Philosophy of Science*, 135–87 (Dordrecht: Kluwer).

—— (1995b), 'Logic and Mathematical Existence', in L. Krüger and B. Falkenburg (eds.), *Physik, Philosophie und die Einheit der Wissenschaft. Für Erhard Scheibe*, 33–82 (Heidelberg: Spektrum Akademischer Verlag).

—— (1996a), 'Introduction to Zermelo 1930', in Ewald (1996), 1208–18.

—— (1996b), 'Hilbert on Geometry, Number, and Continuity', unpublished.

—— (2008), 'Reflections on the Purity of Method in Hilbert's *Grundlagen der Geometrie*', in Paolo Mancosu (ed.), *The Philosophy of Mathematical Practice*, 198–255 (Oxford: Oxford University Press).

HARDCASTLE, GARY L. (2003), 'Debabelizing Science: The Harvard Science of Science Discussion Group, 1940–41', in G. L. Hardcastle and A. W. Richardson (eds.), *Logical Positivism in North America*, 170–98 (Minneapolis: University of Minnesota Press).

HASSE, HELMUT (1950), *Vorlesungen über Zahlentheorie* (Berlin: Springer).

HAUSDORFF, FELIX (1904), 'Der Potenzbegriff in der Mengenlehre', *Jahresbericht der Deutschen Mathematiker-Vereinigung*, 13, 569–71.

HAWKINS, THOMAS (2000), *Emergence of the Theory of Lie Groups* (New York, Berlin, and Heidelberg: Spinger Verlag).

HAZEN, ALLEN P. (2004), 'A "constructive" proper extension of ramified type theory (The logic of *Principia Mathematica*, Second edition, Appendix B)', in Godehard Link (ed.), *One Hundred Years of Russell's Paradox*, 449–80 (Berlin: de Gruyter).

HAZEN, ALLEN P. and DAVOREN, JENNIFER (2000), 'Russell's 1925 Logic', *Australasian Journal of Philosophy*, 78, 534–66.

HEGSELMANN, RAINER (1985), 'Die Korrespondenz zwischen Otto Neurath und Rudolf Carnap aus den Jahren 1934 bis 1945—Ein vorläufiger Bericht', in H.-J. Dahms (ed.), *Philosophie, Wissenschaft, Aufklärung*, 276–90 (Berlin: de Gruyter).

HEINZMANN, GERHARD (1985), *Entre Intuition et Analyse. Poincaré et le Concept de Prédicativité* (Paris: Albert Blanchard).

HELLMAN, GEOFFREY (2004), 'Predicativity as a philosophical position', *Revue Internationale de Philosophie*, 229(3), 295–312.

HEMPEL, CARL G. (1935), 'On the Logical Positivists' theory of truth', *Analysis*, 2, 49–59. Also in *Selected Philosophical Essays*, ed. R. Jeffrey (Cambridge University Press, Cambridge, 1999), pp. 21–25.

—— (1982), 'Schlick und Neurath: Fundieung vs. Kohärenz in der wissenschaftlichen Erkenntnis', *Grazer Philosophische Studien*, 16/7, 1–18. Translated in *Selected Philosophical Essays*, ed. R. Jeffrey (Cambridge University Press, Cambridge, 1999), pp. 181–198.

HENDRY, JOHN (1980), 'Weimar Culture and Quantum Causality', *History of Science*, 18, 155–80.

HENKIN, LEON, ADDISON, JOHN, CHANG, CHEN CHUNG, SCOTT, DANA, VAUGHT, ROBERT L., and CRAIG, WILLIAM (eds.) (1974), *Proceedings of the Tarski Symposium: An International Symposium held to honor Alfred Tarski on the occasion of his seventieth birthday*, vol. 25 of *Proceedings of Symposia in Pure Mathematics* (Providence, RI: American Mathematical Society).

HERBRAND, JACQUES (1930), *Recherches sur la théorie de la démonstration*, Doctoral dissertation, University of Paris. English translation in Herbrand (1971), pp. 44–202.

—— (1931a), 'Sur la non-contradiction de l'arithmétique', *Journal für die Reine und Angewandte Mathematik*, 166, 1–8. English translation in van Heijenoort (1967a), pp. 618–28.

—— (1931b), 'Sur le probléme fondamental de la logique mathémathique', *Sprwozdania z posiedzeń Towarzystwa Naukowego Warszawskiego, Wydział III*, 24, 12–56. English translation in Herbrand (1971), pp. 215–71.

—— (1971), *Logical Writings* (Cambridge, Mass.: Harvard University Press).

HERTZ, HEINRICH (1894), *Gesammelte Werke III: Die Prinzipien der Mechanik* (Leipzig: Barth). Translated as *The Principles of Mechanics* (New York: Dover, 1956).

HERTZ, PAUL (1922), 'Über Axiomenssysteme für beliebige Satzmengen. I. Teil', *Mathematische Annalen*, 87, 246–69.

—— (1923), 'Über Axiomensysteme für beliebige Satzmengen. II. Teil', *Mathematische Annalen*, 89, 76–102.

—— (1928), 'Reichen die üblichen syllogistischen Regeln für das Schließen in der positiven Logik elementarer Sätze aus?', *Annalen der Philosophie und philosophischen Kritik*, 7, 272–77.

—— (1929), 'Über Axiomensysteme für beliebige Satzmengen', *Mathematische Annalen*, 101, 457–514.

HESSELING, DENNIS E. (2003), *Gnomes in the Fog. The reception of Brouwer's Intuitionism in the 1920s* (Basel: Birkhäuser).

HESSENBERG, GERHARD (1906), 'Grundbegriffe der Mengenlehre', *Abhandlungen der Fries'schen Schule, Neue Folge*, 1, 479–706.

HEYTING, AREND (1930a), 'Die formalen Regeln der intuitionistischen Logik', *Sitzungsberichte der Preussischen Akademie der Wissenschaften*, 42–56. English translation in Mancosu (1998a), pp. 311–27.

—— (1930b), 'Die formalen Regeln der intuitionistischen Mathematik', *Sitzungsberichte der Preussischen Akademie der Wissenschaften*, 57–71.

——(1930c), 'Die formalen Regeln der intuitionistischen Mathematik', *Sitzungsberichte der Preussischen Akademie der Wissenschaften*, 158–69.

——(1930d), 'Sur la logique intuitionniste', *Académie Royale de Belgique, Bulletin*, 16, 957–963. English translation in Mancosu (1998a), pp. 306–10.

——(1931), 'Die intuitionistische Grundlegung der Mathematik', *Erkenntnis*, 2, 106–15. English translation in Benacerraf and Putnam (1983), pp. 52–61.

——(1934), *Mathematische Grundlagenforschung. Intuitionismus. Beweistheorie* (Berlin: Springer). Extended French translation in Heyting (1955).

——(1955), *Les Fondements des Mathématiques. Intuitionisme, Théorie de la Démonstration* (Paris: Gauthier-Villars).

HILBERT, DAVID (1899), 'Grundlagen der Geometrie', in *Festschrift zur Feier der Enthüllung des Gauss-Weber-Denkmals in Göttingen*, 1–92 (Leipzig: Teubner), 1st edn.

——(1900a), 'Mathematische Probleme', *Nachrichten von der Königlichen Gesellschaft der Wissenschaften zu Göttingen, Math.-Phys. Klasse*, 253–297. Lecture given at the International Congress of Mathematicians, Paris, 1900. Partial English translation in Ewald (1996), pp. 1096–1105.

——(1900b), 'Über den Zahlbegriff', *Jahresbericht der Deutschen Mathematiker-Vereinigung*, 8, 180–84. English translation in Ewald (1996), pp. 1089–96.

——(1902), *Foundations of Geometry* (Chicago: Open Court). 1st English edn.

——(1903), 'Die Grundlagen der Geometrie', *Mathematische Annalen*, 56, 381–422.

——(1905a), 'Logische Principien des mathematischen Denkens', Vorlesung, Sommer-Semester 1905. Lecture notes by Ernst Hellinger. Unpublished manuscript, 277 pp. Bibliothek, Mathematisches Institut, Universität Göttingen.

——(1905b), 'Über die Grundlagen der Logik und der Arithmetik', in A. Krazer (ed.), *Verhandlungen des dritten Internationalen Mathematiker-Kongresses in Heidelberg vom 8. bis 13. August 1904*, 174–85 (Leipzig: Teubner). English translation in von Heijenoort (1967a), pp. 129–38.

——(1918a), 'Axiomatisches Denken', *Mathematische Annalen*, 78, 405–15. Lecture given at the Swiss Society of Mathematicians, 11 September 1917. Reprinted in Hilbert (1935), pp. 146–56. English translation in Ewald (1996), pp. 1105–15.

——(1918b), 'Mengenlehre', Lecture notes by Margarethe Loeb. Sommer-Semester 1917–18. Unpublished typescript. Bibliothek Mathematisches Institut, Universität Göttingen.

——(1918c), 'Prinzipien der Mathematik', Lecture notes by Paul Bernays. Winter-Semester 1917–18. Unpublished typescript. Bibliothek, Mathematisches Institut, Universität Göttingen.

——(1920a), 'Logik-Kalkül', Vorlesung, Winter-Semester 1920. Lecture notes by Paul Bernays. Unpublished typescript. Bibliothek, Mathematisches Institut, Universität Göttingen.

——(1920b), 'Probleme der mathematischen Logik', Vorlesung, Sommer-Semester 1920. Lecture notes by Paul Bernays and Moses Schönfinkel. Unpublished typescript. Bibliothek, Mathematisches Institut, Universität Göttingen.

——(1922a), 'Grundlagen der Mathematik', Vorlesung, Winter-Semester 1921–22. Lecture notes by Helmut Kneser. Unpublished manuscript, three notebooks.

——(1922b), 'Grundlagen der Mathematik', Vorlesung, Winter-Semester 1921–22. Lecture notes by Paul Bernays. Unpublished typescript. Bibliothek, Mathematisches Institut, Universität Göttingen.

——(1922c), 'Neubegründung der Mathematik: Erste Mitteilung', *Abhandlungen aus dem Seminar der Hamburgischen Universität*, 1, 157–77. Series of talks given at the University of Hamburg, July

25–27, 1921. Reprinted with notes by Bernays in Hilbert (1935), pp. 157–77. English translation in Mancosu (1998a), pp. 198–214. and Ewald (1996), pp. 1115–34.

——(1923a), 'Die logischen Grundlagen der Mathematik', *Mathematische Annalen*, 88, 151–65. Lecture given at the Deutsche Naturforscher-Gesellschaft, September 1922. Reprinted in Hilbert (1935), pp. 178–91. English translation in Ewald (1996), pp. 1134–48.

——(1923b), 'Grundsätzliche Fragen der moderner physik', three lectures delivered at the University of Hamburg, in Hilbert (2009: 396–432).

——(1924–25), 'Über das Unendliche', Vorlesung, Winter-Semester 1924–25. Lecture notes by Lothar Nordheim. Unpublished typescript. Bibliothek, Mathematisches Institut, Universität Göttingen.

——(1926), 'Über das Unendliche', *Mathematische Annalen*, 95, 161–90. Lecture given at Münster, 4 June 1925. English translation in van Heijenoort (1967a), pp. 367–92.

——(1928a), 'Die Grundlagen der Mathematik', *Abhandlungen aus dem Seminar der Hamburgischen Universität*, 6, 65–85. English translation in van Heijenoort (1967a), pp. 464–79.

——(1928b), 'Probleme der Grundlegung der Mathematik', in Nicola Zanichelli (ed.), *Atti del Congresso Internazionale dei Matematici. 3–10 September 1928, Bologna*, 135–41.

——(1929), 'Probleme der Grundlegung der Mathematik', *Mathematische Annalen*, 102, 1–9. Lecture given at the International Congress of Mathematicians, 3 September 1928. English translation in Mancosu (1998a), pp. 227–233.

——(1931), 'Die Grundlegung der elementaren Zahlenlehre', *Mathematische Annalen*, 104, 485–94. Reprinted in Hilbert (1935), pp. 192–95. English translation in Ewald (1996), pp. 1148–57 and Mancosu (1998a), pp. 266–274.

——(1935), *Gesammelte Abhandlungen*, vol. 3 (Berlin: Springer). Reprint (Chelsea, New York, 1965).

——(1988), *Wissen und Mathematisches Denken, Vorlesungen, Winter Semester 1922/3 (Nach der Ausarbeitung von W. Ackermann)* (Mathematisches Institut, Göttingen).

——(1992), *Natur und mathematisches Erkennen* (Basel: Birkhäuser). Vorlesungen, 1919–20.

——(2004), *David Hilbert's Lectures on the Foundations of Geometry, 1891–1902* (New York: Springer).

——(2009), *David Hilbert's Lectures on the Foundations of Physics 1915–1927*, ed. Tilman Sauer and Ulrich Majer (eds.) (Berlin: Springer).

HILBERT, DAVID and ACKERMANN, WILHELM (1928), *Grundzüge der theoretischen Logik* (Berlin: Springer).

HILBERT, DAVID and BERNAYS, PAUL (1923a), 'Logische Grundlagen der Mathematik', Winter-Semester 1922–23. Lecture notes by Helmut Kneser. Unpublished manuscript.

—— ——(1923b), 'Logische Grundlagen der Mathematik', Vorlesung, Winter-Semester 1922–23. Lecture notes by Paul Bernays, with handwritten notes by Hilbert. Hilbert-*Nachlaß*, Niedersächsische Staats- und Universitätsbibliothek, Cod. Ms. Hilbert 567.

—— ——(1934), *Grundlagen der Mathematik*, vol. 1 (Berlin: Springer).

—— ——(1939), *Grundlagen der Mathematik*, vol. 2 (Berlin: Springer).

HINTIKKA, JAAKKO (1988), 'On the development of the model-theoretic viewpoint in logical theory', *Synthese*, 77, 1–36.

HODGES, WILFRID (1986), 'Truth in a structure', *Proceedings of the Aristotelian Society*, 86, 135–51.

——(1993), *Model Theory* (Cambridge: Cambridge University Press).

——(2008), 'Tarski's theory of definitions', in Douglas Patterson (ed.), *New Essays on Tarski and Philosophy*, 94–132 (Oxford: Oxford University Press).

HOFFMANN-GRÜNEBERG, FRANZ (1988), *Radikal-empirische Wahrheitstheorie. Eine Studie über Otto Neurath, den Wiener Kreis und das Wahrheitsproblem* (Vienna: Verlag Hölder-Pichler-Tempsky).

HÖLDER, OTTO (1926), 'Der angebliche Circulus Vitiosus und die sogennante Grundlagenkrise in der Analysis', *Sitzungsber. der Leipziger Akademie*, 78, 243–50. English translation in Mancosu (1998a), pp. 143–48.

―― (1929), 'Der indirekte Beweis in der Mathematik', *Berichte der Sächsischen Akademie der Wissenschaften zu Leipzig*, 81, 201–16.

―― (1930), 'Nachtrag zu meinen Aufsatz über den indirekten Beweis', *Berichte der Sächsischen Akademie der Wissenschaften zu Leipzig*, 82, 97–104.

HOWARD, DON (1996), 'Relativity, Eindeutigkeit, and Monomorphism: Rudolf Carnap and the Development of the Categoricity Concept', in Ronald N. Giere and Alan W. Richardson (eds.), *Origins of Logical Empiricism*, Minnesota Studies in the Philosophy of Science 16, 115–64 (Minneapolis: University of Minnesota Press).

HUNTINGTON, EDWARD V. (1902), 'A complete set of postulates for the theory of absolute continuous magnitude', *Transactions of the American Mathematical Society*, 3, 264–79.

―― (1904), 'Sets of independent postulates for the algebra of logic', *Transactions of the American Mathematical Society*, 5, 288–309.

―― (1905), 'A set of postulates for real algebra, comprising postulates for a one-dimensional continuum and for the theory of groups', *Transactions of the American Mathematical Society*, 6, 17–41.

―― (1906–07), 'The fundamental laws of addition and multiplication in elementary algebra', *Annals of Mathematics*, 8, 1–44.

―― (1911), 'The fundamental propositions of algebra', in J. W. A. Young (ed.), *Monographs on Topics of Modern Mathematics* (New York: Longmans, Green and Co.).

―― (1913), 'A set of postulates for abstract geometry, expressed in terms of the simple relation of inclusion', *Mathematische Annalen*, 73, 522–99.

―― (1935), 'The inter-deducibility of the new Hilbert-Bernays theory and *Principia Mathematica*', *Annals of Mathematics*, 36, 313–24.

HUSSERL, EDMUND (1913), *Ideen zu einer reinen Phänomenologie und phänomenologischen Philosophie*, vol. 1 (Halle: Max Niemeyer).

―― (1982), *Ideas Pertaining to a Pure Phenomenology and to a Phenomenological Philosophy. First Book* (The Hague: Martinus Nijhoff).

―― (1994a), *Edmund Husserl Briefwechsel*, vol. 7 (Dordrecht: Kluwer).

―― (1994b), *Edmund Husserl Briefwechsel*, vol. 3 (Dordrecht: Kluwer).

HYLTON, PETER W. (1990), *Russell, Idealism, and the Emergence of Analytic Philosophy* (Oxford: Oxford University Press).

JADACKI, J. J. (1986), 'Leon Chwistek-Bertrand Russell's scientific correspondence', *Dialectics and Humanism*, 1, 239–63.

JAHNKE, HANS N. (1990), 'Hilbert, Weyl und die Philosophie der Mathematik', *Mathematisches Semesterbericht*, 157–79.

JANICH, PETER (2002), 'Oskar Becker und die Geometriebegründung', in Annemarie Gethmann-Siefert and Jürgen Mittelstrass (eds.), *Die Philosophie und die Wissenschaften. Zum Werk Oskar Beckers*. 87–108 (Munich: Wilhelm Fink Verlag).

JAŚKOWSKI, STANISŁAW (1936), 'Recherches sur le système de la logique intuitioniste', in *Actes du Congrès International de Philosophie Scientifique*, vol. 4, 58–61 (Paris). English translation in McCall (1967), pp. 259–63.

JÓNSSON, BJARNI (1986), 'The contributions of Alfred Tarski to General Algebra', *Journal of Symbolic Logic*, 51, 883–89.

JOURDAIN, PHILIP (1914), 'Preface', in Luis Couturat, *The Algebra of Logic* (Chicago and London: Open Court), pp. iii–xiii.

KALMÁR, LÁSZLÓ (1933), 'Über die Erfüllbarkeit derjenigen Zählausdrücke, welche in der Normalform zwei benachbarte Allzeichen enthalten', *Mathematische Annalen*, 108, 466–84.

KAMAREDDINE, FAIROUZ, LAAN, TWAN, and NEDERPELT, ROB (2002), 'Types in logic and mathematics before 1940', *The Bulletin of Symbolic Logic*, 8(2), 185–245.

KANAMORI, AKIHIRO (2003), *The Higher Infinite. Large Cardinals in Set Theory from Their Beginnings* (Berlin: Springer), 2nd edn.

—— (2004), 'Zermelo and Set Theory', *The Bulletin of Symbolic Logic*, 10, 487–563.

KAUFMANN, FELIX (1930), *Das Unendliche in der Mathematik und seine Ausschaltung* (Vienna: Deuticke). English translation in Kaufmann (1978).

—— (1978), *The Infinite in Mathematics* (Reidel). Review in Gillies (1980).

KENNY, ANTHONY (1984), 'From the Big Typescript to the Philosophical Grammar', in A. Kenny (ed.), *The Legacy of Wittgenstein*, 24–37 (Oxford: Blackwell).

KERN, ISO (1964), *Husserl und Kant. Eine Untersuchung über Husserls Verhältnis zu Kant und zum Neukantianismus* (The Hague: Martinus Nijhoff).

KEYSER, CASSIUS J. (1918a), 'Concerning the number of possible interpretations of any axiom system of postulates', *Bulletin of the American Mathematical Society*, 24, 391–93.

—— (1918b), 'Doctrinal functions', *Journal of Philosophy*, 15, 262–67.

—— (1922), *Mathematical Philosophy. A Study of Fate and Freedom* (New York: Dutton & Co.).

KITCHER, PHILIP (1976), 'Hilbert's epistemology', *Philosophy of Science*, 43, 99–115.

KLEENE, STEPHEN C. (1935), 'A theory of positive integers in formal logic', *American Journal of Mathematics*, 57, 219–44.

KLEENE, STEPHEN C. and ROSSER, J. BARKLEY (1935), 'The inconsistency of certain formal logics', *Annals of Mathematics*, 36, 630–36.

KLUGE, FRITZ (1935), *Aloys Müller's Philosophie der Mathematik und der Naturwissenschaft* (Leipzig: Hirzel).

KNEALE, WILLIAM and KNEALE, MARTHA (1962), *The Development of Logic* (Oxford: Oxford University Press).

KOCKELMANS, JOSEPH J. (1989), 'Idealization and Projection in the Empirical Sciences: Husserl vs. Heidegger', *History of Philosophy Quarterly*, 6, 365–80.

KÖHLER, ECKEHART (1991), 'Gödel und der Wiener Kreis', in Paul Kruntorad (ed.), *Jour Fixe der Vernunft*, 127–58 (Vienna: Hölder-Pichler-Tempsky).

KÖHLER, ECKEHART (ed.) (2002), *Wahrheit und Beweisbarkeit. Leben und Werk Kurt Gödels, (Bd. 1, Dokumente und historische Analysen; Bd. 2, Kompendium zum Werk)* (Vienna: Hölder-Pichler-Tempsky).

KOKOSZYŃSKA, MARIA (1936a), 'Syntax, Semantik und Wissenschaftslogik', in *Actes du Congrès International de philosophie scientifique*, volume 3, 9–14 (Paris: Hermann).

—— (1936b), 'Über den absoluten Wahrheitsbegriff und einige andere semantische Begriffe', *Erkenntnis*, 6, 143–65.

KOLMOGOROV, ANDREI N. (1925), 'O principe tertium non datur', *Matematiceskij Sbornik*, 32, 646–67. English translation in van Heijenoort (1967a), pp. 416–37.

—— (1932), 'Zur Deutung der intuitionistischen Logik', *Mathematische Zeitschrift*, 35, 58–65. English translation in Mancosu (1998a), pp. 328–34.

KÖNIG, DÉNES (1926), 'Sur les correspondances multivoques des ensembles', *Fundamenta Mathematicae*, 8, 114–34.

——(1927), 'Über eine Schlussweise aus dem Endlichen ins Unendliche', *Acta litterarum ac scientiarum Regiae Universitatis Hungaricae Francisco-Josephinae, Sectio scientiarum mathematicarum*, 3, 121–30.

KÖNIG, JULIUS (1904), 'Zum Kontinuum-Problem', in *Verhandlungen des dritten internationalen Mathematiker-Kongresses in Heidelberg*, 144–47 (Leipzig: Teubner).

——(1914), *Neue Grundlagen der Logik, Arithmetik und Mengenlehre* (Leipzig: Veit).

KORSELT, A. (1913), 'Was ist Mathematik?', *Archiv der Mathematik und Physik (Series 3)*, 21, 371–73.

KREISEL, GEORG (1959), 'Wittgenstein's Remarks on the Foundations of Mathematics', *British Journal for the Philosophy of Science*, 9, 135–58.

——(1980), 'Kurt Gödel: 1906–1978', *Bibliographical Memoirs of Fellows of the Royal Society*, 26, 148–223. Corrections, ibid. 27, 697, and 28, 718.

KRETSCHMANN, ERICH (1915), 'Über der prinzipielle Bestimmbarkeit der berechtigten Bezugssysteme beliebiger Relativitätstheorien (I)', *Annalen der Physik*, 48, 907–82.

KRONECKER, LEOPOLD (1901), *Vorlesungen über Zahlentheorie* (Leipzig: Teubner).

——(2001), 'Über den Begriff der Zahl in der Mathematik (Sur le concept de nombres en mathématique)', *Revue d'histoire des mathématiques*, 7, 207–75. Retranscribed and annotated by J. Boniface and N. Schappacher.

LAMPERT, TIM (2008), 'Wittgenstein on the infinitude of primes', *History and Philosophy of Logic*, 29(1), 63–81.

LANDINI, GREGORY (1998), *Russell's Hidden Substitutional Theory* (Oxford: Oxford University Press).

——(2005), 'Quantification Theory in *8 of Principia Mathematica and the empty domain', *History and Philosophy of Logic*, 26(1), 53–71.

LANGFORD, COOPER HAROLD (1927a), 'Some theorems on deducibility', *Annals of Mathematics*, series 2, 28, 16–40.

——(1927b), 'Some theorems on deducibility', *Annals of Mathematics (Series 2)*, 28, 459–71.

LARGEAULT, JEAN (1993a), *Intuition et Intuitionisme* (Paris: Vrin).

——(1993b), 'L'Intuitionisme des mathématiciens avant Brouwer', *Archives de Philosophie*, 56, 53–68.

LASCAR, DANIEL (1998), 'Perspective historique sur les rapports entre la théorie des modèles et l'algebrè', *Revue d'Histoire des Mathématiques*, 4, 237–60.

LAUENER, H. (1971), 'Veröffentlichungen schweizerischer Wissenschaftstheoretiker I*', *Zeitschrift für Allgemeine Wissenschaftstheorie*, II(2), 340–51.

——(1978), 'Paul Bernays (1888–1977)', *Zeitschrift für Allgemeine Wissenschaftstheorie*, IX(1), 13–20.

LAUGWITZ, DETLEV (1958), 'Über eine Vermutung von Hermann Weyl zum Raumproblem', *Archiv der Mathematik*, 9, 128–33.

——(1965), *Differential and Riemannian Geometry* (New York: Academic Press).

LEWIS, CLARENCE IRVING (1918), *A Survey of Symbolic Logic* (Berkeley and Los Angeles: University of California Press).

LEWIS, CLARENCE IRVING and LANGFORD, COOPER HAROLD (1932), *Symbolic Logic* (Toronto: The Century Company).

LINSKY, BERNARD (1999), *Russell's Metaphysical Logic* (Stanford: CSLI Publications).

——(2004), 'Leon Chwistek on the no-classes theory in *Principia Mathematica*', *History and Philosophy of Logic*, 25, 53–71.

LINSKY, BERNARD AND IMAGUIRE, GUIDO, *On Denoting 1905–2005*, (Munich: Philosophie Verlag).
LOHMAR, DIETRICH (1989), *Phänomenologie der Mathematik* (Dordrecht: Kluwer).
LORENZEN, PAUL (1965), *Differential und Integral. Eine konstruktive Einführung in die klassische Analysis* (Frankfurt am Main: Akademische Verlagsgesellschaft). English translation, *Differential and Integral. A Constructive Introduction to Classical Analysis* (University of Texas Press, Austin, 1971).
LOVETT, EDGAR O. (1900–01), 'Mathematics at the International Congress of Philosophy, Paris, 1900', *Bulletin of the American Mathematical Society*, 7, 157–83.
LÖWENHEIM, LEOPOLD (1915), 'Über Möglichkeiten im Relativkalkül', *Mathematische Annalen*, 447–70. Translated in van Heijenoort (1967a), pp. 228–51.
—— (1946), 'On making indirect proofs direct', *Scripta Mathematica*, XII, 125–39.
—— (2007), 'Funktionalgleichungen im Gebietekalkül und Umformungsmöglichkeiten', *History and Philosophy of Logic*, 28, 305–36.
LÖWY, HEINRICH (1926), 'Die Krisis der Mathematik und ihre philosophische Bedeutung', *Die Naturwissenschaften*, 14, 706–08.
ŁUKASIEWICZ, JAN (1920a), 'O logice trówartościowej (On three-valued logic)', *Ruch Filozoficzny*, 6, 170–71. English translation in McCall (1967), pp. 16–17.
—— (1920b), 'O pojęciu możliwości (On the concept of possibility)', *Ruch Filozoficzny*, 6, 169–170. English translation in McCall (1967), pp. 15–16.
—— (1922), 'O determiniźmie (On determinism)', English translation in McCall (1967), pp. 19–39.
—— (1924), 'Démonstration de la compatibilité des axiomes de la théorie de la déduction (Abstract)', *Annales de la Société Polonaise de Mathématique*, 3, 149. Talk given 13 June 1924.
—— (1930), 'Philosophische Bemerkungen zu mehrwertigen Systemen des Aussagenkalküls', *Comptes Rendus des Sèancs de la Société des Sciences et des Lettres de Varsovie. Classe III*, 23, 51–77. English translation in McCall (1967), pp. 40–65.
ŁUKASIEWICZ, JAN and TARSKI, ALFRED (1930), 'Untersuchungen über den Aussagenkalkül', *Comptes Rendus des Séances de la Société des Sciences et des Lettres de Varsovie. Classe III*, 23, 30–50. English translation in Tarski (1983), pp. 38–59.
MCCALL, STORRS (1967), *Polish Logic 1920–1939* (Oxford: Oxford University Press).
MADDUX, ROGER D. (1991), 'The Origin of Relation Algebras in the Development and Axiomatization of the Calculus of Relations', *Studia Logica*, 50, 421–55.
MAHNKE, DIETRICH (1923), 'Von Hilbert zu Husserl: Erste Einführung in die Phänomenologie, besonders der formalen Mathematik', *Unterrichtsblätter*, 29, 34–37. English translation in Mahnke (1977).
—— (1925a), 'Leibnizens Synthese von Universalmathematik und Individualmetaphysik', *Jahrbuch für Philosophie und phänomenologische Forschung*, 7, 305–612.
—— (1925b), 'Neue Einblicke in die Entdeckunggeschichte der höheren Analysis', *Abhandlungen der Preussischen Akademie, Phys. Math. Klasse*, 1, 1–68.
—— (1926), 'Die Entstehung der Funktionsbegriffe', *Kantstudien*, 31, 426–28.
—— (1927a), 'Leibniz als Begründer der symbolische Mathematik', *Isis*, 279–93.
—— (1927b), 'Review of W. Dilthey, Gesammelte Schriften, VII', *Deutsche Literaturzeitung, Neue Folge*, 4(26), 1246–51.
—— (1927c), 'Untergang der abenländischen Wissenschaft?', *Archiv für Mathematik und Naturwissenschaft*, 10, 216–32.
—— (1977), 'From Hilbert to Husserl: first introduction to phenomenology especially that of formal mathematics', *Studies in History and Philosophy of Science*, 8, 71–84.

MAJER, ULRICH (1988), 'Zu einer bemerkenswerten Differenz zwischen Brouwer und Weyl', in W. Deppert (ed.), *Exact Sciences and their philosophical foundations*, 543–52 (Frankfurt am Main: Lang).

—— (1991), 'Hilbert, Reichenbach und der Neu-Kantianismus', in L. Schäfer L. Danneberg, A. Kamlah (ed.), *Hans Reichenbach und die Berliner Gruppe*, 253–73 (Braunschweig: Vieweg).

—— (1993a), 'Different forms of finitism', in J. Czermak (ed.), *Philosophie der Mathematik. Akten des 15. Internationalen Wittgenstein-Symposiums*, 185–94 (Vienna: Hölder-Pichler-Tempsky).

—— (1993b), 'Hilbert's finitism and the concept of space', in U. Majer and H.-J. Schmidt (eds.), *Semantical Aspects of Spacetime Theories*, 145–57 (Vienna: Wissenschaftsverlag).

—— (1993c), 'Hilberts Methode der idealen Elemente und Kants regulativer Gebrauch der Ideen', *Kant-Studien*, 84, 51–77.

MAJER, ULRICH and SAUER, TILMAN (2006), 'Intuition and the axiomatic method in Hilbert's foundation of physics', in Emily Carson and Renate Huber (eds.), *Intuition and the Axiomatic Method*, 213–33 (Dordrecht: Springer).

MCKINSEY, JOHN CHARLES CHENOWETH (1940), 'Postulates for the Calculus of binary Relations', *The Journal of Symbolic Logic*, 5, 85–97.

MALCEV, ANATOLY IVANOVICH (1936), 'Untersuchungen aus dem Gebiet der mathematischen Logik', *Matematiceskij Sbornik(n.s.)*, 1, 323–36. English translation in Anatoly I. Maltsev, *The Metamathematics of Algebraic Systems: Collected Papers 1936–1967* (North-Holland, Amsterdam, 1971), 1–14.

MANCOSU, PAOLO (1996), *Philosophy of Mathematics and Mathematical Practice in the Seventeenth Century* (Oxford: Oxford University Press).

MANCOSU, PAOLO (ed.) (1998a), *From Brouwer to Hilbert. The Debate on the Foundations of Mathematics in the 1920s* (Oxford: Oxford University Press).

MANCOSU, PAOLO (1998b), 'Hermann Weyl: Predicativity and an intuitionistic excursion', in Mancosu (1998a), pp. 65–85. Reprinted as Chapter 9 in this volume.

—— (1998c), 'Hilbert and Bernays on metamathematics', in Mancosu (1998a), pp. 149–88. Reprinted as Chapter 2 in this volume.

—— (1999a), 'Between Russell and Hilbert: Behmann on the foundations of mathematics', *Bulletin of Symbolic Logic*, 5(3), 303–330. Reprinted as Chapter 3 in this volume.

—— (1999b), 'Between Vienna and Berlin: The immediate reception of Gödel's incompleteness theorems', *History and Philosophy of Logic*, 20, 33–45. Reprinted as Chapter 7 in this volume.

—— (2001), 'Mathematical Explanation: Problems and Prospects', *Topoi*, 20, 97–117.

—— (2002), 'On the constructivity of proofs. A debate among Behmann, Bernays, Gödel, and Kaufmann', in Wilfried Sieg, Richard Sommer, and Carolyn Talcott (eds.), *Reflections on the foundations of mathematics. Essays in honor of Solomon Feferman*, Lecture Notes in Logic 15, 346–68 (Association for Symbolic Logic). Reprinted as Chapter 5 in this volume.

—— (2003), 'The Russellian Influence on Hilbert and his school', *Synthese*, 137, 59–101. Reprinted as Chapter 4 in this volume.

—— (2004), 'Review of Kurt Gödel, *Collected Works*, vols. IV and V, Solomon Feferman, et al., eds. Oxford: Oxford University Press, 2003', *Notre Dame Journal of Formal Logic*, 45, 109–125. Reprinted as Chapter 8 in this volume.

—— (2005), 'Harvard 1940–1941: Tarski, Carnap and Quine on a finitistic language of mathematics for science', *History and Philosophy of Logic*, 26, 327–57. Reprinted as Chapter 13 in this volume.

—— (2006), 'Tarski on models and logical consequence', in Jeremy Gray and José Ferreiros (eds.), *The Architecture of Modern Mathematics*, 209–37 (Oxford: Oxford University Press). Reprinted as Chapter 16 in this volume.

—— (2008), 'Neurath, Tarski and Kokoszyńska on the semantic conception of truth', in D. Patterson (ed.), *New Essays on Tarski and Philosophy*, 192–224 (Oxford: Oxford University Press). Reprinted as Chapter 15 in this volume.

—— (2009), 'Tarski's philosophical engagement', in Sandra Lapointe et al. (ed.), *The Golden Age of Polish Philosophy*, 131–53 (Dordrecht: Springer).

MANCOSU, PAOLO and RYCKMAN, THOMAS (2002), 'Mathematics and phenomenology: The correspondence between O. Becker and H. Weyl', *Philosophia Mathematica*, 10, 130–202. Reprinted as Chapter 10 in this volume.

—— —— (2005), 'Geometry, Physics and Phenomenology: Four letters of O. Becker to H. Weyl', in Volker Peckhaus (ed.), *Oskar Becker and die Philosophie der Mathematik*, Neuzeit & Gegenwart: Philosophie in Wissenschaft und Gesellschaft, 229–243 (Munich: Fink Verlag). Reprinted as Chapter 11 in this volume.

MANCOSU, PAOLO and VAN STIGT, WALTER P. (1998), 'Intuitionistic Logic', in Mancosu (1998a), 275–85.

MANGIONE, C. and BOZZI, S. (1993), *Storia della Logica* (Milan: Garzanti).

MARCHISOTTO, ELENA A. (1995), 'In the Shadow of Giants: The work of Mario Pieri in the foundations of mathematics', *History and Philosophy of Logic*, 16, 107–19.

MARCHISOTTO, ELENA A. and SMITH, JAMES T. (2007), *The Legacy of Mario Pieri in Geometry and Arithmetic* (Boston: Birkhäuser).

MARION, MATHIEU (1995), 'Kronecker's "Safe Haven of Real Mathematics"', in Cohen, R. S. and Marion, M. (eds.), *Quebec Studies in the Philosophy of Science, vol. 2*, 135–87 (Dordrecht: Kluwer).

—— (1998), *Wittgenstein, Finitism, and the Foundations of Mathematics* (Oxford: Oxford University Press).

—— (2004), 'Wittgenstein on Mathematics : Constructivism or Constructivity?', in A. Coliva & E. Picardi (ed.), *Wittgenstein Today*, 201–22 (Padua: Il Poligrafo).

—— (2008), 'Wittgenstein et le Constructivisme', in E. Rigal (ed.), *Wittgenstein: état des lieux*, 261–73 (Paris: Vrin).

MARTIN, RICHARD M. (1964), 'The Philosophical Import of Virtual Classes', *The Journal of Philosophy*, 61, 377–87.

MEDICUS, FRITZ (1926), *Die Freiheit des Willens und ihre Grenzen* (Tübingen: J. C. B. Mohr (Paul Siebeck) Verlag).

MENZLER-TROTT, ECKART (2007), *Logic's Lost Genius: The Life of Gerhard Gentzen* (Providence, RI: American Mathematical Society).

MILNE, PETER (2008), 'Russell's Completeness Proof', *History and Philosophy of Logic*, 29(1), 31–62.

MIRIMANOFF, DMITRY (1917), 'Les Antinomies de Russell et de Burali-Forti et le problème fondamental de la théorie des ensembles', *L'Enseignement Mathématique*, 19, 37–52.

MOL, LIESBETH DE (2006), 'Closing the circle: an analysis of Emil Post's early work', *The Bulletin of Symbolic Logic*, 12, 267–89.

MONK, RAY (1990a), *Ludwig Wittgenstein. The Duty of Genius* (London: Jonathan Cape).

—— (1990b), 'Unphilosophical Investigations', *The Independent*, 6 October, 1990, 31.

MOOIJ, JAN J.A. (1966), *La Philosophie des Mathématiques de Henri Poincaré* (E. Nauwelaerts Nancy-Pulnoy).

MOORE, GREGORY H. (1980), 'Beyond First-order Logic: The Historical Interplay between Mathematical Logic and Axiomatic Set Theory', *History and Philosophy of Logic*, 1, 95–137.

—— (1982), *Zermelo's Axiom of Choice. Its Origins, Development and Influence* (New York and Heidelberg: Springer).
—— (1988), 'The emergence of first-order logic', in William Aspray and Philip Kitcher (eds.), *History and Philosophy of Modern Mathematics*, Minnesota Studies in the Philosophy of Science 11, 95–135 (Minneapolis: University of Minnesota Press).
—— (1994), 'Editor's Introduction', in Russell (1994a).
—— (1997), 'Hilbert and the emergence of modern mathematical logic', *Theoria*, 12(1), 65–90.
MOORE, GREGORY H. and GARCIADIEGO, ALEJANDRO R. (1981), 'Burali-Forti's paradox: a reappraisal of its origins', *Historia Mathematica*, 8, 319–50.
MORICONI, ENRICO (1987), *La Teoria della Dimostrazione di Hilbert* (Naples: Bibliopolis).
—— (2006), 'Gödel's Completeness theorem: some history, some philosophy', in E. Ballo and M. Franchella (eds.), *Logic and Philosophy in Italy*, 203–14 (Monza: Polimetrica).
MORMANN, THOMAS (1999), 'Neurath's opposition to Tarskian semantics', in J. Wolenski and E. Köhler (eds.), *Alfred Tarski and the Vienna Circle*, 165–78 (Dordrecht: Kluwer).
MORRIS, MICHAEL and THORNE, KIP (1988), 'Wormholes in Spacetime and Their Use for Interstellar Travel: A Tool for Teaching General Relativity', *American Journal of Physics*, 56, 395–412.
MOSTOWSKI, ANDRZEJ (1966), *Thirty Years of Fundational Studies* (Oxford: Basil Blackwell).
—— (1967), 'Alfred Tarski', in P. Edwards (ed.), *Encyclopedia of Philosophy*, vol. 8 (New York: MacMillan).
MÜLLER, ALOYS (1922), *Der Gegenstand der Mathematik mit besonderer Beziehung auf die Relativitätstheorie* (Braunschweig: Vieweg).
—— (1923), 'Über Zahlen als Zeichen', *Mathematische Annalen*, 90, 150–59.
MÜLLER, GERT (1978), *Sets and Classes* (Amsterdam: North-Holland).
MULLIGAN, KEVIN, SMITH, BARRY, SIMONS, PETER (1984), 'Truth-Makers', *Philosophy and Phenomenological Research*, 44, 287–321.
MURAWSKI, ROMAN (1998), 'Undefinability of truth. The problem of priority: Tarski vs Gödel', *History and Philosophy of Logic*, 19, 153–60.
MURAWSKI, ROMAN and WOLEŃSKI, JAN (2008), 'Tarski and his Polish predecessors on truth', in Douglas Patterson (ed.), *New Essays on Tarski and Philosophy*, 44–71 (Oxford: Oxford University Press).
NAESS, ARNE (1936), *Erkenntnis und Wissenschaftliches Verhalten* (Oslo: Jacob Dybwad).
—— (1938), ' "Truth" as conceived by those who are not professional philosophers', *Skrifter Videnskaps Akademi, Oslo: Hist. Fil.Kl,*, 1–178.
NAGEL, ERNST and NEWMAN, J. R. (1959), *Gödels' Proof* (London: Routledge & Kegan Paul).
NELSON, LEONARD (1928), 'Kritische Philosophie und Mathematische Axiomatik', *Beilage zu Unterrichtsblätter f. Math. und Naturw.*, XXXIV(4), 1–14. Partial English translation in L. Nelson, *The Socratic Method and Critical Philosophy* (Dover, 1966), pp. 158–84.
NEURATH, OTTO (1931a), *Empirische Soziologie* (Vienna: Springer). English translation in Otto Neurath *Empiricism and Sociology*, ed. M. Neurath and R. S. Cohen (Dordrecht: Reidel, 1973).
—— (1931b), 'Physikalismus', *Scientia*, 50, 417–21. Translated in Neurath (1983), pp. 52–57.
—— (1932/3), 'Protokollsätze', *Erkenntnis*, 3, 204–14. Translated in Neurath (1983), pp. 91–99.
—— (1934), 'Radikaler Physikalismus und 'wirkliche Welt'', *Erkenntnis*, 4, 346–62. Translated in Neurath (1983), pp. 100–14.
—— (1935), *Le Développement du Cercle de Vienne* (Paris: Hermann).
—— (1936), 'Erster Internationaler Kongress für Einheit der Wissenschaft in Paris 1935', *Erkenntnis*, 5, 377–406.

—— (1937a), 'Bemerkungen zur Privatdiskussion', Call number: K.33, Neurath *Nachlaß*.
—— (1937b), ' "Diskussion Paris 1937 Neurath-Carnap" ', Call number: K.32, Neurath *Nachlaß*.
—— (1937c), ' "Fuer Die Privatsitzung, 30 Juli 1937" ', Call number: K.31, Neurath *Nachlaß*.
—— (1937d), 'Wahrheitsbegriff und Empirismus (Vorbemerkungen zu einer Privatdiskussion mit Carnap im Kreis der Pariser Konferenz)', Call number: K.30, Neurath *Nachlaß*.
—— (1983), *Philosophical Papers 1913–1946* (Dordrecht: Reidel).
NEWMAN, ANDREW (2002), *The Correspondence Theory of Truth* (Cambridge: Cambridge University Press).
NEWMAN, MAXWELL H. A. (1957), 'Hermann Weyl', *Bibliographical Memoirs of Fellows of the Royal Society*, 3, 305–28.
NICOD, JEAN G. P. (1916–19), 'A reduction in the number of the primitive propositions of logic', *Proceedings of the Cambridge Philosophical Society*, 19, 32–41.
NYE, MARY JO (1976), 'The nineteenth-century atomic debates and the dilemma of an "indifferent hypothesis" ', *Studies in History and Philosophy of Science*, 7, 245–68.
OBERDAN, THOMAS (1993), *Protocols, Truth and Convention* (Amsterdam: Rodopi).
O'LEARY, DANIEL J. (1988), 'The propositional logic of *Principia Mathematica* and some of its forerunners', *Russell*, 8, 92–115.
O'RAIFEARTAIGH, LOCHLAINN (1997), *The Dawning of Gauge Theory* (Princeton: Princeton University Press).
PADOA, ALESSANDRO (1901), 'Essai d'une théorie algébrique des nombre entiers, precedè d'une introduction logique à une théorie dèductive quelconque', in *Bibliothèque du Congrès international de philosophie, Paris 1900*, 309–365 (Paris: A. Colin). Partially translated in van Heijenoort (1967), 118–23.
—— (1902), 'Un nouveau système irréductible de postulats pour l'algèbre', in *Compte rendu du deuxième congrès international des mathematicians tenu à Paris du 6 au 12 août 1900*, 249–56 (Paris: Gauthiers-Villars).
—— (1903), 'Le Problème No 2 de M. David Hilbert', *L'Enseignement Mathématique*, 5, 85–91.
PARSONS, CHARLES (1998), 'Hao Wang as philosopher and interpreter of Gödel', *Philosophia Mathematica*, 6, 3–24.
PASCH, MORITZ (1918), 'Die Forderung der Entscheidbarkeit', *Jahresbericht der Deutsche Mathematiker-Vereinigung*, 27, 228–32.
PATTERSON, DOUGLAS (2008), *New Essays on Tarski and Philosophy* (Oxford: Oxford University Press).
PAULI, WOLFGANG (1921), 'Relativitätstheorie', in *Encyklopädie der mathematischen Wissenschaften, mit Einschluss ihrer Anwendungen*, vol. 5 (Leipzig: B. G. Teubner). Citations to Engl. trans., The Theory of Relativity (New York: Pergamon Press, 1958).
PEANO, GIUSEPPE (1889), *I Principii di Geometria Logicamente Esposti* (Turin: Fratelli Bocca). Reprinted in Peano (1958), pp. 56–91.
—— (1958), *Opere Scelte*, vol. 2 (Rome: Edizioni Cremonese).
PECKHAUS, VOLKER (1990), *Hilbertprogramm und Kritische Philosophie* (Göttingen: Vandenhoeck und Ruprecht).
—— (1992), 'Hilbert, Zermelo und die Institutionalisierung der mathematischen Logik in Deutschland', *Berichte zur Wissenschaftsgeschichte*, 15, 27–38.
—— (1993), 'Kurt Grelling und der Logische Empirismus', in R. Haller and F. Stadler (eds.), *Wien-Berlin-Prag. Der Aufstieg der wissenschaftlichen Philosophie*, 362–85 (Vienna: Hölder-Pichler-Tempski).

——— (1994a), 'Hilbert's axiomatic programme and philosophy', in E. Knobloch and D. E. Rowe (eds.), *The History of Modern Mathematics*, vol. 3, 91–112 (Boston: Academic Press).

——— (1994b), 'Von Nelson zu Reichenbach: Kurt Grelling in Göttingen und Berlin', in L. Danneberg et al. (ed.), *Hans Reichenbach und die Berliner Gruppe*, 53–86 (Braunschweig: Vieweg Verlag).

——— (1995a), 'The genesis of Grelling's paradox', in I. Max and W. Stelzner (eds.), *Logik und Mathematik. Frege-Kolloquium Jena 1993*, 269–80 (Berlin: de Gruyter).

——— (1995b), 'Hilberts Logik: Von der Axiomatik zur Beweistheorie', *Internationale Zeitschrift für Geschichte und Ethik der Naturwissenschaften, Technik und Medizin*, 3, 65–86.

——— (2004a), 'Calculus Ratiocinator versus Characteristica Universalis? The two traditions in logic revisited', *History and Philosophy of Logic*, 25, 3–14.

——— (2004b), 'Schröder's Logic', in D. M. Gabbay and J. Woods (eds.), *The Rise of Modern Logic: From Leibniz to Frege*, vol. 3, 557–609 (Amsterdam: North-Holland).

——— (2005a), 'Becker und Zermelo', in Volker Peckhaus (ed.), *Oskar Becker und die Philosophie der Mathematik*, 279–88 (Munich: Wilhelm Fink Verlag).

——— (2005b), 'Impliziert Widerspruchsfreiheit Existenz? Oskar Becker Kritik am formalistischen Existenzbegriff', in Volker Peckhaus (ed.), *Oskar Becker und die Philosophie der Mathematik*, 79–100 (Munich: Wilhelm Fink Verlag).

PECKHAUS, VOLKER (ed.) (2005c), *Oskar Becker und die Philosophie der Mathematik* (Munich: Wilhelm Fink Verlag).

PEIRCE, CHARLES SANDERS (1870), 'Description of a notation for the logic of relatives, resulting from an amplification of the conceptions of Boole's calculus on logic', *Memoirs of the American Academy of Arts and Sciences*, 9, 317–78. Reprinted in Peirce (1984), pp. 359–429.

——— (1883), 'The Logic of Relatives', in *Studies in Logic by Members of the Johns Hopkins University*, 187–203 (Philadelphia: John Benjamins). Reprinted in Peirce (1986), pp. 453–66.

——— (1885), 'On the algebra of logic: a contribution to the philosophy of notation', *American Journal of Mathematics*, 7, 180–202. Reprinted in Peirce (1993), pp. 162–90.

——— (1903), 'Nomenclature and Divisions of Dyadic Relations', in Ch. Hartshorne and P. Weiss (eds.), *Collected Papers*, vol. 3, 366–87 (Cambridge, Mass.: Harvard University Press, 1933).

——— (1984), *Writings of Charles S. Peirce*. Vol. 2. *1867–1871* (Bloomington: Indiana University Press).

——— (1986), *Writings of Charles S. Peirce*. Vol. 4. *1879–1884* (Bloomington: Indiana University Press).

——— (1993), *Writings of Charles S. Peirce*. Vol. 5. *1884–1886* (Bloomington: Indiana University Press).

PIERI, MARIO (1901), 'Sur la géométrie envisagée comme un système purement logique', in *Bibliothèque du Congrès international de philosophie, Paris 1900*, 367–404 (Paris: A. Colin). Reprinted in Pieri (1980), pp. 235–72.

——— (1904), 'Circa il teorema fondamentale di Staudt e i principi della geometria projettiva', *Atti della Reale academia delle scienze di Torino*, 39, 313–31. Reprinted in Pieri (1980), pp. 289–307.

——— (1980), *Opere Scelte I: Opere sui fondamenti della matematica* (Bologna: UMI).

POEGGELER, OTTO (1969), 'Oskar Becker als Philosoph', *Kantstudien*, 60, 298–311.

POINCARÉ, HENRI (1902), *La Science et l'Hypothèse* (Paris: Flammarion).

——— (1906), 'Les Mathématiques et la logique', *Revue de métaphysique et de morale*, 14, 294–317. English translation in Ewald (1996), 1038–52.

——— (1908), *Science et Méthode* (Flammarion). English translation: *Science and Method* (New York: Dover, 1952).

——— (1909), 'Réflexions sur les deux notes précédentes', *Acta Mathematica*, 32, 195–200.

POLLARD, STEVEN (2005), 'Property is prior to set: Fichte and Weyl', in G. Sica (ed.), *Essays on the Foundations of Mathematics*, 209–26 (Monza: Polimetrica).

POLLOCK, JOHN L. and CRUZ, JOSEPH (1999), *Contemporary Theories of Knowledge* (Lanham, Md: Rowman and Littlefield), 2nd edn.

POSER, HANS (2005), 'Ontologie der Mathematik im Anschluß an Oskar Becker', in Volker Peckhaus (ed.), *Oskar Becker und die Philosophie der Mathematik*, 59–78 (Munich: Wilhelm Fink Verlag).

POST, EMIL L. (1921), 'Introduction to a general theory of elementary propositions', *American Journal of Mathematics*, 43, 163–85. Reprinted in van Heijenoort (1967a), pp. 264–83.

POSY, CARL J. (1974), 'Brouwer's Constructivism', *Synthese*, 27, 125–59.

POTTER, MICHAEL (2000), *Reason's Nearest Kin* (Oxford: Oxford University Press).

PRAWITZ, DAG (1981), 'Philosophical aspects of proof theory', in G. Fløistad (ed.), *Contemporary Philosophy. A New Survey*, vol. 1, 235–77 (The Hague: Nijhoff).

—— (1993), 'Remarks on Hilbert's Program for the foundations of mathematics', in G. Corsi, M. L. Dalla Chiara, and G. C. Ghirardi (eds.), *Bridging the Gap. Philosophy, Mathematics and Physics. Lectures on the Foundations of Science*, 87–98 (Dordrecht: Kluwer).

PRESBURGER, MOJÈSZ (1930), 'Über die Vollständigkeit eines gewisse Systems der Arithmetik ganzer Zahlen, in welchem die Addition als einzige Operation hervortritt', in F. Leja (ed.), *Comptes-rendus du I Congrès des Mathématiciens des Pays Slaves, Varsovie 1929*, 92–101. Translated as "On the completeness of a certain system of arithmetic of whole numbers in which addition occurs as the only operation," *History and Philosophy of Logic* 12 (1991), 225–33.

PROOPS, IAN (2007), 'Russell and the Universalist Conception of Logic', *Noûs*, 41(1), 1–32.

QUINE, WILLARD VAN ORMAN (1934–38), 'Logic Notes. Mostly 1934–38', Notebook of 300 pp.; Quine archive, Houghton Library, *2002M-5, Box 02, Compositions 2UDC.

—— (1936), 'Towards a calculus of concepts', *The Journal of Symbolic Logic*, 1, 2–25.

—— (1937), 'Nominalism', Lecture delivered at the Philosophy Club, Harvard, October 25, 1937. Quine archive, Hougton Library, Occasional Lectures, 1935–38, MS Storage 299, box 11.

—— (1939a), 'Designation and existence', *Journal of Philosophy*, 36, 701–09.

—— (1939b), 'A Logistical approach to the ontological problem', Preprint. Published in Quine, *The Ways of Paradox and Other Essays* (New York: Random House, 1966).

—— (1939c), 'A Logistical approach to the ontological problem, Sept. 8, 1939 (Unity of Science Congress)', Quine archive, Hougton Library, Occasional Lectures, 1939, MS Storage 299, box 11.

—— (1940), 'Logic, Mathematics and Science', Lecture delivered by Quine at Harvard on Dec. 20, 1940. Quine archive, Hougton Library, Occasional Lectures, 1940–7, MS Storage 299, box 11.

—— (1941), *Elementary Logic* (Boston and New York: Ginn and Company). Rev. edn. (New York: Harper and Row, 1965).

—— (1944), *O Sentido da Nova Logica* (Sao Paulo: Martins).

—— (1946), 'Nominalism', Presented at Harvard Philosophical Colloquium on March 11, 1946. Original in possession of Douglas Boynton Quine. Transcription with editorial comments by P. Mancosu published in *Oxford Studies in Metaphysics*, vol. IV (2008), pp. 3–21.

—— (1951), 'Ontology and Ideology', *Philosophical Studies*, 2, 11–15.

—— (1953a), 'Logic and the reification of universals', in *From a Logical Point of View* (Cambridge, Mass.: Harvard University Press).

—— (1953b), 'Nominalism and Platonism in Modern Logic', Lecture delivered in Amersfoort in September 1953. Quine archive, Hougton Library, Occasional Lectures, 1951–55, MS Storage 299, box 11.

—— (1955a), 'On Frege's way out', *Mind*, 64, 145–59.

—— (1955b), 'A Proof procedure for Quantification Theory', *Journal of Symbolic Logic*, 20, 141–49.

—— (1972), *Methods of Logic* (New York: Holt, Rinehart & Winston), 3rd edn.

—— (1985), *The Time of my Life: An Autobiography* (Cambridge, Mass.: MIT Press).

—— (1987), 'Peano as Logician', *History and Philosophy of Logic*, 8, 15–24.

—— (1988a), 'Autobiography', in L. E. Hahn and P. A. Schilpp (eds.), *The Philosophy of W.V. Quine*, 2–46 (La Salle, Ill.: Open Court).

—— (1988b), 'Reply to Nelson Goodman', in L. E. Hahn and P. A. Schilpp (eds.), *The Philosophy of W.V. Quine*, 162–63 (La Salle, IU.: Open Court).

—— (2008), 'Nominalism', *Oxford Studies in Metaphysics*, 4, 3–21.

QUINE, WILLARD VAN ORMAN and GOODMAN, NELSON (1940), 'Elimination of extra-logical postulates', *The Journal of Symbolic Logic*, 5, 104–9.

RAMSEY, FRANK P. (1925), 'The Foundations of Mathematics', *Proceedings of the London Mathematical Society*, 25, 338–84. Reprinted in Ramsey (1990), pp. 164–224.

—— (1926), 'Mathematical logic', *The Mathematical Gazette*, 13, 185–94. Reprinted in Ramsey (1990), pp. 225–44.

—— (1930), 'On a problem of formal logic', *Proceedings of the London Mathematical Society*, 2nd series, 30, 264–86.

—— (1990), *Philosophical Papers* (Cambridge: Cambridge University Press).

RANG, BERNHARD and THOMAS, WOLFGANG (1981), 'Zermelo's discovery of the "Russell's paradox"', *Historia Mathematica*, 8, 15–22.

RAY, GREG (1996), 'Logical consequence: a defence of Tarski', *Journal of Philosophical Logic*, 25, 617–77.

READ, STEPHEN (1997), 'Completeness and Categoricity: Frege, Gödel and model theory', *History and Philosophy of Logic*, 18, 79–93.

RECK, ERICH (2008), 'Carnap and modern logic', in M. Friedman and R. Creath (eds.), *The Cambridge Companion to Carnap*, 176–99 (Cambridge: Cambridge University Press).

REECH, FERDINAND (1852), *Cours de Méchanique d'après la nature génaralement flexible et élastique des corps* (Paris: Carilian-Goeurget et V. Dalmont).

REID, CONSTANCE (1970), *Hilbert* (New York: Springer).

RESNIK, MICHAEL (1980), *Frege and the Philosophy of Mathematics* (Ithaca, NY and London: Cornell University Press).

RICHARD, JULES (1905), 'Les principes des mathématiques et le problème des ensembles', *Revue Générale des Sciences Pures et Appliquées*, 16, 541. Reprinted as 'Lettre à Monsieur le rédacteur de la revue Générale des Sciences,' *Acta Mathematica* 30 (1906) 295–96. English translation in van Heijenoort (1967a), pp. 142–44.

RIVENC, F. (1993), *Recherches sur l'Universalisme Logique. Russell et Carnap* (Paris: Payot).

RODRIGUEZ-CONSUEGRA, FRANCISCO A. (1991), *The Mathematical Philosophy of Bertrand Russell: Origins and Development* (Basel: Birkhäuser).

—— (2007), 'Two unpublished contributions by Alfred Tarski', *History and Philosophy of Logic*, 28, 257–64.

ROETTI, J. A. (1996), 'El finitismo matematico de Hilbert y la critica de Oskar Becker', *Revista de Filosofía (Argentina)*, 11, 3–20.

ROSE, ALAN and ROSSER, J. BARKLEY (1958), 'Fragments of many-valued statement calculi', *Transactions of the American Mathematical Society*, 87, 1–53.

ROSSER, J. BARKLEY (1936), 'Extensions of some theorems of Gödel and Church', *Journal of Symbolic Logic*, 1, 87–91.

—— (1937), 'Review of Malcev (1936)', *Journal of Symbolic Logic*, 2, 84.

RUSSELL, BERTRAND (1901a), 'Recent work on the principles of mathematics', *International Monthly*, 4, 83–101. Reprinted in Russell (1994a), pp. 363–69.

—— (1901b), 'Sur la logique des relations avec des applications à la théorie des series', *Revue des Mathématiques*, 7, 115–48. Reprinted in Russell (1994a), pp. 618–27.

—— (1902a), 'Letter to Frege, June 16, 1902', in van Heijenoort (1967a), pp. 124–125.

—— (1902b), 'Théorie générale des series bien-ordonnées', *Revue des mathématiques*, 8, 12–43. Reprinted in Russell (1994a), pp. 661–73.

—— (1903), *The Principles of Mathematics* (Cambridge: Cambridge University Press).

—— (1905), 'On denoting', *Mind*, 14, 479–93. Reprinted in Russell (1994b), pp. 414–27.

—— (1906a), 'Les Paradoxes de la Logique', *Revue de Metaphysique et de Morale*, 14, 627–50.

—— (1906b), 'On some difficulties in the theory of transfinite numbers and order types', *Proceedings of the London Mathematical Society*, 4, 29–53.

—— (1908), 'Mathematical logic as based on the theory of types', *Americal Journal of Mathematics*, 30, 222–262. Reprinted in van Heijenoort (1967a), pp. 150–82.

—— (1912), *The Problems of Philosophy* (London: Williams & Norgale).

—— (1919), *Introduction to Mathematical Philosophy* (London: Allen and Unwin).

—— (1940), *An Enquiry into Meaning and Truth* (London: Allen and Unwin).

—— (1967), *The Autobiography of Bertrand Russell*, vol. 1 (London: Allen and Unwin).

—— (1973), *Essays in Analysis* (London: Allen and Unwin).

—— (1989), 'My mental development', in Paul Arthur Schilpp (ed.), *The Philosophy of Bertrand Russell*, 3–20 (La Salle, Ill.: Open Court).

—— (1994a), *Collected Works, Vol. 3. Toward the "Principles of Mathematics", 1900–1902* (London and New York: Routledge).

—— (1994b), *Collected Works, Vol. 4. Foundations of Logic, 1903–1905* (London and New York: Routledge).

RUTTE, HEINER (1991), 'Neurath contra Schlick. On the discussion of truth in the Vienna Circle', in T. E. Übel (ed.), *Rediscovering the Forgotten Vienna Circle*, 169–74 (Dordrecht: Kluwer).

RYCKMAN, THOMAS A. (1994), 'Weyl, Reichenbach and the Epistemology of Geometry', *Studies in History and Philosophy of Modern Physics*, 25, 831–70.

—— (2003), 'The philosophical roots of the gauge principle: Weyl and transcendental phenomenological idealism', in E. Castellani and K. Brading (eds.), *Symmetries in Physics: Philosophical Reflections*, 61–88 (Cambridge: Cambridge University Press).

—— (2005), *The Reign of Relativity: Philosophy in Physics 1915–1925* (New York and Oxford: Oxford University Press).

—— (2009), 'Hermann Weyl and "First Philosophy": Constituting Gauge Invariance', in Pierre Kerzberg Michel Bitbol and Jean Petitot (eds.), *Constituting Objectivity: Transcendental Perspectives on Modern Physics*, 275–93 (Dordrecht: Springer).

SAGÜILLO, JOSÉ (1997), 'Logical consequence revisited', *The Bulletin of Symbolic Logic*, 3, 216–41.

—— (2002), 'Quine on logical truth and consequence', *AGORA, Papeles de Filosofia*, 20, 139–56.

SCANLAN, MICHAEL (1991), 'Who were the American Postulate Theorists?', *Journal of Symbolic Logic*, 56, 981–1002.

—— (2000), 'The Known and Unknown H. M. Sheffer', *The Transactions of the C. S. Peirce Society*, 36, 193–224.

—— (2003), 'American Postulate theorists and Alfred Tarski', *History and Philosophy of Logic*, 24, 307–25.

SCHEIBE, EHRARD (1988), 'Hermann Weyl and the Nature of Spacetime', in W. Deppert (ed.), *Exact Sciences and their philosophical foundations*, 61–82 (Frankfurt am Main: Lang).

SCHLICK, MORITZ (1918), *Allgemeine Erkenntinislehre* (Berlin: Springer).

—— (1934), 'Über das Fundament der Erkenntnis', *Erkenntnis*, 4, 79–99. Translated in Schlick (1979), pp. 370–87.

—— (1935), 'Facts and Propositions', *Analysis*, 2, 65–70. Reprinted in Schlick (1979), pp. 400–04.

—— (1979), *Philosophical Papers*, vol. 2 (Dordrecht: Reidel).

SCHMID, ANNE-FRANÇOISE (1978), *Une philosophie de Savant. Henri Poincaré et la Logique Mathematique* (Paris: Maspero).

SCHOLZ, ERHARD (1994a), 'Hermann Weyl's Contribution to Geometry, 1917–1923', in C. Sasaki et alii (ed.), *The Intersection of History and Mathematics*, 203–30 (Basel, Boston, and Berlin: Birkhäuser).

—— (1994b), 'Hermann Weyl's 'Purely Infinitesimal Geometry'', in *Proceedings of the International Congress of Mathematicians, Zürich 1994*, 1592–1603 (Basel, Boston, and Berlin: Birkhäuser).

—— (1999), 'Weyl and the Theory of Connections', in J. Gray (ed.), *The Symbolic Universe: Geometry and Physics 1890–1930*, 260–84 (Oxford: Oxford University Press).

—— (2000), 'Hermann Weyl on the concept of continuum', in S. A. Pedersen, Vincent Hendricks, and K. F. Jørgensen et al. (eds.), *Proof Theory: History and Philosophical Significance*, 195–207 (Dordrecht: Kluwer).

—— (2001), 'Weyls Infinitesimalgeometrie', in Ehrard Scholz (ed.), *Hermann Weyl's 'Raum-Zeit-Materie' and a General Introduction to his Scientific Work*, 48–104 (Basel, Boston, and Berlin: Birkhäuser).

—— (2004), 'Hermann Weyl's Analysis of the "Problem of Space" and the Origin of Gauge Structures', *Science in Context*, 17, 165–97.

—— (2005), 'Philosophy as a cultural resource and medium of reflection for Hermann Weyl', *Révue de Synthèse*, 126, 331–51.

—— (2006), 'Practice-Related Symbolic Realism', in José Ferreiros and Jeremy Gray (eds.), *The Architecture of Modern Mathematics*, 291–309 (Oxford: Oxford University Press).

SCHOLZ, HEINRICH (1928), 'Review of Becker's "Mathematische Existenz"', *Deutsche Literaturzeitung*, 5, 680–90.

SCHÖNFINKEL, MOSES (1922), 'Zum Entscheidungsproblem der mathematischen Logik', Manuscript, 21 pp. Bernays *Nachlaß*, WHS, ETH Zurich Archive, Hs. 974.282.

—— (1924), 'Über die Bausteine der mathematischen Logik', *Mathematische Annalen*, 92, 305–16. English translation in van Heijenoort (1967a), pp. 355–66.

SCHRÖDER, ERNST (1890), *Vorlesungen über die Algebra der Logik (exakte Logik)*, vol. 1. Repr. (New York: Chelsea, 1966).

—— (1891), *Vorlesungen über die Algebra der Logik*, vol. 2 (Leipzig: Teubner). Reprinted (New York: Chelsea, 1966).

—— (1895), *Vorlesungen über die Algebra der Logik*, vol. 3 (Leipzig: Teubner). Reprinted (New York: Chelsea, 1960).

—— (1898), 'On pasigraphy. Its present state and the pasigraphic movement in Italy', *The Monist*, 9, 246–62, 320.

SCHRÖDER-HEISTER, PETER (2002), 'Resolution and the origins of structural reasoning: early proof-theoretic ideas of Hertz and Gentzen', *Bulletin of Symbolic Logic*, 8, 246–65.

SCHUHMANN, KARL (1977), *Husserl-Chronik* (The Hague: Martinus Nijhoff).

—— (1987), 'Koyré et les phénoménologues allemands', *History and Technology*, 4, 149–67.

SCHÜLER, WOLFGANG (1983), *Grundlegungen der Mathematik in transzendentaler Kritik* (Hamburg: Meiner).

SCHÜTTE, KURT (1934a), 'Über die Erfüllbarkeit einer Klasse von logischen Formeln', *Mathematische Annalen*, 110, 572–603.

—— (1934b), 'Untersuchungen zum Entscheidungsproblem der mathematischen Logik', *Mathematische Annalen*, 109, 572–603.

SELDIN, JONATHAN P. (1980), 'Curry's program', in J. P. Seldin and J. R. Hindley (eds.), *To H. B. Curry: Essays on Combinatory Logic, Lambda calculus and Formalism*, 3–33 (London: Academic Press).

SHEFFER, HENRY MAURICE (1913), 'A set of five independent postulates for Boolean algebras, with application to logical constants', *Transactions of the American Mathematical Society*, 14, 481–88.

SHER, GILA (1991), *The Bounds of Logic* (Cambridge, Mass.: MIT Press).

SIEG, WILFRIED (1984), 'Foundations for analysis and proof theory', *Synthese*, 60, 159–200.

—— (1988), 'Hilbert's program sixty years later', *Journal of Symbolic Logic*, 53, 338–48.

—— (1990a), 'Relative consistency and accounts', *Synthese*, 84, 259–97.

—— (1990b), 'Reflections on Hilbert's Program', in W. Sieg (ed.), *Acting and Reflecting*, 171–82 (Dordrecht: Kluwer).

—— (1994), 'Mechanical Procedures and Mathematical Experience', in A. George (ed.), *Mathematics and Mind*, 71–117 (Oxford: Oxford University Press).

—— (1996), 'Proof Theory', in *Routledge Encyclopedia of Philosophy*.

—— (1999), 'Hilbert's Programs: 1917–1922', *Bulletin of Symbolic Logic*, 5(1), 1–44.

—— (2000), 'Reductive Structuralism', Technical Report Technical Report No. CMU-PHIL-108, Carnegie Mellon University, Pittsburgh.

—— (2005), 'Only two letters: The correspondence between Herbrand and Gödel', *The Bulletin of Symbolic Logic*, 12, 172–84.

SIEROKA, NORMAN (2007), 'Weyl's "agens theory" of matter and the Zurich Fichte', *Studies in History and Philosophy of Science*, 38, 84–107.

—— (2009), 'Geometrisation versus Transcendent Matter: A Systematic Historiography of Theories of Matter Following Weyl', forthcoming in *British Journal for the Philosophy of Science*.

SIERPINSKI, WACŁAW (1918), 'L'Axiome de M. Zermelo et son rôle dans la théorie des ensembles', *Bulletin de l'Académie des Sciences de Cracovie, Série A*, 99–152.

SIGURDSSON, SKULI (1991), *Hermann Weyl, Mathematics and Physics, 1900–1927*, Ph.D. thesis, Harvard University, Cambridge, Mass.

SIMPSON, STEVEN (1988), 'Partial realizations of Hilbert's program', *Journal of Symbolic Logic*, 53, 349–63.

SINACEUR, HOURYA (1993), 'Du formalisme à la constructivité: le finitisme', *Revue Internationale de Philosophie*, 47(4), 251–83.

—— (1995), 'Le Rôle de Poincaré dans la genèse de la métamathématique de Hilbert', in G. Heinzmann, K. Lorenz J. L. Greffe (ed.), *Henri Poincaré. Science et Philosophie*, 493–511 (Berlin and Paris: Akademie Verlag and Albert Blanchard).

—— (1997), 'Les "enfants naturels" de Descartes', in P. Radelet de Grave and J.-F. Stoffel (eds.), *Actes du colloque Descartes (Louvain-la-Neuve, 21–22 Juin 1996)*, 205–21 (Paris: Brepols).

—— (2000), 'Address at the Princeton University Bicentennial Conference on problems of mathematics, by A. Tarski', *The Bulletin of Symbolic Logic*, 6, 1–44.

—— (2001), 'Alfred Tarski: Semantic shift, heuristic shift in metamathematics', *Synthese*, 126, 49–65.

—— (2007), *Fields and Models. From Sturm to Tarski and Robinson* (Basel: Birkhäuser).

—— (2008), 'Tarski's practice and philosophy: between formalism and pragmatism', in Sten Lindström et alii (ed.), *Logicism, Intuitionism and Formalism. What has become of them?*, 357–98 (Dordrecht: Springer).

SKOLEM, THORALF (1912), 'Review of Weyl (1910a)', *Jahrbuch für die Fortschritte der mathematik*, 41(1), 89–90.

—— (1919), 'Untersuchungen über die Axiome des Klassenkalküls und über Produktations- und Summationsprobleme, welche gewisse Klassen von Aussagen betreffen', *Videnskasselskapets skrifter, I. Matematisk-naturvidenskabelig klasse*, 3. Reprinted in Skolem (1970), pp. 66–101.

—— (1920), 'Logisch-kombinatorische Untersuchungen über die Erfüllbarkeit oder Beweisbarkeit mathematischer Sätze nebst einem Theoreme über dichte Mengen', *Videnskasselskapets skrifter, I. Matematisk-naturvidenskabelig klasse*, 4. Reprinted in Skolem (1970), pp. 103–36. English translation of §1 in van Heijenoort (1967a), pp. 252–63.

—— (1922), 'Einige Bemerkungen zur axiomatischen Begründung der Mengenlehre', in *Matematikerkongressen I Helsingfors*, 217–32 (Helsinki: Akademiska Bokhandeln). Reprinted in Skolem (1970), pp. 137–52. English translation in van Heijenoort (1967a), pp. 290–301.

—— (1928), 'Über die mathematische Logik', *Norsk Mathematisk Tidsskrift*, 106, 125–42. Reprinted in Skolem (1970), pp. 189–206.

—— (1929a), 'Über einige Grundlagenfragen der Mathematik', *Skrifter utgitt av Det Norske Videnskaps-Akademi i Oslo, I. Mathematisk-naturvidenskapelig klasse*, 4. Reprinted in Skolem (1970), pp. 227–73.

—— (1929b), 'Über einige Satzfunktionen in der Arithmetik', *Skrifter utgitt av Det Norske Videnskaps-Akademi i Oslo, I. Mathematisk-naturvidenskapelig klasse*, 7, 1–28. Reprinted in Skolem (1970), pp. 281–306.

—— (1933), 'Über die Unmöglichkeit einer vollständigen Charakterisierung der Zahlenreihe mittels eines endlichen Axiomensystems', *Norsk matematisk forenings skrifter, series 2*, 10, 73–82. Reprinted in Skolem (1970), pp. 345–54.

—— (1934), 'Über die Nicht-Charakterisierbarkeit der Zahlenreihe mittels endlich oder abzählbar unendlich vieler Aussagen mit ausschließlich Zahlenvariablen', *Fundamenta Mathematicae*, 23, 150–61. Reprinted in Skolem (1970), pp. 355–66.

—— (1935), 'Über die Erfüllbarkeit gewisser Zählausdrücke', *Skrifter utgitt av Det Norske Videnskaps-Akademi i Oslo, I. Mathematisk-naturvidenskapelig klasse*, 6, 1–12. Reprinted in Skolem (1970), pp. 383–94.

—— (1938), 'Sur la portée du théorème de Löwenheim-Skolem', in Ferdinand Gonseth (ed.), *Les Entretiens de Zurich sur les Fondements et la Méthode des Sciences Mathematiques. 6–9 December 1938* (Zurich: Leemann). Reprinted in Skolem (1970), pp. 455–82.

—— (1970), *Selected Works in Logic* (Oslo: Universitetsforlaget).

SŁUPECKI, JERZY (1936), 'Der volle dreiwertige Aussagenkalkül', *Comptes Rendus des Séances de la Société des Sciences et des Lettres de Varsovie, Classe III*, 29, 9–11. English translation in McCall (1967), pp. 335–37.

SMITH, BARRY and SMITH, DAVID WOODRUFF (eds.) (1995), *The Cambridge Companion to Husserl* (Cambridge and New York: Cambridge University Press).

SMORYŃSKI, CRAIG (1989), 'Hilbert's programme', *CWI Quarterly*, 1, 3–59.

SOAMES, SCOTT (1999), *Understanding Truth* (Oxford: Oxford University Press).

SPECKER, ERNST (1979), 'Paul Bernays', in K. McAloon M. Boffa, D. van Dalen (ed.), *Logic Colloquium 78*, 381–389 (Amsterdam: North-Holland).

SPIEGELBERG, HERBERT and SCHUHMANN, KARL (1982), *The Phenomenological Movement* (The Hague: Martinus Nijhoff), 3rd edn.

STADLER, FRIEDRICH (1997), *Studien zum Wiener Kreis* (Frankfurt am Main: Suhrkamp).

—— (2003), *The Vienna Circle. Studies in the Origins, Development, and Influence of Logical Empiricism* (Dordrecht: Springer).

SUNDHOLM, GÖRAN (1994), 'Existence, Proof and Truth-Making. A Perspective on the Intuitionistic Conception of Truth', *Topoi*, 13, 117–26.

TAIT, WILLIAM (2002), 'Remarks on Finitism', in W. Sieg et al. (ed.), *Reflections on the Foundations of Mathematics: Essays in honor of Solomon Feferman*, 407–416 (Natick: AK Peters).

—— (2005), *The Provenance of Pure Reason* (Oxford: Oxford University Press).

—— (2006), 'Review of Kurt Gödel, Collected Works, vol. IV and vol. V', *Philosophia Mathematica*, 14, 76–114.

TAPPENDEN, JAMIE (1997), 'Metatheory and Mathematical Practice in Frege', *Philosophical Topics*, 25, 213–64.

TARSKI, ALFRED (1924a), 'Sur les ensembles finis', *Fundamenta Mathematicae*, 6, 45–95. Reprinted in Tarski (1986a), vol. I, 67–117.

—— (1924b), 'Sur quelques théorèmes qui èquivalent à l'axiome du choix', *Fundamenta Mathematicae*, 5, 147–54. Reprinted in Tarski (1986a), vol. I, 41–48.

—— (1929a), 'Geschichtliche Entwicklung und gegenwärtiger Zustand der Gleichmässigkeitstheorie und der Kardinalzahlarithmetik', *Sprawozdania z posiedzeń Towarzystwa Naukowego Warszwskiego, wydzial III*, 23, 22–29. Reprinted in Tarski (1986a), vol. I, 311–20. English translation with revisions in Tarski (1983), pp. 30–37.

—— (1929b), 'Les Fondements de la géométrie des corps', *Annales de la Societé Polonaise de Mathématiques*, 29–33. Reprinted in Tarski (1986a), vol. I, 227–31. English translation with substantial modifications in Tarski (1983), pp. 24–29.

—— (1930a), 'Fundamentale Begriffe der Methodologie der deduktiven Wissenschaften, I', *Monatshefte für Mathematik und Physik*, 37, 361–404. Reprinted in Tarski (1986a), vol. I, 345–90. English translation with revisions in Tarski (1983), pp. 60–109.

—— (1930b), 'Über einige fundamentale Begriffe der Metamathematik', *Sprawozdania z posiedzeń Towarzystwa Naukowego Warszwskiego, wydzial III*, 23, 22–29. Reprinted in Tarski (1986a), vol. I, 311–20. English translation with revisions in Tarski (1983), pp. 30–37.

—— (1931), 'Sur les ensembles définissables de nombres réels, I.', *Fundamenta Mathematicae*, 17, 210–39. Reprinted in Tarski (1986a), vol. I, 517–48. English translation with revisions in Tarski (1983), pp. 110–42.

—— (1933a), 'Einige Betrachtungen über die Begriffe der ω-Widerspruchsfreiheit und der ω-Vollständigkeit', *Monatshefte für Mathematik und Physik*, 40, 97–112. Reprinted in Tarski (1986a), vol. I, 619–36. English translation with revisions in Tarski (1983), pp. 279–95.

—— (1933b), 'Pojęcie prawdy w językach nauk dedukcyjnych (The concept of truth in the languages of deductives sciences)', *Prace Towarzystwa Naukowego Warszawskiego, Wydzial III*, 34. German translation in Tarski (1935a). English translation in Tarski (1983), pp. 152–278.

—— (1934–35), 'Einige methodologische Untersuchungen über die Definierbarkeit der Begriffe', *Erkenntnis*, 5, 80–100. Reprinted in Tarski (1986a), vol. I, 637–59. English translation in Tarski (1983), pp. 296–319.

—— (1935a), 'Der Wahrheitsbegriff in den formalisierten Sprachen', *Studia Philosophica (Lemberg)*, 1, 261–405. Reprinted in Tarski (1986a), vol. II, 51–198. English translation in Tarski (1983), pp. 152–278.

—— (1935b), 'Grundzüge des Systemenkalküls. Erster Teil', *Fundamenta Mathematicae*, 25, 503–526. Reprinted in Tarski (1986a), vol. II, 25–50. English translation in Tarski (1983), 342–83.

—— (1935c), 'Zur Grundlegung der Boole'schen Algebra, I', *Fundamenta Mathematicae*, 24, 177–98. Reprinted in Tarski (1986a), vol. II, 1–24. English translation with revisions in Tarski (1983), pp. 320–41.

—— (1936a), 'Grundlegung der wissenschaftlichen Semantik', in *Actes du Congrès International de Philosophie Scientifique*, volume 3, 1–8 (Hermann). Reprinted in Tarski (1986a), vol. II, pp. 259–68. English translation in Tarski (1983), pp. 401–08.

—— (1936b), 'Grundzüge des Systemenkalküls. Zweiter Teil', *Fundamenta Mathematicae*, 26, 283–301. Reprinted in Tarski (1986a), vol. II, pp. 225–43. English translation in Tarski (1983), pp. 342–83.

—— (1936c), 'O pojeciu wynikania logicznego', *Przeglad Filozoficzny*, 39, 58–68. English translation in Tarski 2002.

—— (1936d), 'Über den Begriff der logischen Folgerung', in *Actes du Congrès International de Philosophie Scientifique 7, Actualités Scientifiques et Industrielles*, 1–11 (Paris: Hermann). Reprinted in Tarski (1986a), vol. II, pp. 269–82. English translation with revisions in Tarski (1983), pp. 409–20.

—— (1937a), *Einführung in die mathematische Logik und die Methodologie der Mathematik* (Vienna: Springer).

TARSKI, ALFRED (1937b), 'Sur la méthode deductive', in *Travaux du IXe Congrès International de Philosophie*, vol. 6 of *Actualités Scientifiques et Industrielles*, 95–103 (Paris: Hermann). Reprinted in Tarski (1986a), vol. II, 323–34.

—— (1938a), 'Ein Beitrag zur Axiomatik der Abelschen Gruppen', *Fundamenta Mathematicae*, 30, 253–6. Reprinted in Tarski (1986a), vol. II, 447–50.

—— (1938b), 'Einige Bemerkungen zur Axiomatik der Boole'schen Algebra', *Sprawozdania z posiedzen Towarzystwa Naukowego Warszwskiego, wydzial III*, 31, 33–5. Reprinted in Tarski (1986a), vol. II, 353–55.

—— (1939), 'On undecidable statements in enlarged systems of logic and the concept of truth', *The Journal of Symbolic Logic*, 4, 105–12.

—— (1940), 'On the Completeness and Categoricity of Deductive Systems', Unpublished typescript, Alfred Tarski Papers, Carton 15, Bancroft Library, UC Berkeley. Printed in chapter 18.

—— (1941), 'On The Calculus of Relations', *Journal of Symbolic Logic*, 6, 73–89.

—— (1944), 'The semantic conception of truth and the foundations of semantics', *Philosophy and Phenomenological Research*, 4, 341–76. Reprinted in Tarski (1986a), vol. II, 661–99.

—— (1948), *A Decision Method for Elementary Algebra and Geometry* (Berkeley and Los Angeles: University of California Press). 2nd, rev. edn. 1951.

—— (1956), *Logic, Semantics, Metamathematics* (Oxford: Oxford University Press). 2nd edn. 1983.

—— (1967), *The Completeness of Elementary Algebra and Geometry* (Paris: CNRS, Institut Blaise Pascal). Reprinted in Tarski (1986a), vol. IV, 289–346.

—— (1983), *Logic, Semantics, Metamathematics* (Indianapolis: Hackett), 2nd edn.

—— (1986a), *Collected Papers* (Basel: Birkhäuser). Vols. I–IV.

—— (1986b), 'What are logical notions?', *History and Philosophy of Logic*, 7, 143–54.

—— (1987), ' "A philosophical letter of Alfred Tarski" with a prefatory note by Morton White', *Journal of Philosophy*, 84, 28–32.

—— (1995), 'Some current problems in metamathematics', *History and Philosophy of Logic*, 16, 159–68. Edited by Jan Woleński and Jan Tarski.

—— (2002), 'On the concepts of following logically', *History and Philosophy of Logic*, 23, 155–96. A translation of (1936c).

TARSKI, ALFRED and LINDENBAUM, ADOLF (1926), 'Communication sur les recherches de la théorie des ensembles', *Sprawozdania z posiedzeń Towarzystwa Naukowego Warszwskiego, wydzial III*, 19, 299–330. Reprinted in Tarski (1986a), vol. II, pp. 173–204.

TARSKI, ALFRED and LINDENBAUM, ADOLF (1936), 'Über die Beschränktheit der Ausdrucksmittel deduktiver Theorien', *Ergebnisse eines mathematischen Kolloquiums*, 7, 15–22. Reprinted in Tarski (1986a), vol. II, pp. 203–12. English translation in Tarski (1983), pp. 384–92.

TARSKI, ALFRED, MOSTOWSKI, ANDRZEJ, and ROBINSON, RAPHAEL (1953), *Undecidable Theories* (Amsterdam: North-Holland).

TAUSSKY-TODD, OLGA (1987), 'Remembrances of Kurt Gödel', in *Gödel Remembered*, 29–48 (Naples: Bibliopolis).

THIEL, CHRISTIAN (1972), *Grundlagenkrise und Grundlagenstreit* (Meisenheim: Anton Hain).

—— (1988), 'Die Kontroverse um die intuitionistische Logik vor ihrer Axiomatisierung durch Heyting im Jahre 1930', *History and Philosophy of Logic*, 9, 67–75.

—— (1994), 'Schröders zweiter Beweis für die Unabhängigkeit der zweiten Subsumtion des Distributivgesetzes im logischen Kalkül', *Modern Logic*, 4, 382–91.

—— (2002), 'Gödels Anteil am Streit über Behmanns Behandlung der Antinomien', in E. Köhler et alii (ed.), *Wahrheit und Beweisbarkeit. Leben und Werk Kurt Gödels, Band 1: Dokumente und historische Analysen, Band 2: Kompendium zu Gödels Werk*, 387–94 (Vienna: Hölder-Pichler-Tempsky).

—— (2005), 'Becker und die Zeutensche These zum Existenzbegriff in der antiken Mathematik', in Volker Peckhaus (ed.), *Oskar Becker und die Philosophie der Mathematik*, 35–46 (Munich: Wilhelm Fink Verlag).

—— (2007), 'A short introduction to Löwenheim's life and work and to a hitherto unknown paper', *History and Philosophy of Logic*, 28, 289–302.

THIELE, RÜDIGER (2003), 'Hilbert's twenty-fourth problem', *American Mathematical Monthly*, 1–24.

THORNE, KIP (1994), *Black Holes and Time Warps: Einstein's Outrageous Legacy* (New York: W. W. Norton).

TIESZEN, RICHARD (1989), *Mathematical Intuition* (Dordrecht: Kluwer).

—— (1995), 'Mathematics', in Barry Smith and David Woodruff Smith (eds.), *The Cambridge Companion to Husserl*, 438–62 (Cambridge and New York: Cambridge University Press).

—— (2000), 'The philosophical background of Weyl's mathematical constructivism', *Philosophia Mathematica*, 8(3), 274–301.

—— (2005), *Phenomenology, Logic, and the Philosophy of Mathematics* (Cambridge: Cambridge University Press).

TOEPELL, MICHAEL-MARKUS (1986), *Über die Entstehung von Hilbert's "Grundlagen der Geometrie"* (Göttingen: Vandenhoek und Ruprecht).

TONIETTI, TITO (1988), 'Four Letters of E. Husserl to H. Weyl and their Context', in W. Deppert (ed.), *Exact Sciences and their philosophical foundations*, 343–84 (Frankfurt am Main: Lang).

TROELSTRA, ANNE S. (1982), 'On the origin and development of Brouwer's concept of choice sequence', in Anne S. Troelstra and Dirk van Dalen (eds.), *The L. E. J. Brouwer Centenary Symposium*, 465–86 (Amsterdam: North-Holland).

—— (1990), 'On the early history of intuitionistic logic', in Petio P. Petkov (ed.), *Mathematical Logic*, 3–17 (New York and London: Plenum Press).

TROELSTRA, ANNE S. and VAN DALEN, DIRK (1988), *Constructivism in Mathematics. An Introduction*, vol. 1 (Amsterdam: North-Holland).

TUGENDHAT, ERNST (1960), 'Tarskis semantische Definition der Wahrheit und ihre Stellung innerhalb der Geschichte des Wahrheitsproblems im logischen Positivismus', *Philosophische Rundschau*, 8, 131–59. Reprinted in G. Skirbekk, ed., *Wahrheitstheorien* (Suhrkamp, Frankfurt, 1977), pp. 189–223.

TURING, ALAN M. (1937), 'On computable numbers, with an application to the "Entscheidungsproblem"', *Proceedings of the London Mathematical Society*, 2nd series, 42, 230–65.

ÜBEL, THOMAS (1992), *Overcoming Logical Positivism from within: the emergence of Neurath's Naturalism from the Vienna Circle's Protocol Debate* (Amsterdam: Rodopi).

URBANIAK, RAFAL (2008), *Leśniewski's Systems of Logic and Mereology; History and Re-evaluation.*, Ph.D. thesis, University of Calgary, Calgary.

VAIHINGER, HANS (1911), *Die Philosophie des Als Ob* (Leipzig: Felix Meiner).

VAN ATTEN, M., BOLDINI, P., BOURDEAU, M., and HEINZMANN, G. (eds.) (2008), *One Hundred Years of Intuitionism (1907–2007): The Cerisy Conference* (Basel: Birkhäuser).

VAN ATTEN, MARK (1999), *Phenomenology of choice sequences*, Ph.D. thesis, University of Utrecht. Quaestiones Infinitae, vol. 31 (Utrecht and Leiden: Zeno Institute of Philosophy).

—— (2003), *On Brouwer* (Belmont: Wadsworth).

—— (2005a), 'The correspondence between Oskar Becker and Arend Heyting', in Volker Peckhaus (ed.), *Oskar Becker und die Philosophie der Mathematik*, 119–42 (Munich: Wilhelm Fink Verlag).

—— (2005b), 'On Gödel's awareness of Skolem's Helsinki lecture', *History and Philosophy of Logic*, 26(4), 321–26.

—— (2005c), 'Phenomenology's reception of Brouwer's choice sequences', in Volker Peckhaus (ed.), *Oskar Becker und die Philosophie der Mathematik*, 101–18 (Munich: Wilhelm Fink Verlag).

—— (2007), *Brouwer as never read by Husserl. On the phenomenology of Choice Sequences* (Dordrecht: Springer).

VAN ATTEN, MARK, VAN DALEN, DIRK, and TIESZEN, RICHARD (2002), 'Brouwer and Weyl: The phenomenology and mathematics of the intuitive continuum', *Philosophia Mathematica*, 10, 203–26.

VAN DALEN, DIRK (1984), 'Four letters from Edmund Husserl to Hermann Weyl', *Husserl Studies*, 1, 1–12.

—— (1995), 'Hermann Weyl's Intuitionistic Mathematics', *The Bulletin of Symbolic Logic*, 1, 145–69.

—— (1999), *Mystic, Geometer, Intuitionist. The Life of L. E. J. Brouwer* (Oxford: Oxford University Press).

VAN DALEN, DIRK and EBBINGHAUS, HANS-DIETER (2000), 'Zermelo and the Skolem Paradox', *Bulletin of Symbolic Logic*, 6, 145–61.

VAN HEIJENOORT, JEAN (ed.) (1967a), *From Frege to Gödel. A Source Book in Mathematical Logic, 1897–1931* (Cambridge, Mass.: Harvard University Press).

VAN HEIJENOORT, JEAN (1967b), 'Logic as calculus and logic as language', *Boston Studies in the Philosophy of Science*, 3, 440–46.

VAN STIGT, WALTER P. (1990), *Brouwer's Intuitionism* (Amsterdam: North-Holland).

—— (1998), 'Brouwer's Intuitionist Programme', in Paolo Mancosu (ed.), *From Brouwer to Hilbert. The Debate on the Foundations of Mathematics in the 1920s*, 1–22 (New York: Oxford University Press).

VAN ULSEN, PAUL (2000), *E. W. Beth als Logicus* (Amsterdam: ILLC Dissertation Series).
VAUGHT, ROBERT L. (1974), 'Model Theory before 1945', in Henkin *et al.* (1974), 153–72.
VEBLEN, OSWALD (1904), 'A system of axioms for geometry', *Transactions of the American Mathematical Society*, 5, 343–84.
VEBLEN, OSWALD and YOUNG, JOHN WESLEY (1910), *Projective Geometry* (Boston: Ginn and Company).
VERCELLONI, LUCA (1988), *Filosofia delle Strutture* (Florence: La Nuova Italia).
VISSER, MATT (1996), *Lorentzian Wormholes: From Einstein to Hawking* (Woodbury, NY: American Institute of Physics Press).
VIZGIN, VLADIMIR P. (1994), *Unified Field Theories in the first third of the 20th century* (Basel, Boston, and Berlin: Birkhäuser Verlag). Translated from the Russian by J. B. Barbour.
VOLKERT, KLAUS (1994), 'Zur Rolle der Anschauung in mathematischen Grundlagenfragen: Die Kontroverse zwischen Hans Reichenbach und Oskar Becker über die Apriorität der euclidischen Geometrie', in L. Danneberg et al. (ed.), *Hans Reichenbach und die Berliner Gruppe*, 275–94 (Braunschweig: Vieweg Verlag).
VON NEUMANN, JOHANN (1923), 'Zur Einführung der transfiniten Zahlen', *Acta Litterarum ac Scientiarum Regiae Universitatis Hungaricae*, 1, 199–208. English translation in van Heijenoort (1967a), pp. 346–54.
—— (1925), 'Eine Axiomatisierung der Mengenlehre', *Journal für die reine und angewandte Mathematik*, 154, 219–40. English translation in van Heijenoort (1967a), pp. 393–413.
—— (1927), 'Zur Hilbertschen Beweistheorie', *Mathematische Zeitschrift*, 26, 1–46.
—— (1928), 'Die Axiomatisierung der Mengenlehre', *Mathematische Zeitschrift*, 27, 669–752.
—— (1931), 'Die formalistische Grundlegung der Mathematik', *Erkenntnis*, 2, 116–34. English translation in Benacerraf and Putnam (1983), pp. 61–65.
VON NEUMANN, JOHN (2005), *John von Neumann: Selected Letters* (Washington: American Mathematical Society, London Mathematical Society).
VON PLATO, JAN (2004), 'Kurt Gödel Collected Works IV–V', *The Bulletin of Symbolic Logic*, 10, 558–63.
—— (2007), 'In the Shadows of the Löwenheim-Skolem Theorem: Early Combinatorial Analyses of Mathematical Proofs', *The Bulletin of Symbolic Logic*, (13), 189–225.
—— (2008), 'Gentzen's Proof of Normalization for Natural Deduction', *The Bulletin of Symbolic Logic*, 14, 240–57.
VON WRIGHT, GEORG H. (1993), 'The Wittgenstein Papers', in J. Klagge & A. Nordmann (eds.), *Philosophical Occasions. 1912–1951*, 480–510 (Hackett).
VUILLEMIN, JULES (1968), *Leçons sur la première philosophie de Russell* (Paris: Colin).
WAISMANN, FRIEDRICH (1982), 'Über das Wesen der Mathematik. Der Standpunkt Wittgensteins', in F. Waismann (ed.), *Lectures on the Foundations of Mathematics*, 157–67 (Amsterdam: Rodopi). English translation: 'The Nature of Mathematics: Wittgenstein's Standpoint', in S. G. Shanker (ed.), *Ludwig Wittgenstein: Critical Assessments*, 1986 vol. 3 (London, Croom Helm), pp. 60-67.
WAJSBERG, MORDECHAJ (1931), 'Aksjomatyzacja trjwartościowego rachnuku zdań (Axiomatization of the three-valued propositional calculus)', *Comptes Rendus des Sèances de la Société des Sciences et des Lettres de Varsovie, Classe III*, 24, 126–45.
WANG, HAO (1967), 'Introductory note to Kolmogorov (1925)', in van Heijenoort (1967a), pp. 414–16.
—— (1970), 'A survey of Skolem's work in logic', in Skolem (1970), pp. 17–52.
—— (1974), *From Mathematics to Philosophy* (London: Routledge & Kegan Paul).
—— (1981), 'Some facts about Kurt Gödel', *The Journal of Symbolic Logic*, 46, 653–59.

—— (1987), *Reflections on Kurt Gödel* (Cambridge, Mass.: MIT Press).
—— (1996), *A Logical Journey* (Cambridge, Mass.: MIT Press).
WAVRE, ROLIN (1926), 'Logique formelle et logique empiriste', *Revue de métaphysique et de morale*, 33, 65–75.
WEAVER, GEORGE and GEORGE, BENJAMIN (2003), 'The Fraenkel-Carnap question for Dedekind algebras', *Mathematical Logic Quarterly*, 49, 92–6.
—— (2005), 'Fraenkel-Carnap Properties', *Mathematical Logic Quarterly*, 51, 285–90.
WEBB, JUDSON (1980), *Mechanism, Mentalism, and Metamathematics* (Dordrecht: Reidel).
—— (1995), 'Tracking contradictions in geometry: the idea of model from Kant to Hilbert', in Jaakko Hintikka (ed.), *Essays on the Development of the Foundations of Mathematics*, 1–20 (Dordrecht: Kluwer).
WEITZENBOCK, R. (1920), 'Über die Wirkungsfunktion in der Weyl'schen Physik', *Akademie der Wissenschaften in Wien. Sitzungsberichte. Abteilung IIa. Mathematisch-naturwissenschaftliche Klasse*, 129, 683–708.
WEYL, HERMANN (1908), *Singuläre Integralgleichungen mit besonderer Berücksichtigung des Fourierschen Integraltheorem*, Ph.D. thesis, Göttingen Universität, Göttingen. Reprinted in Weyl (1968b), pp. 1–87.
—— (1910a), 'Über die Definitionen der mathematischen Grundbegriffe', *Mathematisch-naturwissenschaftliche Blätter*, 7, 93–95, 109–13. Reprinted in Weyl (1968b), pp. 298–304.
—— (1910b), 'Über gewöhnliche Differentialgleichungen mit Singularitäten und die zugehörigen Entwicklungen willkürlicher Funktionen', *Mathematische Annalen*, 68, 220–60. Reprinted in Weyl (1968b), pp. 248–97.
—— (1913), *Die Idee der Riemannschen Fläche* (Leipzig: B. G. Teubner).
—— (1918a), *Das Kontinuum* (Leipzig: Veit). English translation in Weyl (1994).
—— (1918b), 'Gravitation und Elektrizität', *Sitzungsberichte, Preußische Akademie der Wissenschaften, Math-Physik. Kl.*, 465–80. Reprinted in Weyl (1968c), pp. 29–42. English translation in O'Raifeartaigh (1997), pp. 24–43.
—— (1918c), *Raum-Zeit-Materie* (Berlin: Springer), 1st edn.
—— (1918d), 'Reine Infinitesimalgeometrie', *Mathematische Zeitschrift*, 2, 384–411. Reprinted in Weyl (1968c), pp. 1–28.
—— (1919a), 'Der circulus vitiosus in der heutigen Begründung der Analysis', *Jahresbericht der Deutschen Mathematiker-Vereinigung*, 28, 85–92. Reprinted in Weyl (1968c), pp. 43–50. English translation in Weyl (1994).
—— (1919b), 'Eine Neue Erweiterung der Relativitätstheorie', *Annalen der Physik*, 59, 101–33. Reprinted in Weyl (1968c), pp. 55–87.
—— (1919c), *Raum-Zeit-Materie* (Berlin: Springer), 3rd edn.
—— (1920a), 'Das Verhältnis der kausalen zur statistischen Betrachtungsweise in der Physik', *Schweizerische Medizinische Wochenschrift*, 50 (19 Aug. 1920), 737–41. Reprinted in Weyl (1968c), pp. 113–22.
—— (1920b), 'Die Diskussion über die Relativitätstheorie auf der Naturforscherversammlung', *Die Umschau*, 24, 609–11.
—— (1921a), 'Feld und Materie', *Annalen der Physik*, 65, 541–631. Reprinted in Weyl (1968c), pp. 237–59.
—— (1921b), *Raum-Zeit-Materie* (Berlin: Springer), 4th edn.
—— (1921c), 'Über die neue Grundlagenkrise der Mathematik', *Mathematische Zeitschrift*, 10, 37–79. Reprinted in Weyl (1968c), 143–80. English translation in Mancosu (1998a), pp. 86–118.
—— (1922), 'Die Einzigartigkeit der Pythgoreischen Maßbestimmung', *Mathematische Zeitschrift*, 12, 114–46. Reprinted in Weyl (1968c), pp. 263–95.

—— (1923a), *Raum-Zeit-Materie* (Berlin: Springer), 5th edn.

—— (1923b), 'Zur Charakterisierung der Drehungsgruppe', *Mathematische Zeitschrift*, 17, 293–320. Reprinted in Weyl (1968c), pp. 345–72.

—— (1923/24), 'Review of M. Schlick's Allgemeine Erkenntnislehre', *Jahrbuch über die Fortschritte der Mathematik*, 46, 59–62.

—— (1924a), *Mathematische Analyse des Raumproblems* (Berlin: Springer).

—— (1924b), 'Randbemerkungen zu Hauptproblemen der Mathematik', *Mathematische Zeitschrift*, 20, 131–50. Reprinted in Weyl (1968c), pp. 433–52.

—— (1924c), *Was ist Materie? Zwei Aufsätze zur Naturphilosophie* (Berlin: Springer). Containing two previously published essays: 'Massenträgheit und Kosmos. Ein Dialog', *Die Naturwissenschaften*, 12 (1924), pp. 197–204, and 'Was ist Materie?', *Die Naturwissenschaften*, 12 (1924), pp. 561–68, 585–93, 604–11. Reprinted in Weyl (1968c), pp. 478–85, 486–510.

—— (1925), 'Die heutige Erkenntnislage in der Mathematik', *Symposion*, 1, 1–32. Reprinted in Weyl (1968c), pp. 511–42. English translation in Mancosu (1998c), pp. 123–42.

—— (1927), *Philosophie der Mathematik und Naturwissenschaft* (Munich: Oldenbourg). Augmented and revised English translation by Olaf Helmer, *Philosophy of Mathematics and Natural Science* (Princeton: Princeton University Press, 1949).

—— (1928a), 'Diskussionsbemerkungen zu dem zweiten Hilbertschen Vortrag über die Grundlagen der Mathematik', *Abhandlungen aus dem Mathematischen Seminar der Universität Hamburg*, 6, 86–88. English translation in van Heijenoort (1967a), pp. 480–84.

—— (1928b), *Gruppentheorie und Quantenmechanik* (Leipzig: S. Hirzel).

—— (1931a), *Die Stufen des Unendlichen* (Jena: Fischer).

—— (1931b), 'Geometrie und Physik (Rouse Ball Lecture at Cambridge University, May, 1930)', *Die Naturwissenschaften*, 19(3), 49–58. Reprinted in: Weyl (1968d), pp. 336–45.

—— (1932), *The Open World* (New Haven: Yale University Press).

—— (1940), 'The Ghost of Modality', in M. Farver (ed.), *Philosophical Essays in Memory of Edmund Husserl*, 278–303 (Cambridge, Mass.: Harvard University Press). Reprinted in Weyl (1968d), pp. 684–709.

—— (1944), 'David Hilbert and his mathematical work', *Bulletin of the American Mathematical Society*, 50, 612–54.

—— (1949), *Philosophy of Mathematics and Natural Science* (Princeton: Princeton University Press).

—— (1955), 'Erkenntnis und Besinnung (Ein Lebensrückblick)', *Studia Philosophica*, XVII, 153–71. Reprinted in Weyl (1968a), pp. 631–49.

—— (1968a), *Gesammelte Abhandlungen*, vol. 4 (Berlin: Springer).

—— (1968b), *Gesammelte Abhandlungen*, vol. 1 (Berlin: Springer).

—— (1968c), *Gesammelte Abhandlungen*, vol. 2 (Berlin: Springer).

—— (1968d), *Gesammelte Abhandlungen*, vol. 3 (Berlin: Springer).

—— (1985), 'Axiomatic versus constructive procedures in mathematics', *The Mathematical Intelligencer*, 7, 10–17. Edited by Tito Tonietti.

—— (1994), *The Continuum* (New York: Dover).

—— (1996), 'In Memoriam Helene Weyl', in I. Lange (ed.), *Arnold Zweig, Beatrice Zweig, Helene Weyl: Komm Her, Wir Lieben Dich–Briefe einer ungewöhnlichen Freundschaft*, 379–90 (Berlin: Aufbau-Verlag).

WHEELER, JOHN (1962), *Geometrodynamics* (New York: Academic Press).

—— (1968), 'Superspace and the Nature of Quantum Geometrodynamics', in C. DeWitt and J. Wheeler (eds.), *Battelle Rencontres, 1967 Lectures in Mathematics and Physics*, 242–307 (New York: W. A. Benjamin).
WHITEHEAD, ALFRED NORTH (1898), *A Treatise on Universal Algebra, with Applications* (Cambridge: Cambridge University Press).
—— (1902), 'On Cardinal Numbers', *American Journal of Mathematics*, 24, 367–94.
—— (1907), 'Introduction logistique à la géométrie', *Revue de Métaphysique et de Morale*, 15, 34–39.
WHITEHEAD, ALFRED NORTH and RUSSELL, BERTRAND (1910), *Principia Mathematica*, vol. 1 (Cambridge: Cambridge University Press).
—— (1912), *Principia Mathematica*, vol. 2 (Cambridge: Cambridge University Press).
—— (1913), *Principia Mathematica*, vol. 3 (Cambridge: Cambridge University Press).
WILLE, MATTHIAS (2005), ' "Dem Unendlichen einen finiten Sinn beilegen". Von Becker und Gentzen zu Lorenzen', in Volker Peckhaus (ed.), *Oskar Becker und die Philosophie der Mathematik*, 325–50 (Munich: Wilhelm Fink Verlag).
—— (2008), *Beweis und Reflexion: Philosophische Untersuchungen über die beweistheoretischer Praxen* (Paderborn: Mentis).
WITTGENSTEIN, LUDWIG (1921), 'Logisch-philosophische Abhandlung', *Annalen für Naturphilosophie*, 14, 198–262. Reprinted with English translation in Wittgenstein (1922).
—— (1922), *Tractatus logico-philosophicus* (London: Kegan Paul).
—— (1974), *Philosophical Grammar* (Oxford: Blackwell).
—— (1975), *Philosophical Remarks* (Oxford: Blackwell).
—— (1979), *Ludwig Wittgenstein and the Vienna Circle* (Oxford: Blackwell).
—— (1994), *Wiener Ausgabe. Philosophische Betrachtungen-Philosophische Bemerkungen*, vol. 2 (Vienna: Springer).
—— (1996), *Wiener Ausgabe. Philosophische Grammatik*, vol. 5 (Vienna: Springer).
—— (2000), *Wiener Ausgabe. The Big Typescript*, vol. 11 (Vienna: Springer).
WOHLTMANN, HANS (1957), 'Dietrich Mahnke 1884–1939', in O. H. May (ed.), *Niedersächsische Lebensbilder* (Hildesheim: August Lax Verlagsbuchhandlung).
WOLEŃSKI, JAN (1989a), *Logic and Philosophy in the Lvov-Warsaw School* (Dordrecht: Kluwer).
—— (1989b), 'The Lvov-Warsaw School and the Vienna Circle', in K. Szaniawski (ed.), *The Vienna Circle and the Lvov-Warsaw School*, pp. 443–53 (Dordrecht: Kluwer).
—— (1993), 'Tarski as a philosopher', in R. Poli F. Coniglione and J. Wolenski (eds.), *Polish Scientific Philosophy*, pp. 319–38 (Amsterdam: Rodopi).
—— (1994), 'Jan Łukasiewicz on the liar paradox, logical consequence, truth, and induction', *Modern Logic*, 4, 392–99.
—— (1995a), 'Mathematical logic in Poland 1900–1939: people, circles, institutions, ideas', *Modern Logic*, 5, 363–405.
—— (1995b), 'On Tarski's background', in J. Hintikka (ed.), *From Dedekind to Gödel*, 331–41 (Dordrecht: Reidel).
WOLEŃSKI, JAN and KÖHLER, ECKEHART (eds.) (1998), *Alfred Tarski and the Vienna Circle* (Dordrecht: Kluwer).
WOLEŃSKI, JAN and SIMONS, PETER (1989), 'De Veritate: Austro-Polish contributions to the theory of truth from Brentano to Tarski', in K. Szaniawski (ed.), *The Vienna Circle and the Lvov-Warsaw School*, 391–442 (Dordrecht: Kluwer).
YANDELL, B. H. (2002), *The Honors Class: Hilbert's Problems and their Solvers* (Natick, Mass.: A. K. Peters).

YANG, CHENG NING (1981), 'Geometry and Physics', in Y. Ne'eman (ed.), *To Fulfill a Vision: Jerusalem Einstein Centennial Symposium on Gauge Theories and Unification of Physical Forces*, 3–11 (Reading, Mass.: Addison-Wesley).

YOUNG, JOHN WESLEY (1917), *Lectures on the Fundamental Concepts of Algebra and Geometry* (New York: MacMillan). First edition 1911.

ZACH, RICHARD (1999), 'Completeness before Post: Bernays, Hilbert, and the development of propositional logic', *Bulletin of Symbolic Logic*, 5(3), 331–66.

—— (2003), 'The practice of finitism. Epsilon calculus and consistency proofs in Hilbert's Program', *Synthese*, 137, 211–59.

—— (2004), 'Hilbert's "Verunglückter Beweis," the first epsilon theorem and consistency proofs', *History and Philosophy of Logic*, 25, 79–94.

—— (2006), 'Hilbert's program then and now', in Dale Jacquette (ed.), *Philosophy of Logic. Handbook of the Philosophy of Science*, vol. 5, 411–47 (Amsterdam: Elsevier).

ZAMBELLI, PAOLA (1999), 'Alexandre Koyré alla scuola di Husserl a Gottinga', *Giornale Critico della Filosofia Italiana*, 19, 303–54.

ZERMELO, ERNST (1904), 'Beweis, daß jede Menge wohlgeordnet werden kann', *Mathematische Annalen*, 59, 514–516. English translation in van Heijenoort (1967a), pp. 139–41.

—— (1908a), 'Mathematische Logik', Vorlesung gehalten von Prof. Dr E. Zermelo zu Göttingen im S.S. 1908. Lecture notes by Kurt Grelling. *Nachlaß* Zermelo, Kapsel 4, Universitätsbibliothek Freiburg im Breisgau.

—— (1908b), 'Neuer Beweis für die Möglichkeit einer Wohlordung', *Mathematische Annalen*, 65, 107–128. English translation in van Heijenoort (1967a), pp. 183–98.

—— (1908c), 'Untersuchungen über die Grundlagen der Mengenlehre I', *Mathematische Annalen*, 65, 261–81. English translation in van Heijenoort (1967a), pp. 199–215.

—— (1909), 'Sur les ensembles fini et le principe de l'induction complète', *Acta Mathematica*, 32, 185–93.

—— (1929), 'Über den Begriff der Definitheit in der Axiomatik', *Fundamenta Mathematicae*, 14, 339–44.

—— (1930), 'Über Grenzzahlen und Mengenbereiche: Neue Untersuchungen über die Grundlagen der Mengenlehre', *Fundamenta Mathematicae*, 16, 29–37.

—— (1931), 'Über die Stufen der Quantifikation und die Logik des Unendlichen', *Jahresbericht der Deutschen Mathematiker-Vereinigung*, 41, 85–88.

ZYGMUNT, JAN (1973), 'On the sources of the notion of reduced product', *Reports on Mathematical Logic*, 1, 53–67.

—— (1990), 'Mojżes Presburger: Life and Work', *History and Philosophy of Logic*, 12, 211–23.

—— (2004), 'Bibliografia prac naukowych Marii Kokoszynskiej-Lutmanowej', *Filozofia Nauki*, 12, 155–66.

INDEX

Aanderaa, Stål 80, 578
Abrusci, Michele 126, 128, 498, 502, 514, 571
Ackermann, Richard Wilhelm, 278, 309, 539
Ackermann, Wilhelm 3, 26, 50, 53, 60–5, 70, 75–9, 81, 99,
  105–6, 108–9, 122, 130, 153, 174, 189–90, 215, 252–3,
  281–2, 287, 349, 498–9, 539, 566, 571, 586
Addison, John 584
Ajdukiewicz, Kazimierz 3, 99, 109, 113, 444, 493, 571, 574
Anderson, C. Anthony 582
Andrade, Jules F. C. 310–11, 547–8, 571
Andrews, Peter B. 80–1, 571, 578
Angoff, Allan 241–2
Antonelli, Aldo 440
Aquinas, Thomas 418
Aristotle 172, 199, 437
Arnauld, Antoine 199
Aspray, William 578, 593
Astekar, Abhay 576
Aust, Bernd Peter 346–7, 572
Avenarius, Richard 155
Avigad, Jeremy 498, 571
Awodey, Steve 12, 14, 111, 469, 472–3, 483–4, 494, 497,
  500, 567, 570–1
Ayer, Alfred J. 416, 571

Bachmann, Friedrich 567, 576
Badesa, Calixto viii, xi, 2, 5, 493, 496, 536, 571
Baire, René-Louis 27–8, 84, 571
Baker, Alan 403, 571
Baker, Gordon 217–18, 571
Balas, Yosseff 246, 536
Baldus, Richard 14, 99, 497, 566, 571
Ballo, Edoardo 593
Barbour, Julian B. 606
Barzin, Marcel 91, 572
Bays, Timothy 440, 442, 446–7, 451, 459, 500, 566–7, 572
Becker, Astrid 277, 282, 308, 346, 348
Becker, Oskar vi–vii, ix, xi, 92–3, 129, 132, 141–4, 147, 149,
  256–7, 276–8, 280–91, 293–313, 323–6, 330–2, 334–44,
  346–56, 500, 503, 539–47, 549, 551–3, 572, 578, 581,
  587, 592, 595–7, 599, 604, 606, 609
Beeson, Michael 254
Behmann, Heinrich vi, xi, 60, 62, 64–5, 105, 122, 156–7,
  159–76, 179–82, 190–205, 207–16, 218–25, 227–8, 232,
  242, 248, 505–15, 518, 520–31, 572–3, 583, 591, 604
Bell, John 257, 572
Bellotti, Luca 566, 572
Beltrami, Eugenio 499, 566, 572
Benacerraf, Paul 408, 530–1, 573, 575, 585, 606

Benary, Wilhelm 283, 287, 541
Benmakhlouf, Ali 579
Bennett, Albert A. 381
Bernays, Paul vi, xi, 25–6, 35, 55–65, 68, 73, 75, 78–9, 81,
  83, 92, 99–100, 102–3, 105, 108–9, 122–7, 130, 133, 135–7,
  139–161, 164, 173–4, 176, 180, 182–8, 190, 198–200, 202,
  211, 213–16, 218–19, 232–3, 237–8, 241–2, 244–53, 269,
  274–5, 291–2, 367, 493, 495, 497–502, 504–5, 507,
  511–12, 514–15, 517–521, 524–6, 528, 531, 535, 543, 554,
  566, 573, 579, 585–7, 589, 591, 599, 602, 610
Bernhard, Peter 159
Bernstein, Felix 160, 162, 177–9, 506, 513–15, 574, 580
Berry, G. G. 19, 22, 31
Beth, Evert W. 10, 358, 387–8, 403–7, 559, 574, 605
Betsch, Christian 504, 574
Betti, Arianna 3, 387, 413, 574
Biggs, Norman L. 526, 574
Billinge, Helene 544, 574
Bimbó, Katalin 3, 574
Bitbol, Michel 598
Blaschke, Wilhelm 548, 574
Blok, Willem J. 566, 574
Blumenthal, Otto 500, 574
Bôcher, Maxime 13, 99, 443, 494, 499, 568, 574
Boffa, Maurice 602
Boi, Luciano 257, 574
Boldini, Pascal 605
Bolzano, Bernard 142, 162, 180, 199
Bolzmann, Ludwig 566, 574
Bondoni, Davide 3, 574
Boniface, Jacqueline 123, 574, 589
BonJour, Laurence 438, 562, 574
Boole, George 7, 16, 36, 129, 162, 175, 179, 493, 595
Boolos, George 250
Boone, William W. 241
Borel, Armand 550, 574
Borel, Émile 27–9, 84, 91, 503, 571, 574
Borga, Marco 493, 574–5, 580
Börger, Egon 498, 575
Born, Max 279
Boskovitz, A. 160, 190–1, 505, 520
Bourdeau, Michel 605
Bouveresse, Jacques 577
Bozzi, Silvio 6, 592
Brading, Katherine 123, 549, 575, 598
Bradley, Francis H. 425
Bredehorn, Uwe 278, 309, 346, 541
Breger, Herbert 538, 575
Brentano, Franz 418, 434, 560, 609

Brouwer, Luitzen Egbertus Jan 3, 71, 84–8, 90–3, 125, 129–132, 134, 137–140, 142, 144–5, 147, 151, 157, 161, 188, 201, 210, 212, 219, 226, 228, 259–260, 271–275, 278–9, 286, 290–3, 295, 297–8, 304–6, 309–10, 313, 331, 336, 350–1, 392, 498, 502–4, 514, 526, 533, 538, 542–5, 572, 575, 580–1, 584, 589, 591, 596, 604–5
Buldt, Bernd 531
Burali-Forti, Cesare 6, 18, 22, 29, 165, 178, 180, 287, 495, 583, 592–3
Burgess, John 408, 469, 478, 575

Cantini, Andrea 495, 575
Cantor, Georg 16, 18, 27, 31, 84, 86, 127, 131, 140, 163–5, 172, 185, 191–2, 194–5, 197, 235, 261, 288–9, 299–301, 305–6, 313, 354, 356, 367, 391–2, 403, 511, 523, 533, 537, 554, 574–5, 577, 580, 583
Cariani, Fabrizio viii
Carnap, Rudolf vi-vii, xii, 26, 99–100, 109, 111, 122–3, 199–201, 210–12, 228, 232–4, 236–9, 241–3, 270–1, 358, 361–388, 392, 395–400, 404, 406, 408, 412, 415–18, 420, 423–439, 444, 462, 469, 472–4, 477, 483, 493, 499–500, 525, 527, 530–4, 536–7, 541, 550, 553–5, 559–560, 562–7, 569, 571, 575–7, 580, 583–4, 587, 591, 594, 597, 607
Carson, Emily 591
Cartan, Henri 549, 576
Carus, Andrew W. 111, 473, 500, 571
Cassirer, Ernst 142, 281, 504, 552, 571, 576
Castellani, Elena 598
Cauchy, Augustin 29, 262, 336, 537
Chandrasekharan, Komaravolu S. 574
Chang, Chen Chung 496, 576, 584
Chern, Shiing-Shen 550, 576
Chevalley, Claude 537, 539, 576
Chihara, Charles S. 262, 495, 578
Christoffel, Elwin Bruno 318–319
Church, Alonzo 3, 65–7, 81, 91, 241, 400, 499, 575–576, 580, 598
Chwistek, Leon 25, 177–8, 233, 238–9, 358, 367, 397, 406, 513–14, 536, 555, 587, 589
Cipolla, Michele 504, 576
Cirera, Ramon 559, 576
Clifford, William K. 332
Cohen, Hermann 288
Cohen, Paul J. 34, 241, 578
Cohen, Robert S. 583, 593
Cohnitz, Daniel 392, 400–2, 576
Coliva, Annalisa 592
Columbus, Cristoforo vi
Colyvan, Mark 403, 576
Compton, Arthur H. 549
Conrad-Martius, Hedwig 284
Cooley, John 361
Coolidge, Julian L. 328, 576
Coniglione, Francesco 609
Corcoran, John 440, 451–2, 479, 567, 570, 576
Corry, Leo 100, 123, 178, 566, 577
Corsi, Giovanna 596
Courant, Richard 150–1, 161, 248, 251, 514, 566

Couturat, Louis 513, 588
Craig, William 584
Creath, Richard 359, 364, 377, 577, 597
Cruz, Joseph 438, 596
Curry, Haskell B. 3, 66, 498, 577, 600
Czelakowski, Janusz 500, 566, 577
Czermak, Johannes 578–9, 591

Dahms, Hans-Jachim 584
Dalla Chiara, Maria Luisa 596
Danneberg, Lutz 591, 595, 606
Da Silva, Jairo 280, 577
Dauben, Joseph W. 495, 577
Daubert, Johannes 284
Davis, Martin 247, 537, 577
Davoren, Jennifer 495, 584
Dawson, John W. 199, 211–12, 232–4, 236–7, 242, 249, 254, 451, 498–9, 527, 531, 536–7, 577, 583
Decock, Lieven 387, 400–1, 577
de Jong, Wim 387
de Morgan, Augustus 36, 159, 175, 493, 496
de Risi, Vincenzo 257, 577
de Rouilhan, Philippe 21–2, 495, 500, 554, 566, 577
de Sitter, Willem 313, 548
Dedekind, Richard 18, 28, 31, 51, 71, 110–11, 128, 132–4, 142, 169, 173, 186–7, 261–4, 266, 268–70, 313, 324, 473–4, 503, 537, 577, 607, 609
Demopoulous, William 494–5, 577–8
Deppert, William 575, 591, 599, 604
Descartes, René 600
Detlefsen, Michael 136–7, 139–40, 500, 503, 578
Dettweiler, Marco viii
Dewey, John 13, 473
DeWitt, C. M. 609
Dickson, Leonard E. 529, 578
Dieudonné, Jean 500, 537, 539, 578
Dilthey, Wilhelm 347, 349, 356, 590
Dingler, Hugo 176, 178–9, 335, 348, 352–4, 506, 514, 578
Dirac, Paul 323
Dirichlet, Johann Peter Gustav Lejeune 220, 529
Dreben, Burton 41, 50, 80, 104, 106, 495, 499, 578
Drucker, Thomas 577
Dugas, René 547, 578
Duhem, Pierre 546
Dumitriu, Anton 6, 578
Dummett, Michael 498, 578

Ebbinghaus, Hans-Dieter 3, 495, 578, 605
Eddington, Arthur S. 319, 399, 549, 578
Edwards, Harold M. 123, 502, 578
Edwards, Paul 593
Einstein, Albert 279, 282, 310, 313–16, 319, 322–3, 328–9, 333, 335, 340–3, 547–9, 574, 578–9, 604, 606, 610
Emrich, Johannes 257, 578
Endress, Martin 199, 232, 278, 309, 540
Errera, Alfred 91, 572
Etchemendy, John 413, 440–1, 446–7, 458, 461–2, 465, 479–80, 500, 568, 578

Euclid 11, 126, 219–20, 487, 529–30
Euler, Leonhard xi, 122–3, 207, 210, 213, 217–9, 220–2, 223–6, 228, 529–30
Ewald, William Brag 502, 577, 579, 583, 585–6, 595

Fang, Joong 500, 579
Falkenburg, Brigitte 583
Farber, Marvin 608
Feferman, Anita 362, 559, 569, 579
Feferman, Solomon viii-ix, xi, 232, 240, 244, 247, 252, 254, 267–8, 276, 277, 294, 308, 361, 362, 376, 387, 413, 415, 440, 495, 500, 505, 524, 536–7, 539, 542, 549, 559, 566, 569, 579, 591, 602
Feigl, Herbert 233, 236
Feist, Richard 257, 572
Fenstad, Jens 254
Ferreiros, José x, 5, 26, 34, 36, 440, 472, 495, 554, 568, 579, 592, 599
Fichte, Johann G. 99, 244, 257, 299, 333, 546, 596, 600
Field, Hartry 408
Finsler, Paul 328–9
Flamm, Ludwig 551, 579
Fløistad, Guttorm 596
Floyd, Juliet 582
Folina, Janet 123, 257, 579
Follett, Wilson 242, 536
Forman, Paul 312, 334, 580
Fraenkel, Adolf Abraham 33–6, 99, 110–11, 157, 238, 282, 349, 469, 472–3, 477, 495, 503–4, 535, 541, 566–7, 580, 607
Franchella, Miriam 498, 580, 593
Frank, Matthew 544, 580
Frank, Philip 420, 561, 580
Frege, Gottlob 7, 16–19, 25, 37, 51, 128–9, 132–4, 162, 169, 173, 177, 180, 182, 261, 305, 403, 493–6, 501, 513, 537, 577–8, 580, 582, 595, 597–8, 602, 605
Freguglia, Paolo 8, 575, 580
Frei, Günther 279, 542, 580
Frewer, Magdalene 513, 515, 580
Freytag, Willy 155, 580
Friedman, Harvey, 250
Friedman, Michael 309, 542, 550–1, 580, 597
Fries, Jakob Friedrich 147, 149–50, 244
Frost-Arnold, Greg 358–9, 361–2, 368, 374, 395, 398, 415, 558, 560, 580

Gabbay, Dov M. 3, 575, 580, 595
Galileo 549
Garciadiego, Alejandro R. 18, 495, 580, 593
Gauss, Carl F. 350, 545, 549, 582, 585
Gauthier, Yvon 123, 502, 580
Geiger, Moritz 142, 281, 284–6, 346, 500, 542, 552, 581
Gentzen, Gerhard 3, 67–9, 77–8, 81, 83–4, 92–3, 178, 216, 498–9, 504, 581, 592, 600, 606
George, Alexander 600
George, Benjamin 484, 607
Gethmann, Carl Friedrich 257, 581
Gethmann-Siefert, Annemarie 257, 307, 581, 587

Ghirardi, Gian Carlo 596
Giaquinto, Marcus viii, 136, 361, 387, 440, 461, 495, 581
Giedymin, Jerzy 571
Giere, Ronald N. 587
Gillies, Donald viii, 276, 581, 588
Gillispie, Charles C. 578
Giugliano, Antonello 257, 581
Givant, Steven 467
Glivenko, Valerii 68, 91–3, 211, 581
Gleason, Andrew M. 294
Glock, Hans-Johann 562, 581
Gödel, Kurt vi-vii, xi, 3, 6, 26, 34–5, 49–51, 58, 61–2, 65, 69, 79–81, 82–4, 88, 92–4, 97–9, 106–11, 118, 122–4, 144, 153, 199–201, 210–14, 216, 218–9, 232–8, 240–54, 257, 260, 370–2, 375, 405–6, 408, 458, 460, 469, 472–5, 487–8, 495–500, 504, 524, 527, 529–37, 556, 566, 569, 577–9, 581–3, 588–9, 591, 593–4, 597–8, 600, 604–6, 609
Gödel, Marianne 241
Goldfarb, Warren D. 43, 81, 111, 152, 248, 254, 474, 495–6, 582, 584
Goldstein, Catherine 123, 582
Gomez-Torrente, Mario 413, 440–1, 442, 445, 447–8, 451–3, 455, 463–8, 479–80, 500, 566, 582
Gomperz, Heinrich 246
Goodman, Nelson 358, 361, 368, 380, 382–4, 388, 397–8, 400–5, 408, 471–5, 558–9, 569, 576, 582, 597
Gosselin, Mia 400, 582
Grädel, Erich 575
Grandjean, Burke 241–2, 246
Granger, Gilles Gaston 500, 582
Grattan-Guinness, Ivor 17, 24, 233, 493, 495, 536, 582
Gray, Jeremy x, 123, 582–3, 592, 599
Greffe, Jean-Louis 600
Grelling, Kurt 150, 160, 162, 177–9, 183, 391, 506, 512–14, 583, 594–5, 610
Grundmann, Thomas 559, 583
Guillaume, Marcel 98, 566, 583
Günther, Gotthard 244
Gurevich, Yuri 575

Haaparanta, Leila, ix
Haas, Gerrit 162, 506, 515, 525, 583
Hacker, Peter 217–18, 571
Hadamard, Jacques 27–8, 571
Hafner, Johannes 5, 199, 346, 415, 440, 469
Hagenmaier, Winfried 278, 309, 541
Hahn, Hans 200–1, 210, 228, 235–6, 532–3, 583
Hahn, Lewis E. 582, 597
Hale, Bob 495, 583
Haller, Rudolf 363, 417–18, 553, 583, 594
Hallett, Michael 123, 128–9, 136–7, 301, 495–6, 501–2, 538, 583
Hardcastle, Gary L. 553, 583
Hartmann, Gudrun 346
Hartshorne, Charles 595
Hasse, Helmut 529, 583
Hausdorff, Felix 317, 583

Hawking, Stephen 606
Hawkins, Thomas 550, 583
Hazen, Allen P. 495, 583–4
Hecke, Erich 291, 542
Hegel, Georg Wilhelm Friedrich 244
Hegselmann, Rainer 560, 584
Heidegger, Martin 142, 257, 281, 285–7, 290, 302, 307, 348, 354–6, 546, 552–3, 581, 588
Heinzmann, Gerhard 262, 495, 502, 539, 584, 600, 605
Heisenberg, Werner 549
Hellinger, Ernst 279, 585
Hellmann, Geoffrey 408, 574, 579, 584
Helmer, Olaf 436, 608
Hemholtz, Hermann Ludwig 311, 328, 330, 338, 340–1, 551
Hempel, Carl G. 200, 233–4, 248, 361, 364, 382, 385, 397, 415–16, 423–5, 431, 434, 436, 438, 525, 531–2, 558–60, 564, 584
Hendricks, Vincent 599
Hendry, John 551, 584
Henkin, Leon 576, 584, 605
Herbrand, Jacques 3, 41, 49, 61, 64–5, 69, 79–81, 94, 123, 154, 174, 216, 219, 232–3, 241, 245, 248–249, 251, 496, 499, 503, 531–2, 537, 571, 578, 582, 584, 600
Herglotz, Gustav 281, 539
Hertz, Heinrich 499, 566, 584
Hertz, Paul 67–8, 176, 178, 498, 506, 514, 566, 571, 584, 600
Hertz Rudolf 514
Hesseling, Dennis E. 499, 542, 584
Hessenberg, Gerhard 150, 512–13, 584
Heyting, Arend 3, 57, 89, 92–3, 228, 235, 237, 242, 282, 499, 502, 531, 533, 539–40, 543, 583–5, 604–5
Hilbert, David vi, ix, xi, 2–3, 6–7, 11–12, 14, 26–7, 30, 34, 51–63, 65–75, 79, 81, 83, 88–9, 98–100, 105–6, 108–9, 111, 117–18, 122–65, 171–84, 187–91, 197–8, 201, 214–16, 218–19, 235, 237–9, 246, 248–53, 256, 259–61, 269–70, 274–5, 278, 281–2, 286–8, 290, 292, 295–300, 304–5, 307, 313, 321, 345, 347, 349–51,354–6, 372–3, 405, 443–4, 472, 493–4, 497–507, 511–18, 520–1, 524, 528, 531–2, 534–5, 537–40, 543–6, 551–2, 566, 571, 573–81, 583, 585–8, 590–1, 593–7, 600, 602, 604–5, 607–10
Hindley, J. Roger 600
Hintikka, Jaakko 495, 586, 607, 609
Hirano, J. 215–16, 528
Hodges, Wilfrid 14, 413, 494, 496, 500, 586
Hoffmann-Grüneberg, Franz 438, 559–60, 587
Hölder, Otto 213, 269–70, 281, 539, 587
Honderich, Ted 241
Howard, Don 500, 587
Hubble, Edwin 548
Huber, Renate 591
Huntington, Edward V. 2, 12–15, 101–2, 443–4, 472, 493, 497, 499, 567, 587
Husserl, Edmund vi-vii, 52, 93, 99, 141–2, 147, 176–7, 228, 244, 256–7, 260, 264, 275, 278–87, 289–290, 293–7, 302–5, 307, 309, 314–16, 321–323, 325, 328, 332, 336–7, 346–56, 500, 513, 538–9, 541–2, 544–6, 549, 551–3, 572, 577, 579, 587–8, 590, 600–1, 604–5, 608, 610

Hylton, Peter W. 21, 495, 587

Imaguire, Guido 3, 590
Irvine, Andrew D. 583
Isaacson, Daniel viii, 361, 440

Jacquette, Dale 610
Jadacki, Jacek J. 513–14, 587
Jaentsch, Erich 348
Jahnke, Hans N. 538, 587
Jané, Ignasi 5, 440
Janich, Peter 257, 587
Jaśkowski, Stanisław 98, 587
Jeffrey, Richard 584
Jeffreys, Harold 371
Jevons, Stanley 493
Joachim, Harold H. 425
Johnson, William E. 6
Johansson, Ingebrigt 89
Jónsson, Bjarni 496, 588
Jørgensen, Klaus Frovin 599
Jørgensen, Jørgen 559
Joseph, Helene See Weyl, Helene
Jourdain, Philip 16, 29, 493, 582, 588
Juhos, Bela 559

Kalmár, László 65, 588
Kamareddine, Farouz 3, 588
Kamlah, Andreas 591
Kanamori, Akihiro 3, 254, 495, 588
Kant, Immanuel 84, 137, 145–7, 149–52, 155, 157–8, 170, 174, 199, 244, 275, 287–9, 329, 333, 336, 351–2, 503–5, 552, 573, 575, 579, 588, 590–1, 595, 607
Kaufmann, Felix vi, xi, 99, 111, 122, 199–203, 207, 210–15, 218–21, 225–8, 232–6, 238, 248, 271, 276, 281–2, 346, 473, 500, 525–33, 535–6, 538, 540, 566, 588, 591
Kaufmann, Walter 228
Kenny, Anthony 221, 588
Kern, Iso 552, 588
Kerzberg, Pierre 574, 598
Keyser, Cassius J. 444, 493–4, 588
Khintchine, Alexandre 91
Kisiel, Theodore 588
Kitcher, Philip 136–7, 500, 578, 588, 593
Klagge, James 606
Kleene, Stephen C. 66–7, 588
Klein, Felix 99, 332, 372–3, 550, 566
Kluge, Fritz 504, 588
Kneale, Martha 6, 106, 190, 497, 588
Kneale, William 6, 106, 190, 497, 588
Kneser, Helmut 585–6
Knobloch, Eberhard 595
Kockelmans, Joseph J. 277, 572, 588
Köhler, Eckehart 232, 536–7, 588, 593, 604, 609
Kokoszyńska, Maria xii, 415–19, 425–34, 436–8, 559–65, 588, 592, 610
Kolmogorov, Andrei N. 88–9, 90, 93, 211, 499, 589, 606
König, Dénes 47, 108, 589
König, Julius 19, 27, 29, 31, 498, 537, 589

König, Robert 279
Korselt, Alwin 37, 444, 493, 589
Kotarbiński, Tadeusz 367, 373–4, 383, 396, 418, 429, 434, 436, 555–6, 563–4
Koyré, Alexandre 281, 513, 600, 610
Krazer, A. 585
Kreisel, Georg 217, 219, 245, 500, 536, 589
Kretschmann, Erich 335, 589
Kronecker, Leopold 122–3, 128, 131–2, 145, 147, 217, 220–2, 226–7, 502, 529–30, 578, 581, 589, 592
Krüger, Lorenz 583
Kruntorad, Paul 588

Laan, Twan 588
Ladd-Franklin, Christine 493
Lampert, Tim 124, 589
Landini, Gregory 3, 17, 22, 25, 495, 589
Landau, Edmund 161
Landgrebe, Ludwig 546
Lange, Ilse 608
Langford, Cooper Harold 112, 444, 449–50, 471, 486, 500, 568, 589
Laplace, Pierre-Simon 425
Lapointe, Sandra 415, 592
Largeault, Jean 498, 537, 589
Lascar, Daniel 496, 589
Lauener, Henri 500, 589
Laugwitz, Detlev 549–550, 589
Lebesgue, Henri 27–8, 84, 571
Leibniz, Gottfried 244, 333, 347, 349–52, 354–6, 543, 545, 575, 577, 590, 595
Leja, F. 596
Leonard, Henry S. 388, 390
Lenard, Philip 547
Leśniewski, Stanisław 3, 358, 367, 383, 388, 390, 405–6, 555, 567, 574, 605
Levi-Civita, Tullio 317, 318, 321
Levy, Paul 91
Lewis, Clarence Irving 26, 59, 444, 449, 499, 589
Lie, Sophus 311, 328–30, 340–1, 548, 574, 583
Lindenbaum, Adolf 97, 112, 372, 450–2, 460, 470, 474–5, 481, 483, 567, 604
Lindström, Per 577
Lindström, Sten 601
Link, Godehard 583
Linnebo, Øystein 254
Linsky, Bernard 3, 5, 495, 589–90
Lionnais, François Le 578
Lipps, Hans 281, 346
Lohmar, Dietrich 544, 590
London, Fritz 346, 500
Longo, Giuseppe 278, 547
Lorenz, Kuno 600
Lorenzen, Paul 276, 539, 590, 609
Lovett, Edgar O. 6, 590
Löwenheim, Leopold 2–3, 6, 36–51, 63–4, 79, 99, 110–12, 213, 215, 376, 405, 459–60, 471, 483, 486, 496–7, 528, 531, 566, 571, 590, 601, 604, 606

Löwy, Heinrich 503, 590
Łukasiewicz, Jan 60, 91, 95–7, 103–4, 418, 429, 497, 499–500, 560, 563, 590, 609
Lusin, Nicolai 236, 533
Lutman, Maria, See Kokoszyńska
Lutman-Kokoszyńska, Maria, See Kokoszyńska

MacColl, Hugh 6
Mach, Ernst 136–7, 147, 155–7, 538
Machover, Moshé 254
MacFarlane, John 361, 440
Maddox, J. R. 578
Maddux, Roger D. 496, 590
Mahnke, Dietrich vi-vii, xi, 256–8, 281–4, 293–7, 300, 346–56, 504, 540–1, 543–6, 552–3, 572, 577, 590, 609
Majer, Ulrich 123, 137, 149, 504–5, 542–3, 546, 586, 591
Malament, David 254
Malcev, Anatoly Ivanovich 111, 591, 598
Malinowski, Grzegorz 500, 566, 577
Mancosu, Paolo ix-x, 25, 65, 71, 102, 123, 144, 159, 174, 235, 247–8, 271, 274, 277, 289–90, 293, 308–9, 362, 370, 372, 395, 398, 462, 466, 472, 493, 497–500, 511, 513, 524–5, 528–9, 537, 539–40, 542–4, 546, 549, 552, 559–60, 569–70, 573, 575, 581, 583–8, 591–2, 596, 605, 607–8
Mangione, Corrado 6, 592
Mannoury, Gerrit 503
Marchisotto, Elena A. 3, 8, 592
Marion, Mathieu viii, xi, 122, 124, 502, 530, 546, 583, 592
Martin, Richard M. 558, 592
Max, Ingolf 595
Maxwell, James C. 317, 322–3, 343, 551
McAloon, Kenneth 602
McCall, Storrs 587, 590, 601
McCleary, John 578
McGuinness, Brian 574
McKinsey, John Charles 496, 591
Medicus, Fritz 333–4, 344, 592
Menger, Karl 233, 241, 248, 429, 563
Menzler-Trott, Eckart 3, 592
Meyerhof, Otto 150
Millsop, Rebecca viii
Milne, Peter 3, 592
Mirimanoff, Dmitry 35–6, 495, 592
Misner, Charles W. 551
Mittelstraß, Jürgen 307, 581, 587
Mol, Liesbeth De 3, 592
Momtchiloff, Peter vii
Monk, Ray 530, 592
Mooij, Jan J. A. 502, 592
Moore, Gregory H. 18, 42, 159, 164, 182–3, 495–6, 525, 571, 592–3
Moriconi, Enrico 3, 5, 145, 500, 593
Mormann, Thomas 432, 560, 565, 593
Morris, Charles William 438, 560
Morris, Michael 551, 593
Mostowski, Andrzej 367, 373–4, 496, 554, 593, 604

Müller, Aloys 134–6, 146–9, 152, 156, 173, 504, 515, 573, 588, 593
Müller, Gert 500, 593
Mulligan, Kevin 562, 593
Murawski, Roman 413, 500, 593

Naess, Arne 418, 431, 434, 436–8, 559–60, 564, 593
Nagayama, Tadashi 240
Nagel, Ernst 242, 425, 536, 559, 593
Natorp, Paul 288
Nederpelt, Rob 588
Ne'eman, Yuval 610
Nef, Frederic 577, 582
Neider, Heinrich 364, 386, 425, 438
Nelson, Leonard 134, 146–51, 161, 177–8, 183, 244, 504–5, 513, 573, 583, 593, 595
Nemeth, Elizabeth 583
Ness, Arne see Naess
Neumann, Robert 235, 533
Neurath, Maria 593
Neurath, Otto vi, xii, 363–4, 386, 412–13, 415–39, 553–4, 559–65, 583–4, 587, 592–4, 598, 605
Newman, Andrew 562, 594
Newman, James R. 242, 536, 593
Newman, Maxwell H. A. 537, 539, 594
Newton, Isaac 549
Nicod, Jean G. P. 26, 59–60, 103, 594
Nietzsche, Friedrich 581
Nöbeling, Georg 248
Noether, Emmy 575
Nollan, Richard 234
Nordheim, Lothar 586
Nordmann, Alfred 606
Nye, Mary Jo 503, 594

Oberdan, Thomas 559, 594
O'Leary, Daniel J. 497, 594
O'Raifeartaigh, Lochlainn 549, 594
Ore, Øystein 100, 566

Padoa, Alessandro 6, 8–11, 442–4, 493–4, 574, 594
Palladino, Dario 575
Parakenings, Brigitte 232, 278, 309, 415, 540, 554
Parsons, Charles 242, 244, 254, 500, 594
Pasch, Moritz 442, 512, 594
Patras, Frederic 574
Patterson, Douglas x, 413, 574, 579–80, 582, 586, 592–4
Pauli, Wolfgang 320, 549, 594
Peano, Giuseppe 6–8, 11, 14–17, 20, 25, 29, 82, 98, 101, 110–12, 118, 140, 162, 177, 179, 234–5, 244–5, 248, 376, 379, 381, 384, 406, 442–3, 458, 474–5, 493–4, 503, 532–3, 575, 594, 597
Peckhaus, Volker viii-ix, 3, 51, 126, 128–9, 150, 159, 177, 199, 232, 257, 493, 495, 500, 502, 506, 513–15, 572, 578, 581, 592, 594–6, 604–5, 609
Peirce, Charles Sanders 16, 36–7, 102, 175, 493, 496, 595, 599
Petitot, Jean 598

Petkov, Petio P. 604
Petzoldt, Joseph 155
Pfänder, Alexander 284–5, 290, 307
Picardi, Eva 592
Pichler, Hans 350
Pier, Jean-Paul 583
Pieri, Mario 3, 6–9, 11, 15, 32, 98, 442–3, 494, 592, 595
Pigozzi, Don 566, 574
Pincock, Chris 5, 176, 387
Plato 246
Poeggeler, Otto 595
Poincaré, Henri 19–21, 29, 71, 123, 129–30, 132, 138, 141–3, 157, 178, 262, 264, 267, 274, 292, 304, 324, 335, 389, 392, 495, 502–3, 537, 551, 577, 579, 584, 592, 595, 599, 600
Poli, Roberto 609
Pollard, Steven 257, 596
Pollock, John L. 438, 596
Popper, Karl 382, 415, 418, 439, 558, 562, 565
Poretsky, Platon 6
Poser, Hans 257, 596
Post, Emil L. 3, 52, 62, 96, 100, 103–5, 471, 486, 493, 497, 592, 596, 610
Posy, Carl J. 498, 596
Potter, Michael 495, 596
Poznanski, Edward 562
Prawitz, Dag 136, 500, 505, 596
Presburger, Mojėsz 486, 500, 596, 610
Proops, Ian 3, 596
Putnam, Hilary 376, 403, 530–1, 573, 575, 585, 606

Quine, Douglas B. 387
Quine, Willard Van Orman vi-vii, xii, 18, 49, 358–9, 361–5, 367–73, 375–405, 407–9, 462, 470–1, 475, 493, 528, 553–9, 569–70, 577, 580, 582, 591, 596–8

Radelet de Grave, Patricia 600
Ramsey, Frank P. 3, 24, 26, 31, 64, 174, 190, 260, 270–1, 597
Rand, Rose 234, 537
Rang, Bernhard 513, 597
Rappaport, Leon 253
Ray, Greg 440–1, 447, 479–80, 500, 568, 597
Read, Stephen 500, 597
Reck, Erich 12, 14, 472–3, 483–4, 494, 497, 500, 570–1, 597
Redéi, Miklós 534
Reech, Ferdinand 310–11, 547
Reichenbach, Hans 228, 232–4, 236–8, 257, 282, 349, 531, 534–5, 540, 577, 591, 595, 598, 606
Reid, Constance 131, 241, 249, 500, 537, 597
Reidemeister, Kurt 583
Reinach, Adolf 161, 284–5
Resnik, Michael 136, 597
Rhees, Rush 221, 228
Richard, Jules 19–20, 22, 597
Richardson, Alan W. 583, 587
Rickert, Heinrich 147–8, 348
Riemann, Bernhard 316–17, 319, 321, 328, 338, 340–1, 529, 548–50

Rigal, Elizabeth 592
Ritter, Jim 582
Rivenc, François 495, 597
Robinson, Abraham 241, 601
Robinson, Raphael 461, 604
Rodriguez-Consuegra, Francisco A. 17, 359, 493, 597
Roetti, Jorge A. 540, 546, 597
Rolf, George 500
Rose, Alan 97, 598
Rosen, Gideon 408, 575
Rosen, Nathan 333
Rossberg, Marcus 387–8, 392, 400–2, 576
Rosser, J. Barkley 66, 83, 97, 500, 588, 598
Rougier, Louis 428, 561–3
Rowe, David E. 578, 595
Russell, Bertrand vi, xi, 2–3, 5–7, 15–25, 31, 37, 53, 65, 71, 79, 100, 102–3, 122, 129, 132, 147, 159–64, 169–70, 172–84, 187–98, 215–6, 235, 238, 260–2, 267, 269, 291, 361–2, 364–5, 371–2, 384, 389, 395–9, 402–4, 407, 444, 493, 495, 497, 499, 503, 505–9, 511–15, 518–24, 533, 535, 562, 566, 572–3, 575, 577, 580, 582–4, 587, 589, 591–4, 596–8, 606, 609
Russo, Elena viii
Rutte, Heiner 559, 598
Ryckman, Thomas A. viii, xi, 123, 256–7, 277, 289–90, 293, 308–9, 346, 361, 544, 548–50, 552, 575, 592, 598

Sagüillo, José 469, 567, 570, 598
Sasaki, Chikara 599
Sattler, Jochen 346–7, 572
Sauer, Tilman 123, 505, 586, 591
Savage, C. Wade 582
Scanlan, Michael 3, 12, 472, 500, 568, 598
Schäfer, Lothar 591
Schappacher, Norbert 589
Scheibe, Erhard 330, 550, 583, 599
Scheler, Max 284–5, 290, 307
Schelling, Friedrich W. J. 244
Schilpp, Paul Arthur 241–3, 253, 402, 576, 582, 597–8
Schlick, Moritz 147, 155, 218, 227, 280, 416, 420–6, 428–9, 431, 439, 530, 532–3, 537, 539, 561–3, 583–4, 598–9, 608
Schmid, Anne-Françoise 502, 599
Schmidt, Erhard 250
Schmidt, Hans-Joachim 591
Schmidt, R. 252
Scholz, Erhard 257, 546, 548–50, 599
Scholz, Heinrich 232, 281, 348, 404, 505–6, 512, 531, 552, 583, 599
Schönfinkel, Moses 62, 64–6, 108–9, 122, 160–1, 174, 188, 190, 499, 505, 514, 520, 573, 585, 599
Schopenhauer, Arthur 170
Schouten, Jan A. 318
Schröder, Ernst 6–7, 15–16, 36–40, 42–4, 52, 57, 102, 129, 162, 175, 179, 493, 496, 512, 595, 599, 604
Schröder-Heister, Peter 498, 600
Schuhmann, Karl 279, 513, 542, 600, 602
Schüler, Wolfgang 500, 600
Schur, Issai 248, 251

Schütte, Kurt 65, 600
Scott, Dana 469, 484, 584
Seldin, Jonathan P. 498, 600
Shakespeare, William 313
Shanker, Stuart G. 577, 606
Shapiro, Stewart 469, 579
Sheffer, Henry Maurice 3, 26, 59–60, 65, 102–4, 190–1, 519–20, 599–600
Sher, Gila 440–1, 566, 600
Shieh, Sanford 582
Sica, Giandomenico 596
Sieg, Wilfried ix, 3, 53, 61, 123, 128, 152–3, 159, 164, 171, 176, 182–3, 188, 198, 232–3, 250–1, 254, 502, 504–7, 512, 515, 520, 524, 531, 537, 591, 600, 602
Sieroka, Norman 257, 600
Sierpinski, Wacław, 34, 600
Sigurdsson, Skuli 550, 600
Simons, Peter 560, 593, 609
Simpson, Steven 505, 600
Sinaceur, Hourya 123, 145, 359, 500, 502, 566, 600
Skirbekk, Gunnar 605
Skolem, Thoralf 3, 14, 28, 30, 33–5, 37, 41–3, 45, 49–51, 63–4, 79, 99, 106, 110–13, 243, 260, 291, 366, 374, 376, 405, 459–60, 471, 483, 494–7, 503, 531, 538, 543, 556, 566, 601, 605–6
Słupecki, Jerzy, 97, 499, 601
Smith, Barry 545, 593, 601, 604
Smith, David Woodruff 545, 601, 604
Smith, James T. 3, 592
Smoryński, Craig 503, 602
Soames, Scott 441–2, 602
Solovay, Robert 484
Sommer, Richard ix, 591
Specker, Ernst 500, 602
Spiegelberg, Herbert 542, 602
Stadler, Friedrich ix, 233, 532–3, 583, 594, 602
Stammbach, Urs 279, 542, 580
Stattler, Jochen 572
Stebbing, Susan 425, 438
Steiner, Mark 503
Stelzner, Werner 595
Stemmler, Elke 162, 506, 515, 525, 583
Stoffel, Jean-François 600
Strauss, Martin 437–9, 560
Strohal, Richard 502
Sturm, Jacques C. F. 601
Sundholm, Göran 540, 602
Szaniawski, Klemens 609

Tait, William 5, 123–4, 500, 602
Talcott, Carolyn ix, 591
Tappenden, Jamie 495, 499, 566, 602
Tarski, Alfred vi-vii, x, xii, 3, 5, 10, 14, 26, 38, 97, 99, 103, 109, 111–119, 241, 246–7, 358–9, 361–70, 372–88, 394–400, 403–9, 412–13, 415–20, 426–34, 436–42, 444–72, 474–85, 493, 496, 500, 531, 537, 553–70, 572, 574, 576–80, 582–4, 586, 588, 590–4, 597, 599, 601–5, 609
Tarski, Jan iv, x, 470–1, 569, 604

Tarski, Marja 387
Taussky-Todd, Olga 233, 604
Thiel, Christian 3, 159, 199, 232, 257, 262, 278, 495, 499, 525, 537, 604
Thiele, Rüdiger 123, 604
Thomas, Wolfgang 513, 597
Thorne, Kip 551, 593, 604
Tieszen, Richard 257, 277–8, 280, 308–309, 544–5, 604–5
Toepell, Michael-Markus 500, 604
Tonietti, Tito 280, 604, 608
Troelstra, Anne S. 92, 278, 309, 498–9, 540, 604
Tugendhat, Ernst 605
Turing, Alan M. 65, 81, 245, 605
Twardowski, Kasimierz 418, 559–60

Übel, Thomas E. 415, 423, 427–8, 559, 598, 605
Uhlemann, Brigitte See Parakenings, Brigitte
Urbaniak, Rafal 3, 605

Vaihinger, Hans 170–1, 605
van Atten, Mark 3, 5, 92, 257, 278, 309, 346, 387, 498–9, 540, 542–4, 546, 574, 605
van Dalen, Dirk 144, 274–5, 278, 280, 292, 309, 495, 498, 537–9, 542–3, 550, 602, 604–5
van Heijenoort, Jean 8, 10, 41, 47, 50–1, 104, 106, 246–8, 493, 495–6, 499, 503, 573, 575, 578, 580, 584–6, 588, 590, 594, 596–9, 601, 605–6, 608, 610
van Stigt, Walter P. 132, 260, 271, 498–9, 538, 540, 542, 592, 605
van Ulsen, Paul 387, 404, 605
Vaught, Robert L. 42, 111, 496, 566, 584, 605
Veblen, Oswald 12–14, 110, 450–2, 472–3, 480–2, 490, 499, 566, 568, 606
Venn, John 493
Vercelloni, Luca 100, 566, 606
Vernant, Danis 577, 582
Visser, Henk 387
Visser, Matt 551, 606
Vizgin, Vladimir P. 548, 550, 606
Vögeli, Yvonne 199, 232, 277, 308, 541
Volkert, Klaus 540, 606
von Neumann, Johann 26, 35–6, 70, 75, 78–9, 82–3, 99, 110, 123, 130, 139, 144, 153, 228, 232–8, 241, 245, 248–53, 281–2, 306, 495, 498, 503, 531–5, 537, 539–40, 566, 583, 606
von Plato, Jan 3, 606
von Wright, Georg H. 530, 606
Vuillemin, Jules 17, 606

Waismann, Friedrich 200–1, 210–11, 221, 225–8, 233, 236, 420–1, 527, 529–30, 561, 606
Wajsberg, Mordechaj 97, 606
Wang, Hao 42, 47, 50–1, 89–90, 241–3, 246, 249, 253, 496–8, 536, 594, 606

Wavre, Rolin 90–1, 607
Weaver, George 484, 607
Webb, Judson 501–2, 566, 607
Weierstrass, Karl 131, 202, 214, 313
Weil, André 537, 539, 576
Weiss, Paul 595
Weitzenbock, R. 548, 607
Weyl, Helene 279, 539, 608
Weyl, Hermann vi–vii, xi, 14, 28, 30, 32–4, 71, 99–100, 110–11, 125, 130–2, 134, 137, 144, 147, 151, 157, 160–1, 176–7, 180, 186–8, 202, 214, 226, 237–8, 256–7, 259, 260–84, 286–300, 302–19, 321–34, 338–45, 346–9, 351–5, 389, 392, 495, 499–500, 503, 507, 512–4, 520, 524, 534–5, 537–9, 541–5, 549–51, 553, 566, 572–80, 587, 589, 591–2, 594, 596, 598–601, 604–5, 607–8
Wheeler, John 333, 551, 608–9
White, Morton 364, 369, 604
Whitehead, Alfred North vi, 6, 15, 18, 21, 24–5, 53, 71, 79, 100–1, 129, 132, 160–4, 172, 175, 179–81, 189–94, 215, 371, 388, 390, 493, 506, 508–9, 511, 515, 518, 523, 567, 572, 609
Wiener, Norbert 176–7, 390
Wille, Matthias 3, 123, 257, 609
Wigner, Eugene 279
Wittgenstein, Ludwig xi, 3, 104, 122–4, 201, 210, 217–18, 220–31, 244, 260, 419–21, 429, 433, 529–30, 561–3, 571, 578–9, 588–9, 591–2, 606
Wohltmann, Hans 541, 552, 609
Woleński, Jan 97, 362, 413, 417, 470–1, 499–500, 555, 559–60, 566, 569, 593, 604, 609
Woods, John 3, 575, 580, 595
Woodger, Joseph H. vi, 365, 376–7, 386, 395–6, 402–3
Wright, Crispin 495, 583
Wundheiler, Alexander 372–3, 555, 562

Yandell, B. H. 123, 609
Yang, Cheng Ning 549, 610
Young, John Wesley 499, 566, 587, 606, 610
Young, William Henry 178

Zach, Richard viii, xi, 2, 5, 70, 78, 123, 159, 183, 176, 190, 199, 232, 253, 277, 308–9, 346, 493–4, 497–8, 526, 571, 610
Zambelli, Paola 513, 610
Zanichelli, Nicola 586
Zermelo, Ernst 2–3, 7, 19, 26–36, 52, 85–6, 99, 110, 129, 140, 160, 176–80, 187, 196, 215, 233, 238, 241, 243, 247, 257, 259–60, 261–4, 269, 278, 281–2, 291, 309, 366–7, 371, 403–4, 407, 493, 495–6, 498–9, 503–4, 513, 518, 536–9, 541, 554, 566–7, 577–8, 583, 588, 593–5, 597, 600, 605, 610
Zimmerman, Dean x, 387
Zweig, Arnold 608
Zweig, Beatrice 608
Zygmunt, Jan 113, 500, 560, 610

Printed in Great Britain
by Amazon